Quantum Field Theory of Many-body Systems

Quantum Field Theory of Many-body Systems

From the Origin of Sound to an Origin of Light and Electrons

Xiao-Gang Wen
Department of Physics, MIT

OXFORD
UNIVERSITY PRESS

Great Clarendon Street, Oxford OX2 6DP

Oxford University Press is a department of the University of Oxford.
It furthers the University's objective of excellence in research, scholarship,
and education by publishing worldwide in

Oxford New York

Auckland Bangkok Buenos Aires Cape Town Chennai
Dar es Salaam Delhi Hong Kong Istanbul Karachi Kolkata
Kuala Lumpur Madrid Melbourne Mexico City Mumbai Nairobi
São Paulo Shanghai Taipei Tokyo Toronto

Oxford is a registered trade mark of Oxford University Press
in the UK and in certain other countries

Published in the United States
by Oxford University Press Inc., New York

First published 2004

A catalogue record for this title is available from the British Library

Library of Congress Cataloging in Publication Data
(Data available)

ISBN 0 19 853094 3 (Hbk)

10 9 8 7 6 5 4 3 2 1

Printed in Great Britain
on acid-free paper by
Biddles Ltd. King's Lynn

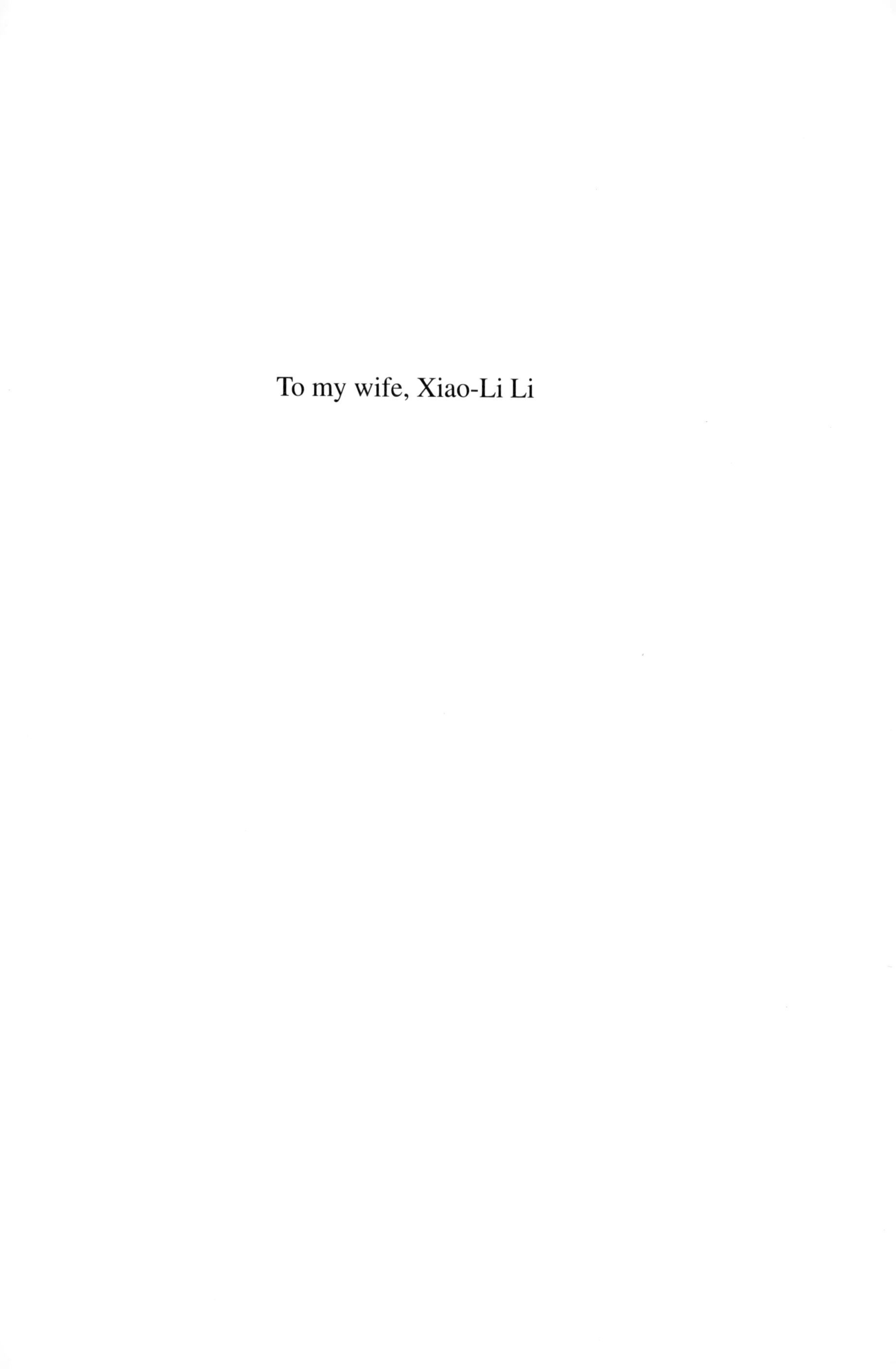

To my wife, Xiao-Li Li

PREFACE

The quantum theory of condensed matter (i.e. solids and liquids) has been dominated by two main themes. The first one is band theory and perturbation theory. It is loosely based on Landau's Fermi liquid theory. The second theme is Landau's symmetry-breaking theory and renormalization group theory. Condensed matter theory is a very successful theory. It allows us to understand the properties of almost all forms of matter. One triumph of the first theme is the theory of semiconductors, which lays the theoretical foundation for electronic devices that make recent technological advances possible. The second theme is just as important. It allows us to understand states of matter and phase transitions between them. It is the theoretical foundation behind liquid crystal displays, magnetic recording, etc.

As condensed matter theory has been so successful, one starts to get a feeling of completeness and a feeling of seeing the beginning of the end of condensed matter theory. However, this book tries to present a different picture. It advocates that what we have seen is just the end of the beginning. There is a whole new world ahead of us waiting to be explored.

A peek into the new world is offered by the discovery of the fraction quantum Hall effect (Tsui *et al.*, 1982). Another peek is offered by the discovery of high-T_c superconductors (Bednorz and Mueller, 1986). Both phenomena are completely beyond the two themes outlined above. In last twenty years, rapid and exciting developments in the fraction quantum Hall effect and in high-T_c superconductivity have resulted in many new ideas and new concepts. We are witnessing an emergence of a new theme in the many-body theory of condensed matter systems. This is an exciting time for condensed matter physics. The new paradigm may even have an impact on our understanding of fundamental questions of nature.

It is with this background that I have written this book.[1] The first half of this book covers the two old themes, which will be called traditional condensed matter theory.[2] The second part of this book offers a peek into the emerging new theme, which will be called modern condensed matter theory. The materials covered in the second part are very new. Some of them are new results that appeared only a few months ago. The theory is still developing rapidly.

---

[1] When I started to write this book in 1996, I planned to cover some new and exciting developments in quantum many-body theory. At that time it was not clear if those new developments would become a new theme in condensed matter theory. At the moment, after some recent progress, I myself believe that a new theme is emerging in condensed matter theory. However, the theory is still in the early stages of its development. Only time will tell if we really do get a new theme or not.

[2] Some people may call the first theme traditional condensed matter theory and the second theme modern condensed matter theory.

After reading this book, I hope, instead of a feeling of completeness, readers will have a feeling of emptiness. After one-hundred years of condensed matter theory, which offers us so much, we still know so little about the richness of nature. However, instead of being disappointed, I hope that readers are excited by our incomplete understanding. It means that the interesting and exciting time of condensed matter theory is still ahead of us, rather than behind us. I also hope that readers will gain a feeling of confidence that there is no question that cannot be answered and no mystery that cannot be understood. Despite there being many mysteries which remain to be understood, we have understood many mysteries which initially seemed impossible to understand. We have understood some fundamental questions that, at the beginning, appeared to be too fundamental to even have an answer. The imagination of the human brain is also boundless.[3]

This book was developed when I taught the quantum many-body physics course between 1996 and 2002 at MIT. The book is intended for graduate students who are interested in modern theoretical physics. The first part (Chapters 2–5) covers traditional many-body physics, which includes path integrals, linear responses, the quantum theory of friction, mean-field theory for interacting bosons/fermions, symmetry breaking and long-range order, renormalization groups, orthogonality catastrophe, Fermi liquid theory, and nonlinear σ-models. The second part (Chapters 6–10) covers topics in modern many-body physics, which includes fractional quantum Hall theory, fractional statistics, current algebra and bosonization, quantum gauge theory, topological/quantum order, string-net condensation, emergent gauge-bosons/fermions, the mean-field theory of quantum spin liquids, and two- or three-dimensional exactly soluble models.

Most of the approaches used in this book are based on quantum field theory and path integrals. Low-energy effective theory plays a central role in many of our discussions. Even in the first part, I try to use more modern approaches to address some old problems. I also try to emphasize some more modern topics in traditional condensed matter physics. The second part covers very recent work. About half of it comes from research work performed in the last few years. Some of the second part is adapted from my research/review papers (while some research papers were adapted from parts of this book).

The book is written in a way so as to stress the physical pictures and to stress the development of thoughts and ideas. I do not seek to present the material in a neat and compact mathematical form. The calculations and the results are presented in a way which aims to expose their physical pictures. Instead of sweeping ugly assumptions under the rug, I try to expose them. I also stress the limitations of some common approaches by exposing (instead of hiding) the incorrect results obtained by those approaches.

[3] I wonder which will come out as a 'winner', the richness of nature or the boundlessness of the human imagination.

Instead of covering many different systems and many different phenomena, only a few simple systems are covered in this book. Through those simple systems, we discuss a wide range of physical ideas, concepts, and methods in condensed matter theory. The texts in smaller font are remarks or more advanced topics, which can be omitted in the first reading.

Another feature of this book is that I tend to question and expose some basic ideas and pictures in many-body physics and, more generally, in theoretical physics, such as 'what are fermions?', 'what are gauge bosons?', the idea of phase transition and symmetry breaking, 'is an order always described by an order parameter?', etc. Here, we take nothing for granted. I hope that those discussions will encourage readers to look beyond the nice mathematical formulations that wrap many physical ideas, and to realize the ugliness and arbitrariness of some physical concepts.

As mathematical formalisms become more and more beautiful, it is increasingly easy to be trapped by the formalism and to become a 'slave' to the formalism. We used to be 'slaves' to Newton's laws when we regarded everything as a collection of particles. After the discovery of quantum theory,[4] we become 'slaves' to quantum field theory. At the moment, we want to use quantum field theory to explain everything and our education does not encourage us to look beyond quantum field theory.

However, to make revolutionary advances in physics, we cannot allow our imagination to be trapped by the formalism. We cannot allow the formalism to define the boundary of our imagination. The mathematical formalism is simply a tool or a language that allows us to describe and communicate our imagination. Sometimes, when you have a new idea or a new thought, you might find that you cannot say anything. Whatever you say is wrong because the proper mathematics or the proper language with which to describe the new idea or the new thought have yet to be invented. Indeed, really new physical ideas usually require a new mathematical formalism with which to describe them. This reminds me of a story about a tribe. The tribe only has four words for counting: one, two, three, and many-many. Imagine that a tribe member has an idea about two apples plus two apples and three apples plus three apples. He will have a hard time explaining his theory to other tribe members. This should be your feeling when you have a truly new idea. Although this book is entitled *Quantum field theory of many-body systems*, I hope that after reading the book the reader will see that quantum field theory is not everything. Nature's richness is not bounded by quantum field theory.

I would like to thank Margaret O'Meara for her proof-reading of many chapters of the book. I would also like to thank Anthony Zee, Michael Levin, Bas Overbosch, Ying Ran, Tiago Ribeiro, and Fei-Lin Wang for their comments and

[4] The concept of a classical particle breaks down in quantum theory. See a discussion in Section 2.2.

suggestions. Last, but not least, I would like to thank the copy-editor Dr. Julie Harris for her efforts in editing and polishing this book.

Lexington, MA Xiao-Gang Wen
October, 2003

CONTENTS

1

INTRODUCTION

1.1 More is different

- The collective excitations of a many-body system can be viewed as particles. However, the properties of those particles can be very different from the properties of the particles that form the many-body system.
- Guessing is better than deriving.
- Limits of classical computing.
- Our vacuum is just a special material.

A quantitative change can lead to a qualitative change. This philosophy is demonstrated over and over again in systems that contain many particles (or many degrees of freedom), such as solids and liquids. The physical principles that govern a system of a few particles can be very different from the physical principles that govern the collective motion of many-body systems. New physical concepts (such as the concepts of fermions and gauge bosons) and new physical laws and principles (such as the law of electromagnetism) can arise from the correlations of many particles (see Chapter 10).

Condensed matter physics is a branch of physics which studies systems of many particles in the 'condensed' (i.e. solid or liquid) states. The starting-point of current condensed matter theory is the Schrödinger equation that governs the motion of a number of particles (such as electrons and nuclei). The Schrödinger equation is mathematically complete. In principle, we can obtain all of the properties of any many-body system by solving the corresponding Schrödinger equation.

However, in practice, the required computing power is immense. In the 1980s, a workstation with 32 Mbyte RAM could solve a system of eleven interacting electrons. After twenty years the computing power has increased by 100-fold, which allows us to solve a system with merely two more electrons. The computing power required to solve a typical system of 10^{23} interacting electrons is beyond the imagination of the human brain. A classical computer made by all of the atoms in our universe would not be powerful enough to handle the problem.[5] Such an impossible computer could only solve the Schrödinger equation for merely about 100

[5] It would not even have enough memory to store a single state vector of such a system.

particles.[6] We see that an generic interacting many-body system is an extremely complex system. Practically, it is impossible to deduce all of its exact properties from the Schrödinger equation. So, even if the Schrödinger equation is the correct theory for condensed matter systems, it may not always be helpful for obtaining physical properties of an interacting many-body system.

Even if we do get the exact solution of a generic interacting many-body system, very often the result is so complicated that it is almost impossible to understand it in full detail. To appreciate the complexity of the result, let us consider a tiny interacting system of 200 electrons. The energy eigenvalues of the system are distributed in a range of about 200 eV. The system has at least 2^{200} energy levels. The level spacing is about $200\,\text{eV}/2^{200} = 10^{-60}\,\text{eV}$. Had we spent a time equal to the age of the universe in measuring the energy, then, due to the energy–time uncertainty relation, we could only achieve an energy resolution of order 10^{-33} eV. We see that the exact result of the interacting many-body system can be so complicated that it is impossible to check its validity experimentally in full detail.[7] To really understand a system, we need to understand the connection and the relationship between different phenomena of a system. Very often, the Schrödinger equation does not directly provide such an understanding.

As we cannot generally directly use the Schrödinger equation to understand an interacting system, we have to start from the beginning when we are faced with a many-body system. We have to treat the many-body system as a black box, just as we treat our mysterious and unknown universe. We have to guess a low-energy effective theory that directly connects different experimental observations, instead of deducing it from the Schrödinger equation. We cannot assume that the theory that describes the low-energy excitations bears any resemblance to the theory that describes the underlying electrons and nuclei.

This line of thinking is very similar to that of high-energy physics. Indeed, the study of strongly-correlated many-body systems and the study of high-energy physics share deep-rooted similarities. In both cases, one tries to find theories that connect one observed experimental fact to another. (Actually, connecting one observed experimental fact to another is almost the definition of a physical theory.) One major difference is that in high-energy physics we only have one 'material' (our vacuum) to study, while in condensed matter physics there are many different materials which may contain new phenomena not present in our vacuum (such as fractional statistics, non-abelian statistics, and gauge theories with all kinds of gauge groups).

---

[6] This raises a very interesting question—how does nature do its computation? How does nature figure out the state of 10^{23} particles one second later? It appears that the mathematics that we use is too inefficient. Nature does not do computations this way.

[7] As we cannot check the validity of the result obtained from the Schrödinger equation in full detail, our belief that the Schrödinger equation determines all of the properties of a many-body system is just a *faith*.

1.2 'Elementary' particles and physics laws are emergent phenomena

- Emergence—the first principle of many-body systems.
- Origin of 'elementary' particles.
- Origin of the 'beauty' of physics laws. (Why nature behaves reasonably.)

Historically, in our quest to understand nature, we have been misled by a fundamental (and incorrect) assumption that the vacuum is empty. We have (incorrectly) assumed that matter placed in a vacuum can always be divided into smaller parts. We have been dividing matter into smaller and smaller parts, trying to discover the smallest 'elementary' particles—the fundamental building block of our universe. We have been believing that the physics laws that govern the 'elementary' particles must be simple. The rich phenomena in nature come from these simple physics laws.

However, many-body systems present a very different picture. At high energies (or high temperatures) and short distances, the properties of the many-body system are controlled by the interaction between the atoms/molecules that form the system. The interaction can be very complicated and specific. As we lower the temperature, depending on the form of the interaction between atoms, a crystal structure or a superfluid state is formed. In a crystal or a superfluid, the only low-energy excitations are collective motions of the atoms. Those excitations are the sound waves. In quantum theory, all of the waves correspond to particles, and the particle that corresponds to a sound wave is called a phonon.[8] Therefore, at low temperatures, a new 'world' governed by a new kind of particle—phonons—emerges. The world of phonons is a simple and 'beautiful' world, which is very different from the original system of atoms/molecules.

Let us explain what we mean by 'the world of phonons is simple and beautiful'. For simplicity, we will concentrate on a superfluid. Although the interaction between atoms in a gas can be complicated and specific, the properties of emergent phonons at low energies are simple and universal. For example, all of the phonons have an energy-independent velocity, regardless of the form of the interactions between the atoms. The phonons pass through each other with little interaction despite the strong interactions between the atoms. In addition to the phonons, the superfluid also has another excitation called rotons. The rotons can interact with each other by exchanging phonons, which leads to a dipolar interaction with a force proportional to $1/r^4$. We see that not only are the phonons emergent, but even the physics laws which govern the low-energy world of the phonons and rotons are emergent. The emergent physics laws (such as the law of the dipolar interaction and the law of non-interacting phonons) are simple and beautiful.

---

[8] A crystal has three kinds of phonons, while a superfluid has only one kind of phonon.

I regard the law of $1/r^4$ dipolar interaction to be beautiful because it is not $1/r^3$, or $1/r^{4.13}$, or one of billions of other choices. It is precisely $1/r^4$, and so it is fascinating to understand why it has to be $1/r^4$. Similarly, the $1/r^2$ Coulomb law is also beautiful and fascinating. We will explain the emergence of the law of dipolar interaction in superfluids in the first half of this book and the emergence of Coulomb's law in the second half of this book.

If our universe itself was a superfluid and the particles that form the superfluid were yet to be discovered, then we would only know about low-energy phonons. It would be very tempting to regard the phonon as an elementary particle and the $1/r^4$ dipolar interaction between the rotons as a fundamental law of nature. It is hard to imagine that those phonons and the law of the $1/r^4$ dipolar interaction come from the particles that are governed by a very different set of laws.

We see that in many-body systems the laws that govern the emergent low-energy collective excitations are simple, and those collective excitations behave like particles. If we want to draw a connection between a many-body system and our vacuum, then we should connect the low-energy collective excitations in the many-body system to the 'elementary' particles (such as the photon and the electron) in the vacuum. But, in the many-body system, the collective excitations are not elementary. When we examine them at short length scales, a complicated non-universal atomic/molecular system is revealed. Thus, in many-body systems we have collective excitations (also called quasiparticles) at low energies, and those collective excitations very often do not become the building blocks of the model at high energies and short distances. The theory at the atomic scale is usually complicated, specific, and unreasonable. The simplicity and the beauty of the physics laws that govern the collective excitations do not come from the simplicity of the atomic/molecular model, but from the fact that those laws have to allow the collective excitations to survive at low energies. A generic interaction between collective excitations may give those excitations a large energy gap, and those excitations will be unobservable at low energies. The interactions (or physics laws) that allow gapless (or almost gapless) collective excitations to exist must be very special—and 'beautiful'.

If we believe that our vacuum can be viewed as a special many-body material, then we have to conclude that there are no 'elementary' particles. All of the so-called 'elementary' particles in our vacuum are actually low-energy collective excitations and they may not be the building blocks of the fundamental theory. The fundamental theory and its building blocks at high energies[9] and short distances are governed by a different set of physical laws. According to the point of view of emergence, those laws may be specific, non-universal, and complicated.

---

[9] Here, by high energies we mean the energies of the order of the Planck scale $M_P = 1.2 \times 10^{19}$ GeV.

The beautiful world and reasonable physical laws at low energies and long distances emerge as a result of a 'natural selection': the physical laws that govern the low-energy excitations should allow those excitations to exist at low energies. In a sense, the 'natural selection' explains why our world is reasonable.

Someone who knows both condensed matter physics and high-energy physics may object to the above picture because our vacuum appears to be very different from the solids and liquids that we know of. For example, our vacuum contains Dirac fermions (such as electrons and quarks) and gauge bosons (such as light), while solids and liquids seemingly do not contain these excitations. It appears that light and electrons are fundamental and cannot be emergent. So, to apply the picture of emergence in many-body systems to elementary particles, we have to address the following question: can gauge bosons and Dirac fermions emerge from a many-body system? Or, more interestingly, can gauge bosons and Dirac fermions emerge from a many-boson system?

The fundamental issue here is where do fermions and gauge bosons come from? What is the origin of light and fermions? Can light and fermions be an emergent phenomenon? We know that massless (or gapless) particles are very rare in nature. If they exist, then they must exist for a reason. But what is the reason behind the existence of the massless photons and nearly massless fermions (such as electrons)? (The electron mass is smaller than the natural scale—the Planck mass—by a factor of 10^{22} and can be regarded as zero for our purpose.) Can many-body systems provide an answer to the above questions?

In the next few sections we will discuss some basic notions in many-body systems. In particular, we will discuss the notion that leads to gapless excitations and the notion that leads to emergent gauge bosons and fermions from local bosonic models. We will see that massless photons and massless fermions can be emergent phenomena.

1.3 Corner-stones of condensed matter physics

- Landau's symmetry-breaking theory (plus the renormalization group theory) and Landau's Fermi liquid theory form the foundation of traditional condensed matter physics.

The traditional many-body theory is based on two corner-stones, namely Landau's Fermi liquid theory and Landau's symmetry-breaking theory (Landau, 1937; Ginzburg and Landau, 1950). The Fermi liquid theory is a perturbation theory around a particular type of ground state—the states obtained by filling single-particle energy levels. It describes metals, semiconductors, magnets, superconductors, and superfluids. Landau's symmetry-breaking theory points out that the reason that different phases are different is because they have different symmetries. A phase transition is simply a transition that changes the symmetry. Landau's symmetry-breaking theory describes almost all of the known phases, such as solid phases, ferromagnetic and anti-ferromagnetic phases, superfluid phases, etc., and all of the phase transitions between them.

Instead of the origin of light and fermions, let us first consider a simpler problem of the origin of phonons. Using Landau's symmetry-breaking theory, we can understand the origin of the gapless phonon. In Landau's symmetry-breaking theory, a phase can have gapless excitations if the ground state of the system has a special property called spontaneous breaking of the continuous symmetry (Nambu, 1960; Goldstone, 1961). Gapless phonons exist in a solid because the solid breaks the continuous translation symmetries. There are precisely three kinds of gapless phonons because the solid breaks three translation symmetries in the x, y, and z directions. Thus, we can say that the origin of gapless phonons is the translational symmetry breaking in solids.

It is quite interesting to see that our understanding of a gapless excitation—phonon—is rooted in our understanding of the phases of matter. Knowing light to be a massless excitation, one may perhaps wonder if light, just like a phonon, is also a Nambu–Goldstone mode from a broken symmetry. However, experiments tell us that a gauge boson, such as light, is really different from a Nambu–Goldstone mode in $3 + 1$ dimensions.

In the late 1970s, we felt that we understood, at least in principle, all of the physics about phases and phase transitions. In Landau's symmetry-breaking theory, if we start with a purely bosonic model, then the only way to get gapless excitations is via spontaneous breaking of a continuous symmetry, which will lead to gapless *scalar bosonic* excitations. It seems that there is no way to obtain gapless gauge bosons and gapless fermions from symmetry breaking. This may be the reason why people think that our vacuum (with massless gauge bosons and nearly-gapless fermions) is very different from bosonic many-body systems (which were believed to contain only gapless scalar bosonic collective excitations, such as phonons). It seems that there does not exist any order that gives rise to massless light and massless fermions. Due to this, we put light and fermions into a different category to phonons. We regard them as elementary and introduce them by hand into our theory of nature.

However, if we really believe that light and fermions, just like phonons, exist for a reason, then such a reason must be a certain order in our vacuum that protects their masslessness.[10] Now the question is what kind of order can give rise to light and fermions, and protect their masslessness? From this point of view, the very

[10] Here we have already assumed that light and fermions are not something that we place in an empty vacuum. Our vacuum is more like an 'ocean' which is not empty. Light and fermions are collective excitations that correspond to certain patterns of 'water' motion.

existence of light and fermions indicates that our understanding of the states of matter is incomplete. We should deepen and expand our understanding of the states of matter. There should be new states of matter that contain new kinds of orders. The new orders will produce light and fermions, and protect their masslessness.

1.4 Topological order and quantum order

- There is a new world beyond Landau's theories. The new world is rich and exciting.

Our understanding of this new kind of order starts at an unexpected place—fractional quantum Hall (FQH) systems. The FQH states discovered in 1982 (Tsui *et al.*, 1982; Laughlin, 1983) opened a new chapter in condensed matter physics. What is really new in FQH states is that we have lost the two cornerstones of the traditional many-body theory. Landau's Fermi liquid theory does not apply to quantum Hall systems due to the strong interactions and correlations in those systems. What is more striking is that FQH systems contain many *different* phases at zero temperature which have the *same* symmetry. Thus, those phases cannot be distinguished by symmetries and cannot be described by Landau's symmetry-breaking theory. We suddenly find that we have nothing in the traditional many-body theory that can be used to tackle the new problems. Thus, theoretical progress in the field of strongly-correlated systems requires the introduction of new mathematical techniques and physical concepts, which go beyond the Fermi liquid theory and Landau's symmetry-breaking principle.

In the field of strongly-correlated systems, the developments in high-energy particle theory and in condensed matter theory really feed upon each other. We have seen a lot of field theory techniques, such as the nonlinear σ-model, gauge theory, bosonization, current algebra, etc., being introduced into the research of strongly-correlated systems and random systems. This results in a very rapid development of the field and new theories beyond the Fermi liquid theory and Landau's symmetry-breaking theory. This book is an attempt to cover some of these new developments in condensed matter theory.

One of the new developments is the introduction of quantum/topological order. As FQH states cannot be described by Landau's symmetry-breaking theory, it was proposed that FQH states contain a new kind of order—topological order (Wen, 1990, 1995). Topological order is new because it cannot be described by symmetry breaking, long-range correlation, or local order parameters. None of the usual tools that we used to characterize a phase apply to topological order. Despite this, topological order is not an empty concept because it can be characterized by a new set of tools, such as the number of degenerate ground states (Haldane

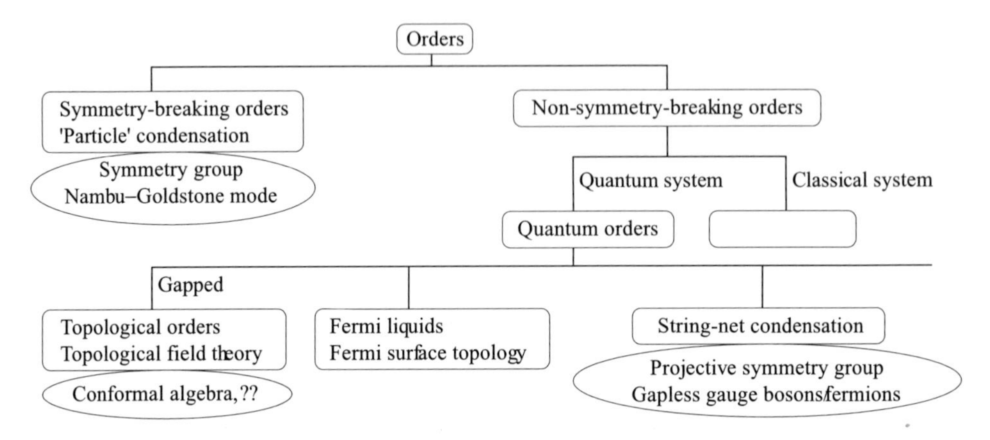

FIG. 1.1. A classification of different orders in matter (and in a vacuum).

and Rezayi, 1985), quasiparticle statistics (Arovas *et al.*, 1984), and edge states (Halperin, 1982; Wen, 1992).

It was shown that the ground-state degeneracy of a topologically-ordered state is robust against *any* perturbations (Wen and Niu, 1990). Thus, the ground-state degeneracy is a universal property that can be used to characterize a phase. The existence of topologically-degenerate ground states proves the existence of topological order. Topological degeneracy can also be used to perform fault-tolerant quantum computations (Kitaev, 2003).

The concept of topological order was recently generalized to quantum order (Wen, 2002c) to describe new kinds of orders in gapless quantum states. One way to understand quantum order is to see how it fits into a general classification scheme of orders (see Fig. 1.1). First, different orders can be divided into two classes: symmetry-breaking orders and non-symmetry-breaking orders. The symmetry-breaking orders can be described by a local order parameter and can be said to contain a condensation of point-like objects. The amplitude of the condensation corresponds to the order parameter. All of the symmetry-breaking orders can be understood in terms of Landau's symmetry-breaking theory. The non-symmetry-breaking orders cannot be described by symmetry breaking, nor by the related local order parameters and long-range correlations. Thus, they are a new kind of order. If a quantum system (a state at zero temperature) contains a non-symmetry-breaking order, then the system is said to contain a non-trivial quantum order. We see that a quantum order is simply a non-symmetry-breaking order in a quantum system.

Quantum orders can be further divided into many subclasses. If a quantum state is gapped, then the corresponding quantum order will be called the topological order. The low-energy effective theory of a topologically-ordered state will be a topological field theory (Witten, 1989). The second class of quantum orders appears in Fermi liquids (or free fermion systems). The different quantum orders in Fermi liquids are classified by the Fermi surface topology (Lifshitz, 1960). The third class of quantum orders arises from a condensation of nets of strings (or simply string-net condensation) (Wen, 2003a; Levin and Wen, 2003; Wen, 2003b). This class of quantum orders shares some similarities with the symmetry-breaking orders of 'particle' condensation.

We know that different symmetry-breaking orders can be classified by symmetry groups. Using group theory, we can classify all of the 230 crystal orders in three dimensions. The symmetry also produces and protects gapless collective excitations—the Nambu–Goldstone bosons—above the symmetry-breaking ground state. Similarly, different string-net condensations (and the corresponding quantum orders) can be classified by mathematical object called projective symmetry group (PSG) (Wen, 2002c). Using PSG, we can classify over 100 different two-dimensional spin liquids that all have the same symmetry. Just like the symmetry group, the PSG can also produce and protect gapless excitations. However, unlike the symmetry group, the PSG produces and protects gapless gauge bosons and fermions (Wen, 2002a,c; Wen and Zee, 2002). Because of this, we can say that light and massless fermions can have a unified origin; they can emerge from string-net condensations.

In light of the classification of the orders in Fig. 1.1, this book can be divided into two parts. The first part (Chapters 3–5) deals with the symmetry-breaking orders from 'particle' condensations. We develop the effective theory and study the physical properties of the gapless Nambu–Goldstone modes from the fluctuations of the order parameters. This part describes 'the origin of sound' and other Nambu–Goldstone modes. It also describes the origin of the law of the $1/r^4$ dipolar interaction between rotons in a superfluid. The second part (Chapters 7–10) deals with the quantum/topological orders and string-net condensations. Again, we develop the effective theory and study the physical properties of low-energy collective modes. However, in this case, the collective modes come from the fluctuations of condensed string-nets and give rise to gauge bosons and fermions. So, the second part provides 'an origin of light and electrons', as well as other gauge bosons and fermions. It also provides an origin of the $1/r^2$ Coulomb law (or, more generally, the law of electromagnetism).

1.5 Origin of light and fermions

- The string-net condensation provides an answer to the origin of light and fermions. It unifies gauge interactions and Fermi statistics.

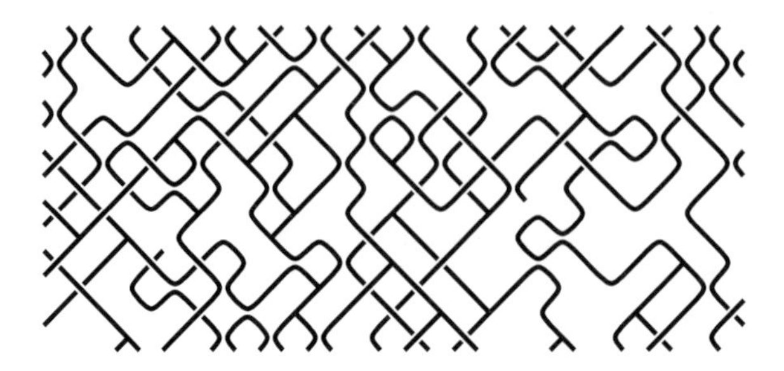

FIG. 1.2. Our vacuum may be a state filled with a string-net. The fluctuations of the string-net give rise to gauge bosons. The ends of the strings correspond to electrons, quarks, etc.

We used to believe that, to have light and fermions in our theory, we have to introduce by hand a fundamental $U(1)$ gauge field and anti-commuting fermion fields, because at that time we did not know of any collective modes that behave like gauge bosons and fermions. However, due to the advances over the last twenty years, we now know how to construct *local bosonic systems* that have emergent *unconfined* gauge bosons and/or fermions (Foerster *et al.*, 1980; Kalmeyer and Laughlin, 1987; Wen *et al.*, 1989; Read and Sachdev, 1991; Wen, 1991a; Moessner and Sondhi, 2001; Motrunich and Senthil, 2002; Wen, 2002a; Kitaev, 2003; Levin and Wen, 2003). In particular, one can construct ugly bosonic spin models on a cubic lattice whose low-energy effective theory is the beautiful quantum electrodynamics (QED) and quantum chromodynamics (QCD) with emergent photons, electrons, quarks, and gluons (Wen, 2003b).

This raises the following issue: do light and fermions in nature come from a fundamental $U(1)$ gauge field and anti-commuting fields as in the $U(1) \times SU(2) \times SU(3)$ standard model, or do they come from a particular quantum order in our vacuum? Is Coulomb's law a fundamental law of nature or just an emergent phenomenon? Clearly, it is more natural to assume that light and fermions, as well as Coulomb's law, come from a quantum order in our vacuum. From the connections between string-net condensation, quantum order, and massless gauge/fermion excitations, we see that string-net condensation provides a way to unify light and fermions. It is very tempting to propose the following possible answers to the three fundamental questions about light and fermions.

What are light and fermions?

Light is the fluctuation of condensed string-nets (of arbitrary sizes). Fermions are ends of condensed strings.

Where do light and fermions come from?

Light and fermions come from the collective motions of string-nets that fill the space(see Fig. 1.2).

Why do light and fermions exist?
Light and fermions exist because our vacuum happens to have a property called string-net condensation.

Had our vacuum chosen to have 'particle' condensation, then there would be only Nambu–Goldstone bosons at low energies. Such a universe would be very boring. String-net condensation and the resulting light and fermions provide a much more interesting universe, at least interesting enough to support intelligent life to study the origin of light and fermions.

1.6 Novelty is more important than correctness

- The Dao that can be stated cannot be eternal Dao. The Name that can be named cannot be eternal Name. The Nameless is the origin of universe. The Named is the mother of all matter.[11]
- What can be stated cannot be novel. What cannot be stated cannot be correct.

In this introduction (and in some parts of this book), I hope to give the reader a sense of where we come from, where we stand, and where we are heading in theoretical condensed matter physics. I am not trying to summarize the generally accepted opinions here. Instead, I am trying to express my personal and purposely exaggerated opinions on many fundamental issues in condensed matter physics and high-energy physics. These opinions and pictures may not be correct, but I hope they are stimulating. From our experience of the history of physics, we can safely assume that none of the current physical theories are completely correct. (According to Lao Zi, the theory that can be written down cannot be the eternal theory, because it is limited by the mathematical symbols that we used to write down the theory.) The problem is to determine in which way the current theories are wrong and how to fix them. Here we need a lot of imagination and stimulation.

[11] These are the first four sentences of *Dao de jing* written by a Chinese philosopher Lao Zi over 2500 years ago. The above is a loose direct translation. Dao has meanings of 'way', 'law', 'conduct', etc. There are many very different translations of *Dao de jing*. It is interesting to search the Web and compare those different translations. The following is a translation in the context of this book. 'The physical theory that can be formulated cannot be the final ultimate theory. The classification that can be implemented cannot classify everything. The unformulatable ultimate theory does exist and governs the creation of the universe. The formulated theories describe the matter we see every day.'

1.7 Remarks: evolution of the concept of elementary particles

- As time goes by, the status of elementary particles is downgraded from the building blocks of everything to merely collective modes of, possibly, a lowly bosonic model.

The Earth used to be regarded as the center of the universe. As times went by, its status was reduced to merely one of the billions of planets in the universe. It appears that the concept of elementary particles may have a similar fate.

At the beginning of human civilization, people realized that things can be divided into smaller and smaller parts. Chinese philosophers theorized that the division could be continued indefinitely, and hence that there were no elementary particles. Greek philosophers assumed that the division could not be continued indefinitely. As a result, there exist ultimate and indivisible particles—the building blocks of all matter. This may be the first concept of elementary particles. Those ultimate particles were called *atomos*. A significant amount of scientific research has been devoted to finding these *atomos*.

Around 1900, chemists discovered that all matter is formed from a few dozen different kinds of particles. People jumped the gun and named them atoms. After the discovery of the electron, people realized that elementary particles are smaller than atoms. Now, many people believe that photons, electrons, quarks, and a few other particles are elementary particles. Those particles are described by a field theory which is called the $U(1) \times SU(2) \times SU(3)$ standard model.

Although the $U(1) \times SU(2) \times SU(3)$ standard model is a very successful theory, now most high-energy physicists believe that it is not the ultimate theory of everything. The $U(1) \times SU(2) \times SU(3)$ standard model may be an effective theory that emerges from a deeper structure. The question is from which structure may the standard model emerge?

One proposal is the grand unified theories in which the $U(1) \times SU(2) \times SU(3)$ gauge group is promoted to $SU(5)$ or even bigger gauge groups (Georgi and Glashow, 1974). The grand unified theories group the particles in the $U(1) \times SU(2) \times SU(3)$ standard model into very nice and much simpler structures. However, I would like to remark that I do not regard the photon, electron, and other elementary particles to be emergent within the grand unified theories. In the grand unified theories, the gauge structure and the Fermi statistics were fundamental in the sense that the only way to have gauge bosons and fermions was to introduce vector gauge fields and anti-commuting fermion fields. Thus, to have the photon, electron, and other elementary particles, we had to introduce by hand the corresponding gauge fields and fermion fields. Therefore, the gauge bosons and fermions were added by hand into the grand unified theories; they did not emerge from a simpler structure.

The second proposal is the superstring theory (Green *et al.*, 1988; Polchinski, 1998). Certain superstring models can lead to the effective $U(1) \times SU(2) \times SU(3)$ standard model plus many additional (nearly) massless excitations. The gauge bosons and the graviton are emergent because the superstring theory itself contains no gauge fields. However, the Fermi statistics are not emergent. The electron and quarks come from the anti-commuting fermion fields on a $(1+1)$-dimensional world sheet. We see that, in the superstring theory, the gauge bosons and the gauge structures are not fundamental, but the Fermi statistics and the fermions are still fundamental.

Recently, people realized that there might be a third possibility—string-net condensation. Banks *et al.* (1977) and Foerster *et al.* (1980) first pointed out that light can emerge as low-energy collective modes of a local bosonic model. Levin and Wen (2003) pointed out that even three-dimensional fermions can emerge from a local bosonic model as the ends of condensed strings. Combining the two results, we find that the photon, electron, quark, and gluon (or, more precisely, the QED and the QCD part of the $U(1) \times SU(2) \times SU(3)$ standard model) can emerge from a local bosonic model (Wen, 2002a, 2003b) if the bosonic model has a string-net condensation. This proposal is attractive because the gauge bosons and fermions have a unified origin. In the string-net condensation picture, neither the gauge structure nor the Fermi statistics are fundamental; all of the elementary particles are emergent.

However, the third proposal also has a problem: we do not yet know how to produce the $SU(2)$ part of the standard model due to the chiral fermion problem. There are five deep mysteries in nature, namely, identical particles, Fermi statistics, gauge structure, chiral fermions, and gravity. The string-net condensation only provides an answer to the first three mysteries; there are two more to go.

2

PATH INTEGRAL FORMULATION OF QUANTUM MECHANICS

This book is about the quantum behavior of many-body systems. However, the standard formulation of the quantum theory in terms of wave functions and the Schrödinger equation is not suitable for many-body systems. In this chapter, we introduce a semiclassical picture and path integral formalism for quantum theory. The path integral formalism can be easily applied to many-body systems. Here, we will use one-particle systems as concrete examples to develop the formalism. We will also apply a path integral formalism to study quantum friction, simple quantum circuits, etc.

2.1 Semiclassical picture and path integral

- A semiclassical picture and a path integral formulation allow us to visualize quantum behavior. They give us a global view of a quantum system.

When we are thinking of a physical problem or trying to understand a phenomenon, it is very important to have a picture in our mind to mentally visualize the connection between different pieces of a puzzle. Mental visualization is easier when we consider a classical system because the picture of a classical system is quite close to what we actually see in our everyday life. However, when we consider a quantum system, visualization is much harder. This is because the quantum world does not resemble what we see every day. In the classical world, we see various objects. With a little abstraction, we view these objects as collections of particles. The concept of a particle is the most important concept in classical physics. It is so simple and plain that people take it for granted and do not bother to formulate a physical law to state such an obvious truth. However, it is this obvious truth which turns out to be false in the quantum world. The concept of a particle with a position and velocity simply does not exist in the quantum world. So, the challenge is how to visualize anything in the quantum world where the concept of a particle (i.e. the building block of objects) does not exist.

The path integral formalism is an attempt to use a picture of the classical world to describe the quantum world. In other words, it is a bridge between the classical world (where we have our experiences and pictures) and the quantum world (which represents realities). The path integral formalism will help us to visualize quantum behavior in terms of the pictures of corresponding classical systems.

2.1.1 Propagator of a particle

- Concept of a propagator.
- The pole structure of a propagator in ω space.

Consider a particle in one dimension:

$$H = \frac{1}{2m}p^2 + V(x) \tag{2.1.1}$$

The time evolution operator

$$U(t_b, t_a) = \mathrm{e}^{-\mathrm{i}\hbar^{-1}(t_b - t_a)H} = \mathrm{e}^{-\mathrm{i}(t_b - t_a)H}$$

completely determines the behavior and properties of the system. In this section we will always assume that $t_b > t_a$. Throughout this book we shall set $\hbar = 1$.

The matrix elements of U in the coordinate basis are

$$\mathrm{i}G(x_b, t_b, x_a, t_a) \equiv \langle x_b | U(t_b, t_a) | x_a \rangle$$

which represent the probability amplitude of finding a particle at position x_b at time t_b if the particle was at x_a at time t_a. Here $G(x_b, t_b, x_a, t_a)$ is called the propagator and it gives a full and complete description of our one-particle quantum system.

It is clear that the propagator G satisfies the Schrödinger equation

$$\mathrm{i}\partial_t G(x, t, x_a, t_a) = HG(x, t, x_a, t_a)$$

with the initial condition

$$G(x, t_a, x_a, t_a) = (-\mathrm{i})\delta(x - x_a)$$

Solving the Schrödinger equation, we find the free-particle propagator for $V(x) = 0$

$$G(x_b, x_a, t) = (-\mathrm{i}) \left(\frac{m}{2\pi \mathrm{i} t} \right)^{1/2} \exp\left[\frac{\mathrm{i}m(x_b - x_a)^2}{2t} \right] \tag{2.1.2}$$

where $t = t_b - t_a$.

We can also use the energy eigenstates $|n\rangle$ (with energy ϵ_n) to expand U. The propagator in the new basis has the following simpler form:

$$G_E(n_b, t_b, n_a, t_a) \equiv (-\mathrm{i})\langle n_b|U(t_b, t_a)|n_a\rangle = (-\mathrm{i})\mathrm{e}^{-\mathrm{i}\epsilon_n(t_b-t_a)}\delta_{n_b,n_a}$$

In the frequency space, it is given by

$$\begin{aligned} G_E(n_b, n_a, \omega) &\equiv \int_0^\infty \mathrm{d}t\, G_E(n_b, t_a + t, n_a, t_a)\mathrm{e}^{\mathrm{i}t\omega - 0^+ t} \\ &= \frac{1}{\omega - \epsilon_{n_a} + \mathrm{i}0^+}\delta_{n_b,n_a} \end{aligned}$$

which has a simple pole at each energy eigenvalue. Here 0^+ is an infinitely small, positive number which was introduced to make the integrals converge. The propagator in coordinate space has a similar structure:

$$\begin{aligned} G(x_b, x_a, t) &= \sum_n \langle x_b|n\rangle\langle n|x_a\rangle G_E(n, n, t) = \sum_n \langle x_b|n\rangle\langle n|x_a\rangle(-\mathrm{i})\mathrm{e}^{-\mathrm{i}\epsilon_n t} \\ G(x_b, x_a, \omega) &= \sum_n \frac{\langle x_b|n\rangle\langle n|x_a\rangle}{\omega - \epsilon_n + \mathrm{i}0^+} \end{aligned}$$

We see that, by analyzing the pole structure of $G(x_b, x_a, \omega)$, we can obtain information about the energy eigenvalues.

To re-obtain the propagator in time

$$G_E(n_b, n_a, t) = \int \frac{\mathrm{d}\omega}{2\pi} \frac{1}{\omega - \epsilon_{n_a} + \mathrm{i}0^+}\delta_{n_b,n_a}\mathrm{e}^{-\mathrm{i}t\omega}$$

we need to choose the contour in the lower half of the complex ω-plane if $t > 0$, and in the upper half of the complex ω-plane if $t < 0$, so that the integral along the contour is finite. As 0^+ makes the pole appear slightly below the real axis of the complex ω-plane, we find that

$$G_E(n_b, n_a, t) = (-\mathrm{i})\mathrm{e}^{-\mathrm{i}\epsilon_n(t_b-t_a)}\delta_{n_b,n_a}\Theta(t)$$

where

$$\Theta(t) = \begin{cases} 1, & t > 0 \\ 0, & t < 0 \end{cases} \tag{2.1.3}$$

Problem 2.1.1.
Show that, in the frequency space, we have

$$G(x_b, x_a, \omega) = \sum_n \frac{\psi_n(x_b)\psi_n^\dagger(x_a)}{\omega - \epsilon_n}$$

where ψ_n are the energy eigenfunctions. The propagator for a harmonic oscillator has the form

$$G(x_b, t, x_a, 0) = (-\mathrm{i}) \left(\frac{m\omega_0}{2\pi \mathrm{i} \sin(t\omega_0)} \right)^{1/2} \exp\left(\frac{\mathrm{i} m\omega_0}{2\pi \sin(t\omega_0)} [(x_b^2 + x_a^2)\cos(\omega_0 t) - 2x_b x_a] \right)$$

Study and explain the pole structure of $G(0, 0, \omega)$ for the harmonic oscillator. (Hint: Try to expand $G(0, t, 0, 0)$ in the form $\sum C_n \mathrm{e}^{-\mathrm{i}t\epsilon_n}$.)

Problem 2.1.2.
Find the propagator $G_E(k_b, k_a, \omega)$ for a free particle in three dimensions in the wave vector and frequency space.

2.1.2 Path integral representation of the propagator

- The path integral is a particular representation of the law of superposition in quantum physics.

The spirit of the path integral formulation of a quantum system is very simple. Consider that a particle propagates from position $|x_a\rangle$ to position $|x_b\rangle$ via an intermediate state $|n\rangle$, $n = 1, 2, ...$ Let the amplitude from $|x_a\rangle$ to $|x_b\rangle$ via state $|n\rangle$ be given by A_n; then the total amplitude from $|x_a\rangle$ to $|x_b\rangle$ is $\sum_n A_n$. Now imagine that we divide the propagation from $|x_a\rangle$ to $|x_b\rangle$ into many time slices. Each time slice has its own set of intermediate states. The propagation from $|x_a\rangle$ to $|x_b\rangle$ can be viewed as the sum of the paths that go through the intermediate states on each time slice. This leads to a picture of the path integral representation of the propagation amplitude.

To find a mathematical representation of the above picture, we note that the amplitude from $|x_a\rangle$ to $|x_b\rangle$ is described by the matrix elements of the time evolution operator U in position space. Here U satisfies

$$U(t_b, t_a) = U(t_b, t)U(t, t_a)$$

Thus, the propagator satisfies

$$\mathrm{i}G(x_b, t_b, x_a, t_a) = \int \mathrm{d}x \; \mathrm{i}G(x_b, t_b, x, t)\mathrm{i}G(x, t, x_a, t_a) \tag{2.1.4}$$

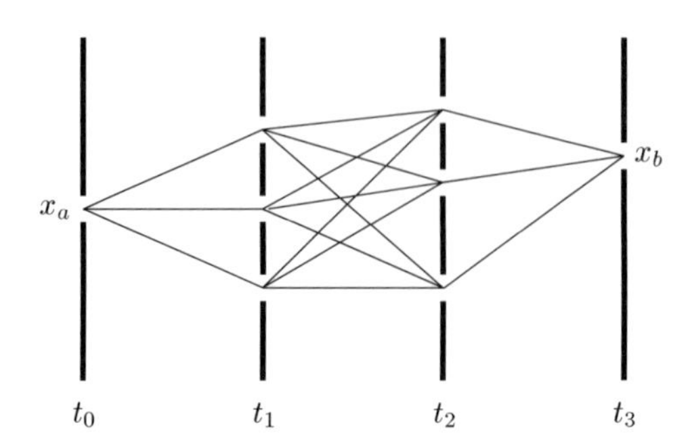

FIG. 2.1. The total amplitude is the sum of all amplitudes associated with the paths connecting x_a and x_b.

Let us divide the time interval $[t_a, t_b]$ into N equal segments, each of length $\Delta t = \frac{t_b - t_a}{N}$. By using eqn (2.1.4) $N - 1$ times, we find that

$$\begin{aligned} \mathrm{i}G(x_b, t_b, x_a, t_a) &= \int \mathrm{d}x_1 ... \mathrm{d}x_{N-1} \prod_{j=1}^{N} \mathrm{i}G(x_j, t_j, x_{j-1}, t_{j-1}) \\ &= A^N \int \prod_i \mathrm{d}x_i \; \exp\left(\mathrm{i} \sum \Delta t L(t_j, \frac{x_j + x_{j-1}}{2}, \frac{x_j - x_{j-1}}{\Delta t}) \right) \\ &\equiv \int \mathcal{D}[x(t)] \; \mathrm{e}^{\,\mathrm{i} \int \mathrm{d}t \; L(t, x, \dot{x})} \end{aligned} \tag{2.1.5}$$

where $(t_0, x_0) = (t_a, x_a)$, $(t_N, x_N) = (t_b, x_b)$, and

$$\mathrm{i} \Delta t L(t_i, \frac{x_i + x_{i-1}}{2}, \frac{x_i - x_{i-1}}{\Delta t}) = \ln[\mathrm{i}G(x_i, t_i, x_{i-1}, t_{i-1})/A] \tag{2.1.6}$$

Assuming that, with the proper choice of A, the definition of L in eqn (2.1.6) makes sense in the $\Delta t \to 0$ limit, then eqn (2.1.5) is the path integral representation of the propagator.

In the path integral representation, each path is assigned an amplitude $\mathrm{e}^{\,\mathrm{i} \int \mathrm{d}t \; L}$. The propagator is then just the sum of all of the amplitudes associated with the paths connecting x_a and x_b (Fig. 2.1). Such a summation is an infinite-dimensional integral, and eqn (2.1.5) defines one way in which to evaluate such an infinite-dimensional integral.

Now let us calculate $L(t, x, \dot{x})$ for our one-particle system (2.1.1). For a free particle system, we can use eqn (2.1.2) to calculate $L(t, x, \dot{x})$. In the following we will calculate L for a more general one-particle system described by $H =$

$\frac{p^2}{2m} + V(x)$. Note that, for small Δt, we have

$$\exp\left(-\mathrm{i}[\frac{p^2}{2m} + V(x)]\Delta t\right) = \exp\left(-\mathrm{i}\frac{p^2}{2m}\Delta t\right)\exp\left(-\mathrm{i}V(x)\Delta t\right) + O(\Delta t^2)$$

Thus,

$$\begin{aligned}
&\mathrm{i}G(x_{i-1}, t_{i-1}, x_i, t_i) \\
&= \langle x_i|\exp\left(-\mathrm{i}\frac{p^2}{2m}\Delta t\right)\exp\left(-\mathrm{i}V(x)\Delta t\right)|x_{i-1}\rangle + O(\Delta t^2) \\
&= \langle x_i|\exp\left(-\mathrm{i}\frac{p^2}{2m}\Delta t\right)|x_{i-1}\rangle\exp\left(-\mathrm{i}V(x_{i-1})\Delta t\right) + O(\Delta t^2)
\end{aligned}$$

Inserting $\int |p\rangle\frac{\mathrm{d}p}{2\pi}\langle p| = 1$ and using $\langle p|x\rangle = \exp(-\mathrm{i}px)$, we find that

$$\begin{aligned}
&\mathrm{i}G(x_{i-1}, t_{i-1}, x_i, t_i) \\
&= \int \frac{\mathrm{d}p_i}{2\pi}\exp\left(\mathrm{i}[p_i\frac{x_i - x_{i-1}}{\Delta t} - \frac{p_i^2}{2m} - V(x_i)]\Delta t\right) + O(\Delta t^2) \qquad (2.1.7) \\
&= \left(\frac{m}{2\pi\mathrm{i}\Delta t}\right)^{1/2}\exp\left(\mathrm{i}[\frac{m}{2}\frac{(x_i - x_{i-1})^2}{\Delta t^2} - V(\frac{x_i + x_{i-1}}{2})]\Delta t\right) \qquad (2.1.8)
\end{aligned}$$

From eqn (2.1.6), we see that

$$\begin{aligned}
L(x, \dot{x}) &= \frac{m}{2}\dot{x}^2 - V(x) \qquad (2.1.9) \\
A &\equiv \left(\frac{m}{2\pi\mathrm{i}\Delta t}\right)^{1/2}
\end{aligned}$$

Thus, the path integral representation of the propagator is given by

$$\begin{aligned}
\mathrm{i}G(x_b, t_b, x_a, t_a) &= \int \mathrm{d}x_1 ... \mathrm{d}x_{N-1}\prod_{j=1}^{N}\mathrm{i}G(x_j, t_j, x_{j-1}, t_{j-1}) \\
&= \left(\frac{m}{2\pi\mathrm{i}\Delta t}\right)^{N/2}\int \mathrm{d}x_1 ... \mathrm{d}x_{N-1}\ \mathrm{e}^{\mathrm{i}\int \mathrm{d}t\ [\frac{m}{2}\dot{x}^2 - V(x)]} \qquad (2.1.10)
\end{aligned}$$

The expression for $L(x, \dot{x})$ calculated above is simply the corresponding Lagrangian of the classical system. Thus, in principle, once we know a classical system (described through a Lagrangian), then the corresponding quantum system can be obtained through a path integral using the same Lagrangian.

We may also insert eqn (2.1.7) into eqn (2.1.5) to obtain

$$\begin{aligned} &\mathrm{i}G(x_b, t_b, x_a, t_a) \\ &= \left(\frac{1}{2\pi}\right)^N \int \mathrm{d}x_1 ... \mathrm{d}x_{N-1} \mathrm{d}p_1 ... \mathrm{d}p_N \ \exp\left(\mathrm{i}\sum_{i=1}^{N} [p_i \frac{x_i - x_{i-1}}{\Delta t} - \frac{p_i^2}{2m} - V(x_i)]\right) \\ &\equiv \int \mathcal{D}[x(t)]\mathcal{D}[p(t)] \ \mathrm{e}^{\mathrm{i}\int \mathrm{d}t \ [p\dot{x} - H(x,p)]} \end{aligned} \tag{2.1.11}$$

which is a path integral in phase space. Here $H(x, p)$ is the classical Hamiltonian. If we define the phase-space measure to be

$$\mathcal{D}[x(t)]\mathcal{D}[p(t)] \equiv \prod_1^{N-1} \mathrm{d}x_i \prod_1^{N} \frac{\mathrm{d}p_i}{2\pi} \tag{2.1.12}$$

then the Hamiltonian H is identical to the classical one. It appears that the phase-space path integral is more fundamental than the coordinate-space path integral, because the annoying coefficient A (see eqn (2.1.9)) does not appear in the measure of the path integral. We also note that the above phase-space measure has $(N - 1)$ $\mathrm{d}x$ integrals and N $\mathrm{d}p$ integrals.

We know that adding a total time derivative term to the Lagrangian, namely $L \to L + \frac{\mathrm{d}f(x)}{\mathrm{d}t}$, does not affect classical dynamics. However, such a change does affect the quantum propagator, but in a trivial way:

$$G(x_b, t_b, x_a, t_a) \to G(x_b, t_b, x_a, t_a)\mathrm{e}^{\mathrm{i}[f(x_b) - f(x_a)]}$$

Problem 2.1.3.

Coherent-state path integral:

Consider a harmonic oscillator $H = \omega\hat{a}^\dagger\hat{a}$. A coherent state $|a\rangle$ labeled by a complex number a is an eigenstate of $\hat{a}$: $\hat{a}|a\rangle = a|a\rangle$. We may define a coherent-state propagator by

$$\mathrm{i}G(a_b, t_b, a_a, t_a) = \langle a_b|U(t_b, t_a)|a_a\rangle$$

Using the completeness of the coherent state

$$\int \frac{\mathrm{d}^2 a}{\pi} |a\rangle\langle a| = 1$$

(where $\mathrm{d}^2 a = \mathrm{dRe}(a) \ \mathrm{dIm}(a)$) and the inner product

$$\langle a|a'\rangle = \mathrm{e}^{-|a|^2/2}\mathrm{e}^{-|a'|^2/2}\mathrm{e}^{a^* a'}$$

show that the path integral representation of G is

$$\mathrm{i}G(a_b, t_b, a_a, t_a) = \int \pi^{-N+1}\mathcal{D}^2[a(t)] \ \mathrm{e}^{\mathrm{i}\int_{t_a}^{t_b} \mathrm{d}t \ [\mathrm{i}\frac{1}{2}(a^\dagger\dot{a} - a\dot{a}^\dagger) - \omega a^\dagger a]}$$

where $\mathcal{D}^2[a(t)] = \prod_{j=1}^{N-1} \mathrm{d}^2 a_j$. Show that the Lagrangian in the coherent-state path integral

$$L = \mathrm{i}\frac{1}{2}(a^\dagger \dot{a} - a\dot{a}^\dagger) - \omega a^\dagger a$$

is simply the Lagrangian in the phase-space path integral

$$L = p\dot{x} - H$$

up to a total time derivative term.

2.1.3 Path integral representation of the partition function

• The path integral can also be used to study statistical systems.

The partition function is very useful when we consider statistical properties of our quantum system at finite temperatures. The partition function of our one-particle system (2.1.1) is given by

$$Z(\beta) = \mathrm{Tr}(\mathrm{e}^{-\beta H}) = \int \mathrm{d}x\, \mathcal{G}(x, x, \beta)$$

where $\beta = \frac{1}{T}$ is the inverse of the temperature and

$$\mathcal{G}(x_b, x_a, \beta) \equiv \langle x_b | \mathrm{e}^{-\beta H} | x_a \rangle$$

can be regarded as a propagator in imaginary time. The calculation in the last section can be repeated, and we find the following path integral representation of $\mathcal{G}$:

$$\mathcal{G}(x_b, x_a, \tau) = A_\tau^N \int \mathcal{D}[x(\tau)]\ \mathrm{e}^{-\int_0^\tau [\frac{m}{2}\left(\frac{\mathrm{d}x}{\mathrm{d}\tau}\right)^2 + V(x)] d\tau'} \tag{2.1.13}$$

$$A_\tau \equiv \left(\frac{m}{2\pi\Delta\tau}\right)^{1/2} \tag{2.1.14}$$

The corresponding phase-space path integral is

$$\mathcal{G}(x_b, x_a, \tau) = \int \mathcal{D}[x(\tau)]\mathcal{D}[p(\tau)]\ \mathrm{e}^{-\int_0^\tau [\frac{p^2}{2m} + V(x) - \mathrm{i}p\frac{\mathrm{d}x}{\mathrm{d}\tau}]\mathrm{d}\tau'} \tag{2.1.15}$$

The partition function can now be written as

$$\begin{aligned} Z(\beta) &= \int \mathcal{D}[x(\tau)]\mathcal{D}[p(\tau)]\ \mathrm{e}^{-\oint [\frac{p^2}{2m} + V(x) - \mathrm{i}p\frac{\mathrm{d}x}{\mathrm{d}\tau}]\mathrm{d}\tau'} \\ &= A_\tau^N \int \mathcal{D}[x(\tau)]\ \mathrm{e}^{-\oint [\frac{m}{2}\left(\frac{\mathrm{d}x}{\mathrm{d}\tau}\right)^2 + V(x)]\mathrm{d}\tau'} \end{aligned}$$

with $x(0) = x(\beta)$ and $p(0) = p(\beta)$, and slightly different measure $\mathcal{D}[x(\tau)] = \prod_1^N dx_j$ and $\mathcal{D}[x(\tau)]\mathcal{D}[p(\tau)] = \prod_1^N \frac{\mathrm{d}x_j\,\mathrm{d}p_j}{2\pi}$. If we compactify the imaginary-time direction, then the partition function will be just a sum of weighted loops wrapping around the imaginary-time direction.

The path integrals (2.1.13) and (2.1.15) are often called imaginary-time path integrals because they can be mapped into the real-time path integrals (2.1.5) and (2.1.11) through an *analytic continuation*:

$$\tau \to \mathrm{e}^{\mathrm{i}\theta} t \to \mathrm{i}t$$

as we continuously change θ from 0 to $\frac{\pi}{2}$. After replacing τ by $\mathrm{e}^{\mathrm{i}\theta} t$ in the imaginary-time path integral, we have

$$\mathcal{G}(x_b, x_a, t\mathrm{e}^{\mathrm{i}\theta}) = \left(\frac{m}{2\pi \mathrm{e}^{\mathrm{i}\theta}\Delta t}\right)^{N/2} \int \mathcal{D}[x(t)]\ \mathrm{e}^{-\int_0^t [\frac{m}{2}(\frac{\mathrm{d}x}{\mathrm{d}t})^2 \mathrm{e}^{-\mathrm{i}\theta} + V(x)\mathrm{e}^{\mathrm{i}\theta}]\mathrm{d}t'}.$$

One finds that the integrals always converge if $-\frac{\pi}{2} < \theta < \frac{\pi}{2}$.

As a result of analytic continuation, the real-time and the imaginary-time propagators are related as follows:

$$\boxed{\mathcal{G}(x_b, x_a, \tau)|_{\tau = \mathrm{i}t} = \mathrm{i}G(x_b, x_a, t)}$$

The analytic continuation often allows us to obtain a real-time propagator through a calculation of the corresponding imaginary-time propagator, which is often a simpler calculation.

Problem 2.1.4.
Derive the path integral representation of the imaginary-time propagator given in eqn (2.1.13) and eqn (2.1.15).

2.1.4 Evaluation of the path integral

- When quantum fluctuations are not strong, stationary paths and quadratic fluctuations around the stationary paths control the path integral.
- Stationary paths are solutions of classical equations of motion.

In the above, we derived a path integral representation of a quantum propagator. Here we will directly evaluate the path integral and confirm that the path integral indeed reproduces the quantum propagator. This calculation also allows us to gain some intuitive understanding of the path integral and the quantum propagator.

Let us consider the semiclassical limit where the action of a typical path is much larger than $\hbar = 1$. In this limit, as we integrate over a range of paths, the phase $\mathrm{e}^{\mathrm{i}S}$ changes rapidly and contributions from different paths cancel with each other. However, near a stationary path $x_c(t)$ that satisfies

$$S[x_c(t) + \delta x(t)] = S[x_c(t)] + O[(\delta x)^2] \tag{2.1.16}$$

all of the paths have the same phase and a constructive interference. Thus, the stationary path and the paths near it have dominant contributions to the path integral in the semiclassical limit. We also note that the stationary path is simply the path of classical motion for the action $S[x(t)]$. Thus, the path integral representation allows us to clearly see the relationship between classical motion and the quantum propagator: quantum propagation of a particle essentially follows the path determined by the classical equation of motion with some 'wobbling'. This wobbling is called quantum fluctuation. The path integral representation allows us to estimate the magnitude of the quantum fluctuations. More precisely, the fluctuation $\delta x(t)$ around the stationary path $x_c(t)$ that satisfies

$$|S[x_c(t) + \delta x(t)] - S[x_c(t)]| \sim \pi/2$$

will have a large contribution to the propagator, and hence represents a typical quantum fluctuation.

Let us first evaluate the path integral (2.1.5) for a free particle. In this case, eqn (2.1.5) becomes

$$\begin{aligned} &\mathrm{i}G(x_b, t_b, x_a, t_a) \\ &= \left(\frac{m}{2\pi \mathrm{i}\Delta t}\right)^{N/2} \int \mathrm{d}x_1 ... \mathrm{d}x_{N-1} \prod_{i=1}^{N} \exp\left(\mathrm{i}[\frac{m}{2}\frac{(x_i - x_{i-1})^2}{\Delta t^2}]\Delta t\right) \end{aligned} \tag{2.1.17}$$

The classical path (or the stationary path) for the free particle is given by

$$x_c(t) = x_a + \frac{x_b - x_a}{t_b - t_a}(t - t_a)$$

with a classical action

$$S_c(x_b, t_b, x_a, t_a) = \frac{m(x_b - x_a)^2}{2(t_b - t_a)}$$

Introducing the fluctuation around the classical path given by

$$\delta x_j = x_j - x_c(t_j), \qquad t_j = t_a + j\Delta t$$

the path integral (2.1.17) can be rewritten as

$$\begin{aligned} &\mathrm{i}G(x_b, t_b, x_a, t_a) \\ &= \left(\frac{m}{2\pi \mathrm{i}\Delta t}\right)^{N/2} \mathrm{e}^{\mathrm{i}S_c} \int \mathrm{d}x_1 ... \mathrm{d}x_{N-1} \exp\left(\mathrm{i} \sum_{j,k=1,N-1} \delta x_j M_{jk} \delta x_k\right) \\ &= \left(\frac{m}{2\pi \mathrm{i}\Delta t}\right)^{N/2} \sqrt{\frac{\pi^{N-1}}{\mathrm{Det}(-\mathrm{i}M)}} \mathrm{e}^{\mathrm{i}S_c} \end{aligned}$$

where the $N-1$ by $N-1$ matrix is given by

$$M = \frac{m}{2\Delta t}\begin{pmatrix} 2 & -1 & 0 & \cdots \\ -1 & 2 & -1 & \cdots \\ 0 & -1 & 2 & \cdots \\ \cdots & \cdots & \cdots & \cdots \end{pmatrix}$$

In the above example we have used the following important formula for a Gaussian integral:

$$\int \mathrm{d}x_1 ... \mathrm{d}x_{N-1} \exp\left(\mathrm{i} \sum_{j,k=1,N-1} x_j M_{jk} x_k \right) = \frac{\sqrt{\pi}^{N-1}}{\sqrt{\mathrm{Det}(-\mathrm{i}M)}}$$

We see that the propagator is given by the classical action $\mathrm{e}^{\mathrm{i}S_c}$ multiplied by a coefficient that is given by a Gaussian integral (or the corresponding determinant). The coefficient is determined by the quantum fluctuations δx around the classical path.

We notice that only the $\mathrm{e}^{\mathrm{i}S_c}$ term depends on x_a and x_b, while all of the other terms depend only on $t_b - t_a \equiv t$. Thus, we have

$$G(x_b, t_b, x_a, t_a) = A(t_b - t_a)\mathrm{e}^{\mathrm{i}S_c(x_b,t_b,x_a,t_a)} \tag{2.1.18}$$

From the normalization condition

$$\int \mathrm{d}x_b\, G(x_b, t_b, x_a, t_a)G^*(x_b, t_b, x'_a, t_a) = \delta(x_a - x'_a)$$

we find that

$$A(t) = \left(\frac{m}{2\pi \mathrm{i}t}\right)^{1/2} \mathrm{e}^{\mathrm{i}\phi}$$

which agrees with the standard result if we set the phase $\phi = 0$.

To calculate $A(t)$ directly we need to calculate $\mathrm{Det}(M)$. In the following, we will discuss some tricks which can be used to calculate the determinants. Let us first consider the following $N \times N$ matrix:

$$M_N = \begin{pmatrix} 2\cosh(u) & -1 & 0 & \cdots \\ -1 & 2\cosh(u) & -1 & \cdots \\ 0 & -1 & 2\cosh(u) & \cdots \\ \cdots & \cdots & \cdots & \cdots \end{pmatrix}$$

One can show that

$$\mathrm{Det}(M_N) = 2\cosh(u)\mathrm{Det}(M_{N-1}) - \mathrm{Det}(M_{N-2}),$$

$$\mathrm{Det}(M_1) = 2\cosh(u), \qquad \mathrm{Det}(M_2) = 4\cosh^2(u) - 1.$$

The above difference equation can be solved by using the ansatz $\text{Det}(M_N) = a\mathrm{e}^{nu} + b\mathrm{e}^{-nu}$. We find that

$$\text{Det}(M_N) = \frac{\sinh((N+1)u)}{\sinh(u)}$$

When $u = 0$, we have

$$\text{Det}(M_N) = N + 1$$

which implies that $\text{Det}(M) = \left(\frac{m}{2\Delta t}\right)^{N-1} N$. Putting everything together, we find that

$$G(x_b, t_b, x_a, t_a) = \left(\frac{m}{2\pi \mathrm{i} t}\right)^{1/2} \mathrm{e}^{\,\mathrm{i} S_c(x_b,t_b,x_a,t_a)}$$

which is precisely the quantum propagator for a free particle.

The path integral can be evaluated exactly for actions that are quadratic in x and $\dot{x}$. The most general quadratic action has the form

$$L = \frac{1}{2} m\dot{x}^2 - bx - \frac{1}{2} m\omega^2 x^2$$

One can easily check that G is given by eqn (2.1.18) with

$$A(t) = -\mathrm{i}\left(\frac{m}{2\pi \mathrm{i} \Delta t}\right)^{N/2} \int \mathcal{D}[x(t')] \exp\left(\mathrm{i} \int_0^t \mathrm{d}t' \left[\frac{1}{2} m\delta\dot{x}^2 - \frac{1}{2} m\omega^2 \delta x^2\right]\right)$$

where $\delta x(0) = \delta x(t) = 0$. The above path integral can be calculated using the formula for $\text{Det}(M_N)$, and one finds that

$$\mathrm{i} A(t) = \left(\frac{m\omega}{2\pi \mathrm{i} \sin(t\omega)}\right)^{1/2} \tag{2.1.19}$$

Problem 2.1.5.
Consider a particle subjected to a constant force f (with potential $V(x) = -fx$) in one dimension. Show that the propagator is given by

$$\mathrm{i} G(x_b, t, x_a, 0) = \left(\frac{m}{2\pi \mathrm{i} t}\right)^{1/2} \exp\left[\mathrm{i}\left(\frac{m(x_b - x_a)^2}{2t} + \frac{1}{2} f t (x_b + x_a) - \frac{f^2 t^3}{24m}\right)\right]$$

Problem 2.1.6.
Show that the propagator of a harmonic oscillator has the form

$$G(x_b, t, x_a, 0) = A(t) \exp\left(\frac{\mathrm{i} m\omega_0}{2\sin(t\omega_0)} [(x_b^2 + x_a^2)\cos(\omega_0 t) - 2x_b x_a]\right)$$

Use the normalization condition or the path integral to show that

$$A(t) = \left(\frac{m\omega_0}{2\pi \mathrm{i} \sin(t\omega_0)}\right)^{1/2} \mathrm{e}^{\,\mathrm{i}\phi(t)}$$

Problem 2.1.7.
Use the imaginary-time path integral to calculate the propagator of a free particle in imaginary time. Perform a proper analytic continuation to recover the real-time propagator.

Problem 2.1.8.
Use the imaginary-time path integral to calculate the propagator $\mathcal{G}(0, 0, \tau)$ of a harmonic oscillator in imaginary time. From the decay of $\mathcal{G}(0, 0, \tau)$ in the $\tau \to \infty$ limit, obtain the ground-state energy.

2.2 Linear responses and correlation functions

- Linear responses (such as susceptibility) are easy to measure. They are the windows which allow us to see and study properties of matter.

To experimentalists, every physical system is a black box. To probe the properties of a physical system, experimentalists perturb the system and see how it responds. This is how we learn about the internal structure of a hydrogen atom, although no one can see it. We hit the atom (i.e. excite it with light, electrons, or atom beams), and then we listen to how it rings (i.e. observe the emission spectrum of an excited atom). As most perturbations are weak, experimentalists usually observe linear responses in which the responses are proportional to the perturbations. There are many types of experiment that measure linear responses, such as measurements of elasticity, magnetic susceptibility, conductivity, etc. Neutron scattering, nuclear magnetic resonance, X-ray diffraction, etc. also measure the linear responses of a system. Thus, to develop a theory for a system, it is very important to calculate various linear responses from the theory, because these results are usually the easiest to check by experiment. In this section, we will consider the linear response of a very simple system. Through this study, we will obtain a general theory of a linear response.

Experiments are very important in any theory and are the only things that matter. In fact, the goal of every theoretical study is to understand the experimental consequences. However, in this book, we will not discuss experiments very much. Instead, we will discuss a lot about linear responses, or, more generally, correlation functions. Due to the close relationship between experiments and correlation functions, we will view correlation functions as our 'experimental results'. The discussions in this book are centered around those 'experimental results'. When we discuss a model, the purpose of the discussion is to understand and calculate various correlation functions. In this book, we will treat the correlation functions as the physical properties of a system.

A philosophical question arises from the above discussion: what is reality? Imagine a world where the only things that can be measured are linear responses. Then, in that world, linear responses will represent the whole reality, and physical theory will be a theory of linear responses. In our world, we can measure things beyond linear responses. However, it appears that we cannot go much further than that. All that we can measure in our world appears to be correlation functions. Thus, it is tempting to define the physical theory of our world as a theory of correlation functions, and correlation functions may represent the reality of our world.

This point is important because many concepts and building blocks in our physical theory do not represent reality. These concepts may change when new theories are developed. For example, the point particle (with position and velocity) is a fundamental concept of Newton's classical theory. We used to believe that everything is formed by particles and that particles are the building blocks of reality. After the development of quantum theory, we now believe that particles do not represent reality; instead, quantum states in linear Hilbert space represent reality. However, quantum states are not things that we can directly measure. There is no guarantee that the concept of a quantum state will not have the same fate as the concept of a particle when a new theory beyond quantum theory is developed. (I hope the reader agrees with me that none of our present physical theories are ultimately correct and represent the final truth.)

Physics is a science of measurement. Unfortunately, physical theories usually contain many things that do not represent reality (such as the concept of a particle in Newton's classical mechanics). It is possible to have two completely different theories (which may even be based on different concepts) to describe the same reality. Thus, it is very important to always keep an eye on the realities (such as correlation functions and measurable quantities) through the smoke-screen of formalism and unreal concepts in the theory. Going to the extreme in this line of thinking, we can treat quantum states and quantum operators as unreal objects and regard them as merely mathematical tools used to calculate correlation functions that can be measured in experiments.

2.2.1 Linear responses and response functions

- Linear responses can be calculated from one type of correlation function—the response functions—of the corresponding operators.

To understand the general structure of linear response theory, let us first study the linear response of a simple quantum system—the polarization of a harmonic oscillator

$$H_0 = \frac{p^2}{2m} + \frac{1}{2}m\omega_0^2 x^2.$$

An electric field $\mathcal{E}$ couples to the dipole operator ex:

$$H_1 = -ex\mathcal{E}$$

The induced dipole moment is given by

$$d = \langle\psi|ex|\psi\rangle$$

where $|\psi\rangle$ is the ground state of $H_0 + H_1$. To the first order of the perturbation theory,

$$|\psi\rangle = |0\rangle + \sum_{n=1,2,\ldots} |n\rangle \frac{\langle n|H_1|0\rangle}{E_0 - E_n}$$

and we have

$$d = -2\mathcal{E} \sum_{n=1,2,\dots} \frac{\langle 0|ex|n\rangle\langle n|ex|0\rangle}{E_0 - E_n} = 2e^2 \frac{\langle 0|x^2|0\rangle}{\omega_0}\mathcal{E} \tag{2.2.1}$$

The susceptibility χ is given by $\chi = 2e^2 \frac{\langle 0|x^2|0\rangle}{\omega_0}$.

The above result can be expressed in terms of correlation functions. To understand the general relationship between linear response and correlation functions, we consider a general quantum system described by H_0. We turn the perturbation on and off slowly and calculate the linear response using time-dependent perturbation theory.

After including a time-dependent perturbation $f(t)O_1$, the total Hamiltonian is

$$H(t) = H_0 + f(t)O_1$$

We assume that the perturbation is turned on at a finite time. That is, $f(t) = 0$ for t less than a starting time t_{start}. To obtain the response of an H_0 eigenstate $|\psi_n\rangle$ under the perturbation $f(t)O_1$, we start with $|\psi_n\rangle$ at $t = t_{-\infty} = -\infty$. At a finite time, the state $|\psi_n\rangle$ evolves into

$$|\psi_n(t)\rangle = T\left(\mathrm{e}^{-\mathrm{i}\int_{t_{-\infty}}^{t} \mathrm{d}t'\ H(t)}\right)|\psi_n\rangle \tag{2.2.2}$$

where $T(...)$ is a time-ordering operator:

$$T(O(t_1)O(t_2)) = \begin{cases} O(t_1)O(t_2), & \text{if } t_1 > t_2 \\ O(t_2)O(t_1), & \text{if } t_2 > t_1 \end{cases}$$

We can expand $|\psi_n(t)\rangle$ to first order in O_1 as follows:

$$|\psi_n(t)\rangle = \mathrm{e}^{-\mathrm{i}\int_{t_{-\infty}}^{t} \mathrm{d}t'\ H_0}|\psi_n(t)\rangle + \delta|\psi_n(t)\rangle,$$

where

$$\begin{aligned} \delta|\psi_n(t)\rangle &= -\mathrm{i}\int_{t_{-\infty}}^{t} \mathrm{d}t'\ f(t')\mathrm{e}^{-\mathrm{i}H_0(t-t')}O_1\mathrm{e}^{-\mathrm{i}H_0(t'-t_{-\infty})}|\psi_n\rangle \\ &= -\mathrm{i}\int_{t_{-\infty}}^{t} \mathrm{d}t'\ f(t')\mathrm{e}^{-\mathrm{i}H_0(t-t_{-\infty})}O_1(t')|\psi_n\rangle \end{aligned}$$

In the above, we have introduced $O_1(t) = \mathrm{e}^{\mathrm{i}H_0(t-t_{-\infty})}O_1\mathrm{e}^{-\mathrm{i}H_0(t-t_{-\infty})}$. To obtain the change of the physical quantity O_2 in the response to the perturbation $f(t)O_1$, we calculate

$$\begin{aligned}
&\langle\psi_n(t)|O_2|\psi_n(t)\rangle - \langle\psi_n|\mathrm{e}^{\mathrm{i}H_0(t-t_{-\infty})}O_2\mathrm{e}^{-\mathrm{i}H_0(t-t_{-\infty})}|\psi_n\rangle\\
\equiv&\delta\langle\psi_n(t)|O_2|\psi_n(t)\rangle\\
=&-\mathrm{i}\int_{-\infty}^{t}\mathrm{d}t'\ f(t')\langle\psi_n|[O_2(t),O_1(t')]|\psi_n\rangle + O(f^2)\\
=&\int_{-\infty}^{\infty}\mathrm{d}t'\ D(t,t')f(t') + O(f^2)
\end{aligned} \tag{2.2.3}$$

where $D(t,t')$ is the response function defined by

$$D(t,t') = -\mathrm{i}\Theta(t-t')\langle\psi_n|[O_2(t),O_1(t')]|\psi_n\rangle \tag{2.2.4}$$

and

$$\Theta(t) = \begin{cases} 1 & \text{for } t>0\\ 0 & \text{for } t<0\end{cases}$$

At zero temperature, we should take $|\psi_n\rangle$ to be the ground state, and obtain the following zero-temperature response function:

$$D(t,t') = -\mathrm{i}\Theta(t-t')\langle\psi_0|[O_2(t),O_1(t')]|\psi_0\rangle \tag{2.2.5}$$

As H_0 is independent of t, we can show that $D(t,t')$ is a function of $t-t'$ only. We will write $D(t,t')$ as $D(t-t')$. The response $\delta\langle O_2\rangle$ for the ground state is given by

$$\delta\langle O_2\rangle = \int_{-\infty}^{\infty}\mathrm{d}t'\ D(t-t')f(t') \tag{2.2.6}$$

To see if eqn (2.2.6) reproduces the result (2.2.1) for the harmonic oscillator, we note that $-e\mathcal{E}$ plays the role of $f(t)$ and $O_1 = O_2 = ex$. To apply eqn (2.2.6), we need to calculate the response function

$$D(t-t') = -\mathrm{i}\Theta(t-t')\langle 0|[x(t),x(t')]|0\rangle = -2\langle 0|x^2|0\rangle\Theta(t-t')\sin(\omega(t-t')). \tag{2.2.7}$$

To obtain the induced dipole moment from eqn (2.2.6), we need to assume that the electric field is turned on and off slowly, namely $\mathcal{E}(t) = \mathcal{E}\,\mathrm{e}^{-0^+|t|}$. We find that

$$d = \int_{-\infty}^{\infty}\mathrm{d}t'\ e^2 D(t-t')\mathrm{e}^{0^+t'}(-\mathcal{E}) = -e^2 D_{\omega=0}\mathcal{E} = \frac{2e^2\langle 0|x^2|0\rangle}{\omega_0}\mathcal{E} \tag{2.2.8}$$

where $D_\omega = \int \mathrm{d}t\ D(t)\mathrm{e}^{\mathrm{i}\omega t}$.

We see that the linear response of an electric dipole to an electric field is related to the correlation of the dipole operator ex. In fact, all of the other linear responses have a similar structure. The coefficients of the linear responses can be calculated from the correlation functions of appropriate operators. For example, the conductivity can be calculated from the correlation function of the current operators.

2.2.2 Time-ordered correlation functions and the path integral

- Response functions can be expressed in terms of time-ordered correlation functions.
- Time-ordered correlation functions can be calculated through the path integral.

In this section, we would like to consider how to calculate the response function using the path integral. We first note that, for $t > t'$, the response function $D(t,t')$ contains two terms $\langle 0|O_2(t)O_1(t')|0\rangle$ and $\langle 0|O_1(t')O_2(t)|0\rangle$. The first term can be written as

$$\begin{aligned} \mathrm{i}G(t,t') &= \langle 0|O_2(t)O_1(t')|0\rangle \\ &= \langle 0|U^\dagger(t,-\infty)O_2(t)U(t,t')O_1(t')U(t',-\infty)|0\rangle \\ &= \frac{\langle 0|U^\dagger(\infty,t)O_2(t)U(t,t')O_1(t')U(t',-\infty)|0\rangle}{\langle 0|U^\dagger(\infty,-\infty)|0\rangle} \end{aligned} \tag{2.2.9}$$

where $U(t,t')$ is given by $\mathrm{e}^{-\mathrm{i}(t-t')H_0}$.

We know that the time evolution operator $U(t,t')$ can be expressed in terms of the path integral. But this is not enough. To calculate $G(t,t')$ using the path integral, we also need to know the ground state $|0\rangle$. In the following, we will discuss a trick which allows us to calculate $G(t,t')$ without knowing the ground-state wave function $|0\rangle$. First, we would like to generalize eqn (2.2.9) to complex time by introducing the modified evolution operator

$$U^\theta(t_b,t_a) \equiv \mathrm{e}^{-\mathrm{i}(t_b-t_a)H\mathrm{e}^{-\mathrm{i}\theta}}$$

with a small positive θ (we have assumed that $t_b - t_a > 0$). Note that, for large $t_b - t_a$, only the ground state $|0\rangle$ contributes to $\langle\psi|U^\theta(t_b,t_a)|\psi\rangle$, where $|\psi\rangle$ is an arbitrary state. Thus, the operator $U^\theta(t_b,t_a)$ performs a projection to the ground state. Therefore, we may write G_x as

$$\mathrm{i}G(t_b,t_a) = \frac{\langle\psi|U^\theta(\infty,t_b)O_2U^\theta(t_b,t_a)O_1U^\theta(t_a,-\infty)|\psi\rangle}{\langle\psi|U^\theta(\infty,-\infty)|\psi\rangle} \tag{2.2.10}$$

If we choose $|\psi\rangle = \delta(x)$, then eqn (2.2.10) can be readily expressed in terms of path integrals:

$$\mathrm{i}G(t_b,t_a) \equiv \frac{\int \mathcal{D}x\; O_2(t_b)O_1(t_a)\mathrm{e}^{\mathrm{i}\int_{-\infty}^{+\infty}\mathrm{d}t\; L(x,\dot{x})}}{\int \mathcal{D}x\; \mathrm{e}^{\mathrm{i}\int_{-\infty}^{+\infty}\mathrm{d}t\; L(x,\dot{x})}} \tag{2.2.11}$$

where it is understood that $x(t) = 0$ at the boundaries $t = \pm\infty$, and the time t contains a small imaginary part, namely $t \to t\mathrm{e}^{-\mathrm{i}\theta}$. We note that the 'nasty' coefficient in the path integral representation of the propagator cancels out when we calculate the correlation functions. This greatly simplifies the path integral calculations.

We would like to remark that the correlation $G(t_2, t_1)$ defined in the path integral (2.2.11) is a so-called *time-ordered* correlation function. This is because, for $t_2 > t_1$, it is equal to

$$\mathrm{i}G(t_2, t_1) = \frac{\langle 0|U(\infty, t_2)O_2U(t_2, t_1)O_1U(t_1, -\infty)|0\rangle}{\langle 0|U(\infty, -\infty)|0\rangle}$$

while, for $t_2 < t_1$, it is equal to

$$\mathrm{i}G(t_2, t_1) = \frac{\langle 0|U(\infty, t_1)O_1U(t_1, t_2)O_2U(t_2, -\infty)|0\rangle}{\langle 0|U(\infty, -\infty)|0\rangle}$$

The above expressions allow us to show that $G(t_2, t_1)$ only depends on $t_2 - t_1$ and can be written as $G(t_2, t_1) = G(t_2 - t_1)$. Introducing $O_{1,2}(t) = U^\dagger(t, -\infty)O_{1,2}U(t, -\infty)$, the time-ordered correlation function can be written in the following more compact form (in the Heisenberg picture):

$$G(t_2, t_1) = -\mathrm{i}\langle 0|T(O_2(t_2)O_1(t_1))|0\rangle = \begin{cases} -\mathrm{i}\langle 0|O_2(t_2)O_1(t_1)|0\rangle & \text{for } t_2 > t_1 \\ -\mathrm{i}\langle 0|O_1(t_1)O_2(t_2)|0\rangle & \text{for } t_1 > t_2 \end{cases} \tag{2.2.12}$$

As an application, let us use the path integral to calculate the time-ordered correlation $G(t_2, t_1) = -\mathrm{i}\langle 0|T(x(t_2)x(t_1))|0\rangle$ in a harmonic oscillator $H_0 = \frac{p^2}{2m} + \frac{1}{2}m\omega_0^2x^2$. After including the small imaginary part for time, namely $t \to t\mathrm{e}^{-\mathrm{i}\theta}$ with $\theta = 0^+$, the action of the oscillator becomes $S = \int \mathrm{d}t\, \frac{\mathrm{e}^{-\mathrm{i}\theta}}{2} x(-m\mathrm{e}^{2\mathrm{i}\theta}(\frac{\mathrm{d}}{\mathrm{d}t})^2 - m\omega_0^2)x$ and eqn (2.2.11) becomes

$$\mathrm{i}G_x(t_2, t_1) \equiv \frac{\int \mathcal{D}x\, x(t_2)x(t_1)\mathrm{e}^{\mathrm{i}\int_{-\infty}^{+\infty} \mathrm{d}t\, \frac{\mathrm{e}^{-\mathrm{i}\theta}}{2} x(-m\mathrm{e}^{2\mathrm{i}\theta}(\frac{\mathrm{d}}{\mathrm{d}t})^2 - m\omega_0^2)x}}{\int \mathcal{D}x\, \mathrm{e}^{\mathrm{i}\int_{-\infty}^{+\infty} \mathrm{d}t\, \frac{\mathrm{e}^{-\mathrm{i}\theta}}{2} x(-m\mathrm{e}^{2\mathrm{i}\theta}\frac{\mathrm{d}}{\mathrm{d}t})^2 - m\omega_0^2)x}} \tag{2.2.13}$$

We can use the following formula (see Problem 2.2.1) to evaluate the above ratio:

$$\frac{\int \prod_i \mathrm{d}x_i\, x_{i_1}x_{i_2}\, \mathrm{e}^{-\frac{1}{2}x_iA_{ij}x_j}}{\int \prod_i \mathrm{d}x_i\, \mathrm{e}^{-\frac{1}{2}x_iA_{ij}x_j}} = (A^{-1})_{i_1j_2} \tag{2.2.14}$$

The only difference is that the vector x_i and the matrix A_{ij} are labeled by integers in eqn (2.2.14), while the vector $x(t)$ and the matrix $m(\frac{\mathrm{d}}{\mathrm{d}t})^2 + m\omega_0^2$ are functions of the real numbers t in eqn (2.2.13). Comparing eqns (2.2.13) and (2.2.14), we see that $\mathrm{i}G_x(t_2, t_1)$ is just the matrix elements of the operator

$\left(-\mathrm{i}m(\mathrm{e}^{\mathrm{i}\theta}(\frac{\mathrm{d}}{\mathrm{d}t})^2+\omega_0^2\mathrm{e}^{-\mathrm{i}\theta})\right)^{-1}$. In other words, it satisfies

$$\mathrm{e}^{-\mathrm{i}\theta}\Big(-m(\frac{\mathrm{d}}{\mathrm{d}t})^2\mathrm{e}^{2\mathrm{i}\theta}-m\omega_0^2\Big)G_x(t-t')=\delta(t-t') \tag{2.2.15}$$

For $0<\theta\leqslant\frac{\pi}{2}$, eqn (2.2.15) has a solution

$$G_x(t-t')=-\frac{\mathrm{i}}{2m\omega_0}\mathrm{e}^{-\mathrm{i}\,\mathrm{e}^{-\mathrm{i}\theta}|t-t'|\omega_0}$$

which is finite as $t-t'\to\pm\infty$. As $\theta\to 0^+$, the above expression becomes

$$G_x(t)=-\,\mathrm{i}\frac{1}{2m\omega_0}\mathrm{e}^{-\mathrm{i}|t|\omega_0(1-\mathrm{i}0^+)} \tag{2.2.16}$$

In frequency space the correlation is given by

$$G_x(\omega)\equiv\int\mathrm{d}t\;G_x(t,0)\mathrm{e}^{\mathrm{i}\omega t}=\frac{m^{-1}}{\omega^2-\omega_0^2+\mathrm{i}0^+} \tag{2.2.17}$$

To summarize,

> $G_x(\omega)$ is just the inverse of the operator $-m\frac{\mathrm{d}^2}{\mathrm{d}t^2}-m\omega_0^2$ that appears in the action $S=\int\mathrm{d}t\;\frac{1}{2}x(t)(-m\frac{\mathrm{d}^2}{\mathrm{d}t^2}-m\omega_0^2)x(t)$.

The lengthy calculation above and the trick of replacing t by $t\mathrm{e}^{-\mathrm{i}\theta}$ only tell us how to introduce the infinitely small, but important, 0^+.

The second term in the response function $D(t,t')$, namely $\langle 0|O_1(t')O_2(t)|0\rangle$, has an incorrect time order because $t>t'$. However, this can be fixed by noting that $\langle 0|O_1(t')O_2(t)|0\rangle=\langle 0|O_2^\dagger(t)O_1^\dagger(t')|0\rangle^*$. Therefore, the response function can be expressed in terms of time-ordered correlation functions:

$$\mathrm{i}D(t,t')=\Theta(t-t')\left(\langle 0|T(O_2(t)O_1(t'))|0\rangle-\langle 0|T(O_2^\dagger(t)O_1^\dagger(t'))|0\rangle^*\right)$$

So, we can use path integrals to calculate the response function through the time-ordered correlation functions.

When $O_{1,2}$ are hermitian, we have

$$D(t)=2\Theta(t)\mathrm{Re}G(t) \tag{2.2.18}$$

When $O_1=O_2=O_1^\dagger$, we have $G(t)=G(-t)$. We can show that, in ω space,

$$\mathrm{Re}D_\omega=\mathrm{Re}G(\omega),\qquad \mathrm{Im}D_\omega=\mathrm{sgn}(\omega)\mathrm{Im}G(\omega) \tag{2.2.19}$$

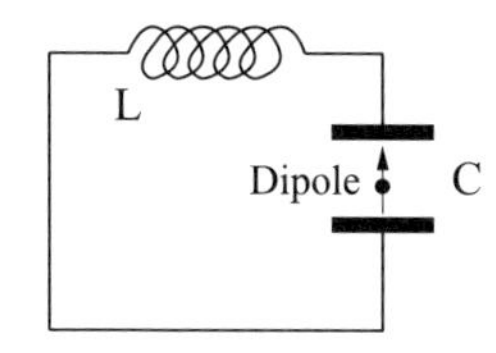

FIG. 2.2. A CL circuit with a dipole in the capacitor.

Problem 2.2.1.
Prove eqn (2.2.14). (Hint: Calculate $\int \prod_i \mathrm{d}x_i \; \mathrm{e}^{J_i x_i} \mathrm{e}^{-\frac{1}{2} x_i A_{ij} x_j}$ and expand the result to second order in J.)

Problem 2.2.2.
(a) Use the coherent-state path integral and the generating functional $\frac{Z[\eta(t),\bar\eta(t)]}{Z[0,0]}$, where

$$Z[\eta(t),\bar\eta(t)] = \int \mathcal{D}^2[a(t)] \mathrm{e}^{\mathrm{i}\int \mathrm{d}t\; [\mathrm{i}\frac{1}{2}(a^* \frac{\mathrm{d}}{\mathrm{d}t} a - a\frac{\mathrm{d}}{\mathrm{d}t} a^*) - \omega_0 a^* a - \bar\eta a - \eta a^*]}$$

to calculate the correlation $\mathrm{i}G(t) = \langle a(t) a^*(0)\rangle$ in real and frequency space. (Hint: Use the complex time $t \to t\mathrm{e}^{-\mathrm{i}0^+}$.)
(b) Compare your result with the corresponding time-ordered correlation function $\langle 0|T(\hat a(t)\hat a^\dagger(0))|0\rangle$ for both $t > 0$ and $t < 0$. Also pay attention to the limits $t \to 0^\pm$.
(c) Now we want to calculate $\langle 0|\hat a^\dagger \hat a|0\rangle$ and $\langle 0|\hat a \hat a^\dagger|0\rangle$ using the coherent-state path integral. The problem is that both operators became $a^* a$ in the path integral. How do we resolve this problem in light of the above calculation? Describe how to use the path integral to calculate the following correlation: $\langle 0|(\hat a\hat a^\dagger)_{t_1}(\hat a\hat a^\dagger)_{t_2}|0\rangle$.

2.2.3 Effective theory

- Effective theory is a very important concept. If you only want to remember one thing from field theory, then remember effective theory.

We know that the electric dipole moment couples to the electric field $\mathcal{E}$. To see how the coupling affects the dynamics of the electric field, let us consider a CL circuit (see Fig. 2.2). In the absence of a dipole moment, the dynamics of the electric field in the capacitor are described by the action $S_0(\mathcal{E}) = \int \mathrm{d}t\; \frac{1}{2g}(\dot{\mathcal{E}}^2 - \omega_{CL}^2 \mathcal{E}^2)$, where ω_{CL} is the oscillation frequency of the CL circuit. The coupled system is described by

$$Z = \int \mathcal{D}x\mathcal{D}\mathcal{E}\; \mathrm{e}^{\mathrm{i}S_0(\mathcal{E}) + \mathrm{i}\int \mathrm{d}t[L(x,\dot x) + ex\mathcal{E}]}$$

where $L(x,\dot x)$ describes the dynamics of the dipole and $ex\mathcal{E}$ describes the coupling between the dipole and the electric field.

Let us assume $\omega_0 \gg \omega_{CL}$ and integrate out the high frequency motion $x(t)$. We obtain the path integral that contains $\mathcal{E}(t)$ alone:

$$Z = \int \mathcal{D}\mathcal{E}\ \mathrm{e}^{\,\mathrm{i}S_0(\mathcal{E})+\mathrm{i}S_I(\mathcal{E})},$$

$$\mathrm{e}^{\,\mathrm{i}S_I} = \frac{Z[\mathcal{E}]}{Z[0]}, \qquad Z[\mathcal{E}] = \int \mathcal{D}x\ \mathrm{e}^{\,\mathrm{i}\int \mathrm{d}t[\frac{m}{2}\dot{x}^2 - \frac{m\omega_0^2}{2}x^2 + ex\mathcal{E}]}$$

We see that the dynamics of the electric field of the coupled system are described by a new action $S_{\text{eff}} = S_0 + S_I$, which will be called the effective action. After performing a Gaussian integral, we find that

$$\begin{aligned} S_I &= \int \mathrm{d}t\ \frac{1}{2m}\mathcal{E}(t)\Big((\frac{\mathrm{d}}{\mathrm{d}t})^2 + \omega_0^2\Big)^{-1}\mathcal{E}(t) \\ &= -\int \mathrm{d}t\mathrm{d}t'\ \frac{e^2}{2}\mathcal{E}(t)G_x(t-t')\mathcal{E}(t') = -\int \frac{\mathrm{d}\omega}{2\pi}\frac{e^2}{2}\mathcal{E}_{-\omega}G_x(\omega)\mathcal{E}_\omega \end{aligned}$$

which in turn gives us the effective Lagrangian for the electric field:

$$L_{\text{eff}} = \frac{1}{2g}(\dot{\mathcal{E}}^2 - \omega_{CL}^2\mathcal{E}^2) + \frac{e^2}{2m}\mathcal{E}\Big((\frac{\mathrm{d}}{\mathrm{d}t})^2 + \omega_0^2\Big)^{-1}\mathcal{E}.$$

In the limit $\omega_0 \gg \omega_{CL}$, L_{eff} can be simplified as

$$L_{\text{eff}} = \frac{1}{2g}(\dot{\mathcal{E}}^2 - (\omega_{CL}^*)^2\mathcal{E}^2), \qquad \omega_{CL}^* = \sqrt{\omega_{CL}^2 - \frac{ge^2}{m\omega_0^2}}.$$

We see that after integrating out the high frequency $x(t)$, the low frequency electric field is described by a simple effective Lagrangian. The coupling to the dipole shifts the oscillation frequency of the electric field to a lower value ω_{CL}^*.

Problem 2.2.3.
Low-energy effective theory of coupled harmonic oscillators: Consider a system of two coupled harmonic oscillators:

$$Z = \int \mathcal{D}x\mathcal{D}X\,\mathrm{e}^{\,\mathrm{i}\int \mathrm{d}t L(x,X)}$$

where

$$L(x,X) = \frac{1}{2}m\dot{x}^2 - \frac{1}{2}m\omega^2x^2 + \frac{1}{2}M\dot{X}^2 - \frac{1}{2}M\Omega^2X^2 - gxX$$

Here we assume that $\Omega \gg \omega$, $g \sim m\omega^2$, and $m \sim M$. In this case, X is a high-energy degree of freedom which oscillates around $X = 0$. Find the low-energy effective theory $L_{\text{eff}}(x)$ that describes the low-energy dynamics of the soft degree of freedom x by integrating out X. (You should include at least the leading $\frac{1}{\Omega}$ corrections.) In particular, find the effective mass m^* and the effective spring constant $m^*(\omega^*)^2$.

2.2.4 Time-dependent response and dissipation

- The response function gives the correct real and imaginary parts of the susceptibility, while the time-ordered correlation function only gives the correct real part of the susceptibility.

We can also use the path integral to calculate the response to a time-dependent electric field. We note that the average dipole moment at time t can be expressed as

$$d(t) = \langle ex(t) \rangle = e \frac{\int \mathcal{D}x \; x(t) \mathrm{e}^{\mathrm{i} \int \mathrm{d}t' \; [L(x,\dot{x}) + ex\mathcal{E}]}}{\int \mathcal{D}x \; \mathrm{e}^{\mathrm{i} \int \mathrm{d}t' \; [L(x,\dot{x}) + ex\mathcal{E}]}}$$

$$= - \int_{-\infty}^{+\infty} \mathrm{d}t' \; e^2 G_x(t - t') \mathcal{E}(t') + O(\mathcal{E}^2)$$

where

$$\mathrm{i} G_x(t_1 - t_2) = \frac{\int \mathcal{D}x \; x(t_1) x(t_2) \mathrm{e}^{\mathrm{i} \int \mathrm{d}t \; L(x,\dot{x})}}{\int \mathcal{D}x \; \mathrm{e}^{\mathrm{i} \int \mathrm{d}t' \; L(x,\dot{x})}}$$

In frequency space, the above expression becomes $d_\omega = -e^2 G_x(\omega) E_\omega$. We find that the finite-frequency susceptibility $\chi(\omega)$ is given by

$$\chi(\omega) = -e^2 G_x(\omega) = -\frac{e^2 m^{-1}}{\omega^2 - \omega_0^2 + \mathrm{i}0^+} \tag{2.2.20}$$

The imaginary part $\mathrm{i}0^+$ was calculated in Section 2.2.2.

The above path integral calculation of the finite-frequency susceptibility belongs to a type of calculation called formal calculation. Using the path integral we can 'formally' calculate many things. However, the results obtained from the 'formal' calculation have to be taken 'with a pinch of salt'. These results are usually more or less correct, but may not be exactly correct. As we will see in Section 2.4.3, $\mathrm{Im}\chi(\omega)\omega$ has a physical meaning. It represents the friction coefficient at the frequency ω. As $\mathrm{Im}G(\omega)$ is always negative, we find that $\mathrm{Im}\chi(\omega)\omega > 0$ if $\omega > 0$ and $\mathrm{Im}\chi(\omega)\omega < 0$ if $\omega < 0$. A negative friction coefficient for $\omega < 0$ is an unphysical and incorrect result. Thus, eqn (2.2.20) is not completely correct and it does not agree with the previous result (2.2.8).

It turns out that the correct result (2.2.8) can be obtained from the incorrect one by replacing the time-ordered correlation G_x by the response function D:

$$d(t) = - \int_{-\infty}^{+\infty} \mathrm{d}t' \; e^2 D(t - t') \mathcal{E}(t') + O(\mathcal{E}^2)$$

$$\chi(\omega) = - e^2 D(\omega) = -\frac{e^2 m^{-1}}{\omega^2 - \omega_0^2 + \mathrm{i}\,\mathrm{sgn}(\omega) 0^+}$$

In this case, $\mathrm{Im}\chi(\omega)\omega$ is always positive.

In the above path integral calculation, we made an error by confusing the path integral average $\frac{\int \mathcal{D}x\, x(t)\mathrm{e}^{\,\mathrm{i}S}}{\int \mathcal{D}x\,\,\mathrm{e}^{\,\mathrm{i}S}}$ with the quantum mechanical average $\langle\psi(t)|x|\psi(t)\rangle$. In order for the path integral calculation to be valid, we need to assume that the two averages are the same in the presence of an electric field. That is,

$$\langle 0|U^\dagger(t,-\infty)xU(t,-\infty)|0\rangle = \frac{\langle 0|U(\infty,t)xU(t,-\infty)|0\rangle}{\langle 0|U(\infty,-\infty)|0\rangle} \tag{2.2.21}$$

where the right-hand side is the path integral average and the left-hand side is the quantum mechanical average. This assumption is correct only when we turn on and turn off the electric field slowly such that the ground state $|0\rangle$ evolves into the same ground state from $t=-\infty$ to $t=\infty$:

$$|0\rangle = \mathrm{e}^{\,\mathrm{i}\theta}U(\infty,-\infty)|0\rangle \tag{2.2.22}$$

In this case,

$$\langle 0|U^\dagger(t,-\infty) = \langle 0|U(-\infty,\infty)U(\infty,t) = \mathrm{e}^{\,\mathrm{i}\theta}\langle 0|U(\infty,t)$$

and we see that eqn (2.2.21) is indeed correct under the condition (2.2.22). However, if $E(t)$ changes rapidly and excites the oscillator to higher and higher excited states, then eqn (2.2.22) will no longer be correct and, even worse, $\langle 0|U(\infty,-\infty)|0\rangle$ may vanish when we have dissipation.

Here we have learnt two lessons. Firstly, the path integral results may be incorrect in the presence of dissipation and, secondly, the problem can be fixed by replacing the time-ordered correlation by the response function.

Problem 2.2.4.
Consider a harmonic oscillator. Calculate the (time-ordered) current correlation function $\mathrm{i}G_j(t) = \langle j(t)j(0)\rangle$ in frequency space, namely $G_j(\omega) = \int \mathrm{d}t\; G_j(t)\mathrm{e}^{\,\mathrm{i}\omega t}$, where the current operator is $j = e\dot{x}$. Calculate the finite-frequency conductance $\sigma(\omega)$ (defined by $j(\omega) = \sigma(\omega)E_\omega$) using, say, classical physics. (You may add a small amount of friction to obtain the dissipation term.) Show that, for $\omega > 0$, $\sigma(\omega)$ has the form $\sigma(\omega) = C\frac{G_j(\omega)}{\omega}$ and find the coefficient C. Show that $\sigma(\omega)$ has a nonzero real part only when $\omega = \omega_0$. (Hint: Keep the $\mathrm{i}0^+$ term in G_j.) At this frequency the oscillator can absorb energies by jumping to higher excited states. We see that, in order to have finite real d.c. conductance, we need to have gapless excitations.

2.2.5 Correlation functions at finite temperatures

At finite temperatures, a linear response is an average response of all excited states, weighted by the Boltzmann factor. As the response function of a state $|\psi_n\rangle$ is given by eqn (2.2.4), the finite temperature response function is given by

$$D^\beta(t-t') = \sum_n -\mathrm{i}\Theta(t-t')\langle\psi_n|[O_2(t), O_1(t')]|\psi_n\rangle \frac{\mathrm{e}^{-\epsilon_n/T}}{Z}$$

We see again that linear responses can be described by the corresponding correlation functions. In frequency space,

$$\langle O_2\rangle_\omega = D^\beta_\omega f_\omega$$

for finite temperatures. Thus,

$$D^\beta_\omega = \int \mathrm{d}t\, D^\beta(t)\mathrm{e}^{\,\mathrm{i}t\omega}$$

can be regarded as the finite-frequency susceptibility at finite temperatures.

Similarly, we define the time-ordered finite-temperature correlation function

$$\mathrm{i}G^\beta(t) = \sum_n \langle\psi_n|T(O_2(t)O_1(0))|\psi_n\rangle \frac{\mathrm{e}^{-\epsilon_n\beta}}{Z}$$

With such a definition, eqn (2.2.18) can be generalized to finite temperatures:

$$D^\beta(t) = 2\Theta(t)\mathrm{Re}G^\beta(t) \tag{2.2.23}$$

2.2.6 Relation between correlation functions

- Response functions, time-ordered correlation functions, and spectral functions at finite temperatures can be calculated from finite-temperature imaginary-time correlation functions through proper analytic continuation.

We have introduced several kinds of correlation function. The response functions are directly related to various susceptibilities measured in linear response experiments. The time-ordered correlation functions are easy to calculate through the path integral. In this section, we will discuss the relationships between these correlation functions. These relations allow us to obtain response functions from time-ordered correlation functions. The calculations in this section are quite formal, but the results are useful. You may skip this section if you do not like formal calculations. You only need to be aware of the relations in the boxed equations.

To begin with, let us study the relationship between the finite-temperature correlations of two operators O_1 and O_2 in real and imaginary time: $\mathrm{i}G^\beta(t) = \langle T[O_2(t)O_1(0)]\rangle$ and $\mathcal{G}^\beta(\tau) = \langle T_\tau[O_2(\tau)O_1(0)]\rangle$. To obtain more general results that will be used later for fermion systems, we generalize our definition of $G^\beta(t)$ and $\mathcal{G}^\beta(\tau)$ as follows:

$$\mathrm{i}G^\beta(t)|_{t>0} = \langle O_2(t)O_1(0)\rangle \qquad \mathrm{i}G^\beta(t)|_{t<0} = \eta\,\langle O_1(0)O_2(t)\rangle$$
$$\mathcal{G}^\beta(\tau)|_{\tau>0} = \langle O_2(\tau)O_1(0)\rangle \qquad \mathcal{G}^\beta(\tau)|_{\tau<0} = \eta\,\langle O_1(0)O_2(\tau)\rangle$$

where $\eta = \pm$. When $\eta = +$ the correlations are called bosonic and when $\eta = -$ they are called fermionic. Note that here we do not assume that $O_{1,2} = O_{1,2}^\dagger$. Using

$$\begin{aligned}\langle\psi_n|O_2(t)O_1(0)|\psi_n\rangle &= \frac{\langle\psi_n|O_2U(t,0)O_1|\psi_n\rangle}{\langle\psi_n|U(t,0)|\psi_n\rangle}\\ &= \sum_m \langle\psi_n|O_2|\psi_m\rangle\langle\psi_m|O_1|\psi_n\rangle\,\mathrm{e}^{-\mathrm{i}(\epsilon_m-\epsilon_n)t}\end{aligned}$$

we find the spectral representation of $G^\beta(t)$ to be

$$\begin{aligned}\mathrm{i}G^\beta(t)|_{t>0} &= \sum_{m,n}\langle\psi_n|O_2|\psi_m\rangle\langle\psi_m|O_1|\psi_n\rangle\frac{\mathrm{e}^{-\epsilon_n\beta-\mathrm{i}(\epsilon_m-\epsilon_n)t}}{Z}\\ \mathrm{i}G^\beta(t)|_{t<0} &= \eta\sum_{m,n}\langle\psi_n|O_1|\psi_m\rangle\langle\psi_m|O_2|\psi_n\rangle\frac{\mathrm{e}^{-\epsilon_n\beta+\mathrm{i}(\epsilon_m-\epsilon_n)t}}{Z}\end{aligned} \tag{2.2.24}$$

If we introduce the finite-temperature spectral function as follows:

$$\begin{aligned}A_+^\beta(\nu) &= \sum_{m,n}\delta[\nu-(\epsilon_m-\epsilon_n)]\langle\psi_n|O_2|\psi_m\rangle\langle\psi_m|O_1|\psi_n\rangle\frac{\mathrm{e}^{-\epsilon_n\beta}}{Z}\\ A_-^\beta(\nu) &= \sum_{m,n}\delta[\nu+(\epsilon_m-\epsilon_n)]\langle\psi_n|O_1|\psi_m\rangle\langle\psi_m|O_2|\psi_n\rangle\frac{\mathrm{e}^{-\epsilon_n\beta}}{Z}\end{aligned}$$

we can rewrite G^β as

$$\mathrm{i}G^\beta(t)|_{t>0} = \int \mathrm{d}\nu\; A_+^\beta(\nu)\mathrm{e}^{-\mathrm{i}\nu t}, \qquad \mathrm{i}G^\beta(t)|_{t<0} = \eta\int \mathrm{d}\nu\; A_-^\beta(\nu)\mathrm{e}^{-\mathrm{i}\nu t} \tag{2.2.25}$$

To understand the meaning of the spectral functions, let us consider a zero-temperature spectral function $A_+(\nu) = \sum_m \delta[\nu-(\epsilon_m-\epsilon_0)]\langle 0|O_2|\psi_m\rangle\langle\psi_m|O_1|0\rangle$ We see that, when O_1 and O_2 act on the ground state $|0\rangle$, they create the excited states $O_1|0\rangle = \sum_m C_{1,m}|\psi_m\rangle$ and $O_2|0\rangle = \sum_m C_{2,m}|\psi_m\rangle$, with amplitudes $C_{1,m}$ and $C_{2,m}$, respectively. The function $A_+(\nu)$ is just a product of $C_{1,m}$ and $C_{2,m}$: $A_+(\nu) = \sum_m \delta[\nu-(\epsilon_m-\epsilon_0)]C_{2,m}^\dagger C_{1,m}$. If $A_+(\nu)\neq 0$, then an excited state of energy $\epsilon_0+\nu$ is created by both O_1 and O_2. When $O_1 = O_2$, $A_+(\nu)$ is simply the weight of the states with energy ν created by O_1.

We note that the finite-temperature spectral function can be rewritten as

$$\begin{aligned}A_+^\beta(\omega) &= \sum_{m,n}\langle\psi_n|O_2|\psi_m\rangle\langle\psi_m|O_1|\psi_n\rangle\frac{\mathrm{e}^{-\epsilon_n\beta}}{Z}\delta\left(\omega-(\epsilon_m-\epsilon_n)\right)\\ &= \mathrm{e}^{\beta\omega/2}\sum_{m,n}\langle\psi_n|O_2|\psi_m\rangle\langle\psi_m|O_1|\psi_n\rangle\frac{\mathrm{e}^{-(\epsilon_m+\epsilon_n)\beta/2}}{Z}\delta\left(\omega-(\epsilon_m-\epsilon_n)\right)\end{aligned} \tag{2.2.26}$$

$$A_-^\beta(\omega) = \sum_{m,n} \langle\psi_n|O_1|\psi_m\rangle\langle\psi_m|O_2|\psi_n\rangle \frac{e^{-\epsilon_n\beta}}{Z}\delta(\omega + (\epsilon_m - \epsilon_n))$$
$$= e^{-\beta\omega/2}\sum_{m,n} \langle\psi_n|O_1|\psi_m\rangle\langle\psi_m|O_2|\psi_n\rangle \frac{e^{-(\epsilon_m+\epsilon_n)\beta/2}}{Z}\delta(\omega + (\epsilon_m - \epsilon_n))$$
$$= e^{-\beta\omega/2}\sum_{m,n} \langle\psi_n|O_2|\psi_m\rangle\langle\psi_m|O_1|\psi_n\rangle \frac{e^{-(\epsilon_m+\epsilon_n)\beta/2}}{Z}\delta(\omega - (\epsilon_m - \epsilon_n)) \quad (2.2.27)$$

In the last line of eqn (2.2.27) we have interchanged m and n. In this form, we can see a very simple relationship between A_+^β and A_-^β at finite temperatures:

$$A_+^\beta(\omega) = e^{\beta\omega}A_-^\beta(\omega) \quad (2.2.28)$$

We also note that A_+^β almost vanishes for $\omega \lesssim -T$ and A_-^β almost vanishes for $\omega \gtrsim -T$. At zero temperature, we have

$$A_+(\omega < 0) = 0, \qquad A_-(\omega > 0) = 0$$

To relate the spectral function $A_\pm^\beta$ of G^β in frequency space we need to introduce

$$G_+^\beta(t) = \Theta(t)G^\beta(t), \qquad G_-^\beta(t) = \Theta(-t)G^\beta(t)$$

In the frequency space, we have

$$G_+^\beta(\omega) = \sum_{m,n} \frac{\langle\psi_n|O_2|\psi_m\rangle\langle\psi_m|O_1|\psi_n\rangle}{\omega - (\epsilon_m - \epsilon_n) + i0^+}\frac{e^{-\epsilon_n\beta}}{Z}$$
$$G_-^\beta(\omega) = \eta\sum_{m,n} \frac{\langle\psi_n|O_1|\psi_m\rangle\langle\psi_m|O_2|\psi_n\rangle}{-\omega - (\epsilon_m - \epsilon_n) + i0^+}\frac{e^{-\epsilon_n\beta}}{Z}$$

We find that

$$G_+^\beta(\omega) = \int d\nu\, \frac{A_+^\beta(\nu)}{\omega - \nu + i0^+} \qquad G_-^\beta(\omega) = -\eta\int d\nu\, \frac{A_-^\beta(\nu)}{\omega - \nu - i0^+} \quad (2.2.29)$$

and $G^\beta(\omega) = G_+^\beta(\omega) + G_-^\beta(\omega)$.

Next, let us consider the response function $D^\beta(t)$ which is generalized to be

$$D^\beta(t - t') = \sum_n -i\Theta(t - t')\langle\psi_n|O_2(t)O_1(t') - \eta O_1(t')O_2(t)|\psi_n\rangle\frac{e^{-\epsilon_n/T}}{Z} \quad (2.2.30)$$

It has the following spectral expansion:

$$D^\beta(\omega) = \sum_{m,n} \frac{\langle\psi_n|O_2|\psi_m\rangle\langle\psi_m|O_1|\psi_n\rangle}{\omega - (\epsilon_m - \epsilon_n) + i0^+}\frac{e^{-\epsilon_n\beta}}{Z}$$
$$+\eta\sum_{m,n} \frac{\langle\psi_n|O_1|\psi_m\rangle\langle\psi_m|O_2|\psi_n\rangle}{-\omega - (\epsilon_m - \epsilon_n) - i0^+}\frac{e^{-\epsilon_n\beta}}{Z} \quad (2.2.31)$$

We find that

$$D^\beta(\omega) = \int d\nu \, \frac{A_+^\beta(\nu)}{\omega - \nu + i0^+} - \eta \int d\nu \, \frac{A_-^\beta(\nu)}{\omega - \nu + i0^+} \tag{2.2.32}$$

We see that we can determine both the time-ordered correlation G^β and the response function D^β from the spectral function $A_\pm^\beta$.

One way to calculate the spectral function from a path integral is to use the time-ordered imaginary-time correlation function

$$\mathcal{G}^\beta(\tau_2, \tau_1) = Z^{-1}\mathrm{Tr}\left(T_\tau(O_2(\tau_2)O_1(\tau_1))\, e^{-\beta H}\right)$$

where $0 < \tau_{1,2} < \beta$. The function $\mathcal{G}^\beta(\tau_2, \tau_1)$ can be also be written as

$$\mathcal{G}^\beta(\tau_2, \tau_1) = \begin{cases} \dfrac{\mathrm{Tr}(e^{-(\beta-\tau_2)H} O_2 \, e^{-(\tau_2-\tau_1)H} O_1 \, e^{-\tau_1 H})}{\mathrm{Tr}(e^{-\beta H})}, & \tau_2 > \tau_1 \\ \eta \dfrac{\mathrm{Tr}(e^{-(\beta-\tau_1)H} O_1 \, e^{-(\tau_1-\tau_2)H} O_2 \, e^{-\tau_2 H})}{\mathrm{Tr}(e^{-\beta H})}, & \tau_1 > \tau_2 \end{cases}$$

which allows us to show that

$$\mathcal{G}^\beta(\tau_2, \tau_1) = \mathcal{G}^\beta(\tau_2 - \tau_1, 0) \equiv \mathcal{G}^\beta(\tau_2 - \tau_1), \qquad \mathcal{G}^\beta(\tau) = \eta \mathcal{G}^\beta(\tau + \beta)$$

The function $\mathcal{G}^\beta(\tau)$ also admits a spectral representation. For $0 < \tau < \beta$, we have

$$\mathcal{G}^\beta(\tau) = \begin{cases} \sum_{m,n} \langle \psi_n | O_2 | \psi_m \rangle \langle \psi_m | O_1 | \psi_n \rangle \dfrac{e^{-\epsilon_n \beta - (\epsilon_m - \epsilon_n)\tau}}{Z}, & \tau > 0 \\ \eta \sum_{m,n} \langle \psi_n | O_1 | \psi_m \rangle \langle \psi_m | O_2 | \psi_n \rangle \dfrac{e^{-\epsilon_n \beta + (\epsilon_m - \epsilon_n)\tau}}{Z}, & \tau < 0 \end{cases} \tag{2.2.33}$$

Comparing eqns (2.2.24) and (2.2.33), we see that G^β and $\mathcal{G}^\beta$ are related as follows:

$$iG^\beta(t) = \mathcal{G}^\beta(it + \mathrm{sgn}(t)0^+) \tag{2.2.34}$$

The real-time Green's function can be calculated from the imaginary-time Green's function even at finite temperatures.

As $\mathcal{G}^\beta(\tau)$ is (anti-)periodic, its frequency is discrete: $\omega_l = 2\pi T \times$ integer for $\eta = +$ and $\omega_l = 2\pi T \times$ (integer $+ \frac{1}{2}$) for $\eta = -$. In frequency space, we have naively

$$\begin{aligned} \mathcal{G}^\beta(\omega_l) &= \int_0^\beta d\tau \, \mathcal{G}^\beta(\tau) e^{i\omega_l \tau} \\ &= -\sum_{m,n} \frac{\langle \psi_n | O_2 | \psi_m \rangle \langle \psi_m | O_1 | \psi_n \rangle}{i\omega_l - \epsilon_m + \epsilon_n} \frac{e^{-\epsilon_n \beta} - \eta e^{-\epsilon_m \beta}}{Z} \end{aligned}$$

However, when $\eta = 1$ and $\omega_l = 0$, the terms with $m = n$ become undefined. Those terms should be $\langle\psi_n|O_2|\psi_n\rangle\langle\psi_n|O_1|\psi_n\rangle\beta e^{-\epsilon_n\beta}/Z$. So, we really have

$$\mathcal{G}^\beta(\omega_l) = -\sum_{m,n} \frac{\langle\psi_n|O_2|\psi_m\rangle\langle\psi_m|O_1|\psi_n\rangle}{i\omega_l - \epsilon_m + \epsilon_n + T\delta} \frac{e^{-\epsilon_n\beta} - (\eta+\delta)e^{-\epsilon_m\beta}}{Z} \tag{2.2.35}$$
$$= -\sum_{m,n} \left(\frac{\langle\psi_n|O_2|\psi_m\rangle\langle\psi_m|O_1|\psi_n\rangle}{i\omega_l - (\epsilon_m - \epsilon_n) + T\delta} + (\eta+\delta)\frac{\langle\psi_n|O_1|\psi_m\rangle\langle\psi_m|O_2|\psi_n\rangle}{-i\omega_l - (\epsilon_m - \epsilon_n) - T\delta} \right) \frac{e^{-\epsilon_n\beta}}{Z}$$

where δ is a small complex number. This allows us to obtain, for $\omega \neq 0$,

$$\frac{1}{2i}[\mathcal{G}^\beta(i\omega_l \to \omega + i0^+) - \mathcal{G}^\beta(i\omega_l \to \omega - i0^+)] \tag{2.2.36}$$
$$= (e^{\beta\omega/2} - \eta e^{-\beta\omega/2}) \sum_{m,n} \langle\psi_n|O_2|\psi_m\rangle\langle\psi_m|O_1|\psi_n\rangle \frac{e^{-(\epsilon_m+\epsilon_n)\beta/2}}{Z} \pi\delta(\omega - (\epsilon_m - \epsilon_n)).$$

Note that, when $\omega \neq 0$, δ has no effect and can be dropped. Finally, we can express $A^\beta_{+,-}$ in terms of $\mathcal{G}^\beta(\omega_l)$ by comparing eqn (2.2.36) with eqn (2.2.26) and eqn (2.2.27). For $\omega \neq 0$, we have

$$A^\beta_+(\omega) = [1 + \eta n_\eta(\omega)](\pi)^{-1} \frac{1}{2i}[\mathcal{G}^\beta(i\omega_l \to \omega + i0^+) - \mathcal{G}^\beta(i\omega_l \to \omega - i0^+)]$$
$$A^\beta_-(\omega) = n_\eta(\omega)(\pi)^{-1} \frac{1}{2i}[\mathcal{G}^\beta(i\omega_l \to \omega + i0^+) - \mathcal{G}^\beta(i\omega_l \to \omega - i0^+)] \tag{2.2.37}$$

where $n_+(\omega) = n_B(\omega) = \frac{1}{e^{\beta\omega}-1}$ and $n_-(\omega) = n_F(\omega) = \frac{1}{e^{\beta\omega}+1}$ are boson and fermion occupation numbers, respectively. This allows us to determine G^β and D^β from $\mathcal{G}^\beta$. In particular, comparing eqns (2.2.31) and (2.2.35), we see that $D^\beta(\omega)$ and $\mathcal{G}^\beta(\omega_l)$ have a very simple relationship:

$$D(\omega) = -\mathcal{G}^\beta(\omega_l)|_{i\omega_l \to \omega + i0^+}, \qquad \text{for } \omega \neq 0. \tag{2.2.38}$$

Again δ has been dropped because $\omega \neq 0$.

It is interesting to see that the Fourier transformation and the analytic continuation of $\mathcal{G}^\beta(\tau)$ do not commute. If we perform the analytic continuation first then we obtain the time-ordered correlation function $G^\beta(t)$. If we perform the Fourier transformation first then we obtain the response function $D^\beta(\omega)$. One may wonder about choosing a different analytic continuation $i\omega_n \to \omega(1 + i0^+)$, which is actually more natural. From our spectral representations, one can see that the new analytic continuation still cannot transform $\mathcal{G}^\beta(\omega_n)$ into $G^\beta(\omega)$. However, at zero temperature, we indeed have

$$G(\omega) = -\mathcal{G}(\omega)|_{i\omega \to \omega(1+i0^+)} \tag{2.2.39}$$

All of the above results have been obtained for a very general situation in which $O_{1,2}$ may not be even hermitian. The results can have a simpler form when $O_1 = O_1^\dagger$ and $O_2 = O_2^\dagger$. At $T = 0$ and for $\omega \neq 0$, we have

$$
\begin{aligned}
&D(t) = 2\Theta(t)\text{Re}G(t), && \text{i}G(t) = \mathcal{G}(\text{i}t + \text{sgn}(t)0^+) \\
&D(\omega) = \text{Re}G(\omega) + \text{i}\,\text{sgn}(\omega)\text{Im}G(\omega), && D(\omega) = -\mathcal{G}(\omega)|_{\text{i}\omega\to\omega+\text{i}0^+}
\end{aligned}
\tag{2.2.40}
$$

For $T \neq 0$ and $\omega \neq 0$, we have

$$
\begin{aligned}
&D^\beta(t) = 2\Theta(t)\text{Re}G^\beta(t), && \text{i}G^\beta(t) = \mathcal{G}^\beta(\text{i}t + \text{sgn}(t)0^+) \\
&D^\beta(\omega) = G_+^\beta(\omega) + \left(G_+^\beta(-\omega)\right)^*, && D^\beta(\omega) = -\mathcal{G}^\beta(\omega_l)|_{\text{i}\omega_l\to\omega+\text{i}0^+}
\end{aligned}
\tag{2.2.41}
$$

There is no simple relationship between $D^\beta(\omega)$ and $G^\beta(\omega)$.

When $O_1 = O_2^\dagger$, the functions $A_\pm^\beta$ are real and positive, and $\frac{1}{2\text{i}}[\mathcal{G}^\beta(\text{i}\omega_l \to \omega + \text{i}0^+) - \mathcal{G}^\beta(\text{i}\omega_l \to \omega - \text{i}0^+)] = \text{Im}\mathcal{G}^\beta(\text{i}\omega_l \to \omega + \text{i}0^+)$. We have

$$
\begin{aligned}
A_+^\beta(\omega) &= [1 + \eta n_\eta(\omega)](\pi)^{-1}\text{Im}\mathcal{G}^\beta(\text{i}\omega_l \to \omega + \text{i}0^+) \\
&= -[1 + \eta n_\eta(\omega)](\pi)^{-1}\text{Im}D^\beta(\omega) \\
&= -\frac{1}{1 + \eta \text{e}^{-\beta\omega}}(\pi)^{-1}\text{Im}G^\beta(\omega)
\end{aligned}
\tag{2.2.42}
$$

We see that the spectral functions $A_\pm^\beta$ can be determined from one of G^β, D^β, or $\mathcal{G}^\beta$, which in turn determines the other correlation functions.

As an example, let us calculate the finite-temperature susceptibility for our oscillator problem. In frequency space the imaginary-time path integral takes the form

$$
Z = \int \mathcal{D}[x(\tau)]\ \text{e}^{-\int_0^\beta \text{d}\tau\ [\frac{m}{2}\dot{x}^2 + \frac{m\omega_0^2}{2}x^2]} = \int \mathcal{D}x_\omega\ \text{e}^{-\frac{1}{2}\sum_{\omega_n=-\infty}^{\infty} x_{-\omega_n}[m\omega_n^2 + m\omega_0^2]x_{\omega_n}}
$$

where $x_{\omega_n} = \int_0^\beta \text{d}\tau\ x(\tau)\beta^{-1/2}\text{e}^{\text{i}\tau\omega_n}$. We find that

$$
\mathcal{G}_x^\beta(\omega_n) = \langle x_{\omega_n} x_{-\omega_n}\rangle = \frac{m^{-1}}{\omega_n^2 + \omega_0^2}
\tag{2.2.43}
$$

The finite-temperature susceptibility is then given by

$$
\chi(\omega) = -D(\omega) = \mathcal{G}_x^\beta(-\text{i}(\omega + \text{i}0^+)) = \frac{m^{-1}}{\omega_0^2 - \omega^2 - \text{i}\,\text{sgn}(\omega)0^+}
$$

which is temperature independent, as expected.

Problem 2.2.5.
Use the spectral representation to prove eqn (2.2.39).

Problem 2.2.6.
Prove eqn (2.2.35) from the definition of $\mathcal{G}^\beta(\tau)$.

Problem 2.2.7.
Prove eqn (2.2.43). Calculate $\langle x^2 \rangle = \mathcal{G}_x^\beta(\tau)|_{\tau=0}$ at finite temperatures. Check the $T \to 0$ and $T \to \infty$ limits. (Hint: The following trick, which is used frequently in

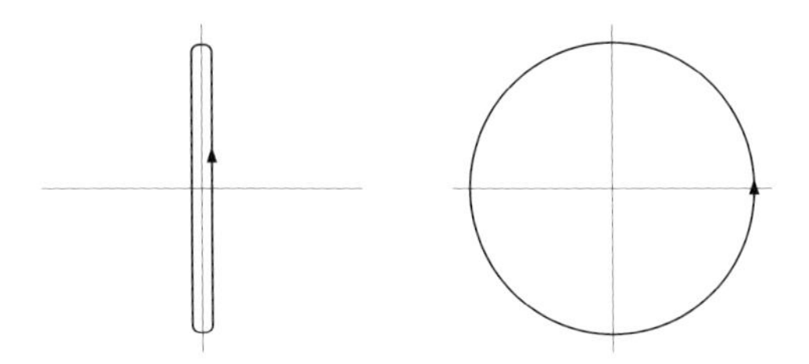

FIG. 2.3. Contours C_1 and C_2.

finite-temperature calculations, may be useful:

$$\sum_{n=-\infty}^{\infty} \frac{T}{\omega_n^2+\omega_0^2} = \oint_{C_1} \frac{\mathrm{d}z}{2\pi\mathrm{i}} \frac{1}{(-\mathrm{i}z)^2+\omega_0^2} \frac{1}{\mathrm{e}^{\beta z}-1}, \quad \oint_{C_2} \frac{\mathrm{d}z}{2\pi\mathrm{i}} \frac{1}{(-\mathrm{i}z)^2+\omega_0^2} \frac{1}{\mathrm{e}^{\beta z}-1} = 0$$

where C_1 is the contour around the imaginary-z axis and C_2 is the contour of an infinite circle around $z = 0$, see Fig. 2.3.)

2.3 Quantum spin, the Berry phase, and the path integral

2.3.1 The path integral representation of a quantum spin

- The Berry phase comes from the coherent-state representation of a family of quantum states. The Berry phase is a geometric phase determined by overlaps (inner products) between the coherent states.
- The action of the coherent path integral contains a Berry phase term.

Consider a spin-$S = \frac{1}{2}$ system described by the Hamiltonian $H = 0$. Can we have a path integral formulation of such a spin system? Firstly, the spin system with only two zero-energy states $|s_z\rangle$, $s_z = \pm\frac{1}{2}$, seems too simple to have a path integral formulation. Secondly, the path, as a time-dependent spin $s_z(t)$, is not a continuous function. So, to obtain a path integral formulation, we first make the simple spin system more complicated. Instead of using $|s_z\rangle$, we use the coherent states $|\boldsymbol{n}\rangle$ to describe different spin states.

The coherent states $|\boldsymbol{n}\rangle$ are labeled by a unit vector $\boldsymbol{n}$. It is an eigenstate of the spin operator in the $\boldsymbol{n}$ direction, namely $\boldsymbol{n}\cdot\boldsymbol{S}|\boldsymbol{n}\rangle = S|\boldsymbol{n}\rangle$, and is given by

$$|\boldsymbol{n}\rangle = |z\rangle = \begin{pmatrix} z_1 \\ z_2 \end{pmatrix}$$

$$\boldsymbol{n} = z^\dagger \boldsymbol{\sigma} z, \qquad |z_1|^2 + |z_2|^2 = 1$$

where $\boldsymbol{\sigma}$ are Pauli matrices. The relation $\boldsymbol{n} = z^\dagger\boldsymbol{\sigma} z$ is also called the spinor representation of the vector. Note that, for a given vector $\boldsymbol{n}$, the total phase of the spinor

z is not determined, i.e. z and $e^{i\theta}z$ give rise to the same $\boldsymbol{n}$. However, we may pick a phase and choose

$$z = \begin{pmatrix} e^{-i\phi}\cos\frac{\theta}{2} \\ \sin\frac{\theta}{2} \end{pmatrix} \tag{2.3.1}$$

where (θ, ϕ) are the spherical angles of $\boldsymbol{n}$.

As the state $|\boldsymbol{n}\rangle$ has both a vanishing potential energy and kinetic energy, one may expect the Lagrangian $L(\dot{\boldsymbol{n}}, \boldsymbol{n})$ in the coherent-state path integral to be simply zero. It turns out that the Lagrangian is quite non-trivial.

To calculate the Lagrangian, we note that the coherent states $|\boldsymbol{n}\rangle$ are complete:

$$\int \frac{d^2\boldsymbol{n}}{2\pi} |\boldsymbol{n}\rangle\langle\boldsymbol{n}| = I_{2\times 2} \tag{2.3.2}$$

(The factor $\frac{1}{2\pi}$ can be obtained by taking a trace on both sides of the equation and noting that $\mathrm{Tr}(|\boldsymbol{n}\rangle\langle\boldsymbol{n}|) = 1$.)

Inserting many $\int \frac{d^2\boldsymbol{n}}{2\pi} |\boldsymbol{n}\rangle\langle\boldsymbol{n}|$ into the time interval $[0, t]$, we find a path integral representation of the propagator $\langle\boldsymbol{n}_2|U(t,0)|\boldsymbol{n}_1\rangle$:[12]

$$\begin{aligned} iG(\boldsymbol{n}_2, \boldsymbol{n}_1, t) &= \langle\boldsymbol{n}_2|U(t,0)|\boldsymbol{n}_1\rangle \\ &= \int \prod_{i=1}^{N} \frac{d^2\boldsymbol{n}(t_i)}{2\pi}] \lim_{N\to\infty} \langle\boldsymbol{n}(t)|\boldsymbol{n}(t_N)\rangle ... \langle\boldsymbol{n}(t_2)|\boldsymbol{n}(t_1)\rangle\langle\boldsymbol{n}(t_1)|\boldsymbol{n}(0)\rangle \\ &= \int \mathcal{D}^2[\frac{\boldsymbol{n}(t)}{2\pi}]\, e^{iS[\boldsymbol{n}(t)]} \\ S[\boldsymbol{n}(t)] &= \int_0^t dt\, i z^\dagger \dot{z} \end{aligned} \tag{2.3.3}$$

The phase term $e^{i\int_0^t dt\, i z^\dagger \dot{z}}$ is purely a quantum effect and is called the Berry phase (Berry, 1984). Note that the term $a^*\dot{a}$ that appears in the coherent-state path integral is also a Berry phase. Thus, Berry's phase appears quite commonly in coherent-state path integrals. We can make the following three observations.

(a) $U(t, 0)$ is the identity operator. Thus, eqn (2.3.3) calculates the matrix elements of an identity operator!

(b) Despite $H = 0$, the action is nonzero. However, because the action contains only first-order time derivative terms, the nonzero action is consistent with $H = 0$. If the spin has a finite energy E, then the action will contain a term which is proportional to time: $S = -tE$. The Berry phase is of order t^0 and is different for the energy term.

(c) The path integral of the action $S(z) = S(\theta, \phi) = \int dt\, [-\frac{1}{2}\cos(\theta)\dot{\phi} -$

[12] Note that $\langle\boldsymbol{n}(\delta t)|\boldsymbol{n}(0)\rangle = z^\dagger(\delta t)z(0) = 1 - z^\dagger(\delta t)[z(\delta t) - z(0)] \approx e^{-z^\dagger \dot{z}\delta t}$.

$\frac{1}{2}\dot{\phi}]$ is a phase-space path integral (see eqn (2.1.11)). Thus, (θ, ϕ), just like (q, p), parametrizes a two-dimensional phase space. However, unlike (q, p) which parametrizes a flat phase space, (θ, ϕ) parametrizes a curved phase space (which is a two-dimensional sphere).

2.3.2 The Berry phase as the extra phase in an adiabatic evolution

Let us look at the Berry phase from another angle. Consider a spin in a constant magnetic field $\boldsymbol{B} = -B\boldsymbol{n}$. The evolution of the ground state is given by

$$\langle \boldsymbol{n}| \mathrm{e}^{-\mathrm{i} \int_0^T \mathrm{d}t\, \boldsymbol{B} \cdot \boldsymbol{S}} |\boldsymbol{n}\rangle = \mathrm{e}^{-\mathrm{i} E_0 T}$$

where E_0 is the ground-state energy. We call $-E_0 T$ the action of the evolution. Now we allow the orientation of $\boldsymbol{B}$ to slowly change in time, namely $\boldsymbol{B} = -B\boldsymbol{n}(t)$. Under such adiabatic motion, the ground state evolves as $|\boldsymbol{n}(t)\rangle$. The action S of such an evolution is given by

$$\langle \boldsymbol{n}| \mathrm{e}^{-\mathrm{i} \int_0^T \mathrm{d}t\, \boldsymbol{B}(t) \cdot \boldsymbol{S}} |\boldsymbol{n}\rangle = \mathrm{e}^{\mathrm{i} S}$$

One may guess that $S = -E_0 T$ because the ground-state energy (at any given time) is still given by E_0. By inserting many identities (2.3.2) into $[0, T]$, we see that actually

$$\langle \boldsymbol{n}| \mathrm{e}^{-\mathrm{i} \int_0^T \mathrm{d}t\, \boldsymbol{B}(t) \cdot \boldsymbol{S}} |\boldsymbol{n}\rangle = \mathrm{e}^{-\mathrm{i} E_0 T} \mathrm{e}^{\mathrm{i} \int_0^T \mathrm{d}t\, \mathrm{i} \langle \boldsymbol{n}(t)| \frac{\mathrm{d}}{\mathrm{d}t} |\boldsymbol{n}(t)\rangle}$$

The action is given by

$$S = -E_0 T + \int_0^T \mathrm{d}t\, \mathrm{i} \langle \boldsymbol{n}(t)| \frac{\mathrm{d}}{\mathrm{d}t} |\boldsymbol{n}(t)\rangle = -E_0 T + \int_0^T \mathrm{d}t\, \mathrm{i} z^\dagger \dot{z} \tag{2.3.4}$$

There is an extra Berry phase.

2.3.3 The Berry phase and parallel transportation

- The Berry phase is well defined only for closed paths.
- The Berry phase represents the frustration of assigning a common phase to all of the coherent states.

The Berry phase has some special properties. The Berry phase associated with the path $\boldsymbol{n}(t)$ connecting $\boldsymbol{n}(0)$ and $\boldsymbol{n}(T)$ is given by $\mathrm{e}^{\mathrm{i}\theta_B} = \mathrm{e}^{-\int_0^T \mathrm{d}t\, \langle \boldsymbol{n}(t)| \frac{\mathrm{d}}{\mathrm{d}t} |\boldsymbol{n}(t)\rangle} = \mathrm{e}^{\mathrm{i} \int_0^T \mathrm{d}t\, \mathrm{i} z^\dagger \dot{z}}$. This expression tells us that the Berry phase for an open path is not well defined. This is because the state $|\boldsymbol{n}\rangle$ that describes a spin in the $\boldsymbol{n}$ direction can have any phase, and such a phase is unphysical. If we change the phases of the states, namely $|\boldsymbol{n}(t)\rangle \to \mathrm{e}^{\mathrm{i}\phi(t)} |\boldsymbol{n}(t)\rangle$, then the

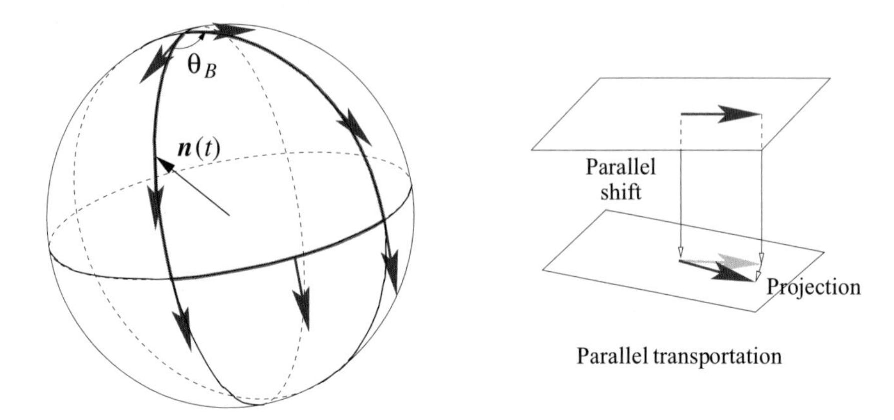

FIG. 2.4. The geometrical representation of the parallel transportation of the phase of the spin-1 coherent state $|\boldsymbol{n}(t)\rangle$. The phase of $|\boldsymbol{n}(t)\rangle$ can be represented by a two-dimensional unit vector in the tangent plane at the point $\boldsymbol{n}(t)$ on the unit sphere. In this representation the parallel transportation can be broken into the two geometrical operations of a parallel shift of the two-dimensional vector in three-dimensional space, and a projection back into the tilted tangent plane. A parallel transportation of a two-dimensional tangent vector around a closed loop will rotate the vector. The angle of the rotation is the Berry phase for the spin-1 state. (The angle of the rotation for a small loop divided by the area enclosed by the loop is the curvature of the sphere. According to Einstein's theory of general relativity, curvature of space = gravity.)

Berry phase is also changed, namely $\theta_B \to \theta_B + \phi(0) - \phi(T)$. In other words, a given path $\boldsymbol{n}(t)$ can have different spinor representations $z(t)$. Different spinor representations give rise to different Berry phases.

The above discussion seems to suggest that the Berry phase is unphysical, just like the phase of the spin state $|\boldsymbol{n}\rangle$. It seems that the origin of the Berry phase is our arbitrary choice of the phases for different spin states. So, if we can choose a 'common' phase for all of the spin states, then there will not be a Berry phase.

Let us try to pick a common phase for the spin states $|\boldsymbol{n}(t)\rangle$ along a path $\boldsymbol{n}(t)$, so that we can remove the Berry phase. We first pick a phase for the spin state $|\boldsymbol{n}(0)\rangle$ at the beginning of the path. Then we pick the 'same phase' for a spin state $|\boldsymbol{n}(\Delta t)\rangle$ at the next point. The problem here is what do we mean by the 'same phase' when $\boldsymbol{n}(0)$ and $\boldsymbol{n}(\Delta t)$ are different. As $\boldsymbol{n}(0)$ and $\boldsymbol{n}(\Delta t)$ are nearly parallel, $|\boldsymbol{n}(0)\rangle$ and $|\boldsymbol{n}(\Delta t)\rangle$ are almost the same states. This means that $|\langle \boldsymbol{n}(0)|\boldsymbol{n}(\Delta t)\rangle| \approx 1$. Thus, we can define $|\boldsymbol{n}(0)\rangle$ and $|\boldsymbol{n}(\Delta t)\rangle$ to have the same phase if $\langle \boldsymbol{n}(0)|\boldsymbol{n}(\Delta t)\rangle$ is real and positive. If we pick the phase this way for all of the states along the path $\boldsymbol{n}(t)$, then we have defined a 'parallel transportation' along the path (see Fig. 2.4). For such a path, we have $\langle \boldsymbol{n}(0)|\frac{\mathrm{d}}{\mathrm{d}t}|\boldsymbol{n}(t)\rangle = 0$. The Berry phase vanishes if we choose a common phase for all of the spin states on the path.

However, for a closed path, we may not be able to pick a common phase for all of the states along the path. This is because, for a closed path, $|\boldsymbol{n}(T)\rangle$ and $|\boldsymbol{n}(0)\rangle$ represent the same spin state. Their choice of phases, as determined from the parallel transportation, may not agree. The discrepancy is simply the Berry phase for the closed path. From the change of the Berry phase, namely $\theta_B \to \theta_B + \phi(0) - \phi(T)$, induced by choosing different phases of the spin states, namely $|\boldsymbol{n}(t)\rangle \to \mathrm{e}^{\,\mathrm{i}\phi(t)}|\boldsymbol{n}(t)\rangle$, we see that the Berry phase for a closed path with $\boldsymbol{n}(0) = \boldsymbol{n}(T)$ is well defined (up to a multiple of 2π) because $\mathrm{e}^{\,\mathrm{i}\phi(0)} = \mathrm{e}^{\,\mathrm{i}\phi(T)}$. Thus, it is meaningful to discuss the Berry phase for a closed path, or to compare the difference between the Berry phases for two paths with the same start and end points.

We see that the Berry phase vanishes if we can define a common phase for all of the states. A nonzero Berry phase for a loop implies that it is impossible to choose a common phase for all of the states on the loop. The value of the Berry phase represents the amount of frustration in defining such a common phase. The concepts of parallel transportation and the frustration in parallel transportation are very important. Both the electromagnetic field and the gravitational field are generalized Berry phases. They describe the frustrations in parallel transportation for some more general vectors, instead of a two-dimensional vector (a complex number) as discussed above.

The Berry phase is a geometric phase in the sense that two paths $\boldsymbol{n}_1(t)$ and $\boldsymbol{n}_2(t)$ will have the same Berry phase as long as they trace out the same trajectory on the unit sphere. If the trajectory $\boldsymbol{n}(t)$ on the unit sphere spans a solid angle Ω, then the Berry phase for that path will be given by (see Problem 2.3.4)

$$\theta_B = \frac{\Omega}{2} \tag{2.3.5}$$

2.3.4 The Berry phase and the equation of motion

- The Berry phase can affect the dynamic properties of a system. It can change the equation of motion.

The path integral representation (2.3.3) can be generalized to a spin-S system and to a nonzero Hamiltonian $H = \boldsymbol{B} \cdot \boldsymbol{S}$. The coherent state (the eigenstate of $\boldsymbol{n} \cdot \boldsymbol{S}$ with the largest eigenvalue S), namely $|\boldsymbol{n}\rangle$, can be written as a direct product of the $2S$ spin-$\frac{1}{2}$ coherent states discussed above:

$$|\boldsymbol{n}\rangle = |z\rangle \otimes |z\rangle \otimes \ldots \otimes |z\rangle$$

Thus, the Berry phase is $2S$ times the spin-$\frac{1}{2}$ Berry phase:

$$\theta_B = \oint dt\, 2S\mathrm{i}\, z^\dagger \dot{z} = S\Omega \tag{2.3.6}$$

The path integral representation now becomes

$$\mathrm{i}G(n_2, n_1, t) = \langle n_2|U(t, 0)|n_1\rangle = \int \mathcal{D}[\frac{\boldsymbol{n}(t)}{2\pi}]\ \mathrm{e}^{\mathrm{i}\int \mathrm{d}t\ [2S\mathrm{i}z^\dagger \dot{z} - \boldsymbol{B}\cdot\boldsymbol{n}S]} \tag{2.3.7}$$

Let us treat

$$S[\boldsymbol{n}(t)] = \int \mathrm{d}t\ [2S\mathrm{i}z^\dagger \dot{z} - \boldsymbol{B}\cdot\boldsymbol{n}S] \tag{2.3.8}$$

as a classical action and consider a classical motion of spin governed by such an action. To obtain the classical equation of motion, let us compare $S[\boldsymbol{n}(t)]$ and $S[\boldsymbol{n}(t)+\delta\boldsymbol{n}(t)]$. (Note that the two paths $\boldsymbol{n}(t)$ and $\boldsymbol{n}(t)+\delta\boldsymbol{n}(t)$ have the same start and end points.) From the geometrical interpretation of the Berry phase (2.3.6), we find that the difference between the Berry phases for the two paths is S times the area between the two trajectories: $\delta\theta_B = \int \mathrm{d}t\ [S\boldsymbol{n}\cdot(\dot{\boldsymbol{n}}\times\delta\boldsymbol{n})$. Thus,

$$S[\boldsymbol{n}(t)+\delta\boldsymbol{n}(t)] - S[\boldsymbol{n}(t)] = \int \mathrm{d}t\ [S\boldsymbol{n}\cdot(\dot{\boldsymbol{n}}\times\delta\boldsymbol{n}) - \boldsymbol{B}\cdot\delta\boldsymbol{n}S]$$

which leads to the equation of motion

$$\boldsymbol{n}\times\dot{\boldsymbol{n}} = \boldsymbol{B} - \boldsymbol{n}(\boldsymbol{n}\cdot\boldsymbol{B})$$

(note that $\delta\boldsymbol{n}$ is always perpendicular to $\boldsymbol{n}$) or

$$\dot{\boldsymbol{S}} = \boldsymbol{B}\times\boldsymbol{S} \tag{2.3.9}$$

where $\boldsymbol{S} = S\boldsymbol{n}$ corresponds to the classical spin vector. This is a strange equation of motion in that the velocity (rather than the acceleration) is proportional to the force represented by $\boldsymbol{B}$. Even more strange is that the velocity points in a direction perpendicular to the force. However, this also happens to be the correct equation of motion for the spin. We see that the Berry phase is essential in order to recover the correct spin equation of motion.

Problem 2.3.1.
Show that θ_B defined in Fig. 2.4 is actually the Berry phase for a spin-1 spin.

Problem 2.3.2.
Actually, it is possible to have a path integral formulation using the discontinuous path $s_z(t)$ to describe a spin-$\frac{1}{2}$ system. Assume that $H = a\sigma^1 + b\sigma^3$ for a spin-$\frac{1}{2}$ system. Find the path integral representation of the time evolution operator $\langle s'_z|U(t,0)|s_z\rangle$. (Hint: $\sum_{s_z=\pm 1/2}|s_z\rangle\langle s_z| = 1$.)

Problem 2.3.3.
Calculate the Berry phase for a closed path $\phi = 0 \to 2\pi$ and θ fixed, using the spinor representation (2.3.1).

Calculate the Berry phase for the same path but with the different spinor representation

$$z = \begin{pmatrix} \cos\frac{\theta}{2} \\ \mathrm{e}^{-\mathrm{i}\phi}\sin\frac{\theta}{2} \end{pmatrix}$$

From this example, we see that the Berry phase has 2π ambiguity depending on different spinor representations. Thus, the Berry phase θ_B cannot be expressed directly in terms of the path $\boldsymbol{n}$. This is why we have to introduce a spinor representation to express the Berry phase. However, $\mathrm{e}^{\mathrm{i}\theta_B}$ is well defined and depends only on the path $\boldsymbol{n}(t)$. This is all we need in order to have a well-defined path integral.

Problem 2.3.4.
Prove eqn (2.3.5) for a small closed path $(\theta, \phi) \to (\theta + d\theta, \phi) \to (\theta + d\theta, \phi + d\phi) \to (\theta, \phi + d\phi) \to (\theta, \phi)$.
Prove eqn (2.3.5) for a general large closed path.

Problem 2.3.5.
Spin and a particle on a sphere:

1. Show that the spin action (2.3.8) can be written as

$$S[\boldsymbol{n}(t)] = \int \mathrm{d}t\ (A_\theta \dot{\theta} + A_\phi \dot{\phi})$$

assuming that $\boldsymbol{B} = 0$. Find A_θ and A_ϕ.

2. The above action can be regarded as the $m \to 0$ limit of the following action:

$$S_p[\boldsymbol{n}(t)] = \int \mathrm{d}t\ (A_\theta \dot{\theta} + A_\phi \dot{\phi} + \frac{1}{2} m \dot{\boldsymbol{n}}^2)$$

S_p describes a particle of mass m moving on a unit sphere. The particle also experiences a magnetic field described by (A_ϕ, A_θ). Show that (A_ϕ, A_θ) gives a uniform magnetic field. Show that the total flux quanta on the sphere is $2S$.

3. We know that a particle in a uniform magnetic field has a Landau level structure. All of the states in the same Landau level have the same energy. The Landau levels are separated by a constant energy gap $\hbar\omega_c$ and $\omega_c \to \infty$ as $m \to 0$. Therefore, in the $m \to 0$ limit, only the states in the first Landau level appear in the low-energy Hilbert space. Find the number of states in the first Landau level for a particle on a sphere if the total flux quanta of the uniform magnetic field is N_ϕ.

Problem 2.3.6.
Spin waves: Consider a spin-S quantum spin chain $H = \sum_i J\boldsymbol{S}_i\boldsymbol{S}_{i+1}$. For $J < 0$, the classical ground state is a ferromagnetic state with $\boldsymbol{S}_i = S\hat{z}$. For $J > 0$, it is an anti-ferromagnetic state with $\boldsymbol{S}_i = S\hat{z}(-)^i$.

1. Write down the action for the spin chain.

2. Find the equation of motion for small fluctuations $\delta\boldsymbol{S}_i = \boldsymbol{S}_i - S\hat{z}$ around the ferromagnetic ground state. Transform the equation of motion to frequency and momentum space and find the dispersion relation. Show that $\omega \propto k^2$ for small k.

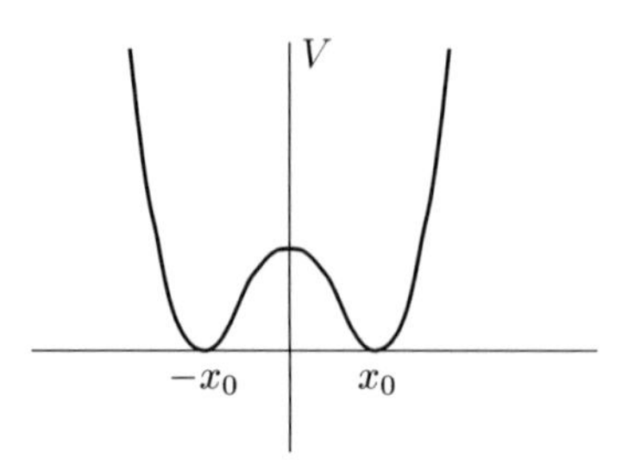

FIG. 2.5. A double-well potential.

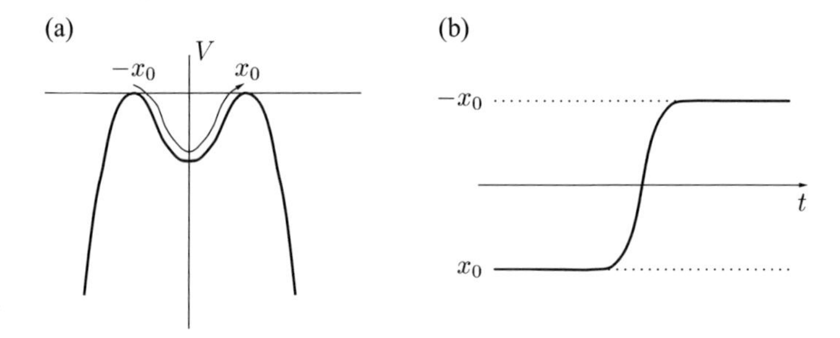

FIG. 2.6. (a) A single instanton solution can be viewed as a motion between the two maxima of the invered potential. (b) The same motion plotted in space-time.

3. Find the equation of motion for small fluctuations $\delta \boldsymbol{S}_i = \boldsymbol{S}_i - S\hat{z}(-)^i$ around the anti-ferromagnetic ground state. Find the dispersion relation. Show that $\omega \propto |k - \pi|$ for k near to π.

It is interesting to see that the spin waves above the ferromagnetic and anti-ferromagnetic states have very different dispersions. In classical statistical physics the ferromagnetic and anti-ferromagnetic states have the same thermodynamical properties. It is the Berry phase that makes them quite different at the quantum level.

2.4 Applications of the path integral formulation

2.4.1 Tunneling through a barrier

- Sometimes the initial and final configurations are connected by several stationary paths. The path with the least action represents the default path and the other paths represent the instanton effect.
- Tunneling through barriers is an instanton effect.

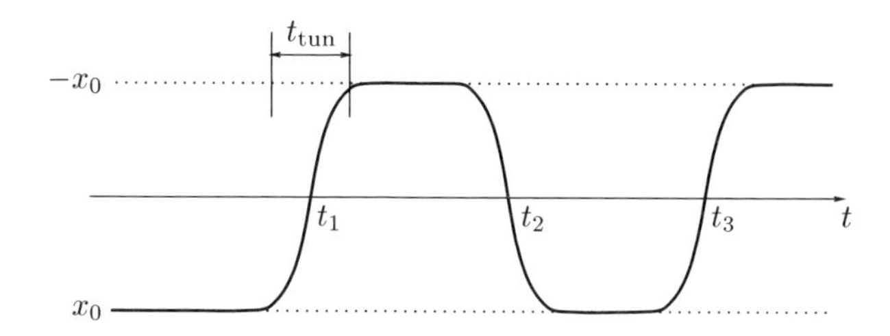

FIG. 2.7. A multi-instanton configuration—an instanton gas.

The path integral not only allows us to calculate perturbative effects, such as linear responses, but it also allows us to calculate non-perturbative effects. Here we would like to study a simple example in order to understand how the path integral can lead to non-perturbative results. We will study tunneling in a double-well potential (Fig. 2.5). In the absence of tunneling, the particle fluctuates near a minimum which can be described by a harmonic oscillator with frequency ω_0. These fluctuations lead to the following propagator:

$$\langle x_0|\mathrm{e}^{-TH}|x_0\rangle = \langle -x_0|\mathrm{e}^{-TH}|-x_0\rangle = \mathrm{e}^{-T\omega_0/2}, \quad \langle x_0|\mathrm{e}^{-TH}|-x_0\rangle = 0.$$

However, the above result is not quite correct due to tunneling. In the presence of tunneling, particles can go back and forth between the two minima and the propagator receives some additional contributions. We note that tunneling is a non-perturbative effect. Performing perturbation around a minimum can never lead to tunneling.

To study the effects of tunneling, we use an imaginary-time path integral and a saddle-point approximation. In addition to the two trivial stationary paths $x(t) = x_0$ and $x(t) = -x_0$ which connect x_0 to x_0 and $-x_0$ to $-x_0$, respectively, there is another stationary path that connects x_0 to $-x_0$. Such a path minimizes the action $\int \mathrm{d}\tau \left[\frac{1}{2}m\dot{x}^2 + V'(x)\right]$ and satisfies $m\frac{\mathrm{d}^2x}{\mathrm{d}\tau^2} = V(x)$. This is Newton's equation for an inverted potential $-V(x)$ (Fig. 2.6(a)) and the solution looks like that shown in Fig. 2.6(b). We see that the switch between $\pm x_0$ happens over a short time interval and this event is called an instanton. The path in Fig. 2.6(b) is not the only stationary path that links x_0 and $-x_0$. The multi-instanton path in Fig. 2.7 is also a stationary path. Similarly, the multi-instanton paths also give rise to other stationary paths that link x_0 to x_0 and $-x_0$ to $-x_0$.

Thus, to calculate the propagator, we need to sum over all of the contributions from the different stationary paths. For the propagator x_0 to x_0, we have

$$\langle x_0|\mathrm{e}^{-TH}|x_0\rangle = \mathrm{e}^{-T\omega_0/2} \sum_{n=\text{even}} \int_0^T \mathrm{d}\tau_n ... \int_0^{\tau_2} \mathrm{d}\tau_1 \left(K\mathrm{e}^{-S_0}\right)^n$$
$$= \mathrm{e}^{-T\omega_0/2} \sum_{n=\text{even}} \frac{T^n}{n!} \left(K\mathrm{e}^{-S_0}\right)^n = \cosh(TK\mathrm{e}^{-S_0})$$

where S_0 is the minimal action of the instanton and K comes from the path integral of fluctuations around a single instanton. Here K should have a dimension of frequency. Its value is roughly given by the frequency at which the particle hits the barrier. Similarly, we also have

$$\langle x_0|\mathrm{e}^{-TH}|-x_0\rangle = \mathrm{e}^{-T\omega_0/2} \sum_{n=\text{odd}} \int_0^T \mathrm{d}\tau_n ... \int_0^{\tau_2} \mathrm{d}\tau_1 \left(K\mathrm{e}^{-S_0}\right)^n$$
$$= \mathrm{e}^{-T\omega_0/2} \sum_{n=\text{odd}} \frac{T^n}{n!} \left(K\mathrm{e}^{-S_0}\right)^n = \sinh(TK\mathrm{e}^{-S_0})$$

Therefore,

$$\langle\psi_\pm|\mathrm{e}^{-TH}|\psi_\pm\rangle = \mathrm{e}^{-T\omega_0/2}[\cosh(TK\mathrm{e}^{-S_0})\pm\sinh(TK\mathrm{e}^{-S_0})] = \mathrm{e}^{-T(\frac{\omega_0}{2}\pm K\mathrm{e}^{-S_0})}$$

where $|\psi_\pm\rangle = \frac{|x_0\rangle\pm|-x_0\rangle}{\sqrt{2}}$. We see that $|\psi_\pm\rangle$ are energy eigenstates with energies $E = \frac{\omega_0}{2} \pm K\mathrm{e}^{-S_0}$.

In order to evaluate S_0, one notes that $\dot{x} = \sqrt{2V(x)/m}$ from energy conservation. Thus,

$$S_0 = \int_{-\infty}^{+\infty} \mathrm{d}\tau\, [\frac{1}{2}m\dot{x}^2 + V(x)] = \int_{-x_0}^{+x_0} \mathrm{d}x\, \sqrt{2mV(x)}$$

Restoring the $\hbar$, we have $\Delta E = 2K\mathrm{e}^{-\hbar^{-1}\int_{-x_0}^{+x_0} \mathrm{d}x\, \sqrt{2mV(x)}}$, which is just the WKB result. However, the path integral does teach us one new thing, namely the tunneling time, t_{tun}, which is the time scale associated with a single instanton. We have a very simple picture of how long it takes for a particle to go through a barrier.

Problem 2.4.1.
Assume that the double-well potential is given by $V(x) = \frac{g}{4}(x^2 - x_0^2)^2$.

1. Estimate S_0 and the tunneling time t_{tun} (i.e. the size of the instantons).
2. Assuming that $K \sim \sqrt{\frac{gx_0^2}{m}}$ (which is the oscillation frequency near one of the minima), estimate the (average) density of instantons in the time direction.

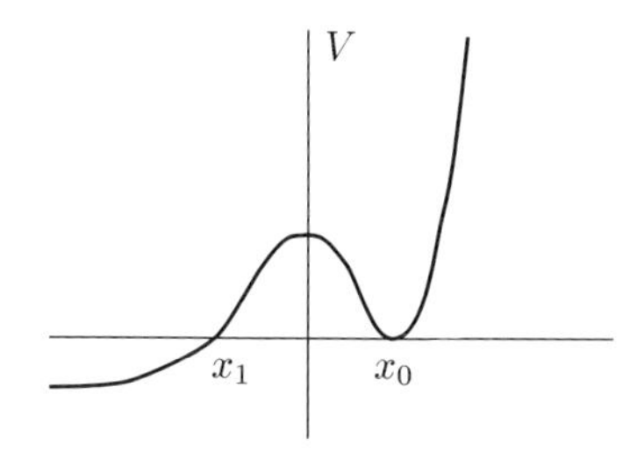

FIG. 2.8. A potential with a meta-stable state.

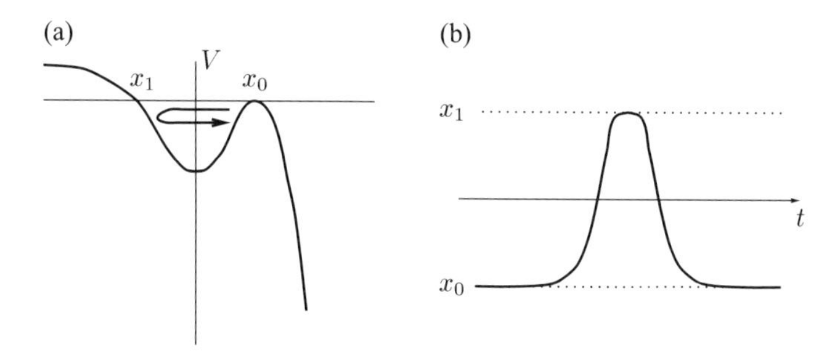

FIG. 2.9. (a) A single-bounce solution can be viewed as a motion in the invered potential. (b) The same motion plotted in space-time.

3. The semiclassical picture is a good description of the tunneling if both $S_0 \gg 1$ and the instantons do not overlap. For which values of g and x_0 are the above two conditions satisfied?

2.4.2 Fate of a meta-stable state

- The decay of a meta-stable state is also an instanton effect.

In this section, we consider a particle in the potential $V(x)$ shown in Fig. 2.8. The particle is initially at x_0. We would like to calculate the decay rate for the particle to escape through the barrier. For this purpose we calculate $\langle x_0|\mathrm{e}^{-TH}|x_0\rangle$. If we only consider small fluctuations around x_0, then we may make the approximation $V(x) = \frac{1}{2}m\omega_0^2(x-x_0)^2$ and find that $\langle x_0|\mathrm{e}^{-TH}|x_0\rangle = \mathrm{e}^{-T\omega_0/2}$ for large T. Analytically continuing to real time, we see that $\langle x_0|\mathrm{e}^{-\mathrm{i}TH}|x_0\rangle = \mathrm{e}^{-\mathrm{i}T\omega_0/2}$, and there is no decay. To find the decay, we need to find a process that makes $|\langle x_0|\mathrm{e}^{-\mathrm{i}TH}|x_0\rangle|$ decrease with time. We will see that the tunneling can cause decay. In the path integral, the tunneling is again described by the instantons.

The imaginary-time Lagrangian $L = \frac{1}{2}m\dot{x}^2 + V(x)$ has a non-trivial stationary path $x_c(t)$ that connects x_0 to x_0. The existence of this stationary path (called a bounce or an instanton) can be seen from the classical motion in the potential $-V(x)$ (Fig. 2.9). Including the bounces and repeating the procedures in Section 2.4.1, we find that

$$\langle x_0|\mathrm{e}^{-TH}|x_0\rangle = \mathrm{e}^{-T\omega_0/2}\sum_n \int_0^T \mathrm{d}\tau_n \int_0^{\tau_n} \mathrm{d}\tau_{n-1}... \int_0^{\tau_2} \mathrm{d}\tau_1 \, \left(K\mathrm{e}^{-S_0}\right)^n$$
$$= \mathrm{e}^{-T(\frac{\omega_0}{2} - K\mathrm{e}^{-S_0})} \tag{2.4.1}$$

where S_0 is the action of the bounce and K is a coefficient due to the fluctuations around the bounce. At a first glance, it appears that the bounce only corrects the ground-state energy $\frac{\omega_0}{2}$ by a small amount $\Delta E = -K\mathrm{e}^{-S_0}$. It is hard to see any decay because the bounce returns back to x_0 after a long time. In fact, we will see that the coefficient K is imaginary, and in real time $\langle x_0|\mathrm{e}^{-\mathrm{i}TH}|x_0\rangle = \mathrm{e}^{-\mathrm{i}T\frac{\omega_0}{2} - T|K|\mathrm{e}^{-S_0}}$. *The bounce precisely describes the decay of the $|x_0\rangle$ state.* This is quite an amazing result.

To understand why K is imaginary, we need to discuss how to calculate K. The discussion here also applies to K in the tunneling amplitude presented in Section 2.4.1.

From the way in which we include the bounces in eqn (2.4.1), we see that $K\mathrm{e}^{-S_0}$ is a ratio of two paths integrals around $x_c(t)$ and $x = x_0$:

$$K\mathrm{e}^{-S_0} = \frac{\int_{x_c} \mathcal{D}\delta x \; \mathrm{e}^{-S}}{\int_{x=x_0} \mathcal{D}\delta x \; \mathrm{e}^{-S}} \tag{2.4.2}$$

For small fluctuations, the Lagrangian can be expanded to quadratic order. We have, near x_c,

$$S = S_0 + \int \mathrm{d}\tau \; \delta x \frac{1}{2}[-m\frac{\mathrm{d}^2}{\mathrm{d}\tau^2} + V''(x_c(\tau))]\delta x$$

and near $x = x_0$ we have

$$S = \int \mathrm{d}\tau \; \delta x \frac{1}{2}[-m\frac{\mathrm{d}^2}{\mathrm{d}\tau^2} + V''(x_0)]\delta x$$

Thus, after performing the Gaussian integral, we have

$$K = \frac{\int_{x_c} \mathcal{D}\delta x \; \mathrm{e}^{-\int \mathrm{d}\tau \; \frac{1}{2}\delta x[-m\frac{\mathrm{d}^2}{\mathrm{d}\tau^2} + V''(x_c(\tau))]\delta x}}{\int_{x=x_0} \mathcal{D}\delta x \; \mathrm{e}^{-\int \mathrm{d}\tau \; \frac{1}{2}\delta x[-m\frac{\mathrm{d}^2}{\mathrm{d}\tau^2} + V''(x_0)]\delta x}}$$
$$= \left(\frac{\mathrm{Det}(-m\frac{\mathrm{d}^2}{\mathrm{d}\tau^2} + V''(x_0))}{\mathrm{Det}(-m\frac{\mathrm{d}^2}{\mathrm{d}\tau^2} + V''(x_c))}\right)^{1/2} \tag{2.4.3}$$

However, eqn (2.4.3) does not make sense because the operator $-m\frac{\mathrm{d}^2}{\mathrm{d}\tau^2} + V''(x_c)$ has a zero eigenvalue. To see this, we note that both $x_c(\tau)$ and $x_c(\tau + \eta)$ are stationary paths and they satisfy the equation of motion

$$-m\frac{\mathrm{d}^2 x_c(\tau)}{\mathrm{d}\tau^2} + V'(x_c(\tau)) = -m\frac{\mathrm{d}^2 x_c(\tau+\eta)}{\mathrm{d}\tau^2} + V'(x_c(\tau+\eta)) = 0$$

Expanding the above to linear order in η, we find that $\eta\dot{x}_c(\tau) = x_c(\tau+\eta) - x_c(\tau)$ satisfies

$$[-m\frac{\mathrm{d}^2}{\mathrm{d}\tau^2} + V''(x_c(\tau))]\dot{x}_c(\tau) = 0$$

Thus, $\dot{x}_c(\tau)$ is an eigenstate of $(-m\frac{\mathrm{d}^2}{\mathrm{d}\tau^2} + V''(x_c))$ with a zero eigenvalue.

From the above discussion, it is also clear that the zero mode $\eta\dot{x}_c(\tau)$ corresponds to the position of the bounce which has already been integrated over in eqn (2.4.1). Thus, eqn (2.4.2) is not quite correct. We actually have

$$\int \mathrm{d}\tau_1\, K\mathrm{e}^{-S_0} = \frac{\int_{x_c} \mathcal{D}\delta x\ \mathrm{e}^{-S}}{\int_{x=x_0} \mathcal{D}\delta x\ \mathrm{e}^{-S}} \tag{2.4.4}$$

To evaluate eqn (2.4.4), we need to define the path integral more carefully. Let $x_n(\tau)$ be the nth normalized eigenstate of the operator $-m\frac{\mathrm{d}^2}{\mathrm{d}\tau^2} + V''(x_c)$. Then δx can be written as $\delta x = \sum c_n x_n(\tau)$ and a truncated 'partition function' can be defined by

$$\begin{aligned} Z_N(x_c) \equiv& A_N \int \prod_1^N \frac{\mathrm{d}c_n}{\sqrt{2\pi}}\ \mathrm{e}^{-\int \mathrm{d}\tau\ \frac{1}{2}\delta x(-m\frac{\mathrm{d}^2}{\mathrm{d}\tau^2}+V''(x_c(\tau))]\delta x} \\ =& A_N\left(\mathrm{Det}_N(-m\frac{\mathrm{d}^2}{\mathrm{d}\tau^2} + V''(x_c))\right)^{-1/2} \end{aligned}$$

where Det_N includes only the products of the first N eigenvalues. The real 'partition function' Z is the $N \to \infty$ limit of Z_N. Due to the zero mode $x_1(\tau)$, the above result needs to be modified. We need to separate the zero mode and obtain

$$Z_N(x_c) = A_N \int \frac{\mathrm{d}c_1}{\sqrt{2\pi}} \left(\mathrm{Det}'_N(-m\frac{\mathrm{d}^2}{\mathrm{d}\tau^2} + V''(x_c))\right)^{-1/2}$$

where Det' excludes zero eigenvalues.

The integration $\int \mathrm{d}c_1$ is actually integration over the position of the bounce, namely $\int \mathrm{d}\tau_1$. It turns out that c_1 and τ_1 are related through $\mathrm{d}\tau_1 = \frac{\mathrm{d}c_1}{\sqrt{S_0/m}}$.[13]

[13] We note that

$$\frac{m}{2}\frac{\mathrm{d}}{\mathrm{d}\tau}(\dot{x}_c)^2 = m\dot{x}_c\ddot{x}_c = \dot{x}_c V'(x_c) = \frac{\mathrm{d}}{\mathrm{d}\tau}V(x_c)$$

Therefore,

$$Z_N(x_c) = A_N \int \mathrm{d}\tau_1 \sqrt{\frac{S_0}{2\pi m}} \left(\mathrm{Det}'_N(-m\frac{\mathrm{d}^2}{\mathrm{d}\tau^2} + V''(x_c)) \right)^{-1/2} \tag{2.4.5}$$

Similarly, we can also introduce $Z_N(x_0)$ for the path integral around $x = x_0$:

$$Z_N(x_0) = A_N \left(\mathrm{Det}_N(-m\frac{\mathrm{d}^2}{\mathrm{d}\tau^2} + V''(x_0)) \right)^{-1/2} \tag{2.4.6}$$

We can now write

$$\int \mathrm{d}\tau\, K = \lim_{N\to\infty} \frac{Z_N(x_c)}{Z_N(x_0)}$$

which leads to

$$\begin{aligned} K &= \sqrt{\frac{S_0}{2\pi m}} \left(\frac{\mathrm{Det}(-m\frac{\mathrm{d}^2}{\mathrm{d}\tau^2} + V''(x_0))}{\mathrm{Det}'(-m\frac{\mathrm{d}^2}{\mathrm{d}\tau^2} + V''(x_c))} \right)^{1/2} \\ &= \sqrt{\frac{S_0}{2\pi}} \left(\frac{\mathrm{Det}(-\frac{\mathrm{d}^2}{\mathrm{d}\tau^2} + m^{-1}V''(x_0))}{\mathrm{Det}'(-\frac{\mathrm{d}^2}{\mathrm{d}\tau^2} + m^{-1}V''(x_c))} \right)^{1/2} \end{aligned} \tag{2.4.7}$$

where we have used the fact that Det contains one more eigenvalue than Det$'$ does. The same fact also tells us that K has dimension τ^{-1}.

All of the eigenvalues of $-\frac{\mathrm{d}^2}{\mathrm{d}\tau^2} + m^{-1}V''(x_0)$ are positive. The operator $-\frac{\mathrm{d}^2}{\mathrm{d}\tau^2} + m^{-1}V''(x_c)$ contains a zero mode $\dot{x}_c(\tau)$ which has one node. Thus, $-\frac{\mathrm{d}^2}{\mathrm{d}\tau^2} + m^{-1}V''(x_c)$ has one, and only one, negative eigenvalue which makes K imaginary.

The same formula (2.4.7) also applies to the tunneling problem considered in Section 2.4.1. However, the zero mode $\dot{x}_c$ in that case has no node, and $-\frac{\mathrm{d}^2}{\mathrm{d}\tau^2} + m^{-1}V''(x_c)$ has no negative eigenvalues. Thus, K is real for the tunneling problem.

We have already mentioned that the coefficient K is due to fluctuations around the stationary path. From the above calculation, we see that K can be expressed in

---

Thus, $\frac{m}{2}(\dot{x}_c)^2 = V(x_c)$ if we assume that $V(x_0) = 0$. As

$$S_0 = \int \mathrm{d}\tau\, [\frac{1}{2}m\dot{x}_c^2 + V(x_c)] = \int \mathrm{d}\tau\, m\dot{x}_c^2$$

the normalized zero mode has the form

$$x_1(\tau) = \frac{\dot{x}_c(\tau)}{\sqrt{S_0/m}}$$

Thus, $\delta x = c_1 x_1$ will displace the bounce by $\delta\tau = \frac{c_1}{\sqrt{S_0/m}}$. This allows us to convert $\mathrm{d}c_1$ to $\mathrm{d}\tau_1$.

terms of determinants of operators if we only include small fluctuations. Coleman (1985) has discussed several ways to calculate the ratio of determinants. I will not go into more detail here.

One may have already realized that the present problem can easily be solved by using the WKB approach. The path integral approach is just too complicated, and one may wonder why we bother to discuss it. The motivation for the present discussion is, firstly, that it demonstrates the calculation steps and subtleties in a typical path integral calculation. Secondly, and more importantly, the same approach can be used to study the decay of a meta-stable state in field theory.

Problem 2.4.2.
The simplest field theory model with a decaying meta-stable state is given by

$$S = \int \mathrm{d}t\mathrm{d}^d\boldsymbol{x} \left[\frac{1}{2}[(\partial_t m)^2 - v^2(\partial_{\boldsymbol{x}} m)^2] - \frac{g}{4}(m^2 - m_0^2)^2 - Bm\right]$$

where $m(t, \boldsymbol{x})$ is a real field which can be regarded as the density of magnetic moments. We will choose the units such that the velocity $v = 1$. When the magnetic field B is zero, the system has two degenerate ground states with $m = \pm m_0$ which are the two minima of the potential $V(m) = \frac{g}{4}(m^2 - m_0^2)^2 + Bm$.

1. Show that, for small B, the system has one ground state with $m = m_+$ and one meta-stable state with $m = m_-$. However, when B is beyond a critical value B_c the meta-stable state can no longer exist. Find the value of B_c.
2. Write down the action in imaginary time. Coleman showed that the action always has a bounce solution (a stationary path) of the form

$$m = f(\sqrt{(\tau - \tau_0)^2 + (\boldsymbol{x} - \boldsymbol{x}_0)^2}) + m_-, \qquad f(\infty) = 0$$

   where $(\tau_0, \boldsymbol{x}_0)$ is the location of the bounce.
3. Assume that without the bounces $\langle m_-|\mathrm{e}^{-TH}|m_-\rangle = \mathrm{e}^{-TE_0}$. After including the bounces, we have

$$\langle m_-|\mathrm{e}^{-TH}|m_-\rangle = \mathrm{e}^{-TE_0} \sum_n \frac{1}{n!} \int \prod_{i=1}^{n} \mathrm{d}\tau_i \mathrm{d}^d\boldsymbol{x}_i\, K^n \mathrm{e}^{-nS_0}$$

   where S_0 is the action of the bounce and $(\tau_i, \boldsymbol{x}_i)$ are the locations of the bounces. (Note that a bounce is now parametrized by its location in both time and space.) If K is imaginary, what is the decay rate of the meta-stable state $|m_-\rangle$? What is the decay rate per unit volume?
4. Estimate the values of S_0, assuming that $B = \frac{B_c}{2}$. (Hint: Rescale $(\tau, \boldsymbol{x})$ and m to rewrite $S = \text{constant} \times S'$ such that all of the coefficients in S' are of order 1.) (One can also estimate the size of a bounce in space–time using this method.)
5. Estimate the values of K, assuming that $B = \frac{B_c}{2}$. (Hint: Consider how K scales as $S \to aS$.)

6. Discuss for which values of g and m_0 the semiclassical tunneling picture is a good description of the decay of the meta-stable state.

2.4.3 Quantum theory of friction

- Dissipation (friction) in a quantum system can be simulated by coupling to a heat bath (a collection of harmonic oscillators).

Friction on a particle is caused by its coupling to the environment. Due to the coupling, the particle can lose its energy into the environment. So, to formulate a quantum theory of friction, we must include coupling to an environment. Here we use a collection of harmonic oscillators to simulate the environment (Caldeira and Leggett, 1981). We need to set up a proper environment that reproduces a finite friction coefficient. This allows us to understand the quantum motion of a particle in the presence of friction.

Our model is described by the following path integral:

$$Z = \text{constant} \times \int \mathcal{D}[x(t)]\mathcal{D}[h_i(t)]\ \mathrm{e}^{\mathrm{i}\int \mathrm{d}t\ \left(\frac{1}{2}m\dot{x}+\sum_i \frac{1}{2}\dot{h}_i^2-\frac{1}{2}(\Omega_i h_i+g_i x)^2\right)}$$

where x is the coordinate of the particle and the h_i describe a collection of oscillators. Note that our system has the translational symmetry $x \to x + \text{constant}$. After integrating out h_i, we get (see Section 2.2.3)

$$Z = \text{constant} \times \int \mathcal{D}[x(t)]\ \mathrm{e}^{\mathrm{i}\int \mathrm{d}t\ \frac{m}{2}\dot{x}^2+\mathrm{i}\int \mathrm{d}t\mathrm{d}t'\ \frac{1}{4}G_{\mathrm{os}}(t-t')(x(t)-x(t'))^2} \tag{2.4.8}$$

where

$$\begin{aligned} G_{\mathrm{os}}(t-t') &= \sum_i \Omega_i^2 g_i^2(-\mathrm{i})\,\langle h_i(t)h_i(0)\rangle = -\mathrm{i}\sum_i \frac{g_i^2\Omega_i}{2}\mathrm{e}^{-\mathrm{i}|t-t'|\Omega_i} \\ &= -\mathrm{i}\int_0^\infty \mathrm{d}\Omega\, n(\Omega)\frac{\Omega g^2(\Omega)}{2}\mathrm{e}^{-\mathrm{i}|t-t'|\Omega}, \end{aligned}$$

$n(\Omega) \equiv \sum_i \delta(\Omega-\Omega_i)$ is the density of states, and $g^2(\Omega) \equiv \sum_i g_i^2\delta(\Omega-\Omega_i)/\sum_i\delta(\Omega-\Omega_i)$. The double-time action in eqn (2.4.8) leads to the following equation of motion:

$$-\omega^2 m x_\omega = -(G_{\mathrm{os}}(\omega)-G_{\mathrm{os}}(0))x_\omega,$$

which can be rewritten as follows:[14]

$$-\omega^2 m^* x_\omega = -(-\mathrm{i}\omega)\gamma x_\omega, \tag{2.4.9}$$

$$m^* = m - \frac{\mathrm{Re}(G_{\mathrm{os}}(\omega) - G_{\mathrm{os}}(0))}{\omega^2}, \qquad \gamma = -\frac{\mathrm{Im}G_{\mathrm{os}}(\omega)}{\omega}$$

If m^* and γ are finite for small ω, then eqn (2.4.9) becomes $m^* \frac{\mathrm{d}^2 x}{\mathrm{d}t^2} = -\gamma \frac{\mathrm{d}x}{\mathrm{d}t}$, which describes a particle of mass m^* whose friction is described by the friction coefficient γ. We see that the coupling to a collection of oscillators can produce friction on the particle. The friction, and hence the dissipation, comes from the imaginary part of the correlation $\mathrm{Im}G_{\mathrm{os}}(\omega)$. The coupling also modifies the mass of the particle from m to m^*. Here m^* is called the effective mass of the particle. The correction $m^* - m$ comes from the real part of the correlation $\mathrm{Re}G_{\mathrm{os}}(\omega)$. Physically, the particle causes deformation of the oscillators. The deformation moves with the particle, which changes the effective mass of the particle.

The above quantum theory of friction has one problem. As $\mathrm{Im}G_{\mathrm{os}}(\omega) < 0$, we see that the friction coefficient $\gamma(\omega) > 0$ for $\omega > 0$, which is correct. However, $\gamma(\omega) < 0$ for $\omega < 0$, which seems to not make any sense. We note that our effective action in eqn (2.4.8) is invariant under the time reversal $t \to -t$. As a consequence, if the $\omega > 0$ solutions correspond to the decaying processes, then the $\omega < 0$ solutions will correspond to the inverse of the decaying processes.

To really describe friction we need to find a time-reversal non-invariant formalism. To achieve this, it is important to notice that in the presence of friction the state at $t = -\infty$, namely $|0_-\rangle$, is very different from the state at $t = \infty$, namely $|0_+\rangle$, because as the particle slows down the oscillators are excited into higher and higher states. This makes our path integral $Z = \langle 0|U(\infty, -\infty)|0\rangle$ a bad starting-point because it is zero. Thus, the problem here is to formulate a path integral for non-equilibrium systems. The closed-time path integral or Schwinger–Keldysh formalism (Schwinger, 1961; Keldysh, 1965; Rammmer and Smith, 1986) is one way to solve these problems. The Schwinger–Keldysh formalism starts with a different path integral $Z_{close} = \langle 0|U(-\infty, \infty)(U(\infty, -\infty)|0\rangle$, which goes from $t = -\infty$ to $t = +\infty$ and back to $t = -\infty$.

The double-time path integral (2.4.8) developed above can only describe the equilibrium properties of a dissipative system, such as fluctuations around the ground state in the presence of friction/dissipation. It cannot describe the non-equilibrium properties, such as the slowing down of the velocity caused by the friction, because this involves the different states $|0_-\rangle$ and $|0_+\rangle$ for $t = \pm\infty$.

The problem encountered here is similar to the one encountered in Section 2.2 when we tried to use path integrals to calculate the imaginary part of the susceptibility, which is also associated with dissipation. There the problem was solved by

[14] Here we have used $\mathrm{Im}G_{\mathrm{os}}(\omega = 0) = 0$. Note that $\mathrm{Im}D_{\mathrm{os}}(\omega) = -\mathrm{Im}D_{\mathrm{os}}(-\omega)$, and thus $\mathrm{Im}D_{\mathrm{os}}(\omega = 0) = 0$. As $\mathrm{Im}G_{\mathrm{os}}(\omega) = \mathrm{sgn}(\omega)\mathrm{Im}D_{\mathrm{os}}(\omega)$, we find that $\mathrm{Im}G_{\mathrm{os}}(\omega = 0) = 0$.

replacing the time-ordered correlation function by the response function. Here the problem can be solved (or, more precisely, fixed by hand) in the same way. If we replace $G_{\mathrm{os}}(\omega)$ by $D_{\mathrm{os}}(\omega)$ (note that $D_{\mathrm{os}}(t) = 2\Theta(t)\mathrm{Re}G_{\mathrm{os}}(t)$) in the *equation of motion* (but not in the double-time path integral), then the friction coefficient γ will be given by

$$\gamma = -\frac{\mathrm{Im}D_{\mathrm{os}}(\omega)}{\omega} \tag{2.4.10}$$

which is always positive for both $\omega > 0$ and $\omega < 0$. With the above restrictions in mind, we may say that the double-time path integral (2.4.8) can describe a system with friction. The friction coefficient is given by eqn (2.4.10).

As

$$G_{\mathrm{os}}(\omega) = \int_0^\infty \mathrm{d}\Omega\, \frac{\Omega^2 n(\Omega) g^2(\Omega)}{\omega^2 - \Omega^2 + i0^+},$$

we find that $\mathrm{Im}G_{\mathrm{os}}(\omega) = -\frac{\pi}{2}|\omega| n(|\omega|) g^2(|\omega|)$. Using the fact that $\mathrm{Im}D_{\mathrm{os}}(\omega) = \mathrm{sgn}(\omega)\mathrm{Im}G_{\mathrm{os}}(\omega)$, we find that

$$\gamma = \frac{\pi}{2} n(|\omega|) g^2(|\omega|) \tag{2.4.11}$$

for small values of ω. Thus, in order for γ to be finite for small ω, we require that $n(\Omega)g^2(\Omega)$ is finite as $\Omega \to 0$.

Let us assume that $n(\Omega)g^2(\Omega) = n_0 g_0^2$ for $\Omega < \Omega_0$ and that $n(\Omega)g^2(\Omega) = 0$ for $\Omega > \Omega_0$. We find that

$$G_{\mathrm{os}}(\omega) = -n_0 g_0^2 \Omega_0 + \frac{n_0 g_0^2 |\omega|}{2} \ln\left|\frac{\Omega_0 + |\omega|}{\Omega_0 - |\omega|}\right| - \mathrm{i}\frac{\pi}{2}|\omega| n_0 g_0^2$$

$$G_{\mathrm{os}}(t) = \mathrm{i}\frac{n_0 g_0^2}{2t^2}\left(1 - (\mathrm{i}|t|\Omega_0 + 1)\mathrm{e}^{-\mathrm{i}|t|\Omega_0}\right)$$

In addition to the finite friction coefficient $\gamma = \frac{\pi}{2} n_0 g_0^2$, we also obtain the effective mass $m^* = m - \frac{n_0 g_0^2}{\Omega_0^2}$.

We would like to point out that the term $(\mathrm{i}|t|\Omega_0 + 1)\mathrm{e}^{-\mathrm{i}|t|\Omega_0}$ in $G_{\mathrm{os}}(t)$ is due to the discontinuity of $n(\Omega)g^2(\Omega)$ at $\Omega = \Omega_0$. If $n(\Omega)g^2(\Omega)$ is a smooth function of Ω, then such a term will vanish exponentially in the large-t limit. For example, if we choose $n(\Omega)g^2(\Omega) = n_0 g_0^2 \mathrm{e}^{-\Omega^2/\Omega_0^2}$, then we will have $G_{\mathrm{os}}(t) = \mathrm{i}\frac{n_0 g_0^2}{2t^2}$ in the large-t limit. The friction coefficient is still given by $\gamma = \mathrm{sgn}(\omega)\frac{\pi}{2} n_0 g_0^2$. Using such a choice of $n(\Omega)g^2(\Omega)$, we find that a particle with a friction of $\gamma = \frac{\pi}{2} n_0 g_0^2$ is described by the following double-time action:

$$S = \int \mathrm{d}t\, \frac{m}{2}\dot{x}^2 + \mathrm{i}\int \mathrm{d}t\mathrm{d}t'\, \frac{\gamma\left(x(t) - x(t')\right)^2}{4\pi(t - t')^2} \tag{2.4.12}$$

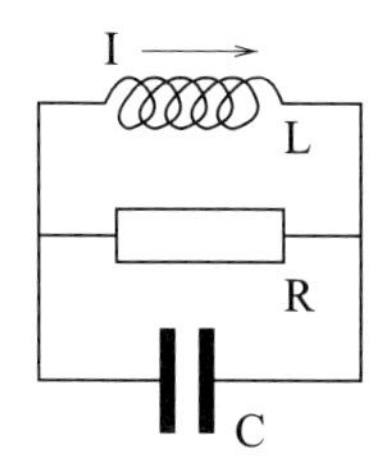

FIG. 2.10. An RCL circuit.

Problem 2.4.3.
Effective mass:

1. Assume that $n(\Omega)g^2(\Omega) \propto \Omega^\alpha$ for small Ω and that $n(\Omega)g^2(\Omega) = 0$ for $\Omega > \Omega_0$. Find the values of α which make the effective mass, m^*, diverge.
2. Consider the two cases of $n(\Omega)g^2(\Omega) = N_0\,\mathrm{e}^{-\Omega/\Omega_0}$ and $n(\Omega)g^2(\Omega) = N_0\,\mathrm{e}^{-(\Omega/\Omega_0)^2}$. Which case has a diverging effective mass?

2.4.4 Quantum theory of an RCL circuit

- The quantum dynamics of an RCL circuit are similar to those of a particle with friction.
- How to include charge quantization in a quantum RCL circuit.

The RCL circuit is a simple circuit. We know how to describe it classically. As everything is believed to be described by quantum theory, the question arises as to how to describe an RCL circuit in terms of quantum mechanics.

Let us first consider an ideal capacitor of capacitance C. The state of the capacitor is described by its charge q, and q is a real number if we ignore the charge quantization. The quantum theory for the ideal capacitor is simple. The Hilbert space is spanned by the states $\{|q\rangle\}$ and the Hamiltonian is $H = \frac{\hat{q}^2}{2C}$, where the charge operator $\hat{q}$ is defined by $\hat{q}|q\rangle = q|q\rangle$. Thus, a quantum capacitor is equivalent to a free particle on a line with q as its 'momentum'. The form of the Hamiltonian $H = \frac{\hat{q}^2}{2C}$ tells us that the mass of the particle is just C. The corresponding classical Lagrangian is $L = \frac{1}{2}C\dot{x}^2$.

To describe a CL circuit, we include a potential term and consider

$$L_{\mathrm{CL}} = \frac{C}{2}\dot{x}^2 - \frac{1}{2L}x^2 \tag{2.4.13}$$

The resulting equation of motion is

$$C\frac{\mathrm{d}^2 x}{\mathrm{d}t^2} = -\frac{x}{L}$$

It is equivalent to the equation of the CL circuit

$$V = \frac{q}{C} = L\frac{\mathrm{d}I}{\mathrm{d}t}, \qquad\qquad I = -\frac{\mathrm{d}q}{\mathrm{d}t},$$

if we interpret x/L to be the current I and L to be the inductance. We see that the CL circuit is equivalent to a harmonic oscillator with coordinate x, mass C, and spring constant $1/L$. Thus, the quantum CL circuit is described by the Hamiltonian

$$H = \frac{1}{2C}\hat{q}^2 + \frac{1}{2L}\hat{x}^2 \tag{2.4.14}$$

where $\hat{q}$ is the charge operator. It is also the momentum operator conjugate to $\hat{x}$, so that $[\hat{q}, \hat{x}] = \mathrm{i}$. The current operator is given by $\hat{I} \equiv \hat{x}/L$.

We can also view q as the coordinate and $-x$ as the corresponding 'momentum'. In this case, the Hamiltonian (2.4.14) leads to the following Lagrangian which is the dual form of eqn (2.4.13):

$$L^d_{\mathrm{CL}} = \frac{L}{2}\dot{q}^2 - \frac{1}{2C}q^2 \tag{2.4.15}$$

After the inclusion of a resistor, as in Fig. 2.10, the classical equation of motion becomes

$$V = \frac{q}{C} = L\frac{\mathrm{d}I}{\mathrm{d}t}, \qquad\qquad I = -\frac{\mathrm{d}q}{\mathrm{d}t} - V/R$$

which leads to

$$C\frac{\mathrm{d}^2x}{\mathrm{d}t^2} = -\frac{x}{L} - \frac{1}{R}\frac{\mathrm{d}x}{\mathrm{d}t} \tag{2.4.16}$$

This is just the equation of motion for a particle with a friction coefficient $\gamma = 1/R$. Such a system can be described by a double-time action (see eqn (2.4.12)):

$$S_{\mathrm{RCL}} = \int \mathrm{d}t\, \left(\frac{1}{2}C\dot{x}^2 - \frac{1}{2L}x^2\right) + \mathrm{i}\int \mathrm{d}t\,\mathrm{d}t'\, \frac{(x(t) - x(t'))^2}{4\pi R(t - t')^2} \tag{2.4.17}$$

Many of the quantum dynamical properties of the RCL circuit can be calculated from the above action.

If we include the charge quantization, then the charge is $q = \mathrm{e}^* \times$ integers. As the charge q is also the momentum of x, the quantization of the momentum implies that the particle lives on a circle. The charge (or the momentum) quantum e^* implies that x and $x + \frac{2\pi}{\mathrm{e}^*}$ should be treated as the same point.

As the charge quantization implies periodicity in x, the action that describes the RCL circuit with the charge quantization must be a periodic function of x.

There are many ways to make eqn (2.4.17) periodic. One of the simplest periodic double-time actions is given by

$$S_{\mathrm{RCL}} = \int \mathrm{d}t \left(\frac{C}{2}\dot{x}^2 - \frac{1-\cos(\mathrm{e}^* x)}{L\mathrm{e}^{*2}} \right) + \mathrm{i} \int \mathrm{d}t\mathrm{d}t' \, \frac{\sin^2\left(\frac{\mathrm{e}^*}{2}[x(t)-x(t')]\right)}{\pi R \mathrm{e}^{*2}(t-t')^2} \tag{2.4.18}$$

We note that eqn (2.4.18) becomes eqn (2.4.17) in the $\mathrm{e}^* \to 0$ limit. Equation (2.4.18) describes the quantum dynamical properties of the RCL circuit with charge quantization.

2.4.5 Relationship between dissipation and fluctuation

- Dissipation and fluctuation always appear together. We can determine one from the other.

The double-time actions (2.4.12) and (2.4.17) are good for calculating equilibrium properties. Let us calculate the equilibrium quantum fluctuations of x and the velocity $v = \dot{x}$. The fluctuations of x are characterized by the noise power spectrum. To define the noise power spectrum, let us first consider a classical fluctuating $x(t)$. The total power (including all of the frequencies) is defined by

$$P_{\mathrm{tot}} \equiv \int_0^{t_\infty} \mathrm{d}t |x(t)|^2 / t_\infty = 2\int_0^\infty \frac{\mathrm{d}\omega}{2\pi} x_{-\omega} x_\omega / t_\infty$$

where $x_\omega = \int_0^{t_\infty} \mathrm{d}t \; x(t) \mathrm{e}^{\mathrm{i}\omega t}$ and $t_\infty \to \infty$. Thus, we can identify $x_{-\omega}x_\omega/t_\infty$ as being the noise power spectrum. The quantum noise power spectrum is then obtained by taking the quantum average of $x_{-\omega}x_\omega/t_\infty$:[15]

$$P(\omega) = 2\langle x_{-\omega}x_\omega\rangle / t_\infty = -2\mathrm{Im}G_\omega \tag{2.4.19}$$

From the following relationship between $\mathrm{Im}D$ and $\mathrm{Im}G$ (see eqn (2.2.42)):

$$\mathrm{Im}G(\omega) = \frac{\mathrm{Im}D}{\tanh(\beta\omega/2)}, \tag{2.4.20}$$

[15] Here we have used

$$\begin{aligned} 2\langle x_{-\omega}x_\omega\rangle / t_\infty &= \int_0^{t_\infty} \mathrm{d}t_1 \mathrm{d}t_2 \left[\langle x(t_1)x(t_2)\rangle + \langle x(t_2)x(t_1)\rangle\right] \mathrm{e}^{\mathrm{i}\omega(t_1-t_2)}/t_\infty \\ &= \int_0^{t_\infty} \mathrm{d}t_1 \mathrm{d}t_2 \left[\langle T(x(t_1)x(t_2))\rangle + \langle T(x(t_1)x(t_2))\rangle^*\right] \mathrm{e}^{\mathrm{i}\omega(t_1-t_2)}/t_\infty \\ &= \int_0^{t_\infty} \mathrm{d}t_1 \mathrm{d}t_2 \left[\langle T(x(t_1)x(t_2))\rangle + \langle T(x(t_2)x(t_1))\rangle^*\right] \mathrm{e}^{\mathrm{i}\omega(t_1-t_2)}/t_\infty \\ &= 2\mathrm{Re}\int_{-\infty}^{+\infty} \mathrm{d}t \; \langle T(x(t)x(0))\rangle \, \mathrm{e}^{\mathrm{i}\omega t} = -2\mathrm{Im}G_\omega \end{aligned}$$

we see that ImG, and hence $P(\omega)$, can be determined from the response function D.

The response function can be calculated from the equation of motion. For a particle with friction, the equation of motion is given by

$$m\ddot{x} = -Kx - \gamma\dot{x} + f(t) \tag{2.4.21}$$

where f is an external force. The solution of the equation of motion can be formally written as

$$x(t) = -\left(\frac{1}{-m\frac{\mathrm{d}^2}{\mathrm{d}t^2} - \gamma\frac{\mathrm{d}}{\mathrm{d}t} - K}\right) f$$

The solution can be interpreted as the response of x to a perturbation $\delta H = -f(t)x$ as follows: $x(t) = \int \mathrm{d}t'\ D(t-t')(-f(t')) \equiv -Df$ (see eqn (2.2.3) with $O_1 = O_2 = x$). Therefore, the equation of motion determines the following response function of x:

$$D = \frac{1}{-m\frac{\mathrm{d}^2}{\mathrm{d}t^2} - \gamma\frac{\mathrm{d}}{\mathrm{d}t} - K}$$

It is now clear that the response function ImD describes the friction (the dissipation) of a moving particle, while the correlation function ImG describes the power spectrum of the fluctuations. Equation (2.4.20) is a direct relationship between the dissipation and the fluctuations.

From the expression for D, we find that the power spectrum of the x fluctuations is given by (see eqn (2.4.19))

$$\begin{aligned} P(\omega) =& -2\left(\tanh(\beta\omega/2)\right)^{-1}\mathrm{Im}D \\ =& \frac{1}{\tanh(\omega\hbar/2T)}\frac{2\gamma\omega\hbar}{(m\omega^2-K)^2+\gamma^2\omega^2} \end{aligned} \tag{2.4.22}$$

The power spectrum of the velocity fluctuations is given by

$$P^v(\omega) = \omega^2 P(\omega) = \frac{1}{\tanh(\omega\hbar/2T)}\frac{2\gamma\omega\hbar}{(m\omega - \frac{K}{\omega})^2 + \gamma^2}$$

We note that the equation of motion (2.4.21) is identical to that of the RCL circuit given in eqn (2.4.16), with (C, L^{-1}, R^{-1}) replaced by (m, K, γ). The voltage is given by $V = \dot{x}$. Thus, the power spectrum of the voltage fluctuations in the

RCL circuit is given by

$$P^V(\omega) = \frac{1}{\tanh(\hbar\beta\omega/2)} \frac{2\hbar R\omega}{(RC\omega - \frac{R}{L\omega})^2 + 1} \tag{2.4.23}$$

When $\hbar\omega \ll T$, we have

$$P^V_c(\omega) = \frac{4TR}{(RC\omega - \frac{R}{L\omega})^2 + 1}$$

which is the classical noise spectrum of the RCL system. When $L = \infty$ and $C = 0$, the above expression becomes the noise spectrum across a resistor, namely $P^V_c(\omega) = 4TR$.

It is interesting to see that the finite-temperature noise spectrum, P^V, and the zero-temperature noise spectrum, P^V_0, are closely related as follows:

$$P^V(\omega) = \frac{1}{\tanh(\hbar|\omega|/2T)} P^V_0(\omega) \tag{2.4.24}$$

$P^V(\omega)$ can also be obtained from the classical noise spectrum $P^V_c(\omega)$:

$$P^V(\omega) = \frac{\hbar\omega}{2T\tanh(\hbar\omega/2T)} P^V_c(\omega)$$

The above relations are very general. They apply to all harmonic systems (see Problem 2.4.5 below).

Problem 2.4.4.
Brownian motion:
(a) A particle of mass m experiences a friction described by γ. If we release the particle at $x = 0$, how far can it have wandered after a time t? (That is, calculate $\bar{x} = \sqrt{\langle x^2 \rangle}$ at time t.) (Hint: Consider the power spectrum of the velocity at $\omega = 0$.)
(b) Find (or estimate) the numerical value of $\bar{x}$ after 1 min for a plastic ball suspended in water. The diameter of the ball is 1 cm. Repeat the calculation for a grain of pollen in water. We model the pollen as a ball made from the same plastic but with a size 10 000 times smaller. The viscosity of water at 20°C is 0.01 poise (1 poise = 1 dyne sec/cm^2).

Problem 2.4.5.
Prove eqn (2.4.24) by directly considering a system of coupled oscillators with arbitrary g_i and Ω_i (see eqn (2.4.8)). (Hint: Consider $\langle x^2 \rangle$ for an oscillator at $T = 0$ and at $T \neq 0$.)

2.4.6 Path integral description of a random differential equation

- The statistical properties of the solutions of a random differential equation can be described by a path integral.

According to the equation of motion (2.4.21), if we release the particle somewhere, then the particle will eventually stop at $x = 0$ after a long time if $f = 0$.

However, the dissipation/fluctuation theorem tells us that the above picture is not correct and the particle does not just stay at $x = 0$. The presence of friction implies the presence of fluctuation, and so the particle will fluctuate around $x = 0$. To have a more correct equation of motion that describes a dissipative system, we need to add a term that generates the fluctuation. One way to do this is to let the force term $f(t)$ have random fluctuations in time. Two questions now arise. Firstly, can a random force term simulate the fluctuations, in particular the quantum fluctuations? If the answer is yes, then, secondly, how do we find the probability distribution of the random force that generates the correct fluctuations? To solve this problem, we need to calculate the average correlation $\langle x(t)x(0)\rangle$ of the random solution of eqn (2.4.21). We then examine whether it is possible to choose a proper distribution of $f(t)$ which will reproduce the power spectrum (2.4.22).

We note that

$$\langle x(t_b)x(t_a)\rangle = \int \mathcal{D}[f(t)]\mathcal{D}[x(t)]\, x(t_b)x(t_a)P[f(t)]\delta[x(t) - x^f(t)]$$

Here $x^f = \mathcal{K}^{-1}f$ is the solution of eqn (2.4.21), where $\mathcal{K} = m\frac{\mathrm{d}^2}{\mathrm{d}t^2} + \gamma\frac{\mathrm{d}}{\mathrm{d}t} + K$, and $P[f(t)]$ is the probability distribution of $f(t)$. We note that $\mathcal{K} = -D^{-1}$. As the operator $\mathcal{K}$ does not depend on f, we have

$$\delta[x(t) - x^f(t)] \propto \delta[\mathcal{K}x(t) - f(t)]$$

Thus,

$$\begin{aligned}\langle x(t_b)x(t_a)\rangle &= \frac{\int \mathcal{D}[f(t)]\mathcal{D}[x(t)]\, x(t_b)x(t_a)P[f(t)]\delta[\mathcal{K}x(t) - f(t)]}{\int \mathcal{D}[f(t)]\mathcal{D}[x(t)]\, P[f(t)]\delta[\mathcal{K}x(t) - f(t)]} \\ &= \frac{\int \mathcal{D}[f(t)]\mathcal{D}[x(t)]\mathcal{D}[\lambda(t)]\, x(t_b)x(t_a)P[f(t)]\mathrm{e}^{\,\mathrm{i}\int \lambda[\mathcal{K}x-f]}}{\int \mathcal{D}[f(t)]\mathcal{D}[x(t)]\mathcal{D}[\lambda(t)]\, P[f(t)]\mathrm{e}^{\,\mathrm{i}\int \lambda[\mathcal{K}x-f]}}\end{aligned}$$

This is quite an interesting result. We have seen that the path integral can describe the Hamiltonian—a linear operator in quantum systems. It can also describe the thermal average in classical and quantum statistical systems. We see from the above that the path integral can even describe a random differential equation.

Let us now assume that the distribution of the random term is Gaussian:

$$P[f(t)] \propto \mathrm{e}^{-\frac{1}{2}\int f\mathcal{V}f}$$

where $\int f\mathcal{V}f \equiv \int \mathrm{d}t\mathrm{d}t'\ f(t)\mathcal{V}(t,t')f(t')$. We can integrate out f and λ in turn as follows:

$$
\begin{aligned}
\langle x(t_b)x(t_a)\rangle &= \frac{\int \mathcal{D}[x(t)]\mathcal{D}[\lambda(t)]\ x(t_b)x(t_a)\mathrm{e}^{\int \mathrm{i}\lambda\mathcal{K}x-\frac{1}{2}\lambda\mathcal{V}^{-1}\lambda}}{\int \mathcal{D}[x(t)]\mathcal{D}[\lambda(t)]\ \mathrm{e}^{\int \mathrm{i}\lambda\mathcal{K}x-\frac{1}{2}\lambda\mathcal{V}^{-1}\lambda}} \\
&= \frac{\int \mathcal{D}[x(t)]\ x(t_b)x(t_a)\mathrm{e}^{-\int \frac{1}{2}x\mathcal{K}^\top\mathcal{V}\mathcal{K}x}}{\int \mathcal{D}[x(t)]\ \mathrm{e}^{-\int \frac{1}{2}x\mathcal{K}^\top\mathcal{V}\mathcal{K}x}} \\
&= (\mathcal{K}^\top\mathcal{V}\mathcal{K})^{-1}(t_b-t_a)
\end{aligned}
$$

Here, the inverse of the operator $\mathcal{K}^\top\mathcal{V}\mathcal{K}(t)$ is interpreted as $\mathcal{G}(t-t')$ such that $\mathcal{K}^\top\mathcal{V}\mathcal{K}(t)\mathcal{G}(t-t') = \delta(t-t')$. We note that $\langle x(t_b)x(t_a)\rangle = \langle x(t_a)x(t_b)\rangle$. So its Fourier transformation is real and equal to one-half of the power spectrum of the x fluctuations (see eqn (2.4.19)). Thus, to reproduce the x fluctuations, we require that

$$
\mathcal{K}^\top\mathcal{V}\mathcal{K} = \frac{\tanh(\omega\hbar/2T)[(-m\omega^2+K)^2+\omega^2\gamma^2]}{\gamma\omega\hbar}
$$

or

$$
\mathcal{V} = \frac{\tanh(\omega\hbar/2T)}{\gamma\omega\hbar}
$$

in frequency space. We see that even quantum noise can be simulated using random classical equations. In particular, we note that the random force term depends only on the temperature T and the friction coefficient γ. A higher temperature or a larger friction will lead to a stronger random force. The classical noise (in the limit $\hbar \to 0$) can be simulated by a simple probability distribution of the random force as follows:

$$
P[f(t)] \propto \mathrm{e}^{-\frac{1}{2}\int \mathrm{d}t\ f\mathcal{V}f}, \qquad \mathcal{V} = \frac{1}{2\gamma T}
$$

which is not correlated in time (i.e. $f(t)$ and $f(t')$ fluctuate independently).

We would like to point out that the theory of random differential equations is a very large field with broad and important applications. Random differential equations are used to describe material growth, chemical and biological processes, and even economical systems. The path integral representation of random differential equations allows us to apply the pictures and concepts developed for the path integral, such as symmetry breaking, phase transition, renormalization group, etc., to random differential equations.

3

INTERACTING BOSON SYSTEMS

The boson system is the simplest system in condensed matter physics. It describes a wide variety of physical phenomena, such as superfluid, magnetism, crystal formations, etc. It is also a model system for phase and phase transition.

3.1 Free boson systems and second quantization

- Second quantization is an operator description of a quantum system. In second quantization a many-body system can be formulated as a field theory.

First quantization is a description of a quantum system using wave functions. Second quantization is another description of a quantum system using operators. In second quantization, we do not need to write down a wave function explicitly. For example, in the operator description of a harmonic oscillator, the ground state is defined through operators for which $\hat{a}|0\rangle = 0$.

Now let us consider a free d-dimensional boson system (in a box of volume $\mathcal{V} = L^d$) which contains N identical bosons. In first quantization, the wave function of the system is a symmetric function of the N variables. These N-boson states form a Hilbert space denoted by $\mathcal{H}_N$. To obtain a second quantization description of the boson system and to avoid writing the complicated N-variable symmetric functions, we combine the Hilbert spaces with all the different numbers of bosons together to form a total Hilbert space:

$$\mathcal{H} = \mathcal{H}_0 \oplus \mathcal{H}_1 \oplus \mathcal{H}_2 \oplus ... \oplus \mathcal{H}_N \oplus ...$$

The total Hilbert space $\mathcal{H}$ has the following base

$$|n_{\boldsymbol{k}_1}\ n_{\boldsymbol{k}_2}\ ...\rangle$$

where $n_{\boldsymbol{k}} = 0, 1, 2, ...$ is the number of bosons in a single-particle momentum eigenstate $|\boldsymbol{k}\rangle$. The total energy of the state $|n_{\boldsymbol{k}_1}\ n_{\boldsymbol{k}_2}\ ...\rangle$ is

$$E = \sum_{\boldsymbol{k}} \frac{\boldsymbol{k}^2}{2m} n_{\boldsymbol{k}}$$

The total Hilbert space $\mathcal{H}$ can also be viewed as a Hilbert space of a collection of harmonic oscillators labeled by a vector $\boldsymbol{k} \equiv (k_x, k_y, ...) = \frac{2\pi}{L}(n_x, n_y, ...)$, and

$n_{\boldsymbol{k}}$ is simply the integer that labels the energy levels of the oscillator $\boldsymbol{k}$. From the total energy of the bosons, we see that the oscillators are independent, and the oscillator $\boldsymbol{k}$ is described by the following Hamiltonian:

$$H_{\boldsymbol{k}} = \epsilon_{\boldsymbol{k}} a^\dagger_{\boldsymbol{k}} a_{\boldsymbol{k}}, \qquad \epsilon_{\boldsymbol{k}} = \frac{\boldsymbol{k}^2}{2m} \tag{3.1.1}$$

Thus, in second quantization, the free boson system is described by a collection of harmonic oscillators with the Hamiltonian

$$H = \sum_{\boldsymbol{k}} \epsilon_{\boldsymbol{k}} a^\dagger_{\boldsymbol{k}} a_{\boldsymbol{k}}$$

The total boson number operator is

$$\hat{N} = \sum_{\boldsymbol{k}} a^\dagger_{\boldsymbol{k}} a_{\boldsymbol{k}}$$

The ground state of the oscillators satisfying

$$a_{\boldsymbol{k}}|0\rangle = 0$$

is a state with no boson. An N-boson state with the ith boson carrying a momentum $\boldsymbol{k}_i$ is created by N of the $a^\dagger_{\boldsymbol{k}}$ operators:

$$|\boldsymbol{k}_1\, \boldsymbol{k}_2\, ...\boldsymbol{k}_N\rangle = C(\boldsymbol{k}_1, ..., \boldsymbol{k}_N) a^\dagger_{\boldsymbol{k}_1} ... a^\dagger_{\boldsymbol{k}_N}|0\rangle$$

where $C(\boldsymbol{k}_1, ..., \boldsymbol{k}_N)$ is a normalization coefficient. Here $C(\boldsymbol{k}_1, ..., \boldsymbol{k}_N) = 1$ only if all of the $\boldsymbol{k}_i$s are distinct from each other. Note that $a^\dagger_{\boldsymbol{k}}$ ($a_{\boldsymbol{k}}$) increases (decreases) the boson number by 1. Thus, $a^\dagger_{\boldsymbol{k}}$ is called the creation operator of the boson, while $a_{\boldsymbol{k}}$ is called the annihilation operator. The creation and annihilation operators do not appear in the first-quantization description of the boson system because there we only consider states with a fixed boson number.

We can introduce

$$a(\boldsymbol{x}) = \sum_{\boldsymbol{k}} a_{\boldsymbol{k}} \mathcal{V}^{-1/2} \mathrm{e}^{\mathrm{i}\boldsymbol{k}\boldsymbol{x}}$$

or

$$a(\boldsymbol{x}) = \int \frac{\mathrm{d}^d\boldsymbol{k}}{(2\pi)^d} a_{\boldsymbol{k}} \mathrm{e}^{\mathrm{i}\boldsymbol{k}\boldsymbol{x}}$$

for the infinite system. We can see that $a^\dagger(\boldsymbol{x})$ creates a boson at $\boldsymbol{x}$. The wave function for the state $|\boldsymbol{k}_1\, \boldsymbol{k}_2\, ...\boldsymbol{k}_N\rangle$ can now be calculated as follows:

$$\psi(\boldsymbol{x}_1, ..., \boldsymbol{x}_N) = \langle \boldsymbol{x}_1 ... \boldsymbol{x}_N | \boldsymbol{k}_1 ... \boldsymbol{k}_N\rangle, \qquad |\boldsymbol{x}_1 ... \boldsymbol{x}_N\rangle = a^\dagger(\boldsymbol{x}_1) ... a^\dagger(\boldsymbol{x}_N)|0\rangle$$

The Hamiltonian can be written in real space as

$$H = \int \mathrm{d}^d \boldsymbol{x}\, a^\dagger(\boldsymbol{x}) \left(-\frac{1}{2m} \frac{\mathrm{d}^2}{\mathrm{d}\boldsymbol{x}^2} \right) a(\boldsymbol{x})$$

The operator $a^\dagger(\boldsymbol{x})a(\boldsymbol{x})$ counts the number of bosons at position $\boldsymbol{x}$, and thus the boson density operator is

$$\rho(\boldsymbol{x}) = a^\dagger(\boldsymbol{x})a(\boldsymbol{x})$$

The Hamiltonian, together with the operators corresponding to physical quantities, allows us to calculate the physical properties of the boson systems by calculating the correlation functions of the physical operators.

The oscillator description allows us to calculate all of the correlation functions of the boson system. In particular, for $t > 0$,

$$\mathrm{i}G(\boldsymbol{x}_f - \boldsymbol{x}_i, t) = \left\langle a(\boldsymbol{x}_f)\mathrm{e}^{-\mathrm{i}tH} a^\dagger(\boldsymbol{x}_i) \right\rangle = \left\langle a(\boldsymbol{x}_f, t) a^\dagger(\boldsymbol{x}_i, 0) \right\rangle$$

is just the propagator of the free particle discussed in Section 2.1.1 and

$$\mathrm{i}G(\boldsymbol{k}, t) = \left\langle a_{-\boldsymbol{k}} \mathrm{e}^{-\mathrm{i}tH} a^\dagger_{\boldsymbol{k}} \right\rangle = \mathrm{e}^{-\mathrm{i}\epsilon_{\boldsymbol{k}} t}$$

is the corresponding propagator in momentum space.

To calculate the correlations of many boson operators, such as the ρ–ρ correlation, we need to use the Wick theorem. The Wick theorem is very useful for performing normal ordering and calculating correlations.

Theorem. (Wick's theorem) *Let O_i be a linear combination of a_k and $a^\dagger_k$: $O_i = \int \mathrm{d}k\; (u_i(k)a_k + v_i(k)a^\dagger_k)$. (Note that, for* free *boson systems, $a(t) = U^\dagger(t,0)aU(t,0)$ is such an operator.) Let $W = \prod_{i=1}^N O_i$, $W_{i_1,i_2} = \prod_{i=1, i\neq i_1, i\neq i_2}^N O_i$, etc. Let $:W:$ be the normal ordered form of the operator W (i.e. putting all of the $a^\dagger_k$ to the left of the a_k) and let $|0\rangle$ be the state annihilated by a_k (i.e. $a_k|0\rangle = 0$ for all k). Then the Wick theorem states that*

$$\begin{aligned} W = {} & :W: + \sum_{(i_1,j_1)} :W_{i_1,j_1}: \langle 0|O_{i_1} O_{j_1}|0\rangle \\ & + \sum_{\substack{(i_1,j_1),(i_2,j_2) \\ (i_1,j_1)\neq(i_2,j_2)}} :W_{i_1,j_1,i_2,j_2}: \langle 0|O_{i_1} O_{j_1}|0\rangle \langle 0|O_{i_2} O_{j_2}|0\rangle + \dots \end{aligned} \tag{3.1.2}$$

where (i_1, j_1) is an ordered pair with $i_1 < j_1$.

To illustrate the power of the Wick theorem, we consider the following correlation:

$$\langle 0|A_1A_2A_3A_4|0\rangle$$

where the A_i are linear combinations of a and $a^\dagger$. As $\langle 0| : A_iA_j : |0\rangle = 0$, the Wick theorem implies that

$$\begin{aligned}&\langle 0|A_1A_2A_3A_4|0\rangle\\&=\langle 0|A_1A_2|0\rangle\langle 0|A_3A_4|0\rangle+\langle 0|A_1A_3|0\rangle\langle 0|A_2A_4|0\rangle+\langle 0|A_1A_4|0\rangle\langle 0|A_2A_3|0\rangle\end{aligned}$$

Thus, a four-operator correlation can be expressed in terms of two-operator correlations.

Problem 3.1.1.
Wick's theorem:

1. Perform normal ordering for the operator $O = aaa^\dagger a^\dagger$ and show that the Wick theorem produces the correct result.
2. Use the Wick theorem to calculate $\langle 0|aa^\dagger aa^\dagger|0\rangle$.

Problem 3.1.2.
Consider a one-dimensional free boson system with N bosons. Let $|\Phi_0\rangle$ be the ground state.

1. Calculate, for $N = 0$ and finite N, the time-ordered propagator

   $$\mathrm{i}G(x,t) = \langle\Phi_0|T(a(x,t)a^\dagger(0,0))|\Phi_0\rangle$$

   where $a(x,t) = \mathrm{e}^{\mathrm{i}Ht}a(x)\mathrm{e}^{-\mathrm{i}Ht}$. Show that $\mathrm{i}G(x,0^+) - \mathrm{i}G(x,0^-) = [a(x), a^\dagger(0)] = \delta(x)$. Show that, for finite N, we have $\mathrm{i}G(x,t) \neq 0$ in the $x \to \infty$ limit, indicating (off-diagonal) long-range order. The off-diagonal long-range order exists only in the boson condensed state.
2. Calculate the time-ordered density correlation function

   $$\langle\Phi_0|T(\rho(x,t)\rho(0,0))|\Phi_0\rangle$$

   Show that $a(x,t)$ is a linear combination of a_k and $a_k^\dagger$ and that one can use the Wick theorem. (Hint: You may want to separate the a_0 operator.)

3.2 Mean-field theory of a superfluid

- The mean-field theory is obtained by ignoring quantum fluctuations of some operators and replacing them by c-numbers (which are usually equal to the averages of the operators).

- The mean-field ground state can be viewed as a trial wave function. The excitation spectrum obtained in mean-field theory can be qualitatively wrong. One needs to justify the correctness of the mean-field spectrum through other means before using the spectrum.

At zero temperature, a free N-boson system is in its ground state

$$|\Phi_0\rangle = (N!)^{-1/2}(a_0^\dagger)^N|0\rangle$$

and all of the bosons are in the $\boldsymbol{k} = 0$ state. This state is called the boson condensed state. The boson condensed state of a free boson system has many special properties, which are only true for non-interacting bosons. To understand the zero-temperature state of a real boson system we need to study the interacting boson system described by the following Hamiltonian:

$$H = \sum_{\boldsymbol{k}}(\epsilon_{\boldsymbol{k}} - \mu)a_{\boldsymbol{k}}^\dagger a_{\boldsymbol{k}} + \int \mathrm{d}^d\boldsymbol{x}\mathrm{d}^d\boldsymbol{x}' \, \frac{1}{2}\rho(\boldsymbol{x})V(\boldsymbol{x}-\boldsymbol{x}')\rho(\boldsymbol{x}')$$

where $V(\boldsymbol{x})$ is a density–density interaction and μ is the chemical potential. The value of μ is chosen such that there are on average N bosons in the system. Due to the inclusion of the chemical potential term $-\mu N$, H is an operator of thermal potential instead of an operator of energy. In momentum space, we have

$$\begin{aligned}
H &= \sum_{\boldsymbol{k}}(\epsilon_{\boldsymbol{k}} - \mu)a_{\boldsymbol{k}}^\dagger a_{\boldsymbol{k}} + \mathcal{V}\sum_{\boldsymbol{q}}\frac{1}{2}\rho_{-\boldsymbol{q}}V_{\boldsymbol{q}}\rho_{\boldsymbol{q}} \\
&= \sum_{\boldsymbol{k}}(\epsilon_{\boldsymbol{k}} - \mu)a_{\boldsymbol{k}}^\dagger a_{\boldsymbol{k}} + \mathcal{V}^{-1}\frac{1}{2}\sum_{\boldsymbol{k},\boldsymbol{k}',\boldsymbol{q}}(a_{\boldsymbol{k}'-\boldsymbol{q}}^\dagger a_{\boldsymbol{k}'})V_{\boldsymbol{q}}(a_{\boldsymbol{k}+\boldsymbol{q}}^\dagger a_{\boldsymbol{k}}) \\
&= \sum_{\boldsymbol{k}}(\epsilon_{\boldsymbol{k}} - \mu + V(0))a_{\boldsymbol{k}}^\dagger a_{\boldsymbol{k}} + \mathcal{V}^{-1}\frac{1}{2}\sum_{\boldsymbol{k},\boldsymbol{k}',\boldsymbol{q}} V_{\boldsymbol{q}}a_{\boldsymbol{k}'-\boldsymbol{q}}^\dagger a_{\boldsymbol{k}+\boldsymbol{q}}^\dagger a_{\boldsymbol{k}'}a_{\boldsymbol{k}}
\end{aligned} \tag{3.2.1}$$

where

$$a_{\boldsymbol{k}} = \int \mathrm{d}\boldsymbol{x}\, a(x)\mathcal{V}^{-1/2}\mathrm{e}^{\,\mathrm{i}\boldsymbol{k}\cdot\boldsymbol{x}}, \quad \rho_{\boldsymbol{k}} = \int \mathrm{d}\boldsymbol{x}\, \rho(x)\mathrm{e}^{\,\mathrm{i}\boldsymbol{k}\cdot\boldsymbol{x}}, \quad V_{\boldsymbol{k}} = \int \mathrm{d}\boldsymbol{x}\, V(\boldsymbol{x})\mathrm{e}^{\,\mathrm{i}\boldsymbol{k}\cdot\boldsymbol{x}}$$

In the last line of eqn (3.2.1), the interaction term has been written in a normal ordered form, i.e. $a^\dagger$ always appears to the left of a. Also, $V_{\boldsymbol{k}}$ is real, and $V_{\boldsymbol{k}} = V_{-\boldsymbol{k}}$ because $V(\boldsymbol{x})$ is symmetric, i.e. $V(\boldsymbol{x}) = V(-\boldsymbol{x})$. In the following, we will absorb $V(0)$ into μ and drop $V(0)$. The interacting Hamiltonian is not quadratic in $a_{\boldsymbol{k}}$ and is very hard to solve.

In the following, we shall use a mean-field approach to find an approximate ground state. The basic idea in mean-field theory is to identify some operators with weak quantum fluctuations and replace them by the corresponding classical value (a c-number). Hopefully, this replacement will reduce the full Hamiltonian to a quadratic one.

Based upon our knowledge of the free boson ground state, we may assume that, in the ground state of the interacting boson system, there are a macroscopic number of particles occupying the $\boldsymbol{k} = 0$ state:

$$\langle \Phi_0 | a_0^\dagger a_0 | \Phi_0 \rangle = N_0$$

while the occupation numbers for the other states, namely $n_{\boldsymbol{k}} = \langle \Phi_0 | a_{\boldsymbol{k}}^\dagger a_{\boldsymbol{k}} | \Phi_0 \rangle$, are small. We also note that

$$\langle \Phi_0 | a_0^\dagger a_0 a_0^\dagger a_0 | \Phi_0 \rangle = N_0^2$$
$$\langle \Phi_0 | a_0^\dagger a_0^\dagger a_0 a_0 | \Phi_0 \rangle = N_0(N_0 - 1) \approx N_0^2$$

Thus, to order N_0^2, a_0 and $a_0^\dagger$ appear to commute with each other and behave like ordinary numbers (c-numbers). This is what we mean when we say that the operator a_0 has weak quantum fluctuations. In this case, we may replace a_0 by its classical value

$$a_0 = a_0^\dagger = \sqrt{N_0}$$

The interacting Hamiltonian can now be written as

$$\begin{aligned} H_{mean} &= \sum_{\boldsymbol{k}} (\epsilon_{\boldsymbol{k}} - \mu) a_{\boldsymbol{k}}^\dagger a_{\boldsymbol{k}} + O((a_{\boldsymbol{k} \neq 0})^3) + \frac{\mathcal{V}}{2} \rho_0^2 V_0 \\ &\quad + \frac{\rho_0}{2} \sum_{\boldsymbol{k} \neq 0} \left(2 a_{\boldsymbol{k}}^\dagger a_{\boldsymbol{k}} V_0 + 2 a_{\boldsymbol{k}}^\dagger a_{\boldsymbol{k}} V_{\boldsymbol{k}} + V_{\boldsymbol{k}} a_{-\boldsymbol{k}} a_{\boldsymbol{k}} + V_{\boldsymbol{k}} a_{-\boldsymbol{k}}^\dagger a_{\boldsymbol{k}}^\dagger \right) \\ &= \sum_{\boldsymbol{k}} (\epsilon_{\boldsymbol{k}} - \mu'_{\boldsymbol{k}}) a_{\boldsymbol{k}}^\dagger a_{\boldsymbol{k}} - \frac{1}{2} \rho_0^2 V_0 + \rho_0 \frac{1}{2} \sum_{\boldsymbol{k} \neq 0} V_{\boldsymbol{k}} \left(a_{-\boldsymbol{k}} a_{\boldsymbol{k}} + a_{-\boldsymbol{k}}^\dagger a_{\boldsymbol{k}}^\dagger \right) \end{aligned} \tag{3.2.2}$$

where $\mu'_{\boldsymbol{k}} = \mu - \rho_0 V_0 - \frac{1}{2} \rho_0 (V_{\boldsymbol{k}} + V_{-\boldsymbol{k}})$, $\rho_0 = N_0/\mathcal{V}$, and we have ignored $(a_{\boldsymbol{k} \neq 0})^3$ and higher-order terms. After the two approximations of firstly replacing a_0 by a c-number, which is accurate in the $N \to \infty$ limit, and secondly dropping the a_k^3 terms, which are accurate in the $V \to 0$ or $\rho \to 0$ limit, we obtain a quadratic mean-field Hamiltonian.

Due to the $a_{-\boldsymbol{k}} a_{\boldsymbol{k}}$ terms, it is clear that the state with all N bosons in the $\boldsymbol{k} = 0$ state is no longer the ground state of the interacting boson system. To find the new (mean-field) ground state, we introduce, for $\boldsymbol{k} \neq 0$,

$$\alpha_{\boldsymbol{k}} = u_{\boldsymbol{k}} a_{\boldsymbol{k}} + v_{\boldsymbol{k}} a^{\dagger}_{-\boldsymbol{k}}$$

We choose $u_{\boldsymbol{k}}$ and $v_{\boldsymbol{k}}$ such that H_{mean} takes the form $\sum_{\boldsymbol{k}} \alpha^{\dagger}_{\boldsymbol{k}} \alpha_{\boldsymbol{k}} E_{\boldsymbol{k}} +$ constant and $\alpha_{\boldsymbol{k}}$ and $\alpha^{\dagger}_{\boldsymbol{k}}$ have the same boson commutation relation

$$[\alpha_{\boldsymbol{k}}, \alpha^{\dagger}_{\boldsymbol{k}'}] = \delta_{\boldsymbol{k},\boldsymbol{k}'}$$

The latter condition requires that $|u^2_{\boldsymbol{k}}| - |v_{\boldsymbol{k}}|^2 = 1$. One can show that the following choice satisfies our requirements:

$$E_{\boldsymbol{k}} = \sqrt{(\epsilon_{\boldsymbol{k}} - \mu'_{\boldsymbol{k}})^2 - \rho_0^2 V_{\boldsymbol{k}}^2} \tag{3.2.3}$$

$$u_{\boldsymbol{k}} = \sqrt{\frac{\epsilon_{\boldsymbol{k}} - \mu'_{\boldsymbol{k}}}{2E_{\boldsymbol{k}}} + \frac{1}{2}} \tag{3.2.4}$$

$$v_{\boldsymbol{k}} = \frac{V_{\boldsymbol{k}}}{|V_{\boldsymbol{k}}|} \sqrt{\frac{\epsilon_{\boldsymbol{k}} - \mu'_{\boldsymbol{k}}}{2E_{\boldsymbol{k}}} - \frac{1}{2}} \tag{3.2.5}$$

and

$$\sum_{\boldsymbol{k}\neq 0} \left((\epsilon_{\boldsymbol{k}} - \mu'_{\boldsymbol{k}}) a^{\dagger}_{\boldsymbol{k}} a_{\boldsymbol{k}} + \frac{V_{\boldsymbol{k}} \rho_0}{2} (a_{-\boldsymbol{k}} a_{\boldsymbol{k}} + h.c.) \right) = \sum_{\boldsymbol{k}\neq 0} E_{\boldsymbol{k}} \alpha^{\dagger}_{\boldsymbol{k}} \alpha_{\boldsymbol{k}} - \sum_{\boldsymbol{k}\neq 0} E_{\boldsymbol{k}} |v_{\boldsymbol{k}}|^2$$

Including the $\boldsymbol{k} = 0$ part, we find that

$$H_{mean} = \sum_{\boldsymbol{k}\neq 0} E_{\boldsymbol{k}} \alpha^{\dagger}_{\boldsymbol{k}} \alpha_{\boldsymbol{k}} + \Omega_g$$

$$\Omega_g = \mathcal{V}(-\mu\rho_0 + \frac{1}{2}\rho_0^2 V_0) - \sum_{\boldsymbol{k}\neq 0} \frac{1}{2}(\epsilon_{\boldsymbol{k}} - \mu'_{\boldsymbol{k}} - E_{\boldsymbol{k}})$$

The mean-field ground state is given by

$$\alpha_{\boldsymbol{k}} |\Phi_{mean}\rangle = 0$$

and Ω_g is the mean-field ground-state energy or, more correctly, the thermal potential for the mean-field ground state. Within mean-field theory, the excitations above the ground state are described by a collection of harmonic oscillators $\alpha_{\boldsymbol{k}}$ which correspond to the waves in the boson fluid. The dispersion relation of the waves is given by $E_{\boldsymbol{k}}$.

The two parameters N_0 and μ remain to be determined. The excitation spectrum

$$E_{\boldsymbol{k}} = \sqrt{(\epsilon_k - \mu + \rho_0 V_0 + \rho_0 V_{\boldsymbol{k}})^2 - (\rho_0 V_k)^2}$$

depends on μ in a sensitive way. If $\mu = \rho_0 V_0$, then $E_0 = 0$ and the excitations are gapless (here we have assumed that $\epsilon_0 = 0$). If $\mu < \rho_0 V_0$, then the excitations have a finite gap. If $\mu > \rho_0 V_0$, then E_0 is imaginary and this suggests an instability.

Firstly, the value of N_0 is obtained by minimizing the thermal potential Ω_g (with μ fixed):

$$\frac{\partial \Omega_g}{\partial N_0} = 0 \tag{3.2.6}$$

After obtaining $N_0 = N_0(\mu)$ and noticing that

$$a_{\boldsymbol{k}} = u^*_{\boldsymbol{k}} \alpha_{\boldsymbol{k}} - v^*_{\boldsymbol{k}} \alpha^\dagger_{-\boldsymbol{k}}$$

we find that

$$N = \langle \Phi_{mean} | \sum_{\boldsymbol{k}} a^\dagger_{\boldsymbol{k}} a_{\boldsymbol{k}} | \Phi_{mean} \rangle = N_0(\mu) + \sum_{\boldsymbol{k}} |v_{\boldsymbol{k}}|^2 \tag{3.2.7}$$

The above equation determines the value of $\mu = \mu(N)$ that gives us the total N bosons.

Let us try to carry out the above calculation in the low-density or weak-coupling limit. Retaining the terms of linear order in $V_{\boldsymbol{k}}$ (note that $\epsilon_{\boldsymbol{k}} - \mu'_{\boldsymbol{k}} - E_{\boldsymbol{k}} = O(V^2_{\boldsymbol{k}})$), we have

$$\Omega_g = \mathcal{V}(-\mu\rho_0 + \frac{1}{2}\rho_0^2 V_0)$$

We see that, to minimize Ω_g, we require that $N_0 = 0$ for $\mu < 0$, and $N_0 = \mathcal{V}\mu/V_0$ for $\mu > 0$. Once we know N_0, the total number of bosons can be calculated through eqn (3.2.7). If we know N, then μ is determined from $N = \mathcal{V}\mu/V_0 + \sum_{\boldsymbol{k}} |v_{\boldsymbol{k}}|^2$.

When the boson density is nonzero, we find that $\mu = \rho_0 V_0 > 0$ and the excitation spectrum

$$E_{\boldsymbol{k}} = \sqrt{\epsilon_{\boldsymbol{k}}(\epsilon_{\boldsymbol{k}} + 2\rho_0 V_{\boldsymbol{k}})} \tag{3.2.8}$$

is gapless and linear for small $\boldsymbol{k}$, regardless of the value of N and $V_{\boldsymbol{k}}$:

$$E_{\boldsymbol{k}} = v|\boldsymbol{k}|, \qquad v^2 = \frac{\rho_0 V_0}{m} - \frac{\rho_0^2}{2} \tag{3.2.9}$$

This result is very reasonable because the excitation created by $\alpha^\dagger_{\boldsymbol{k}}$ should correspond to density waves (the Nambu–Goldstone modes) in the compressible boson fluid and should always have a gapless and linear dispersion. The linear gapless excitation is also called a phonon.

If we were to stop here then everything would be fine. We can say that we have a mean-field theory for the ground state and excitations in an interacting boson

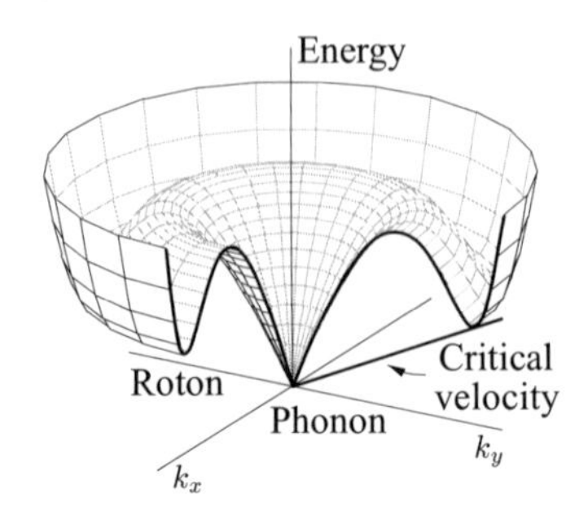

FIG. 3.1. The phonon excitations are near $\boldsymbol{k} = 0$ and the roton excitations are near to the local minimum at finite $\boldsymbol{k}$. The critical velocity of the superfluid flow is also marked (see Section 3.7.3).

system which produces reasonable results. However, if we want to do better, then things may get worse. Within mean-field theory, the vanishing of the energy gap is due to the 'miracle' that μ is exactly $\rho_0 V_0$, which causes the constant term in $E_{\boldsymbol{k}}$ to cancel out. If we keep the higher-order terms of $V_{\boldsymbol{k}}$ appearing in Ω_g, then the more accurate μ may not be equal to $\rho_0 V_0$. In this case, we may have the unreasonable result that the energy gap is not zero. Certainly, when we include higher-order terms in Ω_g, we should at the same time include higher-order terms in E_k, and we may still have a cancellation which leads to a zero energy gap.

However, one thing is clear. The mean-field theory does not guarantee a zero energy gap. To obtain a zero energy gap we have to rely on careful calculations and hope for the miracle. However, from a physical point of view, the existence of a gapless excitation is a general principle which is independent of any details of the boson system. Regardless of the form of the interaction, the boson system in free space is always compressible and always contains a gapless density wave. We see that mean-field theory did not capture the principle behind the gapless excitations. This problem with mean-field theory is not restricted to boson systems. In general, all mean-field theories have the problem that some general principles of the original theory are not captured by the mean-field theory. One needs to keep this in mind and be very careful when using mean-field theory to study excitation spectrums.

In a real superfluid, the excitation spectrum has the form shown in Fig. 3.1. The excitations near the minimum at finite $\boldsymbol{k}$ are called rotons. We note that the roton can have a vanishing group velocity at a proper $\boldsymbol{k}$.

Problem 3.2.1.
Determine N_0 and μ for a one-dimensional interacting boson system with density $\rho = N/\mathcal{V}$. Show that the interacting bosons cannot condense in one dimension, even at zero temperature. (Note that non-interacting bosons can condense in one dimension at zero temperature.)

Problem 3.2.2.
(a) Find a $V_{\boldsymbol{k}}$ such that the roton minimum in $E_{\boldsymbol{k}}$ dips below zero. Plot the $V(\boldsymbol{x})$ in real space.
(b) When the roton minimum does dip below zero, what happens to the superfluid state?

Problem 3.2.3.
A mean-field theory for a spin chain. Consider a spin chain

$$H = J \sum_i \boldsymbol{S}_i \boldsymbol{S}_{i+1}$$

where $J < 0$ and $\boldsymbol{S}$ is the spin operator that carries spin S. When $S \gg 0$, the quantum fluctuations of $\boldsymbol{S}_i$ are weak and we may replace one $\boldsymbol{S}$ by its average $\boldsymbol{s} = \langle \boldsymbol{S}_i \rangle$ (which is assumed to be independent of i) and obtain the mean-field Hamiltonian.

1. Write down the mean-field Hamiltonian and find the mean-field ground state.
2. Determine the value of $\boldsymbol{s}$ by calculating $\langle \boldsymbol{S}_i \rangle$ using the mean-field Hamiltonian.
3. Find the ground-state energy using the mean-field approximation.
4. Find the energies for excitations above the mean-field ground state.

We see here that mean-field theory completely fails to reproduce gapless spin wave excitations. The path integral approach and its classical approximation discussed in Section 2.4 produce much better results.

3.3 Path integral approach to interacting boson systems

3.3.1 Path integral representation of interacting boson systems

As mentioned in Section 3.1, a free boson system can be described by a collection of independent harmonic oscillators with the Hamiltonian

$$H_0 = \sum_{\boldsymbol{k}} (\epsilon_{\boldsymbol{k}} - \mu) a^\dagger_{\boldsymbol{k}} a_{\boldsymbol{k}}$$

The oscillators (and the free boson system) admit a (coherent-state) path integral representation

$$\int \mathcal{D}^2[a_{\boldsymbol{k}}(t)]\ \mathrm{e}^{\mathrm{i}\int \mathrm{d}t\ \sum_{\boldsymbol{k}}[\mathrm{i}\frac{1}{2}(a^*_{\boldsymbol{k}}\dot{a}_{\boldsymbol{k}} - a_{\boldsymbol{k}}\dot{a}^*_{\boldsymbol{k}}) - (\epsilon_{\boldsymbol{k}} - \mu)a^*_{\boldsymbol{k}} a_{\boldsymbol{k}}]}$$

In real space, we have

$$\int \mathcal{D}^2 a\ \mathrm{e}^{\mathrm{i}\int \mathrm{d}^d\boldsymbol{x}\mathrm{d}t\ [\mathrm{i}\frac{1}{2}(a^*(\boldsymbol{x},t)\partial_t a(\boldsymbol{x},t) - a(\boldsymbol{x},t)\partial_t a^*(\boldsymbol{x},t))] - \frac{1}{2m}\partial_{\boldsymbol{x}} a^*(\boldsymbol{x},t)\partial_{\boldsymbol{x}} a(\boldsymbol{x},t) - \mu a^*(\boldsymbol{x},t)a(\boldsymbol{x},t)]}$$

We can now easily include the interaction by adding a term

$$\int \mathrm{d}^d\boldsymbol{x}\,\mathrm{d}^d\boldsymbol{y}\ \frac{1}{2} a^*(\boldsymbol{x},t)a(\boldsymbol{x},t)V(\boldsymbol{x}-\boldsymbol{y})a^*(\boldsymbol{y},t)a(\boldsymbol{y},t)$$

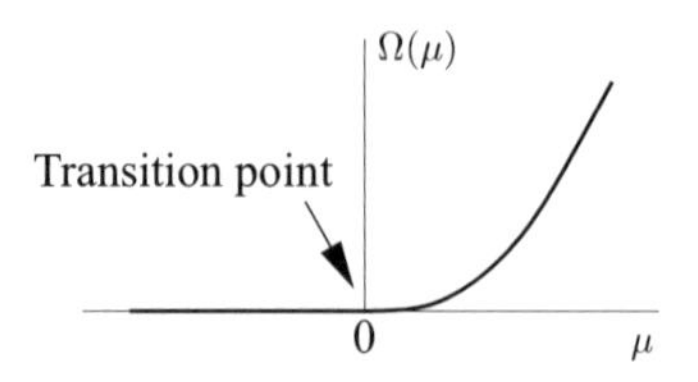

FIG. 3.2. The discontinuity in the second derivative of the thermal potential $\Omega(\mu)$ indicates a second-order phase transition.

to the Hamiltonian and the path integral. Here, we will treat a simpler interaction $V(\boldsymbol{x}) = V_0\delta(\boldsymbol{x})$ whose Fourier component is $V_{\boldsymbol{k}} = V_0$. The path integral for interacting bosons now has the form

$$\int \mathcal{D}^2[\varphi(\boldsymbol{x},t)]\ \mathrm{e}^{\mathrm{i}\int \mathrm{d}^d\boldsymbol{x}\mathrm{d}t\ [\mathrm{i}\frac{1}{2}(\varphi^*\partial_t\varphi-\varphi\partial_t\varphi^*)]-\frac{1}{2m}\partial_{\boldsymbol{x}}\varphi^*\partial_{\boldsymbol{x}}\varphi+\mu|\varphi|^2-\frac{V_0}{2}|\varphi|^4]}$$

where, following convention, we have renamed $a(\boldsymbol{x},t)$ by $\varphi(\boldsymbol{x},t)$.

Having the path integral representation, we can study the physical properties of the quantum interacting boson system by treating it as a classical field theory described by the action

$$S = \int \mathrm{d}^d\boldsymbol{x}\mathrm{d}t\ [\mathrm{i}\frac{1}{2}(\varphi^*\partial_t\varphi - \varphi\partial_t\varphi^*) - \frac{1}{2m}\partial_{\boldsymbol{x}}\varphi^*\partial_{\boldsymbol{x}}\varphi + \mu|\varphi|^2 - \frac{V_0}{2}|\varphi|^4] \quad (3.3.1)$$

Here we are performing a semiclassical approximation (instead of the mean-field approximation performed in Section 3.2), and the leading term is simply the classical theory. The energy (or, more precisely, the thermal potential) of the classical system is

$$\Omega = \int \mathrm{d}^d\boldsymbol{x}\ [\frac{1}{2m}\partial_{\boldsymbol{x}}\varphi^*\partial_{\boldsymbol{x}}\varphi + \mu|\varphi|^2 + \frac{V_0}{2}|\varphi|^4] \quad (3.3.2)$$

3.3.2 Phase transition and spontaneous symmetry breaking

- Zero-temperature phase transitions happen at the non-analytic point of the ground-state energy.
- Spontaneous symmetry breaking happens when an asymmetric ground state emerges from a symmetric system.
- Continuous phase transitions, spontaneous symmetry breaking, and long-range order are closely related. This is the heart of Landau's symmetry-breaking theory for phases and phase transitions.

The classical ground state is translationally invariant and is characterized by a complex constant φ_0: $\varphi(\boldsymbol{x},t) = \varphi_0$. The energy density (or, more precisely, the

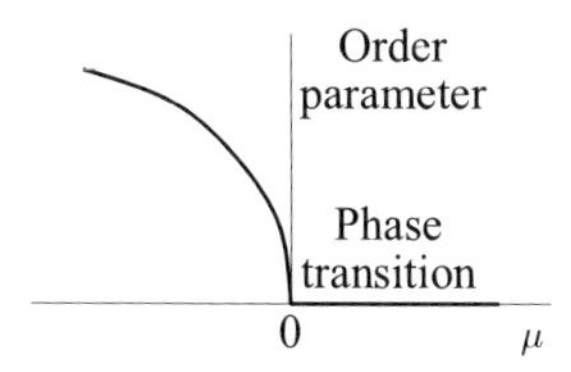

FIG. 3.3. The order parameter φ_0 and the superfluid transition.

density of the thermal potential) of the classical ground state is

$$\frac{\Omega_0}{\mathcal{V}} = -\mu|\varphi_0|^2 + \frac{V_0}{2}|\varphi_0|^4$$

By minimizing the energy, we see that for $\mu < 0$ the ground state is characterized by $\varphi_0 = 0$, which corresponds to a state with no bosons. For $\mu > 0$ the ground state is degenerate and is characterized by the following amplitude of the boson field:

$$\varphi_0 = \sqrt{\frac{\mu}{V_0}} \mathrm{e}^{\mathrm{i}\theta} \tag{3.3.3}$$

with arbitrary θ. Such a state has a boson density $\rho_0 = \mu/V_0$. The thermal potential as a function of μ is given by

$$\Omega_0(\mu) = \begin{cases} 0, & \text{for } \mu < 0 \\ \frac{\mu^2}{V_0}, & \text{for } \mu > 0 \end{cases}$$

We see that, as a function of μ, the ground-state energy $\Omega_0(\mu)$ has a singularity at $\mu = 0$. The singularity signals a second-order phase transition at $\mu = 0$ because $\Omega_0''(\mu)$ has a jump at that point (see Fig. 3.2). This phase transition is called the superfluid transition.[16]

The superfluid phase is characterized by a nonzero boson field $\varphi(\boldsymbol{x}) \neq 0$, and the no-boson phase by a vanishing boson field $\varphi(\boldsymbol{x}) = 0$. The superfluid transition is signaled by the change $\varphi_0 = 0$ to $\varphi_0 \neq 0$ (see Fig. 3.3). For this reason, we call φ_0 the order parameter for the superfluid.

The superfluid phase is also characterized by symmetry breaking. We see that our Hamiltonian or Lagrangian has a $U(1)$ symmetry because they are both invariant under the transformation

$$\varphi \to \mathrm{e}^{\mathrm{i}\theta}\varphi \tag{3.3.4}$$

regardless of which phase we are in. However, the (classical) ground states in the superfluid phase (see eqn (3.3.3)) do not respect the $U(1)$ symmetry, in the sense that they are not invariant under the $U(1)$ transformation (3.3.4).

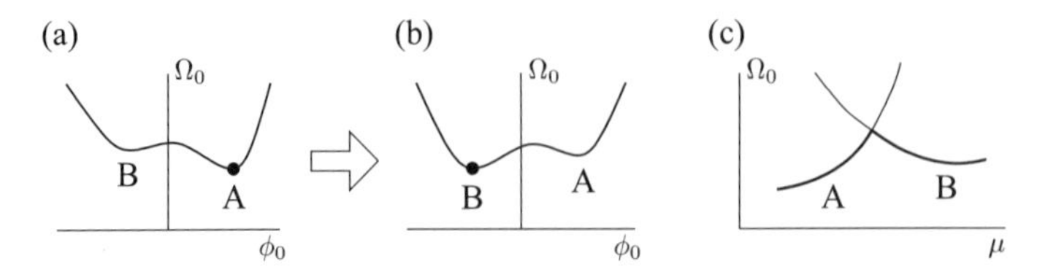

FIG. 3.4. (a) and (b) Switching between different minima causes (c) a kink in Ω_0 which represents a first-order phase transition.

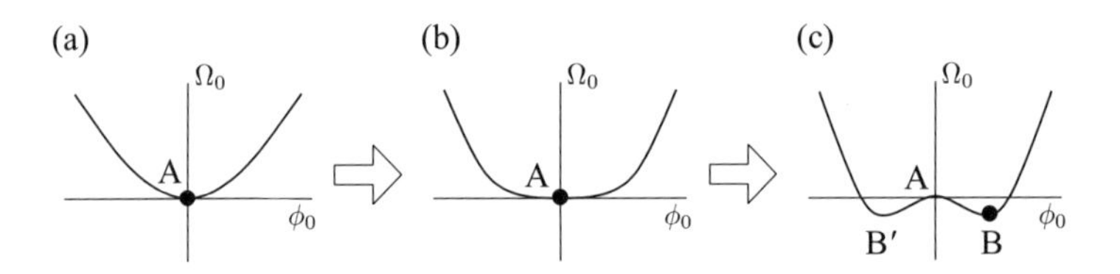

FIG. 3.5. In the presence of $\phi_0 \to -\phi_0$ symmetry, switching of the minima can be continuous and causes a second-order phase transition. (a) The ground state is symmetric under the condition $\phi_0 \to -\phi_0$. (b) The transition point. (c) The ground state breaks the $\phi_0 \to -\phi_0$ symmetry.

The superfluid transition demonstrates a deep principle: a continuous phase transition characterized by a non-analytic point of the thermal potential Ω is related to a change in the symmetry of the state. To appreciate this point, let us start from the beginning. We know that a phase transition is caused by the singularity in the energy or free energy of the system. So, to understand the phase transition, we need to understand how the (free) energy develops a singularity.

For the interacting boson system, the energy function $\Omega_0(\mu; \phi_0)$ as a function of ϕ_0 and μ is an analytic function with no singularities. The ground-state energy $\Omega_0(\mu)$ is the minimal value of $\Omega_0(\mu; \phi_0)$ with respect to ϕ_0. In general, $\Omega_0(\mu)$ is also an analytic function. However, sometimes $\Omega_0(\mu; \phi_0)$ has multiple local minima as a function of ϕ_0. As we change μ, the global minimum may switch from one local minimum to another (see Fig. 3.4). In this case, the resulting $\Omega_0(\mu)$ will have a singularity at the switch point. Such a singularity represents a first-order phase transition because $\Omega_0'(\mu)$ has a jump (see Fig. 3.4(c)).

However, when the energy function $\Omega_0(\mu, \phi_0)$ has symmetry, say $\Omega_0(\mu, \phi_0) = \Omega_0(\mu, \mathrm{e}^{\mathrm{i}\theta}\phi_0)$, then the switching between minima can have a very different behavior and can be continuous (see Fig. 3.5). Due to the symmetry, new global minima are degenerate. The system faces a tough choice to pick one minimum at which to stay. After the system makes the choice, the new phase no longer has $\phi_0 \to \mathrm{e}^{\mathrm{i}\theta}\phi_0$ symmetry. This phenomenon of an asymmetric state emerging from a symmetric

[16] We will discuss superfluidity in Section 3.7.3.

system is called spontaneous symmetry breaking. We see that the second-order phase transition is intimately related to a change of symmetry in the ground state. This is the heart of Landau's symmetry-breaking theory for phase and phase transition (Landau, 1937). This simple idea has a very wide application. In practice, almost all continuous phase transitions are characterized by some kind of spontaneous symmetry breaking.

In general, spontaneous symmetry breaking can be characterized by an order parameter. The order parameter can be chosen to be the expectation value of an operator that transforms non-trivially under the symmetry transformation. In the above, we have chosen $\langle\varphi\rangle$ as the order parameter. Due to its transformational property $\langle\varphi\rangle \to \mathrm{e}^{\mathrm{i}\theta}\langle\varphi\rangle$, $\langle\varphi\rangle = 0$ in the symmetric phase. In contrast, $\langle\varphi\rangle = \varphi_0 \neq 0$ in the symmetry-breaking phase.

In the symmetry-breaking phase, if we know the phase of the φ field at one point, then we know the phase at any other point. This property is called long-range order. The long-range correlation in the φ field can be mathematically expressed as

$$\left\langle \varphi(t,\boldsymbol{x})^\dagger \varphi(0,0) \right\rangle \Big|_{(t,\boldsymbol{x})\to\infty} = |\varphi_0|^2 \neq 0$$

If the phases of φ are random and uncorrelated from one point to another point, then the above correlation will be zero. Certainly, here we only consider the classical ground state without quantum fluctuations. In this case, we see that the nonzero order parameter and the long-range order in the φ operator are closely related. The symmetry-breaking phase can be characterized by either a nonzero order parameter or a long-range correlation. In the next few sections, we will include quantum fluctuations and see how they may affect order parameters and long-range order.

Problem 3.3.1.
Derive the classical equation of motion from the action (3.3.1). Consider only the small fluctuations around the ground state and derive the linearized equation of motion for both the symmetric phase (with $\mu < 0$) and the symmetry-breaking phase (with $\mu > 0$). Show that fluctuations around the symmetric ground state have a finite energy gap, while fluctuations around the symmetry-breaking ground state contain one mode with zero gap. This mode has a linear dispersion $\omega \propto k$.

Problem 3.3.2.
A water–vapor transition can be described by the following Gibbs energy function:

$$G(T,P;n) = h(T,P)n + a(T,P)n^2 + c(T,P)n^3 + b(T,P)n^4$$

where n is the density of the water molecules. The Gibbs energy of the system $G(T,P)$ is given by the minimal value of $G(T,P;n)$ with respect to n. The system has a line of first-order transitions that ends at a critical point. Find the equations that determine the first-order-transition line and the critical point. (Hint: First consider the case with $c = 0$. Such a Gibbs energy has the same form as the free energy of the Ising model in a magnetic field $\delta\rho$, if we regard n as the density of the spin moments.)

3.3.3 Low-energy effective theory

- The effective Lagrangian that describes the slow (i.e. low-energy) fluctuations can be obtained by integrating out the fast (i.e. high-energy) fluctuations.

To study excitations above the classical ground state, let us consider the following small fluctuations around the classical ground state:

$$\varphi = \varphi_0 + \delta\varphi$$

The dynamics of $\delta\varphi$ are governed by the same action (3.3.1). However, we can now assume that $\delta\varphi$ is small and keep only the quadratic terms in $\delta\varphi$. For the symmetric phase $\varphi_0 = 0$, we have

$$S = \int \mathrm{d}^d\boldsymbol{x}\mathrm{d}t\, [\mathrm{i}\frac{1}{2}(\varphi^*\partial_t\varphi - \varphi\partial_t\varphi^*) - \frac{1}{2m}\partial_{\boldsymbol{x}}\varphi^*\partial_{\boldsymbol{x}}\varphi + \mu\varphi^*\varphi] \tag{3.3.5}$$

In the symmetric phase, the equation of motion is

$$[\mathrm{i}\partial_t - \frac{1}{2m}(-\mathrm{i}\partial_{\boldsymbol{x}})^2 + \mu]\varphi = 0$$

which leads to the following dispersion for the fluctuations:

$$\omega = \frac{1}{2m}k^2 - \mu \tag{3.3.6}$$

In the next section, we will see that those fluctuations correspond to particle-like excitations. The frequency and the wave vector of the fluctuations become the energy and the momentum of the particles, respectively. We see that excitations have a finite energy gap $\Delta = -\mu$ (note that $\mu < 0$), which is just the energy needed to create an excitation. In fact, the particles described by the dispersion (3.3.6) just correspond to the bosons that form our system.

To understand the low-energy excitations in the symmetry-breaking phase, we would like to derive a low-energy effective theory that only contains modes related to the low-energy excitations. As to which fluctuations have low energies, we note that, in the symmetry-breaking phase, different phases of the condensate, θ, lead to degenerate ground states. If we change the phase of the condensate in a large but finite region, then, locally, the system is still in one of its degenerate ground states. The only way for the system to know that it is in an excited state is through the gradient of θ. If the phase changes slowly in space, then the corresponding excitations will have small energies. Thus, the low-energy excitations correspond to fluctuations among degenerate ground states and are described by the θ field. Based on this picture, we write

$$\varphi = \sqrt{\rho_0 + \delta\rho}\,\mathrm{e}^{\mathrm{i}\theta} \tag{3.3.7}$$

where $\rho_0 = \varphi_0^2 = \mu/V_0$ is the density of the ground state and $\delta\rho$ is the density fluctuation. In this form, θ describes the low-energy slow fluctuations and

$\delta\rho$ the high-energy fast fluctuations. Substituting eqn (3.3.7) into eqn (3.3.1) and expanding to quadratic order in θ and $\delta\rho$, we obtain

$$S = \int \mathrm{d}^d\boldsymbol{x}\,\mathrm{d}t\,[-(\rho_0+\delta\rho)\partial_t\theta - \frac{\rho_0(\partial_{\boldsymbol{x}}\theta)^2}{2m} - \frac{(\partial_{\boldsymbol{x}}\delta\rho)^2}{8m\rho_0} - \frac{V_0}{2}\delta\rho^2] \tag{3.3.8}$$

Here we are facing a very typical problem in many-body physics. We have two fields, one describing the slow, low-frequency (i.e. low-energy) fluctuations and the other describing the fast, high-frequency fluctuations. If we only care about low-energy physics, how do we remove the fast field to obtain a simple, low-energy effective theory?

One way to obtain the low-energy effective theory is to drop the $\delta\rho$ field by setting $\delta\rho = 0$. This leads to the following low-energy effective action for θ:

$$S = \int \mathrm{d}^d\boldsymbol{x}\,\mathrm{d}t\,[-\varphi_0^2\partial_t\theta - \frac{1}{2m}\varphi_0^2(\partial_{\boldsymbol{x}}\theta)^2]$$

However, this effective action is not quite correct because the $\partial_t\theta$ is a total derivative. It does not enter into the equation of motion. We need to obtain the $(\partial_t\theta)^2$ term to understand the low-energy dynamical properties of θ. To obtain the $(\partial_t\theta)^2$ term, one needs to be more careful in removing the $\delta\rho$ field.

A better way to obtain the low-energy effective theory is to integrate out the fast field $\delta\rho$ in the path integral. This can be easily achieved here because the action (3.3.8) is quadratic in $\delta\rho$. Performing the Gaussian integral, we find that

$$\begin{aligned} Z &= \int \mathcal{D}\theta\; \mathrm{e}^{\,\mathrm{i}\int \mathrm{d}^d\boldsymbol{x}\,\mathrm{d}t\,[-\frac{\rho_0}{2m}(\partial_{\boldsymbol{x}}\theta)^2 + \frac{1}{2}\partial_t\theta\frac{1}{V_0 - \frac{1}{4m\rho_0}\partial_{\boldsymbol{x}}^2}\partial_t\theta]} \\ &\approx \int \mathcal{D}\theta\; \mathrm{e}^{\,\mathrm{i}\int \mathrm{d}^d\boldsymbol{x}\,\mathrm{d}t\,[-\frac{\rho_0}{2m}(\partial_{\boldsymbol{x}}\theta)^2 + \frac{1}{2V_0}(\partial_t\theta)^2]} \end{aligned} \tag{3.3.9}$$

where we have assumed that the field θ varies slowly in space and we have dropped the $\frac{1}{4m\rho_0}(\partial_{\boldsymbol{x}})^2$ term. We have also dropped the total time derivative term in the action.[17] We obtain the following simple, low-energy effective action for θ:

$$S_{\text{eff}} = \int \mathrm{d}^d\boldsymbol{x}\,\mathrm{d}t\,[\frac{1}{2V_0}(\partial_t\theta)^2 - \frac{\rho_0}{2m}(\partial_{\boldsymbol{x}}\theta)^2] \tag{3.3.10}$$

Note that the θ field really lives on a circle, and so θ and $\theta + 2\pi$ describe the same point. To be more accurate, we may introduce $z = \mathrm{e}^{\mathrm{i}\theta}$ and rewrite the above as

$$S_{\text{eff}} = \int \mathrm{d}^d\boldsymbol{x}\,\mathrm{d}t\,[\frac{1}{2V_0}|\partial_t z|^2 - \frac{\rho_0}{2m}|\partial_{\boldsymbol{x}} z|^2] \tag{3.3.11}$$

The model described by eqn (3.3.10) or eqn (3.3.11) is called the XY-model.

[17] The total time derivative term is important in the presence of vortices and should not then be dropped. See Section 3.6.

The classical ground state is given by a uniform θ field: $\theta(\boldsymbol{x},t) = \text{constant}$. The excitations are described by the fluctuations of θ which satisfy the equation of motion

$$(-\frac{1}{2V_0}\partial_t^2 + \frac{\rho_0}{2m}\partial_{\boldsymbol{x}}^2)\theta = 0$$

The above equation is a wave equation. It describes a wave with a linear dispersion

$$\omega = v|\boldsymbol{k}|, \qquad v^2 = \frac{\rho_0 V_0}{m} \tag{3.3.12}$$

where v is the wave velocity.

In many cases, it is not enough to just know the low-energy effective action. It is also important to know how the fields (or operators) in the original theory are represented by the fields (or operators) in the low-energy effective theory. In the following, we will use the density operator as an example to illustrate the representation of the physical operators (or fields) in the original theory by the fields in the effective theory. Firstly, we add a source term $-A_0\rho = -A_0(\rho_0 + \delta\rho)$ that couples to the density in the original Lagrangian. Then, we carry through the same calculation to obtain the effective theory. We find that the effective theory contains an additional term (to linear order in A_0) $-A_0(\rho_0 - \frac{\partial_t\theta}{V_0})$. Thus, the density operator is represented by

$$\rho = \rho_0 - \frac{\partial_t\theta}{V_0} \tag{3.3.13}$$

in the effective theory. Similarly, we find that the boson current density $\boldsymbol{j} = \text{Re}\varphi^\dagger \frac{\partial_{\boldsymbol{x}}}{im}\varphi$ becomes

$$\boldsymbol{j} = \frac{\rho_0}{m}\partial_{\boldsymbol{x}}\theta \tag{3.3.14}$$

in the effective theory. Equations (3.3.13), (3.3.14), and (3.3.10) allow us to use path integrals to calculate the density and current correlations within the simple low-energy effective theory. Those correlations can then be compared with experimental results, such as compressibility and conductivity. The relationship between ρ and θ also tells us that the low-energy fluctuations described by θ are simply the density–current fluctuations.

Problem 3.3.3.

(a) There is another way to obtain the low-energy effective theory for θ. We express $\delta\rho$ in terms of $\partial_t\theta$, $\partial_{\boldsymbol{x}}\theta$, etc. by solving the equation of motion for the $\delta\rho$ field. Then, we substitute $\delta\rho$ back into the action to obtain an effective action that contains only θ. Show that this method produces the same XY-model action (3.3.10).

(b) If we substitute $\delta\rho$ into the density and current operators, then we will reproduce eqn (3.3.13) and eqn (3.3.14), respectively. Show that ρ and $\boldsymbol{j}$ satisfy the conservation law $\partial_t\rho + \partial_{\boldsymbol{x}} \cdot \boldsymbol{j} = 0$.

(c) Use the expression for the density operator given in eqn (3.3.13) to calculate the superfluid density–density correlation function in momentum–frequency space.

3.3.4 Waves are particles

- The emergence of a new kind of particle—quasiparticles—in a superfluid state. The quasiparticles are bosonic and non-interacting. The quasiparticles also have a linear dispersion.

We have viewed the fluctuations of θ around a classical ground state (or the $\delta\varphi$ in the symmetric phase) as waves. Due to the particle–wave duality in quantum physics, the wave can also be viewed as particles. These particles are called quasiparticles. In the following, we will quantize the low-energy effective theory and obtain the quantum theory of the quasiparticles. The quantum theory turns out to be a theory of bosons, which tells us that the quasiparticles are bosons.

We first write the low-energy effective Lagrangian (3.3.10) in the $\boldsymbol{k}$ space as follows:

$$L_{\text{eff}} = \sum_{\boldsymbol{k}} [\frac{A_{\boldsymbol{k}}}{2} \dot{\theta}_{-\boldsymbol{k}} \dot{\theta}_{\boldsymbol{k}} - \frac{B_{\boldsymbol{k}}}{2} \theta_{-\boldsymbol{k}} \theta_{\boldsymbol{k}}], \tag{3.3.15}$$

where

$$A_{\boldsymbol{k}} = V_0^{-1}, \qquad B_{\boldsymbol{k}} = \frac{\rho_0 \boldsymbol{k}^2}{m}, \tag{3.3.16}$$

$\theta_{\boldsymbol{k}} = \mathcal{V}^{-1/2} \int \mathrm{d}^d\boldsymbol{x}\ \theta(\boldsymbol{x}) \mathrm{e}^{-\mathrm{i}\boldsymbol{k}\cdot\boldsymbol{x}}$, and $\mathcal{V}$ is the total volume of the system. The above action describes a collection of decoupled harmonic oscillators.

To obtain the quantum Hamiltonian that describes the quantized theory, we first calculate the classical Hamiltonian. The canonical momentum for $\theta_{\boldsymbol{k}}$ is

$$\pi_{\boldsymbol{k}} = \frac{\partial L_{\text{eff}}}{\partial \dot{\theta}_{\boldsymbol{k}}} = A_{\boldsymbol{k}} \dot{\theta}_{-\boldsymbol{k}}$$

The classical Hamiltonian $H = \sum_{\boldsymbol{k}} \pi_{\boldsymbol{k}} \dot{\theta}_{\boldsymbol{k}} - L_{\text{eff}}$ is given by

$$H = \sum_{\boldsymbol{k}} [\frac{1}{2A_{\boldsymbol{k}}} \pi_{-\boldsymbol{k}} \pi_{\boldsymbol{k}} + \frac{B_{\boldsymbol{k}}}{2} \theta_{-\boldsymbol{k}} \theta_{\boldsymbol{k}}]$$

The quantum Hamiltonian is obtained by simply treating $\pi_{\boldsymbol{k}}$ and $\theta_{\boldsymbol{k}}$ as operators that satisfy the following canonical commutation relation:

$$[\theta_{\boldsymbol{k}}, \pi_{\boldsymbol{k}'}] = i\delta_{\boldsymbol{k}\boldsymbol{k}'}$$

We now introduce

$$\alpha_{\boldsymbol{k}} = u_{\boldsymbol{k}} \theta_{\boldsymbol{k}} + \mathrm{i} v_{\boldsymbol{k}} \pi_{-\boldsymbol{k}}, \quad u_{\boldsymbol{k}} = \frac{1}{\sqrt{2}} (A_{\boldsymbol{k}} B_{\boldsymbol{k}})^{1/4}, \quad v_{\boldsymbol{k}} = \frac{1}{\sqrt{2}} (A_{\boldsymbol{k}} B_{\boldsymbol{k}})^{-1/4}. \tag{3.3.17}$$

We find that the $\alpha_{\boldsymbol{k}}$ satisfy the algebra of the lowering operator of an oscillator:

$$[\alpha_{\boldsymbol{k}}, \alpha_{\boldsymbol{k}'}^\dagger] = \delta_{\boldsymbol{k}\boldsymbol{k}'}, \qquad [\alpha_{\boldsymbol{k}}, \alpha_{\boldsymbol{k}'}] = 0.$$

In terms of the lowering and raising operators $(\alpha_{\boldsymbol{k}}, \alpha_{\boldsymbol{k}}^\dagger)$, the Hamiltonian takes the standard form for the oscillators:

$$H = \sum_{\boldsymbol{k}} \epsilon_{\boldsymbol{k}} \alpha_{\boldsymbol{k}}^\dagger \alpha_{\boldsymbol{k}} + \text{constant}, \qquad \epsilon_{\boldsymbol{k}} = \sqrt{\frac{B_{\boldsymbol{k}}}{A_{\boldsymbol{k}}}} = \sqrt{\frac{V_0 \rho_0}{m}} |\boldsymbol{k}| = v|\boldsymbol{k}| \quad (3.3.18)$$

This is just the free boson Hamiltonian (3.1.1) with single-boson energy $\epsilon_{\boldsymbol{k}} = v|\boldsymbol{k}|$.

We see that the collective fluctuations of θ give rise to bosonic quasiparticles with a linear dispersion. As the fluctuations of θ correspond to the density wave in the superfluid, we can also say that density waves become discrete quasiparticles in quantum theory. This picture applies to more general situations. In fact, any wave-like fluctuations of a quantum ground state correspond to discrete quasiparticles after the quantization.[18] The sound waves in a solid become phonons described by the Hamiltonian (3.3.18). For this reason, we will also call the quasiparticles in the superfluid phonons. The phonon velocity (or the velocity of the density wave) $v = \sqrt{V_0 \rho_0 / m}$ agrees with the mean-field result (3.2.9).

We would like to stress that the bosonic quasiparticles—the phonons—are very different from the original bosons that form the superfluid. The original bosons have a quadratic dispersion $\epsilon_{\boldsymbol{k}} = \boldsymbol{k}^2/2m$, while the phonons have a linear dispersion $\epsilon_{\boldsymbol{k}} = v|\boldsymbol{k}|$. The original bosons are interacting, while the phonons are free. In fact, a phonon corresponds to a collective motion of many original bosons. This is an example of a new type of boson—non-interacting phonons with a linear dispersion—emerging from an interacting boson system.

3.3.5 Superfluid as a toy universe

- If we view the superfluid as a toy universe, then the phonons will be the massless 'elementary particles' in the toy universe.
- The rotons (the vortex rings) can interact by exchanging phonons. This leads to a $1/r^4$ dipolar interaction between two rotons.

To appreciate our universe and to appreciate the elementary particles in it, let us imagine a toy universe that is formed by a superfluid. Let us assume that the bosons that form the superfluid are too small to be seen by the inhabitants in the toy universe. The question is what does the toy universe look like to its inhabitants?

[18] A sound wave in air does not correspond to any discrete quasiparticle. This is because the sound wave is not a fluctuation of any quantum ground state. Thus, it does not correspond to any excitation above the ground state.

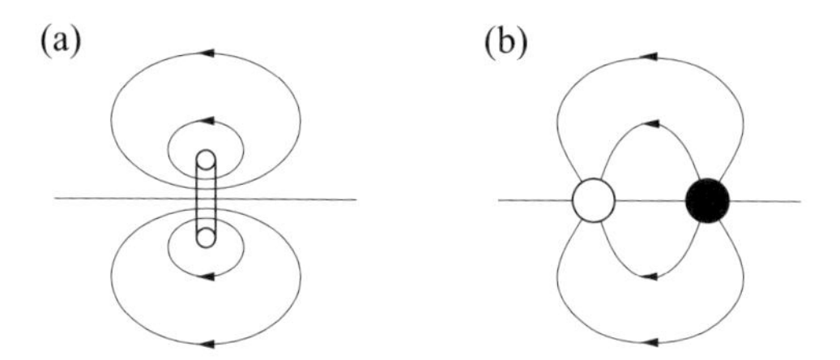

FIG. 3.6. The boson flow pattern around (a) a roton (a vortex ring), and (b) a dipole of the phonon charge.

The toy universe contains a massless excitation[19]—the phonon. To the inhabitants in the toy universe, the phonon is just a kind of particle-like excitation, because nobody can see the bosons that form the superfluid and nobody knows where the phonons come from. Thus, the inhabitants in the toy universe call the phonon an 'elementary particle'.

We see that the toy universe is quite similar to our universe. Our universe also has a massless excitation—the photon. We also call the photon an 'elementary particle'. Just like phonons, photons do not interact with each other either. However, photons are responsible for the $1/r^2$ Coulomb law between electric charges. Do phonons induce a similar interaction between their 'charges'?

The answer to the above question is yes. However, first we need to explain what the 'charges' are for the phonons. We know that the electric charge generates conserved flux—electric field. Similarly, the 'charges' for the phonons also generate conserved flux. However, for phonons the flux is the flux of the bosons in the superfluid. A positive 'phonon charge', by definition, is a source of bosons, where the bosons are created at a constant rate at a certain point. A negative phonon charge is a drain of bosons, where the bosons are annihilated at a constant rate at a certain point.

To understand the interactions between the phonon charges, we minimize the energy $\int \mathrm{d}^3\boldsymbol{x}\ \frac{\rho_0}{2m}(\partial_{\boldsymbol{x}}\theta)^2$ with respect to θ that statisfies the constraint $\partial_{\boldsymbol{x}} \cdot \boldsymbol{j} - J^0 = 0$. Here $\boldsymbol{j} = \frac{\rho_0}{m}\partial_{\boldsymbol{x}}\theta$ is the boson current, $J^0(\boldsymbol{x}, t) = \sum_i q_i \delta(\boldsymbol{x} - \boldsymbol{x}_i)$ is the density of the phonon charges and q_i is the value of the ith phonon charge. $q_i > 0$ corresponds to a source and $q_i < 0$ corresponds to a drain. We find that the potential between two phonon charges q_1 and q_2 is propotional to $q_1 q_2/r$. The interaction between the phonon charges is very similar to the Coulomb law. The force is proportional to $1/r^2$. The like charges repel and the opposite charges attract.

In a real superfluid, the bosons are conserved. As a result, the source and the drains, and hence the phonon charges, are not allowed. The only allowed excitations are the low-energy phonons and the high-energy rotons (see Fig. 3.1). It should be noted that the roton can be regarded as a small ring formed by a vortex line.[20] The flow pattern of the bosons around the roton can be viewed as a flow pattern generated be a collection of the phonon charges if we perform the multipole expansion. The conservation of the number of

---

[19] A relativistic particle of mass m has a dispersion $\epsilon_{\boldsymbol{k}} = \sqrt{m^2c^4 + c^2\boldsymbol{k}^2}$, where c is the speed of light. The phonon dispersion $\epsilon_{\boldsymbol{k}} = v|\boldsymbol{k}|$ can be viewed as the massless limit of $\sqrt{m^2c^4 + c^2\boldsymbol{k}^2}$, where the phonon velocity v plays the role of the speed of light.

[20] The vortex in $(2+1)$-dimensional superfluid is defined in Section 3.4.2. The definition can be generalized to any dimensions.

bosons requires that the total of the phonon charges in the multipole expansion is zero. The symmetry of the vortex ring allows a finite dipole moment in the multipole expansion (see Fig. 3.6), and the dipole moment is in general nonzero for the rotons. Therefore, the rotons have a $1/r^4$ dipolar interaction between them.

To summarize, the toy universe contains particles with a dipole moment. The long-range dipolar interaction is caused by exchanging the massless phonons.

3.3.6 Spontaneous symmetry breaking and gapless excitations

- The concept of universal properties.
- Spontaneous breaking of a continuous symmetry always results in gapless bosonic excitations.

We would like to stress that, in the superfluid, the three properties of the phonons, namely the vanishing interaction, the gapless linear dispersion, and the bosonic statistics, are universal properties that do not depend on the details of the interaction between the bosons. In the above, we have assumed that the bosons have a δ-function interaction. However, the universal properties remain unchanged if the bosons have a more general interaction $L_{int} = -\int \mathrm{d}\boldsymbol{x}\mathrm{d}\boldsymbol{x}'\ \frac{1}{2}|\varphi(\boldsymbol{x})|^2 V(\boldsymbol{x} - \boldsymbol{x}')|\varphi(\boldsymbol{x}')|^2$, as long as the interaction $V(\boldsymbol{x}-\boldsymbol{x}')$ is short-ranged. As the gaplessness of the quasiparticles is a universal property independent of the details of the starting Lagrangian, there should be a general understanding of the gaplessness that does not depend on those details. It turns out that the existence of the gapless excitations in the superfluid phase is intimately related to the $U(1)$ symmetry of the Lagrangian and the spontaneous symmetry breaking of the ground state. In the following, we will explain this relation.

The Lagrangian (3.3.1) has a $U(1)$ symmetry because it is invariant under the $U(1)$ transformation $\varphi \to \mathrm{e}^{\mathrm{i}\eta}\varphi$. If we explicitly break the symmetry, that is to make the Lagrangian not respect the $U(1)$ symmetry, then the low-lying mode will obtain a finite energy gap. For example, if we add a term $C\mathrm{Re}\varphi$ to the Lagrangian to explicitly break the $U(1)$ symmetry, then a term of the form $\cos(\theta)$ will be induced in the XY-model: $\mathcal{L} = \frac{1}{2V_0}(\partial_t\theta)^2 - \frac{\rho_0}{2m}(\partial_{\boldsymbol{x}}\theta)^2 + g\cos(\theta)$. The classical ground state is given by $\theta = 0$. For small fluctuations around the ground state, the Lagrangian can be simplified as follows: $\mathcal{L} = \frac{1}{2V_0}(\partial_t\theta)^2 - \frac{\rho_0}{2m}(\partial_{\boldsymbol{x}}\theta)^2 - \frac{g}{2}\theta^2$. The resulting wave equation $(-\frac{1}{2V_0}\partial_t^2 + \frac{\rho_0}{2m}\partial_{\boldsymbol{x}}^2 - g)\theta = 0$ tells us that the wave always has a finite frequency $\omega = \sqrt{2V_0 g + \frac{V_0\rho_0}{m}\boldsymbol{k}^2}$, regardless of the wave vector. Thus, the potential term $\cos(\theta)$ opens up an energy gap for the θ quasiparticles.

If both the Lagrangian and the ground state are invariant under the $U(1)$ transformation, then the symmetry is not broken. From eqn (3.3.6), we see that the symmetric ground state, in general, also has a finite energy gap.

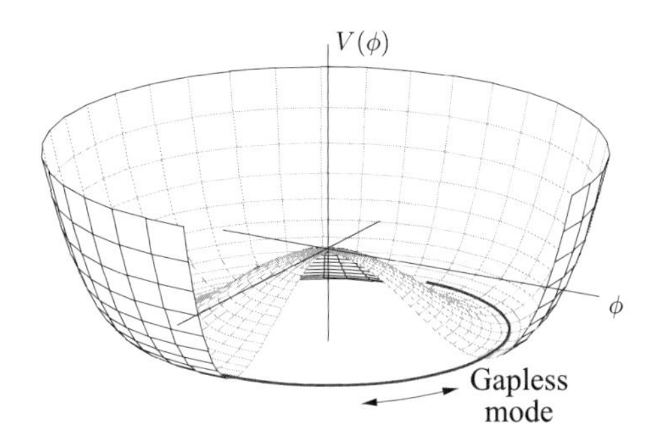

FIG. 3.7. The gapless Nambu–Goldstone mode is a fluctuation between degenerate ground states.

The superfluid phase spontaneously breaks the $U(1)$ symmetry; the Lagrangian has the $U(1)$ symmetry but the ground state does not (see Section 3.3.2). Only in this case do the gapless phonon excitations exist.

Nambu and Goldstone have proved a general theorem, which states that if a continuous symmetry is *spontaneously* broken in a phase, then the phase must contain gapless excitations (Nambu, 1960; Goldstone, 1961). These gapless excitations are usually called the Nambu–Goldstone modes. The gapless phonon in the superfluid is a Nambu–Goldstone mode.

Intuitively, if a symmetry (continuous or discrete) is spontaneously broken, then the ground states must be degenerate, because the Hamiltonian has the symmetry and the ground state does not. The different ground states are related by the symmetry transformations. Furthermore, breaking *continuous* symmetry gives rise to a manifold of degenerate ground states that can be parametrized by the *continuous* variables θ_i. The fluctuations among these degenerate ground states correspond to gapless excitations (see Fig. 3.7). One can write down a low-energy effective action in terms of the $\theta_i(\boldsymbol{x}, t)$ fields to describe the dynamics of the gapless mode. This low-energy effective theory is called the nonlinear σ-model. The XY-model is the simplest nonlinear σ-model. We will see more examples of the nonlinear σ-model later.

3.3.7 Understanding spontaneous symmetry breaking in finite systems

- Quantum fluctuations can restore the symmetry. The real ground state of a finite system can never break $U(1)$ symmetry. However, for a large system, one can form an approximate symmetry-breaking ground state from many closely degenerate ground states.

- For an infinite system, the state characterized by the order parameter $\varphi_0 \mathrm{e}^{\mathrm{i}\theta}$ forms its own 'universe'. It takes an infinitely long time to fluctuate to another state characterized by a different order parameter.

In Section 3.3.2, we mentioned that the superfluid phase (or the symmetry-breaking phase) is characterized by a nonzero order parameter φ_0. In quantum theory, the order parameter is the expectation value of the boson field on the ground state $\varphi_0 = \langle\Phi_0|\varphi(\boldsymbol{x})|\Phi_0\rangle = \langle\Phi_0|a(\boldsymbol{x})|\Phi_0\rangle$ (remember that we have renamed the boson annihilation operator $a(\boldsymbol{x})$ to $\varphi(\boldsymbol{x})$). For a finite quantum system, the order parameter is always zero and there is no spontaneous symmetry breaking. This is because at quantum levels the $U(1)$ symmetry is generated by the operator $W = \mathrm{e}^{-\mathrm{i}\hat{N}\theta}$, where $\hat{N}$ is the total boson number operator. One can easily check that

$$W a(x) W^\dagger = \mathrm{e}^{\mathrm{i}\theta} a(x) \tag{3.3.19}$$

As $[H, W] = 0$, the ground state $|\Phi_0\rangle$ of a finite system is always an eigenstate of W. Therefore,

$\langle\Phi_0|a(x)|\Phi_0\rangle = \langle\Phi_0|W a(x) W^\dagger|\Phi_0\rangle$, and the order parameter $\langle\Phi_0|a(x)|\Phi_0\rangle = \langle\Phi_0|\varphi(x)|\Phi_0\rangle = 0$, regardless of the value of μ.

In contrast, the classical calculation in Section 3.3.2 does imply a phase transition and degenerate ground states that break the $U(1)$ symmetry, even for a finite system. (In fact, all of the calculations in the last section were performed for a finite system of volume $\mathcal{V}$.) The reason that the $U(1)$ symmetry can break in a finite classical system is that the uniform part of the φ field, φ_0, has no fluctuations. We will see below that, if we treat φ_0 as a quantum variable, then its quantum fluctuations will restore the $U(1)$ symmetry for a finite system. We will also see that, as the system volume $\mathcal{V} \to \infty$, the quantum fluctuations of φ_0 approach zero. Thus, the $U(1)$ symmetry breaking can emerge in the $\mathcal{V} \to \infty$ limit of a quantum system.

We know that low-energy physics is governed by the XY-model. Here we are only interested in the uniform fluctuations. Thus, we set $\theta(x,t) = \theta_0(t)$ in eqn (3.3.10). The effective theory that governs the dynamics of θ_0 is given by

$$L = \mathcal{V}[\frac{1}{2V_0}(\partial_t\theta_0)^2 - \rho_0\partial_t\theta_0 + \mu\rho_0 - \frac{1}{2}V_0\rho_0^2] \tag{3.3.20}$$

where we have put back a total time derivative term and the constant terms. eqn (3.3.20) describes a simple system of a particle on a circle parametrized by $\theta_0 \in [0, 2\pi]$. The mass of the particle is $M = \mathcal{V}/V_0$ and the momentum is

$$p = \partial_{\dot{\theta}_0} L = M\dot{\theta}_0 - \frac{\mu\mathcal{V}}{V_0} \tag{3.3.21}$$

where we have used $\mu = \rho_0 V_0$. The Hamiltonian is

$$H = p\dot{\theta}_0 - L = \frac{p^2}{2M} + \frac{\mu\mathcal{V}p}{V_0}$$

If we treat H as a classical system, then the particle is in the ground state if it has momentum $p = -\mu\mathcal{V}p/V_0$, or velocity $\dot{\theta} = 0$. The position of the particle can be arbitrary. Thus, all of the states $|\theta_0\rangle$ describing the particle at different θ_0 are degenerate ground states and the $U(1)$ symmetry is broken.

However, the quantum theory provides a very different picture. In the quantum theory, the momentum p is quantized as an integer. The state with momentum $p = -n$ has energy

$$E_n = \frac{n^2}{2M} - \frac{\mu \mathcal{V} n}{V_0 M} = \frac{V_0 n^2}{2\mathcal{V}} - \mu n \tag{3.3.22}$$

Thus, the state with $-p = n = \mu\mathcal{V}/V_0$ has minimal energy and corresponds to the true ground state

$$|\Phi_0\rangle = \int \mathrm{d}\theta_0 \ \mathrm{e}^{-\mathrm{i}\frac{\mu\mathcal{V}}{V_0}\theta_0}|\theta_0\rangle$$

The ground state has no degeneracy. The order parameter can now be shown explicitly to vanish:

$$\langle\Phi_0|\varphi_0 \mathrm{e}^{\,\mathrm{i}\theta_0}|\Phi_0\rangle = 0$$

Another way to understand why there is no $U(1)$ symmetry breaking is to note that the total boson number is given by $N = \mathcal{V}(\rho_0 - \frac{\partial_t \theta_0}{V_0})$ (see eqn (3.3.13)). From eqn (3.3.21), we see that $-N = p$ is the canonical momentum of θ_0. Thus, in the quantum world, there is an uncertainty relation between them:

$$[\hat{N}, \theta_0] = \mathrm{i}, \qquad \Delta N \Delta\theta_0 \sim 1/2$$

Therefore, a finite system with a fixed number of particles cannot have the definite phase θ_0. In order to have a symmetry-breaking state, we must allow particle numbers to fluctuate easily.[21]

Although the quantum fluctuations lift the degeneracy and restore the $U(1)$ symmetry for a finite system, for large systems the low-lying states are nearly degenerate. The energy gap is only of order $V_0/\mathcal{V}$, which approaches 0 as $\mathcal{V} \to \infty$. If the energy gap is below all energy scales of interest (for example, below the energy resolution of measurements), then those low-lying states can be thought to be degenerate for all practical purposes. In this case, the finite system can be thought to have $U(1)$ symmetry breaking. In Problem 3.3.4 we will see that, if we prepare a state with a phase θ_0, then it can take a very long time for the phase to change to other values if the system is large. Thus, in a short time interval we may

[21] As a result, if it costs a finite energy to add or to remove a boson, then the system cannot be in the $U(1)$-symmetry-breaking phase. As an application of this understanding, let us consider a lattice boson system with strong on-site repulsion

$$H = \sum_{<ij>} t_{ij} a_i^\dagger a_j + \sum_i U(a_i^\dagger a_i)^2, \qquad U \gg t_{ij}$$

The system cannot be in a superfluid phase if the boson density happens to be an integer number of bosons per site, because at that density it costs finite energy to add or remove a boson.

treat the phase as a constant which characterizes the symmetry-breaking ground state.

At the end of Section 3.3.2, we discussed order parameters and long-range correlations. In this section, we discuss a special kind of quantum fluctuation—uniform phase fluctuations of φ. We see that, for a finite system, such fluctuations destroy the order parameter even when $\mu > 0$. However, it is also clear that the long-range correlation present for $\mu > 0$ is not destroyed by uniform phase fluctuations. Thus, we can use long-range correlations to characterize symmetry-breaking phase transitions in large finite systems.

In the above discussion, we considered only the uniform fluctuations of θ and ignored the spatially-dependent fluctuations. The question here is when is it correct to do so? We note that the spatially-dependent fluctuations correspond to sound waves with finite momentum. The smallest energy gap for these sound wave excitations is given by $2\pi v/L$, where L is the linear size of the system. Thus, below the energy $2\pi v/L$, we can ignore spatially-dependent θ fluctuations and consider only uniform fluctuations. The discussion in this section is valid only below $2\pi v/L$.

A symmetry breaking state with a fixed angle $|\theta_0\rangle$ is given by $|\theta_0\rangle = \sum_n \mathrm{e}^{\mathrm{i}\theta_0 n}|n\rangle$. To construct such a state within the above approximation, we need to fit an infinite number of uniform fluctuations below the energy $2\pi v/L$. Beyond one dimension, the energy gap for the uniform fluctuations is $V_0/\mathcal{V}$, which is much less than the energy gap for the sound waves. In the $L \to \infty$ limit, there are infinite energy levels of the uniform fluctuations below $2\pi v/L$. Therefore, it is possible to construct a symmetry-breaking state beyond one dimension. In one dimension, the number of the uniform states below $2\pi v/L$ is finite and we have trouble to construct the symmetry breaking state. In Section 3.3.8, we will see that the superfluid phase in one dimension cannot have true long-range order.

Problem 3.3.4.
Consider a $1\,\mathrm{cm}^3$ superfluid He_4 at zero temperature. We prepare a 'ground' state with phase $\langle\theta_0\rangle = 0$. Assume that the spread of the phase is $\sqrt{\langle\theta_0^2\rangle} = \pi/10$. How long does it take for the spread of the phase to reach 2π, so that the phase of the ground state can no longer be defined? (Hint: You need to make a guess about the sound wave velocity in He_4 in order to estimate V_0, which happens to be the inverse of the compressibility.)

3.3.8 Superfluid phase in low dimensions

- Quantum fluctuations associated with the Nambu–Goldstone modes can destroy long-range order and superfluid phases in low dimensions.
- The effects of quantum fluctuations on the quantum critical point and a concept of the upper critical dimension.

At the end of Section 3.3.2, we showed the existence of long-range order in a symmetry-breaking phase. However, that calculation is almost cheating, because, by considering only the classical ground state, we do not allow spatially-dependent phase fluctuations. As a result, the phases at different positions are always correlated. Here, we will study the effects of spatially-dependent phase fluctuations and see when long-range correlations survive the spatially-dependent phase fluctuations.

First, let us consider the boson system at zero temperature. We want to calculate $\langle \varphi^\dagger(t, \boldsymbol{x}) \varphi(0,0) \rangle$. As we are only interested in low-lying fluctuations, we will use the effective XY-model (3.3.10), which can be rewritten as

$$L = \frac{\chi}{2} \left((\partial_t \theta)^2 - v^2 (\partial_{\boldsymbol{x}} \theta)^2 \right) \tag{3.3.23}$$

with $\chi = 1/V_0$ being the compressibility of the superfluid. In the path integral approach, the correlation function can be written as

$$\left\langle \varphi(t, \boldsymbol{x})^\dagger \varphi(0,0) \right\rangle = |\varphi_0|^2 \left\langle \mathrm{e}^{-\mathrm{i}\theta(t,\boldsymbol{x})} \mathrm{e}^{\mathrm{i}\theta(0,0)} \right\rangle = \frac{\int \mathcal{D}\theta \; \mathrm{e}^{-\mathrm{i}\theta(t,\boldsymbol{x})} \mathrm{e}^{\mathrm{i}\theta(0,0)} \mathrm{e}^{\mathrm{i}S}}{\int \mathcal{D}\theta \; \mathrm{e}^{\mathrm{i}S}}$$

Introducing

$$Z[f(t, \boldsymbol{x})] = \int \mathcal{D}\theta \; \mathrm{e}^{\mathrm{i}S} \mathrm{e}^{\mathrm{i}\int \mathrm{d}t \mathrm{d}^d \boldsymbol{x} \; f(t,\boldsymbol{x})\theta(t,\boldsymbol{x})}$$

we see that

$$\left\langle \mathrm{e}^{-\mathrm{i}\theta(t_0,\boldsymbol{x}_0)} \mathrm{e}^{\mathrm{i}\theta(0,0)} \right\rangle = \frac{Z[-\delta(t - t_0, \boldsymbol{x} - \boldsymbol{x}_0) + \delta(t, \boldsymbol{x})]}{Z[0]}$$

Here, $Z[f(t, \boldsymbol{x})]/Z[0]$ can be calculated easily using the Gaussian integral as follows:

$$\frac{Z[f(t, \boldsymbol{x})]}{Z[0]} = \mathrm{e}^{-\mathrm{i}\frac{1}{2} \int \mathrm{d}t_1 \, \mathrm{d}\boldsymbol{x}_1 \, \mathrm{d}t_2 \, \mathrm{d}\boldsymbol{x}_2 f(t_1,\boldsymbol{x}_1) G_\theta(t_1 - t_2, \boldsymbol{x}_1 - \boldsymbol{x}_2) f(t_2, \boldsymbol{x}_2)}$$

where G_θ is the inverse of $\chi(-\partial_t^2 + v^2 \partial_{\boldsymbol{x}}^2)$. It is also the correlation function of the θ field:

$$\mathrm{i}G_\theta(t, \boldsymbol{x}) = \langle T[\theta(t, \boldsymbol{x})\theta(0,0)] \rangle$$

We find that

$$\left\langle \mathrm{e}^{-\mathrm{i}\theta(t,\boldsymbol{x})} \mathrm{e}^{\mathrm{i}\theta(0,0)} \right\rangle = \mathrm{e}^{\mathrm{i}G_\theta(t,\boldsymbol{x})} \mathrm{e}^{-\mathrm{i}G_\theta(0,0)} \propto \mathrm{e}^{\langle (-\mathrm{i}\theta(t,\boldsymbol{x}))(\mathrm{i}\theta(0,0)) \rangle}$$

3.3.8.1 Superfluid phase in $1+1$ dimensions

The correlation function in $\boldsymbol{k}$–ω space is given by

$$G_\theta(\omega,\boldsymbol{k}) = \frac{\chi^{-1}}{\omega^2 - v^2\boldsymbol{k}^2 + \mathrm{i}0^+}.$$

We can use this to calculate $G_\theta(t,\boldsymbol{x})$ in $1+1$ dimensions as follows:

$$\begin{aligned}
G_\theta(t,x) &= \int \frac{\mathrm{d}k\,\mathrm{d}\omega}{(2\pi)^2}\, G_\theta(\omega,k)\mathrm{e}^{\,\mathrm{i}(-\omega t+kx)} \\
&= \chi^{-1}\int \frac{\mathrm{d}k\,\mathrm{d}\omega}{(2\pi)^2}\,\frac{1}{2v|k|}\left(\frac{1}{\omega - v|k| + \mathrm{i}0^+} - \frac{1}{\omega + v|k| - \mathrm{i}0^+}\right)\mathrm{e}^{\,\mathrm{i}(\omega t - kx)} \\
&= \begin{cases} \chi^{-1}\displaystyle\int \frac{\mathrm{d}k}{(2\pi)^2}\,\frac{-\mathrm{i}\pi}{v|k|}\mathrm{e}^{\,\mathrm{i}(-v|k|t+kx)}, & t>0 \\ \chi^{-1}\displaystyle\int \frac{\mathrm{d}k}{(2\pi)^2}\,\frac{-\mathrm{i}\pi}{v|k|}\mathrm{e}^{\,\mathrm{i}(+v|k|t+kx)}, & t<0 \end{cases} \\
&= \chi^{-1}\int \frac{\mathrm{d}k}{(2\pi)^2}\,\frac{-\mathrm{i}\pi}{v|k|}\mathrm{e}^{\,\mathrm{i}(-v|k||t|+kx)}
\end{aligned}$$

For a finite system, k is quantized: $k = \frac{2\pi}{L}\times$ integer. After replacing $\int \mathrm{d}k/2\pi$ by $L^{-1}\sum_k$, we get

$$\begin{aligned}
G_\theta(t,x) &= \chi^{-1}L^{-1}\sum_k \frac{-\mathrm{i}}{2v|k|}\mathrm{e}^{\,\mathrm{i}(-v|k||t|+kx)} \\
&= \frac{\mathrm{i}}{4\pi v\chi}\left(\ln(1-\mathrm{e}^{2\pi\mathrm{i}\frac{-v|t|+x}{L}-0^+}) + \ln(1-\mathrm{e}^{2\pi\mathrm{i}\frac{-v|t|-x}{L}-0^+})\right)
\end{aligned}$$

The two log terms arise from the $k>0$ and $k<0$ summations. The $k=0$ term is a constant and is dropped. If $v|t|, x \ll L$, then

$$G_\theta(t,x) = \frac{\mathrm{i}}{4\pi v\chi}\ln 4\pi^2\frac{x^2 - v^2t^2 + \mathrm{i}0^+}{L^2} \tag{3.3.24}$$

We see that, in $1+1$ dimensions,

$$\begin{aligned}
\left\langle \mathrm{e}^{-\mathrm{i}\theta(t,x)}\mathrm{e}^{\,\mathrm{i}\theta(0,0)}\right\rangle &= \mathrm{e}^{-\mathrm{i}G_\theta(0,0)}\left(\frac{L^2}{4\pi^2(x^2-v^2t^2+\mathrm{i}0^+)}\right)^{1/4\pi v\chi} \\
&= \mathrm{e}^{-\mathrm{i}G_\theta(0,0)}\mathrm{e}^{-\mathrm{i}\frac{\pi}{4\pi v\chi}\Theta(v^2t^2-x^2)}\left(\frac{L^2}{4\pi^2|x^2-v^2t^2|}\right)^{1/4\pi v\chi}
\end{aligned}$$

From eqn (3.3.24), one has $G_\theta(0,0) = -\mathrm{i}\infty$, which seems to imply that $\langle \mathrm{e}^{-\mathrm{i}\theta(t,x)}\mathrm{e}^{\,\mathrm{i}\theta(0,0)}\rangle = 0$. We note that the XY-model is only a low-energy effective theory. It breaks down at short distances. The divergence of $G_\theta(0,0)$ should be cut off by a short distance scale l. Thus, we may replace $G_\theta(0,0)$ by $G_\theta(0,l)$, which leads to

$$\left\langle \mathrm{e}^{-\mathrm{i}\theta(t,x)}\,\mathrm{e}^{\,\mathrm{i}\theta(0,0)}\right\rangle = \left(\frac{l^2}{x^2 - v^2t^2 + \mathrm{i}0^+}\right)^{1/4\pi v\chi}$$
$$= \mathrm{e}^{-\mathrm{i}\frac{\pi}{4\pi v\chi}\Theta(v^2t^2-x^2)}\left(\frac{l^2}{|x^2 - v^2t^2|}\right)^{1/4\pi v\chi} \tag{3.3.25}$$

The above result tells us that quantum fluctuations destroy long-range order in $1+1$ dimensions and there is no symmetry breaking. However, the destruction is not complete. The φ field still has 'quasi-long-range' order, where the correlation decays algebraically. This leads to the following phase diagram: for both $\mu < 0$ and $\mu > 0$ the boson ground state does not break $U(1)$ symmetry. However, for $\mu < 0$ there is only short-range correlation in the boson field. When $\mu > 0$, the short-range correlation is promoted to 'quasi-long-range' correlation with algebraic decay.

Another important point is that the correlation $\langle \mathrm{e}^{-\mathrm{i}\theta(t,x)}\,\mathrm{e}^{\,\mathrm{i}\theta(0,0)}\rangle$ cannot be expressed in terms of the parameters in the effective theory only. It also depends on a short-distance cut-off scale l. This is not surprising. To obtain the low-energy XY-model we have to throw away a lot of information and structures of the original theory at short distances and high energies. The very definition of many operators depends on these short-distance structures. For example, some operators are products of the fields at the same space point, and the very concept of 'the same space point' is lost, or at least becomes vague, in the low-energy effective theory. Therefore, we should not expect to express all correlation functions in terms of the parameters in the effective theory. In the above example, we find that the correlation $\langle \mathrm{e}^{-\mathrm{i}\theta(t,x)}\,\mathrm{e}^{\,\mathrm{i}\theta(0,0)}\rangle$ does depend on the short-distance structure of the theory. What is surprising is that all of the complicated short-distance structures are summarized into a single parameter l. This demonstrates the universality of the low-energy effective theory.

3.3.8.2 *Superfluid phase in* $2+1$ *dimensions*

To calculate G_θ in $2+1$ dimensions, we will first calculate the imaginary-time correlation

$$\mathcal{G}_\theta(x^i) = \langle \theta(x^i)\theta(0)\rangle$$

Here $x^{1,2}$ are spatial coordinates and x^3 are imaginary times. For simplicity, we will set $v = 1$. The partition function for the XY-model is given by $Z =$

$\int \mathcal{D}\theta \, \exp(-\int d^3x^i \, \frac{\chi}{2}(\partial_i\theta)^2)$. We see that[22]

$$\mathcal{G}_\theta(x^i) = \frac{\chi^{-1}}{-\partial_i^2} = \frac{1}{4\pi\chi r} = \frac{1}{4\pi\chi\sqrt{\boldsymbol{x}^2 + \tau^2}}$$

where $r^2 = x^i x^i$.

From the relationship between $\mathrm{i}G_\theta$ and $\mathcal{G}_\theta$ (see eqn (2.2.34)), we find that

$$\mathrm{i}G_\theta(t, x^1, x^2) = \mathcal{G}_\theta(x^1, x^2, \mathrm{e}^{\mathrm{i}\pi/2}t) = \frac{1}{4\pi\chi}\frac{1}{\sqrt{\boldsymbol{x}^2 - t^2}}$$

However, for $|t| > |\boldsymbol{x}|$ the propagator is ambiguous: $\mathrm{i}G_\theta(t, x^1, x^2) = \pm i\frac{1}{4\pi\chi}\frac{1}{\sqrt{|\boldsymbol{x}^2 - t^2|}}$. To fix the $\pm$ sign we need to perform the analytic continuation more carefully. Assuming that $|t| \gg |\boldsymbol{x}|$ and changing η in $\mathcal{G}_\theta(x^1, x^2, \mathrm{e}^{\mathrm{i}\eta}t)$ continuously from 0 to $\pi/2$, we find that

$$\mathrm{i}G_\theta(t, x^1, x^2) = \frac{1}{4\pi v\chi}\frac{1}{\sqrt{|\boldsymbol{x}^2 - vt^2|}}\mathrm{e}^{-\mathrm{i}\frac{\pi}{2}\Theta(v|t| - |\boldsymbol{x}|)}$$

where we have restored the velocity v.

In any case, $\mathrm{i}G(\boldsymbol{x}, t) \to 0$ as $|t|$ and $|\boldsymbol{x}|$ approach ∞. Therefore, $\langle \mathrm{e}^{-\mathrm{i}\theta(t,x)}\mathrm{e}^{\mathrm{i}\theta(0,0)}\rangle$ approaches to a constant $\mathrm{e}^{-\mathrm{i}G_\theta(0,0)}$ at long distances. However, the value of the constant is different from the classical value (which is 1). Introducing the short-distance cut-off l, we see that

$$\left\langle \mathrm{e}^{-\mathrm{i}\theta(t,x)}\mathrm{e}^{\mathrm{i}\theta(0,0)}\right\rangle \to \mathrm{e}^{-\frac{1}{4\pi v\chi l}} = \left\langle \mathrm{e}^{-\mathrm{i}\theta(t,x)}\right\rangle\left\langle \mathrm{e}^{\mathrm{i}\theta(0,0)}\right\rangle \tag{3.3.26}$$

We find that the phase fluctuations do not destroy long-range order in $2+1$ and higher dimensions; they merely reduce the value of the order parameter.

3.3.8.3 *Short-distance cut-off and nonlinear effects*

In the above, we have discussed the effect of quantum fluctuations on the classical ground state. We find that, in $1+1$ dimensions, quantum fluctuations always have a big effect in that they destroy long-range order. The quantum fluctuations that destroy long-range order come from long wavelengths and low frequencies. They are there no matter how we adjust the parameters in the theory, such as the cut-off scale. In $2+1$ dimensions and above, the long wavelengths and low-frequency fluctuations do not have a diverging effect, and in general long-range order survives the quantum fluctuations. The fact that the order parameter depends sensitively on the cut-off scale implies that the short-distance fluctuations determine the reduction of order parameters. The short-distance fluctuations only quantitatively modify the classical results.

[22] We have used $-\partial_i^2 r^{-1} = 4\pi\delta(x^i)$.

Now a question arises: when are these quantitative corrections small? That is, when can the classical approximation quantitatively describe the physical properties of the system? From our results on the order parameter, see eqn (3.3.26), we see that the fluctuation correction is small if the cut-off scale $l \gg (v\chi)^{-1}$. To obtain the cut-off scale, we notice that we have dropped the $\frac{\varphi_0^2}{2m}\partial_{\boldsymbol{x}}^2$ term in eqn (3.3.9) because we assume that the fluctuations of θ are smooth. However, at the short-distance scale when $\frac{\varphi_0^2}{2m}\partial_{\boldsymbol{x}}^2$ is comparable with the other term $2V_0\varphi_0^4$, we can no longer ignore the gradient term and the XY-model breaks down. Such a crossover length scale is given by

$$\xi = (4mV_0\varphi_0^2)^{-1/2}$$

which is called the coherence length. The coherence length has the following meaning. If we change the h field or the boson density at a point, then such a change will propagate over a distance given by ξ. Setting $l = \xi$, the reduction factor becomes

$$\mathrm{e}^{-\frac{1}{4\pi v\chi l}} = \mathrm{e}^{-\frac{mV_0}{\sqrt{2}\pi}} \tag{3.3.27}$$

We can rewrite the above result in a form that makes more sense. Note that $E_{int} \equiv \frac{1}{2}V_0\varphi_0^2$ has a dimension of energy and it represents the interaction energy per particle. In $2+1$ dimensions, $E_{qua} \equiv \rho_0/m$ also has a dimension of energy. Here E_{qua} is the energy scale (or the temperature scale) below which the boson wave functions start to overlap and we need to treat the bosons as a quantum system. Now we can rewrite the reduction factor as

$$\mathrm{e}^{-\frac{1}{4\pi v\chi l}} = \mathrm{e}^{-\frac{\sqrt{2}E_{int}}{E_{qua}}}$$

We see that in the weak-interaction limit the fluctuation corrections are small and the classical results are good.

However, the above result is not complete because we only considered the effects of short-distance fluctuations within the quadratic approximation. We have not considered the nonlinear effects from higher-order terms. To have a more systematic study of the effects of fluctuations, we would like to perform a dimensional analysis of our boson action

$$S = \int \mathrm{d}^d\boldsymbol{x}\,\mathrm{d}t\,[\mathrm{i}\frac{1}{2}(\varphi^*\partial_t\varphi - \varphi\partial_t\varphi^*) - \frac{1}{2m}\partial_{\boldsymbol{x}}\varphi^*\partial_{\boldsymbol{x}}\varphi - \frac{V_0}{2}|\varphi|^2(|\varphi|^2 - 2\rho_0)] \tag{3.3.28}$$

Here we have chosen $\mu = V_0\rho_0$. We would like to rescale t, $\boldsymbol{x}$, and φ in order to rewrite the action in the form $S = \tilde{S}/g$ such that all of the coefficients in $\tilde{S}$ are of order 1. We find that the following rescaling will suffice:

$$x = \xi \tilde{x}, \qquad t = (V_0 \rho_0)^{-1} \tilde{t}, \qquad \varphi = \sqrt{\rho_0} \tilde{\varphi}$$

which gives us

$$S = g^{-1} \int \mathrm{d}^d \tilde{\boldsymbol{x}} \mathrm{d}\tilde{t} \, [\mathrm{i}\frac{1}{2}(\tilde{\varphi}^* \partial_t \tilde{\varphi} - \tilde{\varphi} \partial_t \tilde{\varphi}^*) - 2\partial_{\boldsymbol{x}} \tilde{\varphi}^* \partial_{\boldsymbol{x}} \tilde{\varphi} - \frac{1}{2}|\tilde{\varphi}|^2(|\tilde{\varphi}|^2 - 2)]$$

with

$$g = N_\xi^{-1}, \qquad N_\xi = \rho_0 \xi^d$$

Here N_ξ is the number of particles in a volume ξ^d. We can also write g as

$$g = \sqrt{\rho_0^{d-2} (4mV_0)^d} \tag{3.3.29}$$

It is now clear that, if g is small in the path integral $\int \mathcal{D}^2 \tilde{\varphi} \mathrm{e}^{\mathrm{i}\frac{\tilde{S}}{g}}$, then the 'potential' is steep and the fluctuations around the potential minimum are small. In this case, the semiclassical approximation is good. In fact, the semiclassical approximation corresponds to an expansion in g. In $2+1$ dimensions $g = 4mV_0$, which happens to agree with the condition (3.3.27) obtained from short-distance fluctuations.

When g is large, the fluctuations can be so large that it is not even clear if the classical picture is qualitatively correct. In fact, it is believed that, when $g \gg 1$, the short-distance fluctuations can destroy long-range order and restore $U(1)$ symmetry in the ground state, by, for example, forming a crystal (which has different symmetry breaking and different long-range order). This is in contrast to what happens in $1+1$ dimensions where long-distance fluctuations destroy long-range order for any value of g.

According to the classical theory, our boson system undergoes a quantum phase transition at $\mu = 0$. The phase transition is continuous and is described by a quantum critical point. Within the classical theory, we can calculate all of the critical exponents. For example, the dynamical exponent z in $\omega \propto k^z$ is $z = 2$. The exponent ν in $\langle \varphi \rangle = \varphi_0 \propto \mu^\nu$ is $\nu = 1/2$. The question is can classical theory correctly describe the quantum critical point? From eqn (3.3.29), we see that if $d > 2$ then the closer we are to the critical point ($\rho_0 \to 0$), the smaller the g, and the better the semiclassical approximation. Therefore, for $d > 2$, classical theory correctly describes the quantum critical point. However, for $d < 2$, g diverges as we approach the critical point. In this case, we have to include quantum fluctuations to obtain the correct critical exponents. This crossover spatial dimension $d_c = 2$ is called the upper critical dimension of the critical point.

Problem 3.3.5.
Consider a boson system with long-range interaction

$$\frac{1}{2}\int \mathrm{d}^d\boldsymbol{x}\,\mathrm{d}^d\boldsymbol{x}'\ |\varphi(\boldsymbol{x})|^2 V(\boldsymbol{x}-\boldsymbol{x}')|\varphi(\boldsymbol{x}')|^2$$

where $V(r) = V_d r^{\epsilon-d}$. If $\epsilon < 0$, then the interaction is effectively short-ranged. Here we assume that $\epsilon > 0$.

1. Derive the modified XY-model.
2. Find the critical spatial dimension d_l below which the phase fluctuations θ always destroy long-range order. Here we treat the dimension as a continuous real number. We know that $d_l = 1$ for short-range interactions.
3. Repeat the discussion at the end of this section to find g (which determines when the quantum fluctuations are small) and the upper critical dimension d_c (beyond which classical theory correctly describes the quantum critical point).

Problem 3.3.6.
Finite temperature correlations in $1+1$ dimensions

1. Calculate the imaginary-time correlation

$$\mathcal{G}(x,\tau) = \left\langle \mathrm{e}^{\,\mathrm{i}\theta(x,\tau)}\mathrm{e}^{-\mathrm{i}\theta(0,0)}\right\rangle$$

   assuming that the one-dimensional space is a finite circle of length L.
2. Calculate the imaginary-time correlation $\mathcal{G}^\beta(x,\tau)$ at finite temperatures, assuming that the one-dimensional space is an infinitely long line. (Hint: You may want to exchange x and τ and use the result in part (a).)
3. Calculate the real-time (time-ordered) correlation $G^\beta(x,t)$ at finite temperatures, assuming that the one–dimensional space is an infinitely long line. Show that

$$G^\beta(0,t) \propto \mathrm{e}^{-\mathrm{i}\frac{\pi}{4\pi v\chi}}\left(\frac{\pi T}{\sinh(\pi T|t|)}\right)^{1/2\pi v\chi}$$

3.4 Superfluid phase at finite temperatures

3.4.1 Path integral at finite temperatures

- The long-distance properties of a $(d+1)$-dimensional quantum system at finite temperatures can be described by the path integral of a d-dimensional system.

In this section, we will study an interacting boson system and its superfluid phase at finite temperatures. We start with an imaginary-time path integral

$$Z = \int \mathcal{D}^2[\varphi(\boldsymbol{x},\tau)]\ \mathrm{e}^{-\int_0^\beta \mathrm{d}^d\boldsymbol{x}\mathrm{d}\tau\ [\frac{1}{2}(\varphi^*\partial_\tau\varphi-\varphi\partial_\tau\varphi^*)+\frac{1}{2m}\partial_{\boldsymbol{x}}\varphi^*\partial_{\boldsymbol{x}}\varphi-\mu|\varphi|^2+\frac{V_0}{2}|\varphi|^4]}$$

which represents the partition function of the system. Note that the Berry phase term remains imaginary in the imaginary-time path integral.

To reduce the above partition function to a familiar statistical model, we first go to the discrete frequency space by introducing

$$\varphi_{\omega_n}(\boldsymbol{x}) = \beta^{-1}\int_0^\beta \mathrm{d}\tau\ \varphi(\boldsymbol{x},\tau)\mathrm{e}^{\,\mathrm{i}\tau\omega_n}$$

Then, we integrate out all of the finite frequency modes:

$$Z = \int \prod_n \mathcal{D}^2[\varphi_{\omega_n}]\ \mathrm{e}^{-S} = \int \mathcal{D}^2[\varphi_c(\boldsymbol{x})]\ \mathrm{e}^{-S_{\text{eff}}}$$

where $\varphi_c(\boldsymbol{x}) = \varphi_{\omega_n=0}(\boldsymbol{x})$ is the zero-frequency mode. This is a difficult step and the resulting effective action S_{eff} can be very complicated. It can even contain non-local terms such as $\int \mathrm{d}^d\boldsymbol{x}\mathrm{d}^d\boldsymbol{x}'\ |\varphi_c(\boldsymbol{x})|^2K(\boldsymbol{x}-\boldsymbol{x}')|\varphi_c(\boldsymbol{x}')|^2$. However, at finite frequency the original action contains a term $\mathrm{i}|\varphi_c(\boldsymbol{x})|^2\omega_n$ that makes the propagator of φ_n have the form $1/(\mathrm{i}\omega_n + ck^2 + \mu)$. The propagator is short-ranged with exponential decay. Thus, the fluctuations of nonzero modes can only mediate short-range interactions and $K(\boldsymbol{x}-\boldsymbol{x}')$ should have an exponential decay: $K(\boldsymbol{x}-\boldsymbol{x}') \sim \mathrm{e}^{-|\boldsymbol{x}-\boldsymbol{x}'|/l_T}$. Here we have a new length scale l_T, beyond which the effective action S_{eff} can be treated as a local action. Although S_{eff} can be very complicated, due to its symmetry, a local action can only take a certain form. At long distances beyond l_T, we can make a gradient expansion

$$S_{\text{eff}} = \beta\int \mathrm{d}^d\boldsymbol{x}\ [\frac{1}{2m^*}|\partial_{\boldsymbol{x}}\varphi_c|^2 + V(|\varphi_c|,T) + ...] \tag{3.4.1}$$

The '...' represent higher-derivative terms. The effective potential has a temperature dependence. When $\mu < 0$, we expect that V always has a single minimum at $\varphi_c = 0$, representing the symmetric state. When $\mu > 0$ and at low temperatures, V should have a circle of minima at $\varphi_c = \varphi_0\mathrm{e}^{\,\mathrm{i}\theta}$, representing the symmetry-breaking state. However, beyond a critical temperature T_c, V is expected to change into one with only a single minimum at $\varphi_c = 0$. Therefore, the temperature dependence of V describes the superfluid transition.

In the superfluid phase, only the phase fluctuations are important at long distances. Setting $\varphi_c(\boldsymbol{x}) = \varphi_0 \mathrm{e}^{\mathrm{i}\theta(\boldsymbol{x})}$, we obtain an XY-model

$$S_{\text{eff}} = \int \mathrm{d}^d\boldsymbol{x}\, \frac{\eta}{2}(\partial_{\boldsymbol{x}}\theta)^2 \tag{3.4.2}$$

where

$$\eta = \frac{\varphi_0^2}{mT} \tag{3.4.3}$$

We note that the energy is given by TS_{eff}. Thus, the value of $T\eta$ determines the energy cost of the phase twist $\partial_{\boldsymbol{x}}\theta \neq 0$. Here $T\eta$ is called the phase rigidity of the XY-model.

Certainly, the above discussions about symmetry breaking are based on the classical picture where we have ignored the fluctuations of φ and θ. (Now those fluctuations correspond to thermal fluctuations.) Again, we can ask whether or not the symmetry-breaking state can survive the thermal fluctuations. As in the last section, we can use the XY-model to address this question. Using those results, we found that, for $d > 2$, the thermal fluctuations do not always destroy long-range order.

For $d < 2$, the thermal fluctuations always destroy long-range order and change it into short-range order. This is because, for $d < 2$, the correlation of θ diverges as a power of $|\boldsymbol{x}|$:

$$\begin{aligned}\langle\theta(\boldsymbol{x})\theta(0)\rangle - \langle\theta(l)\theta(0)\rangle &= C_d\eta^{-1}(L^{2-d} - |\boldsymbol{x}|^{2-d}) - C_d\eta^{-1}(L^{2-d} - l^{2-d}) \\ &= -\,C_d\eta^{-1}|\boldsymbol{x}|^{2-d}\end{aligned}$$

where we have used the Fourier transformation

$$\int \mathrm{d}^d\boldsymbol{k}\, \frac{\mathrm{e}^{\mathrm{i}\boldsymbol{k}\cdot\boldsymbol{x}}}{\eta|\boldsymbol{k}|^2} = C_d\eta^{-1}(L^{2-d} - |\boldsymbol{x}|^{2-d})$$

Therefore,

$$\left\langle \mathrm{e}^{\mathrm{i}n\theta(\boldsymbol{x})}\mathrm{e}^{-\mathrm{i}n\theta(0)} \right\rangle = \mathrm{e}^{-\frac{n^2}{2}\langle(\theta(\boldsymbol{x})-\theta(0))^2} = \mathrm{e}^{n^2\langle\theta(\boldsymbol{x})\theta(0)\rangle - n^2\langle\theta(l)\theta(0)\rangle} = \mathrm{e}^{-C_d n^2\eta^{-1}|\boldsymbol{x}|^{2-d}}$$

The decay length is given by $(\eta/C_d)^{1/(2-d)}$ for $n = 1$.

For $d = 2$, the θ correlation has log divergence $\langle\theta(\boldsymbol{x})\theta(0)\rangle - \langle\theta(l)\theta(0)\rangle = -\frac{1}{2\pi\eta}\ln(|\boldsymbol{x}|/l)$, and we have the following algebraic long-range order:

$$\left\langle \mathrm{e}^{\mathrm{i}n\theta(\boldsymbol{x})}\mathrm{e}^{-\mathrm{i}n\theta(0)} \right\rangle \sim |\boldsymbol{x}|^{-n^2/2\pi\eta} \tag{3.4.4}$$

3.4.2 The Kosterlitz–Thouless transition

- We have seen that spatially-dependent θ fluctuations change long-range order to algebraic long-range order. Here we will see that a new type of fluctuation—vortex fluctuations—can change algebraic long-range order to short-range order.

The above $d = 2$ result is not completely correct. When η is below a critical value η_c (or when the temperature is above a critical temperature T_{KT}), algebraic long-range order cannot survive and there are only short-range correlations (Kosterlitz and Thouless, 1973). To understand this phenomenon, we need to include vortex fluctuations. A vortex configuration is given by

$$\varphi_c(x, y) = f(r)\mathrm{e}^{\mathrm{i}\phi}, \qquad f(0) = 0, \qquad f(\infty) = \varphi_0$$

where $x = r\cos(\phi)$ and $y = r\sin(\phi)$. One can show that $|\partial_{\boldsymbol{x}}\theta| = 1/r$. Therefore, the action of a single vortex is

$$S_v = \int \mathrm{d}^2\boldsymbol{r}\, \frac{1}{2}\eta\frac{1}{r^2} + S_c = \int \pi\,\mathrm{d}(r^2)\, \frac{1}{2}\eta\frac{1}{r^2} + S_c = h\ln\frac{L}{l} + S_c$$

where

$$h = \eta\pi,$$

l is the short-distance cut-off scale, and S_c is the core action (i.e. the contribution from inside the vortex core, where $f(r)$ starts to change from φ_0 to 0). The interaction between a vortex and an anti-vortex separated by a distance r is $2h\ln\frac{r}{l}$.

The problem here is very similar to the instanton gas problem discussed in Sections 2.4.1 and 2.4.3. To calculate the partition function, we must include the vortex contribution. In the presence of a fixed vortex configuration $\varphi_c = \mathrm{e}^{\mathrm{i}\theta_c}$ and using the XY-model, the partition function has the form

$$Z = \int \mathcal{D}\delta\theta\ \mathrm{e}^{-\int\, \mathrm{d}^2\boldsymbol{x}\ \frac{\eta}{2}(\partial_{\boldsymbol{x}}(\delta\theta+\theta_c))^2}$$

As the XY-model is quadratic, the resulting partition function takes a simple factorized form $Z = Z_0\mathrm{e}^{-S_{\text{eff}}(\theta_c)}$, where Z_0 is the partition function with no vortex. As $S_{\text{eff}}(\theta_c)$ is given by the log interaction between the vortices, the total partition function has the form

$$Z = Z_0\sum_n \frac{1}{n!n!}\int\prod_{i=1}^{2n}\mathrm{d}^2\boldsymbol{r}_i\ \mathrm{e}^{-2nS_c}\,\mathrm{e}^{\sum_{i<j}^{2n} 2hq_iq_j\ln\frac{r_{ij}}{l}} \tag{3.4.5}$$

Each term in the summation contains n vortices at $\boldsymbol{r}_1, \ldots\boldsymbol{r}_n$ and n anti-vortices at $\boldsymbol{r}_{n+1}, \ldots, \boldsymbol{r}_{2n}$. Also, $q_i = \pm$ depending on whether the ith vortex is a vortex or an anti-vortex, and r_{ij} is the separation between the ith and jth (anti-)vortices.

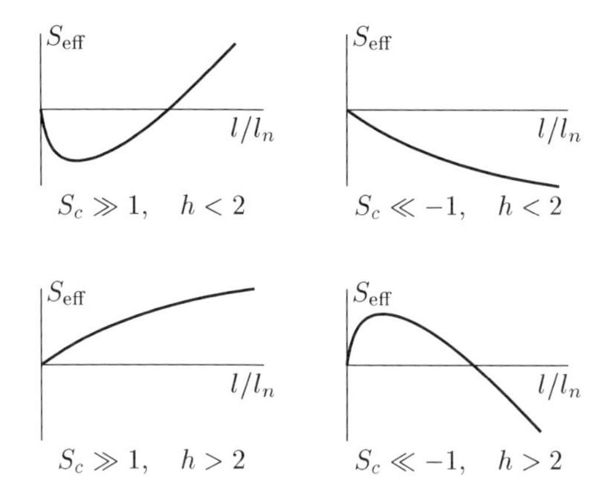

FIG. 3.8. S_{eff} as a function of l/l_n.

To understand the effect of the vortices, let us estimate the partition function for n vortices and n anti-vortices, namely $Z = \mathrm{e}^{-S_{\text{eff}}}$, with

$$S_{\text{eff}} \sim 2n \ln n + n\left(2h \ln \frac{l_n}{l} + 2S_c\right) - 2n \ln \frac{L^2}{l^2} = L^2 \frac{2}{l_n^2}(h-2) \ln \frac{l_n \mathrm{e}^{S_c/(h-2)}}{l} \tag{3.4.6}$$

where L is the linear size of the system and $l_n = L/\sqrt{n}$ is the mean separation between the vortices. Here l_n is always larger than the cut-off, i.e. $l_n > l$. The first term in eqn (3.4.6) comes from $1/(n!)^2$. The second term is the action of n vortex–anti-vortex pairs, each occupying an area of size $\sim l_n$. It contains two contributions: the vortex interaction $2h \ln \frac{l_n}{l}$ and the core action $2S_c$. The third term comes from the integration $\int \prod_{i=1}^{2n} \mathrm{d}^2 \boldsymbol{r}_i$ of the vortex positions (which is the entropy). The behavior of the vortex gas is controlled by the competition between the vortex action, which favors fewer vortices, and the entropy, which favors more vortices.

The number of vortices, n, can be obtained by simply minimizing S_{eff} with respect to l_n in the range $l_n > l$. From the behavior of S_{eff} in Fog. 3.8, we see that, when $S_c \ll -1$ (i.e. when it is cheap to create vortices), S_{eff} is minimized at the boundary $l_n \sim l$. In this case, the vortex fluctuations are important, which changes the algebraic long-range order in eqn (3.4.4) to short-range order:

$$\left\langle \mathrm{e}^{\,\mathrm{i}\theta(\boldsymbol{x})} \mathrm{e}^{-\mathrm{i}\theta(0)} \right\rangle \sim \mathrm{e}^{-|\boldsymbol{x}|/\xi}$$

with the correlation length $\xi \sim l_n \sim l$.

When $S_c \gg 1$ and $h = \pi\eta < 2$, S_{eff} is minimized at $l_n \sim \mathrm{e}^{S_c/(2-h)} l$. Again the vortex density is finite, which again changes the algebraic long-range order to

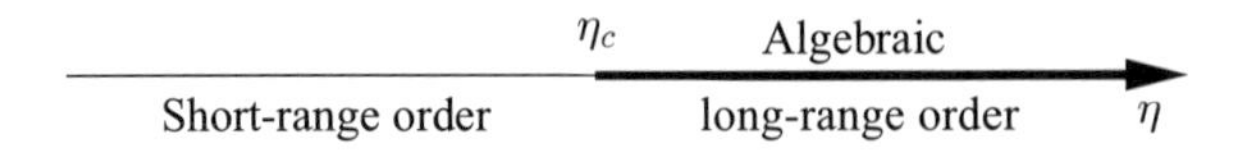

FIG. 3.9. Phase diagram of the two-dimensional XY-model.

short-range order. The correlation length is

$$\xi \sim l_n \sim l e^{S_c/(2-h)} \tag{3.4.7}$$

When $S_c \gg 1$ and $h = \pi\eta > 2$, S_{eff} is minimized at $l_n = \infty$, and the vortex density is zero. In this case, the vortex fluctuations are unimportant at long distances and the algebraic long-range order survives the vortex fluctuations.

From the above discussion, we see that, in the $S_c \gg 1$ limit, our two-dimensional boson system has a finite-temperature phase transition as we change η. The phase transition changes the algebraic long-range order to short-range order. The critical η at the transition is $\eta_c = 2/\pi$. The phase diagram in Fig. 3.9 summarizes our results in the $S_c \gg 1$ limit. The transition at η_c is the famous Kosterlitz–Thouless (KT) transition. From the relation (3.4.3) between η and T, we find the critical temperature

$$T_{KT} = \frac{\pi\varphi_0}{2m}$$

We would like to point out that the KT transition does not change any symmetry, and is a counterexample of Landau's symmetry-breaking theory for phases and phase transitions.

Using $h = \pi\eta = \pi\varphi_0^2/mT$, eqn (3.4.7) can be rewritten as

$$\xi(T) \sim l e^{E_\xi/(T-T_{KT})} \tag{3.4.8}$$

near T_{KT}, where E_ξ is a certain energy scale. Here we would like to point out that eqn (3.4.8) is not correct as $T \to T_{KT}$. The incorrect result is due to an incorrect assumption that the phase rigidity η is not affected by the presence of vortices. In fact, the vortex gas will modify the effective value of η and h. This is because the vortex–anti-vortex pairs can be polarized by the phase twist $\partial_x\theta$, which releases some strain and reduces the effective phase rigidity. Using the effective value h^*, eqn (3.4.7) should be rewritten as

$$\xi \sim l_n \sim l e^{S_c/(2-h^*)} \tag{3.4.9}$$

The temperature dependence of $h^*(T)$ is more complicated than that of $h(T)$. However, by definition $h^*(T_{KT}) = 2$. Based on a renormalization group calculation (see Section 3.5.6), we find that

$$2 - h^*(T_{KT}) \propto (T - T_{TK})^{1/2}$$

which leads to

$$\xi(T) \sim l e^{\left(E_\xi/(T-T_{KT})\right)^{1/2}} \tag{3.4.10}$$

Problem 3.4.1.
Note that, near T_c, the action S_{eff} in eqn (3.4.1) can be approximated by

$$S_{\text{eff}} = \beta \int \mathrm{d}^d\boldsymbol{x} \left[\frac{1}{2m^*} |\partial_{\boldsymbol{x}} \varphi_c|^2 + a(T - T_c)|\varphi_c|^2 + b|\varphi_c|^4 \right] + O(|\varphi_c|^6)$$

because the order parameter φ_c is small near the critical point (or the phase transition point) at $T = T_c$. In the mean-field (or semiclassical) approach to the phase transition and the critical point, we first find the mean-field solution that minimizes the action. We then assume that fluctuations around the mean-field solution are small and expand the action to quadratic order in the fluctuations. The quadratic approximation of S_{eff} can be used to calculate various correlations.

1. Use the mean-field approach to calculate the decay exponent γ in $\langle \varphi_c(\boldsymbol{x}) \varphi_c^*(0) \rangle \sim 1/|\boldsymbol{x}|^\gamma$ at the critical point.
2. The above result is not always valid because the classical theory may break down. Repeat the discussions at the end of Section 3.3.8 (i.e. write S_{eff} in the form $g^{-1}\tilde{S}$ with dimensionless $\tilde{S}$) to see when the mean-field approach can correctly describe the critical point and when the critical point is controlled by strong fluctuations; that is, to find the upper critical point d_c.
3. Here we would like to introduce the concept of relevant versus irrelevant perturbations. We know that above the upper critical dimension the classical theory correctly describes the critical point at the phase transition. Now we add a perturbation $\beta \int \mathrm{d}^d\boldsymbol{x}\, c|\varphi|^\sigma$ to the effective action S_{eff}. If the perturbation is important and modifies the critical exponents, then we say it is a relevant perturbation. If the perturbation becomes vanishingly small near the critical point, then we say it is an irrelevant perturbation. Use the same scalings that you found above to see how the perturbation $\beta \int \mathrm{d}^d\boldsymbol{x}\; c|\varphi|^\sigma$ modifies the scaled action $\tilde{S}$. Determine for what range of σ the perturbation is relevant, and for what range of σ the perturbation is irrelevant.

3.5 Renormalization group

3.5.1 Relevant and irrelevant perturbations

- Relevant perturbations change the long-distance (or low-energy) behavior of a system, while irrelevant perturbations do not.
- We can use the scaling dimension of a perturbation to determine if the perturbation is relevant or irrelevant.

In the above discussion of the KT transition, we note that, when $\mathrm{e}^{-S_c} \ll 1$, the vortex fluctuations are just 'small perturbations'. However, if $h < 2$, then no matter how small e^{-S_c} is, the vortex fluctuations always destroy the algebraic long-range correlation of $\left\langle \mathrm{e}^{\mathrm{i}\theta(x)} \mathrm{e}^{-\mathrm{i}\theta(0)} \right\rangle$. Thus, when $h < 2$, the perturbation of including the vortex fluctuations is called a relevant perturbation. When $h >$

2, the perturbation is called an irrelevant perturbation, and, when $h = 2$, the perturbation is called a marginal perturbation. In the following we would like to discuss relevant/irrelevant/marginal perturbations in a more general set-up.

Consider a theory described by the action

$$S = S_0 + \int \mathrm{d}^d x a O(x)$$

where aO is a perturbation. We assume that S_0 has a Z_2 symmetry and $O \to -O$ under the Z_2 transformation. As a result, $\langle O \rangle = 0$ when $a = 0$. We also assume that, for large x,

$$\langle O(x)O(0) \rangle = \frac{1}{|x|^{2h}} \tag{3.5.1}$$

when $a = 0$. Here h is called the scaling dimension of the operator O (the scaling dimension of $1/x$ is defined as 1). Equation (3.5.1) also defines the normalization of the operator O.

At the second-order perturbation, the partition function is given by

$$Z = Z_0 \int \mathrm{d}^d x \mathrm{d}^d y \, a^2 \langle O(x)O(y) \rangle$$

where Z_0 is the zeroth-order partition function. We see that the second-order perturbation changes the effective action by

$$\Delta S_{\text{eff}} = -\ln Z + \ln Z_0 = -2\ln g + 2h \ln L - 2d \ln L$$

Note that, when $h < d$ and $L > \xi = a^{-1/(d-h)}$, we have $\Delta S_{\text{eff}} < 0$. Thus, the system prefers to have two $O(x)$ insertions. When $L \gg \xi$, the system wants to have two $O(x)$ insertions for each ξ^d volume. We see that, if we are interested in correlation functions at length scales beyond ξ, then the perturbation is always important. We conclude that the perturbation $\int \mathrm{d}^d x \, O(x)$ is relevant if the scaling dimension of $O(x)$ is less than d. In this case, $O(x)$ is called a relevant operator. If the scaling dimension of $O(x)$ is greater than (or equal to) d, then $O(x)$ is called an irrelevant (marginal) operator. An easy way to remember this result is to note that the perturbation $\int \mathrm{d}^d x \, O(x)$ is relevant if $\int \mathrm{d}^d x \, O(x)$ has a dimension less than zero.

The concept of scaling dimension also allows us to use dimensional analysis to estimate the induced $\langle O \rangle$ by a finite perturbation aO. As the scaling dimension of $\delta S = \int \mathrm{d}^d x \, aO$ is zero by definition, the coefficient a has a scaling dimension $[a] = d - [O] = d - h$. When aO is an irrelevant perturbation (i.e. when $h > d$), the induced $\langle O \rangle$ is proportional to a. We have

$$\langle O \rangle \sim a l^{d-2h}$$

where l is the short-distance cut-off. When aO is a relevant perturbation (i.e. when $h < d$), the induced $\langle O \rangle$ is more than $a l^{d-2h}$. By matching the scaling dimensions,

we find that

$$\langle O \rangle \sim a^{h/(d-h)} = a\xi^{d-2h}. \tag{3.5.2}$$

Problem 3.5.1.
The effective action

$$S_{\text{eff}} = \beta \int \mathrm{d}^d \boldsymbol{x} \, \frac{1}{2m^*} |\partial_{\boldsymbol{x}} \varphi|^2$$

describes a critical point. Calculate the scaling dimensions of φ, φ^2, $|\varphi|^2$, and $|\varphi|^4$. Show that, below a spatial dimension d_0, the perturbation $\int \mathrm{d}^d \boldsymbol{x} \, b|\varphi|^4$ becomes a relevant perturbation. Find the value of d_0 and explain why d_0 is equal to the upper critical dimension d_c of

$$S_{\text{eff}} = \beta \int \mathrm{d}^d \boldsymbol{x} \, [\frac{1}{2m^*} |\partial_{\boldsymbol{x}} \varphi|^2 + a(T - T_c)|\varphi|^2 + b|\varphi|^4]$$

3.5.2 The duality between the two-dimensional XY-model and the two-dimensional clock model

- The vortices in the two-dimensional XY-model can be viewed as particles. The field theory that describes those particles is the two-dimensional clock model.

In order to study the vortex fluctuations of the XY-model in more detail, we would like to map the two-dimensional XY-model to the Z_1 two-dimensional clock model. A generic Z_n clock model is defined by

$$S = \int \mathrm{d}^2 \boldsymbol{x} \, \left(\frac{\kappa}{2} (\partial_{\boldsymbol{x}} \theta)^2 - g \cos(n\theta) \right) \tag{3.5.3}$$

When $g = 0$ the clock model is the XY-model at finite temperatures. The action is the energy divided by the temperature: $S = \beta E$. The $g\cos(n\theta)$ term (explicitly) breaks the $U(1)$ rotational symmetry. If we view $(\cos(\theta), \sin(\theta)) = (S_x, S_y)$ as the two components of a spin, then, for $n = 1$, the $g\cos(\theta)$ term is a term induced by a magnetic field in the S_x direction. For general n, the clock model has a Z_n symmetry: $\theta \to \theta + \frac{2\pi}{n}$.

To show the duality relation, we consider the following partition function of eqn (3.5.3) with $n = 1$:

$$\begin{aligned} Z &= \int \mathcal{D}\theta \sum_k \frac{1}{k!} \left(g \int \mathrm{d}^2 \boldsymbol{x} \, \frac{\mathrm{e}^{\mathrm{i}\theta} + \mathrm{e}^{-\mathrm{i}\theta}}{2} \right)^k \mathrm{e}^{-\int \mathrm{d}^2 \boldsymbol{x} \, \frac{\kappa}{2} (\partial_{\boldsymbol{x}} \theta)^2} \\ &= Z_0 \sum_k \frac{1}{k!k!} \int \prod_{i=1}^{2k} \mathrm{d}^2 \boldsymbol{r}_i \, \mathrm{e}^{-2kS_c} \mathrm{e}^{\sum_{i<j}^{2k} 2hq_i q_j \ln \frac{r_{ij}}{l}} \end{aligned} \tag{3.5.4}$$

where Z_0 is the partition function of $S = \int \mathrm{d}^2 \boldsymbol{x} \, \frac{\kappa}{2} (\partial_{\boldsymbol{x}} \theta)^2$. Each term in the summation arises from the correlation $\langle \mathrm{e}^{\mathrm{i}\theta(\boldsymbol{r}_1)} ... \mathrm{e}^{\mathrm{i}\theta(\boldsymbol{r}_k)} \mathrm{e}^{-\mathrm{i}\theta(\boldsymbol{r}_{k+1})} ... \mathrm{e}^{-\mathrm{i}\theta(\boldsymbol{r}_{2k})} \rangle$. Also,

$q_i = 1$ for $1 \leqslant i \leqslant k$, $q_i = -1$ for $k+1 \leqslant i \leqslant 2n$, $\mathrm{e}^{-S_c} = g/2$, and

$$h = \frac{1}{4\pi\kappa}$$

Equation (3.5.4) is identical to the partition function (3.4.5) of the XY-model (3.4.2) if $\frac{1}{4\pi\kappa} = \pi\eta$. So, the Z_1 clock model (3.5.3) is equivalent to the XY-model (3.4.2) (with vortices) if $\frac{1}{2\pi\kappa} = 2\pi\eta$. The vortex in the XY-model is mapped to $\mathrm{e}^{\,\mathrm{i}\theta(\boldsymbol{x})}$ in the clock model. Similarly, the vortex in the clock model is mapped to $\mathrm{e}^{\,\mathrm{i}\theta(\boldsymbol{x})}$ in the XY-model. The vortex in the clock model has a scaling dimension $\pi\kappa$. The $\mathrm{e}^{\,\mathrm{i}\theta(\boldsymbol{x})}$ operator in the XY-model has a scaling dimension $1/4\pi\eta$. The relation $\frac{1}{2\pi\kappa} = 2\pi\eta$ ensures that the two scaling dimensions agree with each other.

We know that the $U(1)$ symmetry in the XY-model does not allow the $\mathrm{e}^{\,\mathrm{i}\theta}$ term to appear in the action. Using the above mapping, we see that the corresponding clock model must not allow vortex fluctuations. Allowing the vortex fluctuations in the clock model corresponds to explicitly breaking the $U(1)$ symmetry in the dual XY-model. We see that there are two different types of clock model, the one with vortex fluctuations and the one without vortex fluctuations. As the corresponding dual models have different symmetries, the two types of clock model have very different properties. We will call the clock model with vortex fluctuations the compact clock model, and the one without vortex fluctuations the non-compact clock model. The XY-model with vortices is mapped to an Z_1 non-compact clock model. Such a mapping allows us to study the KT transition in the XY-model by studying the transition in the corresponding non-compact clock model.

3.5.3 Physical properties of the clock model

- A field theory model is not well defined unless we specify the short-distance cut-off.
- Ginzburg–Landau theory, containing strong vortex fluctuations, cannot describe phase transitions in the non-compact clock model.

In this section, we will discuss possible phase transitions in a generic Z_n clock model. When g is large, the field θ is trapped by one of the minima of the potential $-g\cos(n\theta)$. We believe that, in this case, the model is in a phase that spontaneously breaks the Z_n symmetry. When both κ and g are small, the fluctuation of θ is strong. We expect that the model will be in a Z_n-symmetric phase.

Despite sounding so reasonable, the above statements do not really make sense. This is because g has a dimension. It is meaningless to talk about how large g is. What is worse is that g is the only parameter in the model that has a non-trivial dimension. So, we cannot make a dimensionless combination to determine how large g is.

To understand the importance of the $g\cos(n\theta)$ term in a physical way, we would like to ask how big the $e^{in\theta}$ operator is. One physical way to answer this question is to examine the correlation of $e^{in\theta}$ for the XY-model $S = \int d^2\boldsymbol{x}\, \frac{\kappa}{2}(\partial_{\boldsymbol{x}}\theta)^2$. The correlation is given by (see eqn (3.3.25))

$$\left\langle e^{in\theta(\boldsymbol{x})} e^{-in\theta(0)} \right\rangle = (l/|\boldsymbol{x}|)^{n^2/2\pi\kappa}. \tag{3.5.5}$$

One big surprise is that the correlation depends on the short-distance cut-off l. Thus, the magnitude (or the importance) of the operator $e^{in\theta}$ is not even well defined unless we specify the cut-off l. This illustrates the point that *to have a well-defined field theory we must specify a short-distance cut-off* l. To stress this point, we would like to make the l dependence explicit and write the action as

$$S = \int d^2\boldsymbol{x}\, \left(\frac{\kappa_l}{2}(\partial_{\boldsymbol{x}}\theta_l)^2 - g_l \cos(n\theta_l) \right) \tag{3.5.6}$$

The short-distance cut-off is introduced by requiring that the θ_l field does not contain any fluctuations with wavelengths shorter than l:

$$\theta_l(\boldsymbol{x}) = \int_{|\boldsymbol{k}|<2\pi/l} d^2\boldsymbol{k}\, \theta_{\boldsymbol{k}} e^{i\boldsymbol{x}\cdot\boldsymbol{k}}$$

We see that a well-defined clock model (3.5.6) contains three parameters κ_l, g_l, and l. So, the clock model really contains two dimensionless parameters κ_l and $\bar{g}_l = g_l l^2$.

We can now make sensible statements. When $\bar{g}_l \gg 1$, we believe that the model is in a phase that spontaneously breaks the Z_n symmetry. When both κ_l and $\bar{g}_l$ are much less than 1, we expect the model to be in a Z_n-symmetric phase.

A non-trivial limit is when $\kappa_l \gg 1$ and $\bar{g}_l \ll 1$. Is the model in the Z_n-symmetric phase or in the Z_n-symmetry-breaking phase? The concept of relevant/irrelevant perturbation is very helpful in answering this question. If we treat the $g_l \cos(n\theta)$ term as a perturbation to the XY-model, then, from eqn (3.5.5), we see that the scaling dimension of $e^{in\theta}$ in the XY-model is $[e^{in\theta}] = n^2/4\pi\kappa_l$. Thus, the $g_l\cos(n\theta)$ term is relevant when $n^2/4\pi\kappa_l < 2$ and irrelevant when $n^2/4\pi\kappa_l > 2$.

This result is reasonable. When κ_l is small, the fluctuations of θ are strong. This makes the $g_l\cos(n\theta)$ term average to zero and be less effective. Hence the perturbation $g_l\cos(n\theta)$ is irrelevant. When $g_l\cos(n\theta)$ is irrelevant and $\bar{g}_l$ is small, we can drop the $g_l\cos(n\theta)$ term when we calculate long-range correlations. This suggests that, at long distances, we not only have the Z_n symmetry, but we also have the full $U(1)$ symmetry when both κ_l and $\bar{g}_l$ are small.

When κ is large, the fluctuations of θ are weak. This makes the $g_l\cos(n\theta)$ term a relevant perturbation. The effect of the $g_l\cos(n\theta)$ term becomes important at long distances, no matter how small $\bar{g}_l$ is. Thus, we expect the system to be trapped in one of the n minima of the potential term and the Z_n symmetry is spontaneously broken, even for small $\bar{g}_l$.

After realizing that the clock model can have a Z_n-symmetry-breaking phase and a Z_n-symmetric phase, the next natural question is how do the two phases transform into each other? One way to understand the transition is to introduce a complex order parameter $\varphi(\boldsymbol{x}) = \langle e^{i\theta(\boldsymbol{x})}\rangle$ and write down a Ginzburg–Landau effective theory for the transition

$$S_{GL} = \int d^2\boldsymbol{x}[\frac{\gamma}{2}(\partial_{\boldsymbol{x}}\varphi)^2 + a|\varphi|^2 + b|\varphi|^4 - c\text{Re}\varphi^n] \tag{3.5.7}$$

Note that the $c\text{Re}\varphi^n$ term (explicitly) breaks the $U(1)$ symmetry down to Z_n. When $n > 1$, the Ginzburg–Landau theory describes a symmetry-breaking transition as a changes from a positive value to a negative value.

When $n = 1$, the Ginzburg–Landau theory contains no phase transition because there is no symmetry breaking. This seems to suggest that the Z_1 clock model contains no phase transition and the corresponding XY-model contains no TK transition.

So what is wrong? In the Ginzburg–Landau theory, the order parameter has strong amplitude fluctuations near the transition point. A typical configuration of φ contains many points where $\varphi = 0$. So, there are strong vortex fluctuations. The Ginzburg–Landau theory describes the phase transitions in the compact clock model. The Ginzburg–Landau theory does not apply to a non-compact clock model.

3.5.4 Renormalization group approach to the non-compact clock model

- Through the concept of running coupling constants, the renormalization group (RG) approach allows us to see how a theory evolves as we go to long distances or low energies. It is very useful because it tells us the dynamical properties that emerge at long distances or low energies.
- As an effective theory only evolves into a similar effective theory, we cannot use the renormalization group approach to obtain the emergence of qualitatively new phenomena, such as the emergence of light and fermions from a bosonic model.

In this section, to understand the physical properties of the non-compact clock model, we will work directly with the θ field in the clock model.

We note that, if the fluctuations $O(\boldsymbol{x})$ and $O(\boldsymbol{y})$ at different locations fluctuate independently, then the so-called *connected correlation*

$$G_{conn} = \langle O(\boldsymbol{x})O(\boldsymbol{y})\rangle - \langle O(\boldsymbol{x})\rangle\langle O(\boldsymbol{y})\rangle$$

vanishes. So, the connected correlation measures the correlation between the fluctuations of $O(\boldsymbol{x})$ and $O(\boldsymbol{y})$.

When $g_l = 0$, the non-compact clock model always has an algebraic long-range correlation: $\langle \mathrm{e}^{\,in\theta(\boldsymbol{x})}\mathrm{e}^{-in\theta(0)}\rangle - \langle \mathrm{e}^{\,in\theta(\boldsymbol{x})}\rangle\langle \mathrm{e}^{-in\theta(0)}\rangle \sim \frac{1}{|\boldsymbol{x}|^{n^2/2\pi\kappa_l}}$, regardless of the value of κ_l. The issue here is how the $g_l\cos(n\theta)$ term affects the algebraic long-range correlation.

As discussed in the last section, when $\kappa_l < n^2/8\pi$, $\mathrm{e}^{\,in\theta(\boldsymbol{x})}$ is irrelevant and a small $g_l\cos(n\theta)$ term will not affect the algebraic long-range correlation. When $\kappa_l > n^2/8\pi$, $\mathrm{e}^{\,in\theta(\boldsymbol{x})}$ is relevant. We expect that a $g_l\cos(n\theta)$ term will change the algebraic long-range correlation into a short-ranged one, no matter how small $\bar{g}_l$ is. We see that, for small $\bar{g}_l$, the non-compact clock model has a phase transition at $k_l = n^2/8\pi$. In the following, we will use the RG approach to understand the above phase transition.

We note that the clock model is well defined only after we specify a short-distance cut-off l. The key step in the RG approach is to integrate out the fluctuations between the wavelengths l and λ ($\lambda > l$). This results in a model with a new cut-off λ. To integrate out the short-wavelength θ fluctuations, we first write

$$\theta_l = \theta_\lambda + \delta\theta$$

where $\delta\theta$ only contains fluctuations with wavelengths between l and λ. As the short-wavelength fluctuations $\delta\theta$ are suppressed by the $\kappa(\partial_{\boldsymbol{x}}\delta\theta)^2$ term, we expect $\delta\theta$ to be small and expand the action to second order in $\delta\theta$ as follows:

$$S = \int \mathrm{d}^2\boldsymbol{x}\ \left(\frac{\kappa_l}{2}(\partial_{\boldsymbol{x}}\theta_\lambda)^2 - g_l\cos(n\theta_\lambda) + \frac{\kappa_l}{2}(\partial_{\boldsymbol{x}}\delta\theta)^2\right)$$
$$+ \int \mathrm{d}^2\boldsymbol{x}\ \left(-ng_l\sin(\theta_\lambda)\delta\theta + \frac{n^2}{2}g_l\cos(n\theta_\lambda)(\delta\theta)^2\right)$$

We treat θ_λ as a smooth background field, and integrate out $\delta\theta$ (this approach is called the background-field RG approach). We obtain the effective action

$$S = \int \mathrm{d}^2\boldsymbol{x}\ \left(\frac{\kappa_l}{2}(\partial_{\boldsymbol{x}}\theta_\lambda)^2 - g_l\cos(n\theta_\lambda) + \frac{1}{2}g_l\cos(n\theta_\lambda)K(0)\right)$$
$$- \int \mathrm{d}^2\boldsymbol{x}\mathrm{d}^2\boldsymbol{y}\ \frac{1}{2}(g_l)^2\sin(n\theta_\lambda(\boldsymbol{x}))K(\boldsymbol{x}-\boldsymbol{y})\sin(n\theta_\lambda(\boldsymbol{y}))$$

where $K(\boldsymbol{x}) = n^2 \langle \delta\theta(\boldsymbol{x})\delta\theta(0)\rangle$. We note that the last term can be rewritten as

$$\int \mathrm{d}^2\boldsymbol{x}\mathrm{d}^2\boldsymbol{y}\, \frac{g_l^2}{4}[\sin(n\theta_\lambda(\boldsymbol{x})) - \sin(n\theta_\lambda(\boldsymbol{y}))]^2 K(\boldsymbol{x}-\boldsymbol{y}) - \int \mathrm{d}^2\boldsymbol{x}\, \frac{g_l^2}{2}\sin^2(n\theta_\lambda)\bar{K}$$
$$= \int \mathrm{d}^2\boldsymbol{x}\mathrm{d}^2\boldsymbol{y}\, \frac{n^2}{8}(g_l)^2\cos^2(n\theta_\lambda(\boldsymbol{x}))(\partial_{\boldsymbol{x}}\theta_\lambda(\boldsymbol{x}))^2(\boldsymbol{x}-\boldsymbol{y})^2 K(\boldsymbol{x}-\boldsymbol{y})$$
$$- \int \mathrm{d}^2\boldsymbol{x}\, \frac{1}{2}(g_l)^2(\sin(n\theta_\lambda(\boldsymbol{x})))^2\bar{K}$$

where $\bar{K} = \int \mathrm{d}^2\boldsymbol{x}\ K(\boldsymbol{x})$. We see that terms like $\cos(2n\theta_\lambda)$, $(\partial_{\boldsymbol{x}}\theta_\lambda)^2$, $\cos(2n\theta_\lambda)(\partial_{\boldsymbol{x}}\theta_\lambda)^2$, $(\partial_{\boldsymbol{x}}\theta_\lambda)^4$, etc. are generated. RG flow can generate many new terms that are not in the starting action. In fact, any local terms that do not break the Z_n symmetry can be generated. However, the term $\cos(\theta_\lambda)$ is not generated when $n > 1$ because it breaks the Z_n symmetry. For the time being, let us only keep the terms $(\partial_{\boldsymbol{x}}\theta_\lambda)^2$ and $\cos(\theta_\lambda)$ that are already in our starting action.[23] We find that the action of our model becomes

$$S = \int \mathrm{d}^2\boldsymbol{x}\ \left(\frac{\kappa_\lambda}{2}(\partial_{\boldsymbol{x}}\theta_\lambda)^2 - g_\lambda\cos(n\theta_\lambda)\right) \tag{3.5.8}$$

where λ is the new cut-off. The effective coupling constants depend on the cut-off λ and are called *running coupling constants*. They are given by (assuming that $\frac{\lambda-l}{l} \ll 1$)

$$g_\lambda = g_l - \frac{1}{2}K(0), \qquad \kappa_\lambda = \kappa_l + \frac{n^2}{8}g_l^2 K_2$$

$$K(0) = \int_{2\pi/\lambda<|\boldsymbol{k}|<2\pi/l} \frac{\mathrm{d}^2\boldsymbol{k}}{(2\pi)^2}\frac{n^2}{\kappa_l|\boldsymbol{k}|^2} = \frac{n^2}{2\pi\kappa_l}\ln\frac{\lambda}{l}$$

$$K_2 \equiv \int \mathrm{d}^2\boldsymbol{x}\ |\boldsymbol{x}|^2 K(\boldsymbol{x}) = \int_{2\pi/\lambda<|\boldsymbol{k}|<2\pi/l} \mathrm{d}^2\boldsymbol{x}\frac{\mathrm{d}^2\boldsymbol{k}}{(2\pi)^2}\frac{n^2\boldsymbol{x}^2\,\mathrm{e}^{\,\mathrm{i}\boldsymbol{k}\cdot\boldsymbol{x}-0^+|\boldsymbol{x}|}}{\kappa_l\boldsymbol{k}^2}$$
$$= \frac{\lambda-l}{l}\frac{3n^2l^4}{16\pi^5\kappa_l}\int \mathrm{d}\theta\frac{1}{(\cos\theta + \mathrm{i}0^+)^4} = \frac{\lambda-l}{l}\frac{3n^2l^4}{2\pi^4\kappa_l}$$

[23] It turns out that all of the other terms are irrelevant. If those terms are small at the start of the RG flow, then they will become even smaller after a long flow. This is the reason why we can ignore those terms. Certainly, if those terms are large at the beginning, then they can change everything.

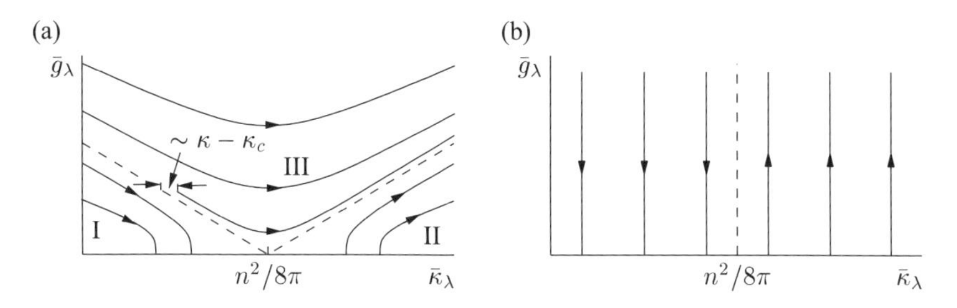

FIG. 3.10. (a) The RG flow of $\bar{g}_\lambda$ and $\bar{\kappa}_\lambda$ as determined by eqn (3.5.9). (b) The RG flow of $\bar{g}_\lambda$ and $\bar{\kappa}_\lambda$ as determined by eqn (3.5.10).

Let $b = \ln \lambda$; then the changes of the coupling constants are described by the following differential equations:

$$\frac{\mathrm{d}g_\lambda}{\mathrm{d}b} = -\frac{n^2}{4\pi\kappa_l} g_\lambda$$
$$\frac{\mathrm{d}\kappa_\lambda}{\mathrm{d}b} = \frac{3n^4 g_\lambda^2 \lambda^4}{16\pi^4 \kappa_\lambda}$$

In terms of the dimensionless couplings $\bar{\kappa}_\lambda = \kappa_\lambda \lambda^0$ and $\bar{g}_\lambda = g_\lambda \lambda^2$, these differential equations can be rewritten as follows:

$$\frac{\mathrm{d}\bar{g}_\lambda}{\mathrm{d}b} = \left(2 - \frac{n^2}{4\pi\bar{\kappa}_\lambda}\right) \bar{g}_\lambda$$
$$\frac{\mathrm{d}\bar{\kappa}_\lambda}{\mathrm{d}b} = \frac{3n^4 \bar{g}_\lambda^2}{16\pi^4 \bar{\kappa}_\lambda} \tag{3.5.9}$$

which are called the RG equations. The flow of $(\bar{\kappa}, \bar{g})$ is illustrated in Fig. 3.10(a).

3.5.5 Renormalization group theory and phase transition

• The concept of a fixed point and effective theory for a fixed point.

To understand the physical implications of the RG flow, let us first ignore the flow of $\bar{\kappa}_\lambda$ and study, instead, the following RG equations:

$$\frac{\mathrm{d}\bar{g}_\lambda}{\mathrm{d}b} = \left(2 - \frac{n^2}{4\pi\bar{\kappa}_\lambda}\right) \bar{g}_\lambda$$
$$\frac{\mathrm{d}\bar{\kappa}_\lambda}{\mathrm{d}b} = 0 \tag{3.5.10}$$

The flow of $(\bar{g}_\lambda, \bar{\kappa}_\lambda)$ is illustrated in Fig. 3.10(b). We find that

$$\bar{g}_\lambda = \bar{g}_l \mathrm{e}^{(2 - \frac{n^2}{4\pi\kappa_l}) \ln(\lambda/l)} = \bar{g}_l (\lambda/l)^{2-h}$$

from the RG equations, where $h = \frac{n^2}{4\pi\kappa_l}$ is the scaling dimension of $\cos(n\theta)$. When $\cos(n\theta)$ is relevant, a very small $\bar{g}_l$ can become as large as one wants for a long enough flow. In particular, $\bar{g}_\lambda = 1$ when $\lambda = l(\bar{g}_l)^{-1/(2-h)}$. At this point, the coupling constants stop flowing because the RG equations (3.5.9) become invalid due to the higher-order $\bar{g}_\lambda$ terms that were ignored in the RG equations. The resulting effective theory has the same form as eqn (3.5.8) and is called fixed-point theory. We can use the fixed-point theory to obtain the long-distance correlations and other long-distance physical properties of the original model.

When $\bar{g}_\lambda = 1$, everything in the renormalized fixed-point theory is of order 1 when measured in units of λ. Thus, if we believe that a large $g_\lambda \cos(n\theta_\lambda)$ will make θ_λ have short-range correlation, then the correlation length ξ must be of order 1 when measured by λ. This way, we find that

$$\xi = l\left(\frac{1}{\bar{g}_l}\right)^{1/(2-h)}$$

which agrees with the general result $\xi = a^{-1/(d-h)}$ obtained in the last section, after realizing that the perturbation O in the last section corresponds to $O = l^{-h}\cos(n\theta)$. (Equation (3.5.1) determines the normalization of O.) Thus $a = gl^h$. In terms of κ, the above result leads to

$$\xi = l\bar{g}_l^{-4\pi\kappa/(8\pi\kappa - n^2)}. \tag{3.5.11}$$

Also, a large $g_\xi \cos(n\theta_\xi)$ potential term at the length scale ξ traps θ_ξ in one of the potential minima. Thus, a relevant perturbation $g_l \cos(n\theta_l)$ always causes a spontaneous Z_n symmetry breaking, no matter how small g_l is at the cut-off scale.

When κ approaches $n^2/8\pi$, the correlation length $\xi \to \infty$. Thus, there is a phase transition at $n^2/8\pi$. When $\kappa < n^2/8\pi$, the perturbation $g_l \cos(n\theta_l)$ is irrelevant. After a long RG flow, we obtain a different fixed-point theory $\frac{\bar{\kappa}_\lambda}{2}(\partial_{\boldsymbol{x}}\theta_\lambda)^2$ because the $\bar{g}_\lambda$ flow goes to zero. This fixed-point theory has full $U(1)$ symmetry! This is a very striking and very important phenomenon called dynamical symmetry restoration. Sometimes a term may explicitly break a certain symmetry (such as the $g_l \cos(n\theta_l)$ term breaks the $U(1)$ symmetry down to the Z_n symmetry). If the term is irrelevant, then, at long distances and/or low energies, the term flows to zero and the symmetry is restored.

To summarize, the non-compact clock model (3.5.3) has Z_n symmetry. When κ is less than a critical value $\kappa_c = n^2/8\pi$, the model is in a phase that does not break the Z_n symmetry. Furthermore, the phase has $U(1)$ symmetry at long distances. The correlation length is infinite. When κ is above the critical value κ_c, the model is in a phase that breaks the Z_n symmetry. The correlation length is finite.

The above discussion is correct and general if there is no marginal operator in the model. In that case, h can be treated as a constant. However, for the XY-model, the operator $(\partial_{\boldsymbol{x}}\theta)^2$ has a dimension exactly equal to 2 and is an exact marginal operator. As a result, κ is a marginal coupling constant. The constant κ, and hence h, can shift their values in an RG flow. This results in the RG flow described by eqn (3.5.9) and shown in Fig. 3.10(a). We note that the RG flow described in Fig. 3.10(a) is quite different from that in Fig. 3.10(b) near the transition point $\bar{\kappa}_\infty = n^2/8\pi$. The result (3.5.11) only applies to the RG flow in Fig. 3.10(b), and is not valid for the RG flow in Fig. 3.10(a) near the transition point $\kappa_c = n^2/8\pi$. In the next section, we will calculate ξ for the RG flow in Fig. 3.10(a).

In the above, we have used the RG approach to discuss the phases and the phase transitions in the non-compact clock model. We can also use the same RG result to discuss the phases and the phase transitions in the compact clock model with vortex fluctuations.

At first sight, one may say that vortices and anti-vortices are always confined due to the potential term $g_l \cos(n\theta_l)$. This is indeed true if $\kappa > \kappa_c$ and $\kappa \cos(n\theta)$ is relevant. When $\kappa < \kappa_c$, $\kappa \cos(n\theta)$ is irrelevant. In this case, the properties of the vortices are just like those in the XY-model. The vortex fluctuations are relevant if $\kappa < 2/\pi$ and irrelevant if $\kappa > 2/\pi$. When vortex fluctuations are relevant, they modify the phase structure of the clock model.

The compact two-dimensional clock model can have several different behaviors depending on the value of n.

1. $n > 4$: The model is in the Z_n-symmetry-breaking phase when $\kappa > n^2/8\pi$. The Z_n order parameter $\mathrm{e}^{\,\mathrm{i}\theta}$ has a long-range order: $\langle \mathrm{e}^{\,\mathrm{i}\theta(\boldsymbol{x})} \mathrm{e}^{-\,\mathrm{i}\theta(0)} \rangle = \text{constant} + C\mathrm{e}^{-|\boldsymbol{x}|/\xi}$. Near the transition point $n^2/8\pi$, we have $\kappa > 2/\pi$ and the vortex fluctuations are irrelevant. Thus, when $n^2/8\pi > \kappa > 2/\pi$, the system is in a Z_n-symmetric phase with emergent $U(1)$ symmetry at long distances. The Z_n order parameter $\mathrm{e}^{\,\mathrm{i}\theta}$ has an algebraic long-range order: $\langle \mathrm{e}^{\,\mathrm{i}\theta(\boldsymbol{x})} \mathrm{e}^{-\,\mathrm{i}\theta(0)} \rangle \sim 1/|\boldsymbol{x}|^{1/2\pi\kappa}$. When $\kappa < 2/\pi$, the vortex fluctuations are relevant, which destroys the algebraic long-range order. The system is in a Z_n-symmetric phase. The Z_n order parameter $\mathrm{e}^{\,\mathrm{i}\theta}$ has a short-ranged correlation: $\langle \mathrm{e}^{\,\mathrm{i}\theta(\boldsymbol{x})} \mathrm{e}^{-\,\mathrm{i}\theta(0)} \rangle \sim \mathrm{e}^{-|\boldsymbol{x}|/\xi}$. As there is no long-range correlation, we cannot even talk about the emergent $U(1)$ symmetry at long distances.

2. $n = 4$: The model is in the Z_4-symmetry-breaking phase when $\kappa > n^2/8\pi = 2/\pi$, and in a Z_4-symmetric phase when $\kappa < 2/\pi$. In the symmetry-breaking phase, $g_l \cos(4\theta)$ is relevant and the vortex is irrelevant. In the Z_4-symmetric phase, $g_l \cos(4\theta)$ is irrelevant and the vortex is relevant. Thus, the Z_4-symmetric phase has no algebraic long-range order and no emergent $U(1)$ symmetry. At the transition point, both $g_l \cos(4\theta)$ and the vortex are marginal.

3. $n < 4$: The model is in the Z_n-symmetry-breaking phase when $\kappa \gg n^2/8\pi$, and in the Z_n-symmetric phase when $\kappa \ll n^2/8\pi$. Near the transition point, both $\mathrm{e}^{\,\mathrm{i}n\theta}$ and the vortices are relevant and fluctuate strongly. The phase transition is described by Ginzburg–Landau theory, see eqn (3.5.7). When $n = 1$, there is no symmetry breaking and no phase transition.

Problem 3.5.2.

Running 'coupling function' Consider a model $S = \int \mathrm{d}^2\boldsymbol{x}\ \left(\frac{\kappa}{2}(\partial_{\boldsymbol{x}}\theta)^2 + V(\theta)\right)$, where $V(\theta)$ is a small periodic function: $V(\theta + 2\pi) = V(\theta)$. Find the RG equations for the flow of the 'coupling function' V. You may ignore the flow of κ because we have assumed that V is small. Discuss the form of V after a long flow if we have started with a very small V.

Problem 3.5.3.

The $n = 1$ clock model (3.5.3) describes a two-dimensional XY-spin system in a magnetic field B_x, where $S_x = \cos(\theta)$, $S_y = \sin(\theta)$, $S_z = 0$, and $B_x = g$. Assume that $\cos(\theta)$ is relevant. Use the RG argument to find the value of S_x induced by a small magnetic field. Now assume that $\cos(\theta)$ is irrelevant. What is the S_x induced by a small magnetic field? Compare your result with eqn (3.5.2). (Hint: You may write the renormalized action in terms of the original coupling constant B_x and use the renormalized action to calculate the induced S_x. You only need to calculate the induced S_x up to an $O(1)$ coefficient.)

3.5.6 The correlation length near the transition point

To understand the behavior of ξ near the transition point for the RG flow in Fig. 3.10(a), let us expand the RG equations (3.5.9) for small $\delta\bar{\kappa}_\lambda \equiv \bar{\kappa}_\lambda - \frac{n^2}{8\pi}$ as follows:

$$\frac{\mathrm{d}\bar{g}_\lambda}{\mathrm{d}b} = \frac{16\pi}{n^2}\delta\bar{\kappa}_\lambda\bar{g}_\lambda$$
$$\frac{\mathrm{d}\delta\bar{\kappa}_\lambda}{\mathrm{d}b} = \frac{3n^2\bar{g}_\lambda^2}{2\pi^3} \tag{3.5.12}$$

We find that

$$\frac{\mathrm{d}\delta\bar{\kappa}_\lambda}{\mathrm{d}\bar{g}_\lambda} = \frac{3n^4}{32\pi^4}\frac{\bar{g}_\lambda}{\delta\bar{\kappa}_\lambda}$$

The differential equation leads to $(\delta\bar{\kappa}_\lambda)^2 = \frac{3n^4}{32\pi^4}\bar{g}_\lambda^2 + C$. Depending on the sign of the constant term C, there are three classes of solutions (see Fig. 3.10(a)). Class I and class II solutions are given by

$$\delta\bar{\kappa}_\lambda = \mathrm{sgn}(\delta\bar{\kappa}_\infty)\sqrt{\frac{3n^4}{32\pi^4}\bar{g}_\lambda^2 + \delta\bar{\kappa}_\infty^2}$$

where $C = \delta\bar{\kappa}_\infty^2 > 0$. Class III solutions have the form

$$\bar{g}_\lambda = \sqrt{\frac{32\pi^4}{3n^4}\delta\bar{\kappa}_\lambda^2 + \bar{g}_{min}^2} \tag{3.5.13}$$

which is for $C < 0$.

Substituting eqn (3.5.13) into the second equation in eqn (3.5.12), we get

$$\frac{\mathrm{d}\delta\bar{\kappa}_\lambda}{\frac{32\pi^4}{3n^4}\delta\bar{\kappa}_\lambda^2 + \bar{g}_{min}^2} = n^2\,\mathrm{d}\ln(\lambda)$$

We can integrate both sides of the above equation from $\lambda = l$ to $\lambda = \xi$ to obtain

$$\int_{\delta\bar{\kappa}_l}^{\delta\bar{\kappa}_\xi} \frac{\mathrm{d}\delta\bar{\kappa}_\lambda}{\frac{32\pi^4}{3n^4}\delta\bar{\kappa}_\lambda^2 + \bar{g}_{min}^2} = n^2\ln(\frac{\xi}{l}) \tag{3.5.14}$$

We know that, at the correlation length ξ, we have $\bar{g}_\xi \sim 1$. Equation (3.5.13) tells us that $\delta\bar{\kappa}_\xi$ is also of order 1. Equations (3.5.13) and (3.5.14) relate $\delta\bar{\kappa}_l$ and $\bar{g}_l$ to ξ and allow us to determine how the correlation length ξ depends on $\kappa_l - \kappa_c$.

Let us first fix g_l and adjust κ_l to make $\delta\bar{\kappa}_l = \kappa_l - \frac{n^2}{8\pi}$ equal to $-\sqrt{\frac{3n^4}{32\pi^4}}\bar{g}_l$. From eqn (3.5.13), we see that $\bar{g}_{min} = 0$. The integral on the left-hand side of eqn (3.5.14) diverges, which implies that $\xi = \infty$. We see that the Z_n-symmetry-breaking transition really happens when $\kappa_l = \kappa_c$, where

$$\kappa_c = \frac{n^2}{8\pi} - \sqrt{\frac{3n^4}{32\pi^4}\bar{g}_l}$$

If κ_l is slightly above κ_c, then we find that

$$\bar{g}_{min}^2 = 2\left(\frac{32\pi^4}{3n^4}\right)^{1/2} \bar{g}_l(\kappa_l - \kappa_c)$$

As $\bar{g}_{min}$ is much less than $|\delta\bar{\kappa}_l|$ and $\delta\bar{\kappa}_\xi$, eqn (3.5.14) becomes

$$\sqrt{\frac{3}{32\pi^2}}\frac{1}{\bar{g}_{min}} = \ln(\frac{\xi}{l})$$

We find that

$$\xi = l\,\mathrm{e}^{\left(C/(\kappa_l - \kappa_c)\right)^{1/2}}, \qquad C = \frac{n^2}{2\pi\bar{g}_l}\left(\frac{3}{32\pi^2}\right)^{3/2}$$

3.5.7 Fixed points and phase transitions

- Fixed points and universal properties.
- A fixed point with no relevant perturbations corresponds to a stable phase. A fixed point with one relevant perturbation corresponds to the transition point between two stable phases.

Running coupling constants and fixed points (or universality classes) are probably the two most important concepts in RG theory. In this section, we are going to discuss them in a general setting. Let us consider a theory with two coupling constants g_1 and g_2. When combined with the cut-off scale l, we can define the dimensionless coupling constants $\tilde{g}_a = g_a l^{\lambda_a}$, $a = 1, 2$. As we integrate out short-distance fluctuations, the dimensionless coupling constants may flow. One of the possible flow diagrams is given in Fig. 3.11(a).

What can we learn from such a flow diagram? First, we note that the flow has two attractive fixed points A and B. If $(\tilde{g}_1, \tilde{g}_2)$ is anywhere below the DCD′ line, then, after a long flow, the system will be described by $(\tilde{g}_1(A), \tilde{g}_2(A))$. So the system is described by the fixed point A at long distances. This picture demonstrates the principle of universality. The long-distance behavior of a system does not depend on the short-distance details of the system. All of the systems below the DCD′ line share a common long-distance behavior described by the fixed-point theory at A. One of the common long-distance properties is the algebraic decay exponent in the correlation function. All of the systems below the DCD′ line have

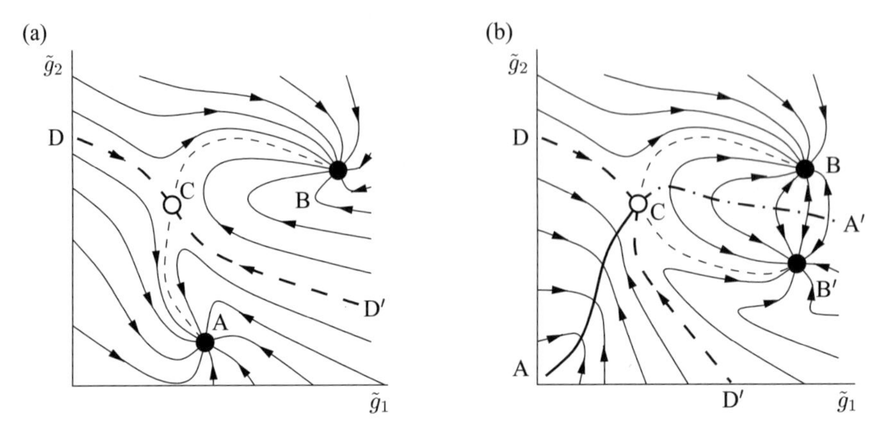

FIG. 3.11. (a) A and B are two stable fixed points representing two phases. C is an unstable fixed point with one relevant operator/direction. The transition between phase A and phase B is continuous. The critical point is described by the unstable fixed point C. (b) The fixed point/line structure of the model (3.5.3). CA is a stable fixed line. B and B′ are two stable fixed points. CA, B, and B' represent three phases. C is the critical point representing the transition between the A phase (with algebraic long-range correlations) and the B/B′ phase (with no long-range correlations). The transition is the KT transition. CA′ is an unstable fixed line, describing the transition between the B phase and the B′ phase.

the same decay exponent in the corresponding correlations. Those common properties are called universal properties. All of the systems that flow to the same fixed point form a universality class.

The systems above the DCD′ line flow to a different fixed point and form a different universality class. Those systems have different universal properties (at long distances). In particular, the decay exponents are different.

The universality classes and phases are closely related. We see that, as $(\tilde{g}_1, \tilde{g}_2)$ moves across the DCD′ line, the long-distance behavior and the long-wavelength fluctuations of the system change suddenly. As a result, the free energy of the system has a singularity at the DCD′ line. Thus, the DCD′ line is a phase transition line that separates two phases. Under this picture, we can say that the systems below the DCD′ line form one phase and the systems above the DCD′ line form the other phase. Phase and universality class mean the same thing here.

Let us start with a system exactly on the fixed point A. We add some perturbations to move the coupling constant $(\tilde{g}_1, \tilde{g}_2)$ away from $(\tilde{g}_1(A), \tilde{g}_2(A))$. As $(\tilde{g}_1, \tilde{g}_2)$ flows back to $(\tilde{g}_1(A), \tilde{g}_2(A))$, the perturbations flow to zero at long distances. Thus, the perturbations are irrelevant perturbations. As all perturbations around a stable fixed point flow to zero, *the effective theory at a stable fixed point contains no relevant or marginal perturbations*.

Now let us consider the long-distance properties of the transition point (or the critical point). If we start anywhere on the DCD$'$ line, then we can see that the system flows to the fixed point C. Thus, the long-distance behavior of the critical point is described by the unstable fixed point C. Here again, we see universality. No matter where we cross the transition line, the long-distance behavior of the transition point is always the same.

The fixed point C has one (and only one) unstable direction. A perturbation in that direction will flow away from the fixed point. Therefore, the fixed-point theory for C has one, and only one, relevant perturbation. In general, *a critical point describing a transition between two phases has one, and only one, relevant perturbation.* If an unstable fixed point has two relevant perturbations, then the fixed point will describe a tri-critical point.

The model (3.5.3) contains a marginal perturbation. Its flow diagram is more complicated. (See Fig. 3.11(b), where $(\tilde{g}_1, \tilde{g}_2)$ corresponds to $(\tilde{\eta}, \tilde{g})$.) The system has three phases. The phase below the DCD$'$ line is controlled by the stable fixed line AC. This phase has algebraic long-range correlations. The exponent of the algebraic long-range correlations depends on the position on the AC line. The phase above the DCA$'$ line is controlled by the stable fixed point B. It has no long-range correlation and is characterized by, say, $\langle\cos(\theta)\rangle < 0$. The phase to the right of the D$'$CA$'$ line is controlled by the stable fixed point B$'$. It has no long-range correlation either, and is characterized by $\langle\cos(\theta)\rangle > 0$. The transition between phase AC and phase B (or phase B$'$) is controlled by the unstable fixed point C, and the transition between phase B and phase B$'$ is controlled by the unstable fixed line CA$'$. The critical exponents depend on the position on the CA$'$ line.

From the above two simple examples, we see that we can learn a lot about the phases and phase transitions from the RG flow diagram of a system. In Section 3.3.2, we discussed phases and phase transitions from the point of view of symmetry breaking. In this section, we see that phases and phase transitions can also be understood based on an RG picture. Here, I would like to point out that the RG picture (although less concrete) is more fundamental than the symmetry-breaking picture. The symmetry-breaking picture assumes that the two stable fixed points in Fig. 3.11(a) have different symmetries and the phase transition line DCD$'$ is a symmetry-breaking transition line. This symmetry-breaking picture is not always true. We can construct explicit examples where the fixed points A and B have the same symmetry and the phase transition line DCD$'$ does not change any symmetry (Coleman and Weinberg, 1973; Halperin *et al.*, 1974; Fradkin and Shenker, 1979; Wen and Wu, 1993; Senthil *et al.*, 1999; Read and Green, 2000; Wen, 2000).

3.6 Boson superfluid to Mott insulator transition

- The Berry phase term (the total derivative term $-\rho_0\partial_t\theta$ in eqn (3.3.20)) is important in the quantum XY-model. It qualitatively changes the properties of vortices.

In $1+1$ dimensions and at zero temperature, the quantum boson superfluid is also described by an XY-model (3.3.10) at low energies. In imaginary time, the quantum XY-model is identical to the two-dimensional XY-model studied in the last section if we set $v = 1$. The vortices in the imaginary-time XY-model correspond to tunneling processes which can change the dynamics of the quantum system. Based on the results obtained in Section 3.4, it appears that when

$$\chi v < 2/\pi$$

the instantons will destroy the long-range order and make all correlations short range in both space and time directions. This means that the instantons will open up an energy gap for all excitations. The above conclusion is obviously wrong. A boson system in free space is always compressible. It at least contains a gapless mode from the density waves.

The mistake in the above argument is a tricky one. To understand the mistake, we need to first discuss the excitation spectrum above the boson superfluid. There are two types of low-energy excitations in a boson superfluid, namely local excitations that correspond to sound waves, and global excitations that correspond to the total number of bosons and the Galileo boost. Let us use a free boson system as an example to illustrate the two types of excitations. The ground state of a free boson system is given by $|k_1, ..., k_N\rangle = |0, ..., 0\rangle$. The local excitations are created by changing a few of the ks to some nonzero values. At low energies, $\sum_i |k_i| \ll \sqrt{\rho}$. The global excitations are created by adding (or removing) a few bosons to the $k = 0$ state, or shifting all of the k_i by the same amount. The latter is called the Galileo boost. For interacting boson systems, a Galileo boost can be obtained by twisting the boundary condition of the bosons from 0 to 2π or $2\pi \times$ integer, i.e. changing the constant $\varphi(x) = \varphi_0$ to $\varphi(x) = \mathrm{e}^{\mathrm{i}2\pi n x/L}\varphi_0$.

Let us consider the energy levels of an interacting N-boson system in the superfluid state. The low-energy and small-momentum excitations are given by the sound waves. The energy levels for these excitations are described in Fig. 3.12(a). The Galileo boost involves only the center of mass motion. The smallest Galileo boost corresponds to shifting all of the k_i by $2\pi/L$, where L is the linear size of the system. Such a Galileo boost gives an excitation of momentum $K_1 = 2\pi N/L = 2\pi\rho$ and energy $E_1 = K_1^2/2mN$. We see that the Galileo boost has very low energy but large momentum. Thus, the Galileo boost cannot be generated by the sound waves. A more general Galileo boost generated by a $2\pi n/L$ shift has momentum $K_n = 2\pi n N/L = 2\pi n\rho$ and energy $E_n = K_n^2/2mN$. Thus, the

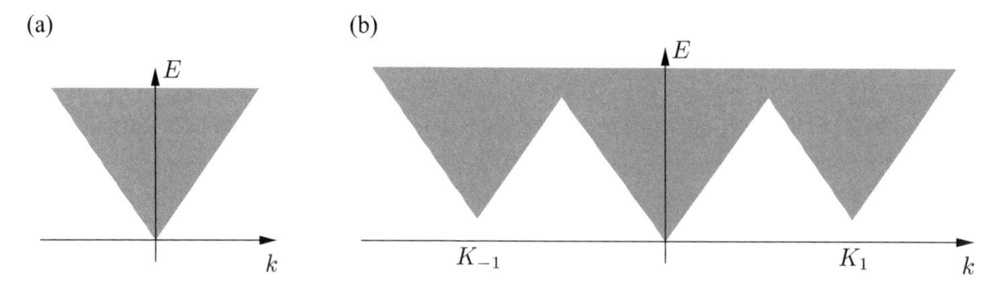

FIG. 3.12. Energy levels of $(1+1)$-dimensional interacting bosons, (a) for $k \sim 0$ and (b) for large k.

low-energy levels of an interacting N-boson system are as given in Fig. 3.12(b). The levels near $k = K_n$ are given by the nth Galileo boost plus the sound waves.

Now the question is can the low-energy XY-model for the superfluid, namely

$$L_{XY} = \frac{\chi}{2}\dot{\theta}^2 - \frac{\rho}{2m}(\partial_x \theta)^2 \tag{3.6.1}$$

reproduce the above spectrum? Let us expand

$$\theta(x,t) = \theta_0(t) + n\frac{2\pi x}{L} + \sum_{k\neq 0} \theta_k L^{-1/2} \mathrm{e}^{\,\mathrm{i}kx}$$

The second term describes the winding of $\mathrm{e}^{\,\mathrm{i}\theta(x)}$ around zero as x goes from $x = 0$ to $x = L$. The action can be rewritten as

$$S_{XY} = \frac{\chi L}{2}\dot{\theta}_0^2 - \frac{K_n^2}{2mN} + \sum_{k>0}[\chi \dot{\theta}_k^\dagger \dot{\theta}_k - \frac{\rho k^2}{2m}\theta_k^\dagger \theta_k]$$

We see that (θ_k, θ_{-k}) describes a two-dimensional oscillator which corresponds to the sound modes. From Section 3.3.7, we see that $\dot{\theta}_0$ describes the boson number fluctuations, which are ignored here. Now it is clear that the winding term $n\frac{2\pi x}{L}$ describes a Galileo boost. This is because the Galileo boost twists the boundary condition of the boson field from $\varphi(L) = \varphi(0)$ to $\varphi(L) = \mathrm{e}^{\,\mathrm{i}2\pi n}\varphi(0)$, which changes $\theta(x) = 0$ to $\theta(x) = 2\pi n x/L$. Such a relationship between the winding of the θ field and the Galileo boost is also consistent with the energy $\frac{K_n^2}{2mN}$ that the winding generates.

A vortex at (x,t) in the $(1+1)$-dimensional superfluid corresponds to an operator $O_v(x,t)$. The above analysis indicates that the operator $O_v(x,t)$ maps the states near $k = 0$ to states near $k = 2\pi\rho$. This is because the vortex changes $\theta(x) = 0$ to $\theta(x) = \frac{2\pi x}{L}$. Thus, the vortex generates a Galileo boost. In imaginary

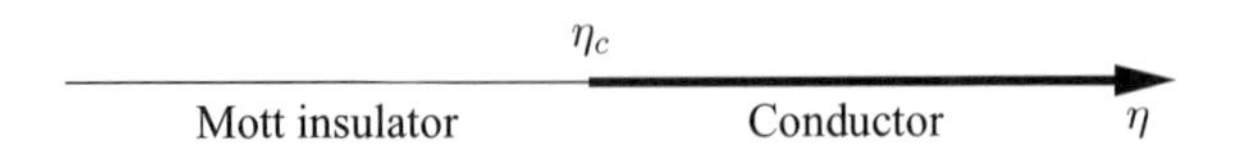

FIG. 3.13. The phase diagram of $(1 + 1)$-dimensional interacting bosons in a weak periodic potential. Here $\eta = \chi v$ and $\eta_c = 2/\pi$.

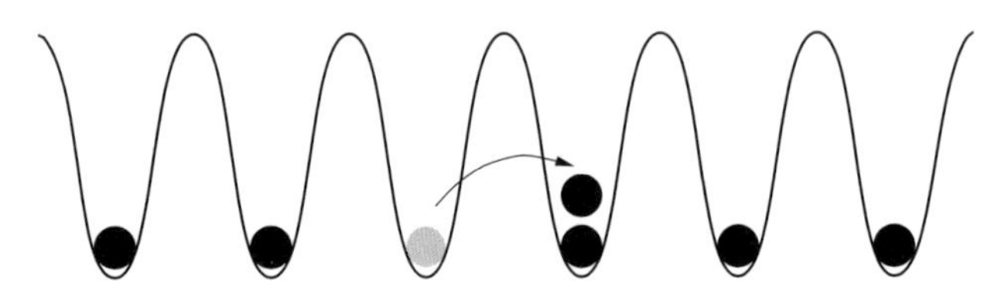

FIG. 3.14. Mott insulator of $(1 + 1)$-dimensional interacting bosons in a strong periodic potential.

time, the partition function with vortices should have the form

$$Z = Z_0 \sum_n \frac{1}{n!n!} \int \prod_{i=1}^{2n} \mathrm{d}^2 \boldsymbol{r}_i \; K^{2n} \mathrm{e}^{\mathrm{i}2\pi\rho \sum_j q_j x_j} \mathrm{e}^{\sum_{i<j}^{2n} q_i q_j 2\pi\eta \ln \frac{|x_i - x_j|}{l}}$$

where Z_0 is the partition function with no vortex. The additional phase term $\mathrm{e}^{\mathrm{i}2\pi\rho \sum_j q_j x_j}$ reflects the large momentum carried by the vortices. Due to the phase term, vortices and anti-vortices have to move in pairs to cancel the phase, in order to have a large contribution. Thus, the phase term confines the vortices and anti-vortices, and the 'Coulomb' gas has no plasma phase. Therefore, the boson system is in the superfluid phase for any values of χ and v, even after we include the vortices.

However, if we add a weak periodic potential whose period is equal to the average boson separation, $a = 1/\rho$, then the phase term can interfere with the periodic potential and on average becomes a constant. Now we can have the KT transition at the critical value $\chi v = 2/\pi$. The phase diagram of $(1 + 1)$-dimensional interacting bosons in a weak periodic potential is plotted in Fig. 3.13. For $\chi v > 2/\pi$, the bosons form a conducting state with algebraic long-range order and gapless sound modes. For $\chi v < 2/\pi$, all excitations will have a finite energy gap and the bosons form an insulator, called the Mott insulator, because the insulating property is caused by interactions rather than energy bands. The Mott insulator in the large-potential limit is easy to understand. In that limit, the ground state has one boson per potential well. Moving a boson to another well causes a finite energy due to the repulsion between bosons (see Fig. 3.14).

The above understanding of vortices also allows us to calculate density correlations at low energies and *large* momenta. Let us consider the spectral expansion of the density correlation function:

$$\begin{aligned}\langle \rho(t,x)\rho(0,0)\rangle &= \frac{\langle 0|\rho(x)U(t,0)\rho(0)|0\rangle}{\langle 0|U(t,0)|0\rangle} \\ &= \sum_{n,k} \langle 0|\rho(x)|n,k\rangle\langle n,k|\rho(0)|0\rangle\, \mathrm{e}^{\,\mathrm{i}kx-\mathrm{i}\epsilon_{n,k}t}\end{aligned}$$

where the state $|n,k\rangle$ has energy $\epsilon_{n,k}$ and momentum k. We have also assumed that the ground state $|0\rangle$ has zero energy. As the low-energy states only appear near K_i, at low energy we may regroup the above summation as follows:

$$\langle \rho(t,x)\rho(0,0)\rangle = \sum_{n,|\delta k|\ll K_1,i} \langle 0|\rho(x)|n,\delta k,i\rangle\langle n,\delta k,i|\rho(0)|0\rangle\, \mathrm{e}^{\,\mathrm{i}(\delta k+K_i)x-\mathrm{i}\epsilon_{n,\delta k,i}t}$$

where the state $|n,\delta k,i\rangle$ has energy $\epsilon_{n,\delta k,i}$ and momentum $\delta k+K_i$. We see that the density correlation at large momentum, say for $k\sim K_i$, is generated by the matrix elements $\langle n,\delta k,i|\rho(0)|0\rangle$. We know that the vortex operator O_v^i ($O_v^i \equiv (O_v^\dagger)^{-i}$ if $i<0$) maps the low-lying states near $k=0$ to low-lying states near $k=K_i$. So here we make the following bold assumption:

$$\langle n,\delta k,i|\rho(t,x)|0\rangle = C_i\langle n,\delta k,i|O_v^i(t,x)|0\rangle$$

The low-energy density operator can now be generalized as

$$\rho(t,x) = \rho_0 - \chi\dot{\theta}(t,x) + \sum_n C_n O_v^n$$

The old result $\rho(t,x) = \rho_0 - \chi\dot{\theta}(t,x)$ only describes the low-energy long-wavelength density fluctuations. The new result describes the low-energy all-wavelength density fluctuations. We find that the correlation function has the form

$$\begin{aligned}&\langle \rho(t,x)\rho(0,0)\rangle \qquad (3.6.2)\\ &= \rho_0^2 - \frac{\chi v}{4\pi}\left(\frac{1}{(x-vt)^2}+\frac{1}{(x+vt)^2}\right) + \sum_n |C_n|^2 \mathrm{e}^{\,\mathrm{i}K_n x}\left(\frac{l^2}{x^2-v^2t^2}\right)^{n^2\pi\chi v} \\ &= \rho_0^2 - \left(\frac{\chi v/4\pi}{(x-vt)^2}+\frac{\chi v/4\pi}{(x+vt)^2}\right) + \sum_n |C_n|^2 \mathrm{e}^{\,\mathrm{i}K_n x}\left(\frac{l^2\,\mathrm{e}^{-\mathrm{i}\pi\Theta(v^2t^2-x^2)}}{|x^2-v^2t^2|}\right)^{n^2\pi\chi v}\end{aligned}$$

where l is a short-distance cut-off. Note that the correlation $\left\langle O_v^\dagger(x,t)O_v(0,0)\right\rangle$ in imaginary time is given by e^{-V}, where V is the potential between the vortex/anti-vortex pair. Then, the real-time correlation is obtained by analytic continuation.

Problem 3.6.1.
Prove eqn (3.6.2).

Problem 3.6.2.
Susceptibility at finite momenta.

1. Assume that the bosons see a weak potential $V(x)$. The potential induces a change in density $\delta\rho$. Express the finite-momentum susceptibility $\chi(k)$ in $\delta\rho_k = \chi(k)V_k$ in terms of a density correlation function in frequency–momentum space.
2. Determine for which values of χ and v the susceptibility $\chi(K_n)$ diverges. (Hint: You may want to do the calculation in imaginary time first.)
3. If we turn on a weak periodic potential with a period $a = \rho^{-1}/n$ (n is an integer), then for which values of χ and v do the bosons form a Mott insulator?

3.7 Superfluidity and superconductivity

3.7.1 Coupling to a gauge field and conserved current

- A theory with global $U(1)$ symmetry contains a conserved charge.
- A theory with global $U(1)$ symmetry can be coupled to a $U(1)$ gauge field. The electromagnetic vector potential is a $U(1)$ gauge field.

A charged boson system couples to an electromagnetic gauge field. In the presence of a nonzero electromagnetic field, the Lagrangian for a charged boson system needs to be modified. The question here is how do we find the modified Lagrangian? We know that the boson Lagrangian

$$L(\varphi) = \mathrm{i}\frac{1}{2}(\varphi^*\partial_t\varphi - \varphi\partial_t\varphi^*) - \frac{1}{2m}\partial_{\boldsymbol{x}}\varphi^*\partial_{\boldsymbol{x}}\varphi + \mu|\varphi|^2 - \frac{V_0}{2}|\varphi|^4 \tag{3.7.1}$$

is invariant under a global $U(1)$ transformation

$$\varphi \to \mathrm{e}^{\mathrm{i}f}\varphi$$

i.e. that $L(\mathrm{e}^{\mathrm{i}f}\varphi) = L(\varphi)$. However, it is not invariant under a local $U(1)$ transformation

$$\varphi(\boldsymbol{x},t) \to \mathrm{e}^{\mathrm{i}f(\boldsymbol{x},t)}\varphi(\boldsymbol{x},t)$$

We have

$$\begin{aligned} L(\mathrm{e}^{\mathrm{i}f(\boldsymbol{x},t)}\varphi) =& \mathrm{i}\frac{1}{2}(\varphi^*(\partial_0 + \mathrm{i}\partial_0 f)\varphi - \varphi(\partial_0 - \mathrm{i}\partial_0 f)\varphi^*) \\ &- \frac{1}{2m}|(\partial_i + \mathrm{i}\partial_i f)\varphi|^2 + \mu|\varphi|^2 - \frac{V_0}{2}|\varphi|^4 \end{aligned} \tag{3.7.2}$$

where the subscript 0 indicates the time direction and the subscript $i = 1, ..., d$ represents the spatial directions. We will use the Greek letters μ, ν, etc. to represent space–time directions. For example, x^μ represents the space–time coordinates. The coupling between the bosons and the electromagnetic gauge field can now be obtained by simply replacing $\partial_\mu f$ by A_μ:

$$L(\varphi, A_\mu) = \mathrm{i}\frac{1}{2}(\varphi^*(\partial_0 + \mathrm{i}A_0)\varphi - \varphi(\partial_0 - \mathrm{i}A_0)\varphi^*)$$
$$- \frac{1}{2m}|(\partial_i + \mathrm{i}A_i)\varphi|^2 + \mu|\varphi|^2 - \frac{V_0}{2}|\varphi|^4 \tag{3.7.3}$$

The resulting Lagrangian has a nice property in that it is gauge invariant:

$$\varphi \to \tilde{\varphi} = \mathrm{e}^{\mathrm{i}f(\boldsymbol{x},t)}\varphi$$
$$A_\mu \to \tilde{A}_\mu = A_\mu - \partial_\mu f$$
$$L(\varphi, A_\mu) \to L(\tilde{\varphi}, \tilde{A}_\mu) = L(\varphi, A_\mu) \tag{3.7.4}$$

The above constructions do not contain much physics. It is simply a trick to obtain a gauge-invariant Lagrangian. Such a trick generates the simplest Lagrangian and is called minimal coupling. We can also write down a different gauge-invariant Lagrangian by, say, replacing $\partial_\mu f$ by $A_\mu + g\partial_\nu F_{\nu\mu}$, where $F_{\mu\nu} = \partial_\mu A_\nu - \partial_\nu A_\mu$ are the field strengths of A_μ. Note that $F_{\mu\nu}$ is invariant under the gauge transformations.

Equation (3.7.3) only describes the coupling between the charge field φ and the gauge field A_μ. The complete gauge-invariant Lagrangian that describes the dynamics of both φ and A_μ is given by

$$L(\varphi, A_\mu) = \mathrm{i}\frac{1}{2}\left(\varphi^*(\partial_0 + \mathrm{i}A_0)\varphi - \varphi(\partial_0 - \mathrm{i}A_0)\varphi^*\right) - \frac{1}{2m}|(\partial_i + \mathrm{i}A_i)\varphi|^2$$
$$+ \mu|\varphi|^2 - \frac{V_0}{2}|\varphi|^4 + \frac{1}{8\pi e^2}\left(\frac{1}{c}\boldsymbol{E}^2 - c\boldsymbol{B}^2\right) \tag{3.7.5}$$

where c is the speed of light and

$$E_i = \partial_0 A_i - \partial_i A_0 = F_{0i}, \qquad B_i = \epsilon_{ijk}\partial_j A_k = \frac{1}{2}\epsilon_{ijk}F_{jk}$$

are the electric field and the magnetic field, respectively, of the $U(1)$ gauge theory.

We know that the model with a global $U(1)$ symmetry has a conserved charge. We can show this easily with the help of the gauge field. The gauged action is gauge invariant:

$$S(\varphi, A_\mu) = S(\mathrm{e}^{\mathrm{i}f(\boldsymbol{x},t)}\varphi, A_\mu - \partial_\mu f)$$

Let $\varphi_c(\boldsymbol{x}, t)$ be a solution of the classical equation of motion. Then

$$S(\mathrm{e}^{\mathrm{i}f(\boldsymbol{x},t)}\varphi_c, A_\mu) = S(\varphi_c, A_\mu) + O(f^2)$$

Therefore

$$S(\varphi_c, A_\mu) = S(\varphi_c, A_\mu - \partial_\mu f) + O(f^2)$$
$$= S(\varphi_c, A_\mu) + \int \mathrm{d}^d\boldsymbol{x}dt\, \partial_\mu f J^\mu(\varphi_c, A_\mu) + O(f^2)$$

where J^μ is the current

$$J^\mu(\varphi, A_\mu) \equiv -\partial_{A_\mu} L(\varphi, A_\mu)$$

We see that, if $\varphi(\boldsymbol{x}, t)$ satisfies the classical equation of motion, then $\int \mathrm{d}^d\boldsymbol{x}\,\mathrm{d}t\, \partial_\mu f J^\mu(\varphi_c, A_\mu) = 0$ for any f and the current $J^\mu(\varphi_c, A_\mu)$ is conserved:

$$\partial_\mu J^\mu(\varphi_c, A_\mu) = \partial_t \rho + \partial_i J^i = 0$$

where $\rho = J^0$ is the density and J^i is the current. Certainly, the above result is also true for a zero A_μ field and we have $\partial_\mu J^\mu(\varphi_c) = 0$, which is the current conservation for neutral systems.

For our boson system, we find that the conserved current is given by

$$J^0 = \rho = \varphi^*\varphi$$
$$J^i = -\frac{\mathrm{i}}{2m}[\varphi^*(\partial_i\varphi) - (\partial_i\varphi^*)\varphi] + A_i|\varphi|^2$$

It is interesting to see that the current in our boson system depends on the gauge potential.

Problem 3.7.1.
Consider a lattice boson system coupled to a gauge field:

$$L = \mathrm{i}\frac{1}{2}\sum_i \left(\varphi_i^*(\partial_0 + \mathrm{i}A_0(i))\varphi_i - \varphi_i(\partial_0 - \mathrm{i}A_0(i))\varphi_i^*\right) + \sum_{\langle ij\rangle}(t_{ij}\varphi_j^*\varphi_i \mathrm{e}^{-\mathrm{i}a_{ij}} + h.c.)$$

where the summation is over all pairs $\langle ij\rangle$, and a_{ij} is the gauge field $a_{ij} = \int_i^j d\boldsymbol{x}\cdot A$.

1. Show that L is invariant under a lattice gauge transformation
$$\varphi_i \to \mathrm{e}^{\mathrm{i}f_i(t)}\varphi_i, \qquad A_0(i) \to A_0(i) - \partial_0 f_i, \qquad a_{ij} \to a_{ij} - f_j + f_i$$
2. Find the equation of motion for $\varphi_i(t)$.
3. Calculate the time derivative of the density, namely $\partial_0\varphi_i^*\varphi_i$. Show that one can introduce a current J_{ij} defined on the links and recover the lattice current conservation relation
$$\partial_0\varphi_i^*\varphi_i + \sum_j J_{ij} = 0$$
4. Show that J_{ij} can be expressed as $-\partial L/\partial a_{ij}$.

3.7.2 Current correlation functions and electromagnetic responses

- Current conservation puts constraints on current correlations. Many important physical quantities, such as compressibility and conductivity, are determined by current correlations.
- To obtain the correct responses, it is important to take the $\boldsymbol{k} \to 0$ and $\omega \to 0$ limits in the right order.

Now let us calculate the response of the system to an external gauge potential. We would like to see how much current J^μ the gauge potential A_μ can generate. Introducing j^μ according to

$$j^0 = \rho, \qquad j^i = -\frac{\mathrm{i}}{2m}[\varphi^*(\partial_i\varphi) - (\partial_i\varphi^*)\varphi]$$

we see that the Lagrangian has the form

$$L(\varphi, A_\mu) = L(\varphi) - A_0 j^0 - A_i j^i - \frac{1}{2m}(A^i)^2\rho$$

Using the linear response theory to calculate $\langle j^\mu(\boldsymbol{x}, t)\rangle$ to leading order in A_μ, we obtain the current $J^\mu \equiv -\partial_{A_\mu} L$

$$\langle J^\mu(\boldsymbol{x}, t)\rangle = \langle j^\mu(\boldsymbol{x}, t)\rangle + (1 - \delta^{\mu 0})A^\mu\rho = \int \mathrm{d}^d\boldsymbol{x}\,\mathrm{d}t\; \Pi^{\mu\nu}(\boldsymbol{x}, t; \boldsymbol{x}', t')A_\nu(\boldsymbol{x}', t')$$

where the response function is given by

$$\Pi^{00}(\boldsymbol{x}, t; \boldsymbol{x}', t') = -\mathrm{i}\Theta(t - t')\left\langle[\rho(\boldsymbol{x}, t), \rho(\boldsymbol{x}', t')]\right\rangle$$

$$\Pi^{0i}(\boldsymbol{x}, t; \boldsymbol{x}', t') = -\mathrm{i}\Theta(t - t')\left\langle[\rho(\boldsymbol{x}, t), j^i(\boldsymbol{x}', t')]\right\rangle$$

$$\Pi^{i0}(\boldsymbol{x}, t; \boldsymbol{x}', t') = -\mathrm{i}\Theta(t - t')\left\langle[j^i(\boldsymbol{x}, t), \rho(\boldsymbol{x}', t')]\right\rangle$$

$$\Pi^{ij}(\boldsymbol{x}, t; \boldsymbol{x}', t') = -\mathrm{i}\Theta(t - t')\left\langle[j^i(\boldsymbol{x}, t), j^j(\boldsymbol{x}', t')]\right\rangle + \delta^{ij}\delta(\boldsymbol{x} - \boldsymbol{x}')\delta(t - t')\frac{\langle\rho\rangle}{m}$$

Due to the A_i dependence of the current, we have an extra contact term $\delta^{ij}\delta(\boldsymbol{x} - \boldsymbol{x}')\delta(t - t')\frac{\langle\rho\rangle}{m}$. If we introduce the correlation function

$$\pi^{\mu\nu}(\boldsymbol{x}, t; \boldsymbol{x}', t') = -\mathrm{i}\Theta(t - t')\left\langle[j^\mu(\boldsymbol{x}, t), j^\nu(\boldsymbol{x}', t')]\right\rangle$$

then

$$\Pi^{\mu\nu} = \pi^{\mu\nu} + \delta^{\mu\nu}(1 - \delta^{0\mu})(1 - \delta^{0\nu})\delta(\boldsymbol{x} - \boldsymbol{x}')\delta(t - t')\langle\rho\rangle \tag{3.7.6}$$

The above result applies for both zero and finite temperatures. We note that

$$\left(\Pi^{\mu\nu}(\boldsymbol{x}, t; \boldsymbol{x}', t')\right)^* = \Pi^{\nu\mu}(\boldsymbol{x}', t'; \boldsymbol{x}, t)$$

or, in the ω–k space,

$$\left(\Pi^{\mu\nu}_{(k_\lambda)}\right)^* = \Pi^{\nu\mu}_{(-k_\lambda)}$$

where $k_0 = \omega$.

Due to current conservation, the components of $\Pi^{\mu\nu}$ are not all independent. We have, in the ω–k space,

$$k_\mu \Pi^{\mu\nu}_{k_\lambda} = 0$$

Thus, the full $\Pi^{\mu\nu}$ can be determined from Π^{ij} as follows:

$$\begin{aligned}
\Pi^{0i}_{k_\lambda} &= (\Pi^{i0}_{(-k_\lambda)})^\dagger = -\frac{k_j}{\omega}\Pi^{ji}_{(k_\lambda)} \\
\Pi^{00}_{(k_\lambda)} &= -\frac{k_j}{\omega}\Pi^{0j}_{(k_\lambda)} = \frac{k_i k_j}{\omega^2}\Pi^{ij}_{(k_\lambda)}
\end{aligned} \tag{3.7.7}$$

For a rotationally invariant system, we can further decompose Π^{ij} and π^{ij} into a longitudinal component $\Pi^{||}_{(k_\lambda)}$ and a transverse component $\Pi^{\perp}_{(k_\lambda)}$ as follows:

$$\begin{aligned}
\Pi^{ij}_{(k_\lambda)} &= \frac{k_i k_j}{\boldsymbol{k}^2}\Pi^{||}_{(k_\lambda)} + (\delta_{ij} - \frac{k_i k_j}{\boldsymbol{k}^2})\Pi^{\perp}_{(k_\lambda)} \\
\pi^{ij}_{(k_\lambda)} &= \frac{k_i k_j}{\boldsymbol{k}^2}\pi^{||}_{(k_\lambda)} + (\delta_{ij} - \frac{k_i k_j}{\boldsymbol{k}^2})\pi^{\perp}_{(k_\lambda)}
\end{aligned} \tag{3.7.8}$$

Equation (3.7.7) can now be written as

$$\Pi^{0i}_{(k_\lambda)} = -k_i \Pi^{||}_{(k_\lambda)}, \qquad \Pi^{00}_{(k_\lambda)} = \frac{\boldsymbol{k}^2}{\omega^2}\Pi^{||}_{(k_\lambda)}$$

and eqn (3.7.6) as

$$\Pi^{||}_{(k_\lambda)} = \pi^{||}_{(k_\lambda)} + \frac{\langle\rho\rangle}{m}, \qquad \Pi^{\perp}_{(k_\lambda)} = \pi^{\perp}_{(k_\lambda)} + \frac{\langle\rho\rangle}{m}, \tag{3.7.9}$$

The response function $\Pi^{\mu\nu}$ is related to many important physical quantities. Let us consider a metal as an example. In the $\omega \to 0$ limit, we have

$$\delta\rho(\boldsymbol{k}) = \Pi^{00}_{(0,\boldsymbol{k})} A_0(\boldsymbol{k})$$

Note that $A_0(\boldsymbol{x})$ is just the external potential and $-\Pi^{00}_{(0,\boldsymbol{k})}$ is simply the compressibility (at wave vector $\boldsymbol{k}$):

$$\chi(\boldsymbol{k}) = -\Pi^{00}_{(0,\boldsymbol{k})}$$

Usually, we expect $\Pi^{00}_{(0,\boldsymbol{k})}$ to be finite for all $\boldsymbol{k}$. Thus, for small ω and in the $\omega \ll \boldsymbol{k}$ limit (note here that we let $\omega \to 0$ first), we have

$$\Pi^{||}_{(k_\lambda)} = -\chi(\boldsymbol{k})\frac{\omega^2}{\boldsymbol{k}^2}$$

From eqn (3.7.9), we see that, in the $\omega \to 0$ limit, $\pi^{||}_{(k_\lambda)}$ must exactly cancel $\frac{\langle\rho\rangle}{m}$ in order for $\Pi^{||}_{(k_\lambda)}$ to vanish like ω^2.

Now let $\boldsymbol{k}$ tend to zero first. In the $|\boldsymbol{k}| \ll \omega$ limit, $-\mathrm{i}\omega A_{i,(\omega,\boldsymbol{k})}$ is an almost uniform electric field, which is expected to generate a current whose direction is given by A_i and whose wave vector is given by $\boldsymbol{k}$:

$$J^i(\omega) = \lim_{\boldsymbol{k}\to 0} \frac{\Pi^{ij}}{-\mathrm{i}\omega}(-\mathrm{i}\omega)A_{j,(\omega,\boldsymbol{k})}$$

From eqn (3.7.8), we see that the limit $\boldsymbol{k} \to 0$ exists only when $\Pi^{||}_{(k_\lambda)} = \Pi^{\perp}_{(k_\lambda)}$ in the limit $\omega \gg |\boldsymbol{k}|$. Suppose that this is the case; then we obtain the conductivity

$$\sigma(\omega) = \frac{\Pi^{||}_{(\omega,0)}}{-\mathrm{i}\omega} = \frac{\Pi^{\perp}_{(\omega,0)}}{-\mathrm{i}\omega}$$

The real part of the conductivity,

$$\mathrm{Re}\sigma(\omega) = -\mathrm{Im}\frac{\Pi^{||}_{(\omega,0)}}{\omega} = -\mathrm{Im}\frac{\Pi^{\perp}_{(\omega,0)}}{\omega}$$

corresponds to dissipation. The imaginary part of the conductivity gives us the dielectric constant

$$\epsilon(\omega) = -\frac{\mathrm{Im}\sigma(\omega)}{\omega} = \mathrm{Re}\frac{\Pi^{||}_{(\omega,0)}}{\omega^2} = \mathrm{Re}\frac{\Pi^{\perp}_{(\omega,0)}}{\omega^2}$$

Thus, in the $\omega \gg |\boldsymbol{k}|$ limit, we may write

$$\Pi^{||}_{(\omega,0)} = \Pi^{\perp}_{(\omega,0)} = \mathrm{i}\omega\mathrm{Re}\sigma(\omega) + \omega^2\epsilon(\omega)$$

We note that the polarization vector $\boldsymbol{P}$ satisfies $\partial_{\boldsymbol{x}} \cdot \boldsymbol{P} = -\delta\rho$ or (assuming that $A_i = 0$)

$$\mathrm{i}k_i P^i = -\delta\rho = -\Pi^{00}A_0 = -\frac{\boldsymbol{k}^2}{\omega^2}\Pi^{||}_{(k_\lambda)}A_0 = \mathrm{i}\frac{\Pi^{||}_{(k_\lambda)}}{\omega^2}k_i E_i$$

Therefore

$$P^i = \frac{\Pi^{||}_{(k_\lambda)}}{\omega^2}E_i$$

and we see again that $\Pi^{||}_{(k_\lambda)}/\omega^2$ is the dielectric constant.

We have considered the $\omega \gg |\boldsymbol{k}|$ limit of $\Pi^{||}$ and $\Pi^{\perp}$, as well as the $\omega \ll |\boldsymbol{k}|$ limit of $\Pi^{||}$. What about the $\omega \ll |\boldsymbol{k}|$ limit of $\Pi^{\perp}$? One natural guess is that it corresponds to the magnetic susceptibility. The magnetic moment density $\boldsymbol{M}$ satisfies $\partial_{\boldsymbol{x}} \times \boldsymbol{M} = -\boldsymbol{j}$. Thus (assuming that $A_0 = 0$), we have

$$\begin{aligned} \mathrm{i}\epsilon^{ijk} k_j M_k &= -j^i = -\Pi^{ij} A_j = -(\delta^{ij}\boldsymbol{k}^2 - k^i k^j) A_j \frac{\Pi^{\perp}_{(k_\lambda)}}{\boldsymbol{k}^2} \\ &= +\epsilon^{ij'k'} k_{j'} \epsilon^{k'i'j} k_{i'} A_j \frac{\Pi^{\perp}_{(k_\lambda)}}{\boldsymbol{k}^2} = -\mathrm{i}\epsilon^{ijk} k_j B_k \frac{\Pi^{\perp}_{(k_\lambda)}}{\boldsymbol{k}^2} \end{aligned}$$

Therefore

$$M_i = -\frac{\Pi^{\perp}_{(k_\lambda)}}{\boldsymbol{k}^2} B_i$$

and $-\Pi^{\perp}_{(k_\lambda)}/\boldsymbol{k}^2$ is the magnetic susceptibility.

Now let us calculate $\Pi^{\mu\nu}$ for the boson superfluid. We start with the boson Lagrangian with the gauge field (3.7.3). We integrate out the amplitude fluctuations and obtain an XY-model with a gauge field. To quadratic order in $(\partial_\mu \theta, A_\mu)$, we have

$$L = \frac{\chi}{2}\left((\partial_0\theta + A_0)^2 - v^2(\partial_i\theta + A_i)^2\right) \tag{3.7.10}$$

We see that

$$j^0 = -\chi\partial_0\theta, \qquad j^i = \chi v^2 \partial_i \theta$$

Therefore

$$\begin{aligned} \pi^{00} &= \chi^2(-\mathrm{i}\omega)(\mathrm{i}\omega)\left\langle \theta_{(\omega,\boldsymbol{k})}\theta_{(-\omega,-\boldsymbol{k})}\right\rangle = \chi\frac{\omega^2}{\omega^2 - v^2\boldsymbol{k}^2 + \mathrm{i}0^+\mathrm{sgn}(\omega)} \\ \pi^{0i} &= \pi^{i0} = \chi^2(-\mathrm{i}\omega)(-\mathrm{i}k_i)\left\langle \theta_{(\omega,\boldsymbol{k})}\theta_{(-\omega,-\boldsymbol{k})}\right\rangle = -\chi v^2\frac{\omega k_i}{\omega^2 - v^2\boldsymbol{k}^2 + \mathrm{i}0^+\mathrm{sgn}(\omega)} \\ \pi^{ij} &= \chi^2(\mathrm{i}k_i)(-\mathrm{i}k_j)\left\langle \theta_{(\omega,\boldsymbol{k})}\theta_{(-\omega,-\boldsymbol{k})}\right\rangle = \chi v^4\frac{k_i k_j}{\omega^2 - v^2\boldsymbol{k}^2 + \mathrm{i}0^+\mathrm{sgn}(\omega)} \end{aligned}$$

Note that the choice $\mathrm{i}0^+\mathrm{sgn}(\omega)$ gives us the response function. The total $\Pi^{\mu\nu}$ is given by

$$\begin{aligned} \Pi^{00} &= \chi\left(\frac{\omega^2}{\omega^2 - v^2\boldsymbol{k}^2 + \mathrm{i}0^+\mathrm{sgn}(\omega)} - 1\right) \\ \Pi^{0i} &= \Pi^{i0} = -\chi v^2\frac{\omega k_i}{\omega^2 - v^2\boldsymbol{k}^2 + \mathrm{i}0^+\mathrm{sgn}(\omega)} \\ \Pi^{ij} &= \chi v^2\left(\delta_{ij} + v^2\frac{k_i k_j}{\omega^2 - v^2\boldsymbol{k}^2 + \mathrm{i}0^+\mathrm{sgn}(\omega)}\right) \end{aligned}$$

and

$$\Pi^{||} = \chi v^2 \left(\frac{v^2 k^2}{\omega^2 - v^2 \boldsymbol{k}^2 + \mathrm{i}0^+ \mathrm{sgn}(\omega)} + 1 \right)$$
$$\Pi^{\perp} = \chi v^2$$

The compressibility $-\Pi^{00}_{(0,\boldsymbol{k})}$ is finite and equal to χ. The magnetic susceptibility $-\Pi^{\perp}/c\boldsymbol{k}^2$ diverges as $\boldsymbol{k} \to 0$. The real part of the conductivity

$$\mathrm{Re}\sigma(\omega) = -\mathrm{Im}\frac{\Pi^{||}_{(\omega,0)}}{\omega} = \mathrm{Im}\frac{\chi v^2}{\omega + \mathrm{i}0^+} = \pi\frac{\rho}{m}\delta(\omega)$$

is zero for finite frequency.

If we choose the Coulomb gauge $\partial_{\boldsymbol{x}}\boldsymbol{A} = 0$, then, from $J^i = \Pi^{ij}A_j$, we find a simple relationship between the current and the *gauge potential*:

$$\boldsymbol{J} = \Pi^{\perp}\boldsymbol{A} = \frac{\rho}{m}\boldsymbol{A} \tag{3.7.11}$$

This is the famous London equation. It is responsible for many novel properties of superconductors, such as the persistent current, the Meisner effect, etc.

3.7.3 Superfluidity and finite-temperature effects

- Excitations reduce superfluid flow, but cannot kill it. This leads to superfluidity.
- Superfluid flow can only be killed by tunneling of vortex.
- The critical velocity of the superfluid flow.

Finally, we are ready to discuss the superfluid property of the symmetry-breaking phase of interacting bosons (Landau, 1941). First, we consider a boson system in free space. The boson system is invariant under a Galileo transformation. Let us assume that the excitations above the symmetry-breaking ground state have a spectrum $\epsilon(\boldsymbol{k})$ and that we ignore the interactions between excitations. The latter assumption is valid at low temperatures when the excitations are dilute.

Consider a single excitation $(\epsilon, \boldsymbol{k})$ above the ground state (at rest). The total energy and the momentum of thc system are $E_{\text{ground}} + \epsilon$ and $\boldsymbol{P} = \boldsymbol{k}$, respectively. If we boost the system by the velocity $\boldsymbol{v}$, then the total energy and the momentum of the system will be $E = E_{\text{ground}} + \epsilon + \frac{1}{2}Nm\boldsymbol{v}^2 + \boldsymbol{v}\cdot\boldsymbol{k}$ and $\boldsymbol{P} = \boldsymbol{k} + Nm\boldsymbol{v}$, respectively. Compared to the energy and the momentum of the boosted ground state, namely $E = E_{\text{ground}} + \frac{1}{2}Nm\boldsymbol{v}^2$ and $\boldsymbol{P} = Nm\boldsymbol{v}$, respectively, we see that the excitations above the boosted ground state have a new spectrum

$$\epsilon_{\boldsymbol{v}}(\boldsymbol{k}) = \epsilon(\boldsymbol{k}) + \boldsymbol{v} \cdot \boldsymbol{k}$$

In the boosted superfluid, the occupation number of the excitations at momentum $\boldsymbol{k}$ is given by $n_B(\epsilon(\boldsymbol{k}))$, where $n_B(\epsilon) = 1/(\mathrm{e}^{\epsilon/T} - 1)$. The energy and the momentum, respectively, for the boosted superfluid can be written as

$$\tilde{E} = E_{\text{ground}} + \sum_{\boldsymbol{k}} \epsilon_{\boldsymbol{v}}(\boldsymbol{k}) n_B(\epsilon(\boldsymbol{k})) + \frac{1}{2} N m \boldsymbol{v}^2$$

$$\tilde{\boldsymbol{P}} = \sum_{\boldsymbol{k}} \boldsymbol{k} n_B(\epsilon(\boldsymbol{k})) + N m \boldsymbol{v} = N m \boldsymbol{v}$$

In the equilibrium state, the excitations should have occupation numbers $n_B(\epsilon_{\boldsymbol{v}}(\boldsymbol{k}))$. We see that the boosted superfluid is not in the equilibrium state. If we let the system relax into the equilibrium state in the laboratory frame by redistributing the excitations, then the energy and the momentum will change to

$$E_{\boldsymbol{v}} = E_{\text{ground}} + \sum_{\boldsymbol{k}} \epsilon_{\boldsymbol{v}}(\boldsymbol{k}) n_B(\epsilon_{\boldsymbol{v}}(\boldsymbol{k})) + \frac{1}{2} N m \boldsymbol{v}^2$$

$$\boldsymbol{P}_{\boldsymbol{v}} = \sum_{\boldsymbol{k}} \boldsymbol{k} n_B(\epsilon_{\boldsymbol{v}}(\boldsymbol{k})) + N m \boldsymbol{v}$$

respectively. For small $\boldsymbol{v}$, the above can be rewritten as[24]

$$E_{\boldsymbol{v}} = E_0 + \frac{1}{2}[Nm - \mathcal{V}(\rho_n m + \delta\rho_n m)]\boldsymbol{v}^2$$

$$\boldsymbol{P}_{\boldsymbol{v}} = (Nm - \mathcal{V}\rho_n m)\boldsymbol{v} \tag{3.7.12}$$

where

$$\rho_n = -\frac{1}{md} \int \frac{\mathrm{d}^d \boldsymbol{k}}{(2\pi)^d} \, \boldsymbol{k}^2 n_B'(\epsilon(\boldsymbol{k})) \tag{3.7.13}$$

$$\delta\rho_n = -\frac{1}{md} \int \frac{\mathrm{d}^d \boldsymbol{k}}{(2\pi)^d} \, \boldsymbol{k}^2 \frac{\mathrm{d}}{\mathrm{d}\epsilon(\boldsymbol{k})} \left(\epsilon(\boldsymbol{k}) n_B'(\epsilon(\boldsymbol{k}))\right) \tag{3.7.14}$$

If $\epsilon(\boldsymbol{k}) \propto |\boldsymbol{k}|$ for small $\boldsymbol{k}$, then $\rho_n \propto T^{d+1}$ for small T.

---

[24] The result for $\boldsymbol{P}_{\boldsymbol{v}}$ is easy to see. To obtain $E_{\boldsymbol{v}}$, we note that

$$E_{\boldsymbol{v}} = E_{\text{ground}} + \sum_{\boldsymbol{k}} \epsilon(\boldsymbol{k}) n_B(\epsilon(\boldsymbol{k})) + \sum_{\boldsymbol{k}} (\boldsymbol{v} \cdot \boldsymbol{k})^2 [n_B'(\epsilon(\boldsymbol{k})) + \frac{1}{2}\epsilon(\boldsymbol{k}) n_B''(\epsilon(\boldsymbol{k}))] + \frac{1}{2} N m \boldsymbol{v}^2$$

$$= E_0 - \frac{1}{2}\mathcal{V} m \rho_n \boldsymbol{v}^2 + \frac{1}{2d}\boldsymbol{v}^2 \sum_{\boldsymbol{k}} \boldsymbol{k}^2 \frac{\mathrm{d}}{\mathrm{d}\epsilon(\boldsymbol{k})} \left[\epsilon(\boldsymbol{k}) n_B'(\epsilon(\boldsymbol{k}))\right] + \frac{1}{2} N m \boldsymbol{v}^2$$

Here we see a striking property of the symmetry-breaking phase: after we let the excitations relax into an equilibrium state, the total momentum $\boldsymbol{P}_{\boldsymbol{v}}$ of the system does not relax to zero. The bosons keep moving forever in the equilibrium state. It is this property that gives the symmetry-breaking phase the name superfluid. To gain a better understanding of this phenomenon, we recall that in the symmetry-breaking phase the boson field can be written as

$$\varphi = \varphi_0 + \delta\varphi$$

where φ_0 describes the condensate and $\delta\varphi$ the excitations above the condensate. Let us assume that the bosons are in a box of size L with a periodic boundary condition. As we boost the system, we twist the boson boundary condition from $\varphi(L) = \varphi(0)$ to $\varphi(L) = \mathrm{e}^{\,imvL}\varphi(0)$. Thus, under the boost, φ is changed to $\mathrm{e}^{\,imvx}\varphi$. In the boosted superfluid, the condensate is twisted: $\varphi_0 \to \mathrm{e}^{\,imvx}\varphi_0$ (see Fig. 3.15(a)). Due to the periodic boundary condition, mvL is quantized as $2\pi\, \times$ integer. Now it is clear that, because $|\varphi_0|$ is fixed, we cannot untwist the condensate without tearing it. Untwisting the condensate requires pushing the condensate φ_0 to zero, which costs large energies. More precisely, untwisting requires the creation of a vortex and the moving of the vortex all the way across the sample (see Fig. 3.15(b)). Thus, untwisting the condensate is a tunneling process with a finite energy barrier. According to this picture, we see that a moving superfluid, although it has a higher energy, cannot relax into the ground state if we ignore the tunneling process. We also learn that the superfluid cannot really flow forever. The tunneling will cause the flow to relax to zero. However, it can take a very, very long time for this to happen. The key to superfluidity is the infinite energy cost of the vortex in an infinite system. The infinite energy cost comes from the finite phase rigidity $\eta \neq 0$ in eqn (3.4.2).

However, the story is very different for the excitations. Here $\delta\varphi$ fluctuates around zero, and can twist and easily change its phase. Thus, the fluctuations (or the excitations) can easily relax to an equilibrium state through their interaction (which can be very weak) with the environment.

The above discussion suggests a two-fluid picture. A boson superfluid contains two components, namely a superfluid component related to the condensate and a normal fluid component related to the excitations. When a boson superfluid flows through a pipe, only the superfluid component flows through without any friction. The normal fluid component cannot flow if the friction is too large. (In a steady state there is no pressure through the pipe and the normal fluid component simply cannot flow.) Certainly, the superfluid velocity cannot be too large. If $\boldsymbol{v}$ is too large, the excitation energy in the laboratory frame $\epsilon_{\boldsymbol{v}}(\boldsymbol{k})$ may become negative, indicating instability and the end of frictionless flow. Therefore, the critical velocity is (see Fig. 3.1)

$$v_c = \mathrm{Min}(\epsilon(\boldsymbol{k})/|\boldsymbol{k}|)$$

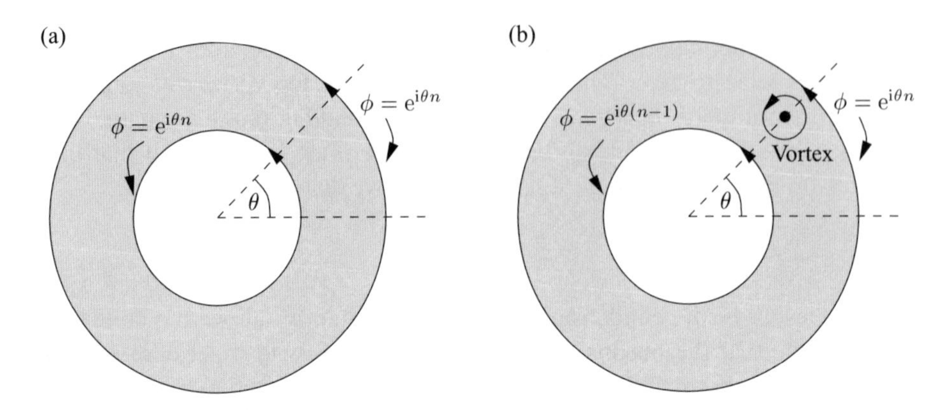

FIG. 3.15. (a) The phase of the condensate for a twist of n times around the circle, which represents a superfluid flow. (b) A vortex can change the phase twist from n times to $n-1$ times.

It is interesting to see that $v_c = 0$ for free bosons, even at zero T. Thus, although the free bosons at zero T have long-range order, they do not form a superfluid because $v_c = 0$.

Let us examine the two-fluid picture in more detail. Note that mv determines the twist of the condensate. Thus, each boson in the superfluid component carries a momentum $m\boldsymbol{v}$, and $\boldsymbol{P_v}/m\boldsymbol{v}$ gives us the number of bosons in the superfluid component. We find the superfluid density to be

$$\rho_s = \frac{\boldsymbol{P_v}}{\mathcal{V}m\boldsymbol{v}} = \rho - \rho_n$$

and ρ_n can be regarded as the normal fluid density. From eqn (3.7.13), we see that at zero temperature $\rho_s = \rho$, despite the fact that the number of bosons in the $\boldsymbol{k} = 0$ state is less than N.

If the bosons are charged and couple to the electromagnetic gauge field A_μ, then the twisted condensate with zero gauge potential $(\boldsymbol{A} = 0, e^{im\boldsymbol{v}\cdot\boldsymbol{x}}\varphi_0)$ is gauge equivalent to an untwisted condensate with a nonzero gauge potential $(\boldsymbol{A} = m\boldsymbol{v}, \varphi_0)$. In this case, eqn (3.7.12) can be interpreted as generating a finite momentum $\boldsymbol{P_v}$ by turning on a constant gauge potential $\boldsymbol{A} = m\boldsymbol{v}$. Due to the Galileo invariance, the momentum density $\boldsymbol{P_v}/\mathcal{V}$ is proportional to the current density $\boldsymbol{j} = \boldsymbol{P_v}/m\mathcal{V}$. Thus, eqn (3.7.12) can be rewritten as

$$\boldsymbol{j} = \frac{\rho_s}{m}\boldsymbol{A}$$

which is simply the London equation. We see that, at finite temperatures, the coefficient in the London equation is given by ρ_s/m and eqn (3.7.13) allows us to calculate its temperature dependence.

Problem 3.7.2.
Galileo non-invariant superfluid. Although the above discussions concentrate on the Galileo invariant superfluid, most of the results obtained above also apply to the Galileo non-invariant superfluid.

1. Derive the XY-model for a boson system with a constant gauge potential $\boldsymbol{A}$, assuming that the condensate $\varphi =$ constant. Find the dispersion $\epsilon_{\boldsymbol{A}}(\boldsymbol{k})$ of low-lying modes and compare your result with $\epsilon_{\boldsymbol{v}}(\boldsymbol{k})$ obtained above.
2. Repeat the above calculation, but now add an extra term $\frac{c}{2}|\dot{\varphi}|^2$ to the boson Lagrangian to break the Galileo invariance.
3. Derive the London equation for the above Galileo non-invariant system. (Hint: You may first derive the London equation at zero temperature. Then consider how excitations at finite T may correct the current. Note that the current of an excitation at $\boldsymbol{k}$ is given by $\frac{\partial}{\partial \boldsymbol{A}}\epsilon_{\boldsymbol{A}}(\boldsymbol{k})$.)

3.7.4 Tunneling and Josephson effects

We have seen that the Green's functions of the boson in a superfluid phase are quite different for different dimensions. In this section, we are going to discuss tunneling between two superfluids or superconductors. We will see that tunneling experiments allow us to measure the properties of the boson Green's functions.

Consider two boson systems described by H_R and H_L which are coupled by a tunneling operator I, so that

$$H = H_R + H_L + \Gamma I + \Gamma I^\dagger$$

where

$$I = \varphi_L \varphi_R^\dagger$$

and Γ describes the tunneling amplitude. In the presence of an electromagnetic gauge field, the total Hamiltonian needs to be rewritten as

$$H = H_R + H_L + \Gamma \mathrm{e}^{-\mathrm{i}a} I + \Gamma \mathrm{e}^{+\mathrm{i}a} I^\dagger$$

where $a = \int_L^R \boldsymbol{A}\,\mathrm{d}\boldsymbol{x}$ is an integration of the vector potential across the tunneling junction. The tunneling current operator is given by

$$j_T = \frac{\partial}{\partial a} H = -\mathrm{i}\Gamma(I\mathrm{e}^{-\mathrm{i}a} - I^\dagger \mathrm{e}^{+\mathrm{i}a})$$

If $\langle j_T \rangle > 0$, then the current flows from L to R. In the $A_0 = 0$ gauge, the voltage difference between the two systems, $V = V_L - V_R$, can be included by setting

$$a(t) = \int_L^R \boldsymbol{A}\,\mathrm{d}\boldsymbol{x} = -Vt$$

The $A_0 = 0$ gauge has the advantage that $H_{R,L}$ are not affected by turning on the voltage.

With the above set-up, we can write down the following expression for the tunneling current using linear response theory (see eqn (2.2.3)):

$$\langle j_T \rangle (t) = -\mathrm{i}\Gamma \int^t \mathrm{d}t' \left\langle [j_T(t), (\mathrm{e}^{-\mathrm{i}a(t')} I(t') + h.c.)] \right\rangle + \langle j_T \rangle_0 (t)$$

$$= -\Gamma^2 \int^t \mathrm{d}t' \, \mathrm{e}^{-\mathrm{i}a(t)} \left(\mathrm{e}^{\,\mathrm{i}a(t')} \langle [I(t), I^\dagger(t')] \rangle + \mathrm{e}^{-\mathrm{i}a(t')} \langle [I(t), I(t')] \rangle + h.c. \right)$$
$$+ \langle j_T \rangle_0 (t)$$

where $\langle j_T \rangle_0 (t)$ is an average in the absence of the tunneling term.

If both boson systems are in a superfluid or superconducting phase, then the main contribution comes from $\langle j_T \rangle_0 (t)$:

$$\langle j_T \rangle (t) = 2\Gamma |\varphi_R \varphi_L| \sin(\theta - a(t))$$

where θ is the phase difference of the two condensates φ_R and φ_L. We see that, even at zero voltage $a(t) = 0$, the tunneling current can be finite if $\theta \neq 0$:

$$\langle j_T \rangle = I_c \sin(\theta), \qquad I_c = 2\Gamma |\varphi_R \varphi_L|$$

where I_c is the maximum tunneling current (which is called the critical tunneling current).

If one of the boson systems is in the normal phase (including the algebraic-decay phase), then $\langle j_T \rangle_0 (t) = 0$ and $\langle [I(t), I(t')] \rangle = 0$. We have

$$\langle j_T \rangle (t) = -\Gamma^2 \int^t \mathrm{d}t' \left(\mathrm{e}^{-\mathrm{i}a(t)+\mathrm{i}a(t')} \left\langle [I(t), I^\dagger(t')] \right\rangle + h.c. \right) \tag{3.7.15}$$

As $\left\langle [I(t), I^\dagger(t')] \right\rangle$ is a correlation in the absence of tunneling, it can be expressed as the product of the Green's functions on two sides of the junction. For example,

$$\left\langle I(t) I^\dagger(t') \right\rangle = \left\langle \varphi_L(t) \varphi_L^\dagger(t') \right\rangle \left\langle \varphi_R^\dagger(t) \varphi_R(t') \right\rangle$$

In $1+1$ dimensions, the time-ordered Green's functions have the following algebraic decay:

$$\left\langle \varphi_{L,R}(t) \varphi_{L,R}^\dagger(t') \right\rangle \sim \mathrm{e}^{-\mathrm{i}\eta_{L,R}\pi/2} |t - t'|^{-\eta_{L,R}}.$$

We find that the tunneling current has the form[25]

$$\langle j_T \rangle = 2\sin(\frac{(\eta_R+\eta_L)\pi}{2}) \left(\frac{l}{v}\right)^{\eta_R+\eta_L}$$
$$\times \Gamma^2 \int^t \mathrm{d}t' \left(\mathrm{i}\,\mathrm{e}^{\mathrm{i}V(t-t')}(t-t')^{-\eta_R-\eta_L} \frac{1}{|t-t'|^{\eta_R+\eta_L}} + h.c. \right)$$

It leads to a nonlinear I–V curve[26]

$$\langle j_T \rangle = 2\left(\frac{l}{v}\right)^{\eta_R+\eta_L} \Gamma^2 \frac{\pi}{\Gamma(\eta_R+\eta_L)} |V|^{\eta_R+\eta_L-1} \mathrm{sgn}(V)$$

We see that the exponent of the algebraic decay $\eta_R + \eta_L$ can be measured in tunneling experiments.

Problem 3.7.3.
Find the expression for the tunneling I–V curve at finite temperatures and show that the differential conductance at $V = 0$ has a temperature dependence $\mathrm{d}I/\mathrm{d}V \propto T^{\eta_R+\eta_L-2}$.

3.7.5 Anderson–Higgs mechanism

- The Anderson–Higgs mechanism combines the gapless Nambu–Goldstone mode and the gapless gauge mode into a mode with a finite energy gap.

In the above discussions, we have treated the electromagnetic gauge field A_μ as a background field, which is fixed and does not respond to the changes of charge

[25] We have used the fact that

$$\left\langle \varphi_{L,R}(t)\varphi^\dagger_{L,R}(t') \right\rangle = \left\langle \varphi^\dagger_{L,R}(t)\varphi_{L,R}(t') \right\rangle,$$
$$\left\langle \varphi^\dagger_{L,R}(t')\varphi_{L,R}(t) \right\rangle = \left\langle \varphi^\dagger_{L,R}(t)\varphi_{L,R}(t') \right\rangle^*$$

to show that

$$\begin{aligned}\left\langle [I(t), I^\dagger(t')] \right\rangle &= \left\langle \varphi^\dagger_R(t)\varphi_R(t') \right\rangle \left\langle \varphi_L(t)\varphi^\dagger_L(t') \right\rangle - \left\langle \varphi_R(t')\varphi^\dagger_R(t) \right\rangle \left\langle \varphi^\dagger_L(t')\varphi_L(t) \right\rangle \\ &= \mathrm{e}^{-\mathrm{i}(\eta_R+\eta_L)\pi/2} \left(\frac{l}{v|t-t'|}\right)^{\eta_R+\eta_L} - \text{h.c.} \\ &= -2\mathrm{i}\sin\left(\frac{(\eta_R+\eta_L)\pi}{2}\right)\left(\frac{l}{v|t-t'|}\right)^{\eta_R+\eta_L}.\end{aligned}$$

[26] We have used the fact that

$$\begin{aligned}\int_{-\infty}^0 \mathrm{d}t\ \mathrm{e}^{\mathrm{i}Vt}|t|^{-a} &= \Big|_{t\to -\mathrm{i}\tau\mathrm{sgn}(V)} - \mathrm{i}\mathrm{sgn}(V)\mathrm{e}^{\mathrm{i}a\pi\mathrm{sgn}(V)/2} \int_0^{+\infty} \mathrm{d}\tau\ \mathrm{e}^{-|V|\tau}\tau^{-a} \\ &= -\mathrm{i}\mathrm{sgn}(V)\mathrm{e}^{\mathrm{i}a\pi\mathrm{sgn}(V)/2}|V|^{a-1}\Gamma(1-a)\end{aligned}$$

and $\Gamma(a)\Gamma(1-a) = \frac{\pi}{\sin(\pi a)}$.

and current. In this section, we will treat A_μ as a dynamical field with its own fluctuations.

The quantum theory of bosons and the gauge field can be defined through the path integral

$$Z = \int \mathcal{D}\varphi\mathcal{D}A_\mu \; \mathrm{e}^{\,\mathrm{i}\int \mathrm{d}^d\boldsymbol{x}\mathrm{d}t \; L(\varphi,A_\mu)+\frac{1}{8\pi e^2}(\boldsymbol{E}^2-\boldsymbol{B}^2)}$$

where we have set the speed of light $c = 1$. Here $L(\varphi, A_\mu)$ is the gauged boson Lagrangian given by eqn (3.7.3). Here we will only consider the classical theory described by

$$L = L(\varphi, A_\mu) + \frac{1}{8\pi e^2}(\boldsymbol{E}^2 - \boldsymbol{B}^2) \tag{3.7.16}$$

The quantum gauge theory will be discussed in Chapter 6.

Because of the gauge invariance (3.7.4), if $(\varphi(\boldsymbol{x}, t), A_\mu(\boldsymbol{x}, t))$ is a solution of the equation of motion, then the gauge transformed fields $(\tilde{\varphi}(\boldsymbol{x}, t), \tilde{A}_\mu(\boldsymbol{x}, t))$ also satisfy the equation of motion. In a gauge theory, we do not view $(\varphi(\boldsymbol{x}, t), A_\mu(\boldsymbol{x}, t))$ and $(\tilde{\varphi}(\boldsymbol{x}, t), \tilde{A}_\mu(\boldsymbol{x}, t))$ as two different motions. We view them as the same motion. In other words, the fields $(\varphi(\boldsymbol{x}, t), A_\mu(\boldsymbol{x}, t))$ are many-to-one labels of physical motions. Two sets of gauge equivalent fields, $(\varphi(\boldsymbol{x}, t), A_\mu(\boldsymbol{x}, t))$ and $(\tilde{\varphi}(\boldsymbol{x}, t), \tilde{A}_\mu(\boldsymbol{x}, t))$, are two labels that describe the same motion.

In the symmetry-breaking phase we have $|\varphi(x, t)| \neq 0$. We can make $\varphi(\boldsymbol{x}, t)$ real, namely $\varphi(\boldsymbol{x}, t) = |\varphi(\boldsymbol{x}, t)| = \phi(\boldsymbol{x}, t)$, through a gauge transformation. This procedure is called fixing the gauge or choosing the gauge. After fixing the gauge, different pairs (φ, A_μ) will describe physically different motions. In the $\varphi(x, t) =$ real gauge, we have

$$L = -A_0\phi^2 - \frac{1}{2m}(\partial_i\phi)^2 - \frac{\phi^2}{2m}(A_i)^2 - V(\phi) + \frac{1}{8\pi e^2}(\boldsymbol{E}^2 - \boldsymbol{B}^2)$$

After integrating out the small fluctuations of $\phi = \varphi_0 + \delta\phi$, we obtain the following low-energy effective theory:

$$L_{\text{eff}} = \frac{1}{2V_0}(A_0)^2 - \frac{\rho}{2m}(A_i)^2 + \frac{1}{8\pi e^2}(\boldsymbol{E}^2 - \boldsymbol{B}^2) \tag{3.7.17}$$

The above can also be obtained from the gauged XY-model (3.7.10) by setting $\theta = 0$. Note that

$$\frac{1}{8\pi e^2}\boldsymbol{E}^2 = \frac{1}{8\pi e^2}(\partial_0 A_i)^2 + \frac{1}{8\pi e^2}(\partial_i A_0)^2 - \frac{1}{4\pi e^2}\partial_i A_0 \partial_0 A_i$$

and A_0 contain no time derivative terms. Thus, A_0 is not dynamical and we integrate it out to give

$$L_{\text{eff}} = \partial_0 A_j \left(\frac{1}{8\pi e^2}\delta_{ij} + \frac{1}{2}\frac{V_0}{(4\pi)^2 e^4}\partial_j\partial_i \right) \partial_0 A_i - \frac{\rho}{2m}A_i^2 - \frac{1}{8\pi e^2}\boldsymbol{B}^2$$

Introducing the transverse and longitudinal components as follows:

$$A_i = \hat{k}_i A^{||} + \hat{n}_a A_a^{\perp} \tag{3.7.18}$$

where $\hat{k}$, $\hat{n}_1$, and $\hat{n}_2$ form a local orthogonal basis, we can rewrite L_{eff} as

$$\begin{aligned} L_{\text{eff}} = & \frac{1}{8\pi e^2}\left((\partial_0 A_a^{\perp})^2 - (\partial_i A_a^{\perp})^2\right) - \frac{\rho}{2m}(A_a^{\perp})^2 \\ & + \frac{1}{8\pi e^2}\partial_0 A^{||}\left(1 + \frac{V_0}{4\pi e^2}(\partial_i)^2\right)\partial_0 A^{||} - \frac{\rho}{2m}(A^{||})^2. \end{aligned}$$

From the equation of motion, we find that all of the three modes described by L_{eff} have the same energy gap $\Delta = e\sqrt{4\pi\rho/m}$. There are no gapless modes. The coupling between the gapless gauge mode and the gapless Nambu–Goldstone mode gives those modes a finite energy gap. This phenomenon is called the Anderson–Higgs mechanism (Anderson, 1963; Higgs, 1964). We can further simplify L_{eff} by replacing ∂_0 by $-\mathrm{i}\Delta$ in the $\partial_i^2\partial_0^2$ term as follows:

$$\begin{aligned} L_{\text{eff}} = & \frac{1}{8\pi e^2}\left((\partial_0 A_a^{\perp})^2 - (\partial_i A_a^{\perp})^2\right) - \frac{\rho}{2m}(A_a^{\perp})^2 \\ & + \frac{1}{8\pi e^2}\left((\partial_0 A^{||})^2 - v^2(\partial_i A^{||})^2\right) - \frac{\rho}{2m}(A^{||})^2 \end{aligned} \tag{3.7.19}$$

(Note that $v^2 = V_0\rho/m$ is the velocity of the XY-model.) Starting from L_{eff}, we can study all of the classical electromagnetic properties of the charged superfluid.

Problem 3.7.4.
Derive the Maxwell equation from the Lagrangian $L = \frac{1}{8\pi e^2}(c^{-1}\boldsymbol{E}^2 - c\boldsymbol{B}^2)$. Show that the gauge fluctuations are gapless and c is the speed of light.

3.8 Perturbative calculation of the thermal potential

3.8.1 Perturbation and Feynman rules

- Feynman diagrams and Feynman rules for interacting bosons.

In this section, we will discuss how to systematically calculate the thermal potential of an interacting boson system. For simplicity, we will assume the boson field to be real. A complex boson field can be treated as comprising two components which are each real boson fields. The discussion below can be easily generalized to a boson field with multiple components. We start with the imaginary-time path integral at finite temperatures. The first step in calculating the thermal potential is to find a classical solution φ_c and expand the action S

around it as follows:

$$S(\varphi) = S(\varphi_c) + \int_0^\beta \mathrm{d}x^\mu \mathrm{d}y^\mu \, \frac{1}{2}\delta\varphi(x^\mu)\mathcal{K}^\beta(x^\mu - y^\mu)\delta\varphi(y^\mu) + \int_0^\beta \mathrm{d}x^\mu V(\delta\varphi)$$

where $V(\delta\varphi) = g_3\delta\varphi^3 + g_4\delta\varphi^4 + ...$ If we ignore the higher-order terms $V(\delta\varphi)$, then we have a free system and its thermal potential Ω_0 can be calculated through a Gaussian integral, as discussed in Section 3.7.5. To calculate the thermal potential of an interacting system, we note that

$$\begin{aligned} Z = A\int \mathcal{D}\varphi \, \mathrm{e}^{-S} &= \mathrm{e}^{-S(\varphi_c)}\mathrm{e}^{-\beta\Omega_0} \frac{\int \mathcal{D}\delta\varphi \, \mathrm{e}^{-S_0 - \int \mathrm{d}x^\mu V(\delta\varphi)}}{\int \mathcal{D}\delta\varphi \, \mathrm{e}^{-S_0}} \\ &= \mathrm{e}^{-S(\varphi_c)}\mathrm{e}^{-\beta\Omega_0} \sum_{n=0} \frac{(-)^n}{n!} \left\langle \left(\int \mathrm{d}x^\mu \, V(\delta\varphi) \right)^n \right\rangle_0 \end{aligned}$$

where $\langle ... \rangle_0$ means the average with weight e^{-S_0} and S_0 is the quadratic part of the action, namely, $S_0 = \int_0^\beta \mathrm{d}x^\mu \mathrm{d}y^\mu \, \frac{1}{2}\delta\varphi(x^\mu)\mathcal{K}^\beta(x^\mu - y^\mu)\delta\varphi(y^\mu)$. If there are several classical solutions (or stationary paths) $\varphi_c^{(a)}$, then we should include all of them as follows:

$$Z = \sum_a \mathrm{e}^{-S(\varphi_c^{(a)})}\mathrm{e}^{-\beta\Omega_0^{(a)}} \frac{\int \mathcal{D}\delta\varphi \, \mathrm{e}^{-S_0^{(a)} - \int \mathrm{d}x^\mu V^{(a)}(\delta\varphi)}}{\int \mathcal{D}\delta\varphi \, \mathrm{e}^{-S_0^{(a)}}}$$

We see that, to calculate the total thermal potential, we need to calculate the multi-point correlation $\langle \prod_{i=1}^n \delta\varphi(i) \rangle_0$. For $n = 2$, it is the Green's function of the boson field $\langle \delta\varphi(1)\delta\varphi(2) \rangle_0 = \mathcal{G}^\beta(1,2)$, where 1 and 2 represent the coordinates x_1^μ and x_2^μ of the first and the second boson fields, respectively. As an operator, $\mathcal{G}^\beta(1,2)$ is the inverse of $\mathcal{K}^\beta(1,2)$, i.e. $\int \mathcal{G}^\beta(1,2)\mathcal{K}^\beta(2,3) = \delta(x_1^\mu - x_3^\mu)$. Here we have also used the abbreviations $\int f(1) \equiv \int \mathrm{d}^d x_1^\mu f(x_1^\mu)$, $\int f(1)g(1,2) \equiv \int \mathrm{d}^d x_1^\mu \mathrm{d}^d x_2^\mu f(x_1^\mu) g(x_1^\mu, x_2^\mu)$, etc. Introducing the generating functional

$$Z[j] = \frac{\int \mathcal{D}\delta\varphi \, \mathrm{e}^{-S_0 - \int \mathrm{d}x^\mu j\delta\varphi}}{\int \mathcal{D}\delta\varphi \, \mathrm{e}^{-S_0}} = \mathrm{e}^{\frac{1}{2}\int j(1)\mathcal{G}^\beta(1,2)j(2)}$$

we find that (for n even)

$$\begin{aligned} \left\langle \prod_{i=1}^n \delta\varphi(i) \right\rangle_0 &= \frac{\delta^n}{\delta j(1)...\delta j(n)} Z[j]\Big|_{j=0} \\ &= \frac{1}{2^{n/2}(n/2)!} \frac{\delta^n}{\delta j(1)...\delta j(n)} \left(\int j\mathcal{G}^\beta j \right)^{n/2} \\ &= \Big(\mathcal{G}^\beta(1,2)\mathcal{G}^\beta(3,4)...\mathcal{G}^\beta(n-1,n) \Big) + \text{all distinct permutations} \end{aligned}$$

The above is another form of the Wick theorem.

FIG. 3.16. Two vertices representing $\int g_4 \delta\varphi^4(1) \int g_4 \delta\varphi^4(2)$.

Using the Wick theorem, we find that, for example, $\langle \int V(\delta\varphi) \int V(\delta\varphi) \rangle_0$ contains many terms:

$$
\begin{aligned}
\left\langle \int V(\delta\varphi) \int V(\delta\varphi) \right\rangle_0 &= \left\langle \int g_4 \delta\varphi^4(1) \int g_4 \delta\varphi^4(2) \right\rangle_0 + \dots \\
= \left[4! g_4^2 \int \left(\mathcal{G}^\beta(1,2) \right)^4 \right] &+ \left[(6^2 \times 2) g_4^2 \int \left(\mathcal{G}^\beta(1,2) \right)^2 \mathcal{G}^\beta(1,1) \mathcal{G}^\beta(2,2) \right] \\
+ \left[3g_4 \int \left(\mathcal{G}^\beta(1,1) \right)^2 \right] &\left[3g_4 \int \left(\mathcal{G}^\beta(2,2) \right)^2 \right] + \dots
\end{aligned} \tag{3.8.1}
$$

The integral factors $4!$, $6^2 \times 2$, etc. arise from the different terms in the Wick expansion which happen to have the same form. For example, in a simpler case, the Wick expansion $\langle ABCD \rangle = \langle AB \rangle \langle CD \rangle + \langle AC \rangle \langle BD \rangle + \langle AD \rangle \langle BC \rangle$ leads to

$$
\begin{aligned}
\langle AABB \rangle &= \langle AA \rangle \langle BB \rangle + \langle AB \rangle \langle AB \rangle + \langle AB \rangle \langle AB \rangle \\
&= \langle AA \rangle \langle BB \rangle + 2 \langle AB \rangle \langle AB \rangle
\end{aligned}
$$

It is more convenient to use Feynman diagrams to directly obtain the result of the Wick expansion (3.8.1) through Feynman rules. Instead of describing Feynman diagrams and Feynman rules in a general setting, we choose to explain them using the particular example given in eqn (3.8.1). I feel that it is easier to understand Feynman diagrams and Feynman rules in action. The Feynman rules that we are going to describe here are the real-space Feynman rules. We will describe momentum-space Feynman rules later in Section 5.4.2.

To calculate $\langle \int g_4 \delta\varphi^4(1) \int g_4 \delta\varphi^4(2) \rangle_0$, we start with the two vertices in Fig. 3.16 that represent $\int g_4 \delta\varphi^4(1) \int g_4 \delta\varphi^4(2)$. To obtain the expectation value $\langle \dots \rangle_0$, we need to connect all of the 'legs' of the vertices with each other. This generates the diagrams in Fig. 3.17. These diagrams are called Feynman diagrams. To construct the result of eqn (3.8.1), we simply use the following three Feynman rules.

(i) Each line in the Feynman diagrams represents a propagator $\mathcal{G}^\beta(a,b)$, where a and b are the two vertices at the ends of the line.

(ii) Each vertex represents $g_4 \int$.

(iii) The value of a connected diagram is given by rules (i) and (ii), and the value of

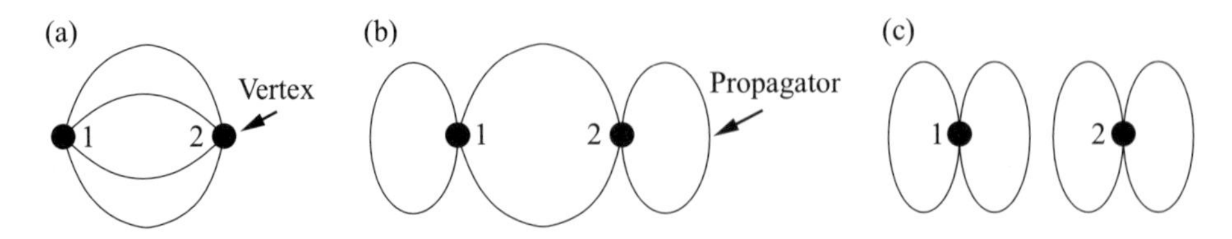

FIG. 3.17. Three Feynman diagrams corresponding to the three terms in eqn (3.8.1). Here (a) and (b) are connected diagrams and (c) is a disconnected diagram which contains two connected diagrams.

the disconnected diagram is given by the product of the connected sub-diagrams. In this way, we obtain the first, second, and third terms in the last line of eqn (3.8.1) from the three Feynman diagrams given in Fig. 3.17(a), (b), and (c), respectively.

Well, not exactly. We neglect the numerical factors such as 4!, $6^2 \times 2$, etc. These numerical factors are the most difficult part of the Feynman rules. Where do they come from? Let us first consider the factor 4! associated with the diagram in Fig. 3.17(a). This factor comes from the fact that each vertex has four legs, and there 4! different ways to connect the four legs of vertex 1 to the four legs of vertex 2. The factor $6^2 \times 2$ associated with the diagram in Fig. 3.17(b) is more complicated, but it still represents different ways to connect the eight legs with each other. First, two legs of vertex 1 connect to each other. There are six ways to choose two legs among the four legs. In this way, we obtain the factor 6 in $6^2 \times 2$. The second factor 6 arises similarly from vertex 2. There are two ways to connect the remaining two legs of vertex 1 to the remaining two legs of vertex 2. We get the final factor of 2. The factor 3 in the third term corresponds to different ways to divide the four legs of a vertex into two groups with two legs in each group. (Note that this is different from the different ways of choosing two legs out of four.)

Let us return to the thermal potential that we want to calculate. After obtaining all of the averages, the correction to the thermal potential due to the interaction can be obtained as follows:

$$\delta\Omega = -T \ln \sum_{n=0} \frac{(-)^n}{n!} \left\langle \left(\int \mathrm{d}x^\mu \, V(\delta\varphi) \right)^n \right\rangle_0$$

There is a linked-cluster theorem which simplifies the above calculation. According to the theorem, $\delta\Omega$ is given by the sum of all of the connected diagrams as follows:

$$\delta\Omega = -T \sum_{n=0} \frac{(-)^n}{n!} \left\langle \left(\int \mathrm{d}x^\mu \, V(\delta\varphi) \right)^n \right\rangle_{0c}$$

where $\left\langle \left(\int \mathrm{d}x^\mu \, V(\delta\varphi) \right)^n \right\rangle_{0c}$ is obtained from $\left\langle \left(\int \mathrm{d}x^\mu \, V(\delta\varphi) \right)^n \right\rangle_0$ by dropping all of the contributions from disconnected diagrams.

3.8.2 Linked-cluster theorem

- The effective action can be calculated directly by summing only the connected Feynman diagrams.

We will derive the linked-cluster theorem by using the replica technique, both for its brevity and to introduce this useful method. The basic idea of the replica method is to evaluate Z^n, for integer n, by replicating the system n times as follows:

$$\left(\frac{Z}{Z_0}\right)^n = \frac{\int \mathcal{D}\delta\varphi_\alpha \; \mathrm{e}^{-\sum_{\alpha=1}^n [S_0(\delta\varphi_\alpha) + \int \mathrm{d}x^\mu V(\delta\varphi_\alpha)]}}{\int \mathcal{D}\delta\varphi_\alpha \; \mathrm{e}^{-\sum_{\alpha=1}^n S_0(\delta\varphi_\alpha)}}$$

Now $(Z/Z_0)^n$ can be calculated through a perturbative expansion. In each Feynman diagram, every propagator carries an index α, and all propagators entering and leaving an interacting vertex have the same index. As we sum α from 1 to n, it is clear that each connected graph is proportional to n, and each disconnected graph is proportional to n^{N_c}, where N_c is the number of connected graphs in the disconnected graph. Therefore,

$$\begin{aligned}\left(\frac{Z}{Z_0}\right)^n &= \mathrm{e}^{n \ln \frac{Z}{Z_0}} = 1 + n \ln \frac{Z}{Z_0} + \sum_{m=2}^{\infty} \frac{(n \frac{Z}{Z_0})^m}{m!} \\ &= 1 + n \sum \text{(all connected graphs)} + O(n^2)\end{aligned}$$

and we have proved the linked-cluster theorem.

Problem 3.8.1.
Consider an anharmonic oscillator $L = \frac{1}{2}m\dot{x}^2 - \frac{1}{2}m\omega_0^2 x^2 - gx^3$. Calculate the finite-temperature free energy of the system to order g^2. (Hint: It may easier to do the calculation in τ space.)
Note that, within perturbation theory, the above anharmonic oscillator is a well-behaved system. The instability comes from the bounce. Show that the decay rate of the ground state has an order of $\omega_0 \,\mathrm{e}^{-\# m^3\omega_0^5/g^2}$ in the small-g limit. Such a term cannot appear in the perturbative calculation around the ground state $x = 0$. However, the bounce as another stationary path does contribute to the free energy. Show that the contribution from the bounce appears as a non-perturbative term proportional to $\mathrm{e}^{-\# m^3\omega_0^5/g^2}$.

4

FREE FERMION SYSTEMS

The fermion system is one of the most important systems in condensed matter physics. Metals, semiconductors, magnets, superconductors, etc. are all fermion systems. Their properties are mainly controlled by the Fermi statistics of electrons. In this chapter, we will study some properties of free many-fermion systems.

4.1 Many-fermion systems

4.1.1 What are fermions?

- Fermions are characterized by the Pauli exclusion principle and a hopping Hamiltonian that generates a π phase shift when two identical fermions are exchanged.
- Fermions are weird because they are non-local objects.
- Fermions can be described by anti-commuting operators.

For a long time, we have felt that we know what fermions are. After reading the next few sections, if you start to feel that you do not understand what fermions really are, then I have achieved my goal. We will give *an* answer to what fermions are later in Chapter 10. Here we will just introduce fermions in the traditional way and show, within the traditional picture, how strange and unnatural fermions are.

Let us first try to describe a system of spinless fermions on a lattice. Due to the Pauli exclusion principle, the Hilbert space of the lattice fermion system can be expanded by the bases $\{|n_{\boldsymbol{i}_1}, n_{\boldsymbol{i}_2}, ...\rangle\}$, where $n_i = 0, 1$ is the number of fermions at site $\boldsymbol{i}$. To have a second quantized description of the free fermion system, we introduce the annihilation operator $\sigma^-_{\boldsymbol{i}}$ and the creation operator $\sigma^+_{\boldsymbol{i}}$ for each site $\boldsymbol{i}$. They have the following matrix form:

$$\sigma^-_{\boldsymbol{i}} = \begin{pmatrix} 0 & 0 \\ 1 & 0 \end{pmatrix} = \frac{1}{2}(\sigma^x - \mathrm{i}\sigma^y) \qquad \sigma^+_{\boldsymbol{i}} = \begin{pmatrix} 0 & 1 \\ 0 & 0 \end{pmatrix} = \frac{1}{2}(\sigma^x + \mathrm{i}\sigma^y)$$

where $\sigma^{x,y,z}$ are the Pauli matrices. The annihilation operator σ^- changes a state with one fermion, $|1\rangle = \binom{1}{0}$, into a state with no fermions, $|0\rangle = \binom{0}{1}$. As $\sigma^\pm_{\boldsymbol{i}}$ creates/annihilates a fermion, we will, for the time being, call them 'fermion' operators.

Using the 'fermion' operators $\sigma_{\boldsymbol{i}}^{\pm}$, we can write down the following Hamiltonian for a fermion system:

$$H_b = \sum_{\langle \boldsymbol{ij} \rangle} (t_{\boldsymbol{ij}} \sigma_{\boldsymbol{i}}^{+} \sigma_{\boldsymbol{j}}^{-} + h.c.)$$

Although, mathematically, H_b is a hermitian operator acting within the fermion Hilbert space $\{|n_{\boldsymbol{i}_1}, n_{\boldsymbol{i}_2}, ...\rangle\}$, a system described H_b is not a fermion system. It is actually a hard-core boson system or a spin-1/2 system.

This is because our fermion Hilbert space can be equally regarded as a hard-core boson Hilbert space, where $|0\rangle$ is the zero-boson state and $|1\rangle$ is the one-boson state.[27] So, from the Hilbert space alone, we cannot determine if the system is a fermion system or a boson system. We have to look at the Hamiltonian to determine if the system is a fermionic or a bosonic system. As the Hamiltonian H_b is written in terms of the $\sigma_{\boldsymbol{i}}^{\pm}$, which commute with each other on different sites, so the system described by H_b is a boson system. We should really call the $\sigma_{\boldsymbol{i}}^{-}$ boson operators.

It is quite amazing to see that a natural local Hamiltonian in a fermion Hilbert space does not describe a fermion system! This raises an interesting question: what makes a many-particle system a fermion system? Obviously, the Pauli exclusion principle alone is not enough. A fermionic system not only has a fermionic Hilbert space that satisfies the Pauli exclusion principle, but it also has a Hamiltonian with a very special property. In fact, a fermion system is described by a highly non-local Hamiltonian

$$H_f = \sum_{\langle \boldsymbol{ij} \rangle} (\hat{t}_{\boldsymbol{ij}}(\{\sigma_{\boldsymbol{i}'}^{z}\}) \sigma_{\boldsymbol{i}}^{+} \sigma_{\boldsymbol{j}}^{-} + h.c.)$$

where $\hat{t}_{\boldsymbol{ij}}$ is a function of the $\sigma_{\boldsymbol{i}}^{z}$ operators, which involve products of many $\sigma_{\boldsymbol{i}}^{z}$ operators. The number of operators $\sigma_{\boldsymbol{i}}^{z}$ is of order N_s^{d-1} for a d-dimensional lattice of N_s sites. If it were not because nature offers us such non-local systems, then no physicist in his/her right mind would want to study them. As such non-local systems do exist in nature, we have to study them. But how?

Here we are extremely lucky. The non-local fermion systems in nature have some special properties which allow us to simplify them. To write down the simplified Hamiltonian, we first order all of the lattice sites in a certain way: $(\boldsymbol{i}_1, \boldsymbol{i}_2, ..., \boldsymbol{i}_a, ...)$. Then we introduce another kind of fermion operator as follows:

$$c_{\boldsymbol{i}_a} = \sigma_{\boldsymbol{i}_a}^{-} \prod_{b<a} \sigma_{\boldsymbol{i}_b}^{z} \tag{4.1.1}$$

[27] The fermion Hilbert space can also be regarded as a spin-1/2 Hilbert space, where $|0\rangle$ is the spin-down state and $|1\rangle$ is the spin-up state.

If we order the sites properly, then H_f will take the following simpler form when written in terms of $c_{\boldsymbol{i}}$:

$$H_f = \sum_{\langle \boldsymbol{ij} \rangle} (t_{\boldsymbol{ij}} c_{\boldsymbol{i}}^\dagger c_{\boldsymbol{j}} + h.c.) \tag{4.1.2}$$

where $t_{\boldsymbol{ij}}$ is independent of $\sigma_{\boldsymbol{i}}^z = 2c_{\boldsymbol{i}}^\dagger c_{\boldsymbol{i}} - 1$. One can check that

$$\begin{aligned} \{c_{\boldsymbol{i}}, c_{\boldsymbol{j}}\} &= \{c_{\boldsymbol{i}}^\dagger, c_{\boldsymbol{j}}^\dagger\} = 0 \\ \{c_{\boldsymbol{i}}, c_{\boldsymbol{j}}^\dagger\} &= \delta_{\boldsymbol{ij}} \end{aligned} \tag{4.1.3}$$

where $\{A, B\} \equiv AB + BA$ is called the anti-commutator. The mapping between the boson operators $\sigma^{x,y,z}$ and the fermion operators $c_{\boldsymbol{i}}$ is the Jordan–Wigner transformation (Jordan and Wigner, 1928). We note that $c_{\boldsymbol{i}}^\dagger c_{\boldsymbol{i}} = \sigma_{\boldsymbol{i}}^+ \sigma_{\boldsymbol{i}}^-$. Here $|0\rangle$ and $|1\rangle$ are the two eigenstates of $c_{\boldsymbol{i}}^\dagger c_{\boldsymbol{i}}$ with eigenvalue 0 and 1, respectively. Thus, $c_{\boldsymbol{i}}^\dagger c_{\boldsymbol{i}}$ is the fermion number operator at the site $\boldsymbol{i}$.

Usually, we do not talk about where fermions come from. We just take eqns (4.1.3) and (4.1.2) as the definition of a free fermion system. If we do ask where fermions come from and if we think that bosons are more fundamental than fermions,[28] then the above discussion indicates that *fermions are non-local excitations*. In fact, fermions in nature do behave like non-local excitations because fermions cannot be created alone. Nature seems to want to keep track of the total number of fermions in our universe to make sure that the number is an even integer. It is impossible for a local excitation to have such a non-local constraint. In Chapter 10, we will see that fermions can be interpreted as ends of condensed strings, and are indeed non-local.

4.1.2 The exact solution of free fermion systems

- All of the eigenstates and the energy eigenvalues of a free fermion system can be obtained from the anti-commuting algebra of the fermion operators.

On a lattice of N_{site} sites, the Hilbert space of the fermion systems has $2^{N_{site}}$ states. The Hamiltonian (4.1.2) is a $2^{N_{site}} \times 2^{N_{site}}$ matrix. Solving the Hamiltonian amounts to finding the eigenvalues and eigenvectors of such a big matrix.

The Hamiltonian (4.1.2) can be solved exactly using the anti-commuting algebra (4.1.3). For simplicity, let us assume that our system has translational symmetry. In this case, $t_{\boldsymbol{ij}}$ only depends on the difference $\boldsymbol{i} - \boldsymbol{j}$, so that $t_{\boldsymbol{i},\boldsymbol{i}+\Delta\boldsymbol{i}} = t_{\Delta\boldsymbol{i}}$. Introducing $c_{\boldsymbol{k}} = \sum_{\boldsymbol{i}} N_s^{-1/2} \mathrm{e}^{-i\boldsymbol{k}\cdot\boldsymbol{i}} c_{\boldsymbol{i}}$, where N_s is the total number of lattice sites,

[28] See Section 10.1 for a definition of boson systems.

we find that

$$H_f = \sum_{\boldsymbol{k}} \epsilon_{\boldsymbol{k}} c^\dagger_{\boldsymbol{k}} c_{\boldsymbol{k}}, \qquad \epsilon_{\boldsymbol{k}} = \sum_{\Delta \boldsymbol{i}} t_{\Delta \boldsymbol{i}} e^{i\boldsymbol{k}\cdot\Delta \boldsymbol{i}},$$

$$\{c_{\boldsymbol{k}}, c_{\boldsymbol{k}'}\} = \{c^\dagger_{\boldsymbol{k}}, c^\dagger_{\boldsymbol{k}'}\} = 0, \qquad \{c_{\boldsymbol{k}}, c^\dagger_{\boldsymbol{k}'}\} = \delta_{\boldsymbol{k}\boldsymbol{k}'}. \tag{4.1.4}$$

One can check that the state $|0\rangle$ satisfying $c_{\boldsymbol{k}}|0\rangle = 0$ is an eigenstate of H_f with zero eigenvalue. Using the anti-commutation relation, we find that the state $|\{n_{\boldsymbol{k}}\}\rangle \equiv \prod_{\boldsymbol{k}} (c^\dagger_{\boldsymbol{k}})^{n_{\boldsymbol{k}}}|0\rangle$, $n_{\boldsymbol{k}} = 0, 1$, is a common eigenstate of $c^\dagger_{\boldsymbol{k}} c_{\boldsymbol{k}}$, i.e. $c^\dagger_{\boldsymbol{k}} c_{\boldsymbol{k}}|\{n_{\boldsymbol{k}}\}\rangle = n_{\boldsymbol{k}}|\{n_{\boldsymbol{k}}\}\rangle$. It is also an eigenstate with energy $E = \sum_{\boldsymbol{k}} n_{\boldsymbol{k}} \epsilon_{\boldsymbol{k}}$. The above form of eigenstates gives us all of the eigenstates of H_f. Here $n_{\boldsymbol{k}}$ is the fermion occupation number at momentum $\boldsymbol{k}$. Adding a fermion to the $\boldsymbol{k}_0$ level increases the total energy by $\epsilon_{\boldsymbol{k}_0}$. Thus, $\epsilon_{\boldsymbol{k}_0}$ can be interpreted as the single-particle energy.

It is remarkable that we can obtain all of the eigenstates of H_f via an algebraic approach without explicitly writing down the eigenvectors. The commuting boson algebra and the anti-commuting fermion algebra are two of only a few algebraic systems that we can solve exactly. Actually, for quite a long time they were the only two algebraic systems that we could solve exactly beyond one dimension. Finding exactly soluble algebraic systems beyond one dimension is very important in our understanding of interacting systems. Chapter 10 will discuss some other algebraic systems discovered recently (Kitaev, 2003; Levin and Wen, 2003; Wen, 2003c) that can be solved beyond one dimension. These algebraic systems lead to exactly soluble interacting models beyond one dimension.

If we include a chemical potential μ, then the Hamiltonian of the free fermion system will become

$$H = \sum_{\boldsymbol{k}} (\epsilon_{\boldsymbol{k}} - \mu) c^\dagger_{\boldsymbol{k}} c_{\boldsymbol{k}} = \sum_{\boldsymbol{k}} \xi_{\boldsymbol{k}} c^\dagger_{\boldsymbol{k}} c_{\boldsymbol{k}}, \qquad \xi_{\boldsymbol{k}} \equiv \epsilon_{\boldsymbol{k}} - \mu \tag{4.1.5}$$

The ground state of H is given by a state $|\Psi_0\rangle$, where all $\boldsymbol{k}$ states with $\xi_{\boldsymbol{k}} < 0$ are filled and all $\boldsymbol{k}$ states with $\xi_{\boldsymbol{k}} > 0$ are empty. Here $|\Psi_0\rangle$ is defined by the following algebraic relation:

$$c_{\boldsymbol{k}}|\Psi_0\rangle = 0, \qquad \text{if } \xi_{\boldsymbol{k}} > 0$$

$$c^\dagger_{\boldsymbol{k}}|\Psi_0\rangle = 0, \qquad \text{if } \xi_{\boldsymbol{k}} < 0$$

The occupation number $n_{\boldsymbol{k}} = c^\dagger_{\boldsymbol{k}} c_{\boldsymbol{k}}$ has a jump at the Fermi surface (see Fig. 4.1).

Problem 4.1.1.
Jordan–Wigner transformation Prove eqn (4.1.3) from eqn (4.1.1).

Problem 4.1.2.
Spectrum of a one-dimensional superconductor A one-dimensional superconductor is

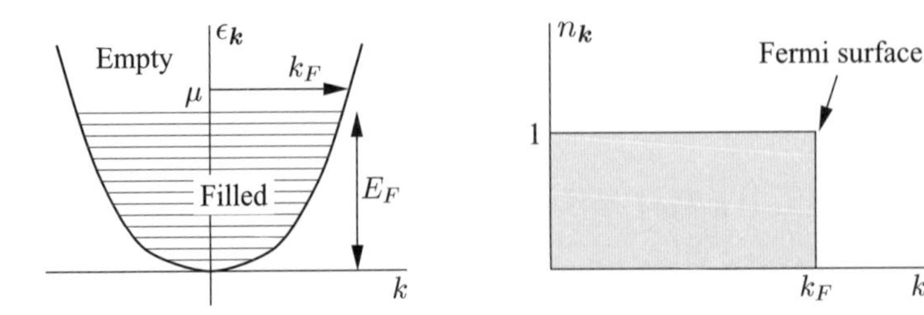

FIG. 4.1. The ground state of a free fermion system is a Fermi sea. Here E_F is called the Fermi energy and k_F is the Fermi momentum. The occupation number $n_{\boldsymbol{k}} = n_F(\xi_{\boldsymbol{k}})$ is discontinuous at the Fermi momentum k_F.

described by the following free fermion Hamiltonian:

$$H = \sum_i (tc_i^\dagger c_{i+1} + \eta c_i^\dagger c_{i+1}^\dagger + h.c.) \tag{4.1.6}$$

Show that, in momentum space, the operator

$$\lambda_k = u_k c_k + v_k c_{-k}^\dagger$$

satisfies the fermion anti-commutation relation $\{\lambda_k, \lambda_{k'}\} = \{\lambda_k^\dagger, \lambda_{k'}^\dagger\} = 0$ and that $\{\lambda_k, \lambda_{k'}^\dagger\} = \delta_{kk'}$ if $|u_k|^2 + |v_k|^2 = 1$. Show that, by properly choosing u_k and v_k, we can rewrite H as

$$H = E_g + \sum_k E_k \lambda_k^\dagger \lambda_k.$$

Find the quasiparticle excitation spectrum E_k and the ground-state energy E_g.

Problem 4.1.3.

Solving the one-dimensional spin-1/2 system (or hard-core boson system) using the Jordan–Wigner transformation Consider a one-dimensional spin-1/2 system (or hard-core boson system) with

$$H = \sum_i \left(J_x \sigma_i^x \sigma_{i+1}^x + J_y \sigma_i^y \sigma_{i+1}^y + B\sigma_i^z \right)$$

1. Use the Jordan–Wigner transformation (with a natural ordering of the one-dimensional lattice) to map the above interacting spin-1/2 (or hard-core boson) system to a free fermion system.
2. Assume that $J_x = J_y = J$ and $B \neq 0$. As we change the value of J from $-\infty$ to $+\infty$, the system experiences several phase transitions. Find the critical values of J for these phase transitions. For each of the phases and each of the critical points, sketch the region in the total energy–momentum space where the system has excitations. You should include all of the low-energy excitations, regardless of their

momentum. You do not need to include high-energy excitations. (Hint: One should first think and guess what the spectrum should look like based on our knowledge of the one-dimensional interacting boson system.)

3. Assume that $J_x = \alpha J_y = J$, $0 < \alpha < 1$, and $B \neq 0$. As we change the value of J from $-\infty$ to $+\infty$, the system experiences several phase transitions. Find the critical values of J for these phase transitions. For each of the phases and each of the critical points, sketch the region in the total energy–momentum space where the system has excitations. You should include all of the low-energy excitations, regardless of their momentum.

4. Discuss why the above two cases, namely $J_x = J_y$ and $J_x \neq J_y$, are qualitatively different.

4.1.3 Majorana fermions

The free fermion system (4.1.2) is an exactly soluble many-body system whose Hilbert space has dimension $2^{N_{site}}$ (i.e. two states per site). There is a smaller exactly soluble system with only $2^{N_{site}/2}$ states in its Hilbert space (assuming that N_{site} is even). The system is called a Majorana-fermion system. It is described by the following Hamiltonian:

$$H = \sum_i (-\mathrm{i}t\lambda_i\lambda_{i+1} + h.c.) \tag{4.1.7}$$

where the λ_i satisfying

$$\{\lambda_i, \lambda_j\} = \delta_{ij}, \qquad (\lambda_i)^\dagger = \lambda_i \tag{4.1.8}$$

are the Majorana-fermion operators. The Hilbert space of the Majorana-fermion system forms a representation of the above anti-commuting algebra.

To construct the Hilbert space, let us consider a one-dimensional periodic system with N_{site} sites. We also assume that N_{site} is even and that the λ_i satisfy the anti-periodic boundary condition $\lambda_{i+N_{site}} = -\lambda_i$. In the k space, the Hamiltonian (4.1.7) takes the form

$$H = -\sum_k \mathrm{i}t\mathrm{e}^{\mathrm{i}k}\lambda(-k)\lambda(k) = \sum_{k>0} t\sin k\,\lambda^\dagger(k)\lambda(k)$$

where $\lambda(k) = N_{site}^{-1/2}\sum_n \mathrm{e}^{-\mathrm{i}kn}\lambda_n$ and $k = \pm\pi/N_{\rm site}, \pm3\pi/N_{\rm site}, \dots, \pm(N_{\rm site}-1)\pi/N_{\rm site}$. One can check that

$$\{\lambda^\dagger(k), \lambda(k')\} = \delta_{k-k'}, \qquad \lambda^\dagger(k) = \lambda(-k)$$

We see that the $\lambda(k)$ for $k > 0$ satisfy the anti-commuting algebra of the complex fermions. Let $|0\rangle$ be the state that satisfies $\lambda(k)|0\rangle = 0$ for $k > 0$. We can regard $|0\rangle$ as a state with no fermions. Let $|\{n_k\}\rangle = \prod_{k>0}(\lambda^\dagger(k))^{n_k}|0\rangle$, where $n_k = 0, 1$. As $\lambda^\dagger(k)\lambda(k)|\{n_k\}\rangle = n_k|\{n_k\}\rangle$, we can view n_k as the occupation number at level k. Note that there are only $N_{site}/2$ levels because the levels are labeled by positive k (see Fig. 4.2). Therefore, the Hilbert space has $2^{N_{site}/2}$ states. It is interesting to see that a Majorana system has $\sqrt{2}$ states per site! The state $|\{n_k\}\rangle$ is an energy eigenstate with energy $\sum_{k>0} 2t\sin(k)n_k$.

For a system with two Majorana fermions and with

$$H = \sum_i (-\mathrm{i}t\lambda_{1,i}\lambda_{1,i+1} - \mathrm{i}t\lambda_{2,i}\lambda_{2,i+1} + h.c.),$$

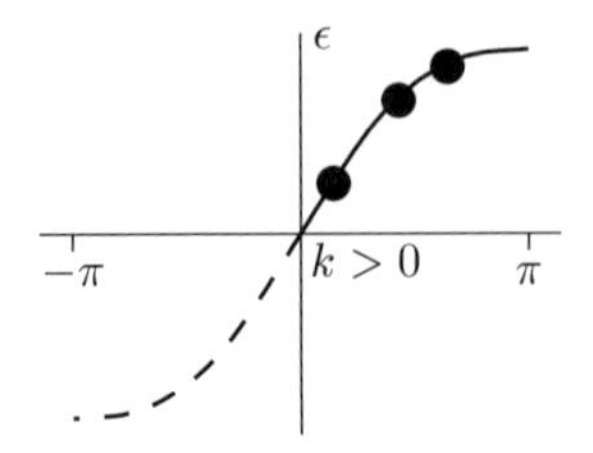

FIG. 4.2. Single-particle energy levels for the Majorana fermion exist only for $k > 0$. A many-body state is described by the occupation numbers n_k on these single-particle levels.

the many-body Hilbert space has 2^{N}_{site} states (or $\sqrt{2} \times \sqrt{2} = 2$ states per site), which is the same as the Hilbert space for one complex fermion. In fact, a system of two Majorana fermions is equivalent to a system with one complex fermion (see Problem 4.1.4).

Problem 4.1.4.
The superconducting Hamiltonian (4.1.6) can also be solved using Majorana-fermion operators. The Majorana-fermion operators are simply the real and imaginary parts of the complex fermion operators $c_{\boldsymbol{i}}$:

$$\lambda_{1,\boldsymbol{i}} = \frac{c_{\boldsymbol{i}} + c_{\boldsymbol{i}}^\dagger}{\sqrt{2}}, \qquad \lambda_{2,\boldsymbol{i}} = \frac{c_{\boldsymbol{i}} - c_{\boldsymbol{i}}^\dagger}{\mathrm{i}\sqrt{2}}$$

Show that the $\lambda_{a\boldsymbol{i}}$ satisfy the anti-commuting algebra of the Majorana fermions. Use the Majorana fermions to find all of the energy eigenstates and their energy eigenvalues. Compare your result with the one obtained in Problem 4.1.2.

4.1.4 Statistical algebra of hopping operators

- The statistics of identical particles is determined by the algebra of their hopping operators.

Usually, a boson system is defined as a system described by commuting operators and a fermion system is defined as a system described by anti-commuting operators. However, these definitions are too formal. To gain a physical understanding of the difference between a boson system and a fermion system, we would like to consider the following many-body hopping system. The Hilbert space is formed by a zero-particle state $|0\rangle$, one-particle states $|\boldsymbol{i}_1\rangle$, two-particle states $|\boldsymbol{i}_1, \boldsymbol{i}_2\rangle$, etc., where $\boldsymbol{i}_n$ labels the sites in a lattice. As an identical particle system, the state $|\boldsymbol{i}_1, \boldsymbol{i}_2, ...\rangle$ does not depend on the order of the indices $\boldsymbol{i}_1, \boldsymbol{i}_2, ...$. For example, $|\boldsymbol{i}_1, \boldsymbol{i}_2\rangle = |\boldsymbol{i}_2, \boldsymbol{i}_1\rangle$. There are no doubly-occupied sites and we assume that $|\boldsymbol{i}_1, \boldsymbol{i}_2, ...\rangle = 0$ if $\boldsymbol{i}_m = \boldsymbol{i}_n$.

A hopping operator $\hat{t}_{\boldsymbol{ij}}$ is defined as follows. When $\hat{t}_{\boldsymbol{ij}}$ acts on the state $|\boldsymbol{i}_1, \boldsymbol{i}_2, ...\rangle$, if there is a particle at site $\boldsymbol{j}$ but no particle at site $\boldsymbol{i}$, then $\hat{t}_{\boldsymbol{ij}}$ moves the particle at site $\boldsymbol{j}$ to the site $\boldsymbol{i}$ and multiplies a complex amplitude $t(\boldsymbol{i}, \boldsymbol{j}; \boldsymbol{i}_1, \boldsymbol{i}_2, ...)$ to the resulting state. Note that the amplitude may depend on the locations of all of the particles and may not be local. Otherwise, the hopping operator $\hat{t}_{\boldsymbol{ij}}$ annihilates the state. The Hamiltonian of our system is

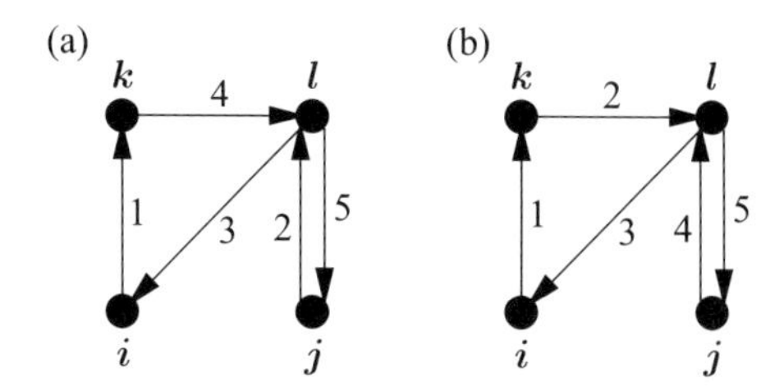

FIG. 4.3. (a) The first way to arrange the five hops swaps the two particles at i and j. (b) The second way to arrange the same five hops does not swap these two particles.

given by

$$H = \sum_{\langle ij \rangle} \hat{t}_{ij}$$

where the sum $\sum_{\langle ij \rangle}$ is over a certain set of pairs $\langle ij \rangle$, such as nearest-neighbor pairs. In order for the above Hamiltonian to represent a local system, we require that

$$[\hat{t}_{ij}, \hat{t}_{kl}] = 0$$

if i, j, k, and l are all different. Now the question is whether the above hopping Hamiltonian describes a hard-core boson system or a fermion system.

Whether the many-body hopping system is a boson system or a fermion system (or even some other statistical system) has nothing to do with the Hilbert space. The fact that the many-body states are labeled by symmetric indices (e.g. $|i_1, i_2\rangle = |i_2, i_1\rangle$) does not imply that the many-body system is a boson system, as we have seen in Section 4.1.1. The statistics are determined by the Hamiltonian H.

Clearly, when the hopping amplitude $t(i, j; i_1, i_2, ...)$ only depends on i and j, i.e. $t(i, j; i_1, i_2, ...) = t(i, j)$, the many-body hopping Hamiltonian will describe a hard-core boson system. The issue is under what condition the many-body hopping Hamiltonian describes a fermion system.

This problem was solved by Levin and Wen (2003). It was found that the many-body hopping Hamiltonian describes a fermion system if the hopping operators satisfy

$$\hat{t}_{lk}\hat{t}_{il}\hat{t}_{lj} = -\hat{t}_{lj}\hat{t}_{il}\hat{t}_{lk} \tag{4.1.9}$$

for any three hopping operators $\hat{t}_{lj}$, $\hat{t}_{il}$, and $\hat{t}_{lk}$, with i, j, k, and l all being different. (Note that the algebra has the structure $\hat{t}_1\hat{t}_2\hat{t}_3 = -\hat{t}_3\hat{t}_2\hat{t}_1$.)

Consider the state $|i, j, ...\rangle$ with two particles at i, j, and possibly other particles further away. We apply a set of five hopping operators $\{\hat{t}_{jl}, \hat{t}_{lk}, \hat{t}_{il}, \hat{t}_{lj}, \hat{t}_{ki}\}$ to the state $|i, j, ...\rangle$, but in different orders (see Fig. 4.3):

$$\hat{t}_{jl}\hat{t}_{lk}\hat{t}_{il}\hat{t}_{lj}\hat{t}_{ki}|i, j, ...\rangle = C_1|i, j, ...\rangle$$
$$\hat{t}_{jl}\hat{t}_{lj}\hat{t}_{il}\hat{t}_{lk}\hat{t}_{ki}|i, j, ...\rangle = C_2|i, j, ...\rangle$$

where we have assumed that there are no particles at sites k and l. We note that after five hops we return to the original state $|i, j, ...\rangle$ with additional phases $C_{1,2}$. However, from Fig. 4.3, we see that the first way to arrange the five hops (Fig. 4.3(a)) swaps the two particles at i and j, while the second way (Fig. 4.3(b)) does not swap these two particles. As the two hopping schemes use the same set of five hops, the difference between C_1 and C_2 is due to exchanging the two particles. Thus, we require $C_1 = -C_2$ in order for the many-body hopping Hamiltonian to describe a fermion system. Noting that the first and the last hops are the same in the two hopping schemes, we find that $C_1 = -C_2$ if the hopping operators satisfy eqn (4.1.9). In fact, eqn (4.1.9) serves as an alternative definition of Fermi statistics if we do not want to use the anti-commuting algebra.

We would like to remark that, on a two-dimensional lattice, the fermion hopping algebra can be generalized to the more general statistical algebra for hopping operators:

$$\hat{t}_{ji}\hat{t}_{ik}\hat{t}_{li} = \mathrm{e}^{\mathrm{i}\theta}\hat{t}_{li}\hat{t}_{ik}\hat{t}_{ji} \tag{4.1.10}$$

In this case, the many-body hopping Hamiltonian describes an anyon system with statistical angle θ.

Problem 4.1.5.
Show that the fermion hopping operator $\hat{t}_{ij} = c_i^\dagger c_j$ satisfies the fermion hopping algebra.

Problem 4.1.6.
We know that the string operator $\hat{W}_{ij} = \hat{t}_{il}\hat{t}_{lm}...\hat{t}_{nj}$ creates a particle at i and annihilates a particle at j. So we may want to write $\hat{W}_{ij} = C_i^+ C_j^-$.
(a) Show that, if the hopping operators $\hat{t}_{ij}$ satisfy the fermion hopping algebra (4.1.9), then the string operators satisfy $\hat{W}_{lk}\hat{W}_{il}\hat{W}_{lj} = -\hat{W}_{lj}\hat{W}_{il}\hat{W}_{lk}$. (You may assume that the three strings only intersect at one point l.)
(b) Show that $C_i^\pm$ and $C_j^\pm$ cannot commute, even when i and j are far apart. However, if $C_i^\pm$ and $C_j^\pm$ anti-commute (for $i \neq j$), then the fermion hopping algebra (4.1.9) can be satisfied.

4.2 Free fermion Green's function

- The algebraic decay of the Green's function over a long time is related to gapless excitations across the Fermi surface. The algebraic decay at long distance is related to the sharp features (discontinuity) in momentum space.
- Electron Green's functions can be measured in tunneling experiments.
- The spectral function of an electron Green's function.

In the next few sections, we will discuss one-body and two-body correlation functions of free fermion systems. One-body correlation functions are important for the understanding of tunneling and photoemission experiments. Two-body correlation functions are even more important, because they are related to all kinds of transport, scattering, and linear response experiments. In these two sections I will include some calculation details. The purpose of the discussion is mostly to introduce mathematical formalisms. Many explicit results are listed for $d = 1, 2, 3$

dimensions. I hope that these results may be useful as references and as concrete examples of various correlations.

4.2.1 Time-ordered correlation functions

- Fermion operators behave like anti-commuting numbers under the time-ordered average.

Consider a free fermion system described by eqn (4.1.5). In the Heisenberg picture, the time-dependent fermion operator is given by

$$c_{\boldsymbol{k}}(t) = U^\dagger(t, -\infty) c_{\boldsymbol{k}} U(t, -\infty), \qquad U(t_1, t_2) = \mathrm{e}^{-\mathrm{i} \int_{t_2}^{t_1} \mathrm{d}t\, H}$$

The fermion propagator (the Green's function) at zero temperature and at finite temperatures is defined as the time-ordered average

$$\begin{aligned} \mathrm{i}G(t_1 - t_2, \boldsymbol{k}_1)\delta_{\boldsymbol{k}_1,\boldsymbol{k}_2} &= \langle 0|T(c_{\boldsymbol{k}_1}(t_1)c^\dagger_{\boldsymbol{k}_2}(t_2))|0\rangle \qquad (4.2.1)\\ &= \begin{cases} +\langle 0|c_{\boldsymbol{k}_1}(t_1)c^\dagger_{\boldsymbol{k}_2}(t_2)|0\rangle, & \text{for } t_1 > t_2 \\ -\langle 0|c^\dagger_{\boldsymbol{k}_2}(t_2)c_{\boldsymbol{k}_1}(t_1)|0\rangle, & \text{for } t_1 < t_2 \end{cases} \qquad \text{for } T = 0 \\ \mathrm{i}G^\beta(t_1 - t_2, \boldsymbol{k}_1)\delta_{\boldsymbol{k}_1,\boldsymbol{k}_2} &= \langle T(c_{\boldsymbol{k}_1}(t_1)c^\dagger_{\boldsymbol{k}_2}(t_2))\rangle \\ &\equiv \frac{\mathrm{Tr}[T(c_{\boldsymbol{k}_1}(t_1)c^\dagger_{\boldsymbol{k}_2}(t_2))\mathrm{e}^{-\beta H}]}{Z} \qquad (4.2.2)\\ &= \begin{cases} +\mathrm{Tr}[c_{\boldsymbol{k}_1}(t_1)c^\dagger_{\boldsymbol{k}_2}(t_2)\mathrm{e}^{-\beta H}], & \text{for } t_1 > t_2 \\ -\mathrm{Tr}[c^\dagger_{\boldsymbol{k}_2}(t_2)c_{\boldsymbol{k}_1}(t_1)\mathrm{e}^{-\beta H}], & \text{for } t_1 < t_2 \end{cases} \qquad \text{for } T > 0 \end{aligned}$$

Note the minus sign in the above definition, and that, by definition,

$$\left\langle T\left(c^\dagger(t)c(t')\right)\right\rangle = -\left\langle T\left(c(t')c^\dagger(t)\right)\right\rangle$$

Knowing the fermion occupation at the $\boldsymbol{k}$ state, namely

$$n_F(\xi_{\boldsymbol{k}}) = \langle c^\dagger_{\boldsymbol{k}} c_{\boldsymbol{k}}\rangle = \frac{1}{\mathrm{e}^{\beta\xi_{\boldsymbol{k}}} + 1}$$

we can calculate the zero-temperature Green's function G and the finite-temperature Green's function G^β in momentum space:

$$\begin{aligned} \mathrm{i}G(t, \boldsymbol{k}) &= +\Theta(t)\Theta(\xi_{\boldsymbol{k}})\mathrm{e}^{-\mathrm{i}t\xi_{\boldsymbol{k}}} - \Theta(-t)\Theta(-\xi_{\boldsymbol{k}})\mathrm{e}^{-\mathrm{i}(-t)(-\xi_{\boldsymbol{k}})}\Big|_{T=0} \\ \mathrm{i}G^\beta(t, \boldsymbol{k}) &= +\Theta(t)(1 - n_F(\xi_{\boldsymbol{k}}))\mathrm{e}^{-\mathrm{i}t\xi_{\boldsymbol{k}}} - \Theta(-t)n_F(\xi_{\boldsymbol{k}})\mathrm{e}^{-\mathrm{i}t\xi_{\boldsymbol{k}}}\Big|_{T>0} \qquad (4.2.3) \end{aligned}$$

In the ω–$\boldsymbol{k}$ space, $G(\omega,\boldsymbol{k}) = \int \mathrm{d}t\ G(t,\boldsymbol{k})\mathrm{e}^{\,\mathrm{i}\omega t}$ and $G^\beta(\omega,\boldsymbol{k}) = \int \mathrm{d}t\, G^\beta(t,\boldsymbol{k})\mathrm{e}^{\,\mathrm{i}\omega t}$ have the following simpler forms:

$$G(\omega,\boldsymbol{k}) = \frac{1}{\omega - \xi_{\boldsymbol{k}} + \mathrm{i}0^+\mathrm{sgn}(\omega)} \qquad \text{for } T=0 \tag{4.2.4}$$

$$G^\beta(\omega,\boldsymbol{k}) = \frac{1-n_F(\xi_{\boldsymbol{k}})}{\omega - \xi_{\boldsymbol{k}} + \mathrm{i}0^+} + \frac{n_F(\xi_{\boldsymbol{k}})}{\omega - \xi_{\boldsymbol{k}} - \mathrm{i}0^+} \qquad \text{for } T>0 \tag{4.2.5}$$

Now let us consider the imaginary-time Green's function at finite temperatures. The time-dependent operator in the Heisenberg picture is now given by

$$c_{\boldsymbol{k}}(\tau) = \mathrm{e}^{H\tau} c_{\boldsymbol{k}} \mathrm{e}^{-H\tau}$$

From the definition ($0<\tau_1,\tau_2<\beta$)

$$\mathcal{G}^\beta(\tau_1,\tau_2,\boldsymbol{k}_1)\delta_{\boldsymbol{k}_1,\boldsymbol{k}_2} = \begin{cases} +\dfrac{\mathrm{Tr}(\mathrm{e}^{-\beta H} c_{\boldsymbol{k}_1}(\tau_1)c^\dagger_{\boldsymbol{k}_2}(\tau_2))}{\mathrm{Tr}(\mathrm{e}^{-\beta H})}, & \text{for } \tau_1>\tau_2 \\ -\dfrac{\mathrm{Tr}(\mathrm{e}^{-\beta H} c_{\boldsymbol{k}_2}(\tau_2)c^\dagger_{\boldsymbol{k}_1}(\tau_1))}{\mathrm{Tr}(\mathrm{e}^{-\beta H})}, & \text{for } \tau_1<\tau_2 \end{cases} \tag{4.2.6}$$

one can show that (see Problem 4.2.1)

$$\begin{aligned} \mathcal{G}^\beta(\tau_1,\tau_2,\boldsymbol{k}) &= \mathcal{G}^\beta(\tau_1-\tau_2,0,\boldsymbol{k}) \equiv \mathcal{G}^\beta(\tau_1-\tau_2,\boldsymbol{k}) \\ \mathcal{G}^\beta(\tau,\boldsymbol{k}) &= -\,\mathcal{G}^\beta(\tau+\beta,\boldsymbol{k}) \end{aligned} \tag{4.2.7}$$

Thus, the fermion Green's function is anti-periodic in the compactified imaginary-time direction. For free fermions, we have

$$\mathcal{G}^\beta(\tau,\boldsymbol{k}) = +\Theta(\tau)(1-n_F(\xi_{\boldsymbol{k}}))\mathrm{e}^{-\tau\xi_{\boldsymbol{k}}} - \Theta(-\tau)n_F(\xi_{\boldsymbol{k}})\mathrm{e}^{-\tau\xi_{\boldsymbol{k}}} \tag{4.2.8}$$

In the ω–$\boldsymbol{k}$ space, we have

$$\mathcal{G}^\beta(\omega_\gamma,\boldsymbol{k}) \equiv \int_0^\beta \mathrm{d}\tau\ \mathcal{G}^\beta(\tau,\boldsymbol{k})\mathrm{e}^{\,\mathrm{i}\omega_\gamma\tau} = \frac{1}{-\mathrm{i}\omega_\gamma + \xi_{\boldsymbol{k}}} \tag{4.2.9}$$

where $\omega_\gamma = 2\pi\gamma T$ with $\gamma = \frac{1}{2} + \text{integer}$.

The time-ordered Green's function in ω–$\boldsymbol{k}$ space can be written as a *time-ordered* average of the fermion operator in ω–$\boldsymbol{k}$ space. For imaginary time, we introduce

$$c_{\omega_\gamma\boldsymbol{k}} = \int_0^\beta \mathrm{d}\tau\ c_{\boldsymbol{k}}(\tau)\beta^{-1/2}\mathrm{e}^{\,\mathrm{i}\omega_\gamma\tau}$$

After a little work, we find that

$$\begin{aligned} \left\langle T(c_{\omega_\gamma\boldsymbol{k}}c^\dagger_{\omega'_\gamma\boldsymbol{k}})\right\rangle &\equiv \beta^{-1}\int_0^\beta \mathrm{d}\tau\mathrm{d}\tau'\ \mathrm{e}^{\,\mathrm{i}\omega_\gamma\tau - \mathrm{i}\omega'_\gamma\tau'}\left\langle T\left(c_{\boldsymbol{k}}(\tau)c^\dagger_{\boldsymbol{k}}(\tau')\right)\right\rangle \\ &= \mathcal{G}^\beta(\omega_\gamma,\boldsymbol{k})\delta_{\omega_\gamma-\omega'_\gamma} \end{aligned} \tag{4.2.10}$$

As $\langle T(c_{\boldsymbol{k}}^\dagger(\tau)c_{\boldsymbol{k}}(\tau'))\rangle = -\langle T(c_{\boldsymbol{k}}(\tau')c_{\boldsymbol{k}}^\dagger(\tau))\rangle$, we have $\langle T(c_{\omega_\gamma \boldsymbol{k}}^\dagger c_{\omega_\gamma \boldsymbol{k}})\rangle = -\langle T(c_{\omega_\gamma \boldsymbol{k}}c_{\omega_\gamma \boldsymbol{k}}^\dagger)\rangle$. Similarly, one can show that, for real time,

$$\left\langle T(c_{\omega\boldsymbol{k}}c_{\omega'\boldsymbol{k}}^\dagger)\right\rangle = -\left\langle T(c_{\omega\boldsymbol{k}}^\dagger c_{\omega'\boldsymbol{k}})\right\rangle = iG^\beta(\omega,\boldsymbol{k})(2\pi)^{-1}\delta(\omega-\omega')$$

This allows us to perform calculations of time-ordered correlations directly in ω–$\boldsymbol{k}$ space. We also note that, within the time-ordered average, c and $c^\dagger$ behave like anti-commuting numbers.

In the above, we discussed time-ordered correlations of two operators. How do we calculate time-ordered correlations of many operators? Here we have the Wick theorem for fermions. Let O_i be *linear* combinations of c and $c^\dagger$. For $W = O_1O_2...O_n$, we have

$$\begin{aligned} W = \ :W: +(-)^{j_1-i_1-1}\sum_{(i_1,j_1)} :W_{i_1,j_1}: \langle 0|O_{i_1}O_{j_1}|0\rangle \\ \pm \sum_{\substack{(i_1,j_1),(i_2,j_2)\\(i_1,j_1)\neq(i_2,j_2)}} :W_{i_1,j_1,i_2,j_2}: \langle 0|O_{i_1}O_{j_1}|0\rangle\langle 0|O_{i_2}O_{j_2}|0\rangle + ... \end{aligned} \tag{4.2.11}$$

where (i_1,j_1) is an ordered pair with $i_1 < j_1$, and $\pm$ is determined by the even or odd number of permutations needed to bring O_{j_1} to the right-hand side of O_{i_1} and O_{j_2} to the right-hand side of O_{i_2}. As $\langle 0| :W: |0\rangle = 0$ by definition, we find that

$$\langle 0|O_1O_2...O_n|0\rangle = \sum(-)^P\langle 0|O_{i_1}O_{i_2}|0\rangle...\langle 0|O_{i_{n-1}}O_{i_n}|0\rangle$$

where $\sum$ is a sum over all of the possible ways to group $1,2,...,n$ into ordered pairs (i_1,i_2), (i_3,i_4),... (i.e. $i_1 < i_2$, $i_3 < i_4$, etc.). Here $(-)^P = 1$ if $1,2,...,n$ and $i_1,i_2,...,i_n$ differ by an even number of permutations, and $(-)^P = -1$ if $1,2,...,n$ and $i_1,i_2,...,i_n$ differ by an odd number of permutations. For example,

$$\begin{aligned} &\langle 0|O_1O_2O_3O_4|0\rangle \\ &= \langle 0|O_1O_2|0\rangle\langle 0|O_3O_4|0\rangle - \langle 0|O_1O_3|0\rangle\langle 0|O_2O_4|0\rangle + \langle 0|O_1O_4|0\rangle\langle 0|O_2O_3|0\rangle \end{aligned}$$

When applied to a time-ordered correlation, we have

$$\langle 0|T(O_1O_2...O_n)|0\rangle = \sum(-)^P\langle 0|T(O_{i_1}O_{i_2})|0\rangle...\langle 0|T(O_{i_{n-1}}O_{i_n})|0\rangle$$

The Wick theorem allows us to express a many-operator correlation in terms of two-operator correlations. The Wick theory also implies that $\langle 0|T(...O_m...O_n...)|0\rangle = -\langle 0|T(...O_n...O_m...)|0\rangle$. Thus, c and $c^\dagger$ behave like anti-commuting numbers in the time-ordered averages of many c and $c^\dagger$ operators

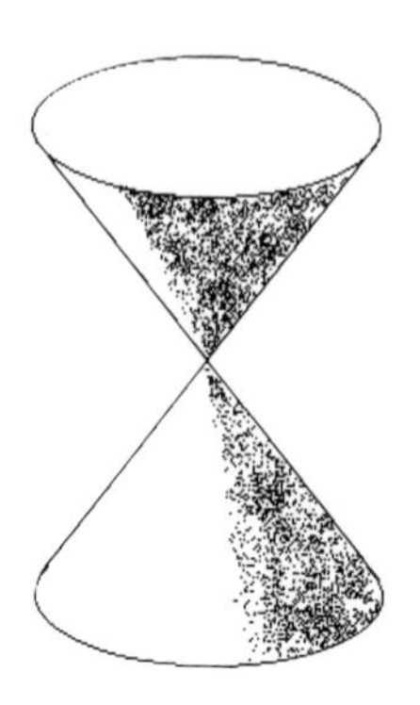

FIG. 4.4. The density of states near the 'Fermi point' vanishes like $N(\xi) \propto |\xi|^{d-1}$.

(for both real time and imaginary time). This fact allows us to construct a path integral formulation of fermion systems (see Section 5.4.1).

Problem 4.2.1.
Prove eqns (4.2.7) and (4.2.9).

Problem 4.2.2.
Prove eqn (4.2.8) from eqn (4.2.6).

Problem 4.2.3.
Prove eqn (4.2.10).

Problem 4.2.4.
Show the validity of the Wick theorem for the cases of $W = c_{k_1} c^\dagger_{k_2} c_{k_3} c^\dagger_{k_4}$ and $W = c_{k_1} c_{k_2} c^\dagger_{k_3} c^\dagger_{k_4}$.

4.2.2 Equal-space Green's function and tunneling

- The algebraic decay of the Green's function over a long time is related to gapless excitations across the Fermi surface.

The Green's function in real space–time is given by $\mathrm{i}G(t, \boldsymbol{x}_1, \boldsymbol{x}_2) = \langle T(c(t, \boldsymbol{x}_1) c(0, \boldsymbol{x}_2)\rangle$. For translation-invariant systems (such as the free fermions systems), the Green's function only depends on $\boldsymbol{x}_1 - \boldsymbol{x}_2$, so that $G(t, \boldsymbol{x}_1, \boldsymbol{x}_2) = G(t, \boldsymbol{x}_1 - \boldsymbol{x}_2)$. For free fermion systems, we have $G(t, \boldsymbol{x}) = \int \frac{\mathrm{d}^d \boldsymbol{k}}{(2\pi)^d} G(t, \boldsymbol{k})$. When $\boldsymbol{x} = 0$ (or when the fermion operators in $\langle T(c(\boldsymbol{x}_1, t_1) c^\dagger(\boldsymbol{x}_2, t_2))\rangle$ are at an equal space point $\boldsymbol{x}_1 = \boldsymbol{x}_2$), we have

$$\mathrm{i}G(t, 0) = \int \frac{\mathrm{d}^d \boldsymbol{k}}{(2\pi)^d} \left(+\Theta(t)\Theta(\xi_{\boldsymbol{k}})\mathrm{e}^{-\mathrm{i}t\xi_{\boldsymbol{k}}} - \Theta(-t)\Theta(-\xi_{\boldsymbol{k}})\mathrm{e}^{-\mathrm{i}t\xi_{\boldsymbol{k}}} \right)$$

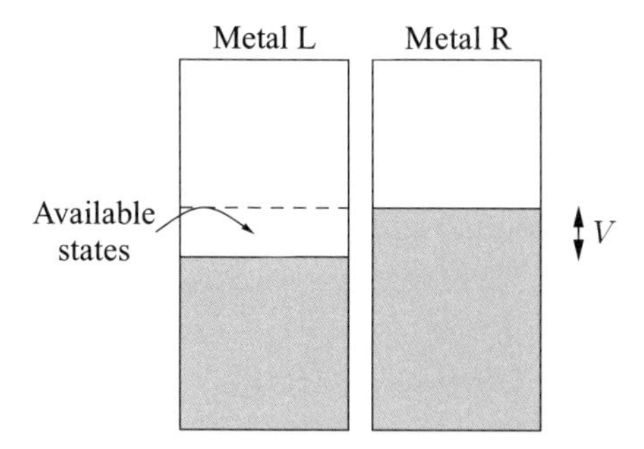

FIG. 4.5. The number of the available states for tunneling is proportional to V

The above can be evaluated by introducing the density of states (number of states per unit volume per unit energy)

$$N(\epsilon) = \int \frac{d^d \boldsymbol{k}}{(2\pi)^d} \delta(\epsilon_{\boldsymbol{k}} - \epsilon)$$

where

$$\begin{aligned}
N(\epsilon) &= \sqrt{\frac{m}{2\pi^2 \epsilon}} && \text{for } d = 1 \\
N(\epsilon) &= \frac{m}{2\pi} && \text{for } d = 2 \\
N(\epsilon) &= \frac{m\sqrt{2m\epsilon}}{2\pi^2} && \text{for } d = 3
\end{aligned} \tag{4.2.12}$$

At $T = 0$ and for large t, only $N(\epsilon)$ near $\epsilon = E_F$ is important. Thus, we may assume that $N(\epsilon) = N(E_F)$ is a constant. After substituting $\int d^d\, \boldsymbol{k}/(2\pi)^d = \int d\xi\, N(\xi + E_F)$, for large t, we find that

$$G(t, 0) = -\frac{N(E_F)}{t - i0^+ \text{sgn}(t)}$$

The algebraic long-time correlation is a consequence of the discontinuity in the density of the occupied states. If the density of states vanishes like $N(\epsilon) \propto |\epsilon - E_F|^g$ (see Fig. 4.4), then the long-time correlation will have a faster decay (assuming that $g > 0$):

$$G(t, 0) \propto -i\,\text{sgn}(t) e^{-i\frac{\pi}{2}(1+g)} \frac{1}{|t|^{1+g}}$$

The fermion Green's function is meaningful only if we can measure it in experiments. To see how to measure the fermion Green's function in experiments, consider tunneling between two metals with densities of states $N_{R,L}(\xi) \propto |\xi|^{g_{R,L}}$. (Note that $g_{R,L} = 0$ for ordinary metals.) Assuming that the tunneling Hamiltonian is given by

$$H_T = \Gamma(\mathrm{e}^{\mathrm{i}Vt} c_R^\dagger c_L + h.c.)$$

the tunneling current from left to right is (see eqn (3.7.15))

$$\begin{aligned}\langle j_T \rangle (t) &= -\Gamma^2 \int^t \mathrm{d}t' \left(\mathrm{e}^{\mathrm{i}V(t-t')} \left\langle [I(t), I^\dagger(t')] \right\rangle + h.c. \right) \\ &= (-\mathrm{i}\Gamma^2 D^\beta(V) + h.c.) = 2\Gamma^2 \mathrm{Im} D^\beta(V)\end{aligned}$$

where $I(t) = c_R^\dagger(t,0)c_L(t,0)$ is the tunneling operator. The correlation of the tunneling operator $I(t) = c_R^\dagger(t,0)c_L(t,0)$ is (assume that $t > 0$ and $T = 0$)

$$\begin{aligned}\left\langle I(t)I^\dagger(0) \right\rangle &= \left\langle c_L(t,0)c_L^\dagger(0,0) \right\rangle \left\langle c_R^\dagger(t,0)c_R(0,0) \right\rangle \\ &= -G_L(t,0)(-)G_R(-t,0) \\ &\propto \mathrm{e}^{-\mathrm{i}\frac{\pi}{2}(2+g_R+g_L)} \frac{1}{t^{2+g_R+g_L}}\end{aligned}$$

We see that the tunneling experiment measures the product of the equal-space Green's functions on the two sides of the junction. The above correlation has the same form as the correlation of the tunneling operator between two $(1+1)$-dimensional interacting boson systems. Repeating the calculation in Section 3.7.4, we find the tunneling current to be

$$I \propto V^{1+g_R+g_L} = V \times V^{g_R} \times V^{g_L}$$

When $g_L = g_R = 0$, it is very easy to see that the available states for tunneling are proportional to V (Fig. 4.5), and hence $I \propto V$. When g_L and g_R are not zero, the available states will have corresponding suppressions, and hence the tunneling current is suppressed by the factor $V^{g_R} \times V^{g_L}$.

We would like to stress that tunneling directly measures the fermion Green's function. Although we only discussed free fermion systems above, the results apply equally well to interacting fermions. Imagine that interactions change the long-time decay exponent g in the fermion Green's function; then such a change can be measured by tunneling experiments.

4.2.3 Fermion spectral function

- The spectral function of the fermion Green's function.
- Fermion Green's functions (or, more precisely, the overlap of fermion spectral functions) can be measured in tunneling experiments.

For interacting electrons, $G_{L,R}(t,0)$ are more complicated. To understand tunneling between two interacting fermion systems, it is useful to introduce the spectral representation of the fermion Green's functions as follows:[29]

$$\mathrm{i}G^{\beta}(t,0) = \begin{cases} \int \mathrm{d}\omega\, A^0_+(\omega)\mathrm{e}^{-\mathrm{i}\omega t}, & \text{for } t > 0 \\ \eta \int \mathrm{d}\omega\, A^0_-(\omega)\mathrm{e}^{-\mathrm{i}\omega t}, & \text{for } t < 0 \end{cases}$$

where $A^0_{+,-}(\omega)$ are the following spectral functions:

$$A^0_+(\nu) = \sum_{m,n} \delta[\nu - (\epsilon_m - \epsilon_n)] \langle \psi_n | c(\boldsymbol{x}) | \psi_m \rangle \langle \psi_m | c^\dagger(\boldsymbol{x}) | \psi_n \rangle \frac{\mathrm{e}^{-\epsilon_n \beta}}{Z}$$

$$A^0_-(\nu) = \sum_{m,n} \delta[\nu + (\epsilon_m - \epsilon_n)] \langle \psi_n | c^\dagger(\boldsymbol{x}) | \psi_m \rangle \langle \psi_m | c(\boldsymbol{x}) | \psi_n \rangle \frac{\mathrm{e}^{-\epsilon_n \beta}}{Z}$$

and $\eta = -1$ for the fermion operators ($\eta = 1$ if we want to introduce a spectral representation of the boson Green's functions). We can also introduce spectral functions in momentum space:

$$\mathrm{i}G^{\beta}(t,\boldsymbol{k}) = \begin{cases} \int \mathrm{d}\omega\, A_+(\omega,\boldsymbol{k})\mathrm{e}^{-\mathrm{i}\omega t} & \text{For } t > 0 \\ \eta \int \mathrm{d}\omega\, A_-(\omega,\boldsymbol{k})\mathrm{e}^{-\mathrm{i}\omega t} & \text{For } t < 0 \end{cases} \tag{4.2.13}$$

where

$$A_+(\nu,\boldsymbol{k}) = \sum_{m,n} \delta[\nu - (\epsilon_m - \epsilon_n)] \langle \psi_n | c_{\boldsymbol{k}} | \psi_m \rangle \langle \psi_m | c^\dagger_{\boldsymbol{k}} | \psi_n \rangle \frac{\mathrm{e}^{-\epsilon_n \beta}}{Z}$$

$$A_-(\nu,\boldsymbol{k}) = \sum_{m,n} \delta[\nu + (\epsilon_m - \epsilon_n)] \langle \psi_n | c^\dagger_{\boldsymbol{k}} | \psi_m \rangle \langle \psi_m | c_{\boldsymbol{k}} | \psi_n \rangle \frac{\mathrm{e}^{-\epsilon_n \beta}}{Z}$$

We see that

$$A^0_\pm(\omega) = \mathcal{V}^{-1} \sum_{\boldsymbol{k}} A_\pm(\omega,\boldsymbol{k})$$

From their definition, we also see that $A^0_{+,-}$ and $A_{+,-}$ are real and positive. In the ω–$\boldsymbol{k}$ space, we find that

$$G^{\beta}_+(\omega,\boldsymbol{k}) \equiv \int \mathrm{d}t\, \Theta(t) G^{\beta}(t,\boldsymbol{k}) \mathrm{e}^{\mathrm{i}\omega t} = \int \mathrm{d}\nu\, \frac{A_+(\nu,\boldsymbol{k})}{\omega - \nu + \mathrm{i}0^+}$$

$$G^{\beta}_-(\omega,\boldsymbol{k}) \equiv \int \mathrm{d}t\, \Theta(-t) G^{\beta}(t,\boldsymbol{k}) \mathrm{e}^{\mathrm{i}\omega t} = -\eta \int \mathrm{d}\nu\, \frac{A_-(\nu,\boldsymbol{k})}{\omega - \nu - \mathrm{i}0^+} \tag{4.2.14}$$

The spectral functions in momentum space $A_{+,-}(\nu,\boldsymbol{k})$ completely determine the Green's function.

[29] The following discussion also applies to boson operators if we choose $\eta = 1$.

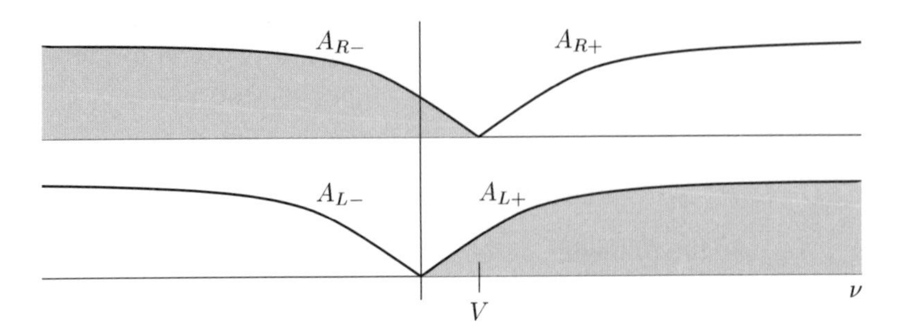

FIG. 4.6. The spectral functions and their overlaps.

The response function $\langle [I(t), I^\dagger(0)] \rangle$ can be expressed in terms of the spectral functions in the R and L metals. We find, for $t > 0$, that

$$\begin{aligned}
&\langle [I(t), I^\dagger(0)] \rangle \\
&= \left\langle c_L(t,0)c_L^\dagger(0,0)\right\rangle \left\langle c_R^\dagger(t,0)c_R(0,0)\right\rangle - \left\langle c_L^\dagger(0,0)c_L(t,0)\right\rangle \left\langle c_R(0,0)c_R^\dagger(t,0)\right\rangle \\
&= (\mathrm{i}G_L(t))(-\mathrm{i}G_R(-t)) - (-\mathrm{i}G_L(-t))^*(\mathrm{i}G_R(t))^* \qquad (4.2.15) \\
&= -\int \mathrm{d}\omega_L \mathrm{d}\omega_R \, \left(A^0_{L+}(\omega_L)A^0_{R-}(\omega_R) - A^0_{L-}(\omega_L)A^0_{R+}(\omega_R)\right) \mathrm{e}^{\,\mathrm{i}(\omega_R-\omega_L)t}
\end{aligned}$$

Thus, the response function $\mathrm{i}D^\beta(t) = \Theta(t)[I(t), I^\dagger(0)]$ in ω space is given by

$$\begin{aligned}
D^\beta(\omega) &= \int \mathrm{d}t \; D^\beta(t)\mathrm{e}^{\,\mathrm{i}\omega t} \qquad (4.2.16) \\
&= \int \mathrm{d}\omega_L \mathrm{d}\omega_R \, \frac{A^0_{L+}(\omega_L)A^0_{R-}(\omega_R) - A^0_{L-}(\omega_L)A^0_{R+}(\omega_R)}{\omega - \omega_L + \omega_R + \mathrm{i}0^+}
\end{aligned}$$

The tunneling current from left to right is determined by the imaginary part:

$$\begin{aligned}
\langle j_T \rangle (t) &= 2\Gamma^2 \mathrm{Im} D^\beta(V) \\
&= 2\pi\Gamma^2 \int \mathrm{d}\nu \, \left(A^0_{L-}(\nu)A^0_{R+}(\nu - V) - A^0_{L+}(\nu)A^0_{R-}(\nu - V)\right)
\end{aligned} \qquad (4.2.17)$$

At zero temperature, $A_+(\omega)$ is nonzero only for $\omega > 0$ and $A_-(\omega)$ is nonzero only for $\omega < 0$. Thus, only one of the two terms contributes, depending on the sign of V (see Fig. 4.6). The contribution is an overlap integral of the spectral functions on the two sides of the tunneling junction.

For free fermions, from the imaginary-time Green's function (4.2.9), we find that (see eqn (2.2.37))

$$A_-(\omega, \boldsymbol{k}) = n_F(\xi_{\boldsymbol{k}})\delta(\omega - \xi_{\boldsymbol{k}}), \qquad A_+(\omega, \boldsymbol{k}) = (1 - n_F(\xi_{\boldsymbol{k}}))\delta(\omega - \xi_{\boldsymbol{k}})$$

(Substituting the above into eqn (4.2.14) allows us to recover the free electron Green's function (4.2.5) at finite temperatures.) After integrating over $\boldsymbol{k}$, we obtain

$$A^0_-(\omega) = n_F(\omega)N(\omega + E_F), \qquad A^0_+(\omega) = (1 - n_F(\omega))N(\omega + E_F) \quad (4.2.18)$$

The tunneling current is

$$\begin{aligned}\langle j_T \rangle (t) = 2\pi\Gamma^2 \int \mathrm{d}\nu \, [&(1 - n_F(\nu))N_L(\nu + E_F)n_F(\nu - V)N_R(\nu - V + E_F) \\ &- n_F(\nu)N_L(\nu + E_F)(1 - n_F(\nu - V))N_R(\nu - V + E_F)] \\ \approx &2\pi\Gamma^2 N_L(E_F)N_R(E_F) \int \mathrm{d}\nu \, (n_F(\nu - V) - n_F(\nu))\end{aligned}$$

which is a very simple result and can be obtained directly using second-order perturbation theory.

Problem 4.2.5.
Calculate the time-ordered fermion Green's function in real space–time for $d = 1$ dimension and in the $(t, x) \to \infty$ limit (but t/x is arbitrary). Check your result in the $t/x \to 0$ and $t/x \to \infty$ limits.

Problem 4.2.6.
Two identical free fermion systems in $d = 1$ dimension are connected at two points located at $x = 0$ and $x = a$. The tunneling Hamiltonian is given by

$$H_T = \Gamma \left[[c_L(0)c_R(0)^\dagger + c_L(a)c_R(a)^\dagger] + h.c. \right]$$

Use the result of the previous problem to find the tunneling conductance as a function of a.

Problem 4.2.7.
Prove the spectral sum rule

$$\int \mathrm{d}\omega \, (A_+(\omega, \boldsymbol{k}) + A_-(\omega, \boldsymbol{k})) = 1$$

for finite temperatures. (Hint: Use $\{c_{\boldsymbol{k}}, c^\dagger_{\boldsymbol{k}}\} = 1$.)

Problem 4.2.8.
Tunneling between parallel two-dimensional electron systems (Eisenstein *et al.*, 1991): Consider two identical fermion systems in two dimensions connected by a vertical area

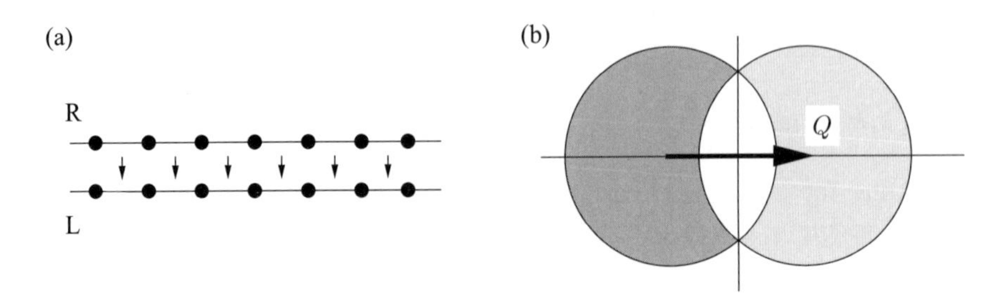

FIG. 4.7. (a) A vertical area tunneling junction, and (b) two shifted Fermi surfaces.

tunneling junction that conserves the two-dimensional momentum (see Fig. 4.7(a)). The tunneling Hamiltonian has the form

$$H_T = A\sum_{\boldsymbol{k}}[c_{L\boldsymbol{k}}c^\dagger_{R\boldsymbol{k}} + h.c.]$$

In the absence of a magnetic field parallel to the two-dimensional layers, the fermions in both layers have the same dispersion relation $\epsilon_{\boldsymbol{k}}$. In the presence of a magnetic field B parallel to the two-dimensional layers, the momenta of the fermions in the two layers have a relative shift (see Fig. 4.7(b)) and the dispersions in the two layers become

$$\epsilon_{R\boldsymbol{k}} = \epsilon_{\boldsymbol{k}-\frac{1}{2}\boldsymbol{Q}}, \qquad \epsilon_{L\boldsymbol{k}} = \epsilon_{\boldsymbol{k}+\frac{1}{2}\boldsymbol{Q}}$$

where $Q \propto B$ and $\boldsymbol{Q}\cdot\boldsymbol{B} = 0$.

1. Show that for free fermions the finite temperature tunneling current is zero for $B = 0$ and a finite voltage $V \neq 0$.
2. Assume that, due to interaction, the spectral functions of the fermions have the forms

$$A_+(\omega,\boldsymbol{k}) = (1-n_F(\omega))\frac{\pi^{-1}\Gamma}{(\omega-\xi_{\boldsymbol{k}})^2+\Gamma^2}, \qquad A_-(\omega,\boldsymbol{k}) = n_F(\omega)\frac{\pi^{-1}\Gamma}{(\omega-\xi_{\boldsymbol{k}})^2+\Gamma^2}$$

   where $\Gamma \ll E_F$ is a decay rate (note that the above $A_\pm$ satisfy eqn (2.2.28)). Find an expression for the tunneling current for small but finite temperatures ($T \ll E_F$).
3. What is the temperature dependence of the tunneling conductance in the $T \to 0$ and $T \gg \Gamma$ limits for $B = 0$? Estimate the tunneling conductance per area at $T = 0$.
4. Assuming that at zero temperature the tunneling conductance for $B = 0$ is σ_0, estimate the tunneling conductance σ_Q for finite Q in terms of σ_0. What is the behavior of σ_Q for Q near 0 and near $2k_F$.

4.2.4 Equal-time Green's function and the shape of the Fermi surface

- The algebraic decay at long distances is related to the sharp features (discontinuity) of the fermion occupation number in momentum space.

Let us turn to the Green's function at equal time. The equal-time correlations are solely determined by ground-state wave functions. It also easy to see from the definition that $G^\beta(0^+, \boldsymbol{x})$ and $G^\beta(-0^+, \boldsymbol{x})$ differ by the average of the anti-commutator (or commutator if the operators are bosonic):

$$G^\beta(0^+, \boldsymbol{x}) - G^\beta(-0^+, \boldsymbol{x}) = \left\langle \{c(\boldsymbol{x}), c^\dagger(0)\} \right\rangle = \delta(\boldsymbol{x})$$

Here we would like to pick a direction $\hat{\boldsymbol{x}}$ and study the large-x behavior of $G(\pm 0^+, x\hat{\boldsymbol{x}})$.

At $T = 0$, $G(-0^+, \boldsymbol{x})$ can be written as

$$\mathrm{i}G(-0^+, \boldsymbol{x}) = -\int \frac{\mathrm{d}^d\boldsymbol{k}}{(2\pi)^d}\, n_F(\xi_{\boldsymbol{k}})\mathrm{e}^{\,\mathrm{i}\boldsymbol{k}\cdot\boldsymbol{x}} = -\int_{-\infty}^{+\infty} \mathrm{d}k\, \tilde{N}(k, \hat{\boldsymbol{x}})\mathrm{e}^{\,\mathrm{i}k|\boldsymbol{x}|}$$

where

$$\tilde{N}(k, \hat{\boldsymbol{x}}) = \int \frac{\mathrm{d}^d\boldsymbol{k}}{(2\pi)^d}\, \Theta(-\xi_{\boldsymbol{k}})\delta(k - \boldsymbol{k}\cdot\hat{\boldsymbol{x}})$$

which can be viewed as the density of occupied states in momentum space.

From Fig. 4.8(a), we see that, in general, $\tilde{N}(k, \hat{\boldsymbol{x}})$ contains two singularities at $k = \boldsymbol{k}_F(\hat{\boldsymbol{x}})\cdot\hat{\boldsymbol{x}}$ and $k = \boldsymbol{k}_F(-\hat{\boldsymbol{x}})\cdot\hat{\boldsymbol{x}}$:

$$\begin{aligned}\tilde{N}(k, \hat{\boldsymbol{x}}) &= c_+\Theta(\boldsymbol{k}_F(\hat{\boldsymbol{x}})\cdot\hat{\boldsymbol{x}} - k)|\boldsymbol{k}_F(\hat{\boldsymbol{x}})\cdot\hat{\boldsymbol{x}} - k|^{(d-1)/2} \\ &\quad + c_-\Theta(-\boldsymbol{k}_F(-\hat{\boldsymbol{x}})\cdot\hat{\boldsymbol{x}} + k)| - \boldsymbol{k}_F(-\hat{\boldsymbol{x}})\cdot\hat{\boldsymbol{x}} + k|^{(d-1)/2}\end{aligned}$$

(For example, in one dimension $\tilde{N}(k, \hat{\boldsymbol{x}})$ has two discontinuous steps.) These two singularities lead to an algebraic long-range equal-time correlation as follows:

$$\mathrm{i}G(-0^+, \boldsymbol{x})|_{\boldsymbol{x}\to\infty} \sim \left(c_+\mathrm{e}^{\,\mathrm{i}\boldsymbol{k}_F(\hat{\boldsymbol{x}})\cdot\boldsymbol{x}}\mathrm{e}^{-\mathrm{i}\,\frac{\pi(d+1)}{4}} + c_-\mathrm{e}^{\,\mathrm{i}\boldsymbol{k}_F(-\hat{\boldsymbol{x}})\cdot\boldsymbol{x}}\mathrm{e}^{\,\mathrm{i}\,\frac{\pi(d+1)}{4}}\right)\frac{1}{|\boldsymbol{x}|^{(d+1)/2}}$$

For a spherical Fermi surface, we have

$$\mathrm{i}G(-0^+, \boldsymbol{x})|_{\boldsymbol{x}\to\infty} \sim \cos\left(k_F|\boldsymbol{x}| - \frac{\pi(d+1)}{4}\right)\frac{1}{|\boldsymbol{x}|^{(d+1)/2}}$$

If the Fermi surface contains a flat piece (see Fig. 4.8(b)), then $\tilde{N}(k, \hat{\boldsymbol{x}})$ will have a discontinuous step in the corresponding direction. In that direction, $G(-0^+, \boldsymbol{x})$ will have a slower algebraic decay. The decaying power is equal to that of a one-dimensional system:

$$\mathrm{i}G(-0^+, \boldsymbol{x})|_{\boldsymbol{x}\to\infty} \sim -\mathrm{i}\,\mathrm{e}^{\,\mathrm{i}\boldsymbol{k}_F(\hat{\boldsymbol{x}})\cdot x}\frac{1}{|\boldsymbol{x}|}$$

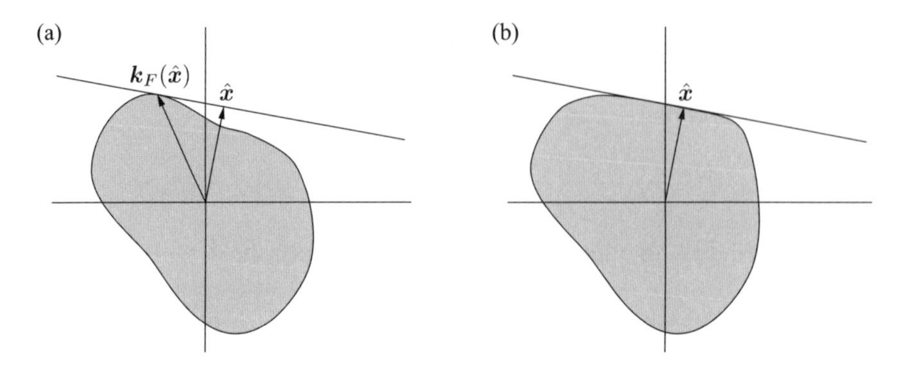

FIG. 4.8. (a) The definition of $\boldsymbol{k}_F(\hat{\boldsymbol{x}})$ where the line is normal to $\boldsymbol{x}$. (b) A Fermi surface with a flat piece.

4.3 Two-body correlation functions and linear responses

- We can calculate linear responses from two-body correlation functions. This allows us to calculate many measurable quantities.
- Two-body correlation functions reveal many potential instabilities of the free fermion ground state.

Two-body correlation functions are the most important correlations. Most quantities measured in experiments are directly related to two-body correlations. These quantities include electrical and thermal conductance, neutron and light scattering cross-sections, elastic constants, dielectric constants, magnetic susceptibility, etc. (see Section 3.7.2). In the following, we will discuss two-body correlations of a spinless free fermion system.

4.3.1 Density–density correlation functions

We first consider one of the two-body correlations—the density correlation function

$$\mathrm{i}P^{00}(t, \boldsymbol{x}) = \langle T(\rho(t, \boldsymbol{x})\rho(0))\rangle$$

where

$$\rho(t, \boldsymbol{x}) = c^\dagger(t, \boldsymbol{x})c(t, \boldsymbol{x}).$$

Density correlation functions are related to compressibility and neutron/light scattering. To calculate $P^{00}(t, \boldsymbol{x})$ we can use the Wick theorem for free fermion

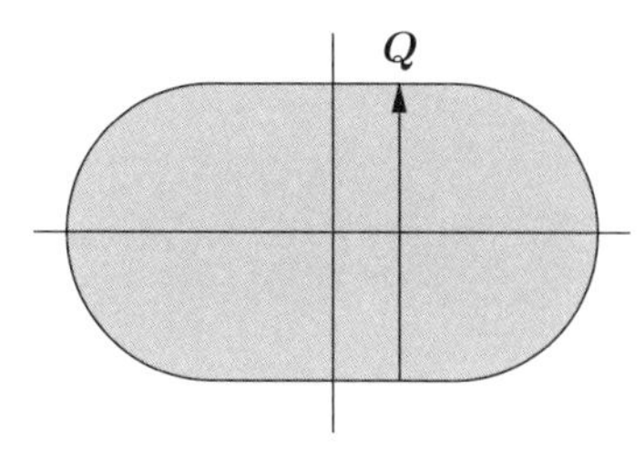

FIG. 4.9. A nested Fermi surface with nesting wave vector $\boldsymbol{Q}$.

operators as follows:

$$\begin{aligned}\mathrm{i}P^{00}(t,\boldsymbol{x}) &= \left\langle T[c^\dagger(t+0^+,\boldsymbol{x})c(t,\boldsymbol{x})c^\dagger(0^+)c(0)]\right\rangle \\ &= \rho_0^2 - \mathrm{i}G(-t,-\boldsymbol{x})iG(t,\boldsymbol{x})\end{aligned}$$

Using the fermion Green's function calculated previously, we find that, in the $t=0$ limit and for a spherical Fermi surface, we have

$$\mathrm{i}P^{00}(0,\boldsymbol{x}) = \rho_0^2 + C\left(1 - \sin\left(2k_F|\boldsymbol{x}| - \frac{\pi(d+1)}{2}\right)\right)\frac{1}{|\boldsymbol{x}|^{(d+1)}}$$

For the nested Fermi surface in Fig. 4.9, we have

$$\mathrm{i}P^{00}(0,\boldsymbol{x}) = \rho_0^2 + C[1+\sin(Q|\boldsymbol{x}|)]\frac{1}{|\boldsymbol{x}|^2}$$

when $\boldsymbol{x}||\boldsymbol{Q}$.

We know that in a crystal the density correlation keeps oscillating without any decay, i.e. $\mathrm{i}P^{00}(0,\boldsymbol{x}) = \rho_0^2 + C[1+\sin(Q|\boldsymbol{x}|)]$. The oscillating part represents the long-range order in an ordered crystal. We see that a free Fermi ground state has no long-range crystal order, but it has an 'algebraic long-range' crystal order, in particular for the nested Fermi surface (see Fig. 4.10). In a sense, a nested free fermion system is on the verge of becoming a crystal. We will see later in this chapter that, by turning on a small interaction, 'algebraic long-range' crystal order can be promoted to long-range crystal order.

To obtain the linear response of the free fermions, we need to calculate the density response function Π^{00}. In the t–$\boldsymbol{k}$ space the density response function can

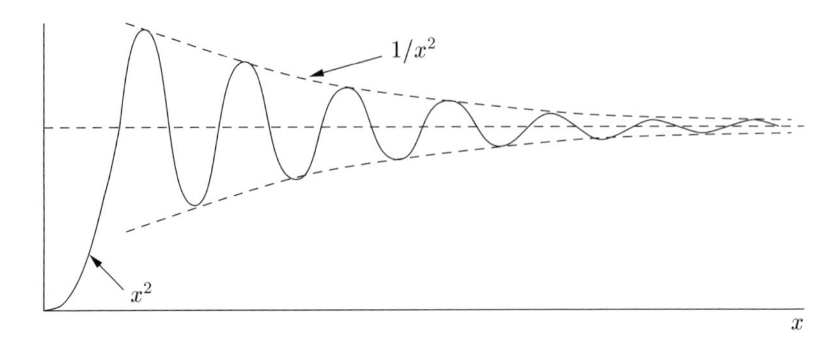

FIG. 4.10. The 'algebraic long-range' crystal order in the density correlation $\mathrm{i}P^{00}(0,\boldsymbol{x})$ for a nested Fermi surface.

be found using Wick's theorem as follows:

$$\begin{aligned}
\mathrm{i}\Pi^{00}(t,\boldsymbol{k}) &= \Theta(t)\mathcal{V}^{-1}\left\langle [\sum_{\boldsymbol{q}}(c^\dagger_{\boldsymbol{q}}c_{\boldsymbol{q}+\boldsymbol{k}})(t), \sum_{\boldsymbol{q}'}(c^\dagger_{\boldsymbol{q}'}c_{\boldsymbol{q}'-\boldsymbol{k}})(0)]\right\rangle \qquad (4.3.1)\\
&= \Theta(t)\mathcal{V}^{-1}\sum_{\boldsymbol{q}}(1-n_F(\xi_{\boldsymbol{q}+\boldsymbol{k}}))n_F(\xi_{\boldsymbol{q}})\mathrm{e}^{-\mathrm{i}t(\xi_{\boldsymbol{q}+\boldsymbol{k}}-\xi_{\boldsymbol{q}})}\\
&\quad - \Theta(t)\mathcal{V}^{-1}\sum_{\boldsymbol{q}}(1-n_F(\xi_{\boldsymbol{q}}))n_F(\xi_{\boldsymbol{q}+\boldsymbol{k}})\mathrm{e}^{+\mathrm{i}t(\xi_{\boldsymbol{q}}-\xi_{\boldsymbol{q}+\boldsymbol{k}})}
\end{aligned}$$

The average is nonzero only when $\boldsymbol{q}' = \boldsymbol{q}+\boldsymbol{k}$. In the first term, we create a fermion at $\boldsymbol{q}+\boldsymbol{k}$ and annihilate one at $\boldsymbol{q}$, which leads to the factor $(1-n_F(\xi_{\boldsymbol{q}+\boldsymbol{k}}))n_F(\xi_{\boldsymbol{q}})$. The second term has a similar structure, where we create a fermion at $\boldsymbol{q}$ and annihilate one at $\boldsymbol{q}+\boldsymbol{k}$.

In the ω–$\boldsymbol{k}$ space, we have

$$\Pi^{00}(\omega,\boldsymbol{k}) = \mathcal{V}^{-1}\sum_{\boldsymbol{q}}\left(\frac{(1-n_F(\xi_{\boldsymbol{q}+\boldsymbol{k}}))n_F(\xi_{\boldsymbol{q}})}{\omega-\xi_{\boldsymbol{q}+\boldsymbol{k}}+\xi_{\boldsymbol{q}}+\mathrm{i}0^+} - \frac{(1-n_F(\xi_{\boldsymbol{q}}))n_F(\xi_{\boldsymbol{q}+\boldsymbol{k}})}{\omega+\xi_{\boldsymbol{q}}-\xi_{\boldsymbol{q}+\boldsymbol{k}}+\mathrm{i}0^+}\right)$$

The imaginary part of $\Pi^{00}(\omega,\boldsymbol{k})$ is nonzero only when $(\omega,\boldsymbol{k})$ corresponds to the energy–momentum of particle–hole excitations (see Fig. 4.11). This feature appears in any two-body correlation function, including current and spin correlations.

In the limit $|\boldsymbol{k}| \ll k_F$ and for finite T, we have

$$\Pi^{00}(\omega,\boldsymbol{k}) = -\int\frac{\mathrm{d}^d\boldsymbol{q}}{(2\pi)^d}\frac{\partial n_F}{\partial\xi}\frac{\boldsymbol{k}\cdot\boldsymbol{v}}{\omega-\boldsymbol{k}\cdot\boldsymbol{v}+\mathrm{i}0^+}$$

where $\boldsymbol{v} = \boldsymbol{q}/m$. In the $T\to 0$ limit, we have $\partial n_F/\partial\xi = -\delta(\xi_{\boldsymbol{k}})$ and the integral becomes an integral over the Fermi surface. Let us first consider the imaginary part

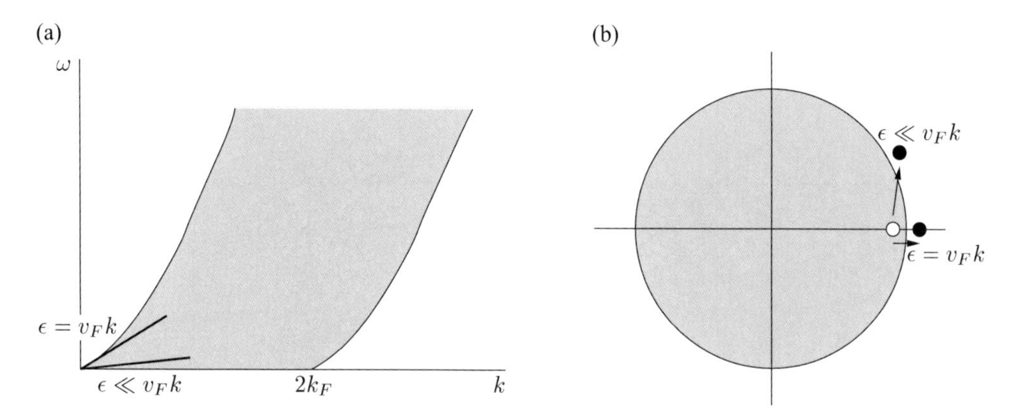

FIG. 4.11. (a) The shaded area represents the energy and momentum of particle–hole excitations (for $d > 1$). $\mathrm{Im}\Pi^{00}(\omega, \boldsymbol{k})$ is nonzero only in the shaded area. The edge at $(\omega, k) = (0, 0)$ has a slope of v_F, as one can see from (b).

of $\Pi^{00}(\omega, \boldsymbol{k})$:

$$\mathrm{Im}\Pi^{00}(\omega, \boldsymbol{k}) = -\frac{\omega}{v_F k}\frac{k_F^{d-1}}{(2\pi)^d v_F}\int \mathrm{d}\Omega\ \pi\delta(\frac{\omega}{v_F k} - \cos(\theta))$$

We find that (see Fig. 4.12)

$$\mathrm{Im}\,\Pi^{00}(\omega, \boldsymbol{k}) = \begin{cases} -(\omega/v_F k)(k_F^2/4\pi v_F)\Theta(v_F k - |\omega|), & d = 3, \\ -(\omega/v_F k)(k_F/2\pi v_F)(1/\sqrt{1-(\omega/v_F k)^2})\Theta(v_F k - |\omega|), & d = 2. \end{cases}$$

To obtain the real part, we note that $\Pi^{00}(\omega, \boldsymbol{k})$ is a sum of functions of the form $1/(\omega - \epsilon + \mathrm{i}0^+)$:

$$\Pi^{00}(\omega, \boldsymbol{k}) = \int \mathrm{d}\epsilon\ \frac{\mathrm{sgn}(\epsilon)A(\epsilon, \boldsymbol{k})}{\omega - \epsilon + \mathrm{i}0^+}$$

where $A(\epsilon, \boldsymbol{k})$ is the spectral function of $\Pi^{00}(\omega, \boldsymbol{k})$. The spectral function is directly related to the imaginary part of $\Pi^{00}(\omega, \boldsymbol{k})$ as follows:

$$\mathrm{Im}\Pi^{00}(\omega, \boldsymbol{k}) = -\mathrm{sgn}(\omega)\pi A(\omega, \boldsymbol{k})$$

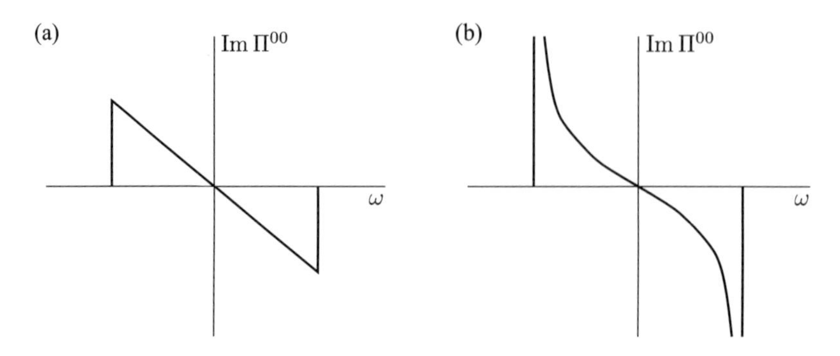

FIG. 4.12. ImΠ^{00} for fixed small k in (a) $d = 3$ dimensions, and (b) $d = 2$ dimensions.

Thus, we can calculate the real part from the imaginary part Im$\Pi^{00}(\omega, \boldsymbol{k})$. We find that

$$
\Pi^{00}(\omega, \boldsymbol{k}) = \begin{cases} -\frac{k_F m}{2\pi^2}\left[1 - \frac{\omega}{2v_F k}\ln\left|\frac{\omega+v_F k}{\omega-v_F k}\right| + \mathrm{i}\frac{\pi\omega}{2v_F k}\Theta(v_F k - |\omega|)\right], & \text{for } d = 3, \\ -\frac{m}{2\pi}\left[1 - \frac{\omega\Theta(|\omega|-v_F k)}{\sqrt{\omega^2-v_F^2 k^2}} + \mathrm{i}\frac{\omega\Theta(v_F k-|\omega|)}{\sqrt{v_F^2 k^2-\omega^2}}\right], & \text{for } d = 2, \\ \frac{1}{\pi}\frac{v_F k^2}{(\omega+\mathrm{i}0^+)^2 - v_F^2 k^2}, & \text{for } d = 1. \end{cases} \quad (4.3.2)
$$

We notice that when Im$\Pi^{00}(\omega, \boldsymbol{k})$ has a discontinuous jump at, say, ω_0, then Re$\Pi^{00}(\omega, \boldsymbol{k})$ will have a log divergence at the same place. We would also like to remark that the time-ordered correlation functions $P^{00}(\omega, \boldsymbol{k})$ are given by the same formula, except that the imaginary part has an extra factor of $\mathrm{sgn}(\omega)$.

The compressibility of the free fermions determines how much change in density can be induced by a change in potential. It is given by $\chi(\boldsymbol{k}) = -\Pi^{00}(0, \boldsymbol{k})$ (at a finite wave vector $\boldsymbol{k}$), and is independent of $\boldsymbol{k}$ for small $\boldsymbol{k}$. From eqn (4.2.12), we find that the compressibility is given by the density of states at the Fermi energy in all dimensions:

$$\chi(\boldsymbol{k}) = N(E_F)$$

as one expected. From

$$\sigma(\omega) = -\lim_{\boldsymbol{k}\to 0} \mathrm{Im}\frac{\omega}{k^2}\Pi^{00}(\omega, \boldsymbol{k})$$

we also see that the optical conductivity vanishes at $\sigma(\omega) = 0$, except at $\omega = 0$. This is also expected. Without any interactions, a uniform oscillating electric field cannot excite any particle–hole excitations, and hence causes no dissipations.

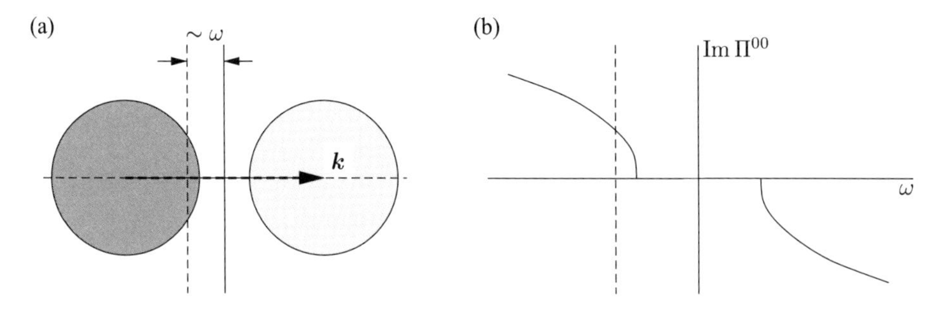

FIG. 4.13. ImΠ^{00} for $|\boldsymbol{k}| > 2k_F$. (a) The intersection between the dashed line and the shaded area determines (b) the value of Im$\Pi^{00}(\omega, \boldsymbol{k})$.

When $\omega \ll v_F k$, Π^{00} has the following form for all $d > 1$ dimensions:

$$\Pi^{00}(\omega, \boldsymbol{k}) = -N(E_F)\left(1 + \mathrm{i}C_d \frac{\omega}{v_F k}\right)$$

where $C_2 = 1$ for $d = 2$ and $C_3 = \pi/2$ for $d = 3$. We see that, when $\boldsymbol{k} \neq 0$, an oscillating electric field causes dissipations. The conductivity has the form

$$\sigma(\omega, \boldsymbol{k}) = N(E_F)C_d \frac{\omega^2}{v_F k^3}$$

in the $\omega \ll v_F k$ limit.

In general, $\Pi^{00}(\omega, \boldsymbol{k})$ is a smooth function of $\boldsymbol{k}$ for small ω, except near $\boldsymbol{k} = 0$ and $|\boldsymbol{k}| = 2k_F$ (for a spherical Fermi surface). We have studied the singular behavior of $\Pi^{00}(\omega, \boldsymbol{k})$ near $\boldsymbol{k} = 0$; next we turn to the behavior of $\Pi^{00}(\omega, \boldsymbol{k})$ near $|\boldsymbol{k}| = 2k_F$. Let us consider the $T = 0$ limit of

$$\Pi^{00}(\omega, \boldsymbol{k}) = \int \frac{\mathrm{d}^2 \boldsymbol{q}}{(2\pi)^d} \left(\frac{n_F(\xi_{\boldsymbol{q}-\frac{\boldsymbol{k}}{2}}) - n_F(\xi_{\boldsymbol{q}+\frac{\boldsymbol{k}}{2}})}{\omega - (\xi_{\boldsymbol{q}+\frac{\boldsymbol{k}}{2}} - \xi_{\boldsymbol{q}-\frac{\boldsymbol{k}}{2}}) + \mathrm{i}0^+} \right)$$

In Figs 4.13, 4.14, and 4.15, $n_F(\xi_{\boldsymbol{q}-\frac{\boldsymbol{k}}{2}}) - n_F(\xi_{\boldsymbol{q}+\frac{\boldsymbol{k}}{2}}) = 1$ in the lightly-shaded area, and $n_F(\xi_{\boldsymbol{q}-\frac{\boldsymbol{k}}{2}}) - n_F(\xi_{\boldsymbol{q}+\frac{\boldsymbol{k}}{2}}) = -1$ in the darkly-shaded area. Also, $\xi_{\boldsymbol{q}+\frac{\boldsymbol{k}}{2}} - \xi_{\boldsymbol{q}-\frac{\boldsymbol{k}}{2}} > 0$ to the right of the y axis, and $\xi_{\boldsymbol{q}+\frac{\boldsymbol{k}}{2}} - \xi_{\boldsymbol{q}-\frac{\boldsymbol{k}}{2}} < 0$ to the left of the y axis. The dashed line is where $\omega = \xi_{\boldsymbol{q}+\frac{\boldsymbol{k}}{2}} - \xi_{\boldsymbol{q}-\frac{\boldsymbol{k}}{2}}$. The intersection between the dashed line and the shaded area determines the value of Im$\Pi^{00}(\omega, \boldsymbol{k})$. From Figs 4.13 and 4.14, it is clear that, for small ω, Im$\Pi^{00}(\omega, \boldsymbol{k}) = 0$ when $\boldsymbol{k}| > 2k_F$, Im$\Pi^{00}(\omega, \boldsymbol{k}) \propto -\omega$ when $\boldsymbol{k}| < 2k_F$, and Im$\Pi^{00}(\omega, \boldsymbol{k}) \propto -\mathrm{sgn}(\omega)|\omega|^{(d-1)/2}$ when $|\boldsymbol{k}| = 2k_F$. These results imply that for $d > 1$ the real part Re$\Pi^{00}(0, \boldsymbol{k})$ is finite near $k = 2k_F$.

However, for a nested Fermi surface, Im$\Pi^{00}(\omega, \boldsymbol{Q})$ has a discontinuous jump at $\omega = 0$ if $\boldsymbol{k}$ is equal to the nesting vector $\boldsymbol{Q}$ (see Fig. 4.15). This leads to a log

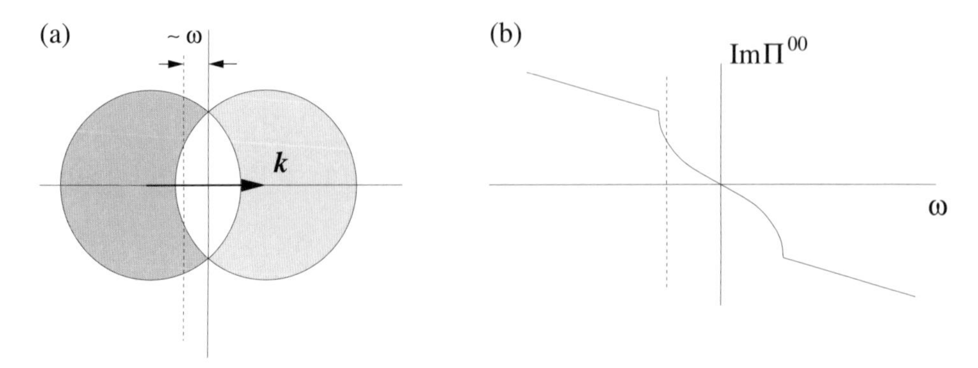

FIG. 4.14. $\mathrm{Im}\Pi^{00}$ for $|\boldsymbol{k}| < 2k_F$.

divergence in the compressibility at the nesting vector $\boldsymbol{Q}$:

$$\chi(\boldsymbol{Q}) \sim N(E_F) \ln \frac{E_F}{\omega}\Big|_{\omega \to 0}$$

At finite temperatures, the discontinuous jump at $\omega = 0$ is smeared by T and the log divergence is cut off by T:

$$\chi(\boldsymbol{Q}) \sim N(E_F) \ln \frac{E_F}{T} \tag{4.3.3}$$

We have seen, from the equal-time density correlation function, that the ground state for a nested Fermi surface contains an 'algebraic long-range' crystal order at the nesting wave vector $\boldsymbol{Q}$. Here we see, from the zero-frequency density correlation function, that turning on a small periodic potential with wave vector $\boldsymbol{Q}$ induces a large density wave. This is again a property similar to a crystal, where an infinite small periodic potential pins the crystal and induces a finite density wave. Later, we will see that, because a nested Fermi surface is so close to a crystal order, turning on an infinitely small interaction can actually cause a spontaneous symmetry breaking and changes the Fermi liquid state into a crystal (which is, more precisely, called a charge-density-wave (CDW) state).

Problem 4.3.1.
Free fermions and interacting bosons in one dimension Calculate the $T = 0$ time-ordered density correlation $\langle T\rho(x,t)\rho(0)\rangle$ in a one-dimensional free fermion system for large (x,t). (Hint: Use the results of Problem 4.2.5). Your result should be a special case of the density correlation for one-dimensional interacting bosons in eqn (3.6.2). Determine the values of χ, v, K_n, and C_n in eqn (3.6.2) so that it reproduces the time-ordered density correlation of the free fermions.

Problem 4.3.2.
Calculate, for a two-dimensional free fermion system, the zero-temperature time-ordered

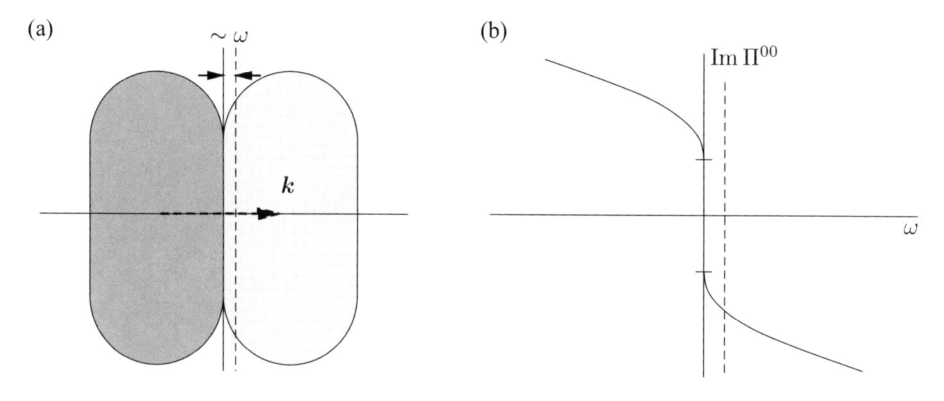

FIG. 4.15. ImΠ^{00} for a nested Fermi surface at $\boldsymbol{k} = \boldsymbol{Q}$.

density correlation $\mathcal{P}^{00}(\omega, k)$ in imaginary time for small (ω, k). Perform the analytic continuation to obtain Π^{00} for a real frequency.

4.3.2 Current operator

- The current operator for particles with a generic dispersion.

To calculate the current response functions π^{ij} and Π^{ij} (see Section 3.7.2), we first need to know what the current operator $\boldsymbol{j}$ is. We start with the time derivative of the density operator $\mathrm{d}\rho/\mathrm{d}t = \mathrm{d}[c^\dagger(\boldsymbol{x},t)c(\boldsymbol{x},t)]/\mathrm{d}t$ and obtain $\boldsymbol{j}$ from the conservation law $\mathrm{d}\rho/\mathrm{d}t + \boldsymbol{\partial}\cdot\boldsymbol{j} = 0$. Assume that the fermion system is described by

$$H = \sum_{\langle \boldsymbol{ij}\rangle}(t_{\boldsymbol{ij}}c_{\boldsymbol{i}}^\dagger c_{\boldsymbol{j}} + h.c.)$$

We find that $\mathrm{i}\partial_t c = \sum_{\boldsymbol{j}}(t_{\boldsymbol{ij}} + t_{\boldsymbol{ji}}^*)c_{\boldsymbol{j}}$ and

$$\partial_t\rho = \sum_{\boldsymbol{j}} j_{\boldsymbol{ij}}, \qquad j_{\boldsymbol{ij}} = -\mathrm{i}c_{\boldsymbol{i}}^\dagger(t_{\boldsymbol{ij}} + t_{\boldsymbol{ji}}^*)c_{\boldsymbol{j}} + \mathrm{i}c_{\boldsymbol{j}}^\dagger(t_{\boldsymbol{ij}}^* + t_{\boldsymbol{ji}})c_{\boldsymbol{i}}$$

These results tell us that, on a lattice, the current operator is given by $j_{\boldsymbol{ij}}$, which describes the number of particles that flow from site $\boldsymbol{j}$ to site $\boldsymbol{i}$ per unit time. This is very different from what we are looking for. We are looking for a current operator $\boldsymbol{j}(\boldsymbol{x})$ which is a vector and depends on one coordinate (rather than two). Such a current operator simply does not exist for a lattice model. However, if $\boldsymbol{A}$ is a smooth function of $\boldsymbol{x}$, then the coupling between $\boldsymbol{A}$ and the current can only see the smooth part of the current. In this limit, we can find such a vector-like current operator, even for a lattice model.

The trick is to start with a gauged Hamiltonian. Let us first consider a continuum model with

$$H = \int \mathrm{d}^d\boldsymbol{x}\; c^\dagger(\boldsymbol{x})\epsilon(-\mathrm{i}\partial_{\boldsymbol{x}})c(\boldsymbol{x})$$

where $\epsilon(\boldsymbol{k})$ is the energy spectrum of the fermion. The gauged Hamiltonian is

$$H = \int \mathrm{d}^d\boldsymbol{x}\; c^\dagger(\boldsymbol{x})\epsilon(-\mathrm{i}\partial_i + A_i(\boldsymbol{x}))c(\boldsymbol{x}) \tag{4.3.4}$$

The total current operator $\boldsymbol{J}$ is obtained through

$$J^i(\boldsymbol{x}) = \frac{\delta H}{\delta A_i(\boldsymbol{x})} \tag{4.3.5}$$

The current operator obtained this way satisfies the current conservation law (see Problem 4.3.3). For a quadratic dispersion $\epsilon(\boldsymbol{k}) = \boldsymbol{k}^2/2m$, we have

$$J^i = \frac{1}{2m}[c^\dagger(\boldsymbol{x})(-\mathrm{i}\partial_i)c(\boldsymbol{x}) - (\mathrm{i}\partial_i c^\dagger(\boldsymbol{x}))c(\boldsymbol{x})] + \frac{1}{m}A_i(\boldsymbol{x})c^\dagger(\boldsymbol{x})c(\boldsymbol{x}) \tag{4.3.6}$$

The first term has the form $m^{-1}\boldsymbol{k}\rho$. The second term is present only when $A_i \neq 0$, and has the form $m^{-1}A_i\rho$. As $m^{-1}(k_i + A_i) = v_i$ is the velocity, the current has the form $v_i\rho$, as expected.

For a more general dispersion, the current can be quite complicated because ∂_i and $A_i(\boldsymbol{x})$ do not commute. For example, for a one-dimensional system with $\epsilon(k) = \gamma k^n$, we find the following 'nasty' form:

$$J = \gamma \sum_{m=0}^{n-1} \left[(\mathrm{i}\partial_x + A_x)^m c^\dagger\right]\left[(-\mathrm{i}\partial_x + A_x)^{n-1-m} c\right]$$

However, in the small momentum limit we may ignore the $\partial_i A_j$ term and treat ∂_i and $A_i(\boldsymbol{x})$ as commuting quantities. In this limit, we obtain

$$\begin{aligned} J^i =& \frac{1}{2m}[c^\dagger(\boldsymbol{x})v^i(-\mathrm{i}\partial_{\boldsymbol{x}})c(\boldsymbol{x}) + (v^i(\mathrm{i}\partial_{\boldsymbol{x}})c^\dagger(\boldsymbol{x}))c(\boldsymbol{x})] \\ &+ \frac{1}{2}A_j(\boldsymbol{x})\left[c^\dagger(\boldsymbol{x})K^{ij}(-\mathrm{i}\partial_{\boldsymbol{x}})c(\boldsymbol{x}) + [K^{ij}(\mathrm{i}\partial_{\boldsymbol{x}})c^\dagger(\boldsymbol{x})]c(\boldsymbol{x})\right] + O(\boldsymbol{A}^2) \end{aligned} \tag{4.3.7}$$

where

$$v^i(\boldsymbol{k}) = \frac{\partial\epsilon(\boldsymbol{k})}{\partial k_i}, \qquad K^{ij}(\boldsymbol{k}) = \frac{\partial^2\epsilon(\boldsymbol{k})}{\partial k_i \partial k_j}$$

In momentum space, eqn (4.3.7) can be rewritten as

$$J^i_{\boldsymbol{q}}(t) = \sum_{\boldsymbol{k}} c^\dagger_{\boldsymbol{k}} c_{\boldsymbol{k}+\boldsymbol{q}} \frac{v^i(\boldsymbol{k}) + v^i(\boldsymbol{k}+\boldsymbol{q})}{2} + \mathcal{V}^{-1} \sum_{\boldsymbol{k}\boldsymbol{k}'} c^\dagger_{\boldsymbol{k}} c_{\boldsymbol{k}+\boldsymbol{k}'} A^j_{\boldsymbol{q}-\boldsymbol{k}'} \frac{K^{ij}(\boldsymbol{k}) + K^{ij}(\boldsymbol{k}+\boldsymbol{k}')}{2} \quad (4.3.8)$$

where $c(\boldsymbol{x},t) = \sum_{\boldsymbol{k}} \mathcal{V}^{-1/2} c_{\boldsymbol{k}}(t) \mathrm{e}^{\,\mathrm{i}\boldsymbol{k}\cdot\boldsymbol{x}}$, $A_i(\boldsymbol{x},t) = \mathcal{V}^{-1} \sum_{\boldsymbol{k}} A_{i,\boldsymbol{k}}(t) \mathrm{e}^{\,\mathrm{i}\boldsymbol{k}\cdot\boldsymbol{x}}$, and $J^i(\boldsymbol{x},t) = \mathcal{V}^{-1} \sum_{\boldsymbol{k}} J^i_{\boldsymbol{k}}(t) \mathrm{e}^{\,\mathrm{i}\boldsymbol{k}\cdot\boldsymbol{x}}$. When $\boldsymbol{A} = 0$, the current operator becomes

$$j^i_{\boldsymbol{q}}(t) = \sum_{\boldsymbol{k}} c^\dagger_{\boldsymbol{k}}(t) c_{\boldsymbol{k}+\boldsymbol{q}}(t) \frac{v^i(\boldsymbol{k}) + v^i(\boldsymbol{k}+\boldsymbol{q})}{2} \quad (4.3.9)$$

We would like to stress that the above is only valid in the small-$\boldsymbol{q}$ limit. So, we can also write the current as $j^i_{\boldsymbol{q}}(t) = \sum_{\boldsymbol{k}} c^\dagger_{\boldsymbol{k}} c_{\boldsymbol{k}+\boldsymbol{q}} v^i(\boldsymbol{k} + (\boldsymbol{q}/2))$ which agrees with eqn (4.3.9) up to $O(\boldsymbol{q})$. If we view $\sum_{\boldsymbol{k}}$ as a summation over the Brillouin zone, then eqn (4.3.8) or eqn (4.3.9) can be viewed as the (approximate) current operator for a lattice model.

Problem 4.3.3.
Gauge invariance and current conservation

1. Show that the gauged Hamiltonian (4.3.4) is invariant under the gauge transformation $c(\boldsymbol{x}) \to \mathrm{e}^{-\mathrm{i}\phi(\boldsymbol{x})} c(\boldsymbol{x})$ and $A_i(\boldsymbol{x}) \to A_i(\boldsymbol{x}) + \partial_i \phi(\boldsymbol{x})$.
2. Use the gauged Hamiltonian (4.3.4) in one dimension to calculate $\partial_t(c^\dagger c)$ and show that the current defined in eqn (4.3.5) satisfies $\partial_t \rho + \partial_x J = 0$.

Problem 4.3.4.
Find the current operator J^i for a free fermion system with dispersion $\epsilon_{\boldsymbol{k}} = \frac{\boldsymbol{k}^2}{2m} + \gamma \boldsymbol{k}^4$ in two dimensions.

4.3.3 Current correlation functions

Using eqn (4.3.9), we can now calculate the first part of the current response function $\mathrm{i}\pi^{ij} = \langle [j^i, j^j] \rangle$ (see eqn (3.7.6)). Remember that the density response function contains two contributions with particle–holes at $(\boldsymbol{q}, \boldsymbol{q}+\boldsymbol{k})$ and at $(\boldsymbol{q}+\boldsymbol{k}, \boldsymbol{q})$. When we calculate the current response function we also have the same two contributions, except that both of the contributions are weighted by an additional weighting factor $\frac{1}{2}[v^i(\boldsymbol{q}) + v^i(\boldsymbol{q}+\boldsymbol{k})]\frac{1}{2}[v^j(\boldsymbol{q}) + v^j(\boldsymbol{q}+\boldsymbol{k})]$. Therefore

$$\pi^{ij}(\omega, \boldsymbol{k}) = \int \frac{\mathrm{d}^2 \boldsymbol{q}}{(2\pi)^d} \frac{[n_F(\xi_{\boldsymbol{q}}) - n_F(\xi_{\boldsymbol{q}+\boldsymbol{k}})][v^i(\boldsymbol{q}) + v^i(\boldsymbol{q}+\boldsymbol{k})][v^j(\boldsymbol{q}) + v^j(\boldsymbol{q}+\boldsymbol{k})]}{4[\omega - (\xi_{\boldsymbol{q}+\boldsymbol{k}} - \xi_{\boldsymbol{q}}) + \mathrm{i}0^+]}$$

In the $\boldsymbol{k} \ll k_F$ limit, we have

$$\pi^{ij}(\omega, \boldsymbol{k}) = -\int \frac{\mathrm{d}^d \boldsymbol{q}}{(2\pi)^d} \frac{\partial n_F}{\partial \xi} \frac{\boldsymbol{k} \cdot \boldsymbol{v}(\boldsymbol{q})[v^i(\boldsymbol{q}) + v^i(\boldsymbol{q}+\boldsymbol{k})][v^j(\boldsymbol{q}) + v^j(\boldsymbol{q}+\boldsymbol{k})]}{4[\omega - \boldsymbol{k} \cdot \boldsymbol{v}(\boldsymbol{q}) + \mathrm{i}0^+]}$$

Let us first limit ourselves to $\epsilon = \frac{\boldsymbol{k}^2}{2m}$, $T = 0$, and $d = 2$. We have already calculated Π^{00}, and hence $\Pi^{||}$. To calculate $\Pi^{\perp}$, we assume that $\boldsymbol{k} = (k, 0)$. Then $\pi^{\perp}(\omega, k) = \pi^{22}(k\hat{\boldsymbol{x}}, \omega)$. We find that (see Section 4.3.6.1)

$$\pi^{\perp}(\omega, k) = -\frac{\rho}{m} + 2\frac{\rho}{m} \begin{cases} \frac{\omega^2}{v_F^2 k^2} - \mathrm{i}\frac{\omega}{v_F k}\sqrt{1 - \frac{\omega^2}{v_F^2 k^2}} & \text{for } |\frac{\omega}{v_F k}| < 1 \\ \frac{\omega^2}{v_F^2 k^2} - |\frac{\omega}{v_F k}|\sqrt{\frac{\omega^2}{v_F^2 k^2} - 1} & \text{for } |\frac{\omega}{v_F k}| > 1 \end{cases}$$

For a quadratic dispersion $\epsilon = \frac{\boldsymbol{k}^2}{2m}$, the current has the form (4.3.6) when $\boldsymbol{A} \neq 0$, which implies that $\Pi^{\mu\nu}$ and $\pi^{\mu\nu}$ are related through eqn (3.7.6). We find that

$$\Pi^{\perp}(\omega, k) = \pi^{\perp}(\omega, k) + \frac{\rho}{m} = 2\frac{\rho}{m} \begin{cases} \frac{\omega^2}{v_F^2 k^2} - \mathrm{i}\frac{\omega}{v_F k}\sqrt{1 - \frac{\omega^2}{v_F^2 k^2}} & \text{for } |\frac{\omega}{v_F k}| < 1 \\ \frac{\omega^2}{v_F^2 k^2} - |\frac{\omega}{v_F k}|\sqrt{\frac{\omega^2}{v_F^2 k^2} - 1} & \text{for } |\frac{\omega}{v_F k}| > 1 \end{cases} \tag{4.3.10}$$

where we have used $\rho = k_F^2/4\pi$. We note that the contact term $\frac{\rho}{m}$ exactly cancels the constant term in $\pi^{\perp}$ (what a relief!). Without this cancellation, the free fermion system would behave like a superconductor (see eqn (3.7.11)).

In $d = 3$ dimensions, we find that (see Section 4.3.6.2)

$$\Pi^{\perp}(\omega, k) \tag{4.3.11}$$
$$= \frac{k_F^3}{8\pi^2 m}\left(2\frac{\omega^2}{v_F^2 k^2} - \frac{\omega}{v_F k}(1 - \frac{\omega^2}{v_F^2 k^2})\left(\ln\frac{|\omega - v_F k|}{|\omega + v_F k|} + \mathrm{i}\pi\Theta(v_F k - |\omega|)\right)\right)$$

Problem 4.3.5.
Calculate $\Pi^{\mu\nu}(\omega, k)$ exactly for a one-dimensional free fermion system. Sketch the regions where $\mathrm{Im}\Pi^{00} \neq 0$, together with all of the boundary lines where $\Pi^{00}(\omega, k)$ becomes non-analytic. Identify the particle–hole excitations that correspond to these boundary lines.

4.3.4 Conductivity

- A finite a.c. conductivity is caused by impurities and/or interactions that break the Galileo invariance.
- An approximate calculation of conductivity.

As mentioned before, the uniform electric field only couples to the center-of-mass motion. As a result, the $\boldsymbol{k} = 0$ a.c. conductivity is $\sigma(\omega) = -\mathrm{Im}\Pi^{\perp}(\omega, 0)/\omega = -\lim_{\boldsymbol{k}\to 0} \mathrm{Im}\frac{\omega}{k^2}\Pi^{00}(\omega, \boldsymbol{k}) = 0$. To have a finite a.c. conductivity, we need to break the Galileo invariance. One way to do so is to include impurities. Another way to break the Galileo invariance is to include an interaction term that violates the Galileo invariance.

The Green's function $\mathrm{i}G(t,\boldsymbol{k})$ in eqn (4.2.3) has the following physical meaning. If we create a fermion in the $\boldsymbol{k}$ state at $t=0$, then $\mathrm{i}G(t,\boldsymbol{k})$ is the amplitude of finding the fermion in the $\boldsymbol{k}$ state at time t. For a free fermion system, the fermion that occupies the $\boldsymbol{k}$ state does not go anywhere. As a result $|\mathrm{i}G(t,\boldsymbol{k})|^2=1$. In the presence of impurities, the fermion in the $\boldsymbol{k}$ state can disappear into states with different momenta, as a result of the scattering by the impurities. Similarly, the interaction between fermions can cause a fermion in the $\boldsymbol{k}$ state to decay into a multiparticle state with two particles and one hole. Again, the fermion in the $\boldsymbol{k}$ state disappears into other states. To a certain degree, such an impurity/interaction effect can be simulated by including a decay term in the fermion propagator as follows:[30]

$$\begin{aligned}
\mathrm{i}G(t,\boldsymbol{k}) &= +\Theta(t)\Theta(\xi_{\boldsymbol{k}})\mathrm{e}^{-\mathrm{i}t\xi_{\boldsymbol{k}}-t\Gamma}-\Theta(-t)\Theta(-\xi_{\boldsymbol{k}})\mathrm{e}^{-\mathrm{i}t\xi_{\boldsymbol{k}})-(-t)\Gamma}\Big|_{T=0}\\
\mathrm{i}G^\beta(t,\boldsymbol{k}) &= +\Theta(t)(1-n_F(\xi_{\boldsymbol{k}}))\mathrm{e}^{-\mathrm{i}t\xi_{\boldsymbol{k}}-t\Gamma}-\Theta(-t)n_F(\xi_{\boldsymbol{k}})\mathrm{e}^{-\mathrm{i}t\xi_{\boldsymbol{k}}+t\Gamma}\Big|_{T>0}
\end{aligned} \tag{4.3.12}$$

As $|G(t,\boldsymbol{k})|^2$ decays as $\mathrm{e}^{-2\Gamma t}$, the decay rate is 2Γ and $\tau=1/2\Gamma$ is the relaxation time.[31] In the ω–$\boldsymbol{k}$ space, eqn (4.3.12) becomes

$$G(\omega,\boldsymbol{k})=\frac{1}{\omega-\xi_{\boldsymbol{k}}+\mathrm{i}\Gamma\mathrm{sgn}(\xi_{\boldsymbol{k}})} \tag{4.3.13}$$

$$G^\beta(\omega,\boldsymbol{k})=G^\beta_+(\omega,\ vk)+G^\beta_-(\omega,\boldsymbol{k})=\frac{1-n_F(\xi_{\boldsymbol{k}})}{\omega-\xi_{\boldsymbol{k}}+\mathrm{i}\Gamma}+\frac{n_F(\xi_{\boldsymbol{k}})}{\omega-\xi_{\boldsymbol{k}}-\mathrm{i}\Gamma} \tag{4.3.14}$$

Equation (4.3.14) allows us to obtain the fermion spectral functions $\pi A_\pm(\omega,\boldsymbol{k})=\pm\mathrm{Im}G^\beta_\pm(\omega,\boldsymbol{k})$ as follows:

$$\pi A_+(\omega,\boldsymbol{k})=\Gamma\frac{1-n_F(\xi_{\boldsymbol{k}})}{(\omega-\xi_{\boldsymbol{k}})^2+\Gamma^2},\qquad \pi A_-(\omega,\boldsymbol{k})=\Gamma\frac{n_F(\xi_{\boldsymbol{k}})}{(\omega-\xi_{\boldsymbol{k}})^2+\Gamma^2}.$$

However, the above $A_\pm$ do not satisfy eqn (2.2.28). Thus, the way that we use eqn (4.3.12) to simulate the impurity/interaction effect is not self-consistent. This

---

[30] For a system with impurities, the Green's function $G(t_1,\boldsymbol{x}_1;t_2,\boldsymbol{x}_2)$ is not a function of $\boldsymbol{x}_1-\boldsymbol{x}_2$, because the translational symmetry is broken. In this case, we cannot even define $G(t,\boldsymbol{k})$. However, the impurity-averaged Green's function only depends on $\boldsymbol{x}_1-\boldsymbol{x}_2$, and $G(t,\boldsymbol{k})$ can be defined. So, for a system with impurities, $G(t,\boldsymbol{k})$ should be regarded as an impurity-averaged Green's function.

[31] Using decay to simulate interaction/impurity effects, we violate the fermion number conservation.

problem can be fixed easily by modifying $A_\pm$ to be

$$\pi A_+(\omega, \boldsymbol{k}) = \Gamma \frac{1 - n_F(\omega)}{(\omega - \xi_{\boldsymbol{k}})^2 + \Gamma^2}, \quad \pi A_-(\omega, \boldsymbol{k}) = \Gamma \frac{n_F(\omega)}{(\omega - \xi_{\boldsymbol{k}})^2 + \Gamma^2}. \tag{4.3.15}$$

The modified $A_\pm$ become the $A_\pm$ of the free fermions when $\Gamma = 0^+$. The modified Green's function is no longer given by eqn (4.3.12), but by the one derived from eqn (4.3.15).

To calculate the density response function $\Theta(t)\langle[\rho(t, \boldsymbol{q}), \rho(0, -\boldsymbol{q})]\rangle$, we need to evaluate (see eqn (4.3.1)) $\langle[c^\dagger_{\boldsymbol{k}}(t)c_{\boldsymbol{k}+\boldsymbol{q}}, c^\dagger_{\boldsymbol{k}+\boldsymbol{q}}(0)c_{\boldsymbol{k}}(0)]\rangle$. It is very tempting to use Wick's theorem to evaluate the above correlation. Here we would like to point out that Wick's theorem only applies to correlation functions in free fermion (or boson) systems. It does not apply to interaction fermions or average Green's functions of impurity systems. Here we will use Wick's theorem anyway, but as an approximation:

$$\begin{aligned}
&\langle[c^\dagger_{\boldsymbol{k}}(t)c_{\boldsymbol{k}+\boldsymbol{q}}, c^\dagger_{\boldsymbol{k}+\boldsymbol{q}}(0)c_{\boldsymbol{k}}(0)]\rangle \\
&\approx \left\langle c_{\boldsymbol{k}+\boldsymbol{q}}(t)c^\dagger_{\boldsymbol{k}+\boldsymbol{q}}(0)\right\rangle \left\langle c^\dagger_{\boldsymbol{k}}(t)c_{\boldsymbol{k}}(0)\right\rangle - \left\langle c^\dagger_{\boldsymbol{k}+\boldsymbol{q}}(0)c_{\boldsymbol{k}+\boldsymbol{q}}(t)\right\rangle \left\langle c_{\boldsymbol{k}}(0)c^\dagger_{\boldsymbol{k}}(t)\right\rangle \\
&= (\mathrm{i}G(t, \boldsymbol{k}+\boldsymbol{q}))(-\mathrm{i}G(-t, \boldsymbol{k})) - (-\mathrm{i}G(-t, \boldsymbol{k}+\boldsymbol{q}))^*(\mathrm{i}G(t, \boldsymbol{k}))^*
\end{aligned}$$

Following the calculation from eqn (4.2.15) to eqn (4.2.17), we find that

$$\begin{aligned}
&\mathrm{Im}\Pi^{00}(\omega, \boldsymbol{q}) \\
&= \frac{\pi}{\mathcal{V}} \int \mathrm{d}\nu \sum_{\boldsymbol{k}} (A_-(\nu, \boldsymbol{k}+\boldsymbol{q})A_+(\nu - \omega, \boldsymbol{k}) - A_+(\nu, \boldsymbol{k}+\boldsymbol{q})A_-(\nu - \omega, \boldsymbol{k}))
\end{aligned} \tag{4.3.16}$$

To calculate the current response function $\Theta(t)\langle[j^i(t, \boldsymbol{k}), j^j(0, -\boldsymbol{k})]\rangle$, we just need to put in the weighting factor $\frac{1}{2}[v^i(\boldsymbol{q}) + v^i(\boldsymbol{q}+\boldsymbol{k})]\frac{1}{2}[v^j(\boldsymbol{q}) + v^j(\boldsymbol{q}+\boldsymbol{k})]$. We find that

$$\begin{aligned}
\mathrm{Im}\Pi^{ij}(\omega, \boldsymbol{q}) = \frac{\pi}{4\mathcal{V}} \int \mathrm{d}\nu \sum_{\boldsymbol{k}} &[v^i(\boldsymbol{k}) + v^i(\boldsymbol{q}+\boldsymbol{k})][v^j(\boldsymbol{k}) + v^j(\boldsymbol{q}+\boldsymbol{k})] \times \\
&(A_-(\nu, \boldsymbol{k}+\boldsymbol{q})A_+(\nu - \omega, \boldsymbol{k}) - A_+(\nu, \boldsymbol{k}+\boldsymbol{q})A_-(\nu - \omega, \boldsymbol{k}))
\end{aligned}$$

Setting $\boldsymbol{q} = 0$, we find the a.c. conductivity to be

$$\begin{aligned}
\sigma^{ij}(\omega) &= -\,\mathrm{Im}\frac{\Pi^{ij}(\omega, 0)}{\omega} \\
&= -\int \frac{\mathrm{d}\nu}{\pi} \frac{\mathrm{d}^d\boldsymbol{k}}{(2\pi)^d} \frac{\Gamma^2 v^i(\boldsymbol{k})v^j(\boldsymbol{k})(n_F(\nu) - n_F(\nu - \omega))}{\omega((\nu - \xi_{\boldsymbol{k}})^2 + \Gamma^2)((\nu - \omega - \xi_{\boldsymbol{k}})^2 + \Gamma^2)}
\end{aligned}$$

Taking $\omega \to 0$, we obtain the following very useful formula for d.c. conductivity:

$$\sigma^{ij}(0) = \int \frac{\mathrm{d}\nu}{\pi} \frac{\mathrm{d}^d \boldsymbol{k}}{(2\pi)^d} \frac{\Gamma^2 v^i(\boldsymbol{k}) v^j(\boldsymbol{k})(-\frac{\partial n_F(\nu)}{\partial \nu})}{((\nu - \xi_{\boldsymbol{k}})^2 + \Gamma^2)^2} \approx \int \frac{\mathrm{d}^d \boldsymbol{k}}{(2\pi)^d} \tau v^i(\boldsymbol{k}) v^j(\boldsymbol{k}) \delta(\xi_{\boldsymbol{k}})$$

where we have assumed that T and Γ are small and have used $-\partial n_F(\nu)/\partial \nu \approx \delta(\nu)$ and $\Gamma^2/(\xi_{\boldsymbol{k}}^2 + \Gamma^2)^2 \approx \frac{\pi}{2\Gamma}\delta(\xi_{\boldsymbol{k}})$.

4.3.5 Other two-body correlation functions

When electrons have spin, we can also calculate the spin correlation function and electron-pair correlation function. The spin density operator is given by $s^i = c_\alpha^\dagger \sigma_{\alpha\beta}^i c_\beta/2$, where c_β, with $\beta = 1, 2$, are spin-up and spin-down electron operators, and σ^i is the Pauli matrix. The spin–spin correlation is equal to the density correlation:

$$\langle s^3(\boldsymbol{x}, t) s^3(0) \rangle = \frac{1}{4}\Big[\Big\langle (c_1^\dagger c_1)(c_1^\dagger c_1) \Big\rangle + \Big\langle (c_2^\dagger c_2)(c_2^\dagger c_2) \Big\rangle \Big] = \frac{1}{2\mathrm{i}} (\Pi^{00}(\boldsymbol{x}, t) - \rho_0^2)$$

We see that the spin–spin correlation also has an algebraic long-range order. For a nested Fermi surface,

$$\langle s^i(\boldsymbol{x}, 0) s^j(0) \rangle \propto \delta_{ij} \frac{1 + \sin(Q|\boldsymbol{x}|)}{|\boldsymbol{x}|^2}$$

if $\boldsymbol{x} || \boldsymbol{Q}$.

The electron-pair operator is given by $b(\boldsymbol{x}, t) = c_1(\boldsymbol{x}, t) c_2(\boldsymbol{x}, t)$ and is related to superconductivity. Its correlation is equal to the square of the Green's function of the spinless electron:

$$\Big\langle b(\boldsymbol{x}, t) b^\dagger(0) \Big\rangle = G^2(\boldsymbol{x}, t) \propto \Big|_{t=0} \cos^2 \left(k_F |\boldsymbol{x}| - \frac{\pi(d+1)}{4} \right) \frac{1}{|\boldsymbol{x}|^{(d+1)}}$$

Once again, we have algebraic long-range order in electron-pair correlations.

In summary, we find that many correlations in free fermion systems have a power law decay. This indicates that the free fermion ground state is a kind of critical state. Sometimes, an arbitrary weak interaction can change algebraic long-range order into true long-range order and induce spontaneous symmetry breaking. For example, an infinitely weak attractive interaction will generate long-range order in an electron-pair correlation and produce a superconducting state.

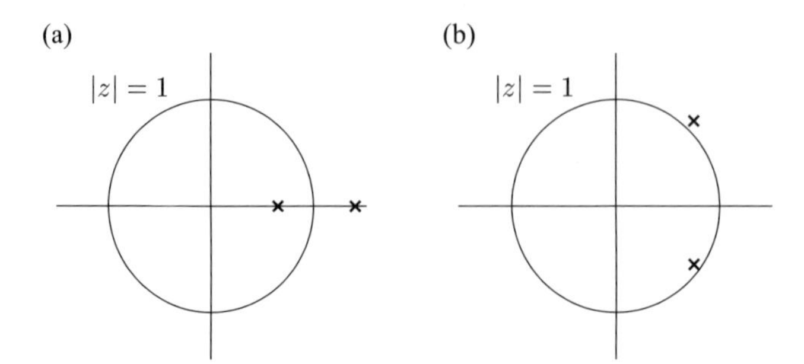

FIG. 4.16. Two poles of $\frac{1}{(z-\alpha-\mathrm{i}0^+)^2+1-(\alpha+\mathrm{i}0^+)^2}$ for (a) $|\alpha| > 1$, and (b) $|\alpha| < 1$. In the absence of the $\mathrm{i}0^+$ term, the two poles for $|\alpha| < 1$ will be on the unit circle $|z| = 1$.

4.3.6 Remarks: some calculation details

4.3.6.1 Calculation of $\pi^\perp$ in two dimensions

We calculate π^{22} as follows:

$$\pi^{22}(k\hat{\boldsymbol{x}},\omega) = \frac{1}{4m^2}\frac{1}{4\pi^2}\int \mathrm{d}(\frac{q^2}{2m})\,\mathrm{d}\theta\;\delta(\epsilon(\boldsymbol{q})-\mu)\frac{\boldsymbol{k}\cdot\boldsymbol{q}}{\omega-\frac{\boldsymbol{k}\cdot\boldsymbol{q}}{m}+\mathrm{i}0^+}\,(2q_2+k_2)^2$$
$$= \frac{1}{m^2}\frac{1}{4\pi^2}\int \mathrm{d}\theta\;\frac{kk_F\cos(\theta)k_F^2\sin^2(\theta)}{\omega-\frac{kk_F\cos(\theta)}{m}+\mathrm{i}0^+} = \frac{k_F^2}{4\pi^2 m}\int \mathrm{d}\theta\;\frac{\cos(\theta)\sin^2(\theta)}{\frac{\omega}{v_F k}-\cos(\theta)+\mathrm{i}0^+}$$

and (see Fig. 4.16)

$$\int \mathrm{d}\theta\;\frac{\cos(\theta)\sin^2(\theta)}{\alpha+\mathrm{i}0^+-\cos(\theta)} = \frac{1}{4}\oint_{|z|=1}\mathrm{d}z\;\frac{1}{\mathrm{i}z^3}\frac{(z^2-1)^2(z^2+1)}{(z-\alpha)^2+1-\alpha^2}$$
$$= \begin{cases} -\pi+2\pi\alpha^2-2\pi\mathrm{i}\alpha\sqrt{1-\alpha^2} & \text{for } |\alpha|<1 \\ -\pi+2\pi\alpha^2-2\pi|\alpha|\sqrt{\alpha^2-1} & \text{for } |\alpha|>1 \end{cases}$$

4.3.6.2 Calculation of $\pi^\perp$ in three dimensions

To calculate $\pi^\perp$, we can take $\boldsymbol{k} = (0,0,k)$ and use

$$\pi^\perp(\omega,k) = \pi^{11}(k\hat{\boldsymbol{z}},\omega) = \frac{1}{4m^2}\frac{1}{8\pi^3}\int q\,\mathrm{d}(\frac{q^2}{2m})\,\mathrm{d}\phi\,\mathrm{d}\theta\;\frac{\cos(\theta)\delta(\epsilon(\boldsymbol{q}))\boldsymbol{k}\cdot\boldsymbol{q}}{\omega-\frac{\boldsymbol{k}\cdot\boldsymbol{q}}{m}+\mathrm{i}0^+}\,(2q_1+k_1)^2$$
$$= \frac{1}{m^2}\frac{1}{8\pi^3}\int \mathrm{d}\phi\,\mathrm{d}\theta\;\frac{k\cos(\theta)k_F^4\sin^3(\theta)\cos^2(\phi)}{\omega-\frac{kk_F\cos(\theta)}{m}+\mathrm{i}0^+} = \frac{k_F^3}{8\pi^2 m}\int_{-1}^{1}\mathrm{d}t\;\frac{t(1-t^2)}{\frac{\omega}{v_F k}-t+\mathrm{i}0^+}$$
$$= \frac{k_F^3}{8\pi^2 m}\left(-\frac{4}{3}+2\frac{\omega^2}{v_F^2k^2}-\frac{\omega}{v_Fk}(1-\frac{\omega^2}{v_F^2k^2})\left(\ln\frac{|\omega-v_Fk|}{|\omega+v_Fk|}+\mathrm{i}\pi\Theta(v_Fk-|\omega|)\right)\right)$$

We also note that $\rho = k_F^3/6\pi^2$.

4.4 Linear responses of insulators and quantized Hall conductance

• The Hall conductance of insulators is a topological quantity and is quantized.

In the last section, we discussed the linear responses of free fermion systems with a Fermi surface. These responses are caused by the particle–hole across the Fermi surface. In this section, we would like to discuss the linear responses of insulators. Here, by 'insulator' we mean any state with a finite energy gap. (Such a state is also called a rigid state.) As there is no low-energy excitation, the imaginary part of the responses is zero for small ω (i.e. there are no low-energy dissipations).

Consider an interacting fermion system coupled to the electromagnetic field: $\mathcal{L}(c, c^\dagger, A_\mu) + \mathcal{L}_{\text{gauge}}(A_\mu)$. The result that we are going to obtain is very general, and we do not even need to know the form of $\mathcal{L}(c, c^\dagger, A_\mu)$. We only assume that the ground state of the fermions has a finite energy gap and that $\mathcal{L}(c, c^\dagger, A_\mu)$ is invariant under a gauge transformation:

$$c(x^\mu) \to \mathrm{e}^{\,\mathrm{i}\phi(x^\mu)} c(x^\mu), \qquad A_\mu \to A_\mu + \partial_\mu \phi(x^\mu).$$

After integrating out the fermions, we obtain an effective action of the gauge field as follows:

$$\begin{aligned}\mathcal{L}_{\text{eff}}(A_\mu) &= \mathcal{L}_{\text{gauge}}(A_\mu) + \delta\mathcal{L}_{\text{eff}}(A_\mu)\\ \delta\mathcal{L}_{\text{eff}}(A_\mu) &= -\frac{1}{2}\int \mathrm{d}x\,\mathrm{d}x'\; A_\mu(x) P^{\mu\nu}(x - x') A_\nu(x')\end{aligned}$$

where $\delta\mathcal{L}_{\text{eff}}(A_\mu)$ is the contribution from the fermions and $P^{\mu\nu}(x - x')$ is the current–current correlation function.

For insulators (or rigid states), $\delta\mathcal{L}_{\text{eff}}$ is local and $P^{\mu\nu}_{k_\mu}$ is a polynomial of k_μ in the frequency–momentum space. This is in sharp contrast to $P^{\mu\nu}_{k_\mu}$ for the metals calculated in the previous section. The singularity in the small-$(\boldsymbol{k}, \omega)$ limits for metals is caused by gapless particle–hole excitations across the Fermi surface.

Here $\delta\mathcal{L}_{\text{eff}}(A_\mu)$ should also be gauge invariant. Thus, $\delta\mathcal{L}_{\text{eff}}(A_\mu)$ should depend on A_μ only via the field strength $\boldsymbol{E}$ and $\boldsymbol{B}$. Therefore, $\delta\mathcal{L}_{\text{eff}}$ has the following form:

$$\delta\mathcal{L}_{\text{eff}} = -\frac{1}{2}(B_i \chi^{ij} B_j + E_i p^{ij} E_j) + \ldots$$

where '...' represents higher-derivative terms. The tensor χ^{ij} is the magnetic susceptibility and p^{ij} represents a correction to the dielectric constant. In general, the above are the only forms of linear responses from an insulator.

However, in $2 + 1$ dimensions, $\delta\mathcal{L}_{\text{eff}}$ may contain a new term—the Chern–Simons term—as follows:

$$\delta\mathcal{L}_{\text{eff}} = \frac{K}{4\pi} A_\mu \partial_\nu A_\lambda \epsilon^{\mu\nu\lambda} - \frac{1}{2}(B_i \chi^{ij} B_j + E_i p^{ij} E_j) + \ldots$$

where $\mu, \nu, \lambda = 0, 1, 2$ and $\epsilon^{\mu\nu\lambda}$ is the total anti-symmetric tensor. Despite the fact that the Chern–Simons term cannot be written in terms of the field strength, it is nevertheless gauge invariant. Under the gauge transformation

$$A_\mu \to A_\mu + \partial_\mu f$$

the action undergoes the following change:

$$S = \int_V \mathrm{d}x \, \mathcal{L}_{\text{eff}} \to S + \oint_S \mathrm{d}S_\mu \, f \frac{K}{4\pi} \partial_\nu A_\lambda \epsilon^{\mu\nu\lambda}$$

where V is the space–time volume and S is the surface of V. We see that, if our system lives on a closed space–time that has no boundary, then the Chern–Simons term is gauge invariant.

The linear response of the Chern–Simons term is

$$J^i = -\frac{\delta S}{\delta A_i} = \frac{K}{2\pi} \partial_0 A_i \epsilon^{ij} = \frac{K}{2\pi} E_i \epsilon^{ij}$$

We see that the Chern–Simons term gives rise to a Hall conductance $\sigma_{xy} = K/2\pi$ (or $\sigma_{xy} = Ke^2/h$ if we put back e and $\hbar$).

In the following, we are going to show that the Hall conductance σ_{xy} or K is quantized (Thouless *et al.*, 1982; Avron *et al.*, 1983). First, let us consider a periodic system and a special gauge field configuration as follows:

$$A_1 = \frac{\theta_1(t)}{L_1}, \qquad A_2 = \frac{\theta_2(t)}{L_2}, \qquad A_0 = 0 \tag{4.4.1}$$

where $L_{1,2}$ are the sizes of the system in the x and y directions. The many-body ground state of the fermions, $|\boldsymbol{\theta}\rangle$, is parametrized by $\boldsymbol{\theta} = (\theta_1, \theta_2)$. If we integrate out the fermions, then we will obtain an effective Lagrangian $\delta L_{\text{eff}} = \int \mathrm{d}^2\boldsymbol{x} \, \delta\mathcal{L}_{\text{eff}}$:

$$\delta L_{\text{eff}} = \frac{K}{4\pi} \theta_i \dot{\theta}_j \epsilon^{ij} + \dot{\theta}_i p^{ij} \dot{\theta}_j$$

where the first term comes from the Chern–Simons term and the second term from the $\boldsymbol{E}^2$ term. If we view $\boldsymbol{\theta}$ as the coordinates of a particle, then δS describes a two-dimensional charged particle in a uniform 'magnetic' field b:

$$\delta L_{\text{eff}} = a_i(\boldsymbol{\theta})\dot{\theta}_i + \dot{\theta}_i p^{ij} \dot{\theta}_j$$

$$a_1 = -\frac{K}{4\pi}\theta_2, \qquad a_2 = \frac{K}{4\pi}\theta_1, \qquad b = \partial_{\theta_1} a_2 - \partial_{\theta_2} a_1 = \frac{K}{2\pi}$$

In Section 4.4.1, we will show that the fermion states $|\theta_1, \theta_2\rangle$, $|\theta_1 + 2\pi, \theta_2\rangle$, and $|\theta_1, \theta_2 + 2\pi\rangle$ are related by a gauge transformation. Physically, this means that $|\theta_1, \theta_2\rangle$, $|\theta_1 + 2\pi, \theta_2\rangle$, and $|\theta_1, \theta_2 + 2\pi\rangle$ are actually the same physical state (see the discussions in Section 6.1). Thus, δL_{eff} actually describes a particle on a torus parametrized by $0 < \theta_1 < 2\pi$ and $0 < \theta_2 < 2\pi$.

Let us move the particle along the loop C: $(0,0) \to (2\pi,0) \to (2\pi,2\pi) \to (0,2\pi) \to (0,0)$. The phase accumulated by such a move is given by the 'magnetic' flux enclosed by C:

$$\oint_C \mathrm{d}\boldsymbol{\theta} \cdot \boldsymbol{a} = \int_D \mathrm{d}^2\boldsymbol{\theta}\, b = 2\pi K$$

where D is the area enclosed by the loop C. On any closed surface, such as a sphere or torus, any loop will enclose two surfaces on the two sides of the loop (see Fig. 6.5). Here D is simply one of the two surfaces enclosed by C. The other surface D' enclosed by C has zero area (because D covers the whole area of the torus). So the phase $\oint_C \mathrm{d}\boldsymbol{\theta} \cdot \boldsymbol{a}$ can also be written as $\int_{D'} \mathrm{d}^2\boldsymbol{\theta}\, b = 0$. To be consistent, $\int_D \mathrm{d}^2\boldsymbol{\theta}\, b$ and $\int_D' \mathrm{d}^2\boldsymbol{\theta}\, b$ should represent the same phase. Hence

$$\int_D \mathrm{d}^2\boldsymbol{\theta}\, b = 2\pi \times \text{integer}$$

Mathematically, one can prove that the total magnetic flux through any closed surface must be quantized as an integer. This is why the magnetic charge of a monopole is quantized. For our case, this means that K is quantized as an integer (see also Section 4.4.1).

We would like to stress that the quantization of K is very general. It applies to any interacting system of fermions and/or bosons. The only assumption is that the ground state $|\boldsymbol{\theta}\rangle$ is not degenerate. As a result, *the Hall conductance of a system is quantized as integer* $\times\, e^2/h$ *if the many-body ground state on a torus is not degenerate and has a finite energy gap.* We would like to point out that, if the many-body ground states have degeneracies (see Section 8.2.1), then K and the Hall conductance can be a *rational* number (Niu *et al.*, 1985).

The quantization of the Hall conductance has some interesting consequences. Let us consider a non-interacting electron system in a uniform magnetic field. If n Landau levels are filled, then the system will have a finite energy gap and the Hall conductance will be ne^2/h (or $K = n$). Now let us turn on an interaction, periodic potential, or even a random potential. As long as the gap never closes in the process, the Hall conductance cannot change! As the Hall conductance is robust against any perturbations, we call it a *topological* quantum number. The only way to change the Hall conductance is to close the energy gap, which induces a quantum phase transition.

To explicitly calculate K, we note that $\int \mathrm{d}t\, \delta L_{\text{eff}}$ is the action for the adiabatic evolution $|\boldsymbol{\theta}(t)\rangle$. Thus, we have $\int \mathrm{d}t\, \delta L_{\text{eff}} = \int \mathrm{d}t\, \mathrm{i}\langle\boldsymbol{\theta}(t)|\frac{\mathrm{d}}{\mathrm{d}t}|\boldsymbol{\theta}(t)\rangle$ or (see eqn (2.3.4))

$$\frac{K}{4\pi}\theta_i\dot{\theta}_j\epsilon^{ij} = \mathrm{i}\langle\boldsymbol{\theta}(t)|\frac{\mathrm{d}}{\mathrm{d}t}|\boldsymbol{\theta}(t)\rangle \tag{4.4.2}$$

Here we would like to remark that, as the ground state of the fermions, the phase of $|\boldsymbol{\theta}(t)\rangle$ is not fixed. If we redefine the phase of $|\boldsymbol{\theta}\rangle$: $|\boldsymbol{\theta}\rangle \to \mathrm{e}^{\,\mathrm{i}\varphi(\boldsymbol{\theta})}|\boldsymbol{\theta}\rangle$, then the Berry phase term will change by a total derivative term:

$$\langle\boldsymbol{\theta}(t)|\frac{\mathrm{d}}{\mathrm{d}t}|\boldsymbol{\theta}(t)\rangle \to \langle\boldsymbol{\theta}(t)|\frac{\mathrm{d}}{\mathrm{d}t}|\boldsymbol{\theta}(t)\rangle + \mathrm{i}\,\frac{\mathrm{d}\varphi}{\mathrm{d}t}$$

Therefore, the relationship between the Chern–Simons term and the Berry phase (4.4.2) cannot be correct, because the Berry phase for an open path is not even well defined (see Section 2.3). The correct relationship should be an integration of eqn (4.4.2) along a closed loop:

$$\oint \mathrm{d}t\, \frac{K}{4\pi}\theta_i\dot{\theta}_j\epsilon^{ij} = \mathrm{i}\oint \mathrm{d}t\, \langle\boldsymbol{\theta}(t)|\frac{\mathrm{d}}{\mathrm{d}t}|\boldsymbol{\theta}(t)\rangle$$

In particular, along the loop C, the Berry phase is given by

$$2\pi K = \frac{\sigma_{xy}\hbar}{e^2} = \oint_C \mathrm{d}\boldsymbol{\theta}\cdot\langle\boldsymbol{\theta}|\mathrm{i}\partial_{\boldsymbol{\theta}}|\boldsymbol{\theta}\rangle \tag{4.4.3}$$

Let us use the Berry phase (4.4.3) to calculate the quantized Hall conductance of a band insulator, for which

$$H = \sum_{\boldsymbol{ij}} c_{\boldsymbol{i}}^\dagger t_{\boldsymbol{ij}} c_{\boldsymbol{j}} = \sum_{\boldsymbol{k}} c_{\boldsymbol{k}}^\dagger M(\boldsymbol{k}) c_{\boldsymbol{k}} \tag{4.4.4}$$

where $c_{\boldsymbol{i}}$ has n components, $t_{\boldsymbol{ij}}$ are $n\times n$ matrices that only depend on $\boldsymbol{i}-\boldsymbol{j}$, and $M(\boldsymbol{k})$ is the Fourier transformation of $t_{\boldsymbol{ij}}$. After coupling to a uniform gauge potential $\boldsymbol{A}$ which has the form given in eqn (4.4.1), we obtain

$$H = \sum_{\boldsymbol{ij}} c_{\boldsymbol{i}}^\dagger t_{\boldsymbol{ij}} \mathrm{e}^{\,\mathrm{i}\boldsymbol{A}\cdot(\boldsymbol{i}-\boldsymbol{j})} c_{\boldsymbol{j}} = \sum_{\boldsymbol{k}} c_{\boldsymbol{k}}^\dagger M^{\boldsymbol{\theta}}(\boldsymbol{k}) c_{\boldsymbol{k}}$$

Then let $\psi^{\boldsymbol{\theta}}_{a,\boldsymbol{k}}$ be the eigenvectors of $M^{\boldsymbol{\theta}}(\boldsymbol{k})$. Here $\psi^{\boldsymbol{\theta}}_{a,\boldsymbol{k}}$ are labeled by the crystal momentum $\boldsymbol{k}$, and $a = 1, ..., n$, where a labels the ath eigenvector. Note that $\psi^{\boldsymbol{\theta}}_{a,\boldsymbol{k}}$ is an n-dimensional complex vector.

Let us assume that the ground state $|\boldsymbol{\theta}\rangle$ is obtained by filling the $a = 1$ band. We have

$$|\boldsymbol{\theta}\rangle = \otimes_{\boldsymbol{k}}|\psi^{\boldsymbol{\theta}}_{1,\boldsymbol{k}}\rangle$$

Using the additive relation of the Berry phase, namely

$$\oint_C \mathrm{d}\boldsymbol{\theta}\cdot\langle\boldsymbol{\theta},0|\mathrm{i}\partial_{\boldsymbol{\theta}}|\boldsymbol{\theta},0\rangle = \oint_C \mathrm{d}\boldsymbol{\theta}\cdot\langle\boldsymbol{\theta},1|\mathrm{i}\partial_{\boldsymbol{\theta}}|\boldsymbol{\theta},1\rangle + \oint_C \mathrm{d}\boldsymbol{\theta}\cdot\langle\boldsymbol{\theta},2|\mathrm{i}\partial_{\boldsymbol{\theta}}|\boldsymbol{\theta},2\rangle$$

if $|\boldsymbol{\theta},0\rangle = |\boldsymbol{\theta},1\rangle\otimes|\boldsymbol{\theta},2\rangle$, we find that

$$\oint_C \mathrm{d}\boldsymbol{\theta}\cdot\langle\boldsymbol{\theta}|\mathrm{i}\partial_{\boldsymbol{\theta}}|\boldsymbol{\theta}\rangle = \sum_{\boldsymbol{k}}\oint_C \mathrm{d}\boldsymbol{\theta}\cdot(\psi^{\boldsymbol{\theta}}_{1,\boldsymbol{k}})^\dagger\,\mathrm{i}\partial_{\boldsymbol{\theta}}\psi^{\boldsymbol{\theta}}_{1,\boldsymbol{k}} \tag{4.4.5}$$

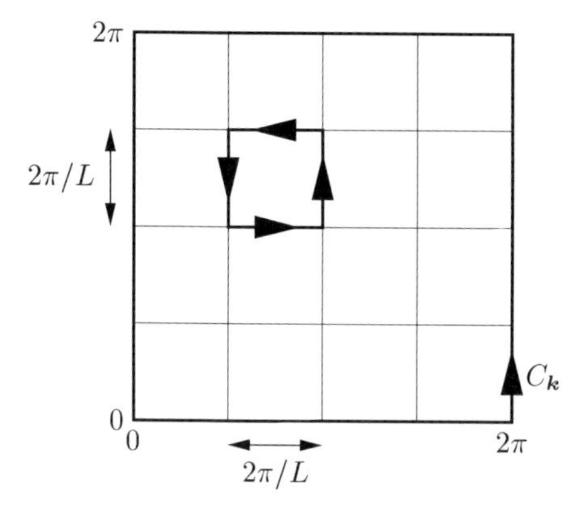

FIG. 4.17. The small loop traced out by $\boldsymbol{k}' = \boldsymbol{k} + \boldsymbol{\theta}L^{-1}$ for a fixed $\boldsymbol{k}$. The small loops for all different $\boldsymbol{k}$s cover the whole Brillouin zone. The loop integral along all of the small loops corresponds to the loop integral along the loop C_k because the contributions from the internal lines cancel each other.

How do we calculate the Berry phase for $\psi^{\boldsymbol{\theta}}_{1,\boldsymbol{k}}$? Here we note that $M^{\boldsymbol{\theta}}(\boldsymbol{k}) = M(\boldsymbol{k} + \boldsymbol{\theta}L^{-1})$, where, for simplicity, we have assumed that $L_1 = L_2 = L$ in eqn (4.4.1). Thus, $\psi^{\boldsymbol{\theta}}_{1,\boldsymbol{k}}$ is an eigenvector of $M(\boldsymbol{k} + \boldsymbol{\theta}L^{-1})$. Let $\psi_{a,\boldsymbol{k}'}$ be the eigenvectors of $M(\boldsymbol{k}')$; then we have $\psi^{\boldsymbol{\theta}}_{1,\boldsymbol{k}} = \psi_{1,\boldsymbol{k}'}$ if $\boldsymbol{k}' = \boldsymbol{k} + \boldsymbol{\theta}L^{-1}$. As $\boldsymbol{\theta}$ moves along the loop C, for a fixed $\boldsymbol{k}$, $\boldsymbol{k}'$ moves along a small loop (see Fig. 4.17). As the discrete $\boldsymbol{k}$ has the form $(n_1\frac{2\pi}{L}, n_2\frac{2\pi}{L})$, we see that all of the small loops for different $\boldsymbol{k}$s cover the whole Brillouin zone without overlap. Thus, the sum of the Berry phases for the $\psi^{\boldsymbol{\theta}}_{1,\boldsymbol{k}}$ in eqn (4.4.5) is equal to the Berry phase for $\psi_{1,\boldsymbol{k}'}$ along a large loop $C_{\boldsymbol{k}}$ that encloses the whole Brillouin zone:

$$\begin{aligned} 2\pi K &= \oint_{C_{\boldsymbol{k}}} \mathrm{d}\boldsymbol{k} \cdot \psi^\dagger_{1,\boldsymbol{k}} \mathrm{i}\partial_{\boldsymbol{k}} \psi_{1,\boldsymbol{k}} \\ &= \int \mathrm{d}^2\boldsymbol{k}\ \mathrm{i}[(\partial_{k_x}\psi^\dagger_{1,\boldsymbol{k}})(\partial_{k_y}\psi_{1,\boldsymbol{k}}) - (\partial_{k_y}\psi^\dagger_{1,\boldsymbol{k}})(\partial_{k_x}\psi_{1,\boldsymbol{k}})] \end{aligned} \tag{4.4.6}$$

If several bands are filled, then we need to sum over the contributions from each band.

As a concrete example, consider the following spin-dependent hopping Hamiltonian on a square lattice:

$$H = \sum_{\boldsymbol{i}} c^\dagger_{\boldsymbol{i}} t_{\boldsymbol{ij}} c_{\boldsymbol{j}}$$

where the 2×2 hopping matrices are given by

$$t_{\boldsymbol{i},\boldsymbol{i}+\boldsymbol{x}} = t\sigma^x, \qquad t_{\boldsymbol{i},\boldsymbol{i}+\boldsymbol{y}} = t\sigma^y, \qquad t_{\boldsymbol{i},\boldsymbol{i}+\boldsymbol{x}+\boldsymbol{y}} = t'\sigma^z,$$

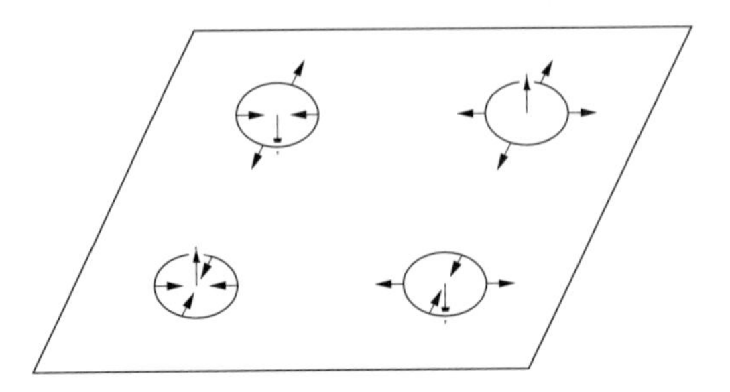

FIG. 4.18. The arrows represent the direction of $\boldsymbol{B}$. The winding number receives contributions only from the neighborhood of four points $\boldsymbol{k} = (\pm\pi/2, \pm\pi/2)$. Around each point, $\boldsymbol{B}$ covers half of the sphere and contributes a winding number of $1/2$. The total winding number is 2.

In momentum space, the Hamiltonian becomes

$$H_{\boldsymbol{k}} = 2t\cos(k_x)\sigma^x + 2t\cos(k_x)\sigma^y + 2t'\cos(k_x + k_y)\sigma^z = -\boldsymbol{B}(\boldsymbol{k}) \cdot \sigma$$

We see that our system has two bands. Here $H_{\boldsymbol{k}}$ is just like the Hamiltonian of a spin-1/2 spin in a magnetic field $\boldsymbol{B}(\boldsymbol{k})$. The lower band corresponds to spin in the $\boldsymbol{B}$ direction. Moving along a loop in $\boldsymbol{k}$ space, the spin in the $\boldsymbol{B}(\boldsymbol{k})$ direction traces out a closed loop on the unit sphere in the spin space and gives rise to a Berry phase. If we fill the lower band, then the Hall conductance can be calculated from the above spin-1/2 Berry phase if we choose the loop to enclose the entire Brillouin zone. From the expression for $\boldsymbol{B}(\boldsymbol{k})$ in the small-t' limit, we see that $\boldsymbol{B}$ is in the x–y plane, except near the four points $\boldsymbol{k} = (\pm\pi/2, \pm\pi/2)$. By examining the spin near $\boldsymbol{k} = (\pm\pi/2, \pm\pi/2)$ (see Fig. 4.18), we find that, as $\boldsymbol{k}$ goes over the Brillouin zone, $\boldsymbol{B}/|\boldsymbol{B}|$ goes over the unit sphere twice. Or, more precisely, $\boldsymbol{B}(\boldsymbol{k})/|\boldsymbol{B}(\boldsymbol{k})|$ maps the Brillouin zone $S^1 \times S^1$ to the unit sphere S^2 with winding number 2. The total Berry phase is thus $2 \times 2\pi$. We find that $K = 2$ and the Hall conductance of the half-filled hopping system is $\sigma_{xy} = 2e^2/h$.

Problem 4.4.1.
For the Hamiltonian (4.4.4), if the $a = 1$ band is partially filled, then show that the Hall conductance is given by an integration of the filled levels as follows:

$$\sigma_{xy} = \frac{e^2}{h}\frac{1}{2\pi}\int_{filled} \mathrm{d}^2\boldsymbol{k}\ \mathrm{i}[(\partial_{k_x}\psi^\dagger_{1,\boldsymbol{k}})(\partial_{k_y}\psi_{1,\boldsymbol{k}}) - (\partial_{k_y}\psi^\dagger_{1,\boldsymbol{k}})(\partial_{k_x}\psi_{1,\boldsymbol{k}})]$$

At finite temperatures, we have

$$\sigma_{xy} = \frac{e^2}{h}\frac{1}{2\pi}\int \mathrm{d}^2\boldsymbol{k}\ n_F(\epsilon_{\boldsymbol{k}})\mathrm{i}[(\partial_{k_x}\psi^\dagger_{1,\boldsymbol{k}})(\partial_{k_y}\psi_{1,\boldsymbol{k}}) - (\partial_{k_y}\psi^\dagger_{1,\boldsymbol{k}})(\partial_{k_x}\psi_{1,\boldsymbol{k}})]$$

(Note that, for the partially filled band, K is not quantized because $|\boldsymbol{\theta}\rangle$ does not have 2π periodicity, see eqn (4.4.9).)

4.4.1 Remarks: periodic structure of $|\theta\rangle$ and quantization of K

First, let us study a periodic structure of the fermion ground state $|\boldsymbol{\theta}\rangle$ in the $\boldsymbol{\theta}$ space. Here $|\boldsymbol{\theta}\rangle$ is the ground state of the Hamiltonian

$$H_{\boldsymbol{\theta}} = \int \mathrm{d}^2\boldsymbol{x} \left(-\frac{\hbar^2}{2m} \sum_{j=1,2} c^\dagger \left(\partial_j - \mathrm{i} A_j - \frac{\mathrm{i}\theta_j}{L_j} \right)^2 c + V(c^\dagger c) \right)$$

We note that the $U(1)$ gauge transformations

$$c(\boldsymbol{x}) \to \mathrm{e}^{2\,\mathrm{i}\,\pi x_1 L_1^{-1}} c(\boldsymbol{x})$$
$$(A_1, A_2) \to (A_1 + \frac{2\pi}{L_1}, A_2)$$

and

$$c(\boldsymbol{x}) \to \mathrm{e}^{2\,\mathrm{i}\,\pi x_2 L_2^{-1}} c(\boldsymbol{x})$$
$$(A_1, A_2) \to (A_1, A_2 + \frac{2\pi}{L_2})$$

do not change the periodic boundary conditions of the fermion operator $c(\boldsymbol{x})$ in the x and y directions. However, the gauge transformations generate the following shift: $(\theta_1, \theta_2) \to (\theta_1 + 2\pi, \theta_2)$ and $(\theta_1, \theta_2) \to (\theta_1, \theta_2 + 2\pi)$. Or, more precisely, we have

$$H_{\theta_1+2\pi,\theta_2} = W_1 H_{\theta_1,\theta_2} W_1^\dagger \qquad W_1 = \mathrm{e}^{\,\mathrm{i} \int \mathrm{d}^2\boldsymbol{x}\ 2\pi x_1 L_1^{-1} c^\dagger c}$$
$$H_{\theta_1,\theta_2+2\pi} = W_2 H_{\theta_1,\theta_2} W_2^\dagger \qquad W_2 = \mathrm{e}^{\,\mathrm{i} \int \mathrm{d}^2\boldsymbol{x}\ 2\pi x_2 L_2^{-1} c^\dagger c}$$

Therefore, if $|\boldsymbol{\theta}\rangle$ is the many-body ground state for $\boldsymbol{\theta}$, then $W_1|\boldsymbol{\theta}\rangle$ is the ground state for the shifted $\boldsymbol{\theta}$. Hence

$$|\theta_1 + 2\pi, \theta_2\rangle = \mathrm{e}^{\,\mathrm{i}\, f_1(\boldsymbol{\theta})} W_1 |\theta_1, \theta_2\rangle \tag{4.4.7}$$

for a choice of $f_1(\boldsymbol{\theta})$. Similarly, we have

$$|\theta_1, \theta_2 + 2\pi\rangle = \mathrm{e}^{\,\mathrm{i}\, f_2(\boldsymbol{\theta})} W_2 |\theta_1, \theta_2\rangle \tag{4.4.8}$$

The two unitary operators $\mathrm{e}^{\,\mathrm{i}\, f_1(\boldsymbol{\theta})} W_1$ and $\mathrm{e}^{\,\mathrm{i}\, f_2(\boldsymbol{\theta})} W_2$ generate the gauge transformations that relate $|\theta_1, \theta_2\rangle$, $|\theta_1 + 2\pi, \theta_2\rangle$, and $|\theta_1, \theta_2 + 2\pi\rangle$.

We can always redefine the phase of $|\boldsymbol{\theta}\rangle$: $|\boldsymbol{\theta}\rangle \to \mathrm{e}^{\,\mathrm{i}\,\phi(\boldsymbol{\theta})}|\boldsymbol{\theta}\rangle$ to make $f_1 = 0$ if we choose ϕ to satisfy

$$\phi(\theta_1 + 2\pi, \theta_2) - \phi(\theta_1, \theta_2) = f_1(\theta_1, \theta_2).$$

Then, from

$$|\theta_1 + 2\pi, \theta_2 + 2\pi\rangle = \mathrm{e}^{\,\mathrm{i}\, f_2(\theta_1+2\pi,\theta_2)} W_2 |\theta_1 + 2\pi, \theta_2\rangle$$

we find that

$$W_1 |\theta_1, \theta_2 + 2\pi\rangle = \mathrm{e}^{\,\mathrm{i}\, f_2(\theta_1+2\pi,\theta_2)} W_2 W_1 |\theta_1, \theta_2\rangle$$

As $[W_1, W_2] = 0$, we have

$$|\theta_1, \theta_2 + 2\pi\rangle = \mathrm{e}^{\,\mathrm{i}\, f_2(\theta_1+2\pi,\theta_2)} W_2 |\theta_1, \theta_2\rangle$$

Comparing the above equation to eqn (4.4.8), we find that

$$\mathrm{e}^{\,\mathrm{i}\,f_2(\theta_1+2\pi,\theta_2)} = \mathrm{e}^{\,\mathrm{i}\,f_2(\theta_1,\theta_2)}$$

We see that $|\boldsymbol{\theta}\rangle$ is quasi-periodic in θ_1 and θ_2 with period 2π:

$$\begin{aligned} |\theta_1+2\pi,\theta_2\rangle &= W_1|\theta_1,\theta_2\rangle \\ |\theta_1,\theta_2+2\pi\rangle &= \mathrm{e}^{\,\mathrm{i}\,f_2(\theta_1,\theta_2)}W_2|\theta_1,\theta_2\rangle \end{aligned} \tag{4.4.9}$$

where $W_{1,2}$ are unitary operators which are independent of $\boldsymbol{\theta}$.

Using $\langle\boldsymbol{\theta}|\partial_{\boldsymbol{\theta}}|\boldsymbol{\theta}\rangle = \langle\boldsymbol{\theta}|W_{1,2}^{\dagger}\partial_{\boldsymbol{\theta}}W_{1,2}|\boldsymbol{\theta}\rangle$ and eqn (4.4.9), we can reduce the Berry phase (4.4.3) to

$$2\pi K = \int \mathrm{d}\theta_1\, \partial_{\theta_1} f_2(\theta_1,0) = f_2(2\pi,0) - f_2(0,0)$$

As $\mathrm{e}^{\,\mathrm{i}\,f_2(\theta_1,\theta_2)}$ is periodic in θ_1, we find that $f_2(2\pi,0)-f_2(0,0) = 2\pi \times$ integer and K is quantized as an integer.

5

INTERACTING FERMION SYSTEMS

Electron systems in nature have quite strong Coulomb interactions. Many interesting properties of materials, such as magnetism, superconductivity, etc. are due to interactions. In this chapter, we will study the effect of electron–electron interactions. In particular, we will discuss Landau's Fermi liquid theory for metals and symmetry-breaking transitions induced by the interaction.

5.1 Orthogonality catastrophe and X-ray spectrum

- Orthogonality catastrophe is a phenomenon that illustrates how many-body effects affect fermion Green's functions and the associated experimental results.
- Orthogonality catastrophe is a general phenomenon that can appear in many systems, such as an X-ray spectrum, an I–V curve of tunneling into quantum dots, etc.

5.1.1 Physical model

The X-ray spectrum is caused by a transition between a core level and a conduction band of a metal (see Fig. 5.1(a)). An absorption of X-rays removes an electron from a core level and adds it into an unoccupied band state. An emission removes an electron from an occupied band state to an unoccupied core state. The system can be modeled by the following Hamiltonian:

$$H = \sum_{\boldsymbol{k}} \epsilon_{\boldsymbol{k}} c^\dagger_{\boldsymbol{k}} c_{\boldsymbol{k}} - E_C C^\dagger C + \Gamma(\mathrm{e}^{-\mathrm{i}\omega t} c(0) C^\dagger + h.c.) \tag{5.1.1}$$

where C describes a core electron at $\boldsymbol{x} = 0$, and $\Gamma\mathrm{e}^{-\mathrm{i}\omega t}$ is the time-dependent coupling induced by the X-rays. For simplicity, we will assume spinless electrons in this section.

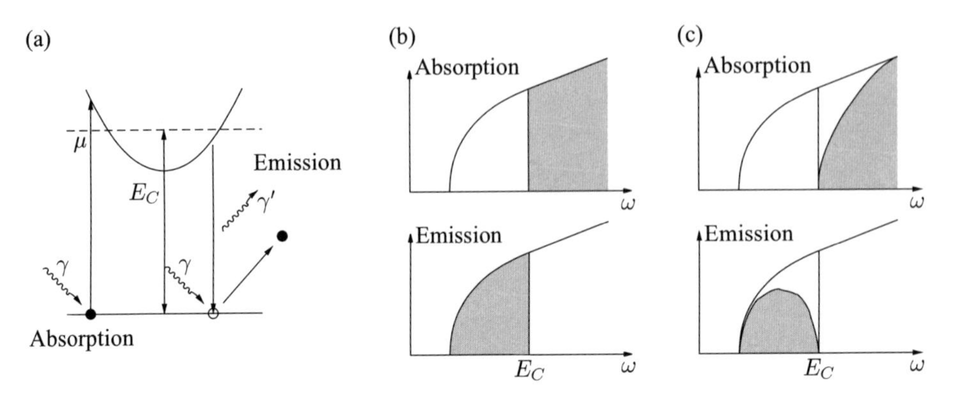

FIG. 5.1. (a) X-ray absorption/emission are caused by transitions between a core level and a conduction band of a metal. Also shown are the X-ray spectra of the transitions (b) with no interaction, and (c) with interaction.

Note that the emission rate is equal to the current from the conduction band to the core state. According to linear response theory, the emission rate is given by

$$A_{em}(\omega) = \langle 0|\mathrm{i}\Gamma(\mathrm{e}^{-\mathrm{i}\omega t}I(t) - h.c.)|0\rangle$$
$$= \Gamma^2 \int^t \mathrm{d}t' \left(\mathrm{e}^{-\mathrm{i}\omega(t-t')}\langle 0|[I(t), I^\dagger(t')]|0\rangle + h.c.\right)$$

where $I(t) = c(0,t)C^\dagger(t)$. We can express the emission rate A_{em} in terms of the spectral function of the core electron and the conduction electron (see eqn (4.2.17)) as follows:

$$A_{em}(\omega) = -2\pi\Gamma^2 \int \mathrm{d}\nu \; \left(A^0_{c-}(\nu)A^0_{C+}(\nu-\omega) - A^0_{c+}(\nu)A^0_{C-}(\nu-\omega)\right)$$

For free fermions, the spectral function is given by the density of states, as in eqn (4.2.18). If we choose our energy to be zero at the Fermi surface, then

$$A^0_{c-}(\omega) = n_F(\omega)N(\omega+E_F), \qquad A^0_{c+}(\omega) = (1-n_F(\omega))N(\omega+E_F)$$
$$A^0_{C-}(\omega) = 0, \qquad A^0_{C+}(\omega) = \delta(\omega+E_C)$$

We note that, before emission, the core state is completely empty. This gives rise to the above non-equilibrium core-electron spectral function. We find that

$$A_{em}(\omega) = 2\pi\Gamma^2 n_F(\omega - E_C)N(E_F - E_C + \omega)$$

and the emission rate directly measures the density of states for free fermions (see Fig. 5.1(b); note that $E_C > 0$). At zero temperature, $A_{em}(\omega)$ has a step-like

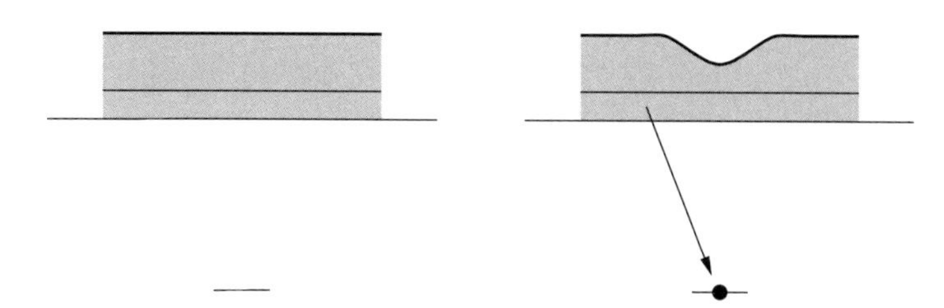

FIG. 5.2. The density of the conduction electrons has to be changed after the tunneling if there is an interaction between the core electron and the conduction electrons.

jump at $\omega = E_C$, indicating the sharpness of the Fermi surface. The above result also applies to interacting electrons if we use the spectral function of interacting electrons.

Problem 5.1.1.
Calculate the absorption rate spectrum $A_{obs}(\omega)$.

5.1.2 The physics of orthogonality catastrophe

- Orthogonality catastrophe is caused by the vanishing overlap between two *many-body* wave functions, which describe the deformed and undeformed Fermi seas.

Now we would like to include an interaction between the core electron and the conduction electrons. We will see that such an interaction can modify the edge singularity of the emission/absorption rate at $\omega = E_C$ in a significant way.

The physics is illustrated in Fig. 5.2. In the absence of the interaction between the core electron and the conduction electrons, the matrix element for the emission at a given frequency involves the matrix element of single-particle wave functions of the conduction electron at a certain energy and the core electron. To generalize to interacting cases, we should view the above matrix element as a matrix element between two many-body eigenstates: one eigenstate has a filled Fermi sea and an empty core state, and the other has a hole in the Fermi sea and a filled core state. In the presence of the interaction, the core electron induces a change in the density of the conduction electrons, and hence a change in the many-body electron wave function. Thus, the matrix element for the emission involves the matrix element between the following two eigenstates. One has a filled but deformed Fermi sea and an empty core state. The other has a hole in the uniform Fermi sea and a filled core state. We can approximate the above matrix element as the product of the matrix element between the single-particle states and the overlap between two many-body states of a uniform Fermi sea and a deformed Fermi sea. If the overlap is finite, then the discussions in the last section remain valid for ω near E_C, except that the tunneling amplitude Γ may be reduced by a finite factor. However, if the overlap is zero, then the results in the last section must be modified in a significant way. In particular, the step-like singularity at $\omega = E_C$ is destroyed. This phenomenon is called infra-red catastrophe or orthogonality catastrophe.

We know that a many-body wave function for free fermions can be constructed from single-body wave functions through the Slater determinant as follows:

$$\Psi(\boldsymbol{x}_1, ..., \boldsymbol{x}_N) = \mathrm{Det}[\psi_i(\boldsymbol{x}_j)]$$

where $\psi_i(\boldsymbol{x})$ is a single-body wave function and $[\psi_i(\boldsymbol{x}_j)]$ is the following matrix:

$$[\psi_i(\boldsymbol{x}_j)] \equiv \begin{pmatrix} \psi_1(\boldsymbol{x}_1) & \psi_1(\boldsymbol{x}_2) & \cdots \\ \psi_2(\boldsymbol{x}_1) & \psi_2(\boldsymbol{x}_2) & \cdots \\ \cdots & \cdots & \ddots \end{pmatrix}$$

The many-body wave function for the uniform Fermi sea, Ψ_0, is constructed from the plane waves $\psi_i(\boldsymbol{x}) = \mathrm{e}^{\mathrm{i}\boldsymbol{k}_i\cdot\boldsymbol{x}}$, where $\boldsymbol{k}_i$ runs through all momenta whose energies are less than the Fermi energy E_F. The many-body wave function for the deformed Fermi sea, Ψ, is constructed from the eigenfunctions in the presence of the potential of the core electron. We can then calculate the overlap $\langle\Psi_0|\Psi\rangle$ to determine if there is an orthogonality catastrophe or not.

To illustrate our point in a simple calculation, we will calculate $\langle\Psi_0|\Psi\rangle$ by treating the fermions as a fluid described by the density $\rho(\boldsymbol{x})$. This treatment give us a correct result in one dimension, but an incorrect result beyond one dimension. Such an approach is called the hydrodynamical approach, or the bosonization of fermion systems. Let us first develop the hydrodynamical approach.

5.1.3 Hydrodynamical approach (bosonization)

- The collective density fluctuations in a metal can be described by a bosonic field theory.
- In one dimension, the boson theory provides a complete description of the particle–hole excitations across the Fermi surface.

The potential energy of a compressible fluid is given by

$$V = \int \mathrm{d}^d\boldsymbol{x}\, \frac{1}{2\chi}\rho^2(\boldsymbol{x})$$

where χ is the compressibility. Here we assume that $\rho(\boldsymbol{x})$ is a fluctuation around the uniform density, ρ_0, of the ground state. The kinetic energy is

$$K = \int \mathrm{d}^d\boldsymbol{x}\, \frac{m\boldsymbol{v}^2(\boldsymbol{x})}{2}\rho_0 = \int \mathrm{d}^d\boldsymbol{x}\, \frac{m\boldsymbol{j}^2(\boldsymbol{x})}{2\rho_0}$$

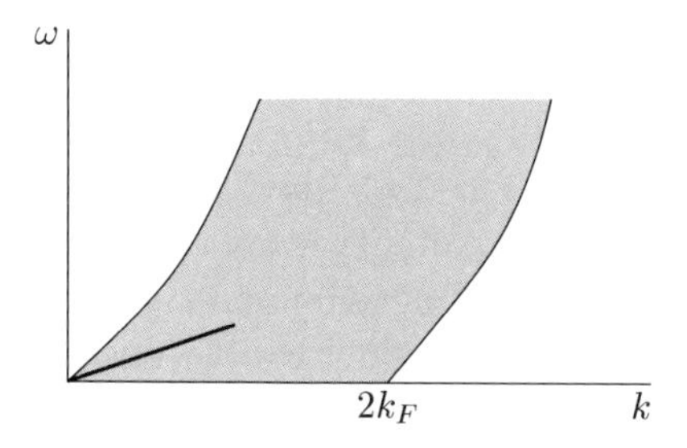

FIG. 5.3. The dispersion of the linear mode always lies within the particle–hole continuum.

where $\boldsymbol{v}(\boldsymbol{x})$ is the velocity and $\boldsymbol{j}(\boldsymbol{x}) = \boldsymbol{v}(\boldsymbol{x})\rho_0$ is the current of the fluid at $\boldsymbol{x}$. Thus, the Lagrangian that describes the dynamics of the fluid is given by

$$L = \int \mathrm{d}^d\boldsymbol{x} \left[\frac{m\boldsymbol{j}^2(\boldsymbol{x})}{2\rho_0} - \frac{1}{2\chi}\rho^2(\boldsymbol{x})\right] \tag{5.1.2}$$

From current conservation $\partial_t\rho + \partial_{\boldsymbol{x}} \cdot \boldsymbol{j} = 0$, we find that $\boldsymbol{j} = -\frac{1}{\partial_{\boldsymbol{x}}}\partial_t\rho$ (assuming that $\partial_{\boldsymbol{x}} \times \boldsymbol{j} = 0$), and the Lagrangian can be written in terms of density only as follows:

$$L = \int \mathrm{d}^d\boldsymbol{x} \left(\frac{m}{2\rho_0}\dot{\rho}\frac{1}{-\partial_{\boldsymbol{x}}^2}\dot{\rho} - \frac{1}{2\chi}\rho^2\right) \tag{5.1.3}$$

$$= \mathcal{V}^{-1}\sum_{\boldsymbol{k}} \left(\frac{m}{2\rho_0}\dot{\rho}_{-\boldsymbol{k}}\frac{1}{\boldsymbol{k}^2}\dot{\rho}_{\boldsymbol{k}} - \frac{1}{2\chi}\rho_{-\boldsymbol{k}}\rho_{\boldsymbol{k}}\right) \tag{5.1.4}$$

where $\rho_{\boldsymbol{k}} = \int \mathrm{d}^d\boldsymbol{x}\ \rho(\boldsymbol{x})\mathrm{e}^{-\mathrm{i}\boldsymbol{k}\cdot\boldsymbol{x}}$ and $\mathcal{V}$ is the volume of the system.

Such a Lagrangian is identical to the phonon Lagrangian (3.3.15) studied previously, with $A_{\boldsymbol{k}} = m/\mathcal{V}\rho_0\boldsymbol{k}^2$ and $B_{\boldsymbol{k}} = \frac{1}{\mathcal{V}\chi}$. The phonons are described by the following quantum Hamiltonian (see eqn (3.3.18)):

$$H = \sum_{\boldsymbol{k}\neq 0} \Omega_{\boldsymbol{k}} \alpha_{\boldsymbol{k}}^\dagger \alpha_{\boldsymbol{k}}, \qquad \Omega_{\boldsymbol{k}} = \sqrt{\frac{B_{\boldsymbol{k}}}{A_{\boldsymbol{k}}}} = \sqrt{\frac{\rho_0}{m\chi}}|\boldsymbol{k}| \tag{5.1.5}$$

For free fermions, the compressibility is equal to the density of states: $\chi = N(E_F)$. We find the velocity of the hydrodynamical mode to be

$$v = v_F \qquad \text{for } d = 1$$

$$v = \frac{1}{\sqrt{2}}v_F \qquad \text{for } d = 2$$

$$v = \frac{1}{\sqrt{3}}v_F \qquad \text{for } d = 3$$

The dispersion of the hydrodynamical mode always lies within the particle–hole continuum (see Fig. 5.3). One can show that the specific heat of the single linear mode is less than the specific heat of the corresponding free fermion system in $d > 1$ dimensions. Thus, for $d > 1$, the hydrodynamical treatment does not reproduce the full low-lying excitations; it only describes some average density fluctuations. However, in $d = 1$ dimension and at low temperatures, the specific heat of the single hydrodynamical mode is exactly equal to the specific heat of the corresponding free fermion system, suggesting that the single boson mode reproduces all low-lying excitations of the free fermion system. This indicates that the single boson mode is a faithful description of the fermion system at low energies, and we can use a boson theory to describe the low-energy properties of a fermion system. The method of using bosons to describe a one-dimensional fermion system is called bosonization (Tomonaga, 1950; Luttinger, 1963; Coleman, 1975).

To compare the hydrodynamical model and the fermion model in more detail, let us consider the many-body low-energy excitations in the quantum Hamiltonian (5.1.5). In one dimension, the hydrodynamical Hamiltonian (5.1.5) reproduces the spectrum of all of the low-energy excitations in the corresponding fermion system. To see this, let us put the system on a circle of size L and assume that the ground state has zero energy. The hydrodynamical model has two eigenstates with energy $E = 2k_0 v$ and momentum $K = 2k_0$, namely $(\alpha^\dagger_{k_0})^2|0\rangle$ and $\alpha^\dagger_{2k_0}|0\rangle$, where $k_0 = 2\pi/L$. The fermion system also has two particle–hole excited states at momentum $2k_0$ which have exactly the same energy $E = 2k_0 v_F = 2k_0 v$. The two particle–hole states are given by $c^\dagger_{k_F+2k_0} c_{k_F}|\psi_0\rangle$ and $c^\dagger_{k_F+k_0} c_{k_F-k_0}|\psi_0\rangle$. Here $|\psi_0\rangle$ is the ground state of the fermion system where the single particle states $|k_F\rangle$, $|k_F - k_0\rangle$, ..., $|-k_F\rangle$ are filled by one fermion. The hydrodynamical model has one eigenstate with energy $E = 2k_0 v$ and momentum $K = 0$, namely $a^\dagger_{-k_0} a^\dagger_{k_0}|0\rangle$. The fermion system also has one particle–hole state with the same momentum and energy, namely $c^\dagger_{-k_F-k_0} c_{-k_F} c^\dagger_{k_F+k_0} c_{k_F}|\psi_0\rangle$. In fact, the excitation spectrum of the hydrodynamical model is identical to the particle–hole excitation spectrum of the fermion model (see Problem 5.1.3). The hydrodynamical model is a faithful description of the fermion system at low energies and small momenta.

Problem 5.1.2.
Write $\alpha_{\boldsymbol{k}}$ in terms of $\rho_{\boldsymbol{k}}$ and $\dot{\rho}_{\boldsymbol{k}}$. Show that $\alpha_{\boldsymbol{k}}$ carries a definite momentum.

Problem 5.1.3.
Show that, in one-dimension, both the hydrodynamical model (with a fixed number of particles) and the fermion system (with a fixed number of fermions) have the following properties.
(a) There are p_n excited states with energy nk_0v_F and momentum nk_0, where $(p_0, p_1, p_2, ...) = (1, 1, 2, 3, 5, 7, ...)$ are the partition numbers. (If you cannot find the general proof, then you may check the result up to p_5.)
(b) There are p_n excited states with energy nk_0v_F and momentum $-nk_0$.

(c) There are $p_{n-m}p_m$ states with energy nk_0v_F and momentum $(n-2m)k_0$.
Thus, the hydrodynamical model and the fermion system have an identical excitation spectrum at low energies and small momenta.

Problem 5.1.4.
(a) From eqn (4.3.2), we see that the compressibility $\chi(\omega, \boldsymbol{k}) = -\Pi^{00}(\omega, \boldsymbol{k})$ contains an imaginary part. Using such a compressibility and the equation of motion of the Lagrangian (5.1.2), find the decay rate of the density mode (the phonons) in $d = 1$, $d = 2$, and $d = 3$ dimensions. Compare the decay rate with the frequency of the mode. Does the density mode gives rise to a well-defined phonon?
(b) In fact, part (a) is not quite right. One should calculate the density response function of the hydrodynamical model using eqn (5.1.2) and then obtain $\chi(\omega, \boldsymbol{k})$ by choosing it to fit the fermion results of eqn (4.3.2). Find $\chi(\omega, \boldsymbol{k})$ using this method. Are the main results in (a) changed by the new $\chi(\omega, \boldsymbol{k})$?

5.1.4 Orthogonality catastrophe from the hydrodynamical approach

- In the hydrodynamical approach, the overlap of many-body wave functions can be calculated easily from the shifted harmonic oscillators.

In the presence of the potential caused by the core electron, the Hamiltonian (5.1.5) contains an additional term (see eqn (3.3.17)):

$$\delta H = \int \mathrm{d}^d\boldsymbol{x}\, \rho(\boldsymbol{x})V(\boldsymbol{x}) = \mathcal{V}^{-1}\sum_{\boldsymbol{k}} \rho_{\boldsymbol{k}} V_{-\boldsymbol{k}} = \mathcal{V}^{-1}\sum_{\boldsymbol{k}} V_{-\boldsymbol{k}} \frac{\alpha_{\boldsymbol{k}} + \alpha^\dagger_{\boldsymbol{k}}}{2u_{\boldsymbol{k}}}$$

$$u_{\boldsymbol{k}} = \frac{1}{\sqrt{2}}(A_{\boldsymbol{k}}B_{\boldsymbol{k}})^{1/4} = \left(\frac{m}{4\mathcal{V}^2\rho_0\chi\boldsymbol{k}^2}\right)^{1/4}$$

Here $H + \delta H$ describes a collection of shifted oscillators. The ground state of the system in the presence of the potential is given by

$$|\Psi\rangle = \frac{\mathrm{e}^{-\mathcal{V}^{-1}\sum_{\boldsymbol{k}} \frac{V_{-\boldsymbol{k}}}{2\Omega_{\boldsymbol{k}} u_{\boldsymbol{k}}}\alpha^\dagger_{\boldsymbol{k}}}|\Psi_0\rangle}{\sqrt{\langle\Psi_0|\mathrm{e}^{-\mathcal{V}^{-1}\sum_{\boldsymbol{k}} \frac{V^*_{-\boldsymbol{k}}}{2\Omega_{\boldsymbol{k}} u_{\boldsymbol{k}}}\alpha_{\boldsymbol{k}}}\,\mathrm{e}^{-\mathcal{V}^{-1}\sum_{\boldsymbol{k}} \frac{V_{-\boldsymbol{k}}}{2\Omega_{\boldsymbol{k}} u_{\boldsymbol{k}}}\alpha^\dagger_{\boldsymbol{k}}}|\Psi_0\rangle}}$$

where $|\Psi_0\rangle$ is the ground state in the absence of the potential. Here $|\Psi\rangle$ corresponds to the deformed state. The overlap between $|\Psi\rangle$ and $|\Psi_0\rangle$ is

$$\langle\Psi_0|\Psi\rangle = \mathrm{e}^{-\mathcal{V}^{-2}\sum_{\boldsymbol{k}} \frac{|V_{\boldsymbol{k}}|^2}{4\Omega^2_{\boldsymbol{k}} u^2_{\boldsymbol{k}}}} = \mathrm{e}^{-\int \frac{\mathrm{d}^d\boldsymbol{k}}{(2\pi)^d} \frac{\rho_0|V_{\boldsymbol{k}}|^2}{2mv^3|\boldsymbol{k}|}}$$

For a short-range potential, $V_{\boldsymbol{k}}$ is nonzero for small $\boldsymbol{k}$. In $d = 1$ dimension, the integral diverges for small $\boldsymbol{k}$ and $\langle\Psi_0|\Psi\rangle = 0$. We have an orthogonality catastrophe. In $d > 1$ dimensions, the integral is finite for small $\boldsymbol{k}$. For large $\boldsymbol{k}$, the integral is cut off by a short distance scale (say $l = 1/k_F$). Therefore, the overlap is finite within the hydrodynamical approach. (This result turns out to be incorrect.)

In the following, we will calculate the correlation of the tunneling operator $I(t)$ to obtain the ω dependence of the emission rate. We approximate the time-ordered imaginary-time correlation as

$$\langle T(I(\tau)I(0))\rangle = -G_c(\tau)G_C(-\tau) \tag{5.1.6}$$

The above is exact only when there are no interactions between the core and the conduction electrons.

It turns out that the influence of the core electron on the conduction electron Green's function is not important and we can approximate $G_c(\tau)$ by the free fermion Green's function. To calculate $G_C(\tau)$, we note that the path of the core electron in space–time is a straight line in the time direction. Thus,

$$G_C(\tau) = \left\langle \mathrm{e}^{-\int_0^\tau \mathrm{d}\tau'\ V_C\rho(\tau',0)} \right\rangle \mathrm{e}^{E_C\tau}$$

where $\rho(\tau)$ is the density operator of the conduction electrons, and $-E_C$ is the energy of the core electron. Within the hydrodynamical approach, we may use the path integral to evaluate the above expression. In Problem 5.1.5 it will be shown that the Lagrangian for the hydrodynamical approach (5.1.2) is equivalent to the standard Lagrangian of the XY-model, namely

$$L = \frac{\chi}{2}\left((\partial_t\theta)^2 - v^2(\partial_{\boldsymbol{x}}\theta)^2\right), \qquad \rho = -\chi\partial_t\theta$$

In imaginary time ($t \to -\mathrm{i}\tau$), we have

$$L = \frac{\chi}{2}\left((\partial_\tau\theta)^2 + v^2(\partial_{\boldsymbol{x}}\theta)^2\right), \qquad \rho = -\mathrm{i}\chi\partial_\tau\theta$$

Therefore

$$\begin{aligned} G_C(\tau)\mathrm{e}^{-E_C\tau} &= \frac{\int \mathcal{D}\theta\ \mathrm{e}^{\mathrm{i}\int_0^\tau \mathrm{d}\tau'\ V_C\chi\partial_\tau\theta(\tau',0) - \int \mathrm{d}\tau'\,\mathrm{d}^d\boldsymbol{x}\ L(\theta)}}{\int \mathcal{D}\theta\ \mathrm{e}^{-\int \mathrm{d}\tau'\,\mathrm{d}^d\boldsymbol{x}\ L(\theta)}} \\ &= \left\langle \mathrm{e}^{\mathrm{i}V_C\chi\theta(\tau,0)}\,\mathrm{e}^{-\mathrm{i}V_C\chi\theta(0,0)} \right\rangle \end{aligned}$$

The long-time behavior of $G_C(\tau)$ depends on the spatial dimension. In $d = 1$ dimension (see eqn (3.3.25)), we have

$$G_C(\tau)\mathrm{e}^{-E_C\tau} = \left(\frac{l^2}{v^2\tau^2}\right)^{\chi V_C^2/4\pi v}$$

where we have assumed that the temperature $T = 0$. In real time, we have

$$G_C(t) = \mathrm{e}^{-\mathrm{i}\frac{\pi\chi V_C^2}{4\pi v}} \left(\frac{l}{v|t|}\right)^{\chi V_C^2/2\pi v} \mathrm{e}^{\mathrm{i}E_C t}$$

We see that the Green's function is very different from the one for the free core electron, namely $G_C(t) = e^{iE_C t}$. The spectral function for the dressed core electron is

$$A_{C+}(\omega) \sim \Theta(\omega + E_C)(\omega + E_C)^{\frac{\chi V_C^2}{2\pi v} - 1}$$

and the emission rate behaves like (see Fig. 5.1(c))

$$A_{em} \sim \Theta(E_C - \omega)(E_C - \omega)^{\chi V_C^2/2\pi v}$$

In one dimension, the exponent is given by

$$\frac{\chi V_C^2}{2\pi v} = \frac{V_C^2 N^2(E_F)}{2}. \tag{5.1.7}$$

In $d > 1$ dimensions, $\langle e^{iV_C\chi\theta(\tau,0)} e^{-iV_C\chi\theta(0,0)} \rangle$ approaches a constant for large τ and $G_C(t) \to e^{iE_C t}$, which is the same as the free core-electron Green's function. Thus, according to the hydrodynamical model, the emission rate should be similar to the one for the non-interacting case. This result is consistent with the above analysis of wave function overlap. However, as we will see in the next section, such a result is incorrect for fermion systems.

Problem 5.1.5.
In the hydrodynamical approach, the energy (Hamiltonian) of the conduction electrons is given by

$$H = \int d^d\boldsymbol{x} \left[\frac{m\boldsymbol{j}^2(\boldsymbol{x})}{2\rho_0} + \frac{1}{2\chi}\rho^2(\boldsymbol{x}) \right]$$

We may replace $\boldsymbol{j}$ by $\partial_{\boldsymbol{x}}\phi$ if we assume that $\partial_{\boldsymbol{x}} \times \boldsymbol{j} = 0$.

1. If we view $\rho(\boldsymbol{x})$ as the canonical coordinates, $q_{\boldsymbol{x}}$, then $C\phi(\boldsymbol{x})$ can be viewed as the canonical momenta, $p_{\boldsymbol{x}}$. (Here $\boldsymbol{x}$ can be regarded as an index which labels different canonical coordinates and momenta.) Show that the current conservation $\dot{\rho} + \partial_{\boldsymbol{x}}\boldsymbol{j} = 0$ can be reproduced by one of the Hamiltonian equations

$$\dot{q}_{\boldsymbol{x}} = \frac{\delta H}{\delta p_{\boldsymbol{x}}}$$

   if we choose the value of C properly. Find the value of C.

2. Show that the Hamiltonian equations

$$\dot{p}_{\boldsymbol{x}} = -\frac{\delta H}{\delta q_{\boldsymbol{x}}}, \qquad \dot{q}_{\boldsymbol{x}} = +\frac{\delta H}{\delta p_{\boldsymbol{x}}}$$

   reproduce the equation of motion of the Lagrangian (5.1.2). Show that the Hamiltonian and the canonical coordinate–momentum pairs $(q, p) = (\rho(\boldsymbol{x}), C\phi(\boldsymbol{x}))$ reproduce the Lagrangian (5.1.2).

3. Now let us view $C\phi(\boldsymbol{x})$ as coordinates and $\rho(\boldsymbol{x})$ as momenta. Show that the Hamiltonian and the canonical coordinate–momentum pairs $(q, p) = (C\phi(\boldsymbol{x}), \rho(\boldsymbol{x}))$ reproduce the Lagrangian of the XY-model.

4. We know that, in the standard XY-model, the θ field is periodic, i.e. θ and $\theta + 2\pi$ describe the same point. Find the periodic condition of the ϕ field. (Hint: Consider the $\boldsymbol{k} = 0$ mode and find the relationship between the quantization of the total particle number and the periodic condition of the ϕ field.) We see that, after a proper scaling, $D\phi$ can be identified as the θ field and our hydrodynamical model can be identified as the XY-model.

5. Compare the density correlation function (say, time-ordered) calculated from eqn (5.1.2) and the XY-model. Comment on your results.

Problem 5.1.6.
Use the relation between the hydrodynamical model and the XY-model obtained in the previous problem, to show that the one-dimensional hydrodynamical model can reproduce the low-energy excitations near $\pm 2k_F$, $\pm 4k_F$, ... in the one-dimensional fermion system (see Section 3.6).

5.1.5 Direct calculation for fermion systems

- The hydrodynamical approach can only produce the corrected overlap of many-body wave functions in one dimension.
- In higher dimensions, the single boson mode is not sufficient for describing particle–hole excitations across the Fermi surface. The particle–hole excitations correspond to many-boson modes, which leads to a smaller overlap.

Let us directly calculate

$$G_C(\tau) = \left\langle \mathrm{e}^{-\int_0^\tau \mathrm{d}\tau'\, V_C \rho(\tau', 0)} \right\rangle \mathrm{e}^{E_C \tau}$$

for $\tau > 0$ in a fermion system. Here we view ρ as the fluctuations around the constant density ρ_0. (The constant part can be absorbed into E_C.) We can remove the constant part from ρ through a normal ordering with respect to the filled Fermi sea, i.e. $\rho(\boldsymbol{x}) =: c^\dagger(\boldsymbol{x}) c(\boldsymbol{x}) :$. Here $: c^\dagger_{\boldsymbol{k}} c_{\boldsymbol{k}} := c^\dagger_{\boldsymbol{k}} c_{\boldsymbol{k}}$ if $\xi_{\boldsymbol{k}} > 0$, and $: c^\dagger_{\boldsymbol{k}} c_{\boldsymbol{k}} := -c_{\boldsymbol{k}} c^\dagger_{\boldsymbol{k}}$ if $\xi_{\boldsymbol{k}} < 0$.

To calculate $G_C(\tau)$, we first expand $\mathrm{e}^{-\int_0^\tau \mathrm{d}\tau'\, V_C \rho(\tau', 0)}$ as follows:

$$G_C(\tau) = \left\langle 1 + \left(-\int_0^\tau \mathrm{d}\tau'\, V_C \rho(\tau', 0) \right) + \frac{1}{2!} \left(-\int_0^\tau \mathrm{d}\tau'\, V_C \rho(\tau', 0) \right)^2 + \ldots \right\rangle$$

Using the linked-cluster theorem, we find that

$$G_C(\tau) = \mathrm{e}^{\left\langle -\int_0^\tau \mathrm{d}\tau'\, V_C \rho(\tau', 0) \right\rangle_c + \frac{1}{2!} \left\langle \left(-\int_0^\tau \mathrm{d}\tau'\, V_C \rho(\tau', 0) \right)^2 \right\rangle_c + \ldots}$$

where $\langle ... \rangle_c$ contains only the connected graphs. The first term $\left\langle -\int_0^\tau \mathrm{d}\tau' \, V_C \rho(\tau', 0) \right\rangle_c$ vanishes, and the second term is the leading term. If we assume that the density fluctuations are small, then we may ignore the higher-order terms. Thus,

$$G_C(\tau) = \mathrm{e}^{\frac{V_C^2}{2} \int_0^\tau \mathrm{d}\tau_1 \, \mathrm{d}\tau_2 \, \langle \rho(\tau_1, 0) \rho(\tau_2, 0) \rangle}$$

Using $\mathcal{G}(\tau, 0) = N(E_F)/\tau$, we find that

$$\langle \rho(\tau, 0) \rho(0, 0) \rangle = \mathcal{G}(\tau, 0)(-)\mathcal{G}(-\tau, 0) = \frac{N^2(E_F)}{\tau^2}$$

and

$$G_C(\tau) = \mathrm{e}^{V_C^2 N^2(E_F)(\frac{\tau}{\tau_l} - \ln \frac{\tau}{\tau_l})} = \left(\frac{\tau_l}{\tau}\right)^\alpha \mathrm{e}^{V_C^2 N^2(E_F)\tau/\tau_l}$$

where $\tau_l \sim 1/E_F$ is a short-time cut-off and

$$\alpha = V_C^2 N^2(E_F)$$

The emission rate behaves like

$$A_{em} \sim \Theta(E_C - \omega)(E_C - \omega)^\alpha$$

We see that the edge singularity is modified as long as the fermions have a finite density of states, regardless of the number of spatial dimensions. The result from the hydrodynamical approach is incorrect for $d > 1$.

In one dimension, the result (5.1.7) from the hydrodynamical approach is similar to the one obtained here. However, if we compare the value of the exponent, then the exponent obtained here is twice as large. So, one may wonder, what can go wrong in our one-dimensional hydrodynamical approach? After all, the hydrodynamical modes reproduce the specific heat of the particle–hole excitations near the Fermi surface in one dimension. It appears that the hydrodynamical modes capture all of the low-lying excitations of the fermion systems in one dimension. Thus, one expects that the one-dimensional hydrodynamical approach should give us the correct result. As we will see below, the one-dimensional hydrodynamical approach indeed gives us a correct result in a certain sense.

To understand the discrepancy, we note that the fermion Green's function at the equal-space point, $\mathcal{G}(\tau, 0) = N(E_F)/\tau$, contains two contributions from the two Fermi points. For a small but finite x, we have

$$\mathcal{G}(\tau, x) = \frac{N(E_F)}{2} \left(\frac{v \, \mathrm{e}^{\mathrm{i} k_F x}}{v\tau + \mathrm{i}x} + \frac{v \, \mathrm{e}^{-\mathrm{i} k_F x}}{v\tau - \mathrm{i}x} \right)$$

Therefore, the density correlation also contains two parts with momenta $k \sim 0$ and $k \sim \pm 2k_F$:

$$\begin{aligned} &\langle \rho(\tau, x) \rho(0, 0) \rangle = \mathcal{G}(\tau, x)(-)\mathcal{G}(-\tau, -x) \\ &= \frac{N^2(E_F)}{4} \left(\frac{v^2}{(v\tau + \mathrm{i}x)^2} + \frac{v^2}{(v\tau - \mathrm{i}x)^2} \right) + \frac{N^2(E_F) \cos(2k_F x)}{2} \frac{v^2}{v^2\tau^2 + x^2} \end{aligned}$$

In the hydrodynamical approach, only the small-momentum part $\frac{v^2}{(v\tau+\mathrm{i}x)^2} + \frac{v^2}{(v\tau-\mathrm{i}x)^2}$ is included, which leads to a smaller exponent.

We know that a free fermion system contains low-lying excitations with momenta near $k = 0, \pm 2k_F, \pm 4k_F, \ldots$. From the above discussion, it is clear that all of these low-lying excitations could contribute to the suppression of the edge singularity. In general, if the core electron induces a potential of finite range, then the coupling between the core and the conduction electrons will have the form $\int \mathrm{d}x\, V(x)\rho(x)$. The correlation of such an operator is given by

$$\left\langle \int \mathrm{d}x\, V(x)\rho(\tau, x) \int \mathrm{d}x\, V(x)\rho(0, x) \right\rangle = V_0^2 \frac{N^2(E_F)}{2\tau^2} + |V_{2k_F}|^2 \frac{N^2(E_F)}{2\tau^2}$$

where $V_k = \int \mathrm{d}x\, V(x)\mathrm{e}^{-\mathrm{i}kx}$. In this case, the exponent is given by

$$\alpha = \frac{1}{2}(V_0^2 + |V_{2k_F}|^2)N^2(E_F)$$

For free fermions, the density correlation contains only $k = 0$ and $k = 2k_F$ singularities. Thus, α only contains contributions from V_0 and V_{2k_F}. In the hydrodynamical approach, the contribution from V_{2k_F} is dropped. This is the source of the discrepancy.

Problem 5.1.7.
Calculate the Green's function of the core electron, $\mathcal{G}_C(\tau)$, in the long-time limit, assuming that the interaction between the core and the *one-dimensional* conduction electrons is given by $\int \mathrm{d}x\, V(x)\rho(x)$, with $V(x) = V/|x|^\gamma$ and $0 < \gamma < 1$. What is the resulting emission rate $A_{em}(\omega)$?

Problem 5.1.8.
Calculate the Green's function of the core electron, $\mathcal{G}_C(\tau)$, in the long-time limit, assuming that the core electron interacts with a one-dimensional interacting boson system through a δ-potential, i.e. $H_I = V_C\rho(0)$, where $\rho(x)$ is the density of the bosons. We may describe the interacting boson system by its effective XY-model (3.3.23). Note that the boson density correlation contains singularities at momenta $k = K_n$ (see eqn (3.6.2)). For a given value of (χ, v), determine which singularity controls the long-time behavior of $\mathcal{G}_C(\tau)$ and calculate this long-time behavior.

5.2 Hartree–Fock approximation

- The Hartree–Fock approximation can be viewed as a variational approach.

When we encounter a condensed matter system, we would like to answer two questions. What are the properties of the ground state and what are the properties of the excitations above the ground state? In this section, we are going to use the Hartree–Fock approximation to answer these two questions for an interacting electron system.

5.2.1 Ground-state energy and ferromagnetic transition

- The Hartree term represents the interaction between the total densities, while the Fock term represents an effective attraction between the like spins.
- Just like the Hund rule in atoms, a strong repulsion between electrons in a metal can lead to a ferromagnetic state.

Let us assume that the electrons are on a lattice and carry spin-1/2. The electrons interact through $\frac{1}{2}\sum_{\boldsymbol{i},\boldsymbol{j}} c_\alpha^\dagger(\boldsymbol{i})c_\alpha(\boldsymbol{i})V(\boldsymbol{i}-\boldsymbol{j})c_\beta^\dagger(\boldsymbol{j})c_\beta(\boldsymbol{j})$. The Hamiltonian is given by (in momentum space)

$$H = \sum_{\boldsymbol{k}} \xi_{\boldsymbol{k}} c_{\alpha\boldsymbol{k}}^\dagger c_{\alpha\boldsymbol{k}} + \frac{1}{2N_{\text{site}}} \sum_{\boldsymbol{k},\boldsymbol{k}',\boldsymbol{q}} c_{\alpha\boldsymbol{k}}^\dagger c_{\alpha,\boldsymbol{k}-\boldsymbol{q}} V_{\boldsymbol{q}} c_{\beta\boldsymbol{k}'}^\dagger c_{\beta,\boldsymbol{k}'+\boldsymbol{q}} \tag{5.2.1}$$

where N_{site} is the number of sites.

To understand the ground-state properties of the above interacting system, let us use the ground state $|\Psi_0\rangle$ of a non-interacting system as a trial wave function. More precisely, $|\Psi_0\rangle$ is simply a state described by a set of occupation numbers $n_{\boldsymbol{k}\alpha}$. Here $n_{\boldsymbol{k}\alpha} = 1$ if the state $\boldsymbol{k}$ is occupied by an α-spin, and $n_{\boldsymbol{k}\alpha} = 0$ otherwise. We determine the occupation number $n_{\boldsymbol{k}\alpha}$ by minimizing the average energy of the trial state.

Using Wick's theorem, we find that $\langle\Psi_0|H|\Psi_0\rangle$ contains three terms as follows:

$$\begin{aligned}\langle\Psi_0|H|\Psi_0\rangle = \langle\Psi_0|H_0|\Psi_0\rangle &+ \frac{1}{2}\sum_{\boldsymbol{i},\boldsymbol{j}} \left\langle c_\alpha^\dagger(\boldsymbol{i})c_\alpha(\boldsymbol{i})\right\rangle V(\boldsymbol{i}-\boldsymbol{j}) \left\langle c_\beta^\dagger(\boldsymbol{j})c_\beta(\boldsymbol{j})\right\rangle \\ &+ \frac{1}{2}\sum_{\boldsymbol{i},\boldsymbol{j}} \left\langle c_\alpha^\dagger(\boldsymbol{i})c_\beta(\boldsymbol{j})\right\rangle V(\boldsymbol{i}-\boldsymbol{j}) \left\langle c_\alpha(\boldsymbol{i})c_\beta^\dagger(\boldsymbol{j})\right\rangle\end{aligned}$$

The first term $\langle\Psi_0|H_0|\Psi_0\rangle$ is the kinetic energy, which is minimized when $N_\uparrow = N_\downarrow$, where $N_\uparrow$ is the number of spin-up electrons and $N_\downarrow$ is the number of spin-down electrons. The second term is called the Hartree term:

$$\frac{1}{2}\sum_{\boldsymbol{i},\boldsymbol{j}} \rho(\boldsymbol{i})V(\boldsymbol{i}-\boldsymbol{j})\rho(\boldsymbol{j})$$

which is independent of $N_\uparrow/N_\downarrow$ if N is fixed. It is clear that the Hartree term is simply the classical potential energy. If we only include the first two terms, then the average energy of the trial state will be minimized at $N_\uparrow = N_\downarrow$. The (trial) ground state is then a spin singlet and does not break spin–rotation symmetry.

The third term in the above equation is the Fock term, which can be rewritten as

$$\frac{1}{2N_{\text{site}}} \sum_{\boldsymbol{k},\boldsymbol{k}',\boldsymbol{q}} V_{\boldsymbol{q}} \left\langle c^\dagger_{\alpha\boldsymbol{k}} c_{\beta,\boldsymbol{k}'+\boldsymbol{q}} \right\rangle \left\langle c_{\alpha,\boldsymbol{k}-\boldsymbol{q}} c^\dagger_{\beta\boldsymbol{k}'} \right\rangle$$

$$= -\frac{1}{2N_{\text{site}}} \sum_{\boldsymbol{k},\boldsymbol{q},\alpha} V_{\boldsymbol{q}} n_{\boldsymbol{k}\alpha} n_{\boldsymbol{k}-\boldsymbol{q},\alpha} + \frac{1}{2} V(0) N$$

where N is the total number of electrons. We see that, if we ignore the constant term $\frac{1}{2}V(0)N$, then the Fock term represents an effective attractive interaction between parallel spins, but no interactions between opposite spins. The contribution from the Fock term becomes more negative as $|N_\uparrow - N_\downarrow|$ increases.

To understand this effective attraction between parallel spins, we note that the average potential energy can be written in terms of density correlation functions as follows:

$$\frac{1}{2} \sum_{\boldsymbol{i},\boldsymbol{j}} V(\boldsymbol{i}-\boldsymbol{j}) \left(\langle \rho_\uparrow(\boldsymbol{i}) \rho_\uparrow(\boldsymbol{j}) \rangle + 2 \langle \rho_\uparrow(\boldsymbol{i}) \rho_\downarrow(\boldsymbol{j}) \rangle + \langle \rho_\downarrow(\boldsymbol{i}) \rho_\downarrow(\boldsymbol{j}) \rangle \right)$$

where $\rho_\uparrow(\boldsymbol{i})$ and $\rho_\downarrow(\boldsymbol{i})$ are the densities of the spin-up and the spin-down electrons, respectively. Comparing this with the Hartree term

$$\frac{1}{2} \sum_{\boldsymbol{i},\boldsymbol{j}} V(\boldsymbol{i}-\boldsymbol{j}) \left(\langle \rho_\uparrow(\boldsymbol{i}) \rangle \langle \rho_\uparrow(\boldsymbol{j}) \rangle + 2 \langle \rho_\uparrow(\boldsymbol{i}) \rangle \langle \rho_\downarrow(\boldsymbol{j}) \rangle + \langle \rho_\downarrow(\boldsymbol{i}) \rangle \langle \rho_\downarrow(\boldsymbol{j}) \rangle \right)$$

we find that the Hartree term correctly reproduces the cross-term because $\rho_\uparrow(\boldsymbol{i})$ and $\rho_\downarrow(\boldsymbol{j})$ are independent and $\langle \rho_\uparrow(\boldsymbol{i}) \rho_\downarrow(\boldsymbol{j}) \rangle = \langle \rho_\uparrow(\boldsymbol{i}) \rangle \langle \rho_\downarrow(\boldsymbol{j}) \rangle$. However, the Hartree term overestimates the interaction between the parallel spins. This is because when site $\boldsymbol{i}$ has an electron, it is less likely to find another electron with the same spin on a nearby site $\boldsymbol{j}$. In fact the probability will be zero if $\boldsymbol{j} = \boldsymbol{i}$ due to the Pauli principle. We find that $\langle \rho_\uparrow(\boldsymbol{i}) \rangle \langle \rho_\uparrow(\boldsymbol{j}) \rangle > \langle \rho_\uparrow(\boldsymbol{i}) \rho_\uparrow(\boldsymbol{j}) \rangle$. The Fock term simply corrects this error and hence represents an effective attraction between parallel spins (assuming that $V(\boldsymbol{i}) > 0$). As a consequence of the attraction between parallel spins, the Fock term may make a ferromagnetic state (with $N_\uparrow \neq N_\downarrow$) have a lower energy than the paramagnetic state (with $N_\uparrow = N_\downarrow$), if the Fock term is larger than the kinetic energy term. This is an example in which an interaction can cause symmetry breaking.

As the trial wave function describes a state of uniform density, the Hartree term is given by

$$\frac{1}{2N_{\text{site}}} V_0 \Big(\sum_{\boldsymbol{k},\alpha} n_{\boldsymbol{k}\alpha} \Big)^2 = \frac{1}{2N_{\text{site}}} V_0 N^2.$$

Thus, the total energy of the trial state is given by

$$\langle\Psi_{\{n_{\boldsymbol{k}\alpha}\}}|H|\Psi_{\{n_{\boldsymbol{k}\alpha}\}}\rangle \tag{5.2.2}$$
$$= \sum_{\boldsymbol{k},\alpha} n_{\boldsymbol{k}\alpha}(\epsilon_{\boldsymbol{k}} - \mu + \frac{V_0}{2}) + \frac{V_0}{2N_{\text{site}}}(\sum_{\boldsymbol{k},\alpha} n_{\boldsymbol{k}\alpha})^2 - \frac{1}{N_{\text{site}}}\sum_{\boldsymbol{k},\boldsymbol{q},\alpha} \frac{V_{\boldsymbol{q}}}{2} n_{\boldsymbol{k}\alpha}n_{\boldsymbol{k}-\boldsymbol{q},\alpha}$$

If we change $n_{\boldsymbol{k}\alpha}$ to $n_{\boldsymbol{k}\alpha} + \delta n_{\boldsymbol{k}\alpha}$, then the change in the energy is given by

$$\delta E = \sum_{\boldsymbol{k},\alpha} \delta n_{\boldsymbol{k}\alpha}[\epsilon_{\boldsymbol{k}} - \mu + \rho_0 V_0 + \frac{1}{2}V(0) + \Sigma_{\boldsymbol{k}\alpha}] \tag{5.2.3}$$

where

$$\Sigma_{\boldsymbol{k}\alpha} = -\frac{1}{N_{\text{site}}}\sum_{\boldsymbol{k},\alpha} V_{\boldsymbol{q}} n_{\boldsymbol{k}-\boldsymbol{q},\alpha}, \tag{5.2.4}$$

$\rho_0 = N/N_{\text{site}}$ is the number of electrons per site, and we have only kept the terms which are linear in $\delta n_{\boldsymbol{k}\alpha}$.

The occupation $n_{\boldsymbol{k}\alpha}$ that minimizes the energy has the property that δE is positive for any $\delta n_{\boldsymbol{k}\alpha}$. Such a set of occupation numbers satisfies

$$\begin{aligned} n_{\boldsymbol{k}\alpha} &= 1, \quad \text{if } \epsilon_{\boldsymbol{k}} - \mu' + \Sigma_{\boldsymbol{k}\alpha} < 0 \\ n_{\boldsymbol{k}\alpha} &= 0, \quad \text{if } \epsilon_{\boldsymbol{k}} - \mu' + \Sigma_{\boldsymbol{k}\alpha} > 0 \end{aligned} \tag{5.2.5}$$

where $\mu' = \mu - \rho_0 V_0 - \frac{1}{2}V(0)$. We see that $\epsilon_{\boldsymbol{k}} - \mu' + \Sigma_{\boldsymbol{k}\alpha} = 0$ at the Fermi surface, where $n_{\boldsymbol{k}\alpha}$ change from 0 to 1. The ground-state occupation numbers $n_{\boldsymbol{k}\alpha}$ are obtained by solving the coupled eqns (5.2.5) and (5.2.4).

5.2.2 Spectrum of excitations in the Hartree–Fock approximation

After obtaining the ground-state occupation numbers $n_{\boldsymbol{k}\alpha}$, we can consider the excitations above the ground state. Let us assume that the trial wave function for an excitation is described by a different set of occupation numbers $n_{\boldsymbol{k}\alpha} + \delta n_{\boldsymbol{k}\alpha}$. The (average) energy for such an excitation has been calculated above:

$$\delta E = \sum_{\boldsymbol{k},\alpha} \delta n_{\boldsymbol{k}\alpha}[\epsilon_{\boldsymbol{k}} - \mu' + \Sigma_{\boldsymbol{k}\alpha}] \tag{5.2.6}$$

We see that, under the Hartree–Fock approximation, the low-lying excitations of the interacting system are described by a free fermion system. In particular, the following statements hold.

1. The many-body eigenstates (the ground state and the low-lying excitations) are labeled by the occupation numbers $n_{\boldsymbol{k}\alpha} = 0, 1$. The number of low-lying excitations is the same as in free fermion theory.

2. The energy of a many-body eigenstate is given by the sum of the energies of the occupied states. To be more precise, we can assign an energy $\xi^*_{\boldsymbol{k}\alpha}$ to each momentum state $(\boldsymbol{k}, \alpha)$ such that the total energy of an eigenstate can be expressed as $\sum_{\boldsymbol{k},\alpha} n_{\boldsymbol{k}\alpha}\xi^*_{\boldsymbol{k}}$ + constant.

3. The energy for each momentum state is modified by the interaction $\xi^*_{\boldsymbol{k}\alpha} = \epsilon_{\boldsymbol{k}} - \mu' + \Sigma_{\boldsymbol{k}\alpha}$. The correction $\Sigma_{\boldsymbol{k}\alpha}$ is called the self-energy of the electron.

The above statements can be summarized by the low-energy effective Hamiltonian

$$H_{\text{eff}} = \sum_{\boldsymbol{k},\alpha}(\epsilon_{\boldsymbol{k}} - \mu + \Sigma_{\boldsymbol{k}\alpha})c^\dagger_{\alpha\boldsymbol{k}}c_{\alpha\boldsymbol{k}} \tag{5.2.7}$$

One can directly derive the above effective Hamiltonian using mean-field theory (see Problem 5.2.2).

For a Coulomb interaction in three dimensions, $V(\boldsymbol{x}) = e^2/|\boldsymbol{x}|$, we find that[32]

$$\Sigma_{\boldsymbol{k}\alpha} = -\frac{e^2 k_{F\alpha}}{\pi}\left(1 + \frac{1-y^2}{2y}\ln\left|\frac{1+y}{1-y}\right|\right), \quad y = \frac{k}{k_{F\alpha}} \tag{5.2.8}$$

We note that the renormalized Fermi velocity

$$v^*_{F\alpha}(k) = v_{F\alpha}(k) + \frac{\partial \Sigma_{k\alpha}}{\partial k}$$

diverges as $-\ln|k - k_{F\alpha}|$ when $k \to k_{F\alpha}$ (or $y \to 1$). However, this divergence has not been observed experimentally, and this particular result from the Hartree–Fock approximation turns out to be incorrect.

Problem 5.2.1.
Consider a one-dimensional electron system on a lattice with an on-site potential interaction, so that

$$H = -t\sum_i (c^\dagger_{\alpha,i}c_{\alpha,i+1} + h.c.) + V\sum_i(\sum_\alpha c^\dagger_{\alpha,i}c_{\alpha,i})^2$$

1. Find the energy ϵ_k of a single-particle state with momentum k for a non-interacting system with $V = 0$.

2. Use the Hartree–Fock approximation to find the ground-state energy $E(N_\uparrow, N_\downarrow)$, assuming that the numbers of spin-up and spin-down electrons are given by $N_\uparrow$ and

[32] As $V_{\boldsymbol{q}} = 4\pi e^2/\boldsymbol{q}^2$, we have

$$\begin{aligned}\Sigma_{\boldsymbol{k}\alpha} &= -\int \frac{\mathrm{d}^3\boldsymbol{q}}{(2\pi)^3}\frac{4\pi e^2}{|\boldsymbol{k}-\boldsymbol{q}|^2}n_{\boldsymbol{q}\alpha} = -\frac{e^2}{\pi}\int_0^{k_{F\alpha}} q^2\,\mathrm{d}q\int_{-1}^{+1}\frac{\mathrm{d}t}{k^2+q^2-2kqt}\\ &= -\frac{e^2}{\pi k}\int_0^{k_{F\alpha}} q\,\mathrm{d}q\,\ln\left|\frac{k+q}{k-q}\right| = -\frac{e^2 k_{F\alpha}}{\pi}\left(1+\frac{1-y^2}{2y}\ln\left|\frac{1+y}{1-y}\right|\right)\end{aligned}$$

$N_\downarrow$, respectively. You may assume that the total number of electrons N is less than the number of lattice sites N_{site}.

3. Consider the energy $E(N_\uparrow, N_\downarrow)$ in the neighborhood of $N_\uparrow - N_\downarrow = 0$ and for fixed N. Show that $E(N_\uparrow, N_\downarrow)$ has a local minimum at $N_\uparrow - N_\downarrow = 0$ if $V < V_0$, and a local maximum at $N_\uparrow - N_\downarrow = 0$ if $V > V_0$. Find the critical value V_0.

4. Calculate the spin susceptibility χ of the spin-unpolarized state for $V < V_0$. What is the behavior of χ as $V \to V_0$?

5. Minimize the ground-state energy obtained above with fixed N. Determine the critical value V_c of V beyond which the system becomes a ferromagnet. Determine the value of the total spin S_{z0} in the minimized ground state, assuming that the spin of the ferromagnet is pointing in the z direction.

6. Calculate $\Sigma_{\boldsymbol{k}\alpha}$ for both $V < V_c$ and $V > V_c$, and determine the energy spectrum of the excitations above the minimized ground state.

7. Find the minimum energy cost to create an excitation of flipping one spin (i.e. change the total spin S_z by 1), for both $V < V_c$ and $V > V_c$. Comment on your result (do you believe it?).

Problem 5.2.2.
Derive eqn (5.2.7) by replacing a pair $c^\dagger_{\alpha\boldsymbol{k}} c_{\beta\boldsymbol{k}'}$ by its average $\left\langle c^\dagger_{\alpha\boldsymbol{k}} c_{\beta\boldsymbol{k}'} \right\rangle$. (This is a mean-field approximation.) There are several contributions from different replacements. Note that $\langle H_{\text{eff}} \rangle$ is different from the ground-state energy $\langle H \rangle$ calculated in the Hartree–Fock approximation. Include a constant term (which can be written in terms of $\left\langle c^\dagger_{\alpha\boldsymbol{k}} c_{\beta\boldsymbol{k}'} \right\rangle$), so that $\langle H_{\text{eff}} \rangle = \langle H \rangle$.

5.3 Landau Fermi liquid theory

- Low-energy excitations of an interacting fermion system are described by free quasiparticles. In other words, interacting fermions $\approx$ free fermions.
- Perturbation theory works even when the interaction is much much larger than the level spacing.

Landau Fermi liquid theory (Landau, 1956, 1959) is one of the two cornerstones of traditional many-body theory. It essentially states that a metal formed by interacting electrons behaves almost like a free fermion system. Landau Fermi liquid theory is very useful because it describes (almost) all known metals. It also forms the foundation of our understanding of many non-metallic states, such as superconductors, anti-ferromagnetic states, etc. These non-metallic states are realized as certain instabilities of a Fermi liquid. On the other hand, Landau Fermi liquid theory is very mysterious because the Coulomb interaction between electrons in ordinary metals is as large as the Fermi energy. It is much larger than the energy-level spacing near the Fermi energy. Perturbation theory (in the usual sense) breaks down for such a strong interaction. It is hard to see why a gapless system with such a strong interaction resembles a non-interacting system.

To appreciate the brilliance of Landau Fermi liquid theory, let us look at the many-body Hamiltonian of interacting electrons, namely

$$H = \sum_i \left(\frac{\hbar^2}{2m} \partial^2_{\boldsymbol{x}_i} + U(\boldsymbol{x}_i) \right) + \sum_{i<j} \frac{e^2}{|\boldsymbol{x}_i - \boldsymbol{x}_j|}$$

It is hopeless for a theorist to solve such a 'nasty' system, not to mention to guess that such a system behaves almost like a free electron system. Certainly, condensed matter physicists did not provide such a bold guess. It is nature itself who hints to us over and over again that metals behave just like a free electron system, despite the strong Coulomb interaction. Even now, I am amazed that so many metals can be described by Landau Fermi liquid theory, and puzzled by the difficulty to find a metal that cannot be described by Landau Fermi liquid theory.

On a technical level, Landau Fermi liquid theory implies that a perturbative expansion in interaction works, despite the interaction being much larger than the level spacing. Due to this, to understand the physical properties of a metal, we can start with a free electron system and use perturbation theory to calculate various quantities. This approach was so successful that Landau Fermi liquid theory became the 'standard model' for interacting fermion systems. The dominant position of Landau Fermi liquid theory was only challenged not so long ago by fractional quantum Hall states in 1982, and by high-T_c superconductors in 1987.

5.3.1 Basic assumptions and their consequences

- The concept of a quasiparticle.

Landau Fermi liquid theory has only one really basic assumption.

> The low-energy eigenstates (including the ground state and the low-lying excitations) are labeled by a set of quantum numbers $n_{\boldsymbol{k}\alpha} = 0, 1$, which are called the 'occupation numbers'.

As the excitations are in one-to-one correspondence with those in the free fermion system, we can use the language of the free fermion systems to describe the excitations in the Landau Fermi liquid, and we call these excitations quasiparticle excitations. The energy of an eigenstate is a function of $n_{\boldsymbol{k}\alpha}$. Expanding around the ground-state occupation numbers $n_{0,\boldsymbol{k}\alpha}$, we may write

$$E_{\{n_{\boldsymbol{k}\alpha}\}} = E_g + \sum_{\boldsymbol{k},\alpha} \delta n_{\boldsymbol{k}\alpha} \xi^*_{\boldsymbol{k}\alpha} + \frac{1}{2\mathcal{V}} \sum_{\boldsymbol{k}\alpha,\boldsymbol{k}'\beta} f(\boldsymbol{k}, \alpha; \boldsymbol{k}', \beta) \delta n_{\boldsymbol{k}\alpha} \delta n_{\boldsymbol{k}'\beta} \tag{5.3.1}$$

where $\xi^*_{\boldsymbol{k}\alpha}$ is the energy of the quasiparticles, and $f(\boldsymbol{k},\alpha;\boldsymbol{k}',\beta)$ represents the interaction between quasiparticles. Here $f(\boldsymbol{k},\alpha;\boldsymbol{k}',\beta)$ is called the Fermi liquid function, which is also very important in determining the low-energy properties of the system. We note that the energy obtained in the Hartree–Fock approximation has the form (5.3.1) (see eqn (5.2.2)). The quadratic term in $\delta n_{\boldsymbol{k}\alpha}$ which we ignored in the last section has the form

$$\frac{1}{2\mathcal{V}}V_0(\sum_{\boldsymbol{k},\alpha}\delta n_{\boldsymbol{k}\alpha})^2 - \frac{1}{2\mathcal{V}}\sum_{\boldsymbol{k},\boldsymbol{q},\alpha}V_{\boldsymbol{q}}\delta n_{\boldsymbol{k}\alpha}\delta n_{\boldsymbol{k}-\boldsymbol{q},\alpha}$$

Thus, we have

$$\begin{aligned}\xi^*_{\boldsymbol{k}\alpha} &= \epsilon_{\boldsymbol{k}} - \mu + \Sigma_{\boldsymbol{k}\alpha}\\ f(\boldsymbol{k},\alpha;\boldsymbol{k}',\beta) &= V_0 - V_{\boldsymbol{k}-\boldsymbol{k}'}\delta_{\alpha\beta}\end{aligned} \tag{5.3.2}$$

within the Hartree–Fock approximation.

Equation (5.3.1) determines the energies of all low-lying excitations, and many low-energy properties of the interacting fermion system can be expressed in terms of the quasiparticle energy ξ^* and the Fermi liquid function f. Creating an excitation above the ground state by changing an occupation number $n_{\boldsymbol{k}\alpha}$ from 0 to 1 (or from 1 to 0) costs an energy $\xi^*_{\boldsymbol{k}\alpha}$. However, in the presence of other excitations (say, due to finite temperatures) the energy cost is different from $\xi^*_{\boldsymbol{k}\alpha}$ due to the interaction between quasiparticles. From eqn (5.3.1), we find the new energy cost to be

$$\epsilon^T_{\boldsymbol{k}\alpha} = \xi^*_{\boldsymbol{k}\alpha} + \frac{1}{\mathcal{V}}\sum_{\boldsymbol{k}'\beta} f(\boldsymbol{k},\alpha;\boldsymbol{k}',\beta)\delta n_{\boldsymbol{k}'\beta}$$

The momentum change due to the creation of such a quasiparticle is the same as the free fermion system, i.e. $\Delta\boldsymbol{K} = \boldsymbol{k}$, and the total momentum of a state described by $n_{\boldsymbol{k}\alpha}$ is

$$\boldsymbol{K} = \sum_{\boldsymbol{k}\alpha}\boldsymbol{k}n_{\boldsymbol{k}\alpha}$$

This result can be obtained from the second assumption of Landau Fermi liquid theory, which is stated as follows.

> As we turn off the interaction, the low-energy eigenstates adiabatically change into the corresponding eigenstates of the free fermion system labeled by the same occupation numbers $n_{\boldsymbol{k}\alpha}$.

As the momenta are quantized, they cannot change during the adiabatic turning on of the interaction.

From the momentum and the energy, we find the velocity of the quasiparticle to be

$$\boldsymbol{v}^T(\boldsymbol{k}) = \partial_{\boldsymbol{k}} \epsilon^T_{\boldsymbol{k}\alpha}$$

At zero temperature, it changes to

$$\boldsymbol{v}^*(\boldsymbol{k}) = \partial_{\boldsymbol{k}} \xi^*_{\boldsymbol{k}\alpha}$$

We note that, near the Fermi surface, the quasiparticle behaves like a particle with mass

$$m^*_\alpha = \frac{k_{F\alpha}}{|\boldsymbol{v}^*(\boldsymbol{k}_{F\alpha})|}$$

Here m^* is called the effective mass of the quasiparticle, which may not be equal to the mass of the original electrons.

We see that the energy of the quasiparticle ϵ^T depends on the presence of other quasiparticles and is not a constant. However, because the sum $\frac{1}{\mathcal{V}} \sum_{\boldsymbol{k}\alpha,\boldsymbol{k}'\beta} f(\boldsymbol{k},\alpha;\boldsymbol{k}',\beta)\delta n_{\boldsymbol{k}'\beta}$ effectively averages over many quasiparticles (assuming that $f(\boldsymbol{k},\alpha;\boldsymbol{k}',\beta)$ is not too singular), its thermal fluctuations are small and ϵ^T can be treated as a constant:

$$\epsilon^T_{\boldsymbol{k}\alpha} = \xi^*_{\boldsymbol{k}\alpha} + \frac{1}{\mathcal{V}} \sum_{\boldsymbol{k}\alpha,\boldsymbol{k}'\beta} f(\boldsymbol{k},\alpha;\boldsymbol{k}',\beta) \langle\langle \delta n_{\boldsymbol{k}'\beta} \rangle\rangle$$

where $\langle\langle ... \rangle\rangle$ represents a thermal average. From this, we obtain the average occupation number

$$\langle\langle n_{\boldsymbol{k}\alpha} \rangle\rangle = n_F(\epsilon^T_{\boldsymbol{k}\alpha}) = \frac{1}{\mathrm{e}^{\epsilon^T_{\boldsymbol{k}\alpha}/T} + 1}$$

Certainly, $\epsilon^T_{\boldsymbol{k}\alpha}$ depends on $\langle\langle n_{\boldsymbol{k}\alpha} \rangle\rangle$, and we have to solve the above equation to obtain $\langle\langle n_{\boldsymbol{k}\alpha} \rangle\rangle$. Given $\langle\langle n_{\boldsymbol{k}\alpha} \rangle\rangle$, all of the thermodynamical properties of the Fermi liquids can be determined.

To calculate the specific heat at low temperatures, we note that the average occupation numbers can be approximated by

$$\langle\langle n_{\boldsymbol{k}\alpha} \rangle\rangle = n_F(\epsilon^T_{\boldsymbol{k}\alpha}) = \frac{1}{\mathrm{e}^{\epsilon^T_{\boldsymbol{k}\alpha}/T} + 1} \approx \frac{1}{\mathrm{e}^{\xi^*_{\boldsymbol{k}\alpha}/T} + 1}$$

because $(\epsilon^T_{\boldsymbol{k}\alpha} - \xi^*_{\boldsymbol{k}\alpha}) \sim \langle\langle \delta n_{\boldsymbol{k}\alpha} \rangle\rangle$ vanishes as T^2 when $T \to 0$ (note that $\delta n_{\boldsymbol{k}\alpha}$ changes its sign across the Fermi surface). Therefore, $\langle\langle n_{\boldsymbol{k}\alpha} \rangle\rangle$ are the same as the average occupation numbers of a mass-m^* free electron system. The specific heat C_V is given by the same formula. In terms of the density of states (see eqn (4.2.12) but with m replaced by m^*), we have $C_V = 2\frac{2\pi^2}{3} N(E_F)T$. The interaction contributes to a term of order T^3/E_F^3. We see that the specific heat depends only on the effective mass m^*. This gives us a way to experimentally measure the effective mass m^* of a Landau Fermi liquid. In Problem 5.3.1, we will see that other low-energy properties, such as the compressibility χ, depend on both the effective mass m^* and the Fermi liquid function f.

We have mentioned that, in Fermi liquid theory, a quasiparticle above the ground state is created by changing $n_{\boldsymbol{k}\alpha}$ from 0 to 1. Can we create such a quasiparticle state $|\Psi_{\boldsymbol{k}}\rangle$ by applying $c^\dagger_{\boldsymbol{k}\alpha}$ to the ground-state wave function $|\Psi_0\rangle$? The answer is no. Although $c^\dagger_{\boldsymbol{k}\alpha}|\Psi_0\rangle$ carries the same momentum as the quasiparticle state, it may not be an energy eigenstate. The state $c^\dagger_{\boldsymbol{k}\alpha}|\Psi_0\rangle$ may be a superposition of a state with one quasiparticle, and a state with two quasiparticles and one quasihole, etc. All of these states can carry the same momentum $\boldsymbol{k}$ but different energies. Therefore, the overlap between $c^\dagger_{\boldsymbol{k}\alpha}|\Psi_0\rangle$ and $|\Psi_{\boldsymbol{k}}\rangle$ given by

$$Z = |\langle\Psi_{\boldsymbol{k}}|c^\dagger_{\boldsymbol{k}\alpha}|\Psi_0\rangle|^2$$

may not be 1. The third assumption of Landau Fermi liquid theory is stated as follows.

> The overlap $Z = |\langle\Psi_{\boldsymbol{k}}|c^\dagger_{\boldsymbol{k}\alpha}|\Psi_0\rangle|^2$ is not zero in the thermodynamic limit (i.e. the $\mathcal{V} \to \infty$ limit) for $\boldsymbol{k}$ near the Fermi surface.

This implies that the electron spectral function $\pi^{-1}\text{Im}G(\boldsymbol{k},\omega)$ at $T = 0$ contains a δ-function: $Z\delta(\omega - \xi^*_{\boldsymbol{k}\alpha})$. The energies of the states with two quasiparticles and one quasihole can spread over a finite range. These states contribute a finite background to the spectral function.

Problem 5.3.1.
Calculate the compressibility χ and the spin susceptibility χ_s of a Landau Fermi liquid. Calculate the ratios between χ, χ_s, and C_V, and compare them with the corresponding ratios for a free electron system.

5.3.2 Boltzmann equation of a Fermi liquid at $T = 0$

In Section 5.1.3, we have tried to use density fluctuations to describe Fermi liquids. The hydrodynamical approach works in one dimension but fails badly beyond one dimension. The reason is that beyond one dimension there are a lot more particle–hole excitations across the Fermi surface than a single density mode can possibly describe. On the other hand, Landau Fermi liquid theory suggests that a uniform Fermi liquid can be completely described by the occupation number $n_{\boldsymbol{k}}$. The density $\sum_{\boldsymbol{k}} n_{\boldsymbol{k}}$ is just one combination of $n_{\boldsymbol{k}}$. This suggests that, if we can generalize Landau theory to non-uniform and time-dependent $n_{\boldsymbol{k}}$, then we can obtain the hydrodynamical theory that provides a complete description of Fermi liquids beyond one dimension. In this section, we are going to develop such a theory. The hydrodynamical theory can also be regarded as bosonization beyond one dimension (Luther, 1979; Haldane, 1992; Houghton and Marston, 1993; Neto and Fradkin, 1994).

Let us consider a spin-1/2 interacting electron system. The ground state is described by the occupation number $n_{0,\boldsymbol{k}\alpha}$. A collective excited state is described by $n_{\boldsymbol{k}\alpha}(\boldsymbol{x},t)$. Let $\delta n_{\boldsymbol{k}\alpha}(\boldsymbol{x},t) = n_{\boldsymbol{k}\alpha}(\boldsymbol{x},t) - n_{0,\boldsymbol{k}\alpha}$. Here we assume that $\delta n_{\boldsymbol{k}\alpha}(\boldsymbol{x},t)$ is a smooth function of $\boldsymbol{x}$ and t, and treat it as a local constant. The quasiparticle energy on the background of a collective excited state is given by

$$\tilde{\epsilon}_{\boldsymbol{k}\alpha}(\boldsymbol{x},t) = \xi^*_{\boldsymbol{k}\alpha} + \frac{1}{\mathcal{V}}\sum_{\boldsymbol{k}'\beta} f(\boldsymbol{k},\alpha;\boldsymbol{k}',\beta)\delta n_{\boldsymbol{k}'\beta}(\boldsymbol{x},t) \tag{5.3.3}$$

Thus, the changes in the position and the momentum of the quasiparticle are given by

$$\dot{\boldsymbol{x}} = \partial_{\boldsymbol{k}}\tilde{\epsilon}_{\boldsymbol{k}\alpha}(\boldsymbol{x},t), \qquad \dot{\boldsymbol{k}} = -\partial_{\boldsymbol{x}}\tilde{\epsilon}_{\boldsymbol{k}\alpha}(\boldsymbol{x},t). \tag{5.3.4}$$

The motion of the quasiparticle causes the following change in $n(\boldsymbol{x},t)$: $\partial n_{\boldsymbol{k}\alpha}/\partial t = -(\partial n_{\boldsymbol{k}\alpha}/\partial \boldsymbol{x})\dot{\boldsymbol{x}} - (\partial n_{\boldsymbol{k}\alpha}/\partial \boldsymbol{k})\dot{\boldsymbol{k}} = -(\partial n_{\boldsymbol{k}\alpha}/\partial \boldsymbol{x}) \cdot (\partial\tilde{\epsilon}/\partial \boldsymbol{k}) + (\partial n_{\boldsymbol{k}\alpha}/\partial \boldsymbol{k}) \cdot (\partial\tilde{\epsilon}/\partial \boldsymbol{x})$. Such a hydrodynamical equation is also written as follows: $\mathrm{d}n/\mathrm{d}t = (\partial n_{\boldsymbol{k}\alpha}/\partial t) + (\partial n_{\boldsymbol{k}\alpha}/\partial \boldsymbol{x}) \cdot (\partial\tilde{\epsilon}/\partial \boldsymbol{k}) - (\partial n_{\boldsymbol{k}\alpha}/\partial \boldsymbol{k}) \cdot (\partial\tilde{\epsilon}/\partial \boldsymbol{x}) = 0$. The Boltzmann equation is obtained by letting $\mathrm{d}n/\mathrm{d}t$ be equal to the redistribution of n due to the additional scattering caused by electron–electron interactions:

$$\frac{\partial n_{\boldsymbol{k}\alpha}}{\partial t} + \frac{\partial n_{\boldsymbol{k}\alpha}}{\partial \boldsymbol{x}} \cdot \frac{\partial\tilde{\epsilon}}{\partial \boldsymbol{k}} - \frac{\partial n_{\boldsymbol{k}\alpha}}{\partial \boldsymbol{k}} \cdot \frac{\partial\tilde{\epsilon}}{\partial \boldsymbol{x}} = I[n_{\boldsymbol{k}\alpha}]$$

The Boltzmann equation describes the dynamics of collective excitations.

As an application of the Boltzmann equation, let us calculate the current carried by a quasiparticle. We know that the velocity of the quasiparticle is $\boldsymbol{v}^* = \partial\xi^*_{\boldsymbol{k}}/\partial\boldsymbol{k}$. So, naively, one expects that the current should be $\mathcal{V}^{-1}\sum_{\boldsymbol{k}}\delta n_{\boldsymbol{k}\alpha}\boldsymbol{v}^*(\boldsymbol{k})$. It turns out that the quasiparticle interaction has a non-trivial correction to the current. The electron number conservation implies that $\sum_{\boldsymbol{k}} I[n] = 0$. Thus, $\sum_{\boldsymbol{k}} \mathrm{d}n_{\boldsymbol{k}\alpha}/\mathrm{d}t = 0$. This allows us to show that

$$\partial_t\mathcal{V}^{-1}\sum_{\boldsymbol{k}} n_{\boldsymbol{k}\alpha}(\boldsymbol{x},t) + \partial_{\boldsymbol{x}}\boldsymbol{J}(\boldsymbol{x},t) = 0$$

where

$$\boldsymbol{J}(\boldsymbol{x},t) = \mathcal{V}^{-1}\sum_{\boldsymbol{k}} n_{\boldsymbol{k}\alpha}(\boldsymbol{x},t)\frac{\partial\tilde{\epsilon}}{\partial\boldsymbol{k}} \tag{5.3.5}$$

As

$$\rho(\boldsymbol{x},t) = \mathcal{V}^{-1}\sum_{\boldsymbol{k}} n_{\boldsymbol{k}\alpha}(\boldsymbol{x},t)$$

is the electron number density, $\boldsymbol{J}$ can be interpreted as the number current density. If we only keep the linear δn term, then $\boldsymbol{J}$ becomes

$$\boldsymbol{J}(\boldsymbol{x},t) = \mathcal{V}^{-1}\sum_{\boldsymbol{k}} \delta n_{\boldsymbol{k}\alpha}(\boldsymbol{x},t)\tilde{\boldsymbol{v}}$$

$$\tilde{\boldsymbol{v}}(\boldsymbol{k}) = \boldsymbol{v}^*(\boldsymbol{k}) + \int \frac{\mathrm{d}^d\boldsymbol{k}'}{(2\pi)^d}\sum_{\beta} f(\boldsymbol{k},\alpha;\boldsymbol{k}',\beta)\boldsymbol{v}^*(\boldsymbol{k}')\delta(\xi^*_{\boldsymbol{k}'}) \tag{5.3.6}$$

The second term in the expression of $\tilde{\boldsymbol{v}}$ is a correction to the free fermion result. It is called the drag term or the back-flow term. To understand the correction, we note that a quasiparticle at $\boldsymbol{k}$ also causes a slight change in the velocity of the quasiparticles in the filled Fermi sea, as a result of interaction between the quasiparticles. Thus, the electrons in the Fermi sea also contribute to the number current density. This is the source of the drag/back-flow term.

In the relaxation-time approximation, we assume that $I[n] = -\tau^{-1}\delta n$. To include an external force $\boldsymbol{F}$ on the quasiparticle, we need to replace $-\partial_{\boldsymbol{x}}\tilde{\epsilon}$ in eqn (5.3.4) by $\boldsymbol{F} - \partial_{\boldsymbol{x}}\tilde{\epsilon}$. The resulting Boltzmann equation becomes

$$\frac{\partial n_{\boldsymbol{k}\alpha}}{\partial t} + \frac{\partial n_{\boldsymbol{k}\alpha}}{\partial \boldsymbol{x}}\cdot\frac{\partial\tilde{\epsilon}}{\partial\boldsymbol{k}} + \frac{\partial n_{\boldsymbol{k}\alpha}}{\partial\boldsymbol{k}}\cdot\left(\boldsymbol{F} - \frac{\partial\tilde{\epsilon}}{\partial\boldsymbol{x}}\right) = -\tau^{-1}\delta n_{\boldsymbol{k}\alpha} \tag{5.3.7}$$

We can use the above equation to calculate many different transport properties.

Within Landau Fermi liquid theory, the quasiparticle is assumed to have infinite lifetime and does not decay into any other states. For a uniform $\delta n_{\boldsymbol{k}\alpha}$ and vanishing $\boldsymbol{F} = 0$, eqn (5.3.7) is reduced to $\frac{\partial n_{\boldsymbol{k}\alpha}}{\partial t} = -\tau^{-1}\delta n_{\boldsymbol{k}\alpha}$. We see that τ is the quasiparticle lifetime. Thus, the collision term $I[n]$ is assumed to be zero in Landau Fermi liquid theory. For a real interacting electron system, $I[n]$ behaves like $|k_F - k|^2\delta n$ (see Section 5.4.4) and can be ignored at low energies.

Problem 5.3.2.
Prove eqns (5.3.5) and (5.3.6). (Hint: $\frac{\partial n}{\partial\boldsymbol{x}}\frac{\partial\tilde{\epsilon}}{\partial\boldsymbol{k}} - \frac{\partial n}{\partial\boldsymbol{k}}\frac{\partial\tilde{\epsilon}}{\partial\boldsymbol{x}} = \frac{\partial}{\partial\boldsymbol{x}}\left(n\frac{\partial\tilde{\epsilon}}{\partial\boldsymbol{k}}\right) - \frac{\partial}{\partial\boldsymbol{k}}\left(n\frac{\partial\tilde{\epsilon}}{\partial\boldsymbol{x}}\right)$. Also, $\partial_{\boldsymbol{k}} n_{0,\boldsymbol{k}} = -\boldsymbol{v}^*(\boldsymbol{k})\delta(\xi^*)$.)

5.3.3 Hydrodynamical theory of a Fermi liquid

In this section, we will obtain a reduced Boltzmann equation, which serves as the classical equation of motion for the hydrodynamical description of a Fermi liquid (Kim *et al.*, 1995). Such a reduced Boltzmann equation generalizes the equation of motion of the density mode discussed in Section 5.1.3.

The key to obtaining the reduced Boltzmann equation is to use the Fermi surface displacement h to describe the collective fluctuations (see Fig. 5.4(a)). One may wonder why we do not use the occupation number $n_{\boldsymbol{k}\alpha}$ to describe the collective fluctuations. One important step in developing a hydrodynamical theory of a

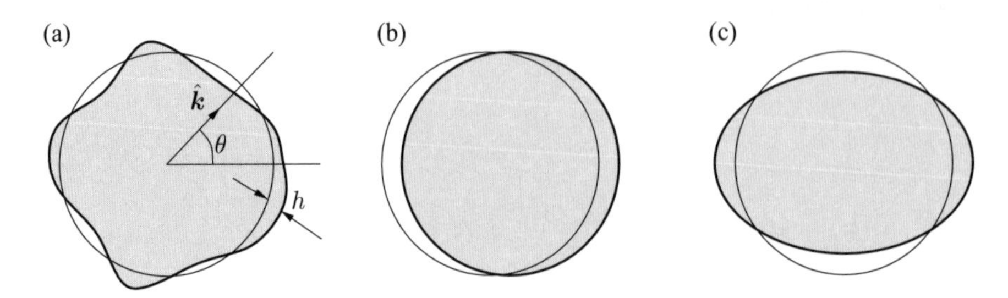

FIG. 5.4. (a) Collective fluctuations in a Fermi liquid can be described by the displacement of the Fermi surface. (b) The $l = 1$ mode corresponds to a dipole fluctuation, and (c) the $l = 2$ mode corresponds to a quadrupole fluctuation.

Fermi liquid is to identify the classical field that we can use to formulate the classical hydrodynamical theory. The uniform occupation numbers $n_{\boldsymbol{k}\alpha}$ can already describe all excitated states. After including the spatial dependence, $n_{\boldsymbol{k}\alpha}(\boldsymbol{x}, t)$ will be an over complete set of variables. It turns out that h, with its spatial dependence, is a proper choice.

For simplicity, we limit our discussion to two-dimensional systems. We will also suppress the spin index. The displacement h and the occupation number $n_{\boldsymbol{k}}$ are related as follows:

$$\tilde{\rho}(\theta) = \int \frac{k\mathrm{d}k}{(2\pi)^2} \delta n_{\boldsymbol{k}=k\hat{\boldsymbol{k}}}, \qquad h(\theta) = (2\pi)^2 \tilde{\rho}(\theta)/k_F$$

where $\hat{\boldsymbol{k}} = \cos(\theta)\boldsymbol{x} + \sin(\theta)\boldsymbol{y}$ is the unit vector in the θ direction (see Fig. 5.4(a)). It turns out that $\tilde{\rho}(\theta)$ is more convenient. So we will use $\tilde{\rho}$ instead of h. From the definition, we see that $\int \mathrm{d}\theta\ \tilde{\rho}(\theta)$ is the fluctuation of the total density of the electrons, and it corresponds to the ρ in eqn (5.1.2).

For small fluctuations, h is close to zero and the quasiparticle energy (5.3.3) can be simplified to give

$$\tilde{\epsilon}_{\boldsymbol{k}}(\boldsymbol{x}, t) = \xi^*_{\boldsymbol{k}} + \int \mathrm{d}\theta'\ f(\theta, \theta')\tilde{\rho}(\theta', \boldsymbol{x}, t) \tag{5.3.8}$$

where θ is the angle of $\boldsymbol{k}$ and $f(\theta, \theta') = f(k_F\hat{\boldsymbol{k}}, k_F\hat{\boldsymbol{k}'})$ is the Fermi function between two points on the Fermi surface. The reduced Boltzmann equation can be obtained from eqn (5.3.7) by performing the integration $\int \frac{\mathrm{d}dk}{(2\pi)^2}$ on both sides of the equation. Assuming rotational symmetry, we find that[33]

$$\frac{\partial \tilde{\rho}}{\partial t} + \frac{\partial \tilde{\rho}}{\partial \boldsymbol{x}} \cdot \frac{\partial \tilde{\epsilon}}{\partial \boldsymbol{k}} + k_F^{-1} \left(\hat{\boldsymbol{k}}_\perp \frac{\partial \tilde{\rho}}{\partial \theta} - \hat{\boldsymbol{k}} \tilde{\rho} - \frac{\hat{\boldsymbol{k}} k_F^2}{(2\pi)^2} \right) \cdot \left(\boldsymbol{F} - \frac{\partial \tilde{\epsilon}}{\partial \boldsymbol{x}} \right) = -\tau^{-1} \tilde{\rho} \tag{5.3.9}$$

where $\hat{\boldsymbol{k}}_\perp$ is the unit vector which is perpendicular to $\hat{\boldsymbol{k}}$ and $\tilde{\rho}$ is a function $\tilde{\rho}(\theta, \boldsymbol{x}, t)$. The above can be viewed as a classical equation of motion for a field in $(1+2)$-dimensional *space* parametrized by $(\theta, \boldsymbol{x})$. The linearized equation has the form

$$\frac{\partial \tilde{\rho}}{\partial t} + v_F^* \hat{\boldsymbol{k}} \cdot \frac{\partial \tilde{\rho}}{\partial \boldsymbol{x}} - \frac{\hat{\boldsymbol{k}} k_F}{(2\pi)^2} \cdot \left(\boldsymbol{F} - \int \mathrm{d}\theta' \, f(\theta, \theta') \frac{\partial \tilde{\rho}(\theta', \boldsymbol{x}, t)}{\partial \boldsymbol{x}} \right) = -\tau^{-1} \tilde{\rho} \tag{5.3.10}$$

When $\boldsymbol{F} = \tau^{-1} = 0$, eqn (5.3.10) can be viewed as an equation of motion for a Fermi liquid which has a form:

$$\frac{\partial \tilde{\rho}}{\partial t} + v_F^* \hat{\boldsymbol{k}} \cdot \frac{\partial \tilde{\rho}}{\partial \boldsymbol{x}} + \frac{k_F}{(2\pi)^2} \int \mathrm{d}\theta' \, f(\theta, \theta') \hat{\boldsymbol{k}} \cdot \frac{\partial \tilde{\rho}(\theta', \boldsymbol{x}, t)}{\partial \boldsymbol{x}} = 0 \tag{5.3.11}$$

The corresponding Fermi liquid energy (or Hamiltonian) (5.3.1) can also be written in terms of $\tilde{\rho}$:

$$\begin{aligned} H &= E_g + \int \mathrm{d}^2\boldsymbol{x} \, \mathrm{d}\theta \, \tilde{\rho} \frac{h v_F^*}{2} + \frac{1}{2} \int \mathrm{d}^2\boldsymbol{x} \, \mathrm{d}\boldsymbol{k} \, \mathrm{d}\boldsymbol{k}' \, f(\boldsymbol{k}, \boldsymbol{k}') \delta n_{\boldsymbol{k}} \delta n_{\boldsymbol{k}'} \\ &= E_g + \int \mathrm{d}^2\boldsymbol{x} \, \mathrm{d}\theta \, \frac{2\pi^2 v_F^*}{k_F} \tilde{\rho}^2 + \frac{1}{2} \int \mathrm{d}^2\boldsymbol{x} \, \mathrm{d}\theta \, \mathrm{d}\theta' \, f(\theta, \theta') \tilde{\rho}(\theta) \tilde{\rho}(\theta') \end{aligned} \tag{5.3.12}$$

The Lagrangian that reproduces the equation of motion and the Hamiltonian is given by

$$\begin{aligned} L = &\int \mathrm{d}^2\boldsymbol{x} \, \mathrm{d}\theta \left(\frac{1}{2k_F} \hat{\boldsymbol{k}} \cdot \partial_{\boldsymbol{x}} \tilde{\varphi} \partial_t \tilde{\varphi} - \frac{v_F^*}{2k_F} (\hat{\boldsymbol{k}} \cdot \partial_{\boldsymbol{x}} \tilde{\varphi})^2 \right) \\ &- \frac{1}{8\pi^2} \int \mathrm{d}^2\boldsymbol{x} \, \mathrm{d}\theta \, \mathrm{d}\theta' \, f(\theta, \theta') \hat{\boldsymbol{k}} \cdot \partial_{\boldsymbol{x}} \tilde{\varphi}(\theta, \boldsymbol{x}, t) \hat{\boldsymbol{k}}' \cdot \partial_{\boldsymbol{x}} \tilde{\varphi}(\theta', \boldsymbol{x}, t) \end{aligned} \tag{5.3.13}$$

where $\tilde{\varphi}$ is related to $\tilde{\rho}$ according to $\tilde{\rho}(\theta, \boldsymbol{x}, t) = (2\pi)^{-1} \hat{\boldsymbol{k}} \cdot \frac{\partial \tilde{\varphi}(\theta, \boldsymbol{x}, t)}{\partial \boldsymbol{x}}$. Equation (5.3.13) is a hydrodynamical description of a two-dimensional Fermi liquid, or,

[33] Note that $\frac{\partial n}{\partial \boldsymbol{k}} = -\hat{\boldsymbol{k}} \delta(|\boldsymbol{k}| - k_F) + \frac{\partial \delta n}{\partial \boldsymbol{k}}$. We find that $\int \frac{k \, \mathrm{d}k}{(2\pi)^2} \hat{\boldsymbol{k}} \delta(|\boldsymbol{k}| - k_F) = \frac{k_F \hat{\boldsymbol{k}}}{(2\pi)^2}$ and

$$\int \frac{k \, \mathrm{d}k}{(2\pi)^2} \left(\hat{\boldsymbol{k}}_\perp k_F^{-1} \frac{\partial \delta n}{\partial \theta} + \hat{\boldsymbol{k}} \frac{\partial \delta n}{\partial k} \right) = \hat{\boldsymbol{k}}_\perp k_F^{-1} \frac{\partial \tilde{\rho}}{\partial \theta} - \int \frac{k \, \mathrm{d}k}{(2\pi)^2} k^{-1} \hat{\boldsymbol{k}} \delta n = k_F^{-1} (\hat{\boldsymbol{k}}_\perp \frac{\partial \tilde{\rho}}{\partial \theta} - \hat{\boldsymbol{k}} \tilde{\rho})$$

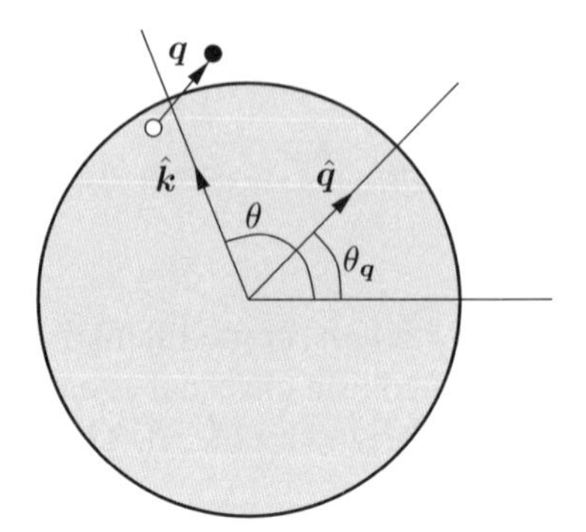

FIG. 5.5. A particle–hole excitation of momentum $\boldsymbol{q}$ at the angle θ.

in other words, a bosonization of a two-dimensional Fermi liquid. Comparing this with the hydrodynamical description discussed in Section 5.1.3, we see that eqn (5.3.13) contains more than just the density mode.

Problem 5.3.3.
In this section, we have derived the reduced Boltzmann equation and the associated Hamiltonian for a two-dimensional rotationally-invariant system. Generalize these results (eqns (5.3.11) and (5.3.12)) to a two-dimensional system with no rotational symmetry.

Problem 5.3.4.
(a) Derive the bosonizated Lagrangian (the one similar to eqn (5.3.13)) for a one-dimensional Fermi liquid.
(b) Consider eqn (5.1.3) in one dimension. Introduce ϕ through $\rho = (2\pi)^{-1}\partial_x\phi$. Write eqn (5.1.3) in terms of ϕ. Find the canonical momentum of ϕ (denoted by $\tilde{\phi}$), and the Hamiltonian H in terms of ϕ and $\tilde{\phi}$. Compare the action $S = \int \mathrm{d}t\ (\tilde{\phi}\dot{\phi} - H)$ with the result in (a).

5.3.4 Application of the hydrodynamical description of a Fermi liquid

As the first application of the bosonized description of a Fermi liquid, let us calculate the spectrum of the collective modes. In $\boldsymbol{q}$ space, the Fermi liquid equation of motion (5.3.11) becomes

$$\mathrm{i}\partial_t\tilde{\rho} = q\cos(\theta_{\boldsymbol{q}} - \theta)\int \mathrm{d}\theta' \left(v_F^*\delta(\theta-\theta') + \frac{k_F}{(2\pi)^2}f(\theta,\theta')\right)\tilde{\rho}(\theta',\boldsymbol{q},t) \quad (5.3.14)$$

where $\theta_{\boldsymbol{q}}$ is the angle of $\boldsymbol{q}$ (see Fig. 5.5).

Equation (5.3.14) can be solved easily if $f(\theta,\theta') = 0$. The eigenmode has the form $\tilde{\rho}(\theta,\boldsymbol{q},t) = \delta(\theta-\theta_0)\delta(\boldsymbol{q}-\boldsymbol{q}_0)\mathrm{e}^{-\mathrm{i}\omega(\boldsymbol{q}_0,\theta_0)t}$, with a frequency $\omega(\boldsymbol{q}_0,\theta_0) = v_F^* q_0\cos(\theta_{\boldsymbol{q}_0} - \theta_0)$. Such an eigenmode describes an excitation with momentum $\boldsymbol{q}_0$. Depending on θ, the energy of such an excitation ranges for 0 to $v_F^* q_0$, which agrees with the energy range of a particle–hole excitation of the same momentum (see Fig. 4.11). In fact, the eigenmode corresponds to the particle–hole excitation in Fig. 5.5.

To find the energies of the collective mode for nonzero $f(\theta, \theta')$, we note that eqn (5.3.14) can be rewritten as

$$iq^{-1}\partial_t \tilde{\rho}(\theta, \boldsymbol{q}, t) = KM\tilde{\rho}(\theta, \boldsymbol{q}, t) \tag{5.3.15}$$

where the two real symmetric operators K and M are given by

$$\begin{aligned} K(\theta, \theta') &= \cos(\theta_{\boldsymbol{q}} - \theta)\delta(\theta - \theta') \\ M(\theta, \theta') &= v_F^* \delta(\theta - \theta') + \frac{k_F}{(2\pi)^2} f(\theta, \theta') \end{aligned} \tag{5.3.16}$$

From the energy of the Fermi liquid, see eqn (5.3.12), we find that the stability of the Fermi liquid requires M to be positive definite. Thus, we can write M as $M = \tilde{M}\tilde{M}^\top$. Letting $u = \tilde{M}^\top \tilde{\rho}$, then eqn (5.3.15) becomes $\mathrm{i}\partial_t u = q\tilde{M}^\top K \tilde{M} u$. The energies of the collective excitations with a fixed momentum $\boldsymbol{q}$ are given by the eigenvalues of the real symmetric matrix $q\tilde{M}^\top K \tilde{M}$.

The rotational symmetry requires that $f(\theta, \theta') = f(\theta - \theta')$. Because the energy spectrum does not depend on the direction of $\boldsymbol{q}$, we can choose $\boldsymbol{q}$ to be in the $\boldsymbol{x}$ direction, i.e. $\boldsymbol{q} = q\boldsymbol{x}$ and $\theta_{\boldsymbol{q}} = 0$. Introducing (see Fig. 5.4)

$$\tilde{\rho}_l(\boldsymbol{q}, t) = \int \mathrm{d}\theta \, \frac{\mathrm{e}^{-\mathrm{i}l\theta}}{\sqrt{2\pi}} \tilde{\rho}(\theta, \boldsymbol{q}, t),$$

eqn (5.3.15) can be written as

$$\begin{aligned} &\mathrm{i}\partial_t \tilde{\rho}_l(q\boldsymbol{x}, t) = qK_{lm}M_{ml'}\tilde{\rho}_{l'}(q\boldsymbol{x}, t), && f(\theta - \theta') = \sum_l f_l \mathrm{e}^{\mathrm{i}l(\theta - \theta')}, \\ &K_{ll'} = \frac{1}{2}(\delta_{l,l'+1} + \delta_{l,l'-1}), && M_{ll'} = \left(v_F^* + \frac{k_F f_l}{2\pi}\right)\delta_{ll'}. \end{aligned}$$

From M, we see that the Fermi liquid is unstable if one of the f_l is less than $-2\pi v_F^*/k_F$. We also see that $\tilde{M}_{ll'} = \sqrt{v_F^* + \frac{k_F f_l}{2\pi}}\delta_{ll'}$ and

$$q(\tilde{M}^\top K \tilde{M})_{ll'} \equiv \tilde{H}_{ll'} = \frac{q}{2}(\delta_{l,l'+1} + \delta_{l,l'-1})\sqrt{v_F^* + \frac{k_F f_l}{2\pi}}\sqrt{v_F^* + \frac{k_F f_{l'}}{2\pi}}$$

The eigenvalues of $\tilde{H}$ give us the frequencies of the collective modes.

Here $\tilde{H}$ describes a particle hopping on a one-dimensional lattice. The nearest-neighbor hopping amplitude between site l and site $l + 1$ is given by

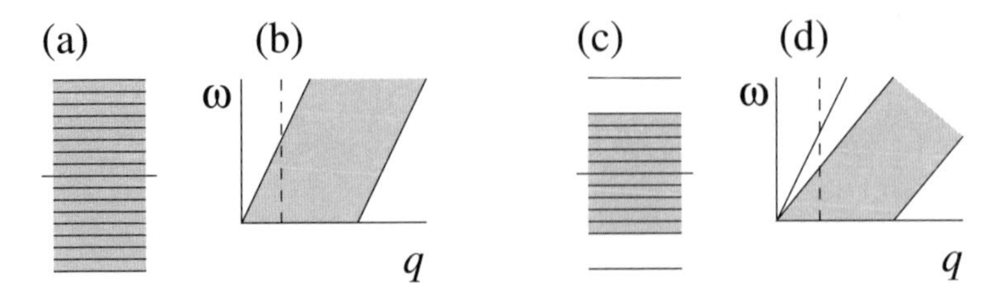

FIG. 5.6. (a) The spectrum of $\tilde{H}$ when $f_l = 0$, and (b) the corresponding particle–hole spectrum. (c) The spectrum of $\tilde{H}$ when $f_l > 0$, and (d) the corresponding particle–hole spectrum.

$\frac{q}{2}\sqrt{\frac{v_F^*}{2\pi} + \frac{k_F f_l}{(2\pi)^2}}\sqrt{\frac{v_F^*}{2\pi} + \frac{k_F f_{l+1}}{(2\pi)^2}}$. When $f_l = 0$, $\tilde{H}$ has a continuous spectrum ranging from $-v_F^* q$ to $v_F^* q$ (see Fig. 5.6(a)). It reproduces the particle–hole continuum of a free fermion system (see Fig. 5.6(b)). However, if $f_l > f_\infty$, then the particle described by $\tilde{H}_{ll'}$ is attracted to the small-l region. Thus, the spectrum of $\tilde{H}$ has isolated eigenvalues corresponding to the bound states in the attractive region, in addition to the continuous spectrum (see Fig. 5.6(c)). The corresponding particle–hole spectrum contains well-defined collective modes (see Fig. 5.6(d)).[34]

In the second application, we would like to use the linearized hydrodynamical equation of motion (5.3.10) to calculate the d.c. conductance of the Fermi liquid. In the presence of a uniform static electric field $\boldsymbol{E}$, the force is given by $\boldsymbol{F} = e\boldsymbol{E}$. The force $\boldsymbol{F}$ will induce a uniform and static $\tilde{\rho}$, i.e. $\partial_t \tilde{\rho} = \partial_{\boldsymbol{x}} \tilde{\rho} = 0$. Equation (5.3.10) becomes

$$\tilde{\rho} = e\tau \frac{k_F}{(2\pi)^2} \hat{\boldsymbol{k}} \cdot \boldsymbol{E}$$

From eqn (5.3.6), we see that the above $\tilde{\rho}$ induces an electric current:

$$e\boldsymbol{J} = e \int \mathrm{d}\theta\, \tilde{\rho} \tilde{\boldsymbol{v}}$$

$$\tilde{\boldsymbol{v}}(\boldsymbol{k}) = \boldsymbol{v}^*(\boldsymbol{k}) + \int \frac{\mathrm{d}^2 \boldsymbol{k}'}{(2\pi)^2} f(\boldsymbol{k}, \boldsymbol{k}') \boldsymbol{v}^*(\boldsymbol{k}') \delta(\xi^*_{\boldsymbol{k}'})$$

[34] A bosonic mode must have a positive energy. A negative energy implies an instability. However, $\tilde{H}$ has both positive and negative eigenvalues, corresponding to positive and negative frequencies. If the positive eigenvalues of $\tilde{H}$ correspond to the positive energies of the bosonic collective modes, then do the negative eigenvalues of $\tilde{H}$ also correspond to the negative energies of the collective modes? Here we would like to point out that so far we have treated eqn (5.3.13) as a classical Lagrangian and only discussed the resulting classical equation of motion. After quantization, the equation of motion becomes an equation for an operator $\hat{\tilde{\rho}}$. It turns out that the modes with positive frequencies correspond to creation operators of the bosonic mode and the modes with negative frequencies correspond to annihilation operators of the bosonic mode. The quantization of a one-dimensional version of eqn (5.3.13) will be discussed in Section 7.4.5, see eqns (7.4.35) and (7.4.36).

Due to the rotational symmetry, $\tilde{\boldsymbol{v}} = |\tilde{\boldsymbol{v}}|\hat{\boldsymbol{k}}$. We find that

$$e\boldsymbol{J} = e^2\tau \frac{|\tilde{\boldsymbol{v}}|k_F}{(2\pi)^2} \int \mathrm{d}\theta\, \hat{\boldsymbol{k}}(\hat{\boldsymbol{k}} \cdot \boldsymbol{E}) = \frac{e^2\tau|\tilde{\boldsymbol{v}}|\rho}{k_F}\boldsymbol{E}$$

The conductance is $\sigma = \frac{e^2\tau|\tilde{\boldsymbol{v}}|\rho}{k_F}$. The above reduces to the Drude result $\sigma = \frac{e^2\tau\rho}{m^*}$ if the Fermi liquid function $f(\boldsymbol{k}, \boldsymbol{k}')$ vanishes.

Problem 5.3.5.
In the presence of a uniform magnetic field $\boldsymbol{B} = B\boldsymbol{z}$, the force $\boldsymbol{F}$ in the two-dimensional reduced Boltzmann equation (5.3.9) has the form $\boldsymbol{F}(\boldsymbol{k}, \boldsymbol{x}) = \frac{e}{c}\boldsymbol{v}^*(\boldsymbol{k}) \times \boldsymbol{B}$.
(a) Show that, after linearization, the reduced Boltzmann equation in the magnetic field has the following form (Kim *et al.*, 1995):

$$(\mathrm{i}\partial_t + \mathrm{i}\omega_c\partial_\theta - v_F^* q\cos(\theta - \theta_{\boldsymbol{q}}))\tilde{\rho} - \frac{qk_F\cos(\theta_{\boldsymbol{q}} - \theta)}{(2\pi)^2} \int \mathrm{d}\theta' f(\theta - \theta')\tilde{\rho}(\theta', \boldsymbol{q}, t) = 0$$

if we set $\tau^{-1} = 0$. Find ω_c.
(b) Show that the energy spectrum of the collective mode with momentum $\boldsymbol{q}$ is determined by the eigenvalues of

$$\tilde{H}_{ll'} = \omega_c l\delta_{ll'} + \frac{q}{2}(\delta_{l,l'+1} + \delta_{l,l'-1})\sqrt{v_F^* + \frac{k_F f_l}{2\pi}}\sqrt{v_F^* + \frac{k_F f_{l'}}{2\pi}}$$

Show that the spectrum of $\tilde{H}$ is symmetric around zero. (Hint: $f_l = f_{-l} =$ real.)
(c) Show that, when $f_l = 0$, the eigenvalues of $\tilde{H}$ are given by $\omega_c \times$ integer. (Hint: Consider $\tilde{\rho} = \exp(\mathrm{i}l\theta - \mathrm{i}\frac{v_F^* q}{\omega_c}\sin(\theta - \theta_{\boldsymbol{q}}))$.) So we can recover the free fermion result using hydrodynamical theory.

Problem 5.3.6.
The a.c. transport of a two-dimensional Fermi liquid
(a) Assume that the force $\boldsymbol{F}$ in eqn (5.3.9) has the form $\boldsymbol{F} = e\boldsymbol{E}\mathrm{e}^{\,\mathrm{i}\omega t - \boldsymbol{q}\cdot\boldsymbol{x}}$. Find $\tilde{\rho}$, assuming that $f(\theta - \theta') = 0$.
(b) Find $\tilde{\rho}$ to the first order in $f(\theta - \theta')$.
(c) Find the induced electric current $e\boldsymbol{J} = \sigma(\omega, \boldsymbol{q})\boldsymbol{E}$, and the optical conductance $\sigma(\omega, \boldsymbol{q})$ to the first order in $f(\theta - \theta')$. Here $\boldsymbol{J}$ is the quasiparticle current given in eqn (5.3.6).

5.3.5 The essence of Fermi liquid theory

It is commonly believed that the essence of Fermi liquid theory is the concept and the existence of well-defined quasiparticles. We adopted this point of view in Section 5.3.1. The existence of well-defined quasiparticles implies that, not only is the total number of quasiparticles conserved, but the number of quasiparticles in a given momentum direction is also conserved. Thus, a Fermi liquid has an infinite number of conserved quantities (Luther, 1979; Haldane, 1992). This allows us to develop a hydrodynamical theory for a Fermi liquid which contains many bosonic modes, one mode for each conserved quantity. Thus, the existence of an infinite number of conserved quantities is also an essence of Fermi liquid theory. In Section 5.3.3, we developed a Fermi liquid theory based on this point of view.

We would like to point out that the two points of view are not equivalent, and the second point of view is more general. The existence of well-defined quasiparticles implies that the number of quasiparticles at each momentum is conserved. In the second point of view, we only assume that the quasiparticles in each momentum direction are conserved. In one dimension, the hydrodynamical theory in Section 5.3.3 actually describes one-dimensional Tomonaga–Luttinger liquids.[35] In higher dimensions, the hydrodynamical theory describes Fermi liquids.

5.4 Perturbation theory and the validity of Fermi liquid theory

In this section, we will test the correctness of Fermi liquid theory. We will develop a perturbation theory to calculate the electron Green's function for an interacting system. We will check if $\text{Im}G(\boldsymbol{k},\omega)$ contains a δ-function. The presence of the δ-function indicates the presence of well-defined Landau quasiparticles and the validity of Landau Fermi liquid theory.

5.4.1 Path integrals and perturbation theory for fermions

- The path integral for fermions is a bookkeeping formalism that allows us to formally express the correlation functions of a fermion system. Unlike the path integral for bosons, the physical meaning of the fermion path integral is unclear.
- The path integral for fermions allows us to systematically develop a perturbation theory for an interacting fermion system.
- The structure of perturbation theory is captured by Feynman diagrams and Feynman rules.

To develop a path integral for fermions, we first note that the time-ordered correlation of fermion operators satisfies $\langle T[c(t)c^\dagger(t')]\rangle = -\langle T[c^\dagger(t')c(t)]\rangle$. Thus, we will use (complex) Grassmann numbers ξ_α to represent the fermion operators. The reader may ask, what do you mean by 'use Grassmann numbers to represent the fermion operators'? Well, I have to say I do not know. I do not know the physical meaning of a Grassmann number. I will treat the following fermion path integral as some kind of bookkeeping device, which allows us to formally pack various fermion correlation functions into a compact formula. The Grassmann numbers are anti-commuting numbers which

[35] See Section 7.4.4 for a discussion of a Tomonaga–Luttinger liquid.

satisfy

$$\xi_\alpha \xi_\beta = -\xi_\beta \xi_\alpha, \qquad \xi_\alpha \xi_\beta^* = -\xi_\beta^* \xi_\alpha, \qquad \xi_\alpha^2 = 0,$$

and

$$(\xi_{\alpha_1} ... \xi_{\alpha_n})^* = \xi_{\alpha_n}^* ... \xi_{\alpha_1}^*$$

Functions of the Grassmann numbers are defined by the following expansion:

$$f(\xi) = f_0 + f_1 \xi, \qquad A(\xi, \xi^*) = a_0 + a_1 \xi + \bar{a}_1 \xi^* + a_{12} \xi^* \xi$$

The derivative is identical to the ordinary derivative, except that, in order for the derivative operator $\frac{\partial}{\partial \xi}$ to act on ξ, the variable ξ has to be anti-commuted through until it is adjacent to $\frac{\partial}{\partial \xi}$. For example, $\frac{\partial}{\partial \xi}(\xi^* \xi) = \frac{\partial}{\partial \xi}(-\xi \xi^*) = -\xi^*$. With these definitions, we have

$$\frac{\partial}{\partial \xi} A(\xi, \xi^*) = a_1 - a_{12} \xi^*$$

$$\frac{\partial}{\partial \xi^*} A(\xi, \xi^*) = \bar{a}_1 + a_{12} \xi$$

$$\frac{\partial}{\partial \xi^*} \frac{\partial}{\partial \xi} A(\xi, \xi^*) = -a_{12} = -\frac{\partial}{\partial \xi} \frac{\partial}{\partial \xi^*} A(\xi, \xi^*)$$

We see that

$$\frac{\partial}{\partial \xi^*} \frac{\partial}{\partial \xi} = -\frac{\partial}{\partial \xi} \frac{\partial}{\partial \xi^*}$$

The Grassmann integration, as a linear mapping, is defined as the Grassmann derivative:

$$\int \mathrm{d}\xi = \frac{\partial}{\partial \xi}$$

We have that

$$\int \mathrm{d}\xi \, 1 = 0, \qquad \int \mathrm{d}\xi \xi = 1$$

The Grassmann integration has the following properties of standard integration:

$$\int \mathrm{d}\xi \, \frac{\partial}{\partial \xi} f(\xi) = 0, \qquad \int \mathrm{d}\xi \, f(\xi + \eta) = \int \mathrm{d}\xi \, f(\xi)$$

From the definition of the determinant, namely

$$\mathrm{Det}(H_0) = \sum_P (-)^P (H_0)_{1P_1} ... (H_0)_{nP_n}$$

where P are permutations $(1, ..., n) \to (P_1, ..., P_n)$, we find the Grassmann Gaussian integral:

$$\int \prod_{i=1}^n \mathrm{d}\eta_i^* \, \mathrm{d}\eta_i \; \mathrm{e}^{-\eta_i^* (H_0)_{ij} \eta_j} = \mathrm{Det}(H_0)$$

and

$$\int \prod_{i=1}^{n} \mathrm{d}\eta_i^* \, \mathrm{d}\eta_i \; \mathrm{e}^{-\eta_i^*(H_0)_{ij}d\eta_j + \zeta_i^*\eta_i + \zeta_i\eta_i^*} = \mathrm{Det}(H_0)\mathrm{e}^{\zeta_i^*(H_0^{-1})_{ij}\zeta_j}$$

Note that, compared to the boson Gaussian integral, the determinant appears in the numerator instead of the denominator.

The fermion Gaussian integral allows us to obtain the Wick theorem for fermion correlations (defined by the path integral):

$$\begin{aligned}
&\langle \psi_1 ... \psi_n \psi_n^* ... \psi_1^* \rangle \\
&= \frac{\int \mathcal{D}\psi^* \mathcal{D}\psi \; \psi_1 ... \psi_n \psi_n^* ... \psi_1^* \mathrm{e}^{-\sum_{ij} \psi_i^* (H_0)_{ij} \psi_j}}{\int \mathcal{D}\psi^* \mathcal{D}\psi \; \mathrm{e}^{-\sum_{ij} \psi_i^* (H_0)_{ij} \psi_j}} \\
&= \sum_P \zeta^P (H_0^{-1})_{P_n, n} ... (H_0^{-1})_{P_1, 1}
\end{aligned} \tag{5.4.1}$$

where $\zeta = -1$. (If we choose $\zeta = +1$, then the above will be the Wick theorem for boson operators.)

Let us apply the above formalism to a free fermion system at zero temperature. We consider the following path integral:

$$Z_0 = \int \mathcal{D}\psi^* \mathcal{D}\psi \; \mathrm{e}^{\mathrm{i}\int_0^\beta \mathrm{d}t \; L_0}$$

$$L_0 = \sum_{\boldsymbol{k}} \mathrm{i}\psi_{\boldsymbol{k}}(t)^* \partial_t \psi_{\boldsymbol{k}}(t) - \psi_{\boldsymbol{k}}(t)^* \xi_{\boldsymbol{k}} \psi_{\boldsymbol{k}}(t)$$

We can calculate the path integral average

$$\mathrm{i}G_{\boldsymbol{k}}(t, t') = \langle \psi_{\boldsymbol{k}}(t) \psi_{\boldsymbol{k}}^*(t') \rangle = \frac{1}{(-\mathrm{i})(\mathrm{i}\partial_t - \xi_{\boldsymbol{k}})}$$

In frequency space, we have

$$G_{\boldsymbol{k},\omega} = \frac{1}{\omega - \xi_{\boldsymbol{k}}}$$

which agrees with the time-ordered average[36] $\langle 0 | T(\psi_{\boldsymbol{k}}(t) \psi_{\boldsymbol{k}}^\dagger(t')) | 0 \rangle$ in eqn (4.2.9). Using the Wick theorem, we can derive the following more general relationship between the path integral average and the time-ordered average:

$$\langle \psi_{i_1} ... \psi_{i_n} \psi_{j_n}^* ... \psi_{j_1}^* \rangle = \frac{\mathrm{Tr}\left(T(\psi_{i_1} ... \psi_{i_n} \psi_{j_n}^\dagger ... \psi_{j_1}^\dagger \, \mathrm{e}^{-\int_0^\beta \mathrm{d}t \; H_0}) \right)}{\mathrm{Tr}\left(T(\mathrm{e}^{-\int_0^\beta \mathrm{d}t \; H_0}) \right)}$$

Thus, we can use path integrals to calculate arbitrary time-ordered correlations.

[36] The calculation presented here is a formal calculation, which fails to reproduce the important 0^+ term. Section 2.2.2 discussed how to produce the 0^+ term.

The above results are for free fermion systems. Let us apply the path integral to an interacting system with

$$H = H_0 + H_I(t)$$

where H_0 describes a free fermion system. Let us assume that $H_I(t)$ is a slowly varying function of t, that $H_I(t) = H_I$ for finite t, and $H_I(\pm\infty) = 0$. Let

$$Z = \langle 0|T(\mathrm{e}^{-\mathrm{i}\int \mathrm{d}t\ (H_0+H_I(t))})|0\rangle$$

be the partition function, where $|0\rangle$ is the ground state for H_0. The partition function Z has the following perturbative expansion:

$$Z = Z_0\left(1 + Z_0^{-1}\langle 0|T\left((-\mathrm{i}\int \mathrm{d}t\ H_I)\mathrm{e}^{-\mathrm{i}\int \mathrm{d}t\ H_0}\right)|0\rangle + ...\right)$$
$$Z_0 = \langle 0|T(\mathrm{e}^{-\mathrm{i}\int \mathrm{d}t\ H_0})|0\rangle.$$

As $Z_0^{-1}\langle 0|T(O_1(t_1)...O_n(t_n)\mathrm{e}^{-\mathrm{i}\int \mathrm{d}t\ H_0})|0\rangle$ in the Schrödinger picture is equal to $\langle 0|T(O_1(t_1)...O_n(t_n))|0\rangle$ in the Heisenberg picture, both of these quantities represent the time-ordered average. We see that the partition function Z for an interacting system can be expressed in terms of the time-ordered averages in the free fermion system H_0. This allows us to express Z in terms of a path integral as follows:

$$Z = \int \mathcal{D}\psi^*\mathcal{D}\psi\ \mathrm{e}^{\mathrm{i}\int_0^\beta \mathrm{d}t\ (L_0+L_I)}, \qquad L_I = -H_I$$

Similarly, using a perturbative expansion, we can show that the time-ordered correlation for an interacting system (in the Schrödinger picture) can be expressed in terms of path integrals as follows:

$$\frac{\langle 0|T(O_1(t_1)...O_n(t_n)\mathrm{e}^{-\mathrm{i}\int \mathrm{d}t\ H})|0\rangle}{\langle 0|T(\mathrm{e}^{-\mathrm{i}\int \mathrm{d}t\ H})|0\rangle} = \frac{\int \mathcal{D}\psi^*\mathcal{D}\psi\ O_1(t_1)...O_n(t_n)\mathrm{e}^{\mathrm{i}\int_0^\beta \mathrm{d}t\ (L_0+L_I)}}{\int \mathcal{D}\psi^*\mathcal{D}\psi\ \mathrm{e}^{\mathrm{i}\int_0^\beta \mathrm{d}t\ (L_0+L_I)}}$$

Now, let us calculate the partition function Z perturbatively using path integrals:

$$Z = Z_0\frac{\int \mathcal{D}\psi^*\mathcal{D}\psi\ \mathrm{e}^{\mathrm{i}\int \mathrm{d}t\ (L_0+L_I)}}{\int \mathcal{D}\psi^*\mathcal{D}\psi\ \mathrm{e}^{\mathrm{i}\int \mathrm{d}t\ L_0}}$$
$$= Z_0\left(1 + \left\langle\left(\mathrm{i}\int \mathrm{d}t\ L_I\right)\right\rangle_0 + \frac{1}{2!}\left\langle\left(\mathrm{i}\int \mathrm{d}t\ L_I\right)^2\right\rangle_0 + ...\right)$$

If we assume that the interaction has the form

$$\int \mathrm{d}t\ L_I = -\int \mathrm{d}x\mathrm{d}x'\ \psi^*(x^\mu)\psi(x^\mu)V(x^\mu, x'^\nu)\psi^*(x'^\nu)\psi(x'^\nu)$$

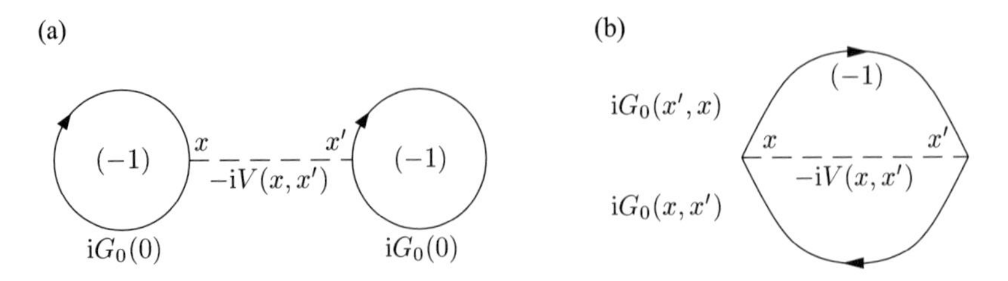

FIG. 5.7. Diagrams that contribute to Z/Z_0 at first order in V.

where $x^\mu = (t, x^1, ..., x^d)$ and $\mathrm{d}x = \mathrm{d}t\mathrm{d}^d\boldsymbol{x}$, then we find that

$$\begin{aligned}\frac{Z}{Z_0} &= 1 + \left\langle \int \mathrm{d}x\mathrm{d}x'\ \psi^*(x^\mu)\psi(x^\mu)[-\mathrm{i}V(x^\mu, x'^\nu)]\psi^*(x'^\nu)\psi(x'^\nu) \right\rangle_0 + ...\\ &= 1 + \int \mathrm{d}x\mathrm{d}x'\ (-)\mathrm{i}G_0(0)[-\mathrm{i}V(x^\mu, x'^\nu)](-)\mathrm{i}G_0(0)\\ &\quad + \int \mathrm{d}x\mathrm{d}x'\ (-)\mathrm{i}G_0(x^\mu - x'^\nu)\mathrm{i}G_0(x'^\nu - x^\mu)[-\mathrm{i}V(x^\mu, x'^\nu)] + ...\end{aligned}$$

The two terms of order V can be represented by the Feynman diagrams shown in Fig. 5.7. The following Feynman rules for fermions are almost identical to those of bosons, expect that each closed loop contributes an extra factor of -1.
(a) Each line in the Feynman diagrams represents a propagator $G_0(x, x')$, where x and x' are the coordinates of the two vertices at the ends of the line.
(b) Each vertex represents an integration over the location of the vertex $\int \mathrm{d}x$.
(c) Each dashed line represents the interaction potential $-\mathrm{i}V(x, x')$.
(d) Each closed fermion loop contributes a factor of -1.
(e) The value of a connected diagram is given by rules (a)–(d) and the value of a disconnected diagram is given by the product of connected sub-diagrams.
Feynman diagrams capture the structure of a perturbative expansion. They not only produce the order-V terms, but they also produce order-V^2 terms and all other higher-order terms.

In the perturbative expansion of Z/Z_0, the path integral averages $\langle ... \rangle_0$ contain both connected and disconnected diagrams. We can use the linked-cluster expansion to rewrite the expansion of Z/Z_0 as

$$\ln \frac{Z}{Z_0} = \sum_{n=0}^{\infty} \frac{1}{n!} \left\langle \left(\mathrm{i} \int \mathrm{d}t\ L_I \right)^n \right\rangle_{0c} = \left\langle \mathrm{e}^{\mathrm{i}\int \mathrm{d}t\ L_I} \right\rangle_{0c} \tag{5.4.2}$$

where the path integral average $\langle ... \rangle_{0c}$ includes only the connected graphs.

Problem 5.4.1.
(a) Calculate the order-V^2 terms in the perturbative expansion of Z/Z_0. Draw the corresponding Feynman diagrams.
(b) Calculate the order-V^2 terms in the perturbative expansion of $\ln(Z/Z_0)$. Draw the corresponding Feynman diagrams.

5.4.2 Self-energy and two-body interactions

Now we are ready to calculate the electron Green's function perturbatively for the following interacting electron system:

$$H = \sum_{\boldsymbol{k}} \xi_{\boldsymbol{k}} c^\dagger_{\alpha\boldsymbol{k}} c_{\alpha\boldsymbol{k}} + \frac{1}{2\mathcal{V}} \sum_{\boldsymbol{k},\boldsymbol{k}',\boldsymbol{q}} c^\dagger_{\alpha\boldsymbol{k}} c_{\alpha,\boldsymbol{k}-\boldsymbol{q}} V_{\boldsymbol{q}} c^\dagger_{\beta\boldsymbol{k}'} c_{\beta,\boldsymbol{k}'+\boldsymbol{q}}$$

The time-ordered Green's function is given by the following path integral average:

$$\mathrm{i}G(\boldsymbol{x},t)\delta_{\alpha\beta} = \frac{\int \mathcal{D}c\mathcal{D}c^*\ c_\alpha(\boldsymbol{x},t)c^*_\beta(0,0)\mathrm{e}^{\mathrm{i}\int \mathrm{d}t\ L}}{\int \mathcal{D}c\mathcal{D}c^*\ \mathrm{e}^{\mathrm{i}\int \mathrm{d}t\ L}} \equiv \left\langle c_\alpha(\boldsymbol{x},t)c^*_\beta(0,0)\right\rangle$$

for interacting electrons, where

$$L = \mathrm{i}c^*_\alpha \partial_t c_\alpha - H$$

is the Lagrangian of the interacting electrons. Note that here we have assumed that $\langle c_\alpha(\boldsymbol{x},t)c^*_\beta(0,0)\rangle$ is spin–rotation invariant. As a result, it has the form $\mathrm{i}G(\boldsymbol{x},t)\delta_{\alpha\beta}$.

To calculate $\mathrm{i}G$, it is convenient to introduce

$$Z[\eta_\alpha,\eta^*_\alpha] = \int \mathcal{D}c\mathcal{D}c^*\ \mathrm{e}^{\mathrm{i}\int \mathrm{d}t\ L+\int \mathrm{d}t\mathrm{d}^d\boldsymbol{x}\ (\eta_\alpha c^*_\alpha+\eta^*_\alpha c_\alpha)}$$

where $\eta(\boldsymbol{x},t)$ and $\eta^*(\boldsymbol{x},t)$, just like $c(\boldsymbol{x},t)$ and $c^*(\boldsymbol{x},t)$, are Grassmann numbers. We see that

$$\mathrm{i}G(\boldsymbol{x},t)\delta_{\alpha\beta} = \frac{\partial}{\partial_{\eta^*_\alpha(\boldsymbol{x},t)}} \frac{\partial}{\partial_{\eta_\beta(0,0)}} \ln Z[\eta_\alpha,\eta^*_\alpha]$$

Using eqn (5.4.2), we obtain

$$\begin{aligned}\mathrm{i}G(\boldsymbol{x},t)\delta_{\alpha\beta} &= \frac{\partial}{\partial_{\eta^*_\alpha(\boldsymbol{x},t)}} \frac{\partial}{\partial_{\eta_\beta(0,0)}} \left\langle \mathrm{e}^{\mathrm{i}\int \mathrm{d}t\ L_I+\int \mathrm{d}t\mathrm{d}^d\boldsymbol{x}\ (\eta_\alpha c^*_\alpha+\eta^*_\alpha c_\alpha)}\right\rangle_{0c} \\ &= \left\langle c_\alpha(\boldsymbol{x},t)c^*_\beta(0,0)\mathrm{e}^{\mathrm{i}\int \mathrm{d}t\ L_I}\right\rangle_{0c} \qquad (5.4.3)\end{aligned}$$

This formula lays the foundation for using Feynman diagrams to calculate the time-ordered Green's function of interacting electrons.

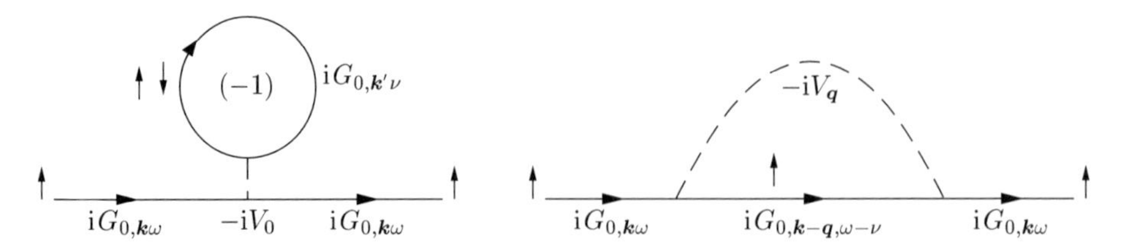

FIG. 5.8. The two Feynman diagrams that contribute to $iG_{\boldsymbol{k}\,\omega}$. The dashed line represents the potential $-iV_{\boldsymbol{q}}$.

To first order in $V_{\boldsymbol{q}}$, we find that

$$iG(\boldsymbol{x},t)\delta_{\alpha\beta} = \left\langle c_\alpha(\boldsymbol{x},t)c^\dagger_\beta(0,0)\right\rangle_0 + \left\langle c_\alpha(\boldsymbol{x},t)c^\dagger_\beta(0,0)(-i)\int dt'\, H_I(t'))\right\rangle_{0c}$$

Using the Wick theorem, we can express the above equation in terms of the free electron Green's function G_0. In $\boldsymbol{k}$–ω space, we find that (see Problem 5.4.2)

$$iG_{\boldsymbol{k}\omega} = iG_{0,\boldsymbol{k}\omega} + (-1)(iG_{0,\boldsymbol{k}\omega})^2(-iV_0)\frac{1}{\mathcal{V}}\sum_{\boldsymbol{k}',\alpha'}\int\frac{d\nu}{2\pi}\, iG_{0,\boldsymbol{k}'\nu}e^{i0^+\nu}$$
$$+ (iG_{0,\boldsymbol{k}\omega})^2\frac{1}{\mathcal{V}}\sum_{\boldsymbol{q}}\int\frac{d\nu}{2\pi}\, iG_{0,\boldsymbol{k}-\boldsymbol{q},\nu}(-iV_{\boldsymbol{q}})e^{-i0^+\nu} \tag{5.4.4}$$

where the free electron Green's function $G_{0,\boldsymbol{k}\omega}$ is given by eqn (4.2.4). The two order-$V_{\boldsymbol{q}}$ terms are summarized by the Feynman diagrams shown in Fig. 5.8. The Feynman rules in momentum space are different from those in real space and are as follows.

(a) Each line in the Feynman diagrams carries an energy–momentum vector, say, $(\boldsymbol{q},\nu)$. It represents a propagator $iG_{\boldsymbol{q}\nu} = i/(\nu - \xi_{\boldsymbol{q}})$.
(b) The energy–momentum flowing into a node is conserved.
(c) Each loop represents the integral $(\beta\mathcal{V})^{-1}\int\frac{dt}{2\pi}\sum_{\boldsymbol{q}}$.
(d) The dashed line represents the interaction $-iV_{\boldsymbol{q}}$, where $\boldsymbol{q}$ is the momentum flowing through the dashed line.
(e) Each closed loop contributes an extra factor of -1 due to the anti-commuting property of the fermion operators. (The momentum space Feynman rules for boson systems do not have such a -1 factor.)

Equation (5.4.4) can be written as

$$iG_{\boldsymbol{k}\omega} = iG_{0,\boldsymbol{k}\omega} + (iG_{0,\boldsymbol{k}\omega})^2(-i\Sigma_{\boldsymbol{k}\omega})$$

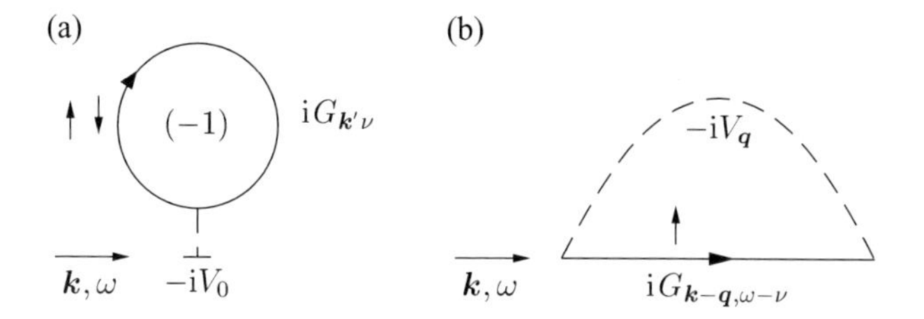

FIG. 5.9. Two diagrams that contribute to the self-energy $-\mathrm{i}\Sigma_{\boldsymbol{k}\,\omega}$.

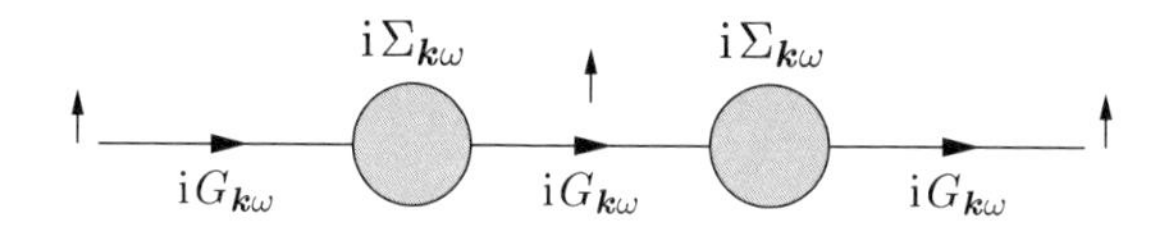

FIG. 5.10. Higher-order diagrams that contribute to $\mathrm{i}G_{\boldsymbol{k}\,\omega}$.

where

$$
\begin{aligned}
&-\mathrm{i}\Sigma_{\boldsymbol{k}\omega}\\
&=(-1)\frac{-\mathrm{i}V_0}{\mathcal{V}}\sum_{\boldsymbol{k}',\alpha'}\int\frac{\mathrm{d}\nu}{2\pi}\mathrm{i}G_{0,\boldsymbol{k}'\nu}\mathrm{e}^{\mathrm{i}0^+\nu}+\frac{1}{\mathcal{V}}\sum_{\boldsymbol{q}}\int\frac{\mathrm{d}\nu}{2\pi}\mathrm{i}G_{0,\boldsymbol{k}-\boldsymbol{q},\nu}(-\mathrm{i}V_{\boldsymbol{q}})\mathrm{e}^{-\mathrm{i}0^+\nu}.
\end{aligned}
$$

The above can be represented by the diagrams in Fig. 5.9. If we include some of the higher-order terms represented by the diagrams in Fig. 5.10, then the Green's function can be written as

$$
\mathrm{i}G_{\boldsymbol{k}\omega}=\mathrm{i}G_{0,\boldsymbol{k}\omega}\sum_{n=0}^{\infty}[(\mathrm{i}G_{0,\boldsymbol{k}\,\omega})(-\mathrm{i}\Sigma_{\boldsymbol{k}\,\omega})]^n
$$

We find that

$$
G_{\boldsymbol{k}\omega}=\frac{1}{\omega-\xi_{\boldsymbol{k}}-\Sigma_{\boldsymbol{k}\omega}+\mathrm{i}\,\mathrm{sgn}(\omega)0^+}
$$

The dispersion relation can be obtained from the position of the pole in $G_{\boldsymbol{k}\omega}$, or from the position of the zero in $\omega-\xi_{\boldsymbol{k}}-\Sigma_{\boldsymbol{k}\omega}$. We see that $\Sigma_{\boldsymbol{k}\omega}$ corrects the energy of the electron (or quasiparticle) and is called the self-energy.

Note that

$$
\int\frac{\mathrm{d}\nu}{2\pi}\mathrm{i}G_{0,\boldsymbol{k}'\nu}\mathrm{e}^{\mathrm{i}0^+\nu}=\mathrm{i}\int\frac{\mathrm{d}\nu}{2\pi}\frac{1}{\nu-\xi_{\boldsymbol{k}'}+\mathrm{i}0^+\mathrm{sgn}(\nu)}\mathrm{e}^{\mathrm{i}0^+\nu}=-\Theta(-\xi_{\boldsymbol{k}'})=-n_{\boldsymbol{k}'}
$$

As we must close the contour in the upper-half complex ν plane, only the poles with negative ν contribute. As $\sum_{\boldsymbol{k}',\alpha}\Theta(-\xi_{\boldsymbol{k}'})$ is simply the total number of

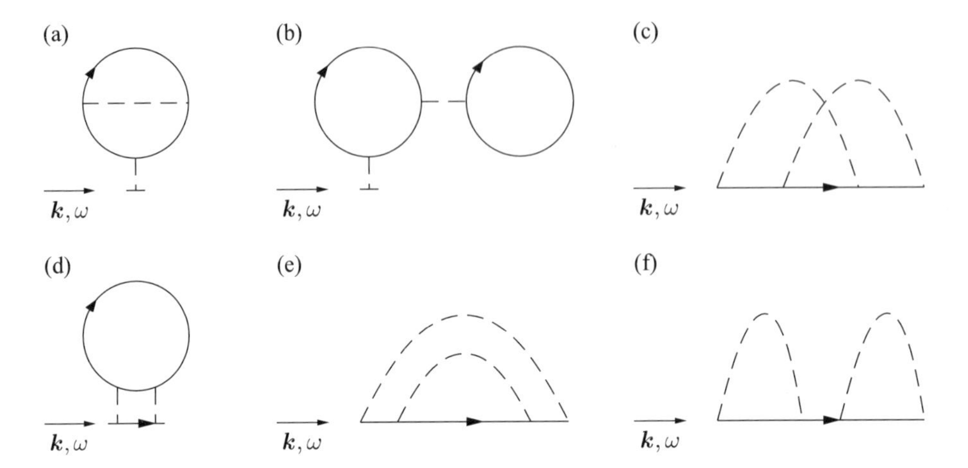

FIG. 5.11. The diagrams (a)–(e) contribute to the self-energy at order $V_{\boldsymbol{q}}^2$. The diagram (f) does not contribute to the self-energy because it is already included in Fig. 5.10.

electrons, the first term in the self-energy is simply the Hartree contribution (Fig. 5.9(a)):

$$\Sigma_{\boldsymbol{k}\omega} = V_0 \rho_0 + ...$$

Similarly, the second term in the self-energy (Fig. 5.9(b)) has the following form:

$$\Sigma_{\boldsymbol{k}\omega} = ... + \frac{1}{\mathcal{V}} \sum_{\boldsymbol{q}} (-n_{\boldsymbol{k}-\boldsymbol{q}}) V_{\boldsymbol{q}}$$

which is just the Fock contribution to the self-energy.

In the above, we have seen how to calculate the self-energy $\Sigma_{\boldsymbol{k}\omega}$ and the quasiparticle dispersion $\xi_{\boldsymbol{k}}^* = \xi_{\boldsymbol{k}} + \Sigma_{\boldsymbol{k}\omega}|_{\omega=\xi_{\boldsymbol{k}}^*}$ from the diagrams. This approach reproduces the Hartree–Fock result to first order in V. However, the diagram approach allows us to calculate higher-order contributions to the self-energy in a systematic way (see Fig. 5.11).

We can also use diagrams to calculate the Fermi liquid function in Fermi liquid theory. We know that the interaction $V_{\boldsymbol{q}}$ can scatter two electrons with momenta $\boldsymbol{k}_1$ and $\boldsymbol{k}_2$ into two electrons with momenta $\boldsymbol{k}_1'$ and $\boldsymbol{k}_2'$. The interaction term $n_{\boldsymbol{k}_1} n_{\boldsymbol{k}_2} V(\boldsymbol{k}_1 - \boldsymbol{k}_2)$ in Fermi liquid theory corresponds to terms that scatter two electrons with momenta $\boldsymbol{k}_1$ and $\boldsymbol{k}_2$ into two electrons with the same momenta. Thus, the Fermi liquid function f receives two contributions, which are illustrated by the two diagrams in Fig. 5.12:

$$-\mathrm{i} f(\boldsymbol{k}_1, \alpha; \boldsymbol{k}_2, \beta) = (-\mathrm{i} V_0) + (-)(-\mathrm{i} V_{\boldsymbol{q}}) \delta_{\alpha\beta}$$

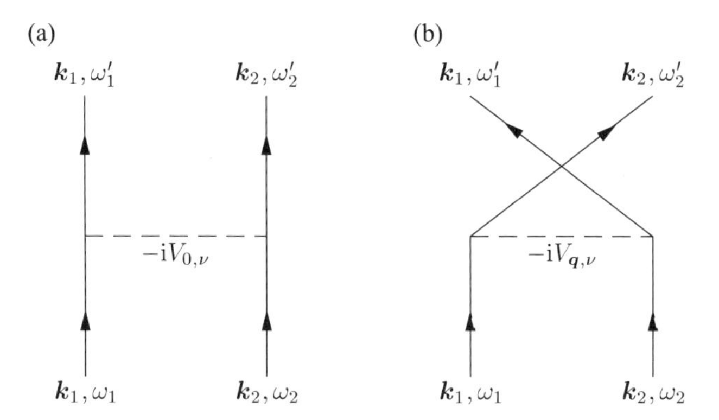

FIG. 5.12. Diagrams that contribute to the Fermi function $f(\boldsymbol{k}_1, \alpha; \boldsymbol{k}_2, \beta)$. (a) Here $\boldsymbol{q} = \boldsymbol{k}_1 - \boldsymbol{k}_1 = 0$ is the momentum transfer and $\nu = \omega_1 - \omega_1'$ is the energy transfer. (b) Here $\boldsymbol{q} = \boldsymbol{k}_1 - \boldsymbol{k}_2$ is the momentum transfer and $\nu = \omega_1 - \omega_2'$ is the energy transfer.

The $(-)$ sign arises from the exchange and $\delta_{\alpha\beta}$ arises from the requirement that the two electrons must have the same spins before and after the scattering. As the interaction considered here is instantaneous, $V_{\boldsymbol{q}}$ does not depend on the energy transfer ν (see Fig. 5.12). The resulting Fermi liquid function f is also instantaneous and independent of the energy transfer ν:

$$f(\boldsymbol{k}_1, \alpha; \boldsymbol{k}_2, \beta) = V_0 - V_{\boldsymbol{k}_1 - \boldsymbol{k}_2} \delta_{\alpha\beta}$$

The above is just the Hartree–Fock result (5.3.2) that we obtained before.

Problem 5.4.2.
Prove eqn (5.4.4) using perturbation theory. We note that $\int \frac{d\nu}{2\pi} \mathrm{i}G_{0,\boldsymbol{k}'\nu} \mathrm{e}^{\mathrm{i}0^+\nu}$ represents $\langle c^\dagger c \rangle$ at equal time, while $\int \frac{d\nu}{2\pi} \mathrm{i}G_{0,\boldsymbol{k}'\nu} \mathrm{e}^{-\mathrm{i}0^+\nu}$ represents $\langle c c^\dagger \rangle$ at equal time.

Problem 5.4.3.
Find the expressions for the order-$V_{\boldsymbol{q}}^2$ terms in the perturbative expansion of the self-energy $\Sigma_{\boldsymbol{k}\omega}$, as described in Fig. 5.11(a)–(e).

5.4.3 Random phase approximation and the effective potential

- The screening of the interaction potential by particle–hole excitations can be included in the random phase approximation.

We have seen that, in three dimensions and for a Coulomb interaction, the Hartree–Fock approximation results in a self-energy that is singular near the Fermi surface. The singularity gives rise to an infinite Fermi velocity. The singularity arises from the long-range potential of the Coulomb interaction. A short-ranged interaction cannot produce any singular self-energy. However, we know that two

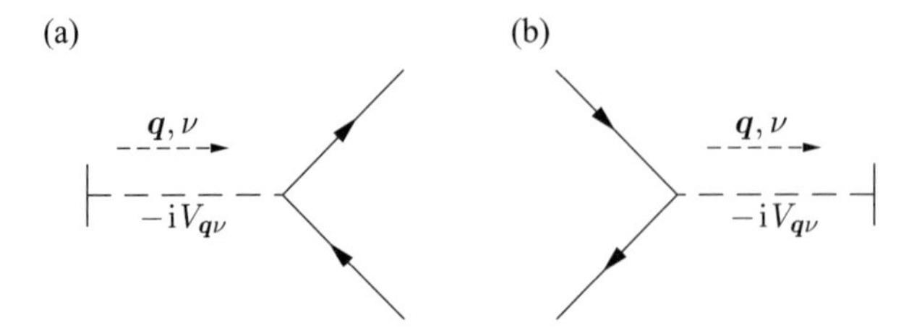

FIG. 5.13. (a) The interaction generates particle–hole excitations, and (b) the particle–hole excitations modify the interaction.

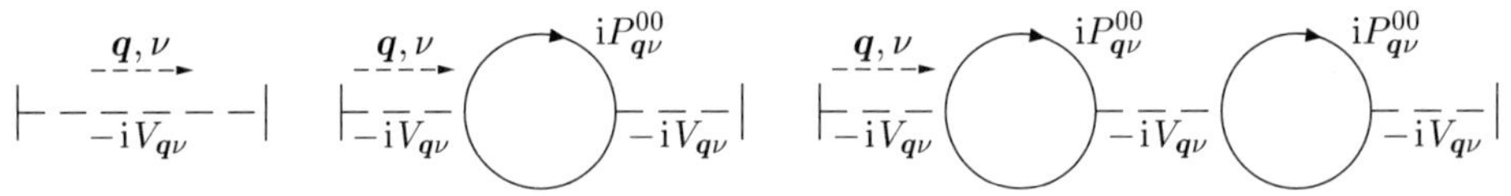

FIG. 5.14. The effective potential in the RPA approximation.

charges in a metal do not experience a long-range interaction due to the screening from other electrons. Therefore, one may doubt the correctness of the Hartree–Fock result about the diverging Fermi velocity; in particular, after realizing that the Hartree–Fock approximation includes only direct interaction and no screening effects.

To go beyond the Hartree–Fock approximation and include the screening effects, we must include the effect that the interaction generates particle–hole excitations (Fig. 5.13(a)), and the particle–hole excitations in turn modifying the potential (Fig. 5.13(b)). In the random phase approximation (RPA), we include the diagrams in Fig. 5.14 to calculate the screened effective potential as follows:

$$-\mathrm{i}V^{\text{eff}}_{\boldsymbol{q}\nu} = -\mathrm{i}V_{\boldsymbol{q}\nu}\left(1 + (-\mathrm{i}V_{\boldsymbol{q}\nu})(iP^{00}_{\boldsymbol{q}\nu}) + (-\mathrm{i}V_{\boldsymbol{q}\nu})^2(iP^{00}_{\boldsymbol{q}\nu})^2 + ...\right)$$

We find that

$$V^{\text{eff}}_{\boldsymbol{q}\nu} = \frac{V_{\boldsymbol{q}\nu}}{1 - V_{\boldsymbol{q}\nu}P^{00}_{\boldsymbol{q}\nu}}$$

In the $\nu \ll v_F q$ limit, $P^{00}_{\boldsymbol{q}\nu}$ is equal to the negative of the compressibility, namely $P^{00}_{\boldsymbol{q}\nu} = -N(E_F)$. We have

$$V^{\text{eff}}_{\boldsymbol{q}\nu} = \frac{V_{\boldsymbol{q}\nu}}{1 + V_{\boldsymbol{q}\nu}N(E_F)}$$

and for the Coulomb interaction we have

$$V^{\text{eff}}_{\boldsymbol{q}} = \frac{4\pi e^2}{\boldsymbol{q}^2 + 4\pi e^2 N(E_F)}$$

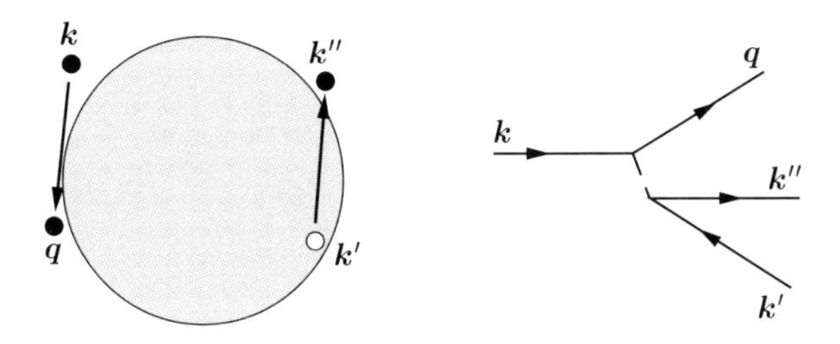

FIG. 5.15. The decay of a quasiparticle.

It is clear that, in the $\nu \ll v_F q$ limit, the effective potential is not singular as $\boldsymbol{q} \to 0$. Such an effective potential represents a screened short-range interaction. The same result can also be obtained through the Thomas–Fermi model.

The relationship between the RPA calculation and the screening can be seen more clearly in the following way. The first term in the RPA calculation (see Fig. 5.14) represents the direct (i.e. unscreened) interaction between two densities (represented by two short vertical bars in Fig. 5.14) at two different places. The second term in the RPA calculation represents the effect that the density at one place induces a particle–hole excitation via the interaction $V_{\boldsymbol{q}\nu}$. The particle–hole excitation, representing a deformed density, interacts in turn with the density at another place. In this way, the second term (and other higher-order terms) modify or screen the interaction between the densities at the two places.

In standard practice, we replace the bare potential $V_{\boldsymbol{q}\nu}$ with the screened potential $V_{\boldsymbol{q}\nu}^{\text{eff}} = V_{\boldsymbol{q}\nu}/(1+V_{\boldsymbol{q}\nu}N(E_F))$, and then use this effective potential to calculate the self-energy and the Fermi liquid function. In this approximation, we ignore the frequency dependence of P^{00}. An instantaneous bare interaction (a frequency-independent interaction) is approximated by an instantaneous effective interaction. If we include the frequency dependence of P^{00}, then an instantaneous bare interaction can result in a frequency-dependent effective interaction. Such a frequency dependence represents the retardation effects.

5.4.4 Justification of Landau Fermi liquid theory

- The interaction causes the quasiparticle to decay, but the decay rate is much less than the quasiparticle energy. Landau Fermi liquid theory remains to be validated.

An electron–electron interaction can cause an electron with momentum $\boldsymbol{k}$ to decay into an electron with momentum $\boldsymbol{q}$ and a particle–hole excitation with momentum $\boldsymbol{k} - \boldsymbol{q}$ (see Fig .5.15). The particle and the hole in the particle–hole excitation can independently have energies in the range between 0 and $\xi_{\boldsymbol{k}}$. The

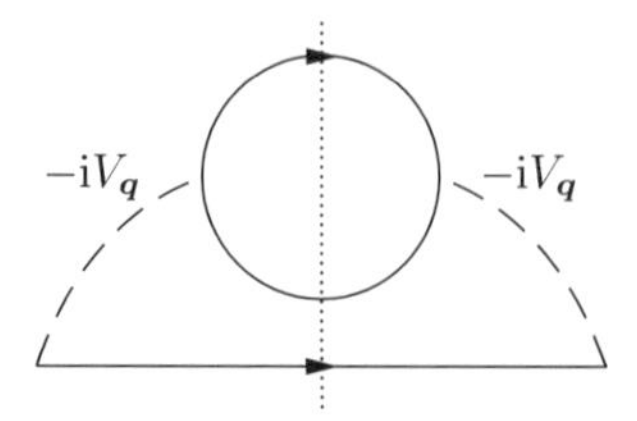

FIG. 5.16. A contribution to the self-energy at order V_q^2. The diagram is formed by two halves, each representing the decay process in Fig. 5.15. The above diagram produces the decay rate, while the one in Fig. 5.15 leads to the decay amplitude.

energy of the particle at $\boldsymbol{q}$ is adjusted accordingly to maintain the energy conservation. Thus, the number of available decay channels is proportional to the square of the initial electron energy $\xi_{\boldsymbol{k}}$. As a result, the decay rate is also proportional to $\xi_{\boldsymbol{k}}^2$. As the decay amplitude is proportional to $V_{\boldsymbol{k}-\boldsymbol{q}}$, we find the decay rate to be

$$\Gamma_{\boldsymbol{k}} \propto |V_{\boldsymbol{k}}|^2 \xi_{\boldsymbol{k}}^2$$

The decay of a quasiparticle can be described by the imaginary part of the self-energy. When $\Sigma_{\boldsymbol{k}\omega}$ has an imaginary part, the pole of the electron Green's function $1/(\omega - \xi_{\boldsymbol{k}} - \Sigma_{\boldsymbol{k}\omega})$ is off the real-ω axis. This leads to a decay in the real-time Green's function:

$$G(t, \boldsymbol{k}) = \int \frac{d\omega}{2\pi} \frac{1}{\omega - \xi_{\boldsymbol{k}} - \Sigma_{\boldsymbol{k}\omega}} e^{-i\omega t} \sim e^{-i\xi_{\boldsymbol{k}}^* t} e^{-|\mathrm{Im}\Sigma_{\boldsymbol{k}\xi_{\boldsymbol{k}}^*}|t}$$

where $\xi_{\boldsymbol{k}}^* = \xi_{\boldsymbol{k}} + \mathrm{Re}\Sigma_{\boldsymbol{k}\xi_{\boldsymbol{k}}^*}$. We see that the imaginary part of the self-energy is given by

$$|\mathrm{Im}\Sigma_{\boldsymbol{k}\xi_{\boldsymbol{k}}^*}| = \Gamma_{\boldsymbol{k}}$$

We would like to mention that such a self-energy can be obtained from the Feynman diagram in Fig. 5.16.

As the decay rate $\Gamma = \mathrm{Im}\Sigma_{\boldsymbol{k}\xi_{\boldsymbol{k}}^*} \propto (\xi_{\boldsymbol{k}}^*)^2$ is much less than the quasiparticle energy $\xi_{\boldsymbol{k}}^*$ in the low-energy limit, we can ignore the decay of the quasiparticle. In this limit, the quasiparticle excitations can be regarded as eigenstates. This is the foundation of Landau Fermi liquid theory. We see that $\mathrm{Im}G_{\boldsymbol{k}\omega}$ is almost a δ-function at $\xi_{\boldsymbol{k}}^*$. The width of the peak is only proportional to $(\xi_{\boldsymbol{k}}^*)^2$.

Near the Fermi surface we may use the expansion

$$\mathrm{Re}\Sigma_{\boldsymbol{k}\omega} = (1 - \frac{1}{Z})\omega + (\frac{v_F^*}{Z} - v_F)(k - k_F)$$

and rewrite the Green's function as follows:

$$G_{\boldsymbol{k}\omega} = \frac{1}{\omega - v_F(k - k_F) - \mathrm{Re}\Sigma_{\boldsymbol{k}\omega}} = \frac{Z}{\omega - v_F^*(k - k_F)}$$

Here Z is the overlap factor introduced near the end of Section 5.3.1. In this way we justify Landau Fermi liquid theory within a perturbative calculation.

5.5 Symmetry-breaking phase and the spin-density-wave state

• Fermi liquids can have many instabilities. Interactions can lead to symmetry-breaking phases.

Interacting fermion systems can have many different states, which lead to materials with so many different properties, such as superconductors, magnets, charge-density-wave states, etc. All of these different states of materials can be understood as certain symmetry-breaking states induced by interactions. In this section, we are going to discuss only one of the symmetry-breaking states—the spin-density-wave state. The picture, the methods, and the results discussed here can be easily generalized to other symmetry-breaking states, such as the superconducting state.

5.5.1 Linear responses and instabilities

• A free fermion system, with so many gapless excitations, has many potential instabilities. Interactions between fermions can turn the potential instabilities into real instabilities. Different interactions induce different instabilities.

Consider an interacting electron system on a lattice with an on-site interaction:

$$\begin{aligned} H &= -\sum_{\langle \boldsymbol{ij}\rangle,\alpha} t(c^\dagger_{\alpha \boldsymbol{j}} c_{\alpha \boldsymbol{i}} + h.c.) - \mu N + U \sum_{\boldsymbol{i}} n_{\boldsymbol{i}}^2 \\ &= -\sum_{\langle \boldsymbol{ij}\rangle,\alpha} t(c^\dagger_{\alpha \boldsymbol{j}} c_{\alpha \boldsymbol{i}} + h.c.) - \mu' N + U \sum_{\boldsymbol{i}} c^\dagger_{\beta \boldsymbol{i}} c^\dagger_{\alpha \boldsymbol{i}} c_{\alpha \boldsymbol{i}} c_{\beta \boldsymbol{i}} \end{aligned} \tag{5.5.1}$$

where $n_{\boldsymbol{i}} = \sum_\alpha c^\dagger_{\alpha \boldsymbol{i}} c_{\alpha \boldsymbol{i}}$ is the number of electrons at site $\boldsymbol{i}$ and $N = \sum_{\boldsymbol{i}} n_{\boldsymbol{i}}$ is the total number of electrons. The above model is called the Hubbard model.

In the following, we will concentrate on the Hubbard model in two dimensions. In momentum space, the two-dimensional Hubbard model can be rewritten as

$$H = \sum_{\boldsymbol{k}} \xi_{\boldsymbol{k}} c^\dagger_{\alpha \boldsymbol{k}} c_{\alpha \boldsymbol{k}} + U \frac{1}{N_{\text{site}}} \sum_{\boldsymbol{k},\boldsymbol{k}',\boldsymbol{q}} c^\dagger_{\alpha \boldsymbol{k}} c^\dagger_{\beta \boldsymbol{k}'} c_{\beta,\boldsymbol{k}'+\boldsymbol{q}} c_{\alpha,\boldsymbol{k}-\boldsymbol{q}}$$

where

$$\xi_{\boldsymbol{k}} = -2t\left(\cos(k_x) + \cos(k_y)\right) - \mu'$$

and N_{site} is the number of lattice sites. We note that, when $\mu' = 0$, the Fermi sea is half-filled ($\langle\langle n_{\boldsymbol{i}} \rangle\rangle = 1$) and the Fermi surface is a square (see Fig. 5.17) which

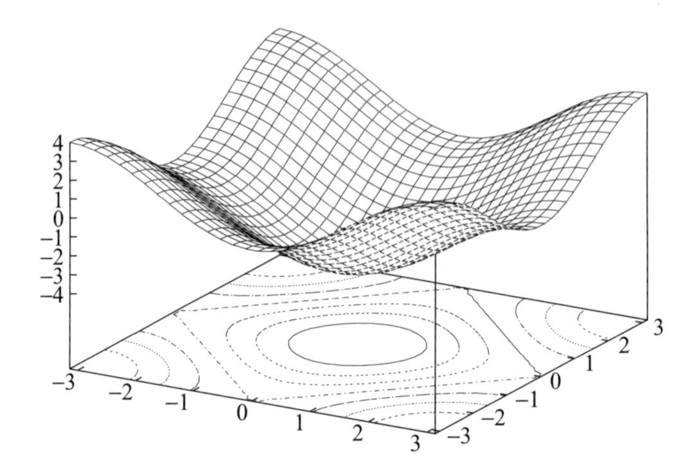

FIG. 5.17. The electron dispersion $\xi_{\boldsymbol{k}}$ in the Hubbard model.

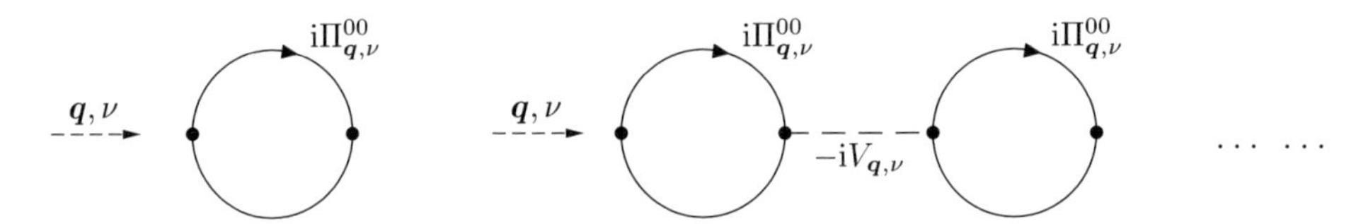

FIG. 5.18. The compressibility and the spin susceptibility in the RPA.

satisfies a nesting condition, with nesting vector $\boldsymbol{Q} = (\pi, \pi)$. From eqn (4.3.3), we see that the free fermion compressibility and the spin susceptibility at the nesting vector $\boldsymbol{Q}$ diverge at low temperatures. As the density of states $N(E_F)$ (now $N(E_F)$ means the number of states per site per unit energy) also has a logarithmic divergence, we have

$$\chi_0(\boldsymbol{Q}) \sim \frac{1}{E_F}(\ln \frac{E_F}{T})^2 \tag{5.5.2}$$

The divergent $\chi_0(\boldsymbol{Q})$ at $T = 0$ implies that the system is on the verge of developing a density order or a spin order. In the following, we will show that, for an interacting system, the susceptibility $\chi(\boldsymbol{Q})$ diverges as T approaches a finite critical temperature T_c. Below T_c, the system spontaneously generates a charge-density-wave (CDW) or a spin-density-wave (SDW) state.

For interacting electrons, the compressibility and the spin susceptibility receive corrections due to the interaction. Within the RPA, we only consider corrections described by the graphs in Fig. 5.18. The one-loop bubble is given by the correlation

$$\langle \rho_{\alpha;\boldsymbol{q}\nu}\rho_{\beta;-\boldsymbol{q},-\nu}\rangle = \mathrm{i}\Pi^{00}_{\alpha\beta;\boldsymbol{q}\nu} = \mathrm{i}\Pi^{00}_{\boldsymbol{q}\nu}\delta_{\alpha\beta}$$

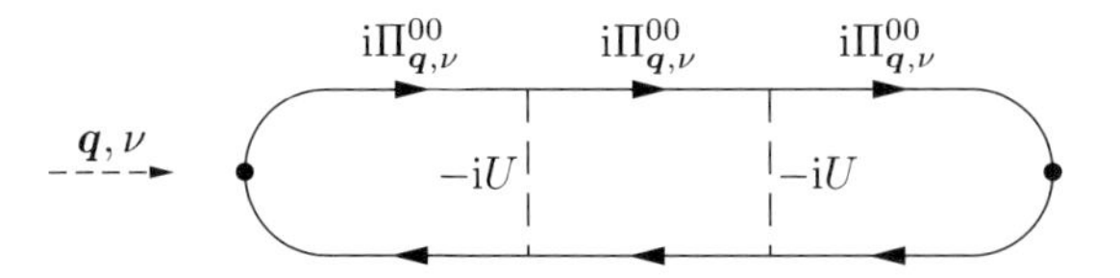

FIG. 5.19. The ladder graphs that contribute to the spin and density correlations.

and the interaction line is given by

$$-\mathrm{i}V_{\alpha\beta;q\nu} = -\mathrm{i}UC_{\alpha\beta}$$

where ρ_α is the density of the spin-α electron and $C = (C_{\alpha\beta})$ is a 2×2 matrix whose elements are all equal to one. We find that, within the RPA,

$$\begin{aligned}
\mathrm{i}\Pi^{\mathrm{RPA}}_{\alpha\beta;q\nu} &= \mathrm{i}\Pi^{00}_{\alpha\beta;q\nu} + \sum_{\gamma,\lambda} \mathrm{i}\Pi^{00}_{\alpha\gamma;q\nu}(-\mathrm{i}V_{\gamma\lambda;q\nu})\mathrm{i}\Pi^{00}_{\lambda\beta;q\nu} + \dots\dots \\
&= \sum_\gamma \mathrm{i}\Pi^{00}_{q\nu}\left(1 - U\Pi^{00}_{q\nu}C\right)^{-1}_{\alpha\beta} = \mathrm{i}\Pi^{00}_{q\nu}\left(\delta_{\alpha\beta} + \frac{U\Pi^{00}_{q\nu}}{1-2U\Pi^{00}_{q\nu}}C_{\alpha\beta}\right)
\end{aligned}$$

where we have used $(a+bC)^{-1} = a^{-1} - b(a^2+2ab)^{-1}C$. Therefore, the density response function is given by

$$\begin{aligned}
\mathrm{i}\Pi^{\mathrm{RPA}}_{\mathrm{den};q\nu} &= \langle(\rho_{\uparrow;q\nu} + \rho_{\downarrow;q\nu})(\rho_{\uparrow;-q,-\nu} + \rho_{\downarrow;-q,-\nu})\rangle \\
&= \mathrm{i}\Pi^{\mathrm{RPA}}_{\uparrow\uparrow;q\nu} + \mathrm{i}\Pi^{\mathrm{RPA}}_{\downarrow\downarrow;q\nu} + 2\mathrm{i}\Pi^{\mathrm{RPA}}_{\uparrow\downarrow;q\nu} = 2\mathrm{i}\Pi^{00}_{q\nu}\frac{1}{1-2U\Pi^{00}_{q\nu}}
\end{aligned}$$

and the spin response function is given by

$$\begin{aligned}
\mathrm{i}\Pi^{\mathrm{RPA}}_{\mathrm{spin};q\nu} &= \left\langle \frac{1}{2}(\rho_{\uparrow;q\nu} - \rho_{\downarrow;q\nu})\frac{1}{2}(\rho_{\uparrow;-q,-\nu} - \rho_{\downarrow;-q,-\nu})\right\rangle \\
&= \frac{1}{4}\mathrm{i}\Pi^{\mathrm{RPA}}_{\uparrow\uparrow;q\nu} + \frac{1}{4}\mathrm{i}\Pi^{\mathrm{RPA}}_{\downarrow\downarrow;q\nu} - \frac{1}{2}\mathrm{i}\Pi^{\mathrm{RPA}}_{\uparrow\downarrow;q\nu} = \frac{1}{2}\mathrm{i}\Pi^{00}_{q\nu}
\end{aligned}$$

However, the above results are not quite complete. For an on-site interaction, the ladder graphs in Fig. 5.19 have similar contributions to the graphs in Fig. 5.18. Including the mixed ladder and bubble graphs, we have

$$\begin{aligned}
\mathrm{i}\Pi^{\mathrm{RPA}^+}_{\alpha\beta;q\nu} &= \mathrm{i}\Pi^{00}_{q\nu}\delta_{\alpha\beta} + \sum_{\gamma\lambda}\left(\mathrm{i}\Pi^{00}_{q\nu}\delta_{\alpha\gamma}(-\mathrm{i}UC_{\gamma\lambda})\mathrm{i}\Pi^{00}_{q\nu}\delta_{\lambda\beta}\right) \\
&\quad + \sum_{\gamma\lambda}\left((-)\mathrm{i}\Pi^{00}_{q\nu}\delta_{\alpha\gamma}(-\mathrm{i}U\delta_{\gamma\lambda})\mathrm{i}\Pi^{00}_{q\nu}\delta_{\lambda\beta}\right) + \dots \\
&= \sum_\gamma \mathrm{i}\Pi^{00}_{q\nu}\left(1 - U\Pi^{00}_{q\nu}C + U\Pi^{00}_{q\nu}\right)^{-1}_{\alpha\beta} \\
&= \mathrm{i}\Pi^{00}_{q\nu}\left(\frac{1}{1+U\Pi^{00}_{q\nu}}\delta_{\alpha\beta} + \frac{U\Pi^{00}_{q\nu}}{1-(U\Pi^{00}_{q\nu})^2}C_{\alpha\beta}\right)
\end{aligned}$$

The new density response function is given by

$$\mathrm{i}\Pi^{\mathrm{RPA}^+}_{\mathrm{den};q\nu} = \mathrm{i}\Pi^{00}_{q\nu}\left(2\frac{1}{1+U\Pi^{00}_{q\nu}} + 2\frac{U\Pi^{00}_{q\nu}}{1-(U\Pi^{00}_{q\nu})^2} + 2\frac{U\Pi^{00}_{q\nu}}{1-(U\Pi^{00}_{q\nu})^2}\right) = 2\frac{\mathrm{i}\Pi^{00}_{q\nu}}{1-U\Pi^{00}_{q\nu}}$$

and the new spin response function is given by

$$\mathrm{i}\Pi^{\mathrm{RPA}^+}_{\mathrm{spin};q\nu} = \mathrm{i}\Pi^{00}_{q\nu}\left(2\frac{1}{1+U\Pi^{00}_{q\nu}} + 2\frac{U\Pi^{00}_{q\nu}}{1-(U\Pi^{00}_{q\nu})^2} - 2\frac{U\Pi^{00}_{q\nu}}{1-(U\Pi^{00}_{q\nu})^2}\right) = \frac{1}{2}\frac{\mathrm{i}\Pi^{00}_{q\nu}}{1+U\Pi^{00}_{q\nu}}$$

We find that the compressibility χ_c and the spin susceptibility χ_s within the RPA are given by

$$\chi_c^{\mathrm{RPA}^+}(\boldsymbol{q}) = \frac{\chi_c^{(0)}(\boldsymbol{q})}{1+\frac{1}{2}U\chi_c^{(0)}(\boldsymbol{q})}$$

$$\chi_s^{\mathrm{RPA}^+}(\boldsymbol{q}) = \frac{\chi_s^{(0)}(\boldsymbol{q})}{1-2U\chi_s^{(0)}(\boldsymbol{q})}$$

where $\chi_c^{(0)}(\boldsymbol{q}) = -2\Pi^{00}_{\boldsymbol{q},0}$ and $\chi_s^{(0)}(\boldsymbol{q}) = -\Pi^{00}_{\boldsymbol{q},0}/2$ are the compressibility and the spin susceptibility for free electrons.

We see that a (short-range) repulsive interaction enhances the spin susceptibility and suppresses the charge compressibility, while an (short-range) attractive interaction suppresses the spin susceptibility and enhances the charge compressibility. For a nested Fermi surface and a repulsive interaction, $\chi_s^{\mathrm{RPA}^+}(\boldsymbol{Q})$ diverges as T approaches a finite critical temperature T_c determined by $1 - U\chi_s^{(0)}(\boldsymbol{Q}) = 0$. Below T_c, the system spontaneously generates an SDW and spontaneously breaks the spin–rotation symmetry and translational symmetry of the lattice. The transition temperature can be estimated from the low-temperature behavior of $\chi_s^{(0)}(\boldsymbol{Q})$ (see eqn (5.5.2)):

$$T_c \sim E_F \mathrm{e}^{-C_1\sqrt{\frac{E_F}{U}}}$$

We see that, for a nested Fermi surface, a repulsive interaction always generates an SDW, no matter how weak the interaction. Similarly, for a nested Fermi surface, an attractive interaction always generates a CDW, no matter how weak the interaction. The transition temperature is

$$T_c \sim E_F \mathrm{e}^{-C_2\sqrt{\frac{E_F}{-U}}}$$

5.5.2 Mean-field approach for the spin-density-wave state

- In the mean-field approach, we replace pairs of fermion operators by their ground-state average to convert an interacting Hamiltonian to a non-interacting Hamiltonian (the mean-field Hamiltonian).

We have seen that a repulsive interaction in a system with a nested Fermi surface may cause an SDW instability. In this and the following sections, we are going study the ground state for such a system and show that the ground state indeed

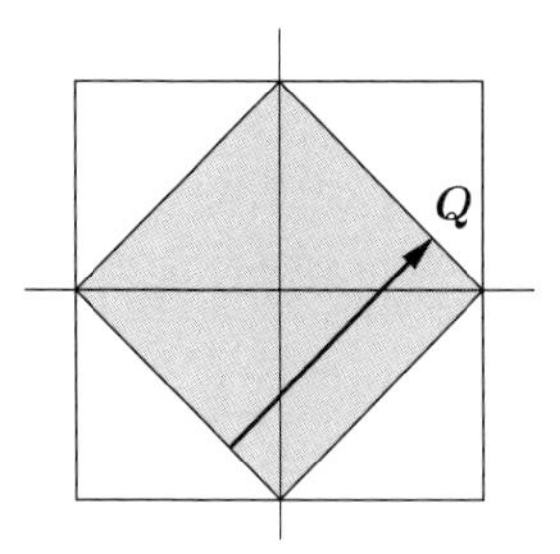

FIG. 5.20. The shaded area is the reduced Brillouin zone. It is also the filled Fermi sea for a half-filled band.

has an SDW order. We are going to use two approaches, namely the mean-field approach and the variational approach. The variational approach is conceptually simpler and more correct. It can also calculate the interaction between quasiparticles. The mean-field approach is mathematically simpler. We first consider the mean-field approach.

To use the mean-field approach to study the SDW state at $T = 0$, we first rewrite the Hamiltonian in a more convenient form. Using the identity

$$Un_{\boldsymbol{i}}^2 = -U(c_{\boldsymbol{i}}^\dagger \sigma^z c_{\boldsymbol{i}})^2 + 2Un_{\boldsymbol{i}}$$

and absorbing the term $2Un_{\boldsymbol{i}}$ into the chemical potential term, we can rewrite the Hubbard model as

$$H = \sum_{\boldsymbol{k},\alpha} \xi_{\boldsymbol{k}} c_{\alpha\boldsymbol{k}}^\dagger c_{\alpha\boldsymbol{k}} - U \sum_{\boldsymbol{i}} (c_{\boldsymbol{i}}^\dagger \sigma^z c_{\boldsymbol{i}})^2$$

Now we can formally obtain a mean-field Hamiltonian by replacing one of the spin $c_{\boldsymbol{i}}^\dagger \sigma^z c_{\boldsymbol{i}}$ by its average $(-)^{\boldsymbol{i}} M$, and replacing $(c_{\boldsymbol{i}}^\dagger \sigma^z c_{\boldsymbol{i}})^2$ by $2M(-)^{\boldsymbol{i}}(c_{\boldsymbol{i}}^\dagger \sigma^z c_{\boldsymbol{i}}) - M^2$. We obtain

$$H_{\text{mean}}^{\text{sub}} = \sum_{\boldsymbol{k},\alpha} \xi_{\boldsymbol{k}} c_{\alpha\boldsymbol{k}}^\dagger c_{\alpha\boldsymbol{k}} - 2U \sum_{\boldsymbol{i}} (-)^{\boldsymbol{i}} M c_{\boldsymbol{i}}^\dagger \sigma^z c_{\boldsymbol{i}} + N_{\text{site}} U M^2$$

Within the mean-field approximation, we say that the physical properties of our system are described by $H_{\text{mean}}^{\text{sub}}$ instead of the original interacting Hamiltonian (5.5.1).

The mean-field Hamiltonian can be solved exactly. In the momentum space, we have

$$\begin{aligned} H_{\text{mean}}^{\text{sub}} &= \sum_{\boldsymbol{k},\alpha} 2t(\cos k_x + \cos k_y) c_{\boldsymbol{k}}^\dagger c_{\boldsymbol{k}} - B \sum_{\boldsymbol{k}} (-)^{\boldsymbol{i}} c_{\boldsymbol{k}+\boldsymbol{Q}}^\dagger \sigma^z c_{\boldsymbol{k}} + \text{constant} \\ &= \sum_{\boldsymbol{k}}{}' \psi_{\boldsymbol{k}}^\dagger M_{\boldsymbol{k}} \psi_{\boldsymbol{k}} + \text{constant} \end{aligned}$$

where $B = 2UM$, $\sum'_{\boldsymbol{k}}$ is a summation over a reduced Brillouin zone (see Fig. 5.20), $(-)^{\boldsymbol{i}} = (-)^{i_x+i_y}$,

$$M_{\boldsymbol{k}} = \begin{pmatrix} \xi_{\boldsymbol{k}} & -B\sigma^z \\ -B\sigma^z & -\xi_{\boldsymbol{k}} \end{pmatrix}, \qquad \xi_{\boldsymbol{k}} = 2t(\cos k_x + \cos k_y) \tag{5.5.3}$$

and

$$\psi_{\alpha\boldsymbol{k}} = \begin{pmatrix} c_{\alpha\boldsymbol{k}} \\ c_{\alpha,\boldsymbol{k}+\boldsymbol{Q}} \end{pmatrix}$$

We can diagonalize the above mean-field Hamiltonian $H^{\text{sub}}_{\text{mean}}$ by introducing

$$\begin{aligned} \eta_{\alpha\boldsymbol{k}} &= -v_{\boldsymbol{k}}\sigma^z_{\alpha\beta}c_{\beta\boldsymbol{k}} + u_{\boldsymbol{k}}c_{\alpha,\boldsymbol{k}+\boldsymbol{Q}} \\ \lambda_{\alpha\boldsymbol{k}} &= u_{\boldsymbol{k}}c_{\alpha\boldsymbol{k}} + v_{\boldsymbol{k}}\sigma^z_{\alpha\beta}c_{\beta,\boldsymbol{k}+\boldsymbol{Q}} \end{aligned}$$

where

$$u_{\boldsymbol{k}} = \frac{1}{\sqrt{2}}\sqrt{1 - \frac{\xi_{\boldsymbol{k}}}{\sqrt{\xi^2_{\boldsymbol{k}} + B^2}}} \qquad v_{\boldsymbol{k}} = \text{sgn}(B)\frac{1}{\sqrt{2}}\sqrt{1 + \frac{\xi_{\boldsymbol{k}}}{\sqrt{\xi^2_{\boldsymbol{k}} + B^2}}}$$

As $u^2_{\boldsymbol{k}} + v^2_{\boldsymbol{k}} = 1$, $\eta_{\alpha\boldsymbol{k}}$ and $\lambda_{\alpha\boldsymbol{k}}$ satisfy the following commutation relations for the fermion operators:

$$\begin{aligned} \{\eta_{\alpha\boldsymbol{k}}, \eta^\dagger_{\alpha\boldsymbol{k}}\} &= \delta_{\boldsymbol{k}\boldsymbol{k}'}\delta_{\alpha\alpha'} \\ \{\lambda_{\alpha\boldsymbol{k}}, \lambda^\dagger_{\alpha\boldsymbol{k}}\} &= \delta_{\boldsymbol{k}\boldsymbol{k}'}\delta_{\alpha\alpha'} \\ \{\eta_{\alpha\boldsymbol{k}}, \lambda^\dagger_{\alpha\boldsymbol{k}}\} &= 0 \end{aligned}$$

In terms of $\eta_{\alpha\boldsymbol{k}}$ and $\lambda_{\alpha\boldsymbol{k}}$, we have

$$H^{\text{sub}}_{\text{mean}} = \sum_{\alpha,\boldsymbol{k}}{}' E_{\boldsymbol{k}}\eta^\dagger_{\alpha\boldsymbol{k}}\eta_{\alpha\boldsymbol{k}} + \sum_{\alpha,\boldsymbol{k}}{}' - E_{\boldsymbol{k}}\lambda^\dagger_{\alpha\boldsymbol{k}}\lambda_{\alpha\boldsymbol{k}} + N_{\text{site}}UM^2$$

and

$$E_{\boldsymbol{k}} = \sqrt{\xi^2_{\boldsymbol{k}} + B^2}$$

We can see that the mean-field ground-state energy is given by

$$E_{\text{mean}} = -2\sum_{\boldsymbol{k}}{}' E_{\boldsymbol{k}} + N_{\text{site}}\frac{B^2}{4U}$$

The spectrum of quasiparticle excitations is given by $\pm E_{\boldsymbol{k}}$ using mean-field theory.

However, the above mean-field results contain an unknown parameter M—the average of the staggered spins. To determine M, we minimize the mean-field ground-state energy E_{mean}. We obtain

$$\frac{1}{4} - UN_{\text{site}}^{-1}{\sum_{\boldsymbol{k}}}' \frac{1}{E_{\boldsymbol{k}}} = 0. \tag{5.5.4}$$

This equation is called the gap equation. We can also obtain the same gap equation by imposing the self-consistency relation

$$M = (-)^{\boldsymbol{i}} \langle \Phi_{\text{mean}} | c_{\boldsymbol{i}}^{\dagger} \sigma^z c_{\boldsymbol{i}} | \Phi_{\text{mean}} \rangle$$

where $|\Phi_{\text{mean}}\rangle$ is the ground state of $H_{\text{mean}}^{\text{sub}}$. That is, the average spin obtained from the mean-field ground state should be equal to the assumed average spin that we used to obtain the mean-field Hamiltonian.

If the gap equation has a solution with $M \neq 0$, then the ground state of our interacting system spontaneously breaks the spin–rotation symmetry and develops an anti-ferromagnetic order. As $\int \frac{\mathrm{d}^2\boldsymbol{k}}{(2\pi)^2} \frac{1}{E_{\boldsymbol{k}}} \to \infty$ as $B \to 0$, we find that eqn (5.5.4) always has $M \neq 0$ solutions as long as $U > 0$. Thus, the positive-U Hubbard model always develops an SDW, regardless of how small U is. This agrees with the result obtained in the last section.

We would like to remark that the above result remains valid as long as $\xi_{\boldsymbol{k}} = -\xi_{\boldsymbol{k}+\boldsymbol{Q}}$. In this case, different points on the Fermi surface (where $\xi_{\boldsymbol{k}} = 0$) are connected by $\boldsymbol{Q}$, or, in other words, the Fermi surface satisfies the nesting condition. For more general $\xi_{\boldsymbol{k}}$, where the Fermi surface does not satisfy the nesting condition, the quasiparticle dispersion is given by

$$\pm\sqrt{\frac{1}{4}(\xi_{\boldsymbol{k}} - \xi_{\boldsymbol{k}+\boldsymbol{Q}})^2 + B^2} + \frac{1}{2}(\xi_{\boldsymbol{k}} + \xi_{\boldsymbol{k}+\boldsymbol{Q}})$$

In this case, the gap equation (5.5.4) no longer applies. It turns out that the SDW is not developed for small U.

Although the mean-field approach (and the variational approach discussed below) correctly reproduces the anti-ferromagnetic ground state of the repulsive Hubbard model, the resulting excitation spectrum is incorrect. This is because the anti-ferromagnetic ground state spontaneously breaks the spin–rotation symmetry and there should be gapless spin wave excitations above the the ground state. The spin wave excitations can be probed through the spin correlation $\mathrm{i}S^{ab}(\boldsymbol{i} - \boldsymbol{j}) = \left\langle (c_{\boldsymbol{i}}^{\dagger}\sigma^a c_{\boldsymbol{i}})(c_{\boldsymbol{j}}^{\dagger}\sigma^b c_{\boldsymbol{j}}) \right\rangle$. Within mean-field theory, we have $S^{ab} \sim \delta_{ab} G_{\text{mean}}^2$, where G_{mean} is the electron Green's function determined from H_{mean}. Thus, $\mathrm{Im}S_{\boldsymbol{k}\omega}^{ab}$ is nonzero only when $|\omega| > 2\Delta$. However, if we go beyond mean-field theory and calculate S^{ab} using the RPA and the ladder graphs in Figs 5.18 and 5.19, then

$\text{Im}S^{ab}_{\boldsymbol{k}\omega}$ will be nonzero down to $\omega = 0$, indicating the presence of a gapless excitation. Also, $S^{ab}_{\boldsymbol{k}\omega}$ has an isolated pole from which we can obtain the dispersion of the spin wave excitations.

5.5.3 Variational approach for the spin-density-wave state—the hard way

- In the variational approach, we use the ground state of a non-interacting Hamiltonian as the trial wave function to approximate the ground state of an interacting Hamiltonian.

In the variational approach, we concentrate on the ground-state wave function, instead of the Hamiltonian. We have seen that the Hubbard model with an on-site repulsive interaction has an SDW instability. To understand the ground-state properties using the variational approach, we just guess a trial wave function for the ground state. One possible choice is to use the ground state $|\Psi_0\rangle$ of $H_0 = \sum_{\boldsymbol{k},\alpha} \xi_{\boldsymbol{k}} c^\dagger_{\alpha\boldsymbol{k}} c_{\alpha\boldsymbol{k}}$ as the trial wave function, as we did in the Hartree–Fock approach. However, $|\Psi_0\rangle$ does not break spin and translational symmetries. The presence of an SDW instability suggests that we should be able to find a different trial wave function which has a lower energy. From symmetry considerations, we would like to use the ground state $|\Psi_B\rangle$ of

$$H_B = \sum_{\boldsymbol{k},\alpha} 2\tilde{t}(\cos k_x + \cos k_y) c^\dagger_{\alpha\boldsymbol{k}} c_{\alpha\boldsymbol{k}} - B\sum_{\boldsymbol{i}} (-)^{\boldsymbol{i}} c^\dagger_{\boldsymbol{i}} \sigma^z c_{\boldsymbol{i}}$$

as the trial wave function. Instead of writing down the many-body trial wave function explicitly, we first diagonalize the 'variational' Hamiltonian H_B, as we did in the mean-field approach. The ground state of H_B satisfies $\lambda^\dagger_{\alpha\boldsymbol{k}}|\Psi_B\rangle = \eta_{\alpha\boldsymbol{k}}|\Psi_B\rangle = 0$. Such an algebraic relation determines our trial wave function.

Note that the Hamiltonian H_B is merely an operator that is used to obtain the trial wave function. Scaling H_B by a constant will not change the trial wave function. Thus, we can set $\tilde{t} = t$, and hence $\tilde{\xi}_{\boldsymbol{k}} = \xi_{\boldsymbol{k}}$. In this case, H_B has the same form as the mean-field Hamiltonian $H^{\text{sub}}_{\text{mean}}$. Here B is a variational parameter that changes the shape of the wave function. We will later vary B to minimize the energy.

To calculate the average energy of our trial state, we note that, for our trial wave function $|\Psi_B\rangle$, we have $\left\langle \eta^\dagger_{\alpha\boldsymbol{k}} \eta_{\alpha\boldsymbol{k}} \right\rangle = 0$ and $\left\langle \lambda^\dagger_{\alpha\boldsymbol{k}} \lambda_{\alpha\boldsymbol{k}} \right\rangle = 1$. From

$$c_{\alpha\boldsymbol{k}} = u_{\boldsymbol{k}} \lambda_{\alpha\boldsymbol{k}} - v_{\boldsymbol{k}} \sigma^z_{\alpha\beta} \eta_{\beta\boldsymbol{k}} \qquad c_{\alpha,\boldsymbol{k}+\boldsymbol{Q}} = v_{\boldsymbol{k}} \sigma^z_{\alpha\beta} \lambda_{\beta\boldsymbol{k}} + u_{\boldsymbol{k}} \eta_{\alpha\boldsymbol{k}}$$

we find that

$$\left\langle c^\dagger_{\alpha\boldsymbol{k}} c_{\beta\boldsymbol{k}} \right\rangle = u^2_{\boldsymbol{k}} \delta_{\alpha\beta} \qquad \left\langle c^\dagger_{\alpha,\boldsymbol{k}+\boldsymbol{Q}} c_{\beta,\boldsymbol{k}+\boldsymbol{Q}} \right\rangle = v^2_{\boldsymbol{k}} \delta_{\alpha\beta}$$

$$\left\langle c^\dagger_{\alpha\boldsymbol{k}} c_{\beta,\boldsymbol{k}+\boldsymbol{Q}} \right\rangle = u_{\boldsymbol{k}} v_{\boldsymbol{k}} \sigma^z_{\alpha\beta} \qquad \left\langle c^\dagger_{\alpha,\boldsymbol{k}+\boldsymbol{Q}} c_{\beta\boldsymbol{k}} \right\rangle = u_{\boldsymbol{k}} v_{\boldsymbol{k}} \sigma^z_{\alpha\beta}$$

In real space, we have

$$\left\langle c_{\alpha i}^{\dagger} c_{\beta i} \right\rangle = \frac{1}{2}\delta_{\alpha\beta} + 2(-)^{i}\sigma_{\alpha\beta}^{z} N_{\text{site}}^{-1} {\sum_{k}}' u_{k} v_{k} = \frac{1}{2}\delta_{\alpha\beta} + (-)^{i}\sigma_{\alpha\beta}^{z} N_{\text{site}}^{-1} {\sum_{k}}' \frac{B}{E_{k}}$$

As expected, our trial state $|\Psi_B\rangle$ has an anti-ferromagnetic order:

$$\langle S_{z,i}\rangle = \frac{1}{2}\left\langle c_{i}^{\dagger}\sigma^{z} c_{i}\right\rangle = (-)^{i} M, \qquad M = N_{\text{site}}^{-1} {\sum_{k}}' \frac{B}{E_{k}}$$

The average energy of our trial state is given by (see Section 5.5.4.1)

$$\langle \Psi_B | H | \Psi_B \rangle = N_{\text{site}} \left(-2N_{\text{site}}^{-1} {\sum_{k}}' \frac{\xi_{k}^{2}}{E_{k}} - U \left(N_{\text{site}}^{-1} {\sum_{k}}' \frac{B}{E_{k}} \right)^{2} + \text{constant} \right)$$

The value of B is obtained by minimizing $\langle \Psi_B | H | \Psi_B \rangle$. We find that such a B satisfies the gap equation (see Section 5.5.4.2):

$$1 - U N_{\text{site}}^{-1} {\sum_{k}}' \frac{1}{\sqrt{\xi_{k}^{2} + B^{2}}} = 0 \tag{5.5.5}$$

We see that eqns (5.5.4) and (5.5.5) do not agree. This illustrates the point that mean-field approaches may lead to different results at the leading order. This is because different mean-field approaches correspond to different expansions, which in general have different leading terms.

The variational approach also allows us study the properties of excitations. The trial wave functions for excitations are created by $\eta_{\alpha k}^{\dagger}$ and $\lambda_{\alpha k}$ from the trial ground state $|\Psi_B\rangle$. For the excited states, we have $\left\langle \eta_{\alpha k}^{\dagger}\eta_{\alpha k}\right\rangle = n_{\alpha k}^{+}$ and $\left\langle \lambda_{\alpha k}^{\dagger}\lambda_{\alpha k}\right\rangle = n_{\alpha k}^{-}$ with $n_{\alpha k}^{+} \neq 0$ and $n_{\alpha k}^{-} \neq 1$. After relating $\langle c^{\dagger} c\rangle$ to $n_{\alpha k}^{+}$ and $n_{\alpha k}^{-}$, we obtain the average energy of our trial excited state (see Section 5.5.4.3) as follows:

$$\langle \Psi_B^{\text{exc}} | H | \Psi_B^{\text{exc}} \rangle = E_{\text{ground}} + {\sum_{k,\alpha}}' E_{k}(\delta n_{\alpha k}^{+} - \delta n_{\alpha k}^{-}) + U\delta N + O(\delta n^{2})$$

where $n_{\alpha k}^{-} = 1 + \delta n_{\alpha k}^{-}$ and $n_{\alpha k}^{+} = \delta n_{\alpha k}^{+}$ describe small fluctuations around the ground state. We see that the quasiparticle excitations have an energy spectrum $\pm E_{k}$. Note that there is a finite energy gap $\Delta = \min E_{k} = |B|$ for quasiparticle excitations. The interaction between the quasiparticles can be easily included through the $(\delta n_{\alpha k}^{\pm})^{2}$ terms.

5.5.4 Remarks: some calculation details

5.5.4.1 *Calculation of the average energy*

We calculate the average energy of the trial state $|\Psi_B\rangle$ as follows:

$$\begin{aligned}
&\langle\Psi_B|H|\Psi_B\rangle \\
&= 2\sum_{\boldsymbol{k}}{}'(u_{\boldsymbol{k}}^2 - v_{\boldsymbol{k}}^2)\xi_{\boldsymbol{k}} + \frac{U}{2}\sum_i \left(\left\langle c_{\alpha i}^\dagger c_{\alpha i}\right\rangle \left\langle c_{\beta i}^\dagger c_{\beta i}\right\rangle + \left\langle c_{\alpha i}^\dagger c_{\beta i}\right\rangle \left\langle c_{\alpha i} c_{\beta i}^\dagger\right\rangle\right) \\
&= 2\sum_{\boldsymbol{k}}{}'(u_{\boldsymbol{k}}^2 - v_{\boldsymbol{k}}^2)\xi_{\boldsymbol{k}} + \frac{U}{2}\sum_i \left(1 + \mathrm{Tr}(\frac{1}{2} + (-)^i M\sigma^z)(\frac{1}{2} - (-)^i M\sigma^z)\right) \\
&= N_{\text{site}}\left(-UM^2 + \frac{3U}{4} + 2N_{\text{site}}^{-1}\sum_{\boldsymbol{k}}{}' - \frac{\xi_{\boldsymbol{k}}}{E_{\boldsymbol{k}}}\xi_{\boldsymbol{k}}\right)
\end{aligned}$$

5.5.4.2 *Calculation of the gap equation*

To obtain the value of B that minimizes the average energy, we calculate $\partial\langle\Psi_B|H|\Psi_B\rangle/\partial B$ and obtain

$$\begin{aligned}
&2N_{\text{site}}^{-1}\sum_{\boldsymbol{k}}{}' - \xi_{\boldsymbol{k}}^2\delta E_{\boldsymbol{k}}^{-1} - 2U\left(N_{\text{site}}^{-1}\sum_{\boldsymbol{k}}{}'\frac{B}{E_{\boldsymbol{k}}}\right)\left(N_{\text{site}}^{-1}\sum_{\boldsymbol{k}}{}'\frac{\delta B}{E_{\boldsymbol{k}}} - \frac{B^2\delta B}{E_{\boldsymbol{k}}^3}\right) \\
&= 2N_{\text{site}}^{-1}\sum_{\boldsymbol{k}}{}' - \xi_{\boldsymbol{k}}^2\delta E_{\boldsymbol{k}}^{-1} - 2U\left(N_{\text{site}}^{-1}\sum_{\boldsymbol{k}}{}'\frac{B}{E_{\boldsymbol{k}}}\right)\left(N_{\text{site}}^{-1}\sum_{\boldsymbol{k}}{}'\frac{\xi_{\boldsymbol{k}}^2\delta B}{E_{\boldsymbol{k}}^3}\right) \\
&= \left(2N_{\text{site}}^{-1}\sum_{\boldsymbol{k}}{}' - \xi_{\boldsymbol{k}}^2\delta E_{\boldsymbol{k}}^{-1}\right) - 2U\left(N_{\text{site}}^{-1}\sum_{\boldsymbol{k}}{}'\frac{B}{E_{\boldsymbol{k}}}\right)\left(N_{\text{site}}^{-1}\sum_{\boldsymbol{k}}{}' - \xi_{\boldsymbol{k}}^2 B^{-1}\delta E_{\boldsymbol{k}}^{-1}\right)
\end{aligned}$$

The requirement that the above vanishes leads to the gap equation (5.5.5).

5.5.4.3 *Calculation of the average energy for excited states*

From the relations

$$\left\langle c_{\alpha\boldsymbol{k}}^\dagger c_{\beta\boldsymbol{k}}\right\rangle = u_{\boldsymbol{k}}^2 n_{\alpha\boldsymbol{k}}^- \delta_{\alpha\beta} + v_{\boldsymbol{k}}^2 n_{\alpha\boldsymbol{k}}^+ \delta_{\alpha\beta}, \quad \left\langle c_{\alpha,\boldsymbol{k}+\boldsymbol{Q}}^\dagger c_{\beta,\boldsymbol{k}+\boldsymbol{Q}}\right\rangle = v_{\boldsymbol{k}}^2 n_{\alpha\boldsymbol{k}}^- \delta_{\alpha\beta} + u_{\boldsymbol{k}}^2 n_{\alpha\boldsymbol{k}}^+ \delta_{\alpha\beta},$$

$$\left\langle c_{\alpha\boldsymbol{k}}^\dagger c_{\beta,\boldsymbol{k}+\boldsymbol{Q}}\right\rangle = u_{\boldsymbol{k}} v_{\boldsymbol{k}}(n_{\alpha\boldsymbol{k}}^- - n_{\alpha\boldsymbol{k}}^+)\sigma_{\alpha\beta}^z, \qquad \left\langle c_{\alpha,\boldsymbol{k}+\boldsymbol{Q}}^\dagger c_{\beta\boldsymbol{k}}\right\rangle = u_{\boldsymbol{k}} v_{\boldsymbol{k}}(n_{\alpha\boldsymbol{k}}^- - n_{\alpha\boldsymbol{k}}^+)\sigma_{\alpha\beta}^z,$$

we find that, in real space,

$$\left\langle c_{\alpha i}^\dagger c_{\beta i}\right\rangle = \delta_{\alpha\beta} N_{\text{site}}^{-1}\sum_{\boldsymbol{k}}{}'(n_{\alpha\boldsymbol{k}}^- + n_{\alpha\boldsymbol{k}}^+) + 2(-)^i\sigma_{\alpha\beta}^z N_{\text{site}}^{-1}\sum_{\boldsymbol{k}}{}' u_{\boldsymbol{k}} v_{\boldsymbol{k}}(n_{\alpha\boldsymbol{k}}^- - n_{\alpha\boldsymbol{k}}^+)$$

This allows us to calculate

$$
\begin{aligned}
&\sum_{i}\left(\left\langle c_{\alpha i}^{\dagger}c_{\alpha i}\right\rangle\left\langle c_{\beta i}^{\dagger}c_{\beta i}\right\rangle+\left\langle c_{\alpha i}^{\dagger}c_{\beta i}\right\rangle\left\langle c_{\alpha i}c_{\beta i}^{\dagger}\right\rangle\right)\\
&=\sum_{i}(n_{\uparrow i}+n_{\downarrow i})^{2}+n_{\uparrow i}(1-n_{\uparrow i})+n_{\downarrow i}(1-n_{\downarrow i})\\
&=\sum_{i}n_{\uparrow i}+n_{\downarrow i}+2n_{\uparrow i}n_{\downarrow i}\\
&=N+2N_{\text{site}}^{-2}\sum_{i}\sum_{k,k'}{}'\left(n_{\uparrow k}^{-}+n_{\uparrow k}^{+}+2(-)^{i}u_{k}v_{k}(n_{\uparrow k}^{-}-n_{\uparrow k}^{+})\right)\times\\
&\qquad\left(n_{\downarrow k'}^{-}+n_{\downarrow k'}^{+}-2(-)^{i}u_{k'}v_{k'}(n_{\downarrow k'}^{-}-n_{\downarrow k'}^{+})\right)\\
&=N+2N_{\text{site}}^{-1}\sum_{k,k'}{}'\left((n_{\uparrow k}^{-}+n_{\uparrow k}^{+})(n_{\downarrow k'}^{-}+n_{\downarrow k'}^{+})-\frac{B^{2}}{E_{k}E_{k'}}(n_{\uparrow k}^{-}-n_{\uparrow k}^{+})(n_{\downarrow k'}^{-}-n_{\downarrow k'}^{+})\right)
\end{aligned}
$$

Using this result, we find that

$$
\begin{aligned}
\langle\Psi_{B}^{\text{exc}}|H|\Psi_{B}^{\text{exc}}\rangle&=\sum_{k,\alpha}{}'(u_{k}^{2}-v_{k}^{2})(n_{\alpha k}^{-}-n_{\alpha k}^{+})\xi_{k} \qquad (5.5.6)\\
&\quad+\frac{U}{2}\sum_{i}\left(\left\langle c_{\alpha i}^{\dagger}c_{\alpha i}\right\rangle\left\langle c_{\beta i}^{\dagger}c_{\beta i}\right\rangle+\left\langle c_{\alpha i}^{\dagger}c_{\beta i}\right\rangle\left\langle c_{\alpha i}c_{\beta i}^{\dagger}\right\rangle\right)\\
&=-\sum_{k,\alpha}{}'\frac{\xi_{k}^{2}}{E_{k}}(n_{\alpha k}^{-}-n_{\alpha k}^{+})+\frac{U}{N_{\text{site}}}\sum_{k,k'}{}'(n_{\uparrow k}^{-}+n_{\uparrow k}^{+})(n_{\downarrow k'}^{-}+n_{\downarrow k'}^{+})\\
&\quad+\frac{U}{N_{\text{site}}}\sum_{kk'}{}'\frac{B^{2}}{E_{k}E_{k'}}(n_{\uparrow k}^{-}-n_{\uparrow k}^{+})(n_{\downarrow k'}^{-}-n_{\downarrow k'}^{+})+\frac{U}{2}N
\end{aligned}
$$

where $N=\sum_{i}n_{\uparrow i}+n_{\downarrow i}$ is the total number of electrons and $n_{\alpha i}$ is the number of spin-α electron at site i.

Expanding $\langle\Psi_{B}^{\text{exc}}|H|\Psi_{B}^{\text{exc}}\rangle$ to linear order in $\delta n_{\alpha k}^{\pm}$, we obtain

$$
\begin{aligned}
\langle\Psi_{B}^{\text{exc}}|H|\Psi_{B}^{\text{exc}}\rangle&=E_{\text{ground}}+\frac{U}{2}\delta N-\sum_{k,\alpha}{}'\frac{\xi_{k}^{2}}{E_{k}}(\delta n_{\alpha k}^{-}-\delta n_{\alpha k}^{+})\\
&\quad+\frac{U}{N_{\text{site}}}\sum_{k,k',\alpha}{}'\left((\delta n_{\alpha k}^{-}+\delta n_{\alpha k}^{+})-\frac{B^{2}}{E_{k}E_{k'}}(\delta n_{\alpha k}^{-}-\delta n_{\alpha k}^{+})\right)\\
&=E_{\text{ground}}+\sum_{k,\alpha}{}'E_{k}(\delta n_{\alpha k}^{+}-\delta n_{\alpha k}^{-})+U\delta N+O(\delta n^{2})
\end{aligned}
$$

where we have used $N=\sum_{k,\alpha}'(n_{\alpha k}^{-}+n_{\alpha k}^{+})$, $\sum_{k}'1=N_{\text{site}}/2$, and $1=\frac{U}{N_{\text{site}}}\sum_{k'}'E_{k'}^{-1}$.

5.6 Nonlinear σ-model

5.6.1 Nonlinear σ-model for the spin-density-wave state

- The nonlinear σ-model describes the dynamics of collective fluctuations around a symmetry-breaking ground state.
- The Hubbard–Stratonovich transformation.

The mean-field approach used in the last section fails to reproduce the gapless spin wave excitations in the SDW state. To obtain the spin wave excitations, we have to go beyond mean-field theory and use, for example, the RPA. However, here we will use the semiclassical approach and the related nonlinear σ-model to study the spin wave excitations. In the semiclassical approach, we need to first derive a low-energy effective Lagrangian for the spin waves.

In our variational approach, we have to include a B field in the 'variational' Hamiltonian to generate a trial wave function with a uniform anti-ferromagnetic spin polarization M. A natural way to include the spin waves is to replace B by a spatially-dependent $\boldsymbol{B}_i$ field in the 'variational' Hamiltonian, which generates a spatially-dependent anti-ferromagnetic spin polarization $\boldsymbol{M}_i$ in the new trial wave function. The energy for the new trial wave function is a functional of $\boldsymbol{M}_i$, i.e. $E[\boldsymbol{M}_i]$, and can be regarded as the energy for the spin waves. However, $E[\boldsymbol{M}_i]$ is the energy for the time-independent spin wave fluctuations and is only the potential energy of the spin wave. To obtain the effective Lagrangian, we also need the kinetic energy for time-dependent spin wave fluctuations. The easiest way to obtain the kinetic energy (as well as the potential energy) of the spin waves is to use the path integral approach developed in Section 5.4.1.

To obtain the low-energy effective theory for spin waves, we note that the Hubbard term $\frac{U}{2}c^\dagger_{\alpha i}c_{\alpha i}c^\dagger_{\beta i}c_{\beta i}$ is equal to 0 for $n_i = 0$, equal to $U/2$ for $n_i = 1$, and equal to $2U$ for $n_i = 2$. Also note that $\boldsymbol{S}_i^2$ is equal to 0 for $n_i = 0$, equal to $3/4$ for $n_i = 1$, and equal to 0 for $n_i = 2$, where

$$\boldsymbol{S}_i = \frac{1}{2}c^\dagger_i \boldsymbol{\sigma} c_i$$

is the total spin operator at site $\boldsymbol{i}$. Therefore

$$\frac{U}{2}c^\dagger_{\alpha i}c_{\alpha i}c^\dagger_{\beta i}c_{\beta i} = Un_i - \frac{2U}{3}\boldsymbol{S}_i^2$$

We will drop the Un_i term by absorbing it into the chemical potential. Thus, the Hubbard model is described by the following path integral:

$$Z = \int \mathcal{D}c^*\mathcal{D}c\ \mathrm{e}^{\mathrm{i}\int \mathrm{d}t\ (L_0+L_I)}$$

where

$$L_0 = \mathrm{i}\sum_i c^*_{\alpha i}\partial_t c_{\alpha i} - \sum_{\langle ij\rangle} -t(c^*_{\alpha i}c_{\alpha j} + c^*_{\alpha j}c_{\alpha i})$$

$$L_I = \frac{U}{6}\sum_i c_i^\dagger \boldsymbol{\sigma} c_i c_i^\dagger \boldsymbol{\sigma} c_i$$

The trick here is to write the interaction term as

$$\mathrm{e}^{\mathrm{i}\int \mathrm{d}t\ \frac{U}{6} c_i^\dagger \boldsymbol{\sigma} c_i c_i^\dagger \boldsymbol{\sigma} c_i} = \text{constant} \times \int \mathcal{D}[\boldsymbol{B}_i(t)]\ \mathrm{e}^{\mathrm{i}\int \mathrm{d}t\ (-)^i \boldsymbol{B}_i(t) c_i^\dagger \boldsymbol{\sigma} c_i - \frac{3}{2U}\boldsymbol{B}_i^2(t)}$$

which converts the fermion quartic term to a fermion quadratic term. Such a transformation is called a Hubbard–Stratonovich transformation. Now the partition function can be rewritten as follows:

$$Z = \int \mathcal{D}c^*\mathcal{D}c\mathcal{D}\boldsymbol{B}\ \mathrm{e}^{\mathrm{i}\int \mathrm{d}t\ L_B - \mathrm{i}\int \mathrm{d}t\ \frac{3}{2U}\sum_i \boldsymbol{B}_i^2} \tag{5.6.1}$$

$$L_B = \mathrm{i}\sum_i c^*_{\alpha i}\partial_t c_{\alpha i} - \sum_{\langle ij\rangle} -t(c^*_{\alpha i}c_{\alpha j} + c^*_{\alpha j}c_{\alpha i}) + \sum_i (-)^i \boldsymbol{B}_i(t) c_i^\dagger \boldsymbol{\sigma} c_i$$

where the fermion Lagrangian is quadratic.

If we integrate out the fermion first, then we obtain

$$Z = \int \mathcal{D}\boldsymbol{B}\ \mathrm{e}^{\mathrm{i}\int \mathrm{d}t\ L_{\text{eff}}(\boldsymbol{B})}$$

where $L_{\text{eff}}(\boldsymbol{B})$ is an effective Lagrangian. As $\boldsymbol{B}$ couples to the electron spin, it describes the spin wave fluctuations.

Let us first study some basic properties of $L_{\text{eff}}(\boldsymbol{B})$. Assuming that $\boldsymbol{B}$ is a constant in space–time, then

$$L_{\text{eff}}(\boldsymbol{B}) = -2{\sum_{\boldsymbol{k}}}'(-E_{\boldsymbol{k}}) - \frac{3}{2U}N_{\text{site}}\boldsymbol{B}^2 = -E_{\text{mean}}(\boldsymbol{B})$$

is the negative of the mean-field energy, where $E_{\boldsymbol{k}} = \sqrt{\xi_{\boldsymbol{k}}^2 + |\boldsymbol{B}|^2}$. Minimizing $E_{\text{mean}}(\boldsymbol{B})$, we obtain the gap equation

$$\frac{3}{2} - UN_{\text{site}}^{-1}{\sum_{\boldsymbol{k}}}'\frac{1}{E_{\boldsymbol{k}}} = 0 \tag{5.6.2}$$

which determines the mean-field value of $\boldsymbol{B}_0$. If $\boldsymbol{B}_0 \neq 0$, then the mean-field ground state is an SDW state. Let us assume that this is the case.

It is clear that the ground states associated with different spin orientations are degenerate. If we have a long-wavelength spin orientation fluctuation, then locally the system is just in one of the many possible ground states. This corresponds to a low-energy fluctuation. To study this kind of low-energy fluctuation, we introduce

$$\boldsymbol{B}_{\boldsymbol{i}}(t) = |\boldsymbol{B}_{\boldsymbol{i}}(t)|\boldsymbol{n}_{\boldsymbol{i}}(t), \qquad |\boldsymbol{n}_{\boldsymbol{i}}(t)| = 1$$

and replace $|\boldsymbol{B}_{\boldsymbol{i}}(t)|$ by $|\boldsymbol{B}_0|$. This gives us a low-energy effective Lagrangian $L_\sigma(\boldsymbol{n}) = L_{\text{eff}}(\boldsymbol{n}\boldsymbol{B}_0)$. In the long-wavelength limit, the effective Lagrangian can be written as

$$L_\sigma(\boldsymbol{n}) = \int \mathrm{d}^d\boldsymbol{x}\, \frac{1}{2g}\left((\partial_t \boldsymbol{n})^2 - v^2(\partial_{\boldsymbol{x}}\boldsymbol{n})^2\right) + \ldots \tag{5.6.3}$$

The terms indicated by '...' are higher-derivative terms. The spin–rotation symmetry forbids the possible potential term (or mass term) $V(\boldsymbol{n})$ and guarantees gapless spin fluctuations. Such an effective Lagrangian is called an $O(3)$ nonlinear σ-model.

We note that the effective Lagrangian is an expansion in powers of $\partial_\mu \boldsymbol{n}$. However, when $\boldsymbol{B} = \boldsymbol{n}B_0$, L_B in eqn (5.6.1) directly depends on $\boldsymbol{n}$. It is not obvious how to obtain an expansion in powers of $\partial_\mu \boldsymbol{n}$. In the following, we discuss a trick to do so. The trick allows us to directly calculate g and v in the nonlinear σ-model.

We introduce $\psi_{\boldsymbol{i}} = U_{\boldsymbol{i}} c_{\boldsymbol{i}}$, where

$$U_{\boldsymbol{i}} = \begin{pmatrix} z_{1\boldsymbol{i}}^* & z_{2\boldsymbol{i}}^* \\ -z_{2\boldsymbol{i}} & z_{1\boldsymbol{i}} \end{pmatrix}, \qquad \psi_{\boldsymbol{i}} = \begin{pmatrix} \psi_{1\boldsymbol{i}} \\ \psi_{2\boldsymbol{i}} \end{pmatrix}$$

$$z_{\boldsymbol{i}} = \begin{pmatrix} z_{1\boldsymbol{i}} \\ z_{2\boldsymbol{i}} \end{pmatrix}, \qquad z_{\boldsymbol{i}}^\dagger z_{\boldsymbol{i}} = 1, \qquad z_{\boldsymbol{i}}^\dagger \boldsymbol{\sigma} z_{\boldsymbol{i}} = \boldsymbol{n}_{\boldsymbol{i}}$$

Here $\psi_{1\boldsymbol{i}}$ corresponds to the spin in the $\boldsymbol{n}_{\boldsymbol{i}}$ direction and $\psi_{2\boldsymbol{i}}$ to the spin in the $-\boldsymbol{n}_{\boldsymbol{i}}$ direction. Now L_B can be rewritten as

$$L_B = \mathrm{i}\sum_{\boldsymbol{i}} \psi_{\boldsymbol{i}}^*(\partial_t + U_{\boldsymbol{i}}\partial_t U_{\boldsymbol{i}}^\dagger)\psi_{\boldsymbol{i}} - \sum_{\langle \boldsymbol{ij}\rangle} -t(\psi_{\boldsymbol{i}}^* u_{ij}\psi_{\boldsymbol{j}} + h.c.) + B_0\sum_{\boldsymbol{i}}(-)^{\boldsymbol{i}}\psi_{\boldsymbol{i}}^\dagger \sigma^z \psi_{\boldsymbol{i}}$$

where

$$u_{ij} = 1 + \begin{pmatrix} z_{\boldsymbol{i}}^\dagger z_{\boldsymbol{j}} - 1 & \frac{1}{2}(z_{\boldsymbol{i}} - z_{\boldsymbol{j}})^\dagger \mathrm{i}\sigma^y (z_{\boldsymbol{i}}^* + z_{\boldsymbol{j}}^*) \\ -\frac{1}{2}(z_{\boldsymbol{i}} - z_{\boldsymbol{j}})^\top \mathrm{i}\sigma^y (z_{\boldsymbol{i}} + z_{\boldsymbol{j}}) & z_{\boldsymbol{j}}^\dagger z_{\boldsymbol{i}} - 1 \end{pmatrix}$$

We see that the B_0 term is independent of $\boldsymbol{n}$ in the new basis. In the new basis, the L_B dependence on $\boldsymbol{n}$ is via the derivative of $\boldsymbol{n}$.

Consider small fluctuations around $\boldsymbol{n} = \hat{\boldsymbol{z}}$ as follows:

$$\boldsymbol{n_i} = \hat{\boldsymbol{z}} + \mathrm{Re}\phi_{\boldsymbol{i}}\hat{\boldsymbol{x}} + \mathrm{Im}\phi_{\boldsymbol{i}}\hat{\boldsymbol{y}} \tag{5.6.4}$$

and the corresponding spinor field

$$z_{\boldsymbol{i}} = \begin{pmatrix} 1 - |\phi_{\boldsymbol{i}}|^2/8 \\ \phi_{\boldsymbol{i}}/2 \end{pmatrix} + O(\phi_{\boldsymbol{i}}^3)$$

Assuming further that ϕ is real, u_{ij} can be simplified as follows:

$$u_{\boldsymbol{ij}} = 1 + \begin{pmatrix} -\frac{1}{8}(\phi_{\boldsymbol{i}} - \phi_{\boldsymbol{j}})^2 & \frac{1}{2}(\phi_{\boldsymbol{j}} - \phi_{\boldsymbol{i}}) \\ -\frac{1}{2}(\phi_{\boldsymbol{j}} - \phi_{\boldsymbol{i}}) & -\frac{1}{8}(\phi_{\boldsymbol{i}} - \phi_{\boldsymbol{j}})^2 \end{pmatrix} = \mathrm{e}^{\mathrm{i}\frac{\phi_{\boldsymbol{i}} - \phi_{\boldsymbol{j}}}{2}\sigma^y}$$

Also, $U_{\boldsymbol{i}}\partial_t U_{\boldsymbol{i}}^\dagger$ can be simplified to

$$U_{\boldsymbol{i}}\partial_t U_{\boldsymbol{i}}^\dagger = \begin{pmatrix} 0 & -\frac{1}{2}\partial_t\phi_{\boldsymbol{i}} \\ \frac{1}{2}\partial_t\phi_{\boldsymbol{i}} & 0 \end{pmatrix}$$

We note that the coupling of the fermions to the spin fluctuations takes the form of a coupling to an $SU(2)$ gauge field. For small spin fluctuations, L_B can be written as

$$L_B = \mathrm{i}\sum_{\boldsymbol{i}} \psi_{\boldsymbol{i}}^* \partial_t \psi_{\boldsymbol{i}} - H_0 - H_I$$

where

$$H_0 + H_I = -t\sum_{\langle \boldsymbol{ij}\rangle} (\psi_{\boldsymbol{i}}^* \mathrm{e}^{\mathrm{i}a_{\boldsymbol{ij}}}\psi_{\boldsymbol{j}} + h.c.) + B_0 \sum_{\boldsymbol{i}} (-)^{\boldsymbol{i}} \psi_{\boldsymbol{i}}^* \sigma^z \psi_{\boldsymbol{i}} + \sum_{\boldsymbol{i}} \psi_{\boldsymbol{i}}^* a_0(\boldsymbol{i})\psi_{\boldsymbol{i}},$$

$a_0(\boldsymbol{i}) = -\frac{1}{2}\partial_t\phi_{\boldsymbol{i}}\sigma^y$, and $a_{\boldsymbol{ij}} = \frac{1}{2}(\phi_{\boldsymbol{i}} - \phi_{\boldsymbol{j}})\sigma^y$.

If $a_0(\boldsymbol{i})$ and $a_{\boldsymbol{ij}}$ are constant in space, then the Hamiltonian $H_0 + H_I$ can be solved in momentum space. Let $a_0(\boldsymbol{i}) = -\frac{1}{2}B_y\sigma^y$ and $a_{\boldsymbol{ij}} = \frac{1}{2}\boldsymbol{a}\cdot(\boldsymbol{i} - \boldsymbol{j})\sigma^y$. Note that $B_y = \partial_t\phi$ and $\boldsymbol{a} = \partial_{\boldsymbol{i}}\phi_{\boldsymbol{i}}$. For small B_y and $\boldsymbol{a}$, the ground-state energy has the form

$$E_0(B_y, \boldsymbol{a}) = -C_1 B_y^2(\boldsymbol{i}) + C_2\boldsymbol{a}^2 \tag{5.6.5}$$

The $a_0^2(\boldsymbol{i})$ term is the $(\partial_t\boldsymbol{n})^2$ term in the effective Lagrangian eqn (5.6.3), i.e. $C_1 a_0^2(\boldsymbol{i}) = \int \mathrm{d}^d\boldsymbol{x}(\partial_t\phi)^2/2g$, and the $a_{\boldsymbol{ij}}^2$ term is the $(\partial_{\boldsymbol{x}}\boldsymbol{n})^2$ term, i.e. $C_2\boldsymbol{a}^2 =$

$\int d^d\boldsymbol{x} v^2(\partial_{\boldsymbol{x}}\phi)^2/2g$. This allows us to obtain g and v in L_σ. We find that

$$g^{-1} = -a^{-d} N_{\text{site}}^{-1} \left. \frac{\partial^2 E_0(B_y)}{\partial B_y^2} \right|_{B_y=0},$$

$$E_0(B_y) = N_{\text{site}}^{-1} \sum_{\boldsymbol{k}}{}' (-E^g_{+,\boldsymbol{k}} - E^g_{-,\boldsymbol{k}}), \tag{5.6.6}$$

$$E^g_{\pm,\boldsymbol{k}} = \sqrt{\left(\xi_{\boldsymbol{k}} \pm \frac{B_y}{2}\right)^2 + B_0^2}$$

and

$$\frac{v^2}{g} = a^{-d} \left. \frac{\partial^2 E_0(\boldsymbol{a})}{\partial \boldsymbol{a}^2} \right|_{\boldsymbol{a}=0},$$

$$E_0(\boldsymbol{a}) = N_{\text{site}}^{-1} \sum_{\boldsymbol{k}}{}' (-E^v_{+,\boldsymbol{k}} - E^v_{-,\boldsymbol{k}}), \tag{5.6.7}$$

$$E^v_{\pm,\boldsymbol{k}} = \left| \sqrt{\frac{1}{4}(\xi_{\boldsymbol{k}+\boldsymbol{a}/2} + \xi_{\boldsymbol{k}-\boldsymbol{a}/2})^2 + B_0^2} \pm \frac{1}{2}(\xi_{\boldsymbol{k}+\boldsymbol{a}/2} - \xi_{\boldsymbol{k}-\boldsymbol{a}/2}) \right|,$$

where a is the lattice constant and $\sum'_{\boldsymbol{k}}$ is the sum over the reduced Brillouin zone (see Fig. 5.20).

Problem 5.6.1.
Prove eqns (5.6.6) and (5.6.7).

5.6.2 Stability of long-range orders

- The nonlinear σ-model allows us to study quantum fluctuations and the stability of the SDW state.
- Stability and the topological term.

Classically, one of the ground states of the nonlinear σ-model (5.6.3) is $\boldsymbol{n}(\boldsymbol{x}) = \hat{z}$, which has a long-range spin order. The small-spin wave fluctuations are described by eqn (5.6.4) with a Lagrangian

$$\frac{1}{2g}\left(|\partial_t\phi|^2 - v^2|\partial_{\boldsymbol{x}}\phi|^2\right)$$

which has a linear dispersion $\epsilon_{\boldsymbol{k}} = v|\boldsymbol{k}|$, just like the fluctuations in the XY-model. The quantity $\langle|\phi(\boldsymbol{x},t) - \phi(0)|^2\rangle$ measures the strength of the long-range transverse spin fluctuations. If $\langle|\phi(\boldsymbol{x},t) - \phi(0)|^2\rangle \gg 1$, then the long-range order cannot exist. This problem is identical to the one in the XY-model. The strength of

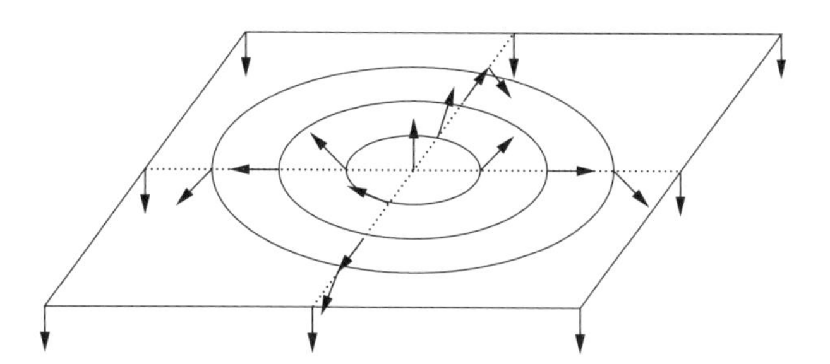

FIG. 5.21. The soliton/instanton solution in the $(1+1)$-dimensional $O(3)$ nonlinear σ-model.

the fluctuations is controlled by g. A small value of g leads to weak fluctuations. When g is too large, the long-range spin order will be destroyed by spin wave fluctuations. However, when $d \leqslant 1$ we find that $\langle |\phi(\boldsymbol{x},t) - \phi(0)|^2 \rangle$ diverges as $(\boldsymbol{x},t) \to \infty$, no matter how small g is. Therefore, there cannot be long-range order in $d \leqslant 1$ dimensions for any values of g. For the XY-model, we can still have algebraic long-range order when $d = 1$ under certain conditions. The question now is whether the $O(3)$ nonlinear σ-model has algebraic long-range order for $d = 1$.

To answer this question, let us consider imaginary time and investigate the following $(1+1)$-dimensional system:

$$\frac{1}{2g}\left(|\partial_\tau \boldsymbol{n}|^2 + v^2|\partial_x \boldsymbol{n}|^2\right)$$

The above action has a non-trivial stationary 'path', which is an instanton solution (see Fig. 5.21). The action is independent of the size of the instanton and is equal to $S_0 = 4\pi v/g$. This is because $S(\boldsymbol{n}(x,\tau)) = S(\boldsymbol{n}(bx, b\tau))$ for any $\boldsymbol{n}(x,\tau)$ and any scaling factor b.

As the action of the instanton is finite, we always have a finite (space–time) density of the instantons. The mean separation of the instantons is of order $l\,\mathrm{e}^{S_0/2}$, where l is a short-distance cut-off. The size of the instantons is also of order $l\,\mathrm{e}^{S_0/2}$ (because that action of the instantons is independent of their size). Thus, the long-range spin order is destroyed by the instantons. The spin–spin correlation is short-ranged, with a correlation length

$$\xi = l\,\mathrm{e}^{S_0/2} = l\,\mathrm{e}^{2\pi v/g} \tag{5.6.8}$$

It is interesting to see that ξ or $1/\xi$ is non-perturbative in g.

We see that the $O(3)$ nonlinear σ-model has a short-ranged correlation in $1+1$ dimensions, and all excitations are gapped. However, this result is only half correct. The $O(3)$ nonlinear σ-model can contain a topological term that changes the low-energy properties of the model.

To understand the origin of the topological term, let us consider the large-U limit of the Hubbard model (5.5.1). When $U \gg t$, the low-energy properties of the

Hubbard model (at half-filling) are described by the Heisenberg model:

$$H = \sum_{\langle ij \rangle} J \boldsymbol{S}_i \cdot \boldsymbol{S}_j \tag{5.6.9}$$

where $\boldsymbol{S}_i = c_i^\dagger \frac{\boldsymbol{\sigma}}{2} c_i$ is the spin operator and $J = 4t^2/U$. The effective Lagrangian of the Heisenberg model is

$$L = \mathrm{i} \sum_i z_i^\dagger \dot{z}_i - \frac{J}{4} \sum_{\langle ij \rangle} \boldsymbol{n}_i \cdot \boldsymbol{n}_j \tag{5.6.10}$$

where $z = \binom{z_1}{z_2}$ and $\boldsymbol{n} = z^\dagger \boldsymbol{\sigma} z$. Now let us assume that the ground state of the Heisenberg model is the anti-ferromagnetic (AF) state $\boldsymbol{n}_i = (-)^i \hat{\boldsymbol{z}}$, and we want to study the small fluctuations around the AF ground state.

The gapless spin wave fluctuations are described by

$$\boldsymbol{n}_i(t) = (-)^i \boldsymbol{n}(x^i, t)$$

It is very tempting to substitute the above into the Lagrangian (5.6.10) so as to obtain a continuum effective theory. However, this simple-minded approach does not lead to a correct result. For one thing, the resulting Lagrangian contains only a first-order time derivative term and no second-order derivative term. One may say that, at low energy, the first-order time derivative term is more important than the second-order time derivative term. It should be alright to drop the second-order time derivative term. However, it turns out that the first-order time derivative term is a total derivative, which has no effect on the equation of motion. We need to include the second-order time derivative term so that the equation of motion has non-trivial dynamics. The situation here is very similar to what we encountered in Section 3.3.3, where we derived the XY-model from an interacting boson model.

To obtain the second-order time derivative term, we consider the following more general fluctuations:

$$\boldsymbol{n}_i(t) = (-)^i \boldsymbol{n}(x^i, t) + \boldsymbol{m}(x^i, t) \tag{5.6.11}$$

where $\boldsymbol{n}(x, t)$ is a smooth function of (x, t) with $|\boldsymbol{n}(x, t)| = 1$, and $\boldsymbol{m}(x, t)$ is also a smooth function of (x, t) but with $|\boldsymbol{m}(x, t)| \ll 1$ and $\boldsymbol{m}(x, t) \cdot \boldsymbol{n}(x, t) = 0$. Here $\boldsymbol{n}$ describes the low-energy AF spin wave fluctuations. Also, $\boldsymbol{m}$ describes the ferromagnetic fluctuations which have high energies and which will be integrated out. As $\int \mathrm{d}t \, 2\mathrm{i} z_i^\dagger \dot{z}_i$ is the solid angle spanned by $\boldsymbol{n}_i$, we have

$$\int \mathrm{d}t \, 2\mathrm{i} z_i^\dagger \dot{z}_i = \int \mathrm{d}t \, 2\mathrm{i} (-)^i z(x^i, t)^\dagger \dot{z}(x^i, t) + \int \mathrm{d}t \, \boldsymbol{n} \cdot (\dot{\boldsymbol{n}} \times \boldsymbol{m})$$

where $\boldsymbol{n}(x,t) = z^\dagger(x,t)\boldsymbol{\sigma} z(x,t)$, or

$$\mathrm{i}\sum_i z_i^\dagger \dot{z}_i = \frac{1}{2}\int \mathrm{d}x \frac{1}{2}\boldsymbol{n}\cdot(\partial_t \boldsymbol{n} \times \partial_x \boldsymbol{n}) + \int \mathrm{d}x \frac{1}{2a}\boldsymbol{n}\cdot(\dot{\boldsymbol{n}} \times \boldsymbol{m})$$

where a is the lattice spacing. The $\frac{J}{4}\sum_{\langle ij\rangle} \boldsymbol{n}_i \cdot \boldsymbol{n}_j$ term induces a term $\boldsymbol{m}^2$. Thus, after integrating out $\boldsymbol{m}$, we obtain

$$S = \int \mathrm{d}t\mathrm{d}x \, \frac{1}{2g}[(\partial_t \boldsymbol{n})^2 - v^2(\partial_x \boldsymbol{n})^2] + \theta W \tag{5.6.12}$$

with $\theta = \pi$. Comparing this with the effective action that we calculated before, see eqn (5.6.3), the above contains an extra term θW with

$$W = \int \mathrm{d}t\mathrm{d}x \, \frac{1}{4\pi}\boldsymbol{n}\cdot(\partial_t \boldsymbol{n} \times \partial_x \boldsymbol{n}) \tag{5.6.13}$$

Such a term is a topological term, because $W(\boldsymbol{n}+\delta\boldsymbol{n}) - W(\boldsymbol{n}) = 0$ for any small variation $\delta\boldsymbol{n}$.[37] Therefore, $W(\boldsymbol{n})$ takes the same value for different $\boldsymbol{n}$s that can continuously deform into each other. Note that $\boldsymbol{n}$ is a mapping from a two-dimensional plane $\mathcal{R}^2$ to a sphere S^2, i.e. $\mathcal{R}^2 \to S^2$. The mapping does have different classes that cannot continuously deform into each other. These different classes are characterized by a winding number—the number of times the $\mathcal{R}^2$ wrap around S^2. Here $W(\boldsymbol{n})$ in eqn (5.6.13) taking only integer values is a mathematical expression of the winding number.

In contrast to g and v, the value of $\theta = \pi$ in (5.6.12) is exact and quantized. It does not receive corrections from high-order terms in the perturbative expansion. To see this, we note that the translation by one lattice space induces a transformation $\boldsymbol{n} \to -\boldsymbol{n}$ in our effective theory (5.6.12). Under such a transformation,

---

[37] We note that

$$W(\boldsymbol{n}+\delta\boldsymbol{n}) - W(\boldsymbol{n}) = \int \mathrm{d}t\,\mathrm{d}x \left[\frac{1}{4\pi}\boldsymbol{n}\cdot(\partial_t\delta\boldsymbol{n} \times \partial_x\boldsymbol{n}) + \frac{1}{4\pi}\boldsymbol{n}\cdot(\partial_t\boldsymbol{n} \times \partial_x\delta\boldsymbol{n})\right]$$

because $\boldsymbol{n}\cdot\delta\boldsymbol{n} = 0$ and $\partial_t\boldsymbol{n} \times \partial_x\boldsymbol{n}$ is parallel to $\boldsymbol{n}$. Also,

$$\begin{aligned}
\int \mathrm{d}t\,\mathrm{d}x\,\boldsymbol{n}\cdot(\partial_t\delta\boldsymbol{n} \times \partial_x\boldsymbol{n}) &= \int \mathrm{d}t\,\mathrm{d}x\,\partial_t\delta\boldsymbol{n}\cdot(\partial_x\boldsymbol{n} \times \boldsymbol{n}) \\
&= -\int \mathrm{d}t\,\mathrm{d}x\,[\delta\boldsymbol{n}\cdot(\partial_t\partial_x\boldsymbol{n} \times \boldsymbol{n}) + \delta\boldsymbol{n}\cdot(\partial_x\boldsymbol{n} \times \partial_t\boldsymbol{n})] \\
&= -\int \mathrm{d}t\,\mathrm{d}x\,\delta\boldsymbol{n}\cdot(\partial_t\partial_x\boldsymbol{n} \times \boldsymbol{n}),
\end{aligned}$$

and similarly

$$\int \mathrm{d}t\,\mathrm{d}x\,\boldsymbol{n}\cdot(\partial_t\boldsymbol{n} \times \partial_x\delta\boldsymbol{n}) = +\int \mathrm{d}t\,\mathrm{d}x\,\delta\boldsymbol{n}\cdot(\partial_t\partial_x\boldsymbol{n} \times \boldsymbol{n}).$$

S should be invariant. As $W \to -W$ under the transformation, only $\theta = 0$ and π are consistent with the symmetry of translation by one lattice space. Thus, as long as the underlying lattice Hamiltonian has the symmetry of translation by one lattice space, then θ can only take the two values 0 and π. Due to the quantization of θ, eqn (5.6.12) not only applies to the Heisenberg model (5.6.9), but it also applies to the AF state of the Hubbard model (5.5.1). We also note that, for the spin-S Heisenberg model, the AF state is also described by the $O(3)$ nonlinear σ-model, but now with $\theta = 2\pi S$. (If we explicitly break the symmetry, say by letting $J_{i,i+1} \neq J_{i+1,i+2}$, then the value of θ is no longer quantized.)

The above discussion about an instanton changing the algebraic long-range order to a short-range one is valid only when $\theta = 0 \bmod 2\pi$. Therefore, the Heisenberg model with integer spin has short-range correlations and a finite energy gap. The ground state is a spin singlet (for a chain with even sites). For the Heisenberg model with half-integer spin, $\theta = \pi$ and the above instanton discussion is not valid. In fact, the spin–spin correlation in this case has algebraic long-range order, i.e. $\langle \boldsymbol{S}_i \cdot \boldsymbol{S}_j \rangle \sim 1/|i-j|$. The ground state is still a spin singlet (for a chain with even sites), but now there are gapless excitations.

The winding number also provides a lower bound for the instanton action S_0, namely $S_0 \geqslant 4\pi v|W|/g$.[38] In fact, the instanton solutions saturate the bound and $S_0 = 4\pi v/g$.

5.6.3 Quantum numbers and low-energy excitations

- Calculation of the crystal momenta of low-energy excitations from the continuum nonlinear σ-model.

Let us assume that $d > 1$ and the system has a true long-range AF order. The low-energy excitations above the AF ground state are described by the $O(3)$ nonlinear σ-model

$$S = \int \mathrm{d}t\,\mathrm{d}x\, \frac{1}{2g}[(\partial_t \boldsymbol{n})^2 - v^2(\partial_x \boldsymbol{n})^2] \tag{5.6.14}$$

Here we want to use S to calculate the low-lying excitations for a finite system.

---

[38] Assume that $v = 1$. For a small square $\delta\tau \times \delta x$ with $\delta\tau = \delta x$, we have

$$\int_{\delta\tau\times\delta x} \mathrm{d}\tau\,\mathrm{d}x\,[(\partial_\tau \boldsymbol{n})^2 + (\partial_x \boldsymbol{n})^2] = (\delta_\tau \boldsymbol{n})^2 + (\delta_x \boldsymbol{n})^2$$

$$\geqslant 2|\boldsymbol{n}\cdot(\delta_\tau \boldsymbol{n} \times \delta_x \boldsymbol{n})| = 2\left|\int_{\delta\tau\times\delta x} \mathrm{d}\tau\,\mathrm{d}x\, \boldsymbol{n}\cdot(\partial_\tau \boldsymbol{n} \times \partial_x \boldsymbol{n})\right|.$$

From the winding number $(4\pi)^{-1}\int \mathrm{d}\tau\,\mathrm{d}x\,\boldsymbol{n}\cdot(\partial_\tau \boldsymbol{n}\times\partial_x \boldsymbol{n}) = \pm 1$, we find that $S_0 \geqslant 4\pi v|W|/g$ (we have put v back).

First, let us consider the uniform fluctuations, namely

$$\boldsymbol{n}(\boldsymbol{x}, t) = \boldsymbol{n}_0(t) \tag{5.6.15}$$

We have assumed that the lattice has an even number of sites in each direction, so that the above ansatz (5.6.15) can be consistent with the AF order. The effective action for $\boldsymbol{n}_0$ is

$$S_0 = \int \mathrm{d}t\, \frac{M}{2}\dot{\boldsymbol{n}}_0^2$$

with $M = \mathcal{V}/g$ and $\mathcal{V}$ being the volume of space. Here S_0 describes a particle of mass M moving on a unit sphere. The energy levels are labeled by the 'angular momentum', which is the total spin (S, S_z) in our system: $E_{S,S_z} = \frac{S(S+1)}{2M}$, $S_z = -S, -S+1, ..., S$. As the lattice has an even number of sites, S is always an integer. We note that E_{S,S_z} scales like $\mathcal{V}^{-1}$, which is much less than the energy of the lowest spin wave excitation, which scales like $\mathcal{V}^{-1/d}$ if $d > 1$.

Now let us consider the momentum for the excited state $|S, S_z\rangle$. Naively, one expects all $|S, S_z\rangle$ states to carry zero momentum because they come from a constant $\boldsymbol{n}_0$. However, $\boldsymbol{n}_0$ is invariant only under the translations that translate the even (odd) lattice points to even (odd) lattice points. Thus, $|S, S_z\rangle$ is only invariant under these translations. This restricts the momenta of $|S, S_z\rangle$ to 0 or $Q = (\pi, \pi, ..., \pi)$. Under the translation by one lattice space, $\boldsymbol{n}_0 \to -\boldsymbol{n}_0$. Thus, the states that are even under $\boldsymbol{n}_0 \to -\boldsymbol{n}_0$ carry momentum 0, and those that are odd under $\boldsymbol{n}_0 \to -\boldsymbol{n}_0$ carry momentum Q. We find that even-S states carry zero momentum and odd-S states carry momentum Q.

Problem 5.6.2.

Calculate the action S in eqn (5.6.12) from eqn (5.6.10) (i.e. determine the value of g and v from J and the lattice spacing a).

6

QUANTUM GAUGE THEORIES

Quantum field theories can be loosely divided into three classes, namely bosonic theory, fermionic theory, and gauge theory. We have discussed quantum field theories of bosons and fermions in the last few chapters. In this chapter, we will discuss gauge theories. Here we will introduce gauge theory formally. A more physical discussion will be given in Chapter 10.

One may wonder why there are three classes of quantum field theories. With which quantum field theories does nature choose to describe itself? According to the $U(1) \times SU(2) \times SU(3)$ standard model, nature chooses fermionic theory and gauge theory. So, gauge theory is important for high energy particle physics. But why does nature choose the more complicated fermionic theory and gauge theory, and skip the simpler bosonic theory? Chapter 10 provides an answer to the above questions. It turns out that we do not have to introduce fermionic theories and gauge theories. They can emerge as effective theories of a local bosonic system. As gauge theory is emergent, it is not surprising that gauge theories also appear in some condensed matter systems. In fact, more varieties of gauge theories could emerge from condensed matter systems than those offered by our vacuum.[39]

6.1 Simple gauge theories

6.1.1 Gauge 'symmetry' and gauge 'symmetry' breaking

- Gauge theory is a theory where we use more than one label to label the same quantum state.[40]
- Gauge 'symmetry' is not a symmetry and can never be broken.

When two different quantum states $|a\rangle$ and $|b\rangle$ (i.e. $\langle a|b\rangle = 0$) have the same properties, we say that there is a symmetry between $|a\rangle$ and $|b\rangle$. If we use two different labels 'a' and 'b' to label the same state, $|a\rangle = |b\rangle$, then $|a\rangle$ and $|b\rangle$ obviously have (or has) the same properties. In this case, we say that there is a

[39] This emergent picture also raises an interesting question: can new classes of quantum field theories, other than the fermionic theories and the gauge theories, emerge from bosonic models?

[40] This notion of gauge theory is quite unconventional, but true. See Section 10.7.4 for a discussion of the historic development of gauge theory.

gauge 'symmetry' between $|a\rangle$ and $|b\rangle$, and the theory about $|a\rangle$ and $|b\rangle$ is a gauge theory (at least formally). As $|a\rangle$ and $|b\rangle$, being the same state, always have (or has) the same properties, the gauge 'symmetry', by definition, can never be broken.

Usually, when the same 'thing' has the same properties, we do not say that there is a symmetry. Thus, the terms 'gauge symmetry' and 'gauge symmetry breaking' are two of the most misleading terms in theoretical physics. In the following, we will not use the above two confusing terms. We will say that there is a gauge structure (instead of a gauge 'symmetry') when we use many labels to label the same state. When we change our labeling scheme, we will say that there is a change of gauge structure (instead of gauge 'symmetry' breaking).

6.1.2 Gauge theory without a gauge field

- The concepts of gauge transformation, gauge-invariant states, and gauge-invariant operators.

The above simple example of gauge theory (containing only one state) is too simple. So, in this section, we will study a more complicated example. Consider a particle moving in a one-dimensional periodic potential, with

$$
\begin{aligned}
H &= \frac{1}{2m}p^2 + V\cos(2\pi x/a) \\
\mathcal{H} &= \{|x\rangle\}
\end{aligned}
\tag{6.1.1}
$$

where $\mathcal{H}$ is the Hilbert space for the possible states in our system. Here we stress that the Hamiltonian alone does not specify the system. It is the Hamiltonian *and* the Hilbert space, $(H, \mathcal{H})$, that specify the system.

The above system has a translational symmetry:

$$
T_a^\dagger H T_a = H, \qquad T_a|x\rangle = |x+a\rangle
$$

We can define a gauge theory by turning the symmetry into a gauge structure. The resulting gauge theory is given by

$$
\begin{aligned}
H &= \frac{1}{2m}p^2 + V\cos(2\pi x/a) \\
\mathcal{H}_a &= \{|\psi\rangle, T_a|\psi\rangle = |\psi\rangle\}
\end{aligned}
\tag{6.1.2}
$$

That is, we define our gauge theory by modifying the Hilbert space while keeping the 'same' Hamiltonian. The new Hilbert space is the invariant subspace under the symmetry transformation. As the Hamiltonian has the symmetry, it acts within the invariant subspace. We call the transformation T_a the *gauge transformation* and the states in the new Hilbert space the *gauge-invariant* states. Our gauge theory simply describes a particle on a circle.

A lot of concepts of gauge theory can be discussed using our 'more complicated' gauge theory. For a particle on a line, the states $|x\rangle$ and $|x+a\rangle$ are different states with the same properties. Thus, there is a translational symmetry. For a particle on a circle, $|x\rangle$ and $|x+a\rangle$ are the same state. Here x and $x+a$ are just two labels which label the same quantum state. Gauge transformations are just transformations between those labels which label the same state. Thus, a gauge transformation, by definition, is a 'do nothing' transformation. All of the physical states are, by definition, gauge invariant. If a state is not gauge invariant (i.e. in our example, a wave function $\psi(x)$ which is not periodic), then such a state does not belong to the physical Hilbert space (i.e. it cannot be a wave function for a particle on a circle). All of the physical operators must also be gauge invariant, i.e. $O = T_a^\dagger O T_a$. Otherwise, the operator will map a physical state into an unphysical state which is outside the physical Hilbert space. We see that it is very important to maintain gauge invariance in a gauge theory. Only gauge-invariant states and gauge-invariant operators are meaningful.

Another familiar gauge theory is the identical-particle system, where the exchange symmetry is changed into a gauge structure. The following two-particle system:

$$H = H(x^1) + H(x^2)$$
$$\mathcal{H} = \{|x^1, x^2\rangle\}$$

has an exchange symmetry $(x^1, x^2) \to (x^2, x^1)$. However, it is not an identical-particle system. If we turn the exchange symmetry into a gauge structure (i.e. by reducing the Hilbert space), then the new system is described by

$$H = H(x^1) + H(x^2)$$
$$\mathcal{H}_b = \{|x^1, x^2\rangle \,|\, |x^1, x^2\rangle = |x^2, x^1\rangle\}.$$

Such a system is an identical-particle system. Note that the identical-particle system does not have exchange *symmetry* because $|x^1, x^2\rangle$ and $|x^2, x^1\rangle$ are simply the same state. All of the special interference phenomena associated with identical particles come from the gauge structure.

6.2 Z_2 lattice gauge theory

In this section, we will study the simplest gauge theory with a gauge field—Z_2 gauge theory. The Z_2 gauge theory is the simplest topological field theory. It appears as an effective theory of several topologically-ordered quantum liquid states which we will discuss in Chapters 9 and 10. So, Z_2 gauge theory is very important for understanding the physical properties of these quantum liquid states.

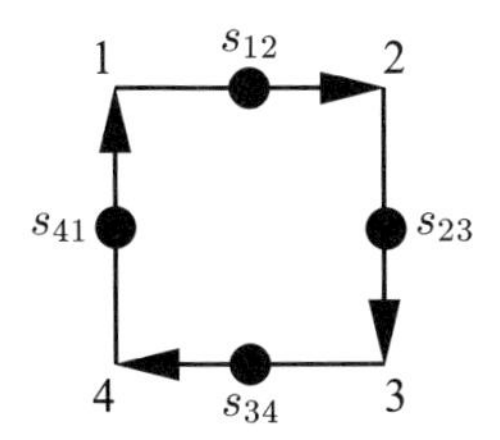

FIG. 6.1. A Z_2 gauge theory on a square.

Just like any quantum theory, to define a Z_2 gauge theory, we need to define the Hilbert space and the Hamiltonian. To understand the basic properties of the Z_2 gauge theory, we need to find the excitations spectrum. This is what we are going to do in the next few sections.

6.2.1 The Hilbert space

- The states in Z_2 gauge theory have a one-to-one correspondence with the gauge-equivalent classes of configurations.

To define a quantum Z_2 lattice gauge theory (Wegner, 1971), we need to first define its Hilbert space. To be concrete, let us consider a two-dimensional square lattice labeled by $\boldsymbol{i}$. On every nearest-neighbor link, we assign a link variable $s_{\boldsymbol{ij}} = s_{\boldsymbol{ji}}$ which can take the two values ± 1. The states in the Hilbert space are labeled by configurations of $s_{\boldsymbol{ij}}$, i.e. $|\{s_{\boldsymbol{ij}}\}\rangle$. If the labeling was one-to-one, then the Hilbert space would be a Hilbert space of a quantum Ising model (with spins on the links). To obtain the Hilbert space of the Z_2 gauge theory, the labeling is not one-to-one: two gauge-equivalent configurations label the same quantum state.

By definition, two configurations $s_{\boldsymbol{ij}}$ and $\tilde{s}_{\boldsymbol{ij}}$ are gauge equivalent if they are related by a Z_2 gauge transformation:

$$\tilde{s}_{\boldsymbol{ij}} = W_{\boldsymbol{i}} s_{\boldsymbol{ij}} W_{\boldsymbol{j}}^{-1} \tag{6.2.1}$$

where $W_{\boldsymbol{i}}$ is an arbitrary function with values ± 1. The gauge transformation defines an equivalence relation. If we group all of the gauge-equivalent configurations together to form a class, then such a class will be called a gauge-equivalent class. We see that the states in the Hilbert space have a one-to-one correspondence with the gauge-equivalent classes.

As an exercise, let us consider a Z_2 gauge theory on a square (see Fig. 6.1). We would like to find out the dimension of its Hilbert space. First, there are 4 links and hence $2^4 = 16$ different $s_{\boldsymbol{ij}}$ configurations. Second, there are 4 sites and hence $2^4 = 16$ different Z_2 gauge transformations. If the number of configurations in a gauge-equivalent class is equal to the number of different gauge transformations,

W_i has a site label which can take two values ± 1

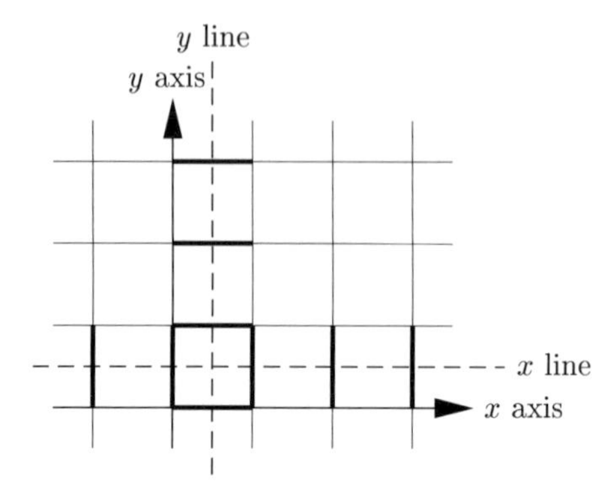

FIG. 6.2. The links cross the x line and/or the y line may receive an additional minus sign.

then there will be only one gauge-equivalent class. However, we note that the two gauge transformations

$$W_{\boldsymbol{i}} = 1, \qquad W_{\boldsymbol{i}} = -1$$

do not change the configurations $s_{\boldsymbol{ij}}$. The two gauge transformations form a group which will be called the invariant gauge group (IGG) (see also Section 9.4.2). As a result, the 16 gauge transformations only induce $16/2 = 8$ gauge-equivalent configurations. The factor of 2 is the number of elements in the IGG. As each gauge-equivalent class contains 8 configurations, there are $16/8 = 2$ gauge-equivalent classes, and hence two states in the Hilbert space.

To find a way to explicitly label the states in the Hilbert space, it is important to find gauge-invariant quantities that do not change under gauge transformations. One of the gauge-invariant quantities is the Wegner–Wilson loop variable (Wegner, 1971; Wilson, 1974), defined for a loop C as

$$U(C) = s_{\boldsymbol{ij}} s_{\boldsymbol{jk}} ... s_{\boldsymbol{li}}$$

where $\boldsymbol{i}$, $\boldsymbol{j}$, $\boldsymbol{k}$, ..., and $\boldsymbol{l}$ are sites on the loop. We will call $U(C)$ the Z_2 flux through the loop. The reader can easily check that $U(C)$ only takes the two values ± 1 and is invariant under the Z_2 gauge transformation. The two states in the Z_2 gauge theory on the square are labeled by $s_{12}s_{23}s_{34}s_{41} = \pm 1$.

Now let us count the number of states in the Z_2 gauge theory on a finite square lattice. We assume that the lattice has a periodic boundary condition in both directions (i.e. the lattice forms a torus). If the lattice has N_{site} sites, then it has $2N_{\text{site}}$ nearest-neighbor links. Thus, there are $2^{2N_{\text{site}}}$ different $s_{\boldsymbol{ij}}$ configurations. There are $2^{N_{\text{site}}}$ different gauge transformations and the IGG still has 2 elements. Thus, there are $\frac{2^{2N_{\text{site}}}}{2^{N_{\text{site}}}/2} = 2 \times 2^{N_{\text{site}}}$ different gauge-equivalent classes which correspond to $2 \times 2^{N_{\text{site}}}$ different states.

To find a way to label the $2 \times 2^{N_{\rm site}}$ states, we consider the Z_2 flux through a plaquette:

$$F_{\boldsymbol{i}} = s_{\boldsymbol{i},\boldsymbol{i}+\boldsymbol{x}} s_{\boldsymbol{i}+\boldsymbol{x},\boldsymbol{i}+\boldsymbol{x}+\boldsymbol{y}} s_{\boldsymbol{i}+\boldsymbol{x}+\boldsymbol{y},\boldsymbol{i}+\boldsymbol{y}} s_{\boldsymbol{i}+\boldsymbol{y},\boldsymbol{i}}$$

As there are N_{site} plaquettes and $F_{\boldsymbol{i}} = \pm 1$, one naively expects that different $\{F_{\boldsymbol{i}}\}$s can provide $2^{N_{\rm site}}$ different labels. However, the $F_{\boldsymbol{i}}$ are not independent. They satisfy

$$\prod_{\boldsymbol{i}} F_{\boldsymbol{i}} = 1 \tag{6.2.2}$$

Thus, the $\{F_{\boldsymbol{i}}\}$ only provide $2^{N_{\rm site}}/2$ different labels. Obviously, we cannot use $\{F_{\boldsymbol{i}}\}$ to label all of the $2 \times 2^{N_{\rm site}}$ states. In fact, each flux configuration $\{F_{\boldsymbol{i}}\}$ corresponds to four different states.

To see this, we consider the following four configurations obtained from one configuration $s^0_{\boldsymbol{ij}}$:

$$s^{(m,n)}_{\boldsymbol{ij}} = f^m_x(\boldsymbol{ij}) f^n_y(\boldsymbol{ij}) s^0_{\boldsymbol{ij}}, \qquad m,n = 0,1 \tag{6.2.3}$$

The functions $f_{x,y}(\boldsymbol{ij})$ take the values -1 or 1, with $f_x(\boldsymbol{ij}) = -1$ if the link $\boldsymbol{ij}$ crosses the x line (see Fig. 6.2) and $f_x(\boldsymbol{ij}) = 1$ otherwise. Similarly, $f_y(\boldsymbol{ij}) = -1$ if the link $\boldsymbol{ij}$ crosses the y line and $f_y(\boldsymbol{ij}) = 1$ otherwise. The four configurations give rise to the same Z_2 flux $F_{\boldsymbol{i}}$ through every plaquette. Despite this, the four configurations are not gauge equivalent (see Problem 6.2.2) because they correspond to inserting a different Z_2 flux through the two holes in the torus (see Fig. 9.6). Therefore, $2^{N_{\rm site}}/2$ different flux configurations $\{F_{\boldsymbol{i}}\}$ plus the four-fold degeneracy allow us to recover $2 \times 2^{N_{\rm site}}$ states.

Problem 6.2.1.
Prove eqn (6.2.2) for the Z_2 gauge theory on a torus.

Problem 6.2.2.
Show that the four configurations in eqn (6.2.3) are not gauge equivalent. (Hint: Consider $U(C)$ with C going all the way around the torus.)

Problem 6.2.3.
The Hilbert space $\mathcal{H}_{Z_2}$ of the Z_2 gauge theory can also be constructed from the Hilbert space $\mathcal{H}_{\text{spin}}$ of a spin-1/2 model which has one spin on each link of the square lattice. Here $\mathcal{H}_{\text{spin}}$ has a subspace $\mathcal{H}_{\text{sub}}$ which is formed by states that are invariant under all of the unitary transformations of the form $\prod_{\langle \boldsymbol{ij}\rangle} (\sigma^1_{\boldsymbol{ij}})^{G_i - G_j}$. Also, $G_{\boldsymbol{i}}$ is an arbitrary function on the sites taking the values $G_{\boldsymbol{i}} = 0, 1$, and $\sigma^1_{\boldsymbol{ij}}$ is the spin operator acting on the spin on the link $\langle \boldsymbol{ij}\rangle$. Show that $\mathcal{H}_{\text{sub}} = \mathcal{H}_{Z_2}$ and $\prod_{\langle \boldsymbol{ij}\rangle} (\sigma^1_{\boldsymbol{ij}})^{G_i - G_j}$ generates the Z_2 gauge transformation.

6.2.2 The Hamiltonian

- The Hamiltonian of a Z_2 gauge theory must be gauge invariant.

So far we have only discussed the Hilbert space of the Z_2 lattice gauge theory. What about the Hamiltonian? We cannot just pick any operator in the Z_2 Hilbert space as our Hamiltonian. The Hamiltonian should have some 'local' properties. One way to construct a local Hamiltonian is to write it as a local function of s_{ij}. However, because s_{ij} is a many-to-one labeling of the physical states, we have to make sure that the equivalent labels give rise to the same energy. That is, the Hamiltonian must be gauge invariant. A simple way to construct a gauge-invariant Hamiltonian is to write the Hamiltonian as a function of gauge-invariant operators, such as $U(C)$ and F_i. In addition to $U(C)$, the σ^1_{ij} operator that flips the sign of s_{ij} on a link σ^1_{ij}, i.e. $s_{ij} \to -s_{ij}$, is also a gauge-invariant operator (see Problem 6.2.4). Thus, a gauge-invariant local Hamiltonian for the Z_2 gauge theory can be written as

$$H_{Z_2} = -g\sum_i F_i - t\sum_{\langle ij\rangle} \sigma^1_{ij} \tag{6.2.4}$$

Problem 6.2.4.
Let $\hat{W}$ generate a gauge transformation $\hat{W}|\{s_{ij}\}\rangle = |\{\tilde{s}_{ij}\}\rangle$, where $\tilde{s}_{ij}$ and s_{ij} are related through eqn (6.2.1). An operator $\hat{O}$ is gauge invariant if $\hat{W}\hat{O}\hat{W}^{-1} = \hat{O}$. Show that σ^1_{ij} is a gauge-invariant operator.

6.2.3 The physical properties

- Despite the finite energy gap, a Z_2 gauge theory at low energies is not trivial. Its low-energy properties on a torus are characterized by the four-fold topologically-degenerate ground states.

When $t = 0$ and $g > 0$, H_{Z_2} has four degenerate ground states on a torus which are characterized by $F_i = +1$. The excitations are created by flipping the signs of F_i (see Fig. 6.3). The excitations behave like local particles and will be called the Z_2 vortices. Due to the constraint (6.2.2), the Z_2 vortices can only be created in pairs on a torus. The energy gap of these excitations is of order g. The t term in H_{Z_2} induces a hopping of the Z_2 vortices. As the σ^1_{ij} operators commute with each other, using the statistical algebra of the hopping operators, see eqn (4.1.10), we find that the Z_2 vortices are bosons.

We would like to point out that the four-fold degeneracy of the ground states is a so-called topological degeneracy. The topological degeneracy, by definition, is a degeneracy that cannot be lifted by any local perturbations of the Hamiltonian. The topological degeneracy is very important. It allows us to define topological orders in Chapter 8.

To understand the robustness of the four-fold degeneracy in the Z_2 gauge theory, let us treat the t term as a perturbation and see how the t term can lift the four-fold-degenerate ground states of the $t = 0$ Hamiltonian. The four degenerate ground states are given by eqn (6.2.3) with $s^0_{ij} = 1$. We can see that the only way

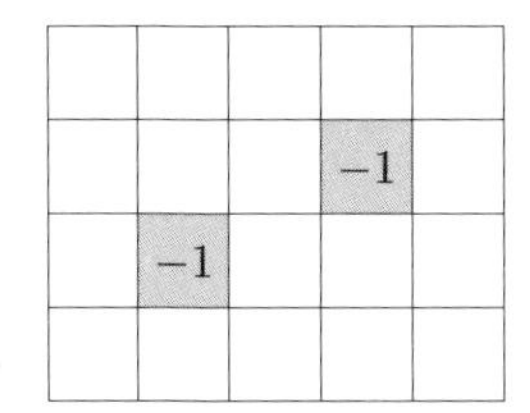

FIG. 6.3. A Z_2 vortex is created by changing F_i from $+1$ to -1.

to change one state $s_{ij}^{(m,n)}$ into a different state $s_{ij}^{(m',n')}$ is to flip the signs of s_{ij}^0 on a line of links all the way around the torus. If the torus is formed by an $L \times L$ lattice, then we need at least L of the σ_{ij}^1 operators to connect different degenerate states. This can happen only beyond the Lth order in perturbation theory. Thus, the t term can lift the four-fold degeneracy. However, the energy splitting ΔE is only of order t^L/g^{L-1}. In the thermodynamical limit, $L \to \infty$ and $\Delta E \to 0$. The above discussion does not depend on any symmetry. It is valid even when t is not uniform. The four-fold degeneracy (in the thermodynamical limit) is a robust property of a phase. The four-fold degeneracy reflects the low-energy dynamics of the Z_2 flux and is the most important characteristic of the Z_2 gauge theory.

When $g = 0$ and $t > 0$, the ground state of H_{Z_2} is given by

$$|\Psi_0\rangle = \sum_{\{s_{ij}\}} |\{s_{ij}\}\rangle \tag{6.2.5}$$

where the summation $\sum_{\{s_{ij}\}}$ is over all configurations $\{s_{ij}\}$. The ground state is non-degenerate and has an energy $-tN_l$, where N_l is the number of links. This result is easy to see if we treat H_{Z_2} as an Ising model with spins on the links. Here s_{ij} can be identified as the eigenvalues of the spin operator σ_{ij}^3 in the z direction, and σ_{ij}^1 is the spin operator in the x direction. In the Ising model picture, the $|\Psi_0\rangle$ state is simply a state with all spins pointing in the x direction.

From the above discussion, we see that H_{Z_2} has two phases. When $g \gg t$, the ground states have four-fold degeneracy and the excitations are Z_2 vortices. We will call this phase the Z_2 deconfined phase. The low-energy properties of this phase are described by a Z_2 gauge theory. When $g \ll t$, the ground state is not degenerate. We will call this phase the Z_2 confined phase. The low-energy properties have no characteristics of a Z_2 gauge theory.

If you feel that the definition of the Z_2 gauge theory is formal and the resulting Z_2 gauge theory is strange, then you get the point. The Z_2 gauge theory is actually a non-local theory, in the sense that its total Hilbert space cannot be expressed as a direct product of local Hilbert spaces. In Section 10.3, we will give a more physical description of Z_2 gauge theory. We will see that Z_2 gauge theory is actually a theory of closed strings.

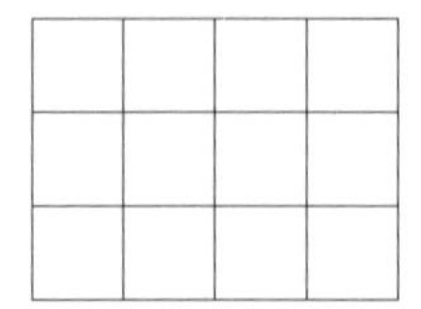

FIG. 6.4. An open two-dimensional square lattice.

Problem 6.2.5.
(a) A Z_2 gauge theory H_{Z_2} (see eqn (6.2.4)) is defined on an open square lattice of $L_x \times L_y$ sites (see Fig. 6.4). Assuming that $t = 0$, find the ground-state energy and degeneracy.
(b) Assume that the lattice is periodic in the y direction and forms a cylinder. Assuming that $t = 0$, find the ground-state energy and degeneracy.
(c) Assume that the lattice is periodic in the both the x and y directions. Assuming that $t = 0$ and $g < 0$, find the ground-state energy and degeneracy for the following three cases: (i) $L_x =$ even and $L_y =$ even; (ii) $L_x =$ even and $L_y =$ odd; and (iii) $L_x =$ odd and $L_y =$ odd.

Problem 6.2.6.
Show that eqn (6.2.5) is the ground state of H_{Z_2} when $g = 0$ and $t > 0$.

Problem 6.2.7.
Consider the two configurations s^1_{ij} and s^2_{ij} on a torus with $L \times L$ sites. The two configurations generate the same Z_2 flux F_i. Show that, if the two configurations are not gauge equivalent, then one needs to flip the signs of s_{ij} for at least L links to change s^1_{ij} to s^2_{ij} (or $s^1_{ij}s^2_{ij} = -1$ on more than L links).

Problem 6.2.8.
Note that the hopping of the Z_2 vortices is generated by the hopping operators $P\sigma^1_{ij}P$, $P\sigma^1_{ij}P\sigma^1_{jk}P$, etc., where P is a projection to a subspace with a fixed number of Z_2 vortices. Using the statistical algebra (4.1.10), show that the Z_2 vortices are bosons.

6.3 $U(1)$ gauge theory and the XY-model in $1+2$ dimensions

We constructed the Lagrangian of the $U(1)$ gauge theory in Section 3.7.1 and studied its classical equation of motion in Section 3.7.5. In this section, we are going to study the quantum $U(1)$ gauge theory. Instead of studying the $U(1)$ gauge theory by itself, we will study it together with the XY-model in $1+2$ dimensions. We will show that there is a duality relation between the XY-model and the $U(1)$ gauge theory. We will use this duality relation to study the instanton effect and confinement in $(1+2)$-dimensional $U(1)$ gauge theory.

6.3.1 Duality between $U(1)$ gauge theory and the XY-model in $1+2$ dimensions

- Gauge fixing and the Coulomb gauge.
- Quantization of the $U(1)$ gauge theory.
- Quantized charge, large gauge transformations, and compact $U(1)$ gauge theory.
- The quantized charge in the $U(1)$ gauge theory is the quantized vortex in the dual XY-model.

The XY-model is described by the path integral

$$Z_{XY} = \int \mathcal{D}\theta \; \mathrm{e}^{\mathrm{i}\int \mathrm{d}t\mathrm{d}^2\boldsymbol{x}\; \mathcal{L}_{XY}}$$

$$\mathcal{L}_{XY} = \frac{\chi}{2}\left(\dot{\theta}^2 - (\partial_{\boldsymbol{x}}\theta)^2\right) \tag{6.3.1}$$

The $U(1)$ gauge theory can also be described by the path integral

$$Z_{U(1)} = \int \mathcal{D}a_\mu \; \mathrm{e}^{\mathrm{i}\int \mathrm{d}t\mathrm{d}^2\boldsymbol{x}\; \mathcal{L}_{U(1)}}, \tag{6.3.2}$$

$$\mathcal{L}_{U(1)} = \frac{1}{2g^2}\left(\boldsymbol{e}^2 - b^2\right), \qquad e_i = \partial_0 a_i - \partial_i a_0, \qquad b = \partial_1 a_2 - \partial_2 a_1,$$

where $\boldsymbol{e}$ is the 'electric' field and b is the 'magnetic' field of a_μ. The two path integrals are so different that it is hard to see any relationship between the two theories. However, we will show below that the two path integrals describe the same physical system.

We note that both theories are quadratic and thus exactly soluble. One way to understand the relationship between the two theories is to quantize the two theories and find all of the low-energy excitations in the two theories.

Let us first try to quantize the $U(1)$ gauge theory, i.e. find the Hilbert space and the Hamiltonian of the $U(1)$ gauge theory. We note that $\mathcal{L}_{U(1)}$ is invariant under the transformation $a_\mu \to a_\mu + \partial_\mu f$:

$$\mathcal{L}_{U(1)}(a_\mu) = \mathcal{L}_{U(1)}(a_\mu + \partial_\mu f)$$

If we treat a_μ and $a_\mu + \partial_\mu f$ as two different paths, then the theory would have a symmetry. However, here we treat a_μ and $a_\mu + \partial_\mu f$ as two different labels of the *same* path. In this case, we get a gauge theory and $a_\mu \to a_\mu + \partial_\mu f$ defines the gauge structure.

To obtain a Hamiltonian description, we first try to find a one-to-one labeling of the paths. Consider a path labeled by a_μ. The gauge transformation tells us that $a'_\mu = a_\mu + \partial_\mu f$ labels the same path. Among so many different labels of the same path, we can always choose the one that satisfies

$$\boldsymbol{\partial} \cdot \boldsymbol{a} = 0 \tag{6.3.3}$$

by adjusting f. So a_μ that satisfies eqn (6.3.3) provides a one-to-one labeling of the different paths. Equation (6.3.3) is called a gauge-fixing condition. There are many different gauge-fixing conditions that can lead to a one-to-one labeling of the paths. The particular gauge-fixing condition (6.3.3) is called the Coulomb gauge.

In the Coulomb gauge, the part of the action that contains a_0 has the form $S = \int \mathrm{d}t\mathrm{d}^2\boldsymbol{x}\, \boldsymbol{e}^2 = \int \mathrm{d}t\mathrm{d}^2\boldsymbol{x}\, (\dot{a}_i^2 + \partial_i a_0 \partial_i a_0)$. The cross-term $\int \mathrm{d}t\mathrm{d}^2\boldsymbol{x}\, \partial_i a_0 \partial_0 a_i$ vanishes because $\partial_i a_i = 0$. We see that a_0 decouples from $\boldsymbol{a}$ and has no dynamics (i.e. no $\dot{a}_0$ terms). Thus, we can drop a_0 (by effectively setting $a_0 = 0$). So, in the Coulomb gauge, the path integral calculates the amplitude of the evolution from $a_i(\boldsymbol{x}, t_1)$ to $a_i(\boldsymbol{x}, t_2)$ that satisfies $\partial_i a_i(\boldsymbol{x}, t) = 0$, $t_1 \leqslant t \leqslant t_2$. Thus, the quantum state of the $U(1)$ gauge theory is described by a wave functional $\Psi[a_i(\boldsymbol{x})]$, where a_i satisfies $\partial_i a_i = 0$. This defines the Hilbert space for the physical states.

Next, we are going to find the eigenstates and their energies, by finding the gauge-fixed action. We first expand a_i as follows:

$$a_i = \frac{X_i(t)}{L_i} + b_0 x^1 \delta_{i,2} + \sum_{\boldsymbol{k}} \mathrm{i} c_{\boldsymbol{k}}(t) \epsilon_{ij} k_j \mathrm{e}^{\mathrm{i}\boldsymbol{k}\cdot\boldsymbol{x}} \tag{6.3.4}$$

where $c_{\boldsymbol{k}} = c^*_{-\boldsymbol{k}}$. The above a_i always satisfies $\partial_i a_i = 0$. Thus, the wave functions of the physical states are functions of b_0, X_i, and $c_{\boldsymbol{k}}$. The gauge-fixed Lagrangian takes the form

$$L = \frac{1}{2g^2}\left(-b_0^2 L_1 L_2 + \frac{L_2}{L_1}\dot{X}_1^2 + \frac{L_1}{L_2}\dot{X}_2^2 + \sum_{\boldsymbol{k}} (|\dot{c}_{\boldsymbol{k}}|^2 - \boldsymbol{k}^2 |c_{\boldsymbol{k}}|^2) L_1 L_2 \boldsymbol{k}^2 \right)$$

We see that the $U(1)$ gauge theory on a torus is described by a collection of oscillators $c_{\boldsymbol{k}}$, plus a free particle in two dimensions described by (X_1, X_2). The energies of the eigenstates are

$$E = \frac{1}{2g^2} b_0^2 L_1 L_2 + \frac{g^2 L_1}{2L_2} P_1^2 + \frac{g^2 L_2}{2L_1} P_2^2 + \sum_{\boldsymbol{k}} |\boldsymbol{k}| n_{\boldsymbol{k}}$$

where $n_{\boldsymbol{k}}$ are the occupation numbers of the oscillators and P_i are the momenta that conjugate to X_i.

There are two kinds of $U(1)$ gauge theory, named compact and non-compact $U(1)$ theories. The gauge transformations in the two $U(1)$ gauge theories are defined differently. As a result, the two theories have different Hilbert spaces and should be distinguished. For the non-compact $U(1)$ theory, $a_\mu \to a_\mu + \partial_\mu f$ is the only allowed gauge transformation. We will see later that the gauge charge is not quantized in the non-compact $U(1)$ theory.

The compact $U(1)$ gauge theory has quantized gauge charge. Only the Wegner–Wilson loop amplitude

$$O_C = \mathrm{e}^{-\mathrm{i}q \oint_C \mathrm{d}x^\mu a_\mu}$$

can be observed. Here q is the charge quantum. By choosing small loops with different orientations, one can show that O_C contains all of the components of the electric field $\boldsymbol{e}$ and the magnetic field b. As O_C is the only observable, any transformations of a_μ that make O_C invariant are gauge transformations. Such transformations have the form $a_\mu \to a_\mu + \partial_\mu f$, but now f may not be a single-valued function of space–time. In particular, the following f on the torus generates a valid gauge transformation:

$$f = \frac{2\pi N_1 x^1}{qL_1} + \frac{2\pi N_2 x^2}{qL_2} \tag{6.3.5}$$

For example, for a loop C that wraps around the torus in the x^1 direction, one can check directly that the Wegner–Wilson loop amplitude transforms as $O_C \to O_C \mathrm{e}^{-\mathrm{i}q \oint_C \mathrm{d}x^\mu \partial_\mu f} = O_C$ under the above gauge transformation.

In general, any multi-valued f that makes $\mathrm{e}^{\mathrm{i}qf}$ single valued will leave O_C invariant. These f will generate valid gauge transformations. We also note that a charge field with charge q transforms as

$$\phi \to \mathrm{e}^{\mathrm{i}qf} \phi.$$

A gauge transformation with a single-valued $\mathrm{e}^{\mathrm{i}qf}$ will leave the charge field single valued.

The gauge transformations in eqn (6.3.5) are called large gauge transformations. These transformations, namely $\mathrm{e}^{\mathrm{i}qf} = \exp(\mathrm{i}q(\frac{2\pi N_1 x^1}{qL_1} + \frac{2\pi N_2 x^2}{qL_2}))$, cannot be continuously deformed into the identity transformation $\mathrm{e}^{\mathrm{i}qf} = 1$. As x^1 moves from 0 to L_1, $\mathrm{e}^{\mathrm{i}qf}$ goes around the unit circle in the complex plane N_1 times. The winding number N_1 cannot be changed continuously.

We note that, under the gauge transformation (6.3.5),

$$X_i \to X_i + 2\pi q^{-1} N_i$$

Therefore, for the compact $U(1)$ theory, (X_1, X_2), $(X_1 + 2\pi q^{-1}, X_2)$, and $(X_1, X_2 + 2\pi q^{-1})$ label the same physical state (because they are related by gauge transformations). The free particle described by X_i lives on a torus of size $\frac{2\pi}{q} \times \frac{2\pi}{q}$. The momenta P_i are quantized as $P_i = n_i q$.

For the non-compact $U(1)$ theory, the large gauge transformations are not allowed. As a result, X_1 and X_2 have no periodic conditions. The free particle described by X_i lives on the plane.

We see that, although their Hamiltonians have the same form, the compact and non-compact $U(1)$ theories have different physical Hilbert spaces. We also see that

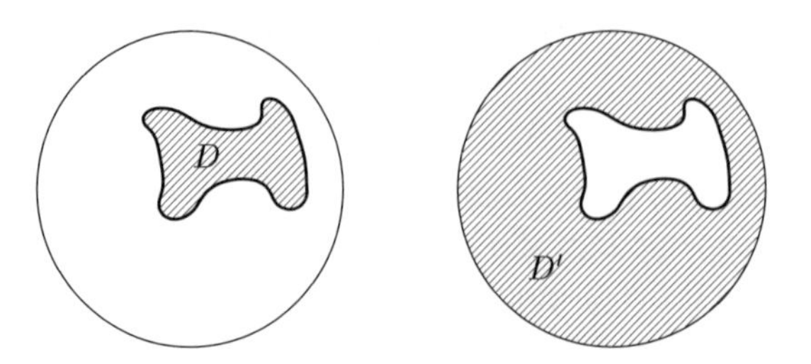

FIG. 6.5. A loop on a unit sphere can be the boundary of two different disks D and D'.

the $q \to 0$ limit of the compact $U(1)$ theory gives rise to the non-compact $U(1)$ theory. This is why we say that the charge is not quantized in the non-compact $U(1)$ theory.

As our system is defined on a torus, every physical quantity must be periodic in both the x^1 and x^2 directions. One may naively apply this requirement to a_i and require a_i to be periodic:

$$a_i(x^1, x^2) = a_i(x^1 + L_1, x^2) = a_i(x^1, x^2 + L_2).$$

However, we note that a_i itself is not physical. Different, but gauge-equivalent, a_i describe the same physical state. Thus, for a_i to describe a physical state on the torus, we only require that $a_i(x^1, x^2)$, $a_i(x^1 + L_1, x^2)$, and $a_i(x^1, x^2 + L_2)$ be gauge equivalent. This can happen only when the total magnetic flux is quantized (see Problem 6.3.1):

$$\int \mathrm{d}^2\boldsymbol{x}\, b = 2\pi N/q, \qquad N = \text{integer} \tag{6.3.6}$$

The above quantization of the total flux is necessary for the loop amplitude $\mathrm{e}^{-\mathrm{i}q\int_C \mathrm{d}x^\mu a_\mu}$ to make sense on a compact surface. This is because, for a loop C in a two-dimensional space S, $\mathrm{e}^{-\mathrm{i}q\int_C \mathrm{d}x^\mu a_\mu} = \mathrm{e}^{\mathrm{i}q\int_D \mathrm{d}^2\boldsymbol{x}\, b}$, where D is a two-dimensional disk with the loop C as its boundary. For a compact two-dimensional space, the disk is not unique (see Fig. 6.5). Two different disks D and D' with the same boundary differ by the total area S of the compact space. Hence $q\int_D \mathrm{d}^2\boldsymbol{x}\, b - q\int_{D'} \mathrm{d}^2\boldsymbol{x}\, b = q\int_S \mathrm{d}^2\boldsymbol{x}\, b$. Therefore, $q\int_S \mathrm{d}^2\boldsymbol{x}\, b/2\pi$ must be quantized as an integer in order for $\mathrm{e}^{-\mathrm{i}q\int_C \mathrm{d}x^\mu a_\mu}$ to be well defined. We also note that, for the non-compact $U(1)$ theory, $q = 0$ and $\int_S \mathrm{d}^2\boldsymbol{x}\, b$ must vanish.

Now the total energy of the compact $U(1)$ gauge theory can be rewritten as

$$E = \frac{(2\pi)^2}{2g^2q^2L_1L_2}N + \frac{g^2q^2L_1}{2L_2}n_1^2 + \frac{g^2q^2L_2}{2L_1}n_2^2 + \sum_{\boldsymbol{k}} |\boldsymbol{k}| n_{\boldsymbol{k}} \tag{6.3.7}$$

All of the eigenstates of the $U(1)$ gauge theory on a torus are labeled by the set of integers $(N, n_i, n_{\boldsymbol{k}})$. In this way, we have solved the compact $U(1)$ gauge theory on a torus.

Next let us turn to the XY-model and its eigenstates. We expand

$$\theta(t, \boldsymbol{x}) = \theta_0(t) + \frac{2\pi}{L_i} m_i x^i + \sum_{\boldsymbol{k}} \lambda_{\boldsymbol{k}}(t) \mathrm{e}^{\,\mathrm{i}\boldsymbol{k}\cdot\boldsymbol{x}}.$$

Here m_1 and m_2 are the winding number of θ in the x^1 and x^2 directions, respectively. The XY-model Lagrangian now takes the form

$$L = \frac{\chi}{2}\left(L_1 L_2 \dot{\theta}_0^2 - \frac{(2\pi)^2 L_2}{L_1} m_1^2 - \frac{(2\pi)^2 L_1}{L_2} m_2^2 + L_1 L_2 \sum_{\boldsymbol{k}} (|\dot{\lambda}_{\boldsymbol{k}}|^2 - \boldsymbol{k}^2 |\lambda_{\boldsymbol{k}}|^2) \right)$$

We see that the XY-model is described by a collection of oscillators (described by $\lambda_{\boldsymbol{k}}$) and a free particle on a circle (described by θ_0). The eigenstates of the XY-model are labeled by $(N, m_i, n_{\boldsymbol{k}})$ with energy

$$E = \frac{1}{2\chi L_1 L_2} N^2 + \frac{\chi(2\pi)^2 L_2}{2L_1} m_1^2 + \frac{\chi(2\pi)^2 L_1}{2L_2} m_2^2 + \sum_{\boldsymbol{k}} |\boldsymbol{k}| n_{\boldsymbol{k}} \tag{6.3.8}$$

We see that the XY-model and the $U(1)$ gauge theory have identical spectra if

$$\chi = \left(\frac{qg}{2\pi}\right)^2$$

and if we identify n_i as $\epsilon_{ij} m_j$ in the labeling of the eigenstates in the $U(1)$ theory and the XY-model. The eigenstates also carry definite total momenta $\boldsymbol{K}$, where

$$K_i = N m_i \frac{2\pi}{L_i} + \sum_{\boldsymbol{k}} k_i n_{\boldsymbol{k}} = N \epsilon_{ij} n_j \frac{2\pi}{L_i} + \sum_{\boldsymbol{k}} k_i n_{\boldsymbol{k}}$$

for the XY-model states $|N, m_i, n_{\boldsymbol{k}}\rangle$ or the $U(1)$ theory states $|N, n_i, n_{\boldsymbol{k}}\rangle$.

From the XY-model, we see that the label N has a physical meaning: it is the total number of bosons minus the number of bosons at equilibrium, namely $N = N_{tot} - N_0$. Therefore, $\frac{1}{\chi}\dot{\theta}$ and $\frac{q}{2\pi} b$ can be regarded as the boson-number density minus the equilibrium density:

$$\rho - \rho_0 = \frac{q}{2\pi} b = \frac{1}{\chi}\dot{\theta}$$

The constraint $\partial_0 b - \epsilon_{ij}\partial_i e_j = 0$ in the $U(1)$ theory can be regarded as a continuous equation for the conserved current, namely $\partial_0 \rho + \partial_i j_i = 0$. Thus, if $\frac{q}{2\pi} b$ is identified as the boson-number density, then the boson-number current density is given by

$$j_i = -\frac{q}{2\pi} \epsilon_{ij} e_j$$

in the $U(1)$ theory.

The equation of motion for the XY-model, namely $\partial_0^2\theta - \partial_{\boldsymbol{x}}^2\theta = 0$, can also be regarded as a continuous equation for the conserved current. We find that the same current is given by

$$j_i = -\frac{1}{\chi}\partial_i\theta$$

in the XY-model. We discover a simple relationship between fields in the $U(1)$ theory and the XY-model:

$$\frac{q}{2\pi}b = \frac{1}{\chi}\dot{\theta}, \qquad \frac{q}{2\pi}\epsilon_{ij}e_j = \frac{1}{\chi}\partial_i\theta. \tag{6.3.9}$$

In fact, substituting eqn (6.3.9) into eqn (6.3.2) will convert the $U(1)$ gauge Lagrangian into the XY-model Lagrangian (6.3.1). This is the simplest way to see the duality relation between the $U(1)$ gauge theory and the XY-model in $2+1$ dimensions.

The $U(1)$ gauge theory can be coupled to charges. Let J_μ be the density and the current density of the $U(1)$ charge. Then the coupled Lagrangian takes the form

$$\mathcal{L} = \frac{1}{2g^2}(\boldsymbol{e}^2 - b^2) - J_\mu a_\mu$$

For a point charge at $\boldsymbol{x} = 0$ we have $J_0 = q\delta(\boldsymbol{x})$ and we find, from the equation of motion, that

$$\frac{1}{g^2}\boldsymbol{\partial}\cdot\boldsymbol{e} = q\delta(\boldsymbol{x})$$

or

$$\boldsymbol{e} = \frac{g^2 q\boldsymbol{x}}{2\pi\boldsymbol{x}^2}$$

The above $\boldsymbol{e}$ corresponds to a circular flow in the XY-model:

$$\partial_i\theta = \frac{\chi q}{2\pi}\frac{g^2 q}{2\pi\boldsymbol{x}^2}\epsilon_{ij}x^j = \frac{\epsilon_{ij}x^j}{\boldsymbol{x}^2}$$

We see that a quantized charge in the $U(1)$ gauge theory is just the quantized vortex in the XY-model.

Problem 6.3.1.
Show eqn (6.3.6). (Hint: You may find the expansion (6.3.4) useful.)

Problem 6.3.2.
Quantize the compact $U(1)$ gauge theory in $1+1$ dimensions, assuming that the space is a ring of length L and the charge quantum is q. Find the set of integers that label all of the energy eigenstates. Find the energies for these eigenstates.

6.3.2 Confinement of the compact $U(1)$ gauge theory in $1+2$ dimensions

- The instanton effect in the $(1+2)$-dimensional $U(1)$ gauge theory gives the gauge boson a finite energy gap and causes a confinement between the $U(1)$ charges.
- The instanton effect can be described by a $\cos(\theta)$ term in the dual XY-model.

Here we consider the $U(1)$ gauge theory in imaginary time:

$$\mathcal{L}_{U(1)} = \frac{1}{2g^2}\left(\boldsymbol{e}^2 + b^2\right) \tag{6.3.10}$$

Formally, the $U(1)$ gauge theory appears to be a free theory with gapless excitations. However, it is not that simple. For the compact $U(1)$ gauge theory with a finite cut-off scale, the theory contains instantons. The instanton effect will make the $U(1)$ gauge theory an interacting theory. We will see that the interaction will affect the low-energy properties of the $U(1)$ gauge theory in a drastic way.

What is the instanton? The instanton at $x^\mu = 0$ is described by the following configuration:

$$b(x^\mu) = \frac{1}{2q}\frac{x^0}{|x|^3}, \qquad e_i(x^\mu) = \frac{1}{2q}\frac{x^i}{|x|^3}$$

The above instanton is designed to change the flux by $2\pi/q$, i.e.

$$\int_{x^0>0} \mathrm{d}^2\boldsymbol{x}\, b - \int_{x^0<0} \mathrm{d}^2\boldsymbol{x}\, b = \frac{2\pi}{q}$$

This is the smallest allowed instanton that is consistent with the charge quantum q. In the presence of a finite cut-off, the path integral should not only include smooth fluctuations of the gauge field, but it should also include instanton fluctuations.

When instanton effects are included, the $U(1)$ gauge theory is no longer a free theory. The problem now is how to understand the low-energy properties of the compact $U(1)$ theory with instantons. One approach is to use the duality between the XY-model and the $U(1)$ gauge theory.

For imaginary time, θ and a_μ are related by

$$\frac{q}{2\pi}b = \frac{1}{\chi}\dot{\theta}, \qquad \frac{q}{2\pi}\epsilon_{ij}e_j = -\frac{1}{\chi}\partial_i\theta \tag{6.3.11}$$

which can be written as

$$\frac{q}{2\pi}\epsilon_{\mu\nu\lambda}\partial_\nu a_\lambda = \frac{1}{\chi}\partial_\mu\theta \tag{6.3.12}$$

After substituting eqn (6.3.12) into eqn (6.3.10), we obtain

$$\mathcal{L} = \frac{\chi}{2}(\partial_\mu\theta)^2$$

with $\chi = \left(\frac{qg}{2\pi}\right)^2$. We note that the instanton creates $2\pi/q$ amount of flux. As we have seen, such an amount of flux corresponds to a single particle in the XY-model. Thus, an instanton creates or annihilates a single particle. In the XY-model, it is the operator $e^{\pm i\theta}$ that creates or annihilates a single particle. We find that an instanton at x^μ corresponds to $e^{i\theta(x^\mu)}$. As both the instanton and the anti-instanton appear with equal weight in the path integral, we can include the instanton effect in the XY-model by adding a term $e^{i\theta} + e^{-i\theta} = 2\cos(\theta)$ as follows:

$$\mathcal{L} = \frac{\chi}{2}(\partial_\mu \theta)^2 - K\cos(\theta)$$

When χ is large, the fluctuations of θ around $\theta = 0$ are small. We may approximate $\mathcal{L}$ as

$$\mathcal{L} = \frac{\chi}{2}(\partial_\mu \theta)^2 + \frac{1}{2}K\theta^2$$

The correlations between $\partial_\mu \theta$ or (b, e_i) become short-ranged after including the instanton effect. We see that the instanton effect opens up an energy gap of the $U(1)$ gauge field in $1 + 2$ dimensions. Here we would like to stress that the $K\cos(\theta)$ term is a relevant perturbation and the $U(1)$ gauge boson will gain an energy gap, no matter how small K is.

If the gauge boson is gapless, then the charged particles will interact through a logarithmic potential in $1 + 2$ dimensions. The potential between a particle of charge q and a particle of charge $-q$ is given by

$$\pi \frac{g^2 q^2}{(2\pi)^2} \ln r = \pi\chi \ln r$$

After including the instanton effect, the gauge field gains an energy gap. But how does the instanton effect affect the long-range interaction between charged particles? We note that a particle of charge q is described by a vortex in the XY-model. In the presence of the instanton effect, that is, in the presence of the $K\cos(\theta)$ term, the potential between a vortex and an anti-vortex grows linearly with the separation of the two vortices (see Fig. 6.6). Thus, the instanton effect changes the logarithmic potential to a linear potential between charges.

We have shown that, in $1 + 2$ dimensions, the XY-model can be described by the compact $U(1)$ gauge theory. We have also shown that the instanton effect in the $U(1)$ gauge theory opens up an energy gap. These two results seem to contradict each other. How can the XY-model with its gapless Nambu–Goldstone mode be equivalent to the $U(1)$ gauge theory, which has an energy gap? However, we realize that an instanton creates flux. As the flux corresponds to particle numbers, the existence of the instanton implies that the particle number in the corresponding XY-model is not conserved. So everything is consistent. In the presence of the instanton, the particle number is not conserved. The compact $U(1)$ gauge theory

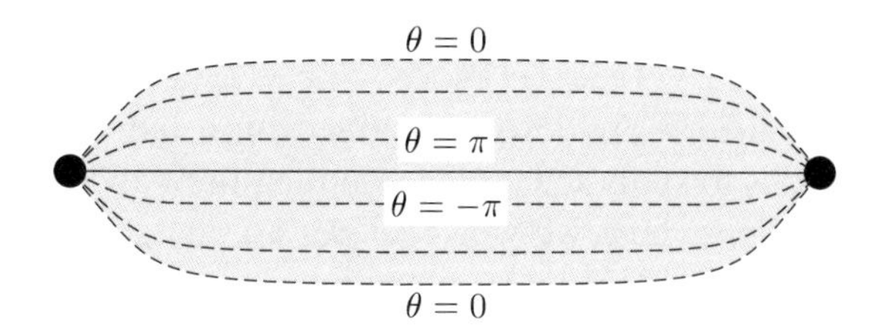

FIG. 6.6. For a vortex–anti-vortex pair in the XY-model separated by a distance l, the θ field must jump by 2π across the line connecting the two vortices. In the presence of the $-K\cos(\theta)$ term, the usual vortex ansatz $\theta(x, y) = \arctan(x/y)$ costs a huge energy which grows like l^2. To minimize the energy, the nonzero θ is confined in the shaded region. (The dashed lines are the equal-value lines for the θ field.) This leads to a linear confining potential between the two vortices.

describes an XY-model that does not have the $U(1)$ symmetry. This explains why the instanton effect corresponds to a potential term $K\cos(\theta)$ in the XY-model. On the other hand, if the XY-model has the $U(1)$ symmetry and conserves particle number, then the instanton is not allowed in the corresponding $U(1)$ gauge theory.

Problem 6.3.3.
Duality between the $(1+2)$-dimensional $U(1)$ gauge theory with instantons and the XY-model with the $\cos(\theta)$ term

1. Calculate the partition function $Z(x^\mu, y^\mu)$ of the $U(1)$ gauge theory with an instanton at x^μ and an anti-instanton at y^μ, up to an overall constant factor.

2. Use the above result to write down an expression for the partition function for the $U(1)$ gauge theory, including contributions from a multi-instanton gas.

3. Calculate the following partition function $Z(x^\mu, y^\mu)$ of the XY-model with an insertion $\mathrm{e}^{\mathrm{i}\theta}$ at x^μ and another insertion $\mathrm{e}^{-\mathrm{i}\theta}$ at y^μ, up to an overall constant factor:

$$Z(x^\mu, y^\mu) = \int \mathcal{D}\theta\ \mathrm{e}^{\mathrm{i}\theta(x^\mu)}\mathrm{e}^{-\mathrm{i}\theta(y^\mu)}\mathrm{e}^{-\int \mathrm{d}^3x^\mu\ \mathcal{L}_{XY}(\theta)}$$

4. Show that the $U(1)$ partition function with instantons agrees with the XY-model partition function with an added $K\cos(\theta)$ term (after expanding in powers of K).

6.4 The quantum $U(1)$ gauge theory on a lattice

6.4.1 The Lagrangian of a lattice $U(1)$ gauge theory

- A lattice $U(1)$ gauge theory is described by variables on the links a_{ij} and on the sites $a_0(\boldsymbol{i})$.
- The compact and non-compact lattice $U(1)$ gauge theories.

• The flux of the electric field e_{ij}.

The $U(1)$ gauge theory can also be defined on a lattice. The $U(1)$ lattice gauge theory allows us to study the strong-coupling limit and confinement in any dimension. For simplicity, here we will consider a $U(1)$ gauge theory on a two-dimensional square lattice. The scalar potential $a_0(\boldsymbol{x})$ is defined at each lattice site $\boldsymbol{i}$ and becomes $a_0(\boldsymbol{i})$. The vector potential $\boldsymbol{a}(\boldsymbol{x})$ is defined on each link and becomes $a_{\boldsymbol{ij}}$. We note that $a_{\boldsymbol{ij}}$ and $a_{\boldsymbol{ji}}$ are not independent; they are related through $a_{\boldsymbol{ij}} = -a_{\boldsymbol{ji}}$. The continuous fields are related to the lattice fields as follows:

$$a_0(\boldsymbol{j}) = a_0(\boldsymbol{x})|_{\boldsymbol{x}=l\boldsymbol{j}}, \qquad a_{\boldsymbol{j}_1\boldsymbol{j}_2} = \int_{l\boldsymbol{j}_1}^{l\boldsymbol{j}_2} \mathrm{d}x^i a_i(\boldsymbol{x}) \tag{6.4.1}$$

where l is the lattice constant. The Lagrangian that describes the $U(1)$ lattice gauge theory is the following function of $a_0(\boldsymbol{i})$ and $a_{\boldsymbol{ij}}$:

$$L = \frac{1}{4J} \sum_{\boldsymbol{i},\boldsymbol{\mu}=\boldsymbol{x},\boldsymbol{y}} [\dot{a}_{\boldsymbol{i},\boldsymbol{i}+\boldsymbol{\mu}} + a_0(\boldsymbol{i}) - a_0(\boldsymbol{i}+\boldsymbol{\mu})]^2 - \frac{g}{2}\sum_{\boldsymbol{p}} \Phi_{\boldsymbol{p}}^2 \tag{6.4.2}$$

where $\Phi_{\boldsymbol{p}}$ is the $U(1)$ flux through the square labeled by $\boldsymbol{p}$:

$$\Phi_{\boldsymbol{p}} = a_{\boldsymbol{i},\boldsymbol{i}+\boldsymbol{x}} + a_{\boldsymbol{i}+\boldsymbol{x},\boldsymbol{i}+\boldsymbol{x}+\boldsymbol{y}} + a_{\boldsymbol{i}+\boldsymbol{x}+\boldsymbol{y},\boldsymbol{i}+\boldsymbol{y}} + a_{\boldsymbol{i}+\boldsymbol{y},\boldsymbol{i}}. \tag{6.4.3}$$

One can check directly that the above Lagrangian is invariant under the lattice $U(1)$ gauge transformation

$$a_{\boldsymbol{ij}}(t) \to a_{\boldsymbol{ij}}(t) + \phi_{\boldsymbol{j}}(t) - \phi_{\boldsymbol{i}}(t), \qquad a_0(\boldsymbol{i},t) \to a_0(\boldsymbol{i},t) + \dot{\phi}_{\boldsymbol{i}}(t) \tag{6.4.4}$$

Equation (6.4.2) defines a non-compact lattice $U(1)$ gauge theory. A compact lattice $U(1)$ gauge theory is defined by regarding $u_{\boldsymbol{ij}} = \mathrm{e}^{\mathrm{i}a_{\boldsymbol{ij}}}$ as link variables. In the compact lattice $U(1)$ gauge theory, $a_{\boldsymbol{ij}}$ and $a'_{\boldsymbol{ij}} = a_{\boldsymbol{ij}} + 2\pi$ are regarded as being equivalent. The Lagrangian for the compact lattice $U(1)$ gauge theory has the form

$$L = \frac{1}{4J} \sum_{\boldsymbol{i},\boldsymbol{\mu}=\boldsymbol{x},\boldsymbol{y}} [\dot{a}_{\boldsymbol{i},\boldsymbol{i}+\boldsymbol{\mu}} + a_0(\boldsymbol{i}) - a_0(\boldsymbol{i}+\boldsymbol{\mu})]^2 + g\sum_{\boldsymbol{p}} \cos(\Phi_{\boldsymbol{p}}) \tag{6.4.5}$$

which is modified from eqn (6.4.2) to be consistent with the periodic condition $a_{\boldsymbol{ij}} \sim a_{\boldsymbol{ij}} + 2\pi$. In the rest of this section, we will consider only the compact $U(1)$ lattice gauge theory.

The classical equation of motion is obtained from $\delta L/\delta a_{ij} = 0$ and $\delta L/\delta a_0(i) = 0$. We find that

$$\frac{1}{2J}\frac{\mathrm{d}e_{i,i+\mu}}{\mathrm{d}t} + g[\sin(\Phi_{p_1}) - \sin(\Phi_{p_2})] = 0, \tag{6.4.6}$$

$$\sum_{\mu=\pm x,\pm y} e_{i,i+\mu} = 0, \qquad e_{ij} = \dot{a}_{ij} + a_0(i) - a_0(j) \tag{6.4.7}$$

where p_1 and p_2 label the two squares on the two sides of the link $(i, i+\mu)$.

By comparing e_{ij} with the e in the continuum model, we see that e_{ij} can be interpreted as the flux of the electric field flowing through the link (i, j). Equation (6.4.7) states that the electric flux is conserved. It corresponds to Gauss's law $\partial_x \cdot e = 0$ in the continuum model. Also, being the flux through the square S_p, Φ_p corresponds to $\Phi_p = \int_{S_p} \mathrm{d}^2 x b$ in the continuum model.

The simplest way to obtain the dynamical properties of the lattice $U(1)$ gauge theory described by eqn (6.4.2) or eqn (6.4.5) is to take the continuum limit. Assuming that a_μ is small and is a smooth function of t and x, we substitute the relation (6.4.1) between the continuous variables and the lattice variables into the lattice Lagrangians. We obtain the following standard $U(1)$ Lagrangian:

$$\mathcal{L} = \frac{1}{8\pi\alpha_{2D}}\left(\frac{1}{c}e^2 - cb^2\right) \tag{6.4.8}$$

From the resulting Maxwell equation, we find that the lattice $U(1)$ gauge theory contains a gapless mode with velocity c. The coupling constant α_{2D} has dimension $1/\text{length}$ in $1+2$ dimensions.

In Section 3.7.1, we used the gauge invariance to construct a coupled theory of charged bosons and the $U(1)$ gauge field in a continuum. The same construction applies to the lattice model. We find that a lattice $U(1)$ gauge theory that couples to a charged boson is described by the following Lagrangian:

$$\begin{aligned} L = &+\sum_{\langle ij\rangle} t_{ij}(\varphi_i^\dagger \varphi_j \mathrm{e}^{-\mathrm{i}a_{ji}} + h.c) + \sum_i \varphi_i^\dagger i[\partial_0 + \mathrm{i}a_0(i)]\varphi_i \\ &+ \frac{1}{4J}\sum_{i,\mu=x,y}[\dot{a}_{i,i+\mu} + a_0(i) - a_0(i+\mu)]^2 + g\sum_p \cos(\Phi_p) \end{aligned} \tag{6.4.9}$$

The Lagrangian is invariant under the following lattice gauge transformation:

$$c_i \to \mathrm{e}^{\mathrm{i}\phi(i,t)}c_i, \quad a_0(i) \to a_0(i) - \partial_0\phi(i,t), \quad a_{ij} \to a_{ij} + \phi(i,t) - \phi(j,t).$$

Problem 6.4.1.

(a) Generalize eqn (6.4.1) to a three-dimensional cubic lattice.

(b) Use eqn (6.4.1) to show that the Lagrangian for the three-dimensional lattice gauge

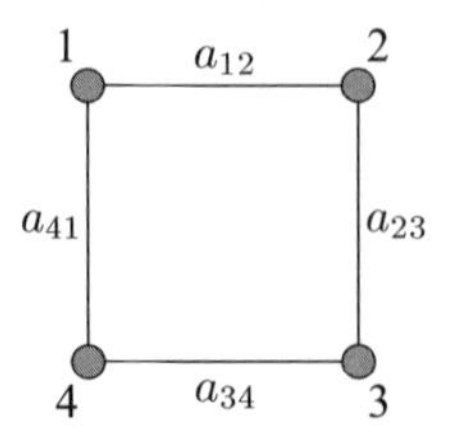

FIG. 6.7. A $U(1)$ gauge theory on a square.

theory becomes $\frac{1}{8\pi\alpha}(c^{-1}\boldsymbol{e}^2 - c\boldsymbol{b}^2)$ in the continuum limit, where $e_i = \partial_0 a_i - \partial_i a_0$ and $b_i = \epsilon_{ijk}\partial_j a_k$. Find the values of α and c in terms of J, g, and the lattice constant l. Here the dimensionless constant α is the fine-structure constant and c is the speed of light.

6.4.2 The Hamiltonian of the lattice $U(1)$ gauge theory

- The Hamiltonian of a lattice $U(1)$ gauge theory can be obtained via gauge fixing or a phase-space path integral.
- The electric flux e_{ij} through a link is the canonical 'momentum' of the vector potential a_{ij} on the same link.

The Lagrangian (6.4.5) describes a classical $U(1)$ lattice gauge theory. In this section, we are going to find the quantum description. For simplicity, we will consider the lattice gauge theory on a single square (see Fig 6.7). The quantum theory is described by the path integral

$$Z = \int \mathcal{D}a\mathcal{D}a_0 \; \mathrm{e}^{\,\mathrm{i}\int \mathrm{d}t\left(\frac{1}{4J}\sum_i(\dot{a}_{i,i+1}+a_0(i)-a_0(i+1))^2+g\cos\Phi\right)}, \tag{6.4.10}$$

where $i = 1, 2, 3, 4$. Here $i = 1$ and $i = 5$ are regarded as the same point. As a gauge theory, $(a_{ij}(t), a_0(i,t))$ is a many-to-one labeling of the path. We can obtain a one-to-one labeling by 'fixing a gauge'. We note that $\sum_{j=i\pm1} a_{ij}$ transforms as $\sum_{j=i\pm1} a_{ij} \to \sum_{j=i\pm1} \tilde{a}_{ij} = \sum_{j=i\pm1}(a_{ij} + \phi_i - \phi_j)$ under the gauge transformation. By tuning ϕ_i, we can always make $\sum_{j=i\pm1} \tilde{a}_{ij} = 0$. Thus, for any path $(a_{ij}(t), a_0(i,t))$, we can always construct a gauge transformation to make $\sum_{j=i\pm1} a_{ij} = 0$. Therefore, we can fix a gauge by choosing a gauge-fixing condition

$$\sum_{j=i\pm1} a_{ij} = 0$$

Such a gauge is called the Coulomb gauge, which has the form $\boldsymbol{\partial} \cdot \boldsymbol{a} = 0$ for a continuum theory. In the Coulomb gauge, our path integral becomes

$$Z = \int \mathcal{D}a\mathcal{D}a_0 \, \mathrm{e}^{-\mathrm{i}\int \mathrm{d}t\left(\frac{1}{4J}\sum_i(\dot{a}_{i,i+1}+a_0(i)-a_0(i+1))^2+g\cos\Phi\right)} \prod_{i,t} \delta(\sum_{j=i\pm1} a_{ij})$$

We note that a coupling between $a_0(i)$ and a_{ij} has the form $a_0(i)\sum_{j=i\pm1}\dot{a}_{ij}$. Thus, for a_{ij} satisfying the constraint $\sum_{j=i\pm1} a_{ij} = 0$, $a_0(i)$ and a_{ij} decouple. As $a_0(i)$ has no dynamics (i.e. no $\dot{a}_0(i)$ terms), we can integrate out $a_0(i)$, which corresponds to simply dropping a_0. The resulting path integral becomes

$$Z = \int \mathcal{D}a \, \mathrm{e}^{-\mathrm{i}\int \mathrm{d}t\left(\frac{1}{4J}\sum_i \dot{a}^2_{i,i+1}+g\cos\Phi\right)} \prod_{i,t} \delta(\sum_{j=i\pm1} a_{ij}).$$

In general, a path integral in the Coulomb gauge can be obtained by the two simple steps of, firstly, inserting the gauge-fixing condition $\prod_{i,t}\delta(\sum_{j=i\pm1} a_{ij})$ and, secondly, dropping the $a_0(i)$ field.

For our problem, the constraint $\prod_{i,t}\delta(\sum_{j=i\pm1} a_{ij})$ requires that $a_{12} = a_{23} = a_{34} = a_{41} \equiv \Phi/4$. Here Φ describes the $U(1)$ flux through the square. The path integral takes the simple form

$$Z = \int \mathcal{D}\theta \, \mathrm{e}^{-\mathrm{i}\int \mathrm{d}t\left(\frac{1}{16J}\dot{\Phi}^2+g\cos\Phi\right)} \tag{6.4.11}$$

We note that the configuration $(a_{12}, a_{23}, a_{34}, a_{41}) = (\pi/2, \pi/2, \pi/2, \pi/2)$ is gauge equivalent to $(a_{12}, a_{23}, a_{34}, a_{41}) = (2\pi, 0, 0, 0)$ (i.e. there is a gauge transformation that transforms $(\pi/2, \pi/2, \pi/2, \pi/2)$ to $(2\pi, 0, 0, 0)$). Also, $a_{12} = 2\pi$ is equivalent to $a_{12} = 0$ because the $a_{i,i+1}$ live on a circle. Thus, $\Phi = 2\pi$ and $\Phi = 0$ correspond to the same physical point. The path integral (6.4.11) describes a particle of mass $(8J)^{-1}$ on a unit circle. The flux energy $-g\cos\Phi$ is the potential experienced by the particle. When $g = 0$, the energy levels are given by $E_n = 4Jn^2$.

There is another way to solve eqn (6.4.10). We first introduce the canonical momentum of a_{ij}, namely

$$\frac{\partial L}{\partial \dot{a}_{ij}} = \frac{1}{2J} e_{ij}$$

and write the path integral (6.4.10) as a phase-space path integral (see Problem 6.4.4)

$$Z = \int \mathcal{D}a\mathcal{D}a_0\mathcal{D}e_{ij} \, \mathrm{e}^{\mathrm{i}\int \mathrm{d}t\left(\sum_i \frac{e_{i,i+1}}{2J}(\dot{a}_{i,i+1}+a_0(i)-a_0(i+1)) - \frac{1}{4J}\sum_i e^2_{i,i+1}+g\cos\Phi\right)}. \tag{6.4.12}$$

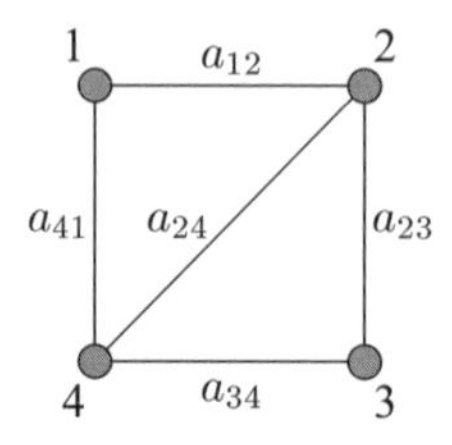

FIG. 6.8. A $U(1)$ gauge theory on a square with a diagonal link.

Then we integrate out a_0 and obtain[41]

$$Z = \int \mathcal{D}a\mathcal{D}e_{ij}\ \mathrm{e}^{\mathrm{i}\int \mathrm{d}t(\sum_i \frac{e_{i,i+1}}{2J}\dot{a}_{i,i+1} - \frac{1}{4J}\sum_i e_{i,i+1}^2 + g\cos\Phi)} \prod_{i,t} \delta(e_{i,i+1} + e_{i,i-1}) \tag{6.4.13}$$

The quantum Hamiltonian of the above system is given by

$$H = \frac{1}{4J}\sum_i e_{i,i+1}^2 - g\cos\Phi, \qquad [a_{i,i+1}, (2J)^{-1}e_{i,i+1}] = \mathrm{i}\hbar.$$

The physical Hilbert space is formed by states that satisfy

$$(e_{i,i+1} + e_{i,i-1})|\text{phy}\rangle = 0$$

As $[e_{i,i+1} + e_{i,i-1}, H] = 0$, H acts within the physical Hilbert space.

To find the energy levels of the system, we note that a_{ij} is periodic with a period 2π and the momentum of a_{ij} is $e_{ij}/2J$. Thus, $e_{ij}/2J$ is quantized as integers, n_{ij}. When $g = 0$, the energy levels of the system are given by $J\sum_i n_{i,i+1}^2$, where the n_{ij} satisfy the constraint $\sum_{j=i\pm 1} n_{ij} = 0$. For our four-site system, the constraint requires that $n_{12} = n_{23} = n_{34} = n_{41} \equiv n$. The energy eigenstates are labeled by n and have energy $E_n = 4Jn^2$, which agrees with the previous calculation.

Problem 6.4.2.
(a) Generalize eqn (6.4.10) to a square with a diagonal link (see Fig. 6.8).
(b) Find the energies of the ground state and the excited states, assuming that $g = 0$.

Problem 6.4.3.
(a) Show that, for a non-compact lattice $U(1)$ gauge theory on a two-dimensional square

[41] Note that the coupling between $a_0(i)$ and e_{ij} has the form $\sum_i a_0(i)(e_{i,i+1} + e_{i,i-1})$. The path integral $\int \mathcal{D}a_0\ \mathrm{e}^{\mathrm{i}\int \mathrm{d}t \sum_i a_0(i)(e_{i,i+1}+e_{i,i-1})}$ is proportional to $\prod_{i,t} \delta(e_{i,i+1}(t) + e_{i,i-1}(t))$.

lattice, the gauge-fixed path integral has the form

$$Z = \int \mathcal{D}a \; \mathrm{e}^{-\mathrm{i}\int \mathrm{d}t\left(\frac{1}{4J}\sum_{i,\mu=x,y} \dot{a}^2_{i,i+\mu} - \frac{1}{2g}\sum_{p}\Phi_p^2\right)} \prod_{i,t} \delta\Big(\sum_{\mu=\pm x,\pm y} a_{i,i+\mu}\Big)$$

(b) Show that the constraint $\sum_{\mu=\pm x,\pm y} a_{i,i+\mu} = 0$ can be solved by introducing a real field ϕ_p defined on the squares, and write the gauge potential on the link as $a_{i,i+\mu} = \phi_{p_1} - \phi_{p_2}$, where p_1 and p_2 are the two squares on the two sides of the link $(i, i+\mu)$. Write the path integral in terms of the ϕ_p field.
(c) Find the dispersion relation of the ϕ_p fluctuations.

Problem 6.4.4.
Prove eqn (6.4.12). (Hint: The phase-space path integral is given by $\int \mathcal{D}p\mathcal{D}q \; \mathrm{e}^{\mathrm{i}\int \mathrm{d}t[p\dot{q} - H(p,q)]}$, where $H(p,q) = p\dot{q} - L$ is the Hamiltonian.)

6.4.3 The Coulomb phase and the confined phase of the lattice $U(1)$ gauge theory

- In $1+3$ dimensions, the compact lattice $U(1)$ gauge theory is in the Coulomb phase when $g/J \gg 1$ and in the confined phase if $g/J \ll 1$.

We know that, in $1+2$ dimensions, the compact $U(1)$ gauge theory is always in the confined phase due to the instanton effect. In this section, we are going to study a compact $U(1)$ gauge theory on a cubic lattice. We will argue that the $U(1)$ gauge theory in $1+3$ dimensions has both the confined phase and the Coulomb phase.

The Lagrangian of the $U(1)$ gauge theory is given by

$$L = \frac{1}{4J} \sum_{i,\mu=x,y,z} [\dot{a}_{i,i+\mu} + a_0(i) - a_0(i+\mu)]^2 + g\sum_{p} \cos(\Phi_p)$$

We can express the action of the above Lagrangian in the following dimensionless form:

$$S = \sqrt{\frac{g}{J}} \int \mathrm{d}\tilde{t} \left(\frac{1}{4} \sum_{i,\mu=x,y} [\partial_{\tilde{t}} a_{i,i+\mu} + \tilde{a}_0(i) - \tilde{a}_0(i+\mu)]^2 + \sum_{p} \cos(\Phi_p) \right)$$

where $\tilde{t} = \sqrt{gJ}t$ is the dimensionless time and $\tilde{a}_0 = a_0/\sqrt{gJ}$ is the dimensionless potential. We see that, when $g/J \gg 1$, the fluctuations of Φ_p are weak and we can assume that $a_{ij} \sim 0$. In this limit, L becomes a non-compact $U(1)$ gauge theory as follows:

$$L = \frac{1}{4J} \sum_{i,\mu=x,y,z} [\dot{a}_{i,i+\mu} + a_0(i) - a_0(i+\mu)]^2 - \frac{g}{2}\sum_{p} \Phi_p^2$$

The quadratic theory can be solved exactly. In the continuum limit, the above becomes the standard Maxwell Lagrangian of the $U(1)$ gauge theory. We find that, at low energies, there are two linearly-dispersing modes $\omega = c_a|\boldsymbol{k}|$ representing the gapless photons.

When $g/J \ll 1$, we can drop the $\cos(\Phi_{\boldsymbol{p}})$ term. The resulting quadratic theory can also be solved exactly, as we did in the last section. The Lagrangian for the phase-space path integral has the form

$$L = \sum_{\boldsymbol{i},\mu=\boldsymbol{x},\boldsymbol{y},\boldsymbol{z}} \frac{e_{\boldsymbol{i},\boldsymbol{i}+\boldsymbol{\mu}}}{2J}[\dot{a}_{\boldsymbol{i},\boldsymbol{i}+\boldsymbol{\mu}}+a_0(\boldsymbol{i})-a_0(\boldsymbol{i}+\boldsymbol{\mu})] - \sum_{\boldsymbol{i},\mu=\boldsymbol{x},\boldsymbol{y},\boldsymbol{z}} \frac{e^2_{\boldsymbol{i},\boldsymbol{i}+\boldsymbol{\mu}}}{4J} + g\sum_{\boldsymbol{p}} \cos(\Phi_{\boldsymbol{p}})$$

After integrating out a_0, we obtain the constraint

$$\sum_{\mu=\pm\boldsymbol{x},\pm\boldsymbol{y},\pm\boldsymbol{z}} e_{\boldsymbol{i},\boldsymbol{i}+\boldsymbol{\mu}} = 0$$

The Hamiltonian is given by

$$H = \sum_{\boldsymbol{i},\mu=\boldsymbol{x},\boldsymbol{y},\boldsymbol{z}} \frac{e^2_{\boldsymbol{i},\boldsymbol{i}+\boldsymbol{\mu}}}{4J} - g\sum_{\boldsymbol{p}} \cos(\Phi_{\boldsymbol{p}})$$

Since $e_{\boldsymbol{ij}}/2J$ is quantized as an integer $n_{\boldsymbol{ij}}$, when $g = 0$, the energy levels are given by $E = J\sum_{\boldsymbol{i},\mu=\boldsymbol{x},\boldsymbol{y},\boldsymbol{z}} n^2_{\boldsymbol{i},\boldsymbol{i}+\boldsymbol{\mu}}$, where $n_{\boldsymbol{i},\boldsymbol{i}+\boldsymbol{\mu}}$ satisfy $\sum_{\mu=\pm\boldsymbol{x},\pm\boldsymbol{y},\pm\boldsymbol{z}} n_{\boldsymbol{i},\boldsymbol{i}+\boldsymbol{\mu}} = 0$. We see that all of the excitations in the $g = 0$ limit are gapped. The gapped excitations are loops formed by nonzero $n_{\boldsymbol{ij}}$s. These loops represent lines of electric flux. In the presence of a pair of positive and negative charges, the two charges are connected by a line of electric flux which causes a linear confining potential between the two charges. Thus, the gapped phase is the confined phase of the $U(1)$ gauge theory.

Problem 6.4.5.

To show that charges interact with a linear potential in the confined phase, we consider the lattice $U(1)$ gauge theory in the presence of charges as follows:

$$L = \sum_{\boldsymbol{i},\mu=\boldsymbol{x},\boldsymbol{y},\boldsymbol{z}} \frac{[\dot{a}_{\boldsymbol{i},\boldsymbol{i}+\boldsymbol{\mu}} + a_0(\boldsymbol{i}) - a_0(\boldsymbol{i}+\boldsymbol{\mu})]^2}{4J} + g\sum_{\boldsymbol{p}} \cos(\Phi_{\boldsymbol{p}}) - \sum a_0(\boldsymbol{i})q_{\boldsymbol{i}}$$

where $q_{\boldsymbol{i}}$ is the $U(1)$ charge at site $\boldsymbol{i}$.
(a) Find the Lagrangian in the phase-space path integral.
(b) Show that integrating out a_0 leads to the following constraint:

$$\sum_{\mu=\pm\boldsymbol{x},\pm\boldsymbol{y},\pm\boldsymbol{z}} e_{\boldsymbol{i},\boldsymbol{i}+\boldsymbol{\mu}} = 2Jq_{\boldsymbol{i}}$$

which is Gauss's law on the lattice.
(c) Find the ground-state energy of the gauge theory with a charge $q = 1$ at $\boldsymbol{i} = 0$ and a charge $q = -1$ at $\boldsymbol{i} = l\boldsymbol{x}$, assuming that $g = 0$. Show that the interaction energy between the two charges grows linearly with l.

7

THEORY OF QUANTUM HALL STATES

- Fractional quantum Hall liquids represent new states of matter which contain extremely rich internal structures.
- Different fractional quantum Hall liquids cannot be characterized by order parameters and long-range orders of local operators. We need a completely new theory to describe fractional quantum Hall liquids.
- Fractional quantum Hall liquids contain gapless edge excitations which form the so-called chiral Luttinger liquids. The chiral Luttinger liquids at the edges also have very rich structures, as a reflection of the rich bulk internal structures.

The fractional quantum Hall (FQH) effect appears in two-dimensional electron systems in a strong magnetic field. Since its discovery in 1982 (Tsui *et al.*, 1982; Laughlin, 1983), experiments on FQH systems have continued to reveal many new phenomena and surprises. The observed rich hierarchical structures (Haldane, 1983; Halperin, 1984) indicate that electron states that demonstrate a fractional quantum Hall effect (these states are called FQH liquids) contain extremely rich internal structures. In fact, FQH liquids represent a whole new state of matter, which is very different from the symmetry-breaking states discussed in the last few chapters. One needs to develop new concepts and new techniques to understand this new kind of state. In this chapter, we will introduce a general theory of FQH states. A detailed discussion of the new orders (the topological orders) in FQH states will be given in Chapter 8.

7.1 The Aharonov–Bohm effect and fractional statistics

7.1.1 The Aharonov–Bohm effect—deflect a particle without touching

- The local phase for contractible loops and the global phase for non-contractible loops.
- The nonzero local phase represents a force, while a nonzero global phase does not correspond to any classical force.

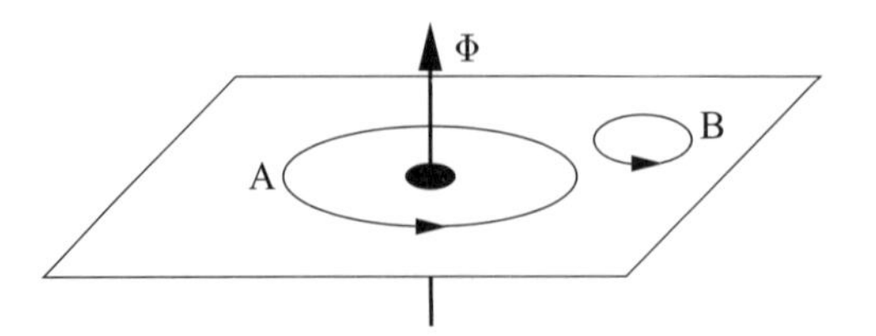

FIG. 7.1. A plane with the $\boldsymbol{r} = 0$ point removed and a flux tube at $\boldsymbol{r} = 0$. The loop A is non-contractible and has a winding number $n_w = 1$. The loop B is contractible.

To prepare for later discussions on the quantum interference in the quantum Hall (QH) state, let us first give a brief discussion of the Aharonov–Bohm (A–B) effects. First, we would like to introduce two concepts, namely local phase and global phase. The local phase is associated with contractible loops. Moving a charged particle slowly around a contractible loop will generate a local phase as follows:

$$\text{local phase} = \exp\left(\mathrm{i}e \oint \boldsymbol{A}\, \mathrm{d}\boldsymbol{x}\right) = \exp\left(\mathrm{i}e \int \boldsymbol{B} \cdot \mathrm{d}\boldsymbol{S}\right).$$

The magnetic field $\boldsymbol{B} = 0$ if and only if the local phases for all contractible loops vanish. Thus, a charged particle that feels no magnetic force has no local phases.

The global phase is related to non-contractible loops. To understand the global phase in a simple setting, let us consider a charged particle moving on a two-dimensional plane. We remove the $\boldsymbol{r} = 0$ point, which changes the topology of the space. We then place an infinite thin flux tube of flux Φ at $\boldsymbol{r} = 0$ (see Fig. 7.1). As the particle goes around $\boldsymbol{r} = 0$, a phase $\mathrm{e}^{\,\mathrm{i}e\Phi}$ is generated. Such a phase is called the global phase. More generally,

$$\text{global phase} = \mathrm{e}^{\,\mathrm{i}e\Phi n_w}$$
$$n_w = \text{winding number.}$$

Clearly, despite the nonzero global phase, the particle feels no magnetic force because there is no magnetic field away from $\boldsymbol{r} = 0$. We see that a free particle (i.e. a particle that feels no force) can have a nonzero global phase.

Global phase exists only when non-contractible loops exist (i.e. when the space is not simply connected). Another example of a non-simply-connected space is a torus (see Fig. 7.2). A free particle on a torus can also have global phases.

The above free particle (with the flux tube at $\boldsymbol{r} = 0$) can be described by the following Hamiltonian:

$$H = -\frac{1}{2m}(\boldsymbol{\partial} - \mathrm{i}\boldsymbol{a})^2, \qquad (a_x, a_y) = \frac{1}{r^2}(-y, -x)\frac{\phi}{2\pi}$$

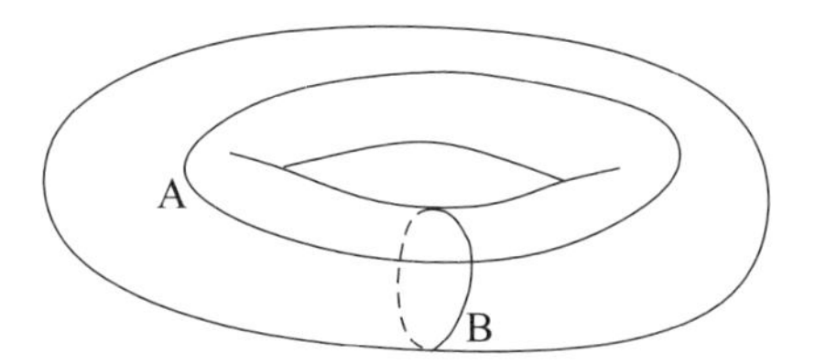

FIG. 7.2. A torus with two non-contractible loops A and B.

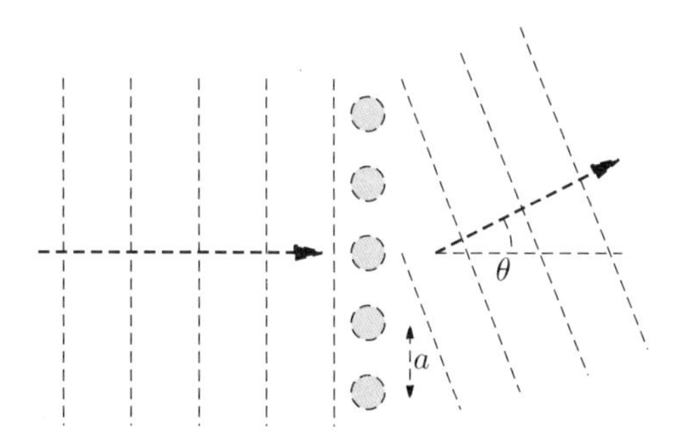

FIG. 7.3. A beam of particles passes through a grid of tubes, with a flux through each tube.

The field strength vanishes, i.e.

$$\boldsymbol{b} = \partial_x a_y - \partial_y a_x = 0 \implies \text{local phase} = 0$$

For a loop around $\boldsymbol{x} = 0$, we have

$$\oint \boldsymbol{a} \cdot \mathrm{d}\boldsymbol{x} = \phi \implies \text{global phase} \neq 0 \ .$$

A non-vanishing local phase will affect the classical equation of motion. The Berry phase for the spin is an example of local phase (see Section 2.3). In contrast, a global phase will not affect the classical equation of motion. However, a global phase will affect the quantum properties of the particle.

Problem 7.1.1.
Deflecting without touching Although a global phase does not produce any classical force, it can still deflect a moving particle. Consider the set-up in Fig. 7.3, where a beam of particles with charge e and momentum $\boldsymbol{k}$ passes through a grid of impenetrable tubes. Calculate the deflecting angle θ if there is a flux Φ through each tube.

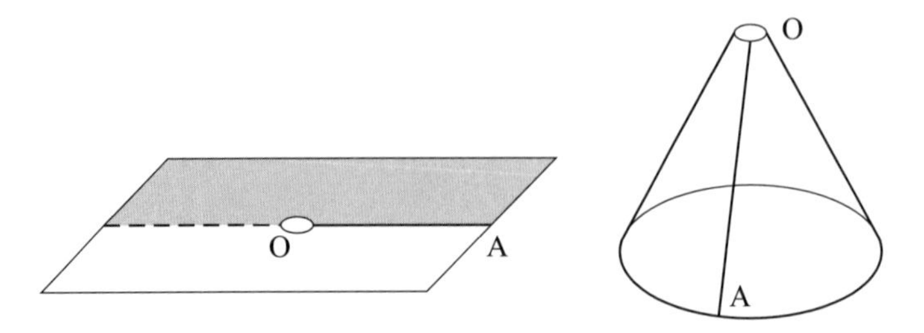

FIG. 7.4. The V_- space, with the $\boldsymbol{r} = 0$ point removed, and the points $\boldsymbol{r}$ and $-\boldsymbol{r}$ identified as the same point.

7.1.2 Particles with a hard-core condition and fractional statistics

- The configuration space for particles with a hard-core condition is not simply connected. The non-trivial topology of the configuration space allows non-trivial statistics.
- Fractional statistics exist and only exist in two spatial dimensions.

One realization of the A–B effect is the fractional statistics in two dimensions (Leinaas and Myrheim, 1977; Wilczek, 1982). Consider two free hard-core identical particles in two dimensions.[42] The configuration space is given by

$$\begin{aligned}\text{configuration space} &= \left\{(\boldsymbol{r}_1, \boldsymbol{r}_2)|_{\boldsymbol{r}_1 \neq \boldsymbol{r}_2, (\boldsymbol{r}_1,\boldsymbol{r}_2)\sim(\boldsymbol{r}_2,\boldsymbol{r}_1)}\right\} \\ &= \left\{(\boldsymbol{r}_+, \boldsymbol{r}_-)|_{\boldsymbol{r}_- \neq 0, (\boldsymbol{r}_+,\boldsymbol{r}_-)\sim(\boldsymbol{r}_+,-\boldsymbol{r}_-)}\right\} = V_+ \otimes V_-\end{aligned}$$

where $\boldsymbol{r}_+ = \frac{1}{2}(\boldsymbol{r}_1 + \boldsymbol{r}_2)$, $\boldsymbol{r}_- = \boldsymbol{r}_1 - \boldsymbol{r}_2$, $V_+ = \{\boldsymbol{r}_+\}$, and $V_- = \{\boldsymbol{r}_-|_{\boldsymbol{r}_- \neq 0, \boldsymbol{r}_- \sim -\boldsymbol{r}_-}\}$ (see Fig. 7.4). Here V_+ is the usual two-dimensional plane, while V_- is only half of the two-dimensional plane because $\boldsymbol{r}_-$ and $-\boldsymbol{r}_-$ are regarded as the same point. Also, the $\boldsymbol{r}_- = 0$ point is removed from V_-.

As pointed out in the last section, 'free' means that the local phase is zero. As the two-particle configuration space is not simply connected (V_- is not simply connected), there exist global phases and we have the freedom to choose the global phases. Note that the non-contractible loops in V_- are characterized by winding numbers n_w (Fig. 7.5). Thus, we can assign the following global phases for non-contractible loops:

$$\text{global phases} = \mathrm{e}^{\,\mathrm{i}\theta n_w}$$

Now it is clear that, in quantum mechanics, there are different kinds of free identical particles, labeled by a parameter θ.

---

[42] Here, by free particles, we mean particles that do not experience any force when separated. However, the particles may be subject to a hard-core condition.

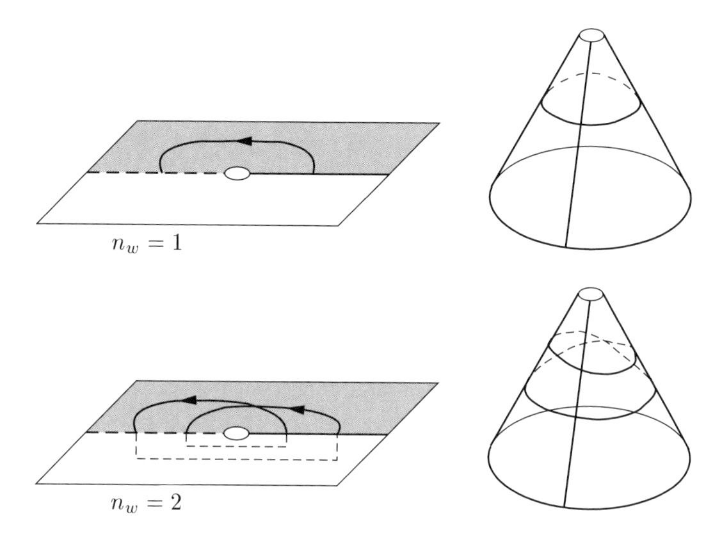

FIG. 7.5. Two loops with winding numbers $n_w = 1$ and $n_w = 2$ in the V_- space.

- The parameter θ describes the flux (the global phase) in many-particle configuration space.
- The parameter θ determines the statistics.

Note that the loop with $n_w = 1$ connects $\boldsymbol{r}_-$ to $-\boldsymbol{r}_-$, or $(\boldsymbol{r}_1, \boldsymbol{r}_2)$ to $(\boldsymbol{r}_2, \boldsymbol{r}_1)$, which corresponds to exchanging the two particles. Thus, the statistics of the particles are as follows:

$$\begin{cases} \theta = 0 \to & \text{boson} \\ \theta = \pi \to & \text{fermion} \\ \theta = \text{other} \to & \text{anyon} \end{cases}$$

The Hamiltonian for the above two free identical particles (i.e. for two anyons with statistics θ) is given by

$$\begin{aligned} H &= -\frac{1}{2m}(\partial_{\boldsymbol{r}_+})^2 - \frac{1}{2}\frac{1}{m}(\partial_{\boldsymbol{r}_-} - \mathrm{i}\boldsymbol{a})^2 \\ (a_x, a_y) &= \frac{\theta}{\pi r^2}(-y, x) \\ &\text{Boundary Conditions}: \ \phi(\boldsymbol{r}_+, \boldsymbol{r}_-) = \phi(\boldsymbol{r}_+, -\boldsymbol{r}_-) \end{aligned} \tag{7.1.1}$$

When $\theta = 0$ and $\boldsymbol{a} = 0$, H obviously describes two bosons. For $\theta = \pi$, we can make a gauge transformation

$$\phi(\boldsymbol{r}_+, \boldsymbol{r}_-) \to \mathrm{e}^{\,\mathrm{i}f(\boldsymbol{r}_-)}\phi(\boldsymbol{r}_+, \mathbf{r}_-), \qquad f(\boldsymbol{r}) \equiv \arctan\left(\frac{x}{y}\right).$$

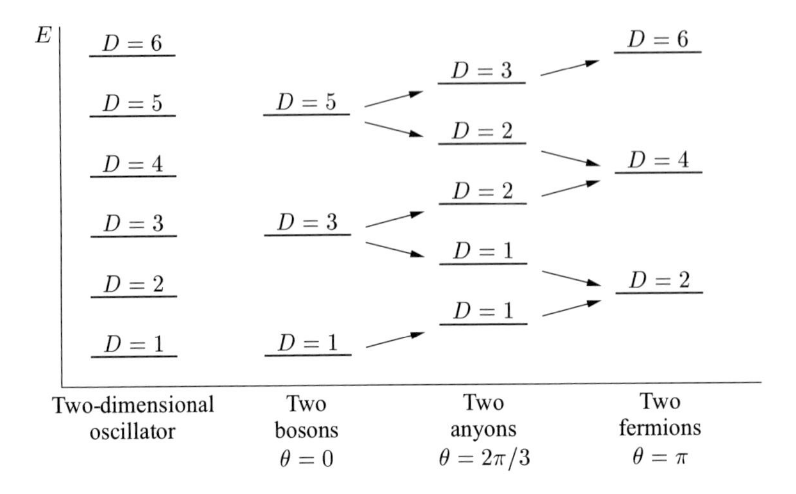

FIG. 7.6. The energy levels E and the degeneracies D (for the relative motion) of two anyons in a two-dimensional harmonic potential well.

As $\nabla f(\boldsymbol{r}) = \frac{1}{r^2}(-y, x)$, we find that $\mathrm{e}^{-\mathrm{i}f}(\partial_{\boldsymbol{r}_-} - \mathrm{i}\boldsymbol{a})^2\mathrm{e}^{\mathrm{i}f} = (\partial_{\boldsymbol{r}_-})^2$ and the Hamiltonian becomes

$$
\begin{aligned}
H &= -\frac{1}{2m}(\partial_{\boldsymbol{r}_+})^2 - \frac{1}{2}\frac{1}{m}(\partial_{\boldsymbol{r}_-})^2 \\
\phi(\boldsymbol{r}_+, \boldsymbol{r}_-) &= -\phi(\boldsymbol{r}_+, \boldsymbol{r}_-) \\
\text{or } \phi(\boldsymbol{r}_1, \boldsymbol{r}_2) &= -\phi(\boldsymbol{r}_2, \boldsymbol{r}_1)
\end{aligned}
$$

which describes two free fermions. In general, we can consistently gauge away $\boldsymbol{a}$, even for N fermions, and describe the Fermi statistics through the 'boundary condition', i.e. the anti-symmetric condition. However, for an N-anyon system, we cannot gauge away $\boldsymbol{a}$ and represent the fractional statistics through the 'boundary condition'.

One can solve the system of two anyons in a two-dimensional harmonic potential well, and find the spectrum shown in Fig. 7.6. One can clearly see how the spectrum changes continuously from the bosonic one to the fermionic one.

An interesting question is do we have an anyon in three dimensions? Two identical free particles in three dimensions have a configuration space $V_+^{(3D)} \otimes V_-^{(3D)}$, with

$$
V_-^{(3D)} = \{\boldsymbol{r}_- |_{\boldsymbol{r}_- \neq 0,\ \boldsymbol{r}_- \sim -\boldsymbol{r}_-}\}
$$

Here $V_-^{(3D)}$ is not simply connected, and a global phase exists. However, the non-contractible loop in $V_-^{(3D)}$ is characterized by Z_2. This is because an exchange in one direction (described by the loop C_1) and the exchange in the opposite direction

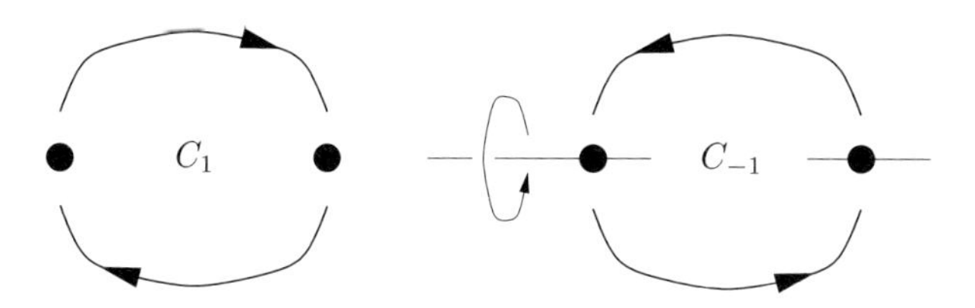

FIG. 7.7. The two exchange loops C_1 and C_{-1} can be continuously changed into each other in three dimensions, by rotating about the axis connecting the two particles.

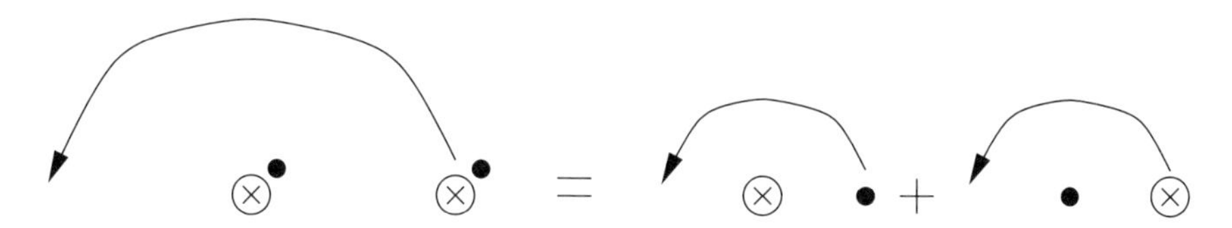

FIG. 7.8. The phase induced by moving a flux–charge bound state around another flux–charge bound state.

(described by the loop C_{-1}) belong to the same class, as the two loops $C_{\pm 1}$ can be continuously changed into each other (see Fig. 7.7). If we assign a global phase $\mathrm{e}^{\mathrm{i}\theta}$ to the loop C_1, then the loop C_{-1} will have a global phase $\mathrm{e}^{-\mathrm{i}\theta}$. As the two loops $C_{\pm 1}$ differ only by a contractible loop which has a vanishing local phase, we also have $\mathrm{e}^{\mathrm{i}\theta} = \mathrm{e}^{-\mathrm{i}\theta}$. Therefore, θ can only take two possible values as follows:

$$\theta = \begin{cases} 0, & \text{boson} \\ \pi, & \text{fermion} \end{cases}$$

The above discussion demonstrates the anyon as a mathematical possibility in quantum mechanics. The next question is how an anyon may appear in some physical models which originally contained no anyons. In fact, an anyon can appear as excitations in boson/fermion systems. In the following, we will consider a model in which an anyon appears as a bound state of charge and flux. Consider a system of charge one boson ϕ. Assume that bosons form a q-boson bound state $\Phi_q = (\phi)^q$, and Φ_q undergoes a boson condensation. The effective Lagrangian is

$$\mathcal{L}_{\text{eff}} = \frac{1}{2} \left|(\partial_\mu - \mathrm{i}qa_\mu)\,\Phi_q\right|^2 - \frac{1}{2}a\left|\Phi_q\right|^2 - \lambda\left|\Phi_q\right|^4 + \frac{1}{2g}\left(f_{\mu\nu}\right)^2 \tag{7.1.2}$$

$$\langle \Phi_q \rangle \neq 0$$

where a_μ is the $U(1)$ gauge field that couples to the charge boson. In the condensed state, the vortex carries a flux $\frac{2\pi}{q}$ (see Problem 7.1.2). Consider a bound state of the vortex $\frac{2\pi}{q}$ and the original boson ϕ. Moving one bound state half-way around the other, we obtain a phase

$$\theta = \frac{1}{2}\frac{2\pi}{q} + \frac{1}{2}\frac{2\pi}{q} = \frac{2\pi}{q}$$

The first term is from the charge going around the flux tube, and the second term is from the flux tube going around the charge (see Fig. 7.8). We see that the bound state is a boson if $q = 1$, a fermion if $q = 2$, and an anyon if $q > 2$.

For BCS superconductors, $q = 2$. However, ϕ is a fermion, and hence the bound state is a boson. The only known condensed matter system that supports anyon excitations is the FQH system.

Problem 7.1.2.
In polar coordinates (r, ϕ), a vortex in eqn (7.1.2) is described by

$$\Phi_q = f(r)\mathrm{e}^{\,\mathrm{i}\phi}$$

where $f(\infty) = \langle \Phi_q \rangle$.
(a) Show that, if $a_\mu = 0$, then the energy of the vortex diverges as $\ln(L)$, where L is the size of the system.
(b) Find $\boldsymbol{a}(r, \phi)$ that makes the vortex have a finite energy.
(c) Show that the total flux of the above $\boldsymbol{a}(r, \phi)$ is $2\pi/q$.

7.2 The quantum Hall effect

7.2.1 The integral quantum Hall effects

- Integral quantum Hall (IQH) states are described by filled Landau levels.
- The many-body wave function of the $\nu = 1$ IQH state.
- The density profile of IQH wave functions.

Let us first discuss classical Hall effects. Consider a 2D gas of charge $q_e = -e$ electrons moving under the influence of an electric field in the plane and a magnetic field normal to the plane. To maintain a static current, the force from the electric field must balance the Lorentz force from the magnetic field, i.e.

$$q_e \boldsymbol{E} = q_e \boldsymbol{v} \times \boldsymbol{B}$$

In this chapter, we will choose a unit such that the speed of light is $c = 1$. As the current $\boldsymbol{j}$ is given by $\boldsymbol{v}n$, where n is the density, $\boldsymbol{E}$ and $\boldsymbol{j}$ are related through

$$\boldsymbol{E} \perp \boldsymbol{j}, \qquad |\boldsymbol{E}| = |\boldsymbol{j}|\frac{B}{nq_e c} = |\boldsymbol{j}|\frac{h}{q_e^2}\frac{B/(hc/q_e)}{n} = |\boldsymbol{j}|\left(\frac{h}{q_e^2}\right)\frac{1}{\nu}$$

where

$$\nu \equiv \frac{nhc}{q_e B} = \frac{\text{number of particles}}{\text{number of flux quanta}}$$

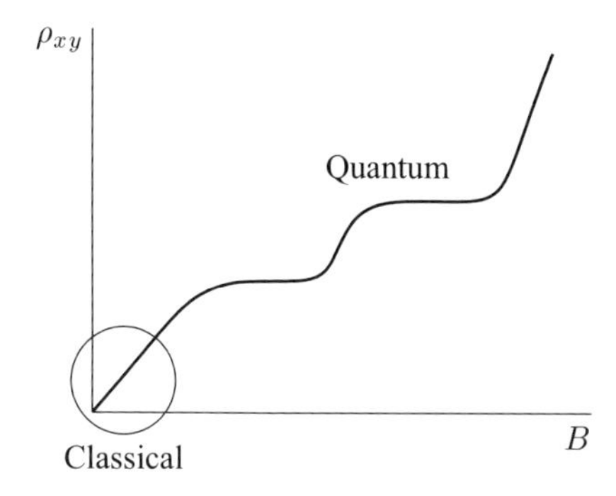

FIG. 7.9. The Hall resistance ρ_{xy} as a function of B.

is the filling fraction. The Hall resistance $\rho_{xy} = \left(\frac{h}{q_e^2}\right)\frac{1}{\nu}$ can be measured directly in experiments. According to the above classical theory, $\rho_{xy} \propto B$ if n is fixed.

Experimentally, one indeed finds that, for a two-dimensional electron gas, $\rho_{xy} \propto B$ at weak fields (see Fig. 7.9). In the early 1980s physicists put a two-dimensional electron gas under strong magnetic fields ($\sim$ 10 Tesla) and cooled it to very low temperatures ($\sim 1K^\circ$). They found that, for strong B, ρ_{xy} develops plateau structures, as shown in Fig. 7.9 (von Klitzing *et al.*, 1980; Tsui *et al.*, 1982). Such a phenomenon is called the QH effect. Physicists soon discovered many amazing properties of QH states. What is most striking is that the plateaus appear exactly at those ρ_{xy} corresponding to

$$\nu = \underbrace{1, 2, 3, \cdots}_{\text{IQH}}, \underbrace{\frac{1}{3}, \frac{2}{3}, \frac{2}{5}, \frac{3}{7}, \cdots}_{\text{FQH}}$$

The value of ρ_{xy} is so accurate and stable that it has become the standard for resistance.

The physics of the plateaus is a little complicated because it involves impurities. In short, we can understand the plateaus at these filling fractions, if the two-dimensional electron gas forms an incompressible liquid[43] at these νs. As we will see below, electrons do form an incompressible state at $\nu = 1, 2, 3, \cdots$, due to the Landau level structure. This allows us to understand the plateaus at integer ν and IQH effects.

Let us ignore the Coulomb interaction and consider a two-dimensional free electron gas in a strong magnetic field B. We will ignore the electron spins or assume that the electron spins are polarized. A single electron in a uniform

[43] That is, all charged excitations have a finite energy gap.

magnetic field is described by

$$H = -\frac{1}{2m}(\partial_i - \mathrm{i}q_e A_i)^2, \qquad \hbar = c = 1 \tag{7.2.1}$$
$$(A_x, A_y) = \frac{B}{2}(-y, x), \qquad \text{(symmetric gauge)}$$
$$B = \partial_x A_y - \partial_y A_x$$

The above Hamiltonian has an interesting symmetry property. It is not invariant under the direct translation $T_{\boldsymbol{a}} = \mathrm{e}^{-\boldsymbol{a}\cdot\partial_{\boldsymbol{x}}}$, i.e. $T_{\boldsymbol{a}} H T_{\boldsymbol{a}}^\dagger \neq H$, despite the uniform magnetic field. However, it is invariant under the translation $T_{\boldsymbol{a}}$ followed by the gauge transformation

$$G_{\boldsymbol{a}} = \mathrm{e}^{\mathrm{i}\frac{1}{2}Bq_e(xa^y - ya^x)}. \tag{7.2.2}$$

or, more precisely,

$$G_{\boldsymbol{a}} T_{\boldsymbol{a}} H (G_{\boldsymbol{a}} T_{\boldsymbol{a}})^\dagger = H$$

The combined transformation $G_{\boldsymbol{a}} T_{\boldsymbol{a}}$ is a symmetry of H and is called the magnetic translation. The magnetic translations in different directions do not commute:

$$(G_{\boldsymbol{a}} T_{\boldsymbol{a}})(G_{\boldsymbol{b}} T_{\boldsymbol{b}}) = \mathrm{e}^{-\mathrm{i}q_e B(a^x b^y - a^y b^x)}(G_{\boldsymbol{b}} T_{\boldsymbol{b}})(G_{\boldsymbol{a}} T_{\boldsymbol{a}}) \tag{7.2.3}$$

Thus, the eigenstates of H cannot be labeled by the momentum vector (k_x, k_y), despite the Hamiltonian having magnetic translation symmetries in both the x and y directions. Note that the phase $q_e B(a^x b^y - a^y b^x)$ is just the phase obtained by moving the charge q_e around the parallelogram $P_{\boldsymbol{ab}}$ spanned by $\boldsymbol{a}$ and $\boldsymbol{b}$, namely $\mathrm{e}^{\mathrm{i}q_e \oint_{P_{\boldsymbol{ab}}} \mathrm{d}\boldsymbol{x}\cdot\boldsymbol{A}}$.

To find the energy levels of eqn (7.2.1), we introduce the complex coordinate $z = x + \mathrm{i}y$, so that

$$\begin{cases} \partial_z = \frac{1}{2}(\partial_x - \mathrm{i}\partial_y), & x = \frac{1}{2}(z + \bar{z}), \\ \partial_{\bar{z}} = \frac{1}{2}(\partial_x + \mathrm{i}\partial_y), & y = \frac{1}{2\mathrm{i}}(z - \bar{z}). \end{cases}$$

In terms of z and $\bar{z}$, we have

$$H = -\frac{1}{m}(D_z D_{\bar{z}} + D_{\bar{z}} D_z)$$

where

$$\begin{cases} D_z = \partial_z - \mathrm{i}q_e A_z, & A_z = \frac{1}{2}(A_x - \mathrm{i}A_y) = \frac{B}{4\mathrm{i}}\bar{z} \\ D_{\bar{z}} = \partial_{\bar{z}} - \mathrm{i}q_e A_{\bar{z}}, & A_{\bar{z}} = \frac{1}{2}(A_x + \mathrm{i}A_y) = -\frac{B}{4\mathrm{i}}z \end{cases}$$

The Hamiltonian H can be simplified through a non-unitary transformation $\mathrm{e}^{+|z|^2/4l_B^2}$, where l_B is the magnetic length defined by

$$l_B^2 = \left|\frac{c\hbar}{q_e B}\right|$$

(Note that $2\pi l_B^2 B = \frac{hc}{q_e} = \Phi_0$ gives us a unit flux quantum.) We find that

$$H = \mathrm{e}^{-|z|^2/4l_B^2} \left[-\frac{1}{m} \left(\tilde{D}_z \tilde{D}_{\bar{z}} + \tilde{D}_{\bar{z}} \tilde{D}_z \right) \right] \mathrm{e}^{+|z|^2/4l_B^2}$$

where

$$\tilde{D}_z = \mathrm{e}^{+|z|^2/4l_B^2} D_z \mathrm{e}^{-|z|^2/4l_B^2} = \partial_z - \frac{q_e B}{4c}\bar{z} - \frac{1}{4l_B^2}\bar{z} = \partial_z - \frac{1}{2l_B^2}\bar{z}$$

$$\tilde{D}_{\bar{z}} = \partial_{\bar{z}} + \frac{q_e B}{4c} z - \frac{1}{4l_B^2} z = \partial_{\bar{z}}$$

Here we have assumed that $q_e B > 0$. We see that the non-unitary transformation $\mathrm{e}^{+|z|^2/4l_B^2}$ can be regarded as a 'gauge transformation' that makes $A_{\bar{z}} = 0$. We can further simplify H as follows:

$$H = \mathrm{e}^{-|z|^2/4l_B^2} \left[-\frac{2}{m} \left(\partial_z - \frac{1}{2l_B^2}\bar{z} \right) \partial_{\bar{z}} \right] \mathrm{e}^{+|z|^2/4l_B^2} + \frac{1}{2}\hbar\omega_c$$

where $\omega_c = q_e B/mc$ is the cyclotron frequency.

Now we can easily write down the eigenstates of an electron in a uniform magnetic field as follows:

$$\Psi_0 = \mathrm{e}^{-|z|^2/4l_B^2} f(z) \qquad E = \frac{1}{2}\hbar\omega_c$$

$$\Psi_1 = \mathrm{e}^{-|z|^2/4l_B^2} \left(\partial_z - \frac{1}{2l_B^2}\bar{z} \right) f(z) \qquad E = \left(\frac{1}{2} + 1 \right) \hbar\omega_c$$

$$\Psi_n = \mathrm{e}^{-|z|^2/4l_B^2} \left(\tilde{D}_z \right)^n f(z) \qquad E = \left(\frac{1}{2} + n \right) \hbar\omega_c$$

which gives us the Landau level structure.

In the above, we have assumed that $q_e B > 0$. The electron wave function in the zeroth Landau level is an analytic function $f(z)$ times $\mathrm{e}^{-|z|^2/4l_B^2}$. If $q_e B < 0$, then the non-unitary transformation $\mathrm{e}^{+|z|^2/4l_B^2}$ will transform A^z to zero. The electron wave function in the zeroth Landau level will be an anti-analytic function $f(z^*)$ times $\mathrm{e}^{-|z|^2/4l_B^2}$.

The wave functions Ψ_0 in the zeroth Landau level can be expanded by the basis states

$$\psi_m = \sqrt{N_m} z^m \mathrm{e}^{-|z|^2/4l_B^2}$$

which carry angular momentum m. The wave function ψ_m has a ring shape. The position of the maximum of $|\psi_m| = \mathrm{e}^{m \ln|z| - |z|^2/4l_B^2}$ is at

$$|z| = r_m = \sqrt{2m} l_B$$

We find that the ring of the mth state has an area of $\pi r_m^2 = 2\pi l_B^2 m$, which encloses m flux quanta (see Fig. 7.10). Therefore, there is one state for every flux quantum,

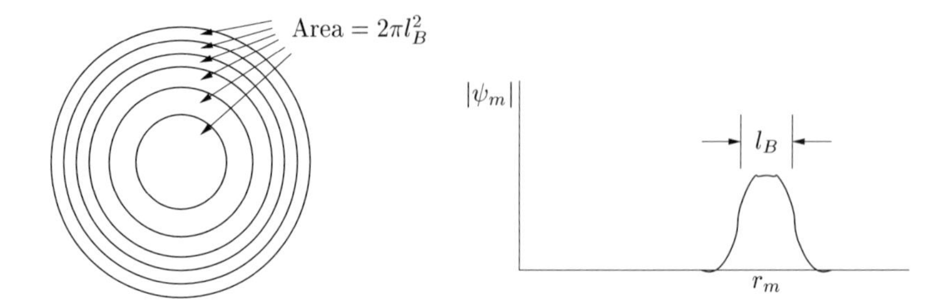

FIG. 7.10. The circular orbits in the zeroth Landau level.

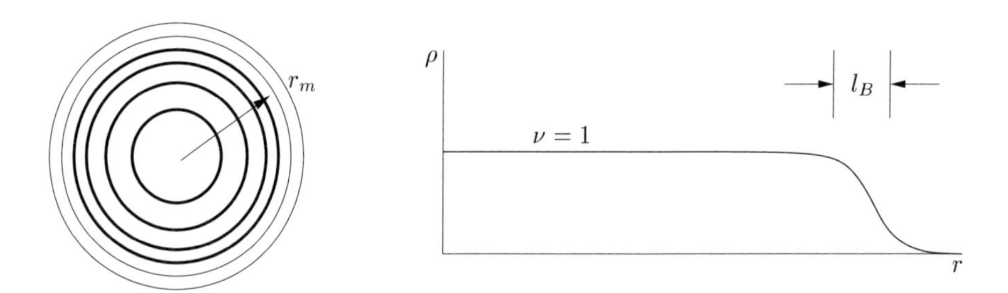

FIG. 7.11. The density profile of the $\nu = 1$ droplet, where the first m levels (represented by the thick lines) are filled.

and the number of states in the zeroth Landau level is equal to the number of flux quanta. So, when $\nu = 1$, every state in the zeroth Landau level is filled by an electron. This gives us a finite gap $\Delta = \hbar\omega_c$ for excitations, and explains the $\nu = 1$ plateau observed in experiments. Incompressibility is due to the Pauli exclusion principle (Fermi statistics).

The many-body wave function of the $\nu = 1$ state has the form

$$\Psi(z_1, \cdots z_N) = \mathcal{A}\left[(z_1)^0 (z_2)^1 (z_3)^2 \cdots\right] \mathrm{e}^{-\sum |z_i|^2/4l_B^2}$$

$$= \det \begin{vmatrix} 1 & 1 & \dots \\ z_1 & z_2 & \dots \\ (z_1)^2 & (z_2)^2 & \dots \\ \vdots & \vdots & \vdots \end{vmatrix} \mathrm{e}^{-\sum |z_i|^2/4l_B^2} = \prod_{i<j} (z_i - z_j) \mathrm{e}^{-\sum |z_i|^2/4l_B^2}$$

where $\mathcal{A}$ is the anti-symmetrization operator. The above wave function describes a circular droplet with uniform density (Fig. 7.11), because every electron occupies the same area $\pi r_{m+1}^2 - \pi r_m^2 = 2\pi l_B^2$. This property is important. A generic wave function of N variables may not have a uniform density in the $N \to \infty$ limit. In this case, the wave function does not have a sensible thermodynamical limit.

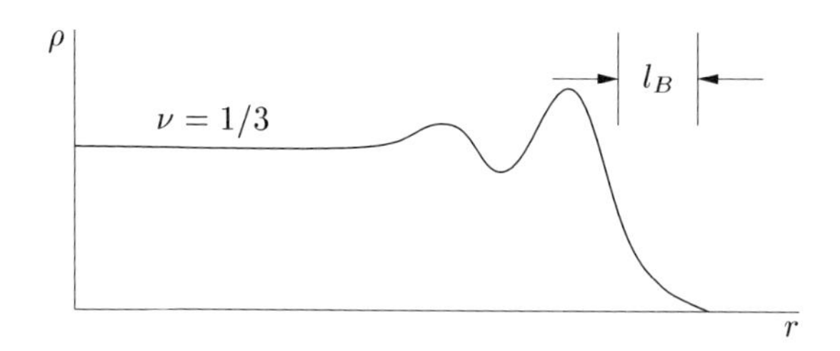

FIG. 7.12. The density profile of the $\nu = 1/3$ droplet.

Problem 7.2.1.
Find the energy levels of an electron in a potential $V = \frac{1}{2}m\omega_0^2 \boldsymbol{r}^2$ and a uniform magnetic field B.

Problem 7.2.2.
(a) Show that the magnetic translation $G_{\boldsymbol{a}}T_{\boldsymbol{a}}$ is a symmetry of the Hamiltonian in the uniform magnetic field (7.2.1) if the gauge transformation $G_{\boldsymbol{a}}$ is given by eqn (7.2.2).
(b) Show that the magnetic translations satisfy the algebra (7.2.3).

7.2.2 Fractional quantum Hall effect

- The Laughlin wave functions of FQH states.
- The plasma analogue and the density profile of Laughlin states.

When $0 < \nu < 1$, the zeroth Landau level is only partially filled. This gives us a huge ground-state degeneracy $\sim \frac{N_\phi!}{N!(N_\phi - N)!}$ for free electrons (where N_ϕ is the number of states in the zeroth Landau level, which is also the number of flux quanta, and N is the number of electrons). Therefore, the experimentally observed $\nu = 1/3$ state must be due to the interaction. Two questions arise. Firstly, how can the interaction produce a gap (incompressibility)? Secondly, why is $\nu = 1/3$ special?

Laughlin answered the above two questions with the trial wave function

$$\Psi_3 = \prod_{i<j}(z_i - z_j)^3 \mathrm{e}^{-\sum |z_i|^2/4l_B^2}$$

The wave function Ψ_3 is good due to the following reasons.

a) Ψ_3 has a third-order zero $(z_i - z_j)^3$ for any pair of electrons. This is good for repulsive interactions.
b) Ψ_3 describes a circular droplet of uniform density (Fig. 7.12), with $\nu = 1/3$.
c) Ψ_3 is an incompressible state.

Due to the rotational invariance of $|\Psi_3|$, it is clear that $|\Psi_3|$ describes a circular droplet. If we assume that the droplet also has a uniform density, then it is easy to understand why it has a filling fraction $\nu = 1/3$. This is because the highest power of z_i (with i fixed) is $3(N-1)$. The orbit at angular momentum $3(N-1)$ has a radius of $r_{\max} = \sqrt{2 \times 3(N-1)} l_B$. Here $r_{\max}$ is also the radius of the droplet, which is $\sqrt{3}$ times the radius $\sqrt{2 \times (N-1)} l_B$ of the $\nu = 1$ droplet. Thus, $\nu = \frac{1}{3}$ for Ψ_3.

To understand why Ψ_3 has a uniform density, let us consider the following joint probability distribution of electron positions:

$$P(z_1 \cdots z_N) \propto |\Psi_m(z_1 \cdots z_N)|^2 = \mathrm{e}^{-\beta V(z_1 \cdots z_N)} \tag{7.2.4}$$

where we have generalized the Ψ_3 state to the Ψ_m state, given by

$$\Psi_m = \prod_{i<j} (z_i - z_j)^m \mathrm{e}^{-\sum |z_i|^2/4l_B^2}$$

We see that m must be an odd integer for Ψ_m to be totally anti-symmetric. Choosing $T = \frac{1}{\beta} = \frac{m}{2}$ in eqn (7.2.4), we can view

$$V = -m^2 \sum_{i<j} \ln|z_i - z_j| + \frac{m}{4l_B^2} \sum_i |z_i|^2$$

as the potential for a two-dimensional plasma of 'charge' m particles and $P(z_1 \cdots z_N)$ as the probability distribution of the plasma. The potential between two particles with 'charges' m_1 and m_2 is $-m_1 m_2 \ln(r)$, and the force is $m_1 m_2/r$. Thus, to understand the density distribution of Ψ_m, we just need to calculate the 'charge' distribution of the plasma.

In V, we see that each particle in the plasma sees a potential $m|z|^2/4l_B^2$, which can be viewed as the potential produced by a background 'charge' distribution. In fact, such a potential is produced by a uniform background 'charge' with 'charge' density $\rho_\phi = 1/2\pi l_B^2$, which is just the density of the flux quanta. To see this, we note that the force of such a background 'charge' acting on a 'charge' m at radius r is a force between 'charge' m and the total 'charge' within a radius r. Such a force is given by $F = m\rho_\phi \pi r^2/r = m\rho_\phi \pi r$. This gives us a potential $\frac{m}{2}\pi\rho_\phi r^2$, which is just the above potential. Due to the complete screening property of plasma, the plasma wants to be 'charge' neutral. The plasma 'charge' density is equal to the background 'charge' density $mn_e = \rho_\phi$, where n_e is the electron density. We find the filling fraction to be $\nu = n_e/\rho_\phi = \frac{1}{m}$.

Why is Ψ_m an incompressible state? Note that the total angular momentum of Ψ_m (the total power of z_i) is $M_0 = m\frac{N(N-1)}{2}$. Here Ψ_m is the state with the minimum total angular momentum, which has mth-order zeros between every pair of electrons. Compressing the droplet reduces the total angular momentum and creates lower-order zeros, which costs energy. We expect that the energy gap is of the order of the Coulomb energy e^2/l_B. There is no mathematical proof of the above statement, but we have plenty of numerical evidence.

We would like to mention that, although Ψ_3 is only an approximate ground state for electrons with a Coulomb interaction, there exists an ideal Hamiltonian for which Ψ_3 is the exact ground state. The ideal Hamiltonian has the form

$$H = \frac{1}{2m} \sum_i (\partial_{\boldsymbol{x}_i} - \mathrm{i} q_e \boldsymbol{A})^2 - \sum_{i<j} UV(\boldsymbol{r}_i - \boldsymbol{r}_j), \qquad V(\boldsymbol{r}) = \partial_{\boldsymbol{r}}^2 \delta(\boldsymbol{r}). \quad (7.2.5)$$

One can show that, for any state ψ, we have $\langle \psi | H | \psi \rangle \geqslant \frac{1}{2} N \hbar \omega_c$, and for Ψ_3 we have $\langle \Psi_3 | H | \Psi_3 \rangle = \frac{1}{2} N \hbar \omega_c$. Thus, Ψ_3 is indeed the exact ground state of H. Numerical calculations also indicate that Ψ_3 is incompressible for the ideal Hamiltonian.

Problem 7.2.3.
Show that $\langle \Psi_3 | H | \Psi_3 \rangle = \frac{1}{2} N \hbar \omega_c$ for the ideal Hamiltonian (7.2.5). Find an ideal Hamiltonian for the $\nu = 1/m$ Laughlin state Ψ_m.

Problem 7.2.4.
A double-layer FQH state is described by the wave function

$$\Psi_{lmn} = \prod_{i<j} (z_i - z_j)^l \prod_{i<j} (w_i - w_j)^m \prod_{i,j} (z_i - w_j)^n \, \mathrm{e}^{-\sum |z_i|^2/4l_B^2} \, \mathrm{e}^{-\sum |w_i|^2/4l_B^2} \quad (7.2.6)$$

where z_i are the coordinates of the electrons in the first layer and w_i in the second layer. Such a state is called the (lmn) state. Let N_1 and N_2 be the numbers of electrons in the two layers. Find the ratio N_1/N_2 such that the sizes of the electron droplets in the two layers are equal. Find the total filling fraction of the electrons in the two layers. (Hint: You may first assume that $n = 0$ to simplify the problem.)

7.2.3 Quasiparticles with fractional charge and fractional statistics

- Fractional charge and fractional statistics from the plasma analogue.

A quasihole excitation above the incompressible ground state Ψ_m is described by the many-body wave function

$$\Psi^h(\xi, \xi^*) = \sqrt{C(\xi, \xi^*)} \prod_i (\xi - z_i) \Psi_m,$$

where ξ is the position of the quasihole and $C(\xi, \xi^*)$ is the normalization coefficient. The simplest way to calculate the charge of the quasihole is to first remove an electron from the Ψ_m state and obtain the wave function $\prod_i (\xi - z_i)^m \Psi_m$, which has a charge e hole at ξ. However, $\prod_i (\xi - z_i)^m \Psi_m$ can also be viewed as the wave function with m quasiholes at ξ. Thus, the quasihole charge is e/m.

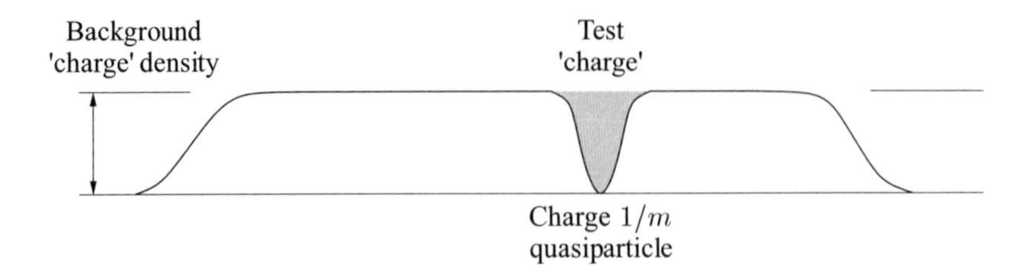

FIG. 7.13. The charge from the electrons and the test charge must neutralize the background charge.

A more direct calculation of the quasihole charge is through the plasma analogue. The electron position distribution in the quasihole state Ψ_h is given by

$$P_\xi^h(\{z_i\}) = \left|\Psi^h(\xi, \xi^*)\right|^2 = \mathrm{e}^{-\beta[V(z_i) - m\ln|\xi - z_i|]}$$

Therefore, the electrons see an addition background potential $-m \ln|\xi - z_i|$. This potential is produced by a unit test charge. The background charge contains an extra unit test charge at ξ. There must be $1/m$ less electrons near ξ to maintain the charge neutrality (see Fig. 7.13). Thus, the quasihole carries a charge e/m.

To understand the fractional statistics of the quasiholes (Arovas *et al.*, 1984), let us start with the following effective Lagrangian for several quasiparticles:

$$Z = \int \mathcal{D}[\boldsymbol{x}_i(t)] \, \mathrm{e}^{\,\mathrm{i}\int \mathrm{d}t \, \sum_i [-E_0 + \frac{1}{2}m\dot{\boldsymbol{x}}_i^2 + \boldsymbol{a}(\boldsymbol{x}_1, \boldsymbol{x}_2, \ldots)\cdot\dot{\boldsymbol{x}}_i]}$$

The effective gauge potential $\boldsymbol{a}$ determines the statistics. Just like the spin systems discussed in Section 2.3, $\boldsymbol{a}$ is determined from the Berry phase of the coherent state $|\Psi^h(\xi(t), \xi^*(t))\rangle$.

Let us first consider only a single quasiparticle. For an adiabatic motion of the quasiparticle $\xi(t)$, the phase change of $\Psi^h(\xi(t), \xi^*(t))$ gives us the Berry phase. For a small Δt, the Berry phase is

$$\langle\Psi^h(\xi(t+\Delta t), \xi^*(t+\Delta t))|\Psi^h(\xi(t), \xi^*(t))\rangle = \mathrm{e}^{\,\mathrm{i}\Delta t \boldsymbol{a}\cdot\dot{\boldsymbol{x}}}.$$

Thus, $\boldsymbol{a}$ is given by

$$\mathrm{i}\langle\Psi^h(\xi(t), \xi^*(t))|\frac{\mathrm{d}}{\mathrm{d}t}|\Psi^h(\xi(t), \xi^*(t))\rangle = \boldsymbol{a}\cdot\dot{\boldsymbol{x}}$$

As Ψ^h depends on t only through $\xi(t)$, we have

$$\begin{aligned}\boldsymbol{a}\cdot\dot{\boldsymbol{x}} &= \mathrm{i}\langle\Psi^h(\xi(t), \xi^*(t))|\frac{\mathrm{d}}{\mathrm{d}t}|\Psi^h(\xi(t), \xi^*(t))\rangle \\ &= \frac{\mathrm{d}\xi}{\mathrm{d}t}\mathrm{i}\langle\Psi^h(\xi, \xi^*)|\frac{\partial}{\partial\xi}|\Psi^h(\xi, \xi^*)\rangle + \frac{\mathrm{d}\xi^*}{\mathrm{d}t}\mathrm{i}\langle\Psi^h(\xi, \xi^*)|\frac{\partial}{\partial\xi^*}|\Psi^h(\xi, \xi^*)\rangle\end{aligned}$$

Thus, if we write $\boldsymbol{a} \cdot d\boldsymbol{x} = a_\xi d\xi + a_{\xi^*} d\xi^*$, then[44]

$$a_\xi = \mathrm{i}\langle \Psi^h(\xi,\xi^*)|\frac{\partial}{\partial \xi}|\Psi^h(\xi,\xi^*)\rangle = -\mathrm{i}\sqrt{C}\frac{\partial}{\partial \xi}\frac{1}{\sqrt{C}}$$

Here $C(\xi,\xi^*) = \langle \Psi(\xi)|\Psi(\xi)\rangle$ and $\Psi(\xi) = \prod(\xi - z_i)\Psi_m$ which is an analytic function of ξ. The key point of the above result is that the normalization of Ψ (an analytic function of ξ) gives us the Berry phase:

$$a_\xi = +\frac{\mathrm{i}}{2}\frac{\partial}{\partial \xi}\ln C, \qquad a_{\xi^*} = -\frac{\mathrm{i}}{2}\frac{\partial}{\partial \xi^*}\ln C$$

The normalization $C(\xi,\xi^*)$ can be calculated from the plasma analogue. We note that, after including a term $\mathrm{e}^{-|\xi|^2/4ml_B^2}$ representing the interaction between the test 'charge' and the background 'charge',[45]

$$\int \prod_i \mathrm{d}^2 z_i \left| \mathrm{e}^{-\frac{1}{4ml_B^2}|\xi|^2} \prod(\xi - z_i) \prod(z_i - z_j)^m \mathrm{e}^{-\frac{1}{4l_B^2}|z_i|^2} \right|^2 \propto \mathrm{e}^{-\beta V(\xi,\xi^*)}$$

gives the total energy of the plasma with a test 'charge' at ξ. Again, due to the complete screening of the plasma, the total force acting on the inserted test 'charge' vanishes; hence the energy of plasma does not depend on ξ, i.e. $V(\xi,\xi^*) = $ constant. We see that

$$C(\xi,\xi^*) \propto \mathrm{e}^{\frac{1}{2l_B^2}\frac{1}{m}|\xi|^2}$$

This gives us

$$a_\xi = \mathrm{i}\frac{1}{m}\frac{1}{4l_B^2}\xi^*, \qquad a_{\xi^*} = -\mathrm{i}\frac{1}{m}\frac{1}{4l_B^2}\xi$$

---

[44] We have used

$$\begin{aligned}
&\mathrm{i}\langle \Psi^h(\xi,\xi^*)|\frac{\partial}{\partial \xi}|\Psi^h(\xi,\xi^*)\rangle = \mathrm{i}\langle \Psi(\xi)|\frac{1}{\sqrt{C(\xi,\xi^*)}}\frac{\partial}{\partial \xi}\frac{1}{\sqrt{C(\xi,\xi^*)}}|\Psi(\xi)\rangle \\
&= \mathrm{i}\int \prod_i \mathrm{d}^2 z_i\, \Psi^*(\xi^*)\frac{1}{\sqrt{C}}\frac{\partial}{\partial \xi}\left(\frac{1}{\sqrt{C}}\Psi(\xi)\right) \\
&= i\frac{\partial}{\partial \xi}\int \prod_i \mathrm{d}^2 z_i\, \Psi^*(\xi^*)\frac{1}{C}\Psi(\xi) - \mathrm{i}\int \prod_i \mathrm{d}^2 z_i\, \Psi^*(\xi^*)\left(\frac{\partial}{\partial \xi}\frac{1}{\sqrt{C}}\right)\frac{1}{\sqrt{C}}\Psi(\xi) \\
&= -\mathrm{i}\sqrt{C}\frac{\partial}{\partial \xi}\frac{1}{\sqrt{C}}
\end{aligned}$$

[45] Note that $\mathrm{e}^{-|\xi|^2/4l_B^2}$ corresponds to the interaction between an electron and the background 'charge'. An electron corresponds to a bound state of m test 'charges'.

The above vector potential $\boldsymbol{a}$ is proportional to the vector potential of the magnetic field:

$$a_\xi = -\frac{1}{m} q_e A_z, \qquad a_{\xi^*} = -\frac{1}{m} q_e A_{z^*}$$

Around a loop, the quasihole picks up an A–B phase $\oint \mathrm{d}\boldsymbol{x} \cdot \boldsymbol{a} = \left(-\frac{q_e}{m}\right) \oint \mathrm{d}\boldsymbol{x} \cdot \boldsymbol{A}$, a phase for a charge $-q_e/m = e/m$ particle.

For two quasiholes, the total energy V of the plasma with two test 'charges' is given by

$$\mathrm{e}^{-\beta V(\xi_1, \xi_1^*, \xi_2, \xi_2^*)}$$
$$= \left| \mathrm{e}^{-\frac{1}{4l_B^2} \frac{1}{m}(|\xi_1|^2 + |\xi_2|^2)} |\xi_1 - \xi_2|^{\frac{1}{m}} \prod_{a,i} (\xi_a - z_i) \prod (z_i - z_j)^m \mathrm{e}^{-\frac{1}{4l_B^2} \sum |z_i|^2} \right|^2$$

where the extra term $|\xi_1 - \xi_2|^{2/m} = \mathrm{e}^{-\beta(-) \ln |\xi_1 - \xi_2|}$ represents the direct interaction between the two test 'chargeis', each carries 'charge' 1. Again, $V(\xi_1, \xi_1^*, \xi_2, \xi_2^*)$ is independent of ξ_1 and ξ_2. Thus,

$$|\Psi(\xi_1, \xi_2)|^2 \propto \mathrm{e}^{\frac{1}{2l_B^2} \frac{1}{m}(|\xi_1|^2 + |\xi_2|^2)} |\xi_1 - \xi_2|^{-\frac{2}{m}}$$

Setting $\xi = \xi_1$ and $\xi_2 = 0$, we find that

$$a_\xi = \mathrm{i} \frac{1}{m} \frac{1}{4l_B^2} \xi^* - \frac{\mathrm{i}}{2m} \frac{\partial}{\partial \xi} \ln |\xi|^2 = \mathrm{i} \frac{1}{m} \frac{1}{4l_B^2} \xi^* - \frac{\mathrm{i}}{2m} \frac{1}{\xi}$$
$$a_{\xi^*} = -\mathrm{i} \frac{1}{m} \frac{1}{4l_B^2} \xi + \frac{\mathrm{i}}{2m} \frac{1}{\xi^*}$$

which gives us $(a_x, a_y) = (-y, x) \frac{1}{r^2} \frac{1}{m} + \ldots$. As we move the quasihole ξ_1 around ξ_2, the quasihole obtains a phase given by $\frac{e}{m} B \times \text{area} + \frac{2\pi}{m}$. The first term is due to the uniform magnetic field. The additional term $\frac{2\pi}{m}$ comes from $(-y, x) \frac{1}{r} \frac{1}{m}$. Such a term gives the quasiholes fractional statistics with a statistical angle $\theta = \frac{\pi}{m}$ (see eqn (7.1.1)).

Problem 7.2.5.
Consider the double-layer (lmn) state eqn (7.2.6). A quasihole in the first layer is described by the wave function

$$\Psi_1^h(\xi_1) = \prod_i (\xi_1 - z_i) \Psi_{lmn}(\{z_i, w_j\})$$

and a quasihole in the second layer by

$$\Psi_2^h(\xi_2) = \prod_i (\xi_2 - w_i) \Psi_{lmn}(\{z_i, w_j\})$$

Find the fractional charge and the fractional statistics of the quasiholes in the first and second layers. Find the mutual statistics between a quasihole in the first layer and a quasihole

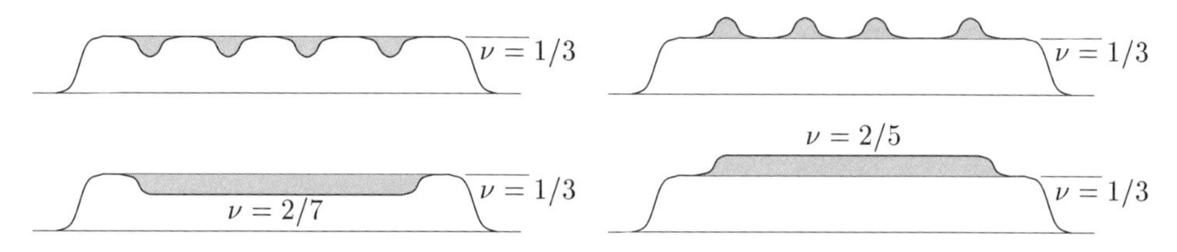

FIG. 7.14. The hierarchical $\nu = 2/7$ and $\nu = 2/5$ states.

in the second layer. The mutual statistics between two non-identical particles are defined by the Berry phase generated by moving one particle all the way around the other.

7.2.4 Hierarchical fractional quantum Hall states—generalization of Laughlin theory

- Hierarchical FQH states are equivalent to Laughlin states for quasiparticles.

To construct more general FQH states at filling fractions other than $1/m$, we may start with a $\nu = 1/m$ Laughlin state. We add quasiholes or quasiparticles to change the average electron density. When its density reaches a certain value, the gas of quasiholes/quasiparticles can also form a Laughlin state by itself. The resulting state is a hierarchical FQH state (see Fig. 7.14).

When the charge $e/3$ quasiholes on top of the $\nu = 1/3$ state condense into a Laughlin state, we obtain the $\nu = 2/7$ FQH state. The wave function has the form

$$\begin{aligned}&\Psi(z_1 \cdots z_N)\\ &= \int \prod_i \mathrm{d}^2\xi_i \prod (z_i - z_j)^3 \mathrm{e}^{-\frac{1}{4l_B^2}\sum |z_i|^2} \prod (\xi_i - z_j)(\xi_i^* - \xi_j^*)^2 \mathrm{e}^{-\frac{1}{4l_B^2}\frac{1}{3}|\xi_i|^2}\end{aligned}$$

When the charge $e/3$ quasiparticles on top of the $\nu = 1/3$ state condense into a Laughlin state, we get the $\nu = 2/5$ FQH state. The wave function is given by

$$\Psi_e(z_i - z_N) = \int \prod_i \mathrm{d}^2\xi_i \prod_{i<j} (\xi_i - \xi_j)^2 (\xi_i^* - 2\partial_{z_i}) \prod (z_i - z_j)^3 \mathrm{e}^{-\frac{1}{4l_B^2}|z_i|}$$

We can calculate the properties of the above two states using the plasma analogue. However, the calculation is complicated. In the next section, we are going to derive the effective field theory for FQH states. We will use the effective theory to calculate the properties of the above hierarchical FQH states. The effective field theory greatly simplifies the calculation. We can easily obtain the filling fractions of the FQH states and the quantum numbers of the quasiparticles.

7.3 Effective theory of fractional quantum Hall liquids

- Low-energy effective theories of FQH states are $U(1)$ Chern–Simons theories, which capture the universal properties of the FQH states.

We have already seen several different FQH states. On one hand, these QH states have different fractional charges and different fractional statistics. This suggests that the different FQH states belong to different quantum phases. On the other hand, these FQH states all have the same symmetry. We cannot use broken symmetry to distinguish them. In fact, FQH states contain a new kind of order—topological order (see Chapter 8). We need to use new tools to characterize topological orders.

One way to systematically study topological orders is to construct a low-energy effective theory for FQH liquids. The effective theory can capture the universal properties of FQH liquids and provide hints on how to characterize and label different topological orders in FQH liquids. In the following, we will introduce a way to construct effective theories that is closely related to the hierarchical construction proposed by Haldane and Halperin (Haldane, 1983; Halperin, 1984).

The first attempt to characterize the internal structures of FQH liquids was proposed by Girvin and MacDonald (1987), where it was shown that a Laughlin state contains an off-diagonal long-range order in a *non-local* operator. Such an observation leads to a description of FQH liquids in terms of Ginzburg–Landau–Chern–Simons effective theory (Read, 1989; Zhang *et al.*, 1989). These developments have led to many interesting and important results, and a deeper understanding of FQH liquids. However, here I will try to describe the internal structures of FQH liquids from a more general point of view. It appears that some internal structures of FQH liquids (especially those in the so-called non-abelian FQH liquids) cannot be described by the Ginzburg–Landau–Chern–Simons effective theory and the associated off-diagonal long-range orders. Thus, we need to develop more general concepts and formulations, such as topological orders and topological field theory, to describe the internal structures in FQH liquids (see Chapter 8). The concept of off-diagonal long-range order, in my opinion, does not capture the essence of the internal structure in FQH liquids. Here we will develop an effective theory without using off-diagonal long-range order (Blok and Wen, 1990a,b). The resulting effective theory contains only pure Chern–Simons terms. The pure Chern–Simons form is more compact and more clearly reveals the internal structure of the hierarchical FQH states.

7.3.1 Effective theory of the Laughlin states

- The hydrodynamics of FQH liquids and effective Chern–Simons theory.

In this section, we will consider only single-layer spin-polarized QH systems. To construct the effective theory for the hierarchical states, we will first try to obtain the effective theory of the Laughlin state. Then we will use the hierarchical construction to obtain the effective theory for the hierarchical states in the next section.

Consider an electron system in a magnetic field. To include more general situations, we will assume that the electrons can have bosonic or fermionic statistics. The Lagrangian of the system has the following form (the first quantized version):

$$\begin{aligned}\mathcal{L} = & -eA^i J^i + \text{kinetic/potential energy} \\ & = eA_i J^i + \text{kinetic/potential energy}\end{aligned} \tag{7.3.1}$$

where

$$\boldsymbol{J}(\boldsymbol{x}) = \sum_i \boldsymbol{v}_i \delta(\boldsymbol{x} - \boldsymbol{x}_i), \quad J^0(\boldsymbol{x}) = \sum_i \delta(\boldsymbol{x} - \boldsymbol{x}_i) \tag{7.3.2}$$

are the current and the density of the electrons, respectively, and $(\boldsymbol{x}_i, \boldsymbol{v}_i)$ are the position and the velocity of the ith electron. The kinetic/potential energy is given by $\sum_i \frac{1}{2} m \boldsymbol{v}_i^2 + \sum_{i<j} V(\boldsymbol{x}_i - \boldsymbol{x}_j)$, whose precise form is not important to us. We have also assumed that the charge of each electron is $-e$ and the speed of light is $c = 1$.

In a hydrodynamical approach, we assume that the low-energy collective modes can be described by the density and the current J^μ, and the low-energy effective theory has the form $\mathcal{L}(A_\mu, J^\nu) = eA_i J^i + \mathcal{L}'(J^\mu)$. From our discussion in Sections 5.1.3 and 5.3.3, we see that sometimes a state may have too many low-lying excitations to be described by single density mode. As the gapped FQH states have even fewer low-energy excitations than a superfluid, it is reasonable to assume that a single density mode is enough to describe the low-energy fluctuations in the FQH states.

At a filling fraction of $\nu = 1/m$, where m is an even integer for a bosonic electron and an odd integer for a fermionic electron, the ground state of the electron system is given by the Laughlin wave function (Laughlin, 1983)

$$\left[\prod (z_i - z_j)^m\right] \mathrm{e}^{-\frac{1}{4l_B^2}\sum |z_i|^2} \tag{7.3.3}$$

(Here we have assumed that $B < 0$, so that $(-e)B > 0$.) To construct the effective theory, we note that the state (7.3.3) is an incompressible fluid, so that its density is tied to the the magnetic field, i.e. $J^0 = (-e)\nu B/2\pi$. When combined with the finite Hall conductance $\sigma_{xy} = \frac{\nu e^2}{h} = \frac{\nu e^2}{2\pi\hbar} = \frac{\nu e^2}{2\pi}$, we find that the electron number current J^μ has the following response to a change of electromagnetic field:

$$-e\delta J^\mu = -\sigma_{xy}\varepsilon^{\mu\nu\lambda}\partial_\nu \delta A_\lambda = -\frac{\nu e^2}{2\pi}\varepsilon^{\mu\nu\lambda}\partial_\nu \delta A_\lambda \tag{7.3.4}$$

We choose the effective Lagrangian $\mathcal{L}(A_\mu, J^\nu)$ in such a way that it produces the equation of motion (7.3.4). It is convenient to introduce a $U(1)$ gauge field a_μ to describe the electron number current:

$$J^\mu = \frac{1}{2\pi}\partial_\nu a_\lambda \varepsilon^{\mu\nu\lambda}$$

The current defined in this way automatically satisfies the conservation law. Then the effective Lagrangian that produces eqn (7.3.4) takes the following form:

$$\mathcal{L} = \left[-m\frac{1}{4\pi}a_\mu\partial_\nu a_\lambda\varepsilon^{\mu\nu\lambda} + \frac{e}{2\pi}A_\mu\partial_\nu a_\lambda\varepsilon^{\mu\nu\lambda}\right] \tag{7.3.5}$$

Equation (7.3.5) describes only the linear response of the ground state to the external electromagnetic fields. To have a more complete description of the topological fluid, such as the FQH liquid, we need to introduce the electron excitations into our effective theory. In particular, we want to make sure that the effective theory contains an excitation that carries the same quantum numbers as the electron.

7.3.2 Electron and quasiparticle excitations in the effective theory

- Effective Chern–Simons theory alone is not a complete description of the FQH state. We need to specify the electron operator to have a complete effective theory.
- The required trivial mutual statistics between the electrons and the quasiparticle/quasihole determine the quantum numbers of the quasiparticles and quasiholes.

Let us introduce a particle that carries an a_μ charge of l. In the effective theory (7.3.5), such a particle corresponds to the following source term:

$$l a_0 \delta(\boldsymbol{x} - \boldsymbol{x}_0) \tag{7.3.6}$$

The source term will create an excitation of charge

$$Q = -le/m. \tag{7.3.7}$$

This can be seen from the equation of motion $\delta\mathcal{L}/\delta a_0 = 0$, with

$$J_0 = \frac{1}{2\pi}\varepsilon_{ij}\partial_i a_j = -\frac{e}{2\pi m}B + \frac{l}{m}\delta(\boldsymbol{x} - \boldsymbol{x}_0)$$

The first term on the right-hand side indicates that the filling fraction $\nu \equiv 2\pi\frac{J_0}{-eB}$ is indeed $\nu = 1/m$, and the second term corresponds to the increase in the electron density associated with the excitation.

We also see that the excitation created by the source term (7.3.6) is associated with l/m units of the a_μ flux. Thus, if we have two excitations carrying a_μ charges of l_1 and l_2, then moving one excitation around the other will induce a phase $2\pi \times$ (number of a_μ-flux quanta) $\times$ (a_μ charge),[46] namely

$$2\pi \times \frac{l_1}{m} \times l_2 \tag{7.3.8}$$

If $l_1 = l_2 \equiv l$, then the two excitations will be identical. Interchanging them will induce half of the phase in eqn (7.3.8), namely

$$\theta = \pi \frac{l^2}{m} \tag{7.3.9}$$

Here θ is simply the statistical angle of the excitation that carries l units of the a_μ charge.

Our electrons carry a charge of $-e$. From the above discussion, we see that a charge $-e$ excitation corresponds to a particle that carries m units of the a_μ charge. Such an excitation has a statistical angle $\theta = \pi m$ (see eqn (7.3.9)); thus, it is a boson if m is even and a fermion if m is odd. Therefore, for the case of a bosonic electron with even m and the case of a fermionic electron with odd m, we can identify the excitations of m units of the a_μ charge as the electrons that form the FQH liquid. The existence of the excitations with the electric charge and the statistics of the electron support the correctness of our Chern–Simons effective theory.

We would like to stress that the identification of the fundamental electrons in the effective theory is very important. It is this identification, together with the effective Lagrangian, that provides a complete description of the topological properties of the FQH liquid. We will see below that this identification allows us to determine the fractional charge and the fractional statistics of the quasiparticle excitations.

A quasihole excitation at a position described by a complex number $\xi = x + iy$ is created by multiplying $\prod_i (\xi - z_i)$ with the ground-state wave function (7.3.3). Note that the phase of the wave function changes by 2π as an electron goes around the quasihole. The 2π phase implies that the electron and the quasihole have trivial mutual statistics. In general, the electron and an allowed excitation must have trivial mutual statistics in order for the excitation to have a single-valued electron wave function.

Now let us try to create an excitation by inserting a source term of l units of the a_μ charge. Moving an electron around such an excitation will induce a phase $2\pi l$

[46] The careful reader may note that the two excitations carry both an a_μ charge and an a_μ flux. So the phase induced by moving one excitation around the other should receive two contributions, one from moving the charge around the flux and the other from moving the flux around the charge, as described in Fig. 7.8. This will lead to a phase $2\pi \times \frac{l_1}{m} \times l_2 + 2\pi \times \frac{l_2}{m} \times l_1$, which is twice as large as the one in eqn (7.3.8). However, a more careful calculation indicates that the picture of Fig. 7.8 does not apply to the charge–flux bound state induced by the Chern–Simons term. The naive result in eqn (7.3.8) happens to be the correct one (see Problem 7.3.1).

(see eqn (7.3.8)). The single-value property of the electron wave function requires such a phase to be a multiple of 2π. So the a_μ charges must be quantized as integers, and only those charges correspond to allowed excitations.

From the charge of the excitations given in eqn (7.3.7), we find that $l = -1$ corresponds to the fundamental quasihole excitation, while $l = 1$ corresponds to the fundamental quasiparticle excitation. The quasiparticle excitation carries an electric charge $-e/m$ and the quasihole carries an electric charge e/m. Both of them have statistics $\theta = \pi/m$, as one can see from eqn (7.3.9). We see that the effective theory reproduces the well-known results for quasiparticles in Laughlin states (Arovas *et al.*, 1984). The full effective theory with quasiparticle excitations is given by

$$\begin{aligned}\mathcal{L} = &\left[-m\frac{1}{4\pi}a_\mu \partial_\nu a_\lambda \varepsilon^{\mu\nu\lambda} + \frac{e}{2\pi}A_\mu \partial_\nu a_\lambda \varepsilon^{\mu\nu\lambda}\right] \\ &+ l a_\mu j^\mu + \text{kinetic/potential energy}\end{aligned} \tag{7.3.10}$$

where j^μ is the current of the quasiparticles, which has the form given in eqn (7.3.2). Equation (7.3.10), together with the quantization condition on l, is a complete low-energy effective theory which captures the topological properties of the $1/m$ Laughlin state.

Problem 7.3.1.

Prove eqn (7.3.8). (Hint: You may start with the following effective theory with two excitations:

$$\mathcal{L} = \left[-\frac{1}{2}\frac{m}{2\pi}a_\mu \partial_\nu a_\lambda \varepsilon^{\mu\nu\lambda} + j^\mu a_\mu\right]$$

where $j^\mu = j_1^\mu + j_2^\mu$ and

$$\begin{aligned} j_1^0(\boldsymbol{x},t) &= l_1\delta(\boldsymbol{x}-\boldsymbol{x}_1(t)), & j_2^0(\boldsymbol{x},t) &= l_2\delta(\boldsymbol{x}-\boldsymbol{x}_2(t)), \\ j_1^i(\boldsymbol{x},t) &= l_1\dot{x}_1^i\delta(\boldsymbol{x}-\boldsymbol{x}_1(t)), & j_2^i(\boldsymbol{x},t) &= l_2\dot{x}_2^i\delta(\boldsymbol{x}-\boldsymbol{x}_2(t)). \end{aligned}$$

Here j^μ is the total current of the two excitations and $\boldsymbol{x}_{1,2}(t)$ are the locations of the two excitations. You can then integrate out a_μ and obtain

$$\int \mathrm{d}^3x\, \frac{1}{2} j^\mu \frac{1}{\frac{m}{2\pi}\partial_\lambda \epsilon^{\mu\lambda\nu}} j^\nu$$

The cross-term can be written as

$$\int \mathrm{d}^3x\, j_1^\mu \frac{1}{\frac{m}{2\pi}\partial_\lambda \epsilon^{\mu\lambda\nu}} j_2^\nu = \int \mathrm{d}^3x\, j_1^\mu f_\mu$$

where f_μ satisfies

$$\frac{m}{2\pi}\partial_\lambda \epsilon^{\mu\lambda\nu} f_\nu = j_2^\mu$$

You can find f_i by assuming that $\boldsymbol{x}_2 = \dot{\boldsymbol{x}}_2 = 0$.)

Problem 7.3.2.
We have been concentrating on the topological properties of the Laughlin states. To have some understanding of the dynamical properties of the Laughlin states, we need to include the Maxwell term in the effective Chern–Simons theory as follows:

$$\mathcal{L} = -m\frac{1}{4\pi}a_\mu \partial_\nu a_\lambda \varepsilon^{\mu\nu\lambda} + \frac{1}{2g_1}\boldsymbol{e}^2 - \frac{1}{2g_2}b^2 \tag{7.3.11}$$

where $\boldsymbol{e}$ and b are the electric field and the magnetic field, respectively, of a_μ, and $e_i = \dot{a}_i - \partial_i a^0$ and $b = \partial_1 a_2 - \partial_2 a_1$. Find the equation of motion for the collective fluctuations described by $\boldsymbol{e}$ and b. Find the energy gap of the collective excitations. Assume that the energy gap in the $1/m$ Laughlin state is of order $e^2/m^2 l_B$ and the size of the quasihole is of order l_B. Estimate the values of g_1 and g_2.

7.3.3 Effective theory of the hierarchical fractional quantum Hall states

- Different hierarchical FQH states can be characterized by an integer K-matrix and a charge vector $\boldsymbol{q}$.
- The topological properties of the FQH state, such as filling fractions and quasiparticle quantum numbers, can be calculated easily from the K-matrix and the charge vector.
- The equivalence relations between different pairs of K-matrix and charge vector.

To obtain an effective theory of the hierarchical FQH states, let us start with a $1/m$ Laughlin state formed by fermionic electrons. Let us increase the filling fraction by creating the fundamental quasiparticles, which are labeled by $l = 1$. Equation (7.3.10) with $l = 1$ describes the $1/m$ state in the presence of these quasiparticles. Now, two equivalent pictures emerge.

(a) In a mean-field theory approach, we may view the gauge field a_μ in eqn (7.3.10) as a fixed background and we do not allow a_μ to respond to the inserted source term j^μ. In this case, the quasiparticle gas behaves like bosons in the 'magnetic' field $b = \partial_i a_j \varepsilon_{ij}$, as one can see from the second term in eqn (7.3.10). These bosons do not carry any electric charge because the quasiparticle number current j^μ does not directly couple with the electromagnetic gauge potential A_μ. When the boson density satisfies

$$j^0 = \frac{1}{p_2}\frac{b}{2\pi}$$

where p_2 is even, the bosons have a filling fraction $\frac{1}{p_2}$. The ground state of the bosons can again be described by a Laughlin state. The final electronic state that we obtained is just the second-level hierarchical FQH state constructed by Haldane (1983).

(b) If we let a_μ respond to the insertion of j^μ, then quasiparticles will be dressed by the a_μ flux. The dressed quasiparticles carry an electric charge of $-e/m$ and statistics of $\theta = \pi/m$. When the quasiparticles have the density

$$j^0 = \frac{1}{(p_2 - \frac{\theta}{\pi})} \frac{(-e)B}{2\pi m}$$

where p_2 is even, the quasiparticle will have a filling fraction $\frac{1}{(p_2 - \frac{\theta}{\pi})}$. In this case, the quasiparticle system can form a Laughlin state described by the wave function

$$\prod_{i<j} (z_i - z_j)^{p_2 - \frac{\theta}{\pi}}$$

The final electronic state obtained this way is again a second-level hierarchical FQH state. This construction was first proposed by Halperin (1984). The two constructions in (a) and (b) lead to the same hierarchical state and are equivalent.

In the following, we will follow Haldane's hierarchical construction to derive the Chern–Simons effective theory of hierarchical FQH states (Blok and Wen, 1990a,b; Read, 1990; Fröhlich and Kerler, 1991; Wen and Zee, 1992a). Notice that, under the assumption (a), the boson Lagrangian (the second term in eqn (7.3.10) with $l = 1$) is just eqn (7.3.1) with an external electromagnetic field eA_μ replaced by a_μ. Thus, we can follow the same steps as from eqn (7.3.3) to eqn (7.3.10) to construct the effective theory of the boson Laughlin state. Introducing a new $U(1)$ gauge field $\tilde{a}_\mu$ to describe the boson current, we find that the boson effective theory takes the form

$$\mathcal{L} = -\frac{p_2}{4\pi} \tilde{a}_\mu \partial_\nu \tilde{a}_\lambda \varepsilon^{\mu\nu\lambda} + \frac{1}{2\pi} a_\mu \partial_\nu \tilde{a}_\lambda \varepsilon^{\mu\nu\lambda} \tag{7.3.12}$$

In eqn (7.3.12), the new gauge field $\tilde{a}_\mu$ describes the density j^0 and the current j^i of the bosons, and is given by

$$j^\mu = \frac{1}{2\pi} \partial_\nu \tilde{a}_\lambda \varepsilon^{\mu\nu\lambda}$$

This reduces the coupling between a_μ and the boson current, $a_\mu j^\mu$, to a Chern–Simons term between a_μ and $\tilde{a}_\mu$ (which becomes the second term in eqn (7.3.12)). The total effective theory (including the original electron condensate) has the form

$$\begin{aligned} \mathcal{L} = & \left[-\frac{p_1}{4\pi} a_\mu \partial_\nu a_\lambda \varepsilon^{\mu\nu\lambda} + \frac{e}{2\pi} A_\mu \partial_\nu a_\lambda \varepsilon^{\mu\nu\lambda} \right] \\ & + \left[-\frac{p_2}{4\pi} \tilde{a}_\mu \partial_\nu \tilde{a}_\lambda \varepsilon^{\mu\nu\lambda} + \frac{1}{2\pi} a_\mu \partial_\nu \tilde{a}_\lambda \varepsilon^{\mu\nu\lambda} \right] \end{aligned} \tag{7.3.13}$$

where $p_1 = m$ is an odd integer. Equation (7.3.13) is the effective theory of a second-level hierarchical FQH state.

The effective theory can be used to determine the physical properties of the hierarchical FQH state. The total filling fraction is determined from the equation of motion $\delta\mathcal{L}/\delta a_0 = \delta\mathcal{L}/\delta\tilde{a}_0 = 0$, with

$$-eB = p_1 b - \tilde{b}, \qquad b = p_2 \tilde{b}$$

We find that

$$\nu = \frac{b}{-eB} = \frac{1}{p_1 - \frac{1}{p_2}}. \tag{7.3.14}$$

Equation (7.3.13) can be written in the following more compact form by introducing $(a_{1\mu}, a_{2\mu}) = (a_\mu, \tilde{a}_\mu)$:

$$\mathcal{L} = -\frac{1}{4\pi} K_{IJ} a_{I\mu} \partial_\nu a_{J\lambda} \varepsilon^{\mu\nu\lambda} + \frac{e}{2\pi} q_I A_\mu \partial_\nu a_{I\lambda} \varepsilon^{\mu\nu\lambda} \tag{7.3.15}$$

where K is the integer matrix

$$K = \begin{pmatrix} p_1 & -1 \\ -1 & p_2 \end{pmatrix}$$

and $\boldsymbol{q}$ is the integer vector $\boldsymbol{q}^\top = (q_1, q_2) = (1, 0)$. Here $\boldsymbol{q}$ will be called the charge vector. The filling fraction (7.3.14) can be rewritten as $\nu = \boldsymbol{q}^\top K^{-1} \boldsymbol{q}$. When $(p_1, p_2) = (3, 2)$, the hierarchical state corresponds to the $\nu = 2/5$ FQH state observed in experiments.

The second-level hierarchical FQH state contains two kinds of quasiparticle. One is the quasihole (or vortex) in the original electron condensate, and the other is the quasihole (or vortex) in the new boson condensate. The two kinds of quasiholes are created by inserting the source terms $-j^\mu a_\mu$ and $-\tilde{j}^\mu \tilde{a}_\mu$, respectively, where $\tilde{j}^\mu$ and j^μ have a similar form to eqn (7.3.2). The first kind of quasihole is created by multiplying $\prod_i (\xi - z_i)$ with the electron wave function, while the second kind is created by multiplying $\prod_i (\eta - \xi_i)$ with the boson Laughlin wave function (here ξ_i are the complex coordinates of the bosons and η is the position of the quasihole).

A generic quasiparticle consists of a number l_1 of quasiparticles of the first kind and a number l_2 of quasiparticles of the second kind, and is labeled by two integers. Such a quasiparticle carries l_1 units of the $a_{1\mu}$ charge and l_2 units of the $a_{2\mu}$ charge, and is described by

$$(l_1 a_{1\mu} + l_2 a_{2\mu}) j^\mu \tag{7.3.16}$$

After integrating out the gauge fields, we find that such a quasiparticle carries $\sum_J K_{IJ} l_J$ units of the $a_{I\mu}$ flux. Hence the statistics of such a quasiparticle are given by

$$\theta = \pi \boldsymbol{l}^\top K^{-1} \boldsymbol{l} = \frac{1}{p_2 p_1 - 1}(p_2 l_1^2 + p_1 l_2^2 + 2 l_1 l_2)$$

and the electric charge of the quasiparticle is

$$Q_q = -e\boldsymbol{q}^\top K^{-1} \boldsymbol{l} = -e\frac{p_2 l_1 + l_2}{p_2 p_1 - 1}$$

For the $\nu = 2/5$ state (i.e. $(p_1, p_2) = (3, 2)$), the quasiparticle with minimal electric charge is labeled by $(l_1, l_2) = (0, 1)$. The minimal electric charge is $-e/5$. Such a quasiparticle has statistics $\theta = \frac{3}{5}\pi$.

We can also construct more general FQH states. The effective theory for these FQH states still has the form given in eqn (7.3.15), but now I runs from 1 to an integer n (n will be called the level of the FQH state). To obtain the form of the matrix K, let us assume that at the level $n-1$ the effective theory is given by eqn (7.3.15) with $a_{I\mu}$, $I = 1, \ldots, n-1$, and $K = K^{(n-1)}$. The quasiparticles carry integer charges of the $a_{I\mu}$ gauge fields. Now consider an nth-level hierarchical state which is obtained by the 'condensation' of quasiparticles with the $a_{I\mu}$ charge $l_I|_{I=1,..,n-1}$. The effective theory of this nth-level hierarchical state will be given by eqn (7.3.15) with n gauge fields. The nth gauge field $a_{n\mu}$ comes from the new condensate. The matrix K is given by

$$K^{(n)} = \begin{pmatrix} K^{(n-1)} & -\boldsymbol{l} \\ -\boldsymbol{l}^\top & p_n \end{pmatrix}$$

with $p_n =$ even. The charge vector $\boldsymbol{q}$ is still given by $(1, 0, 0, ...)$. By iteration, we see that the generalized hierarchical states are always described by integer symmetric matrices, with $K_{II} =$ even, except for $K_{11} =$ odd. The new condensate gives rise to a new kind of quasiparticle, which again carries an integer charge of the new gauge field $a_{n\mu}$. Hence, a generic quasiparticle always carries integral charges of the $a_{I\mu}$ field.

Let us summarize the above results in more general terms. We know that a hierarchical (or generalized hierarchical) FQH state contains many different condensates. The different condensates are not independent. The particles in one condensate behave like a flux tube to the particles in other condensates. To describe such a coupling, it is convenient to use $U(1)$ gauge fields to describe the density and the current $J_{I\mu}$ of the Ith condensate

$$J_I^\mu = \frac{1}{2\pi}\varepsilon^{\mu\alpha\beta}\partial_\alpha a_{I\beta} \tag{7.3.17}$$

In this case, the couplings between different condensates are described by the Chern–Simons term of the gauge fields as follows:

$$\mathcal{L} = -\frac{1}{4\pi}K_{IJ} a_{I\mu}\partial_\nu a_{J\lambda}\varepsilon^{\mu\nu\lambda} + \frac{e}{2\pi}q_I A_\mu \partial_\nu a_{I\lambda}\varepsilon^{\mu\nu\lambda} \tag{7.3.18}$$

When K is taken to be a general integral $n \times n$ matrix with $K_{II}|_{I=1}$ = odd, $K_{II}|_{I>1}$ = even, and $\boldsymbol{q}^\top = (1, 0, ..., 0)$. Equation (7.3.18) describes the most general (abelian) FQH states (Wen and Zee, 1992a). The filling fraction is given by

$$\nu = \boldsymbol{q}^\top K^{-1} \boldsymbol{q} \tag{7.3.19}$$

The quasiparticle excitations can be viewed as vortices in different condensates. A generic quasiparticle is labeled by n integers $l_I|_I = 1, .., n$, and can be generated by the following source term:

$$\mathcal{L} = l_I a_{I\mu} j^\mu \tag{7.3.20}$$

Such a quasiparticle will be denoted by $\psi_{\boldsymbol{l}}$. The j^μ in eqn (7.3.20) has the form

$$\begin{aligned} \boldsymbol{j}(x) &= \dot{\boldsymbol{x}}_0 \delta(\boldsymbol{x} - \boldsymbol{x}_0) \\ j^0(x) &= \delta(\boldsymbol{x} - \boldsymbol{x}_0) \end{aligned} \tag{7.3.21}$$

which creates a quasiparticle at $\boldsymbol{x}_0$.

As we create a quasiparticle $\psi_{\boldsymbol{l}}$, it will induce a change in the densities of all of the condensates, namely δJ_I^0. From the equation of motion of eqn (7.3.18) and eqn (7.3.20), we find that δJ_I^0 satisfies

$$\int \mathrm{d}^2 x\, \delta J_I^0 = l_J (K^{-1})_{JI} \int \mathrm{d}^2 x\, j^0 = (\boldsymbol{l}^\top K^{-1})_I \tag{7.3.22}$$

The charge and the statistics of the quasiparticle $\psi_{\boldsymbol{l}}$ are given by

$$\theta_{\boldsymbol{l}} = \pi \boldsymbol{l}^\top K^{-1} \boldsymbol{l}, \qquad Q_{\boldsymbol{l}} = -e q_I \int \mathrm{d}^2 x\, \delta J_I^0 = -e \boldsymbol{l}^\top K^{-1} \boldsymbol{q} \tag{7.3.23}$$

The result of quasiparticle statistics is obtained by noting that $2\pi \delta J_I^0$ is the flux of the $a_{I\mu}$ and the quasiparticle carries l_I units of $a_{I\mu}$ charge.

A generic electron excitation can be viewed as a special kind of (generic) quasiparticle, $\psi_e = \psi_{\boldsymbol{l}_e}$, where the integer vector $\boldsymbol{l}_e$ is given by

$$l_{eI} = K_{IJ} L_J, \qquad L_I = \text{integers}, \qquad q_I L_I = 1 \tag{7.3.24}$$

We can show that these electron excitations satisfy the following properties: they carry a unit charge (see eqn (7.3.23)); they have fermionic statistics; moving an electron excitation $\psi_e = \psi_L$ around any quasiparticle excitations $\psi_{\boldsymbol{l}}$ always induces a phase of a multiple of 2π; and the excitations defined in eqn (7.3.24) are all of the excitations satisfying the above three conditions.

From the generic effective theory in eqn (7.3.15), we find that the generalized hierarchical states can be labeled by an integer-valued K-matrix and a charge vector $\boldsymbol{q}$. Now we would like to ask the following question: do different $(K, \boldsymbol{q})$s describe different FQH states? Notice that, through a redefinition of the gauge fields $a_{I\mu}$, one can always diagonalize the K-matrix into one with ± 1 as the diagonal elements. Thus, it seems that all K-matrices with the same signature describe the same FQH states, because they lead to the same effective theory after a proper redefinition of the gauge fields. Certainly this conclusion is incorrect. We would like to stress that the effective Lagrangian (7.3.15) alone does not provide a proper description of the internal orders (or topological orders) in the hierarchical states. It is the effective Lagrangian (7.3.15) together with the quantization condition of the $a_{I\mu}$ charges that characterize the topological order. A $U(1)$ gauge theory equipped with a quantization condition on the allowed $U(1)$ charges is called a compact $U(1)$ theory. Our effective theory (7.3.15) is actually a compact $U(1)$ theory with all $U(1)$ charges quantized as integers. Thus, the allowed $U(1)$ charges form an n-dimensional cubic lattice, which will be called the charge lattice. Therefore, when one considers the equivalence of two different K-matrices, one can only use the field redefinitions that keep the charge quantization condition unchanged (i.e. keep the charge lattice unchanged). The transformations that map the charge lattice onto itself belong to the group $SL(n, Z)$ (a group of integer matrices with unit determinant):

$$a_{I\mu} \to W_{IJ} a_{J\mu}, \qquad W \in SL(n, Z) \tag{7.3.25}$$

From the above discussion, we see that the two FQH states described by $(K_1, \boldsymbol{q}_1)$ and $(K_2, \boldsymbol{q}_2)$ are equivalent (i.e. they belong to the same universality class) if there exists a $W \in SL(n, Z)$ such that

$$\boldsymbol{q}_2 = W \boldsymbol{q}_1, \qquad K_2 = W K_1 W^{\top} \tag{7.3.26}$$

This is because, under the transformation (7.3.25), an effective theory described by $(K_1, \boldsymbol{q}_1)$ simply changes into another effective theory described by $(K_2, \boldsymbol{q}_2)$.

We would like to remark that in the above discussion we have ignored another topological quantum number—the spin vector. Because of this, the equivalence condition in eqn (7.3.26) does not apply to rotationally-invariant systems. However, eqn (7.3.26) does apply to disordered FQH systems because the angular momentum is not conserved in disordered systems and the spin vector is not well defined. A discussion of the spin vectors can be found in Wen and Zee (1992c,*d*). Table 7.1 lists the K-matrix, the charge vector $\boldsymbol{q}$, and the spin vector $\boldsymbol{s}$ for some common single-layer spin-polarized FQH states.

Problem 7.3.3.
Prove the four properties listed below eqn (7.3.24).

7.3.4 Effective theory of simple multi-layer fractional quantum Hall states

• The K-matrix and multi-layer FQH wave functions.

TABLE 7.1. Expressions for $(K, \boldsymbol{q}, \boldsymbol{s})$ for some simple single-layer spin-polarized FQH states.

ν	$\boldsymbol{q}$	K	$\boldsymbol{s}$
$1/m$	(1)	(m)	$(m/2)$
$1-1/m$	$\begin{pmatrix}1\\0\end{pmatrix}$	$\begin{pmatrix}1 & 1\\1 & -(m-1)\end{pmatrix}$	$\begin{pmatrix}1/2\\(1-m)/2\end{pmatrix}$
$2/5$	$\begin{pmatrix}1\\0\end{pmatrix}$	$\begin{pmatrix}3 & -1\\-1 & 2\end{pmatrix}$	$\begin{pmatrix}1/2\\1\end{pmatrix}$
$3/7$	$\begin{pmatrix}1\\0\\0\end{pmatrix}$	$\begin{pmatrix}3 & -1 & 0\\-1 & 2 & -1\\0 & -1 & 2\end{pmatrix}$	$\begin{pmatrix}1/2\\1\\1\end{pmatrix}$

So far, we only considered the so called single-layer FQH states where the 2D electron gas lives on an interface of two different semiconductors. We can also make samples with multiple layers of interfaces. The 2D electron gases on different layers can form a multi-layer FQH state when placed under a strong magnetic field.

The same approach used to construct the effective theory of the hierarchical states can also be used to construct the effective theory for the multi-layer FQH states. The connection between the FQH wave function and the K-matrix becomes very transparent for the multi-layer FQH states. In this section, we will concentrate on double-layer FQH states. However, the generalization to the n-layer FQH states is straightforward.

We would like to construct an effective theory for the following simple double-layer FQH state:

$$\prod_{i<j}(z_{1i}-z_{1j})^l \prod_{i<j}(z_{2i}-z_{2j})^m \prod_{i,j}(z_{1i}-z_{2j})^n \mathrm{e}^{-\frac{1}{4l_B^2}(\sum_i |z_{1i}|^2+\sum_j |z_{2j}|^2)}, \tag{7.3.27}$$

where z_{Ii} is the complex coordinate of the ith electron in the Ith layer. Here l and m are odd integers, so that the wave function is consistent with the Fermi statistics of the electrons, while n can be any non-negative integer. The above wave function was first suggested by Halperin as a generalization of the Laughlin wave function (Halperin, 1983). It appears that these wave functions can explain some of the dominant FQH filling fractions observed in double-layer FQH systems.

We start with a single-layer FQH state in the first layer, namely $\prod_{i<j}(z_{1i}-z_{1j})^l \mathrm{e}^{-\sum_i |z_{1i}|^2/4}$, which is a $1/l$ Laughlin state and is described by the effective theory

$$\mathcal{L} = \left[-l\frac{1}{4\pi}a_{1\mu}\partial_\nu a_{1\lambda}\varepsilon^{\mu\nu\lambda} + \frac{e}{2\pi}A_\mu\partial_\nu a_\lambda\varepsilon^{\mu\nu\lambda}\right] \tag{7.3.28}$$

where $a_{1\mu}$ is the gauge field that describes the electron density and current in the first layer.

Examining the wave function in eqn (7.3.27), we see that an electron in the second layer is bounded to a quasihole excitation in the first layer. Such a quasihole excitation is formed by n fundamental quasihole excitations and carries an $a_{1\mu}$ charge of $-n$. A gas of the quasiholes is described by the following effective theory:

$$\mathcal{L} = -na_{1\mu}j^\mu + \text{kinetic/potential energy} \tag{7.3.29}$$

where j^μ has the form given in eqn (7.3.2). As we mentioned before, in the mean-field theory, if we ignore the response of the $a_{1\mu}$ field, then eqn (7.3.29) simply describes a gas of bosons in a magnetic field nb_1, where $b_1 = -\varepsilon_{ij}\partial_i a_{1j}$. Now we would like to attach an electron (in the second layer) to each quasihole in eqn (7.3.29). Such an operation has the following two effects: the bound state of the quasiholes and the electrons can directly couple with the electromagnetic field A_μ because the electron carries the charge $-e$; and the bound state behaves like a fermion. The effective theory for the bound states has the following form:

$$\mathcal{L} = (eA_\mu - na_{1\mu})j^\mu + \text{kinetic/potential energy} \tag{7.3.30}$$

which now describes a gas of fermions (in mean-field theory). These fermions see an effective magnetic field $-eB + nb_1$.

When the electrons (i.e. the fermions in eqn (7.3.30)) in the second layer have a density of $\frac{1}{m}(-eB + nb_1)/2\pi$ (i.e. they have an effective filling fraction $1/m$), they can form a $1/m$ Laughlin state, which corresponds to the $\prod_{i<j}(z_{2i} - z_{2j})^m$ part of the wave function. (Note that here $eB < 0$.) Introducing a new gauge field $j^\mu = \frac{1}{2\pi}\partial_\nu a_{2\lambda}\varepsilon^{\mu\nu\lambda}$ to describe the fermion current j^μ in eqn (7.3.30), the effective theory of the $1/m$ state in the second layer has the form

$$\mathcal{L} = -m\frac{1}{4\pi}a_{2\mu}\partial_\nu a_{2\lambda}\varepsilon^{\mu\nu\lambda} \tag{7.3.31}$$

Putting eqns (7.3.28), (7.3.30), and (7.3.31) together, we obtain the total effective theory of the double-layer state, which has the form given in eqn (7.3.15) with the K-matrix and the charge vector $\boldsymbol{q}$ given by

$$K = \begin{pmatrix} l & n \\ n & m \end{pmatrix}, \qquad \boldsymbol{q} = \begin{pmatrix} 1 \\ 1 \end{pmatrix} \tag{7.3.32}$$

We see that the elements of the K-matrix are simply the exponents in the wave function. The filling fraction of the FQH state is still given by eqn (7.3.19).

There are two kinds of (fundamental) quasihole excitation in the double-layer state. The first kind is created by multiplying $\prod_i(\xi - z_{1i})$ with the ground-state wave function and the second kind is created by $\prod(\xi - z_{2i})$. As the vortices in the two electron condensates in the first and the second layers, a first kind of quasihole is created by the source term $-a_{1\mu}j^\mu$, and the second kind of quasihole is created by $-a_{2\mu}j^\mu$. Thus, a generic quasiparticle in the double-layer state is a

TABLE 7.2. Expressions for $(K, \boldsymbol{q}, \boldsymbol{s})$ for some double-layer FQH states.

ν	$\boldsymbol{q}$	K	$\boldsymbol{s}$
$1/m$	$\begin{pmatrix}1\\1\end{pmatrix}$	$\begin{pmatrix}m & m\\m & m\end{pmatrix}$	$\begin{pmatrix}m/2\\m/2\end{pmatrix}$
$2/5$	$\begin{pmatrix}1\\1\end{pmatrix}$	$\begin{pmatrix}3 & 2\\2 & 3\end{pmatrix}$	$\begin{pmatrix}3/2\\3/2\end{pmatrix}$
$1/2$	$\begin{pmatrix}1\\1\end{pmatrix}$	$\begin{pmatrix}3 & 1\\1 & 3\end{pmatrix}$	$\begin{pmatrix}3/2\\3/2\end{pmatrix}$
$2/3$	$\begin{pmatrix}1\\1\end{pmatrix}$	$\begin{pmatrix}3 & 0\\0 & 3\end{pmatrix}$	$\begin{pmatrix}3/2\\3/2\end{pmatrix}$
$2/3$	$\begin{pmatrix}1\\1\end{pmatrix}$	$\begin{pmatrix}1 & 2\\2 & 1\end{pmatrix}$	$\begin{pmatrix}1/2\\1/2\end{pmatrix}$

bound state of several quasiholes of the first and the second kind, and is described by eqn (7.3.16). The quantum numbers of such quasiparticles are still given by eqn (7.3.23).

In general, a multi-layer FQH state of the type (7.3.27) is described by a K-matrix whose elements are integers and whose diagonal elements are odd integers. The charge vector has the form $\boldsymbol{q}^\top = (1, 1, ..., 1)$. People usually label the double-layer FQH state (7.3.27) by (l, m, n). In Table 7.2, we list the K-matrix, the charge vector $\boldsymbol{q}$, and the spin vector $\boldsymbol{s}$ for some simple double-layer states.

From the above, we see that the (332) double-layer state has the filling fraction $2/5$—a filling fraction that also appears in single-layer hierarchical states. Now the following question arises: do the double-layer $2/5$ state and the single-layer $2/5$ state belong to the same universality class? This question has experimental consequences. We can imagine the following experiment. We start with a (332) double-layer state in a system with very weak inter-layer tunneling. As we make the inter-layer tunneling stronger and stronger, while keeping the filling fraction fixed, the double-layer state will eventually become a single-layer $2/5$ state. The question is whether the transition between the double-layer (332) state and the single-layer $2/5$ state is a smooth crossover or a phase transition. If we ignore the spin vector, then we see that the K-matrices and the charge vectors of the two $2/5$ states are equivalent because they are related by an $SL(2, Z)$ transformation. Therefore, in the absence of rotational symmetry (in which case the spin vector is not well defined), the two $2/5$ states can change into each other smoothly. When we include the spin vector, the two $2/5$ states are not equivalent and, for rotationally-invariant systems, they are separated by a first-order phase transition.

From Table 7.2, we also see that there are two different $2/3$ double-layer states. When the intra-layer interaction is much stronger than the inter-layer interaction (a situation found in real samples), the ground-state wave function prefers to have higher-order zeros between electrons in the same layers. Thus, the (330) state should have lower energy than the (112) state. The K-matrix and the charge vector of the single-layer $2/3$ state are equivalent to those of the (112) state, and are not equivalent to those of the (330) state. Thus, to change a double-layer $2/3$ state (i.e. the (330) state) into the single-layer $2/3$ state, one must go through a phase transition, regardless of the rotational symmetry.

The double-layer (mmm) state is an interesting state because $\det K = 0$. Fertig (1989), Brey (1990), and MacDonald *et al.* (1990) studied the (111) state and discovered a gapless collective mode. Wen and Zee (1992b) studied the more general (mmm) state and pointed out that the (mmm) state (in the absence of inter-layer tunneling) spontaneously breaks a $U(1)$ symmetry and is a neutral superfluid. More detailed discussions (including their experimental implications) can be found in Wen and Zee (1993), Murphy *et al.* (1994), and Yang *et al.* (1994).

Problem 7.3.4.
Let us first ignore the spin vector $\boldsymbol{s}$. Show that $(K, \boldsymbol{q})$ for the single-layer $2/5$ state in Table 7.1 and the double-layer $2/5$ state in Table 7.2 are equivalent. Show that, however, $(K, \boldsymbol{q}, \boldsymbol{s})$ for the two states are not equivalent. (In the presence of the spin vector $\boldsymbol{s}$, the equivalence relation becomes $\boldsymbol{s}_2 = W\boldsymbol{s}_1$, $\boldsymbol{q}_2 = W\boldsymbol{q}_1$, and $K_2 = WK_1W^\top$.)

Problem 7.3.5.
Consider the double-layer (lmn) state (7.2.6). Use the effective theory (7.3.15) and eqn (7.3.32) to find the filling fraction. Also find the fractional charge and the fractional statistics of the quasiholes in the first and the second layers. (You may compare the results with those of Problem 7.2.5.)

Problem 7.3.6.
Assume that the effective theory for a double-layer (mmm) state has the form

$$\mathcal{L} = -\frac{1}{4\pi} K_{IJ} a_{I\mu} \partial_\nu a_{J\lambda} \varepsilon^{\mu\nu\lambda} + \frac{1}{2g_1} \boldsymbol{e}_I \cdot \boldsymbol{e}_I - \frac{1}{2g_2} b_I b_I$$

where $\boldsymbol{e}_I$ and b_I are the 'electric' field and the 'magnetic' field, respectively, of $a_{I\mu}$.

1. Show that the (mmm) state, just like a superfluid, has gapless excitations. Find the velocity of the gapless excitations.
2. Also like the superfluid, the (mmm) state contains vortex excitations with energy $\gamma \ln L$, where L is the linear size of the system. Find the value of γ for the vortex with minimal energy. (Hint: The $a_{I\mu}$ charges of l_I are still quantized as integers in the (mmm) state.)
3. If we do regard the (mmm) state as a superfluid, can you identify the broken symmetry and the order parameter?

7.4 Edge excitations in fractional quantum Hall liquids

- FQH states *always* have gapless edge excitations.
- Edge excitations of an FQH state form a chiral Luttinger liquid.
- The structure of edge excitations is determined by bulk topological orders.

Due to the repulsive interaction and strong correlation between the electrons, a QH liquid is an incompressible state, despite the fact that the first Landau level is only partially filled. All of the bulk excitations in QH states have finite energy gaps. QH states and insulators are very similar in the sense that both states have finite energy gaps and short-ranged electron propagators. Due to this similarity, people were puzzled by the fact that QH systems apparently have very different transport properties to ordinary insulators. Halperin first pointed out that the IQH states contain gapless edge excitations (Halperin, 1982). The non-trivial transport properties of the IQH states come from the gapless edge excitations (Halperin, 1982; Trugman, 1983; MacDonald and Streda, 1984; Streda *et al.*, 1987; Buttiker, 1988; Jain and Kivelson, 1988a,b). For example, a two-probe measurement of an IQH sample can result in a finite resistance only when the source and the drain are connected by the edges. If the source and the drain are not connected by any edge, then the two-probe measurement will yield an infinite resistance at zero temperature, a result very similar to the insulators. Halperin also studied the dynamical properties of the edge excitations of the IQH states and found that the edge excitations are described by a chiral one-dimensional Fermi liquid theory.

Due to the similar transport properties of the FQH and IQH states, it is natural to conjecture that the transport in the FQH states is also governed by the edge excitations (Beenakker, 1990; MacDonald, 1990). However, because the FQH states are intrinsically many-body states, the edge excitations in the FQH states cannot be constructed by filling single-particle energy levels. In other words, the edge excitations of the FQH states should not be described by a Fermi liquid. Thus, we need a completely new approach to understand the dynamical properties of the edge states of FQH liquids (Wen, 1992, 1995).

In the next section, we will use Fermi liquid theory to study the edge excitations of the $\nu = 1$ IQH state. Then we will use the current algebra (or bosonization, see Tomonaga, 1950; Luttinger, 1963; Coleman, 1975) to develop a theory for edge excitations of the FQH states.

7.4.1 Fermi liquid theory of integral quantum Hall edge states

- The ground states of IQH liquids are obtained by filling single-particle energy levels. As a result, the low-energy edge excitations can be obtained by filling these single-particle levels in slightly different ways.

Consider a non-interacting electron gas in a uniform magnetic field B. In the symmetric gauge, the angular momentum is a good quantum number. The common eigenstates of energy and momentum form a ring-like shape (see Fig. 7.10). Therefore, in the presence of a smooth circular potential $V(r)$, the single-particle energy levels are as illustrated in Fig. 7.15. The ground state is obtained by filling

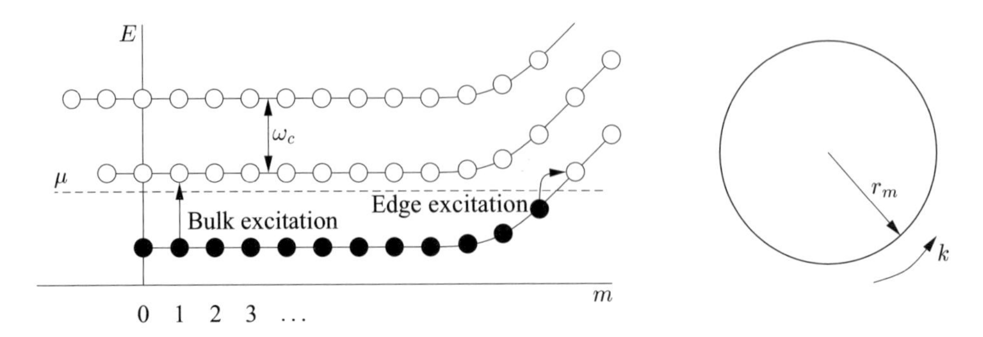

FIG. 7.15. Energy levels in the first three Landau levels in the presence of a smooth potential $V(r)$. Here m is the angular momentum of the levels. The levels below the chemical potential μ are filled by one electron, which gives rise to an IQH state. Modifying the occupation numbers near the edge produces gapless edge excitations of the IQH states.

N lowest energy levels. Such a state correspond to a circular droplet of uniform density (see Fig. 7.11).

In the zeroth Landau level, the energy of the angular momentum m state is given by

$$E_m = \frac{1}{2}\hbar\omega_c + V(r_m)$$

where $r_m = \sqrt{2m}l_B$ is the radius of the m state. From Fig. 7.15, we see that bulk excitations have a finite energy gap $\hbar\omega_c$. An edge excitation can be created by moving an electron from the m state to the $m+1$ state near the edge (see Fig. 7.15). As $r_{m+1} - r_m \to 0$ in the thermodynamical limit $r_m \to \infty$, the edge excitations are gapless.

As the m state has radius r_m, we can also view the angular momentum m as the momentum along the edge $k = m/r_m$. This allows us to regard E_m as an energy–momentum relation

$$E(k) = \frac{1}{2}\hbar\omega_c + V(2l_B^2 k)$$

In term of $E(k)$, our electron system can be described by a non-interacting Hamiltonian

$$H = \sum_k E(k) c_k^\dagger c_k \tag{7.4.1}$$

The above Hamiltonian describes a one-dimensional chiral Fermi liquid. We call it a chiral Fermi liquid because it has only one Fermi point and all of the low-energy excitations propagate in the same direction.[47] We conclude that the edge

[47] In contrast, the usual one-dimensional Fermi liquid has two Fermi points and contains both left and right movers.

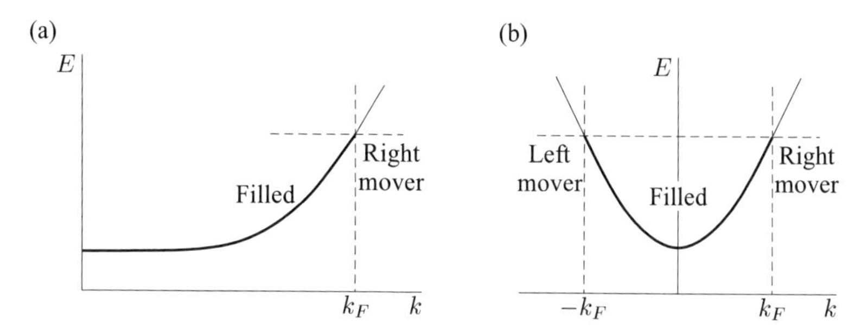

FIG. 7.16. (a) The energy–angular momentum relation E_m can be regarded as an energy–momentum relation $E(k)$. The edge excitations can be regarded as excitations of a one-dimensional chiral Fermi liquid, which contains only right movers. (b) The dispersion relation of a usual one-dimensional Fermi liquid contains both right and left movers.

excitations of the $\nu = 1$ IQH state are described by the one-dimensional chiral Fermi liquid (7.4.1).

Problem 7.4.1.
Show that the velocity of the IQH edge excitations $v = \partial E(k)/\partial k$ is given by cE/B, where c is the speed of light and $E = -\partial V(r)/\partial r$ is the electric field on the edge produced by the confining potential $V(r)$. Find the Fermi momentum k_F for the chiral Fermi liquid, assuming that the $\nu = 1$ IQH droplet contains N electrons.

Problem 7.4.2.
A $\nu = 1$ IQH state is confined by a circular smooth potential $V(r)$ and contains N electrons. Find the total ground-state angular momentum M_0. Find the number of low-energy particle–hole excitations with total angular momentum $M_0 + m$ for $m = -1$ and $m = 1, 2, 3, 4, 5$. Show that the excitations with the same total angular momentum have the same energy in the thermodynamical limit.

Problem 7.4.3.
Consider a one-dimensional non-interacting fermion system described by $H = \sum_k (v_1 k c^\dagger_{1,k} c_{1,k} + v_2 k c^\dagger_{2,k} c_{2,k})$. The mixing term $\sum_k (\gamma c^\dagger_{1,k} c_{2,k} + h.c.)$ is a relevant perturbation. Show that, when the system has both right and left movers (i.e. $v_1 v_2 < 0$), the relevant mixing term opens up a finite energy gap and drastically changes the low-energy properties of the system. However, for a chiral Fermi liquid with all excitations moving in one direction (i.e. $v_1 v_2 > 0$), the relevant mixing term does not induce any energy gap. In fact, we believe that no perturbations can induce an energy gap for one-dimensional chiral Fermi liquids. The gapless excitations in a one-dimensional chiral Fermi liquid are robust against all perturbations and are topologically stable.

7.4.2 The hydrodynamical approach—the $1/m$ Laughlin state

- Edge excitations from the current algebra (the Kac–Moody algebra).

- There is no way to construct FQH edge excitations by filling single-particle energy levels.

As the $1/m$ Laughlin FQH state cannot be obtained by filling single-particle energy levels, we cannot use the picture in the last section to construct edge excitations of the $1/m$ Laughlin state. Not knowing how to derive the FQH edge theory from the Hamiltonian of interacting electrons, in the following we instead try to guess a low-energy effective theory.

The simplest way to guess the dynamical theory of edge excitations is to use the hydrodynamical approach. In this approach, one uses the fact that FQH states are incompressible, irrotational liquids that contain no low-energy bulk excitations. Therefore, the only low-lying excitations (below the bulk energy gap) are surface waves from the deformations of the FQH droplet. These surface waves are identified as edge excitations of the FQH state.

In the hydrodynamical approach (Wen, 1992), we first study the classical theory of the surface wave on the FQH droplet. Then we quantize the classical theory to obtain the quantum description of the edge excitations. It is amazing that the simple quantum description obtained from the classical theory provides a complete description of the edge excitations at low energies and allows us to calculate the electron and the quasiparticle propagators along the edges.

Consider an FQH droplet with a filling fraction ν confined by a potential well. Due to the nonzero conductance, the electric field of the potential well generates a persistent current flowing along the edge, given by

$$\boldsymbol{j} = \sigma_{xy}\hat{z} \times \boldsymbol{E}, \qquad \sigma_{xy} = \nu\frac{e^2}{h}$$

This implies that the electrons near the edge drift along the edge with a velocity

$$v = \frac{E}{B}c$$

where c is the velocity of light. Thus, the edge wave (the deformation of the edge) also propagates with velocity v. Let us use the one-dimensional density $\rho(x) = nh(x)$ to describe the edge wave, where $h(x)$ is the displacement of the edge, x is the coordinate along the edge, and $n = \nu/2\pi l_B^2$ is the two-dimensional electron density in the bulk. (Here $l_B = \sqrt{c/eB}$ is the magnetic length.) We see that the propagation of the edge waves is described by the following wave equation:

$$\partial_t \rho + v\partial_x \rho = 0 \tag{7.4.2}$$

Notice that the edge waves always propagate in one direction; there are no waves that propagate in the opposite direction.

The Hamiltonian (i.e. the energy) of the edge waves is given by

$$H = \int \mathrm{d}x \, \frac{1}{2} e\rho E h = \int \mathrm{d}x \, \pi \frac{v}{\nu} \rho^2 \tag{7.4.3}$$

where we have used $h = \rho/n = \rho \frac{2\pi c}{eB}$. In momentum space, eqns (7.4.2) and (7.4.3) can be rewritten as

$$\dot{\rho}_k = -\mathrm{i}vk\rho_k, \qquad H = 2\pi \frac{v}{\nu} \sum_{k>0} \rho_{-k}\rho_k \tag{7.4.4}$$

where $\rho_k = \int \mathrm{d}x \, \frac{1}{\sqrt{L}} \mathrm{e}^{\mathrm{i}kx} \rho(x)$ and L is the length of the edge. We find that, if we identify $\rho_k|_{k>0}$ as the 'coordinates' and $\pi_k = \mathrm{i}2\pi\rho_{-k}/\nu k$ as the corresponding canonical 'momenta', then the standard Hamiltonian equation

$$\dot{q} = \frac{\partial H}{\partial p}, \qquad \dot{p} = -\frac{\partial H}{\partial q}$$

will reproduce the equation of motion $\dot{\rho}_k = \mathrm{i}vk\rho_k$. This allows us to identify the canonical 'coordinates' and 'momenta'. It is interesting to see that the displacement $h(x)$ contains both the 'coordinates' and the 'momenta'. This is due to the chiral property of the edge wave.

Knowing the canonical coordinates and momenta, it is easy to quantize the classical theory. We simply view ρ_k and π_k as operators that satisfy $[\rho_k, \pi_{k'}] = \mathrm{i}\delta_{kk'}$. Thus, after quantization we have

$$\begin{aligned} [\rho_k, \rho_{k'}] &= \frac{\nu}{2\pi} k \delta_{k+k'}, \qquad k, k' = \text{integer} \times \frac{2\pi}{L} \\ H &= 2\pi \frac{v}{\nu} \sum_{k>0} \rho_{-k}\rho_k \end{aligned} \tag{7.4.5}$$

The above algebra is called the ($U(1)$) Kac–Moody (K–M) algebra (Kac, 1983; Goddard and Olive, 1985, 1986). A similar algebra has also appeared in the Tomonaga model (Tomonaga, 1950; Luttinger, 1963; Coleman, 1975). Notice that eqn (7.4.5) simply describes a collection of decoupled harmonic oscillators (generated by (ρ_k, ρ_{-k})). Thus, eqn (7.4.5) is a one-dimensional free phonon theory (with only a single branch). For $k > 0$, $\rho_k^\dagger = \rho_{-k}$ creates a phonon with momentum k and energy vk, while ρ_k annihilates a phonon. Equation (7.4.5) provides a complete description of the low-lying edge excitations of the Laughlin state.

The edge excitations considered here do not change the total charge of the system and hencc are neutral. In the following, we will discuss the charged excitations and calculate the electron propagator from the K–M algebra (7.4.5).

The low-lying charge excitations obviously correspond to adding (removing) electrons to (from) the edge. These charged excitations carry integer charges and are created by the electron operators $\Psi^\dagger$. The above theory of edge excitations is formulated in terms of a one-dimensional density operator $\rho(x)$. So the central objective is to write the electron operator in terms of the density operator. The electron operator on the edge creates a localized charge and should satisfy

$$[\rho(x), \Psi^\dagger(x')] = \delta(x - x')\Psi^\dagger(x') \tag{7.4.6}$$

As ρ satisfies the K–M algebra (7.4.5), one can show that

$$[\rho(x_1), \rho(x_2)] = \mathrm{i}\frac{\nu}{2\pi}\delta'(x_1 - x_2)$$

or

$$[\rho(x_1), \phi(x_2)] = -\mathrm{i}\nu\delta(x_1 - x_2)$$

where ϕ is given by $\rho = \frac{1}{2\pi}\partial_x\phi$. We see that $\rho(x)$ can be regarded as a functional derivative of ϕ, i.e. $\rho(x) = -\mathrm{i}\nu\frac{\delta}{\delta\phi(x)}$. Using this relation, one can show that the operators that satisfy eqn (7.4.6) are given by (Wen, 1992)

$$\Psi \propto \mathrm{e}^{\mathrm{i}\frac{1}{\nu}\phi} \tag{7.4.7}$$

Equation (7.4.6) only implies that the operator Ψ carries the charge e. In order to identify Ψ as an electron operator, we need to show that Ψ is a fermionic operator. Using the K–M algebra (7.4.5), we find that (see Problem 7.4.8)

$$\Psi(x)\Psi(x') = (-)^{1/\nu}\Psi(x')\Psi(x) \tag{7.4.8}$$

We see that the electron operator Ψ in eqn (7.4.7) is fermionic only when $1/\nu = m$ is an odd integer, in which case the FQH state is a Laughlin state.

In the above discussion, we have made an assumption that is not generally true. We have assumed that the incompressible FQH liquid contains only *one* component of incompressible fluid, which leads to one branch of edge excitations. The above result implies that, when $\nu \neq 1/m$, the edge theory with only one branch does not contain the electron operators and is not self-consistent. Our one-branch edge theory only applies to the $\nu = 1/m$ Laughlin state.

Now let us calculate the electron propagator along the edge of the $\nu = 1/m$ Laughlin state. As ϕ is a free phonon field with a propagator

$$\langle\phi(x,t)\phi(0)\rangle = -\nu\ln(x - vt) + \text{constant}$$

the electron propagator can be easily calculated as

$$G(x,t) = \langle T(\Psi^\dagger(x,t)\Psi(0))\rangle = \exp[\frac{1}{\nu^2}\langle\phi(x,t)\phi(0)\rangle] \propto \frac{1}{(x - vt)^m} \tag{7.4.9}$$

The first thing we see is that the electron propagator on the edge of an FQH state acquires a non-trivial exponent $m = 1/\nu$ that is not equal to one. This implies that the electrons on the edge of the FQH state are strongly correlated and cannot be described by Fermi liquid theory. We will call this type of electron state a chiral Luttinger liquid.

We would like to emphasize that the exponent m is quantized. The quantization of the exponent is directly related to the fact that the exponent is linked to the statistics of the electrons (see eqn (7.4.8)). Thus, the exponent is a topological number that is independent of electron interactions, edge potential, etc. Despite the exponent being a property of the edge states, the only way to change the exponent is through a phase transition in the bulk state. Therefore, the exponent can be regarded as a quantum number that characterizes the topological orders in the *bulk* FQH states.

In momentum space, the electron propagator has the form

$$G(k,\omega) \propto \frac{(vk+\omega)^{m-1}}{\omega - vk + \mathrm{i}0^+\mathrm{sgn}(\omega)}$$

The anomalous exponent m can be measured in tunneling experiments. From $\mathrm{Im}G$, we find that the tunneling density of states of electrons is given by

$$N(\omega) \propto |\omega|^{m-1}$$

This implies that the differential conductance has the form $\frac{\mathrm{d}I}{\mathrm{d}V} \propto V^{m-1}$ for a metal–insulator–FQH junction.

Problem 7.4.4.
Prove eqn (7.4.8). (Hint: First show that $[\phi(x), \phi(y)] = \frac{\pi}{q}\mathrm{sgn}(x-y)$.)

Problem 7.4.5.
Bosonization and fermionization When $m = 1$ the discussion in this section implies that the $\nu = 1$ IQH edge state can be described by the free phonon theory

$$H_b = 2\pi v \sum_{k>0} \rho_k^\dagger \rho_k$$

In Section 7.4.1, we found that the $\nu = 1$ IQH edge state can be described by the free chiral fermion model

$$H_f = \sum_k vk : c_k^\dagger c_k := \sum_{k>0} vk c_k^\dagger c_k - \sum_{k<0} vk c_k c_k^\dagger$$

The normal order is defined as $: c_k^\dagger c_k := c_k^\dagger c_k$ if $k > 0$ and $: c_k^\dagger c_k := -c_k c_k^\dagger$ if $k < 0$. In this problem, we are going to study the direct relationship between the two descriptions. Switching from the fermion description to the boson description is called bosonization, and switching from the boson description to the fermion description is called fermionization.

1. Find the energies of the first five energy levels and their degeneracy for the bosonic model. Compare your result with that from Problem 7.4.2.

2. Let $\rho_f(x) =: c^\dagger(x)c(x) :$. In k space, we have $\rho_{f,k} = \sum'_q : c^\dagger_q c_{q+k} :$, where the summation $\sum'_q$ is limited to the range for which both the momenta of c_{q+k} and c_q are within the range $[-\Lambda, +\Lambda]$. The normal ordered $\rho_f(x)$ has a zero average in the ground state, which agrees with ρ in boson theory. Show that $\rho_{f,k}$ satisfies the same K–M algebra as ρ_k (see eqn (7.4.5) with $\nu = 1$). (Hint: We have $c^\dagger_k c_k = 0$ for large positive k and $c^\dagger_k c_k = 0$ for large negative k.)

3. Show that $H_f = 2\pi v \sum_{k>0} \rho^\dagger_{f,k}\rho_{f,k}$.

7.4.3 A microscopic theory for the edge excitations

- Relationship between the K–M algebra and the Laughlin wave functions.
- Electron equal-time correlation from the plasma analogue.

In this section, we will present a microscopic theory for the edge excitations in the Laughlin states. To be specific, let us consider an electron gas in the first Landau level. We assume that the electrons are described by the ideal Hamiltonian (7.2.5). The $\nu = 1/3$ Laughlin wave function

$$\Psi_3(z_i) = Z^{-1/2} \prod_{i<j} (z_i - z_j)^3 \prod_k \mathrm{e}^{-|z_k|^2/4l_B} \tag{7.4.10}$$

has zero energy[48] and is an exact ground state of the ideal Hamiltonian. In this section, we will assume that the magnetic length $l_B = 1$. In eqn (7.4.10), Z is the normalization factor. However, the Laughlin state (7.4.10) is not the only state with zero energy. One can easily check that the following type of states all have zero energy:

$$\Psi(z_i) = P(z_i)\Psi_3(z_i) \tag{7.4.11}$$

where $P(z_i)$ is a symmetric polynomial of z_i. In fact, the reverse is also true; all of the zero-energy states are of the form (7.4.11). This is because, in order for a fermion state to have zero energy, Ψ must vanish at least as fast as $(z_i - z_j)^3$ when any two electrons i and j are brought together (the possibility $(z_i - z_j)^2$ is excluded by fermion statistics). Because the Laughlin wave function is zero only when $z_i = z_j$, then $P = \Psi/\Psi_3$ is a finite function. As Ψ and Ψ_3 are both anti-symmetric functions in the first Landau level, P is a symmetric holomorphic function that can only be a symmetric polynomial.

Among all of the states in eqn (7.4.11), the Laughlin state describes a circular droplet with the smallest radius. All of the other states are a deformation and/or an inflation of the droplet of the Laughlin state. Thus, the states generated by P correspond to the edge excitations of the Laughlin state.

[48] Here, for convenience, we shift the zero of energy to the zeroth Landau level.

First, let us consider the zero-energy space (i.e. the space of symmetric polynomials). We know that the space of symmetric polynomials is generated by the polynomials $s_n = \sum_i z_i^n$ (through multiplication and addition). Let $M_0 = 3\frac{N(N-1)}{2}$ be the total angular momentum of the Laughlin state (7.4.10). Then the state Ψ will have an angular momentum $M = \Delta M + M_0$, where ΔM is the order of the symmetric polynomial P. As we have only one order-zero and one order-one symmetric polynomial, namely $s_0 = 1$ and $s_1 = \sum_i z_i$, respectively, the zero-energy states for $\Delta M = 0, 1$ are non-degenerate. However, when $\Delta M = 2$, we have two zero-energy states generated by $P = s_2$ and $P = s_1^2$. For general ΔM, the degeneracy of the zero-energy states is given by

$$\begin{array}{rccccccc} \Delta M: & 0 & 1 & 2 & 3 & 4 & 5 & 6 \\ \text{degeneracy}: & 1 & 1 & 2 & 3 & 5 & 7 & 11 \end{array} \tag{7.4.12}$$

Here we would like to point out that the degeneracy in eqn (7.4.12) is exactly what we expected from the macroscopic theory. We know that, for a circular droplet, the angular momentum ΔM can be regarded as the momentum along the edge $k = 2\pi\Delta M/L$, where L is the perimeter of the FQH droplet. According to the macroscopic theory, the (neutral) edge excitations are generated by the density operators ρ_k. One can easily check that the edge states generated by the density operators have the same degeneracies as those in eqn (7.4.12) for every ΔM. For example, the two states at $\Delta M = 2$ are generated by $\rho_{\kappa_0}^2$ and $\rho_{2\kappa_0}$, where $\kappa_0 = 2\pi/L$. Therefore, the space generated by the K–M algebra (7.4.5) and the space of symmetric polynomials are identical.

Now let us ask the following physical question: do the symmetric polynomials generate all of the low-energy states? If this is true, then from the above discussion we see that all of the low-energy excitations of the HQ droplet are generated by the K–M algebra, and we can say that eqn (7.4.5) is a complete theory of the low-lying excitations. Unfortunately, up to now we do not have an analytic proof of the above statement. This is because, although states which are orthogonal to the states generated by the symmetric polynomials have nonzero energies, it is not clear that these energies remain finite in the thermodynamical limit. It is possible that the energy gap approaches zero in the thermodynamical limit. To resolve this problem, we currently have to rely on numerical calculations. In Fig. 7.17 we present the energy spectrum of a system of six electrons in the first twenty-two orbits for the Hamiltonian introduced at the beginning of this section. The degeneracies of the zero-energy states at $M = 45, ..., 51$ (or $\Delta M = 0, ..., 6$) are found to be $1, 1, 2, 3, 5, 7, 11$, respectively, which agrees with eqn (7.4.12). More importantly, we clearly see that a finite energy gap separate all of the other states from the zero-energy states. Thus, the numerical results imply that all of the low-lying edge excitations of the Laughlin state are generated by the symmetric polynomials or the K–M algebra (7.4.5).

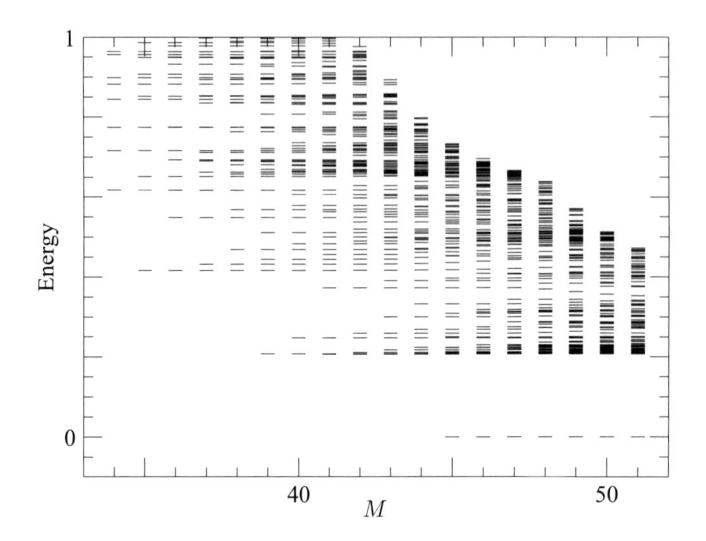

FIG. 7.17. The energy spectrum of a system of six electrons in the first twenty-two orbits for the Hamiltonian H_V. The degeneracies of the zero-energy states at $M = 45, ..., 51$ are found to be $1, 1, 2, 3, 5, 7, 11$, respectively.

In the following, we are going to calculate the equal-time correlation of the quasiparticle/electron from the Laughlin wave function Ψ_m. First, we calculate the norm of $|\xi\rangle = \prod_i (1 - \frac{z_i}{\xi})^n \Psi_m(\{z_i\})$ as follows:

$$\langle \xi | \xi \rangle = |\xi|^{-2Nn} \frac{Z_1}{Z} \tag{7.4.13}$$

$$Z = \int \prod d^2 z_i \ \exp(-\beta \sum_{ij} (-m^2) \ln |z_i - z_j| - \beta \sum_k \frac{m}{4} |z_k|^2)$$

$$Z_1 = \int \prod d^2 z_i \ \exp\left[\beta \sum_{ij} m^2 \ln |z_i - z_j| - \beta \sum_k (\frac{m}{4} |z_k|^2 - nm \ln |\xi - z_k|)\right]$$

where $\beta = 2/m$. Notice that Z is the partition function of a plasma formed by N 'charge' m particles, and Z_1 is the partition function of the plasma interacting with a particle of 'charge' n at ξ. We may write

$$Z = e^{-\beta E}, \qquad Z_1 = e^{-\beta E_1}$$

where E and E_1 are the total energies of the plasma. The change in energy $E_1 - E$ is given by the interaction between the added 'charge' n particle at ξ and N 'charge' m particles at z_i. When $|\xi|$ is large, the N 'charge' m particles form a circular droplet and we may treat them as a point 'charge' mN at $z = 0$. We obtain $E_1 - E = -nmN \ln|\xi|$. For smaller $|\xi|$, the 'charge' n changes the shape of the droplet and

$$E_1 - E = -nmN \ln|\xi| - \frac{n^2}{2}\left[-\ln\left(|\xi| - \frac{R^2}{|\xi|}\right) + \ln(|\xi|)\right] \qquad (7.4.14)$$

where $R = \sqrt{mN} l_B$ is the radius of the droplet. The first term in eqn (7.4.14) is the interaction between the 'charge' n and the *undeformed* droplet. The second term is the correction due to the deformation of the droplet. Since the plasma behaves like a metal, this correction can be represented by the interaction of the 'charge' n with its mirror images at $|z| = R^2/|\xi|$ and $z = 0$.

In the above discussion, we ignore the discreteness of the charges and treat the plasma as a continuous medium. We expect this approximation to give rise to the correct ratio Z_1/Z if ξ is not too close to the droplet, i.e. $|\xi| - R \gg l_B$.

From eqn (7.4.14), we find the norm of $|\psi^n(\xi)\rangle$ to be

$$\langle\xi|\xi\rangle = \left(\frac{\xi\xi^*}{\xi\xi^* - R^2}\right)^{n^2/m} \qquad (7.4.15)$$

As the inner product $\langle\tilde{\xi}|\xi\rangle$ is a holomorphic function of ξ and an anti-holomorphic function of $\tilde{\xi}$, eqn (7.4.15) implies that

$$\langle\tilde{\xi}|\xi\rangle = \left(\frac{\xi\tilde{\xi}^*}{\xi\tilde{\xi}^* - R^2}\right)^{n^2/m}$$

Notice that, when $n = m$,

$$\xi^{Nm}\tilde{\xi}^{*Nm}\langle\tilde{\xi}|\xi\rangle = \xi^{Nm}\tilde{\xi}^{*Nm}\left(\frac{\xi\tilde{\xi}^*}{\xi\tilde{\xi}^* - R^2}\right)^m$$

is proportional to the electron propagator G_e along the edge of an $N_e = N + 1$ electron system at equal time. Choosing $\xi = R\mathrm{e}^{\mathrm{i}\frac{x}{R}}$ and $\tilde{\xi} = R$, we obtain

$$G_e(x) = L^{-m} a^{m-1} \mathrm{e}^{\mathrm{i}m(N_e - \frac{1}{2})\frac{2\pi x}{L}} \sin^{-m}(\pi x/L) \qquad (7.4.16)$$

which reduces to eqn (7.4.9) when x is much less than $L \equiv 2\imath R$. Here a is a length scale of order l_B. Equation (7.4.16) can be expanded as follows:

$$G_e(x) = L^{-m} a^{m-1} \mathrm{e}^{\mathrm{i}m(N_e-1)\frac{2\pi x}{L}} \left[\sum_{n=0}^{\infty} \mathrm{e}^{-\mathrm{i}\frac{2\pi x}{L}n}\right]^m$$

$$= L^{-m} a^{m-1} \mathrm{e}^{\mathrm{i}m(N_e-1)\frac{2\pi x}{L}} \sum_{n=0}^{\infty} C_n^{m+n-1} \mathrm{e}^{-\mathrm{i}\frac{2\pi x}{L}n}$$

$$C_n^{m+n-1} = \frac{(n+m-1)!}{(m-1)!n!} \qquad (7.4.17)$$

From this expansion, we obtain the electron occupation number n_M at the angular momentum M state as follows:

$$n_M = 0, \qquad \delta M = m(N_e - 1) - M < 0$$

$$n_M = \frac{a^{m-1}}{L^{m-1}} C_{\delta M}^{\delta M + m - 1}, \qquad \delta M \geqslant 0 \qquad (7.4.18)$$

We see that the exact position of the Fermi edge is at the last partially occupied single-particle orbit, i.e. at the angular momentum $m(N_e - 1)$ (or $k_F = \sqrt{m(N_e-1)}/l_B$). Note that, when $m(N_e - 1) - M \gg m$, we have $n_M \propto (m(N_e - 1) - M)^{m-1}$. In terms of momentum along the edge, we have $n_k \propto (k_F - k)^{m-1}$. We find that, unlike a Fermi liquid, the occupation number n_k does not have a jump at the Fermi momentum.

We would like to remark that eqn (7.4.16) is correct only when x is much larger than the magnetic length l_B. Therefore, eqn (7.4.18) is valid only when $m(N_e - 1) - M \ll \sqrt{N_e}$.

If the dispersion of the edge excitations is linear, then G_e can only depend on $x - vt$ at low energies. We immediately see that the dynamical electron Green's function is

$$G_e(x,t) = L^{-m} a^{m-1} \mathrm{e}^{\,\mathrm{i} m(N - \frac{1}{2}) \frac{2\pi(x - vt)}{L}} \sin^{-m}[\pi(x - vt)/L]$$

When $m = 1$, the above becomes the electron Green's function of one-dimensional free fermions (for k near a Fermi point).

7.4.4 The hydrodynamical approach—the $2/5$ and $2/3$ states

- The edge states of hierarchical FQH liquids.
- The structure of electron and quasiparticle operators.
- The edge interaction can modify the exponents of electron/quasiparticle propagators if and only if the edge excitations can propagate in opposite directions.
- When $\nu_1 = -\nu_2 = 1$, the discussion here describes the one-dimensional Tomonaga–Luttinger liquid.

In this section, we will use the hydrodynamical approach to study the edge structures of second-level hierarchical states. We will concentrate on the $2/5$ and $2/3$ states as examples. In particular, we will study the structures of the electron and the quasiparticle operators on the edges of the hierarchical states. We will also see that the $2/3$ state contains two edge modes that propagate in opposite directions, which is quite counter-intuitive.

First, let us consider the $\nu = \frac{2}{5}$ FQH state. According to the hierarchical theory, the $\nu = \frac{2}{5}$ FQH state is generated by the condensation of quasiparticles on top of the $\nu = \frac{1}{3}$ FQH state. Thus, the $2/5$ state contains two components of incompressible fluids. To be definite, let us consider a special edge potential such that the FQH state consists of two droplets (see Fig. 7.14); one is the electron condensate with a filling fraction $\nu_1 = \frac{1}{3}$ and radius r_1, and the other is the quasiparticle condensate (on top of the $1/3$ state) with a filling fraction $\nu_2 = \frac{1}{15}$ (note that $\frac{1}{3} + \frac{1}{15} = \frac{2}{5}$) and radius $r_2 < r_1$.

When $r_1 - r_2 \gg l_B$, the two edges are independent. Generalizing the hydrodynamical approach in Section 7.4.2, we can show that there are two branches of the edge excitations whose low-energy dynamics are described by

$$[\rho_{Ik}, \rho_{Jk'}] = \frac{\nu_I}{2\pi} k \delta_{IJ} \delta_{k+k'}$$

$$H = 2\pi \sum_{I,k>0} \frac{v_I}{\nu_I} \rho_{I,-k} \rho_{I,k} \tag{7.4.19}$$

where $I = 1, 2$ labels the two branches and v_I are the velocities of the edge excitations. In eqn (7.4.19), ρ_I are the one-dimensional electron densities. In order for the Hamiltonian to be bounded from below, we require that $\nu_I v_I > 0$. We find that the stability of the $\nu = \frac{2}{5}$ FQH state requires both of the v_I to be positive.

Generalizing the discussion in Section 7.4.2, the electron operators on the two edges are found to be

$$\Psi_I = e^{i \frac{1}{\nu_I} \phi_I(x)}, \qquad I = 1, 2 \tag{7.4.20}$$

with $\partial_x \phi_I = \frac{1}{2\pi} \rho_I$. The electron propagators have the form

$$\langle T(\Psi_I(x,t) \Psi_I^\dagger(0)) \rangle = \frac{e^{i k_I x}}{(x - v_I t)^{-1/|\nu_I|}}, \qquad I = 1, 2$$

where $k_I = r_I / 2 l_B^2$.

According to the hierarchical picture, the $\nu = \frac{2}{3}$ FQH state is also formed by two condensates, an electron condensate with a filling fraction 1 and a hole condensate with a filling fraction $-\frac{1}{3}$. Thus, the above discussion can also be applied to the $\nu = \frac{2}{3}$ FQH state by choosing $(\nu_1, \nu_2) = (1, -\frac{1}{3})$. Again, there are two branches of the edge excitations, but now with *opposite* velocities if the Hamiltonian is positive definite.

As we bring the two edges together ($r_1 - r_2 \sim l_B$), the interaction between the two branches of the edge excitations can no longer be ignored. In this case, the Hamiltonian has the form

$$H = 2\pi \sum_{k>0} V_{IJ} \rho_{I,-k} \rho_{J,k} \tag{7.4.21}$$

The Hamiltonian (7.4.21) can be diagonalized. For $\nu_1\nu_2 > 0$, we may choose

$$\begin{aligned}\tilde\rho_{1k} &= \cos(\theta)\frac{1}{\sqrt{|\nu_1|}}\rho_{1k} + \sin(\theta)\frac{1}{\sqrt{|\nu_2|}}\rho_{2k}\\ \tilde\rho_{2k} &= \cos(\theta)\frac{1}{\sqrt{|\nu_2|}}\rho_{2k} - \sin(\theta)\frac{1}{\sqrt{|\nu_1|}}\rho_{1k}\\ \tan(2\theta) &= 2\frac{\sqrt{|\nu_1\nu_2|}V_{12}}{|\nu_1|V_{11} - |\nu_2|V_{22}}\end{aligned} \tag{7.4.22}$$

One can check that the $\tilde\rho$s satisfy

$$\begin{aligned}[\tilde\rho_{Ik}, \tilde\rho_{Jk'}] &= \frac{\mathrm{sgn}(\nu_I)}{2\pi} k\delta_{IJ}\delta_{k+k'}\\ H &= 2\pi \sum_{I,k>0} \mathrm{sgn}(\nu_I)\tilde v_I \tilde\rho_{I,-k}\tilde\rho_{I,k}\end{aligned} \tag{7.4.23}$$

where the new velocities $\tilde v_I$ of the edge excitations are given by

$$\begin{aligned}\mathrm{sgn}(\nu_1)\tilde v_1 &= \frac{\cos^2(\theta)}{\cos(2\theta)}|\nu_1|V_{11} - \frac{\sin^2(\theta)}{\cos(2\theta)}|\nu_2|V_{22}\\ \mathrm{sgn}(\nu_2)\tilde v_2 &= \frac{\cos^2(\theta)}{\cos(2\theta)}|\nu_2|V_{22} - \frac{\sin^2(\theta)}{\cos(2\theta)}|\nu_1|V_{11}\end{aligned} \tag{7.4.24}$$

We see that there are still two branches of the edge excitations. However, in this case, the edge excitations with a definite velocity are mixtures of those on the inner edge and the outer edge. One can also show that, as long as the Hamiltonian (7.4.21) is bounded from below, the velocities $\tilde v_I$ of the two branches are always positive.

After rewriting the electron operator Ψ_I in eqn (7.4.20) in terms of $\tilde\rho_I$ by inverting eqn (7.4.22), we can calculate their propagators using eqn (7.4.24) as follows:

$$\langle T(\Psi_I(x,t)\Psi_I^\dagger(0))\rangle = \mathrm{e}^{\,\mathrm{i}k_I x}\frac{1}{(x-\tilde v_1 t)^{\alpha_I}}\frac{1}{(x-\tilde v_2 t)^{\beta_I}}$$

where

$$(\alpha_1,\alpha_2) = (\frac{\cos^2\theta}{|\nu_1|}, \frac{\sin^2\theta}{|\nu_2|}), \qquad (\beta_1,\beta_2) = (\frac{\sin^2\theta}{|\nu_1|}, \frac{\cos^2\theta}{|\nu_2|})$$

However, when the two edges are close to each other within the magnetic length, the Ψ_I are no longer the most general electron operators on the edge. The generic electron operator may contain charge transfers between the two edges. For the $\nu = 2/5$ FQH state, the inner edge and the outer edge are separated by the $\nu = \frac{1}{3}$ Laughlin state. Thus, the elementary charge transfer operator is given by

$$\eta(x) = \mathrm{e}^{\,\mathrm{i}(\phi_1 - \frac{\nu_1}{\nu_2}\phi_2)} = (\Psi_1 \Psi_2^\dagger)^{\nu_1}$$

which transfers a $\nu_1 e = e/3$ charge from the outer edge to the inner edge. The generic electron operator then takes the form

$$\Psi(x) = \sum_{n=-\infty}^{+\infty} c_n \psi_n(x)$$
$$\psi_n(x) = \Psi_1(x)\eta^n(x) \tag{7.4.25}$$

To understand this result, we notice that each operator ψ_n always creates a unit localized charge and is a fermionic operator, regardless of the value of the integer n. Therefore, each ψ_n is a candidate for the electron operator on the edge. For a generic interacting system, the electron operator on the edge is expected to be a superposition of different ψ_ns, as represented in eqn (7.4.25). Note that $\Psi_2 = \psi_{-1/\nu_1}$. The propagator of ψ_n can be calculated in a similar way to that outlined above, and is given by

$$\langle T(\psi_n(x,t)\psi_m^\dagger(0))\rangle \propto \delta_{n,m}\mathrm{e}^{\,\mathrm{i}[k_1 + n\nu_1(k_2 - k_1)]x} \prod_I (x - \tilde{v}_I t)^{-\gamma_{In}}$$

where the γ_{In} are

$$\gamma_{1n} = \left[(n + \frac{1}{|\nu_1|})\sqrt{|\nu_1|}\cos\theta - \frac{n\nu_1}{\nu_2}\sqrt{|\nu_2|}\sin\theta \right]^2$$
$$\gamma_{2n} = \left[(n + \frac{1}{|\nu_1|})\sqrt{|\nu_1|}\sin\theta + \frac{n\nu_1}{\nu_2}\sqrt{|\nu_2|}\cos\theta \right]^2 \tag{7.4.26}$$

From eqns (7.4.25) and (7.4.4), we see that the electron propagator has singularities at the discrete momenta $k = k_1 + n\nu_1(k_2 - k_1)$. They are analogous to the $k_F, 3k_F, ...$ singularities of the electron propagator in the interacting one-dimensional electron systems.

For the $\nu = \frac{2}{3}$ FQH state, we have $\nu_1\nu_2 < 0$. In this case, we need to choose

$$\tilde{\rho}_{1k} = \cosh(\theta)\frac{1}{\sqrt{|\nu_1|}}\rho_{1k} + \sinh(\theta)\frac{1}{\sqrt{|\nu_2|}}\rho_{2k}$$
$$\tilde{\rho}_{2k} = \cosh(\theta)\frac{1}{\sqrt{|\nu_2|}}\rho_{2k} + \sinh(\theta)\frac{1}{\sqrt{|\nu_1|}}\rho_{1k}$$
$$\tanh(2\theta) = 2\frac{\sqrt{|\nu_1\nu_2|}V_{12}}{|\nu_1|V_{11} + |\nu_2|V_{22}} \tag{7.4.27}$$

to diagonalize the Hamiltonian. One can check that $\tilde{\rho}_I$ also satisfies the K–M algebra (7.4.23), but now

$$\begin{aligned}\mathrm{sgn}(\nu_1)\tilde{v}_1 &= \frac{\cosh^2(\theta)}{\cosh(2\theta)}|\nu_1|V_{11} - \frac{\sinh^2(\theta)}{\cosh(2\theta)}|\nu_2|V_{22} \\ \mathrm{sgn}(\nu_2)\tilde{v}_2 &= \frac{\cosh^2(\theta)}{\cosh(2\theta)}|\nu_2|V_{22} - \frac{\sinh^2(\theta)}{\cosh(2\theta)}|\nu_1|V_{11}\end{aligned} \tag{7.4.28}$$

Again, as long as the Hamiltonian H is positive definite, the velocities $\tilde{v}_I$ of the edge excitations always have opposite signs. The electron operator still has the form (7.4.25), with $\eta = (\Psi_1\Psi_2^\dagger)^{\nu_1}$. The propagator of ψ_n is still given by eqn (7.4.4), with

$$\begin{aligned}\gamma_{1n} &= \left[(n+\frac{1}{|\nu_1|})\sqrt{|\nu_1|}\cosh\theta + \frac{n\nu_1}{\nu_2}\sqrt{|\nu_2|}\sinh\theta\right]^2 \\ \gamma_{2n} &= \left[(n+\frac{1}{|\nu_1|})\sqrt{|\nu_1|}\sinh\theta + \frac{n\nu_1}{\nu_2}\sqrt{|\nu_2|}\cosh\theta\right]^2\end{aligned} \tag{7.4.29}$$

At equal space, the long-time correlation of the electron operator ψ_n is controlled by the total exponent $g_e^{(n)}$ as follows:

$$\langle\psi_n^\dagger(0,t)\psi_n(0,0)\rangle \sim t^{-g_e^{(n)}}, \qquad g_e^{(n)} = \sum_I \gamma_{In}$$

The equal-space electron correlation determines the I–V curve in the edge tunneling experiments (see Sections 3.7.4 and 4.2.2). The minimum value of the exponents $g_e \equiv \mathrm{Min}(g_e^{(n)})$ controls the scaling properties of the tunneling of electrons between two edges. For example, the tunneling conductance scales at finite temperatures as

$$\sigma \propto T^{2g_e-2}$$

Here $g_e = 3$ for the $\nu = 2/5$ state. For the $\nu = 2/3$ state, g_e depends on the interactions between the two edges.

Problem 7.4.6.

Bosonization of a one-dimensional interacting fermion system Consider a free one-dimensional fermion model

$$H_0 = \sum_k \epsilon_k : c_k^\dagger c_k :$$

The normal order is defined as $: c_k^\dagger c_k := c_k^\dagger c_k$ if $\epsilon_k > 0$ and $: c_k^\dagger c_k := -c_k c_k^\dagger$ if $\epsilon_k < 0$. Here $\epsilon_k = 0$ when $k = \pm k_F$.

1. Let $\rho_{1,k} = \sum'_{q\sim k_F} : c_q^\dagger c_{q+k} :$, where the summation $\sum'_{q\sim k_F}$ is limited to the range for which both the momenta of c_{q+k} and c_q are within the range $[-\Lambda + k_F, +\Lambda + k_F]$.

Let $\rho_{2,k} = \sum'_{q\sim -k_F} : c_q^\dagger c_{q+k} :$, where the summation $\sum'_{q\sim -k_F}$ is such that c_{q+k} and c_q are within the range $[-\Lambda - k_F, +\Lambda - k_F]$. Here $k < \Lambda < k_F$. We note that $\rho_{1,k}$ creates right-moving excitations near k_F, while $\rho_{2,k}$ creates left-moving ones near $-k_F$. Show that $\rho_{I,k}$ satisfies the K–M algebra (7.4.19), with $\nu_1 = 1$ and $\nu_2 = -1$. Show that $H_0 = 2\pi v_F \sum_{k>0}(\rho_{1,k}\rho_{1,-k} - \rho_{2,-k}\rho_{2,k})$. So the one-dimensional free fermion system can be bosonized.

2. We can write the electron operator as $c(x) = \psi_1(x) + \psi_2(x)$, where $\psi_1(x) = \sum_{-\Lambda+k_F}^{\Lambda+k_F} \mathrm{e}^{\,\mathrm{i}kx} c_k$ is the electron operator near k_F, and $\psi_2(x) = \sum_{-\Lambda-k_F}^{\Lambda-k_F} \mathrm{e}^{\,\mathrm{i}kx} c_k$ is the electron operator near $-k_F$. Express $c(x)$ in terms of ϕ_I, where $2\pi^{-1}\partial_x\phi_I(x) = \rho_I(x)$.

3. To verify that a one-dimensional interacting fermion system can also be bosonized, show that an interacting term $H_V = \sum_{q,k_1,k_2} V_q(c_{k_1}^\dagger c_{k_1+q})(c_{k_2}^\dagger c_{k_2+q})$, when limited near to the two Fermi points, becomes $2\pi \sum V_{IJ}\rho_{I,-k}\rho_{J,k}$ + constant. Find V_{IJ}.

4. Calculate the electron Green's function $\langle c^\dagger(x,t)c(0)\rangle$ for the interacting model $H_0 + H_V$.

7.4.5 Bulk effective theory and the edge states

- The structure of the edge states can be directly determined from the bulk topological order characterized by the K-matrix.

In this section, we will directly derive the macroscopic theory of the edge excitations from the Chern–Simons effective theory of the bulk FQH states. In this approach, we do not rely on a specific construction of the FQH states. The relationship between the bulk topological orders and edge states becomes quite clear. I have to warn the reader that the calculations presented in this section are very formal. The correctness of the results is not guaranteed by the formal calculation, but by the comparison with other independent calculations, such as the one in Section 7.4.3.

To understand the relationship between the effective theory and the edge states, let us first consider the simplest FQH state of the filling fraction $\nu = 1/m$ and try to rederive the results of Section 7.4.2 from the bulk effective theory. Such an FQH state is described by the $U(1)$ Chern–Simons theory with the action

$$S = -\frac{m}{4\pi}\int a_\mu \partial_\nu a_\lambda \varepsilon^{\mu\nu\lambda}\,\mathrm{d}^3x \tag{7.4.30}$$

Suppose that our sample has a boundary. For simplicity, we shall assume that the boundary is the x axis and that the sample covers the lower half-plane.

There is one problem with the effective action (7.4.30) for FQH liquids with boundaries. It is not invariant under gauge transformations $a_\mu \to a_\mu + \partial_\mu f$ due to the presence of the boundary, i.e. $\Delta S = -\frac{m}{4\pi}\int_{y=0} \mathrm{d}x\mathrm{d}t\, f(\partial_0 a_1 - \partial_1 a_0)$. To solve this problem, we will restrict the gauge transformations to be zero on the boundary $f(x, y = 0, t) = 0$. Due to this restriction, some degrees of freedom of a_μ on the boundary become dynamical.

We know that the effective theory (7.4.30) is derived only for a bulk FQH state without a boundary. Here we will take eqn (7.4.30) with the restricted gauge transformation as the definition of the effective theory for an FQH state with a boundary. Such a definition is definitely self-consistent. In the following, we will show that such a definition reproduces the results obtained in Section 7.4.2.

One way to study the dynamics of gauge theory is to choose the gauge condition $a_0 = 0$ and regard the equation of motion for a_0 as a constraint. For the Chern–Simons theory, such a constraint becomes $f_{ij} = 0$. Under this constraint, we may write a_i as $a_i = \partial_i \phi$. Substituting this into eqn (7.4.30), one obtains an effective one-dimensional theory on the edge with the action (Elitzur *et al.*, 1989)

$$S_{\text{edge}} = -\frac{m}{4\pi} \int \partial_t \phi \partial_x \phi \, \mathrm{d}x \, \mathrm{d}t \tag{7.4.31}$$

This approach, however, has a problem. It is easy to see that a Hamiltonian associated with the action (7.4.31) is zero and the boundary excitations described by eqn (7.4.31) have zero velocity. Therefore, this action cannot be used to describe any physical edge excitations in real FQH samples. The edge excitations in FQH states always have finite velocities.

The appearance of finite velocities for edge excitations is a boundary effect. The bulk effective theory defined by eqn (7.3.18) does not contain the information about the velocities of the edge excitations. To determine the dynamics of the edge excitations from the effective theory, we must find a way to input the information about the edge velocity. The edge velocities must be treated as the external parameters that are not contained in the bulk effective theory. The problem is how to put these parameters into the theory.

Let us now note that the condition $a_0 = 0$ is not the only choice for the gauge-fixing condition. A more general gauge-fixing condition has the form

$$a_\tau = a_0 + v a_x = 0 \tag{7.4.32}$$

Here a_x is the component of the vector potential parallel to the boundary of the sample and v is a parameter that has the dimension of velocity.

It is convenient to choose new coordinates that satisfy

$$\tilde{x} = x - vt, \qquad \tilde{t} = t, \qquad \tilde{y} = y. \tag{7.4.33}$$

Note that the gauge potential a_μ transforms as $\partial/\partial x^\mu$ under the coordinate transformation. We find that in the new coordinates the components of the gauge field are given by

$$\tilde{a}_{\tilde{t}} = a_t + va_x, \qquad \tilde{a}_{\tilde{x}} = a_x, \qquad \tilde{a}_{\tilde{y}} = a_y. \tag{7.4.34}$$

The gauge-fixing condition becomes the one discussed before in the new coordinates. It is easy to see that the form of the Chern–Simons action is preserved under the transformation given in eqns (7.4.33) and (7.4.34):

$$S = -\frac{m}{4\pi}\int \mathrm{d}^3x\, a_\mu \partial_\nu a_\lambda \varepsilon^{\mu\nu\lambda} = -\frac{m}{4\pi}\int \mathrm{d}^3x\, \tilde{a}_{\tilde{\mu}} \partial_{\tilde{\nu}} \tilde{a}_{\tilde{\lambda}} \varepsilon^{\tilde{\mu}\tilde{\nu}\tilde{\lambda}}$$

Repeating the previous derivation, we find that the edge action is given by

$$S = -\frac{m}{4\pi}\int \mathrm{d}\tilde{t}\,\mathrm{d}\tilde{x}\; \partial_{\tilde{t}}\phi\partial_{\tilde{x}}\phi$$

In terms of the original physical coordinates, the above action acquires the form

$$S = -\frac{m}{4\pi}\int \mathrm{d}t\,\mathrm{d}x\; (\partial_t + v\partial_x)\phi\partial_x\phi \tag{7.4.35}$$

which is a chiral boson theory (Gross *et al.*, 1985; Floreanini and Jackiw, 1988). From the equation of motion $\partial_t\phi + v\partial_x\phi = 0$, we can see that the edge excitations described by eqn (7.4.35) have a nonzero velocity v and move only in one direction.

To obtain the quantum theory for eqn (7.4.35), we need to quantize the chiral boson theory. The quantization can be done in the momentum space:

$$S = \frac{m}{2\pi}\int \mathrm{d}t\, \sum_{k>0}(\mathrm{i}k\dot{\phi}_k\phi_{-k} - vk^2\phi_k\phi_{-k}).$$

One can show that, if we regard ϕ_k with $k > 0$ as 'coordinates', then ϕ_{-k} will be proportional to the corresponding momentum $\pi_k = \partial S/\partial\dot{\phi}_k$. After identifying the 'coordinates' and the 'momentum', we can obtain the Hamiltonian and the commutation relation $[\phi_k, \pi_{k'}] = \mathrm{i}\delta_{kk'}$, which completely define the quantum system. If we introduce $\rho = \frac{1}{2\pi}\partial_x\phi$, then we find that the quantum system is described by

$$[\rho_k, \rho_{k'}] = \frac{k\delta_{k+k'}}{2\pi m}, \qquad H = 2\pi m v \sum_{k>0} \rho_k^\dagger \rho_k \tag{7.4.36}$$

The velocity v of the edge excitations enters our theory through the gauge-fixing condition. Notice that, under the restricted gauge transformations, the gauge-fixing conditions (7.4.32) with different v cannot be transformed into each other. They are physically inequivalent. This agrees with our assumption that v in the gauge-fixing condition is physical and actually determines the velocity of the edge excitations.

The Hamiltonian (7.4.36) is bounded from below only when $vm < 0$. The consistency of our theory requires v and m to have opposite signs. Therefore, the sign of the velocity (the chirality) of the edge excitations is determined by the sign of the coefficient in front of the Chern–Simons terms.

The above results can be easily generalized to the generic FQH states described by eqn (7.3.18) because the matrix K can be diagonalized. The resulting effective edge theory has the form

$$S_{\text{edge}} = \frac{1}{4\pi}\int \mathrm{d}t\,\mathrm{d}x\,\left[K_{IJ}\partial_t\phi_I\partial_x\phi_J - V_{IJ}\partial_x\phi_I\partial_x\phi_J\right] \tag{7.4.37}$$

The Hamiltonian is given by

$$H_{\text{edge}} = \frac{1}{4\pi}\int \mathrm{d}t\,\mathrm{d}x\; V_{IJ}\partial_x\phi_I\partial_x\phi_J$$

Therefore, V must be a positive-definite matrix. Using this result, one can show that a positive eigenvalue of K corresponds to a left-moving branch and a negative eigenvalue corresponds to a right-moving one.

The effective theory of the $\nu = 2/5$ FQH state is given by $K = \begin{pmatrix} 3 & 2 \\ 2 & 3 \end{pmatrix}$. As K has two positive eigenvalues, the edge excitations of the $\nu = 2/5$ FQH state have two branches moving in the same direction. The $\nu = 1 - \frac{1}{n}$ FQH state is described by the effective theory with $K = \begin{pmatrix} 1 & 0 \\ 0 & -n \end{pmatrix}$. The two eigenvalues of K now have opposite signs; hence, the two branches of the edge excitations move in opposite directions.

Problem 7.4.7.
Show that, after quantization, the system (7.4.35) is described by eqn (7.4.36).

7.4.6 Charged excitations and the electron propagator

• Generic electron/quasiparticle operators on a generic FQH edge.

We have studied the dynamics of edge excitations in a few simple FQH liquids. We found that the low-lying edge excitations are described by a free phonon theory. In this section, we will concentrate on the generic charge excitations. In particular, we will calculate the propagators of the electrons and the quasiparticles for the most general (abelian) FQH states. The key point again is to write the electron or the quasiparticle operators in terms of the phonon operator ρ_I. Once we have done so, the propagators can be easily calculated because the phonons are free (at low energies and long wavelength).

We know that, for the FQH state described by eqn (7.3.18), the edge states are described by the action (7.4.37). The Hilbert space of the edge excitations forms a representation of the K–M algebra

$$[\rho_{Ik}, \rho_{Jk'}] = (K^{-1})_{IJ}\frac{1}{2\pi}k\delta_{k+k'}$$
$$k, k' = \text{integer} \times \frac{2\pi}{L} \tag{7.4.38}$$

where $\rho_I = \frac{1}{2\pi}\partial_x\phi_I$ is the edge density of the Ith condensate in the FQH state, $I, J = 1, ..., \kappa$, and κ is the dimension of K. The electron density on the edge is given by

$$\rho_e = -eq_I\rho_I$$

The dynamics of the edge excitations are described by the Hamiltonian

$$H = 2\pi\sum_{IJ} V_{IJ}\rho_{I,-k}\rho_{J,k} \tag{7.4.39}$$

where V_{IJ} is a positive-definite matrix.

Let us first try to write down the quasiparticle operator $\Psi_{\boldsymbol{l}}$ on the edge which creates a quasiparticle labeled by l_I. We know that inserting the quasiparticle on the edge will cause a change $\delta\rho_I$ in the edge density of the Ith condensate (see eqn (7.3.22)). Here $\delta\rho_I$ satisfies

$$\int \mathrm{d}x\, \delta\rho_I = (\boldsymbol{l}^\top K^{-1})_I.$$

Because $\Psi_{\boldsymbol{l}}$ is a local operator that only causes a local change in the density, we have

$$[\rho_I(x), \Psi_{\boldsymbol{l}}(x')] = l_J(K^{-1})_{JI}\delta(x-x')\Psi_{\boldsymbol{l}}(x') \tag{7.4.40}$$

Using the Kac–Moody algebra (7.4.38), one can show that the quasiparticle operators that satisfy eqn (7.4.40) are given by

$$\Psi_{\boldsymbol{l}} \propto \mathrm{e}^{\mathrm{i}\phi_I l_I} \tag{7.4.41}$$

The charge of the quasiparticle $\Psi_{\boldsymbol{l}}$ is determined from the commutator $[\rho_e, \Psi_{\boldsymbol{l}}]$, and is found to be

$$Q_{\boldsymbol{l}} = -e\boldsymbol{q}^\top K^{-1}\boldsymbol{l} \tag{7.4.42}$$

From eqns (7.3.24) and (7.4.41), we see that the electron operator can be written as

$$\Psi_{e,L} \propto \mathrm{e}^{\mathrm{i}\sum_I l_I\phi_I}, \qquad l_I = K_{IJ}L_J, \qquad q_I L_I = 1 \tag{7.4.43}$$

The above operators carry a unit charge, as one can see from eqn (7.4.42). The commutation of $\Psi_{e,L}$ can be calculated as follows:

$$\Psi_{e,L}(x)\Psi_{e,L}(x') = (-)^{\lambda}\Psi_{e,L}(x')\Psi_{e,L}(x)$$
$$\lambda = l_I(K^{-1})_{IJ}l_J = L_I K_{IJ} L_J \tag{7.4.44}$$

In the hierarchical basis, $K_{II}|_{I=1} = \text{odd}$, $K_{II}|_{I>1} = \text{even}$, $\boldsymbol{q}^{\top} = (1, 0, ..., 0)$, and $L_1 = 1$. Using these conditions, we can show that $(-)^{\lambda} = -1$. The electron operators defined in eqn (7.4.43) are indeed fermion operators.

As all of the operators $\Psi_{e,L}$ for different choices of L_I carry a unit charge and are fermionic, each $\Psi_{e,L}$ can be a candidate for the electron operators. In general, the true electron operator is the following superposition of the $\Psi_{e,L}$s:

$$\Psi_e = \sum_L C_L \Psi_{e,L}$$

Here, when we say that there are many different electron operators on the edge, we really mean that the true physical electron operator is a superposition of these operators.

Using the K–M algebra (7.4.38) and the Hamiltonian (7.4.39), we can calculate the propagators of a generic quasiparticle operator

$$\Psi_{\boldsymbol{l}} \propto \mathrm{e}^{\mathrm{i}l_I\phi_I}$$

(which includes the electron operators for suitable choices of $\boldsymbol{l}$). First, we note that the transformation

$$\rho_I \to U_{IJ}\tilde{\rho}_J$$

changes (V, K) to $(\tilde{V}, \tilde{K}) = (U^{\top}VU, U^{\top}KU)$. By choosing a suitable U, K and V can be simultaneously diagonalized;[49] in terms of $\tilde{\rho}_I$, eqns (7.4.38) and (7.4.39) become

$$[\tilde{\rho}_{Ik}, \tilde{\rho}_{Jk'}] = \sigma_I\delta_{IJ}\frac{1}{2\pi}k\delta_{k+k'}$$
$$H = 2\pi \sum_{I,k>0} |v_I|\tilde{\rho}_{I,-k}\tilde{\rho}_{I,k} \tag{7.4.45}$$

where $\sigma_I = \pm 1$ is the sign of the eigenvalues of K. The velocity of the edge excitations created by $\tilde{\rho}_I$ is given by $v_I = \sigma_I|v_I|$.

[49] As V is a positive-definite symmetric matrix, we can find U_1 which transforms $V \to V_1 = U_1^{\top}VU_1 = 1$ and $K \to K_1 = U_1^{\top}KU_1$. Note that K_1 is a symmetric matrix whose eigenvalues have the same sign as the eigenvalues of K (although their absolute values may differ). Then we make an orthogonal transformation to diagonalize K_1, namely $(K_1)_{IJ} \to (K_2)_{IJ} = \sigma_I|v_I|^{-1}\delta_{IJ}$. Under the orthogonal transformation, $V_1 \to V_2 = 1$ is not changed and remains diagonal. Finally, we use a diagonal matrix to scale K_2 to $(K_3)_{IJ} = \sigma_I\delta_{IJ}$. This changes V_2 to $V_3 = |v_I|\delta_{IJ}$. The new (V_3, K_3) lead to eqn (7.4.45).

In terms of $\tilde{\rho}_I$, the operator $\Psi_{\boldsymbol{l}}$ has the form

$$\Psi_{\boldsymbol{l}} \propto e^{i\sum_I \tilde{l}_I \tilde{\phi}_I}, \qquad \tilde{l}_J = l_I U_{IJ} \tag{7.4.46}$$

From eqns (7.4.45) and (7.4.46), we see that the propagator of $\Psi_{\boldsymbol{l}}$ has the following general form:

$$\langle \Psi_{\boldsymbol{l}}^\dagger(x,t)\Psi_{\boldsymbol{l}}(0)\rangle \propto e^{il_I k_I x} \prod_I (x - v_I t + i\sigma_I \delta)^{-\gamma_I}, \qquad \gamma_I = \tilde{l}_I^2 \tag{7.4.47}$$

where $v_I = \sigma_I |v_I|$ are the velocities of the edge excitations. The total exponent from the right movers minus the total exponent from the left movers satisfies the sum rule[50]

$$\sum_I \sigma_I \gamma_I \equiv \lambda_{\boldsymbol{l}} = \boldsymbol{l}^\top K^{-1} \boldsymbol{l} \tag{7.4.48}$$

We see that the difference $\lambda_{\boldsymbol{l}}$ is independent of the edge interaction V and is a topological quantum number. From eqns (7.4.44) and (7.4.48), we see that the difference is directly related to the statistics of $\Psi_{\boldsymbol{l}}$. If $\Psi_{\boldsymbol{l}}$ represents the electron operator (or other fermionic operators), then $\lambda_{\boldsymbol{l}}$ will be an odd integer. From eqn (7.4.47), we also see that the operator $\Psi_{\boldsymbol{l}}$ creates an excitation with momentum near to $\sum_I l_I k_I$.

Problem 7.4.8.
Prove that $\Psi_{\boldsymbol{l}}$ in eqn (7.4.41) satisfies eqn (7.4.40).

7.4.7 Phenomenological consequences of chiral Luttinger liquids

- Three classes of edge states.
- The effects of long-range interactions and impurity scattering.

In the last four sections, we have shown that the electrons at the edges of an FQH liquid form a chiral Luttinger liquid. As one of the characteristic properties of chiral Luttinger liquids, the electron and the quasiparticle propagators obtain anomalous exponents as follows:

$$\langle \Psi_e^\dagger(t, x=0)\Psi_e(0)\rangle \sim t^{-g_e}, \qquad \langle \Psi_q^\dagger(t, x=0)\Psi_q(0)\rangle \sim t^{-g_q}$$

These anomalous exponents can be directly measured by tunneling experiments between the FQH edges.

[50] We have used

$$\lambda_{\boldsymbol{l}} = \sum_{I,J,I'} l_{I'} U_{I'I} \sigma_I (U^\top)_{IJ} l_J, \qquad (K^{-1})_{IJ} = U_{II'} \sigma_{I'} \delta_{I'J'} (U^\top)_{J'J}.$$

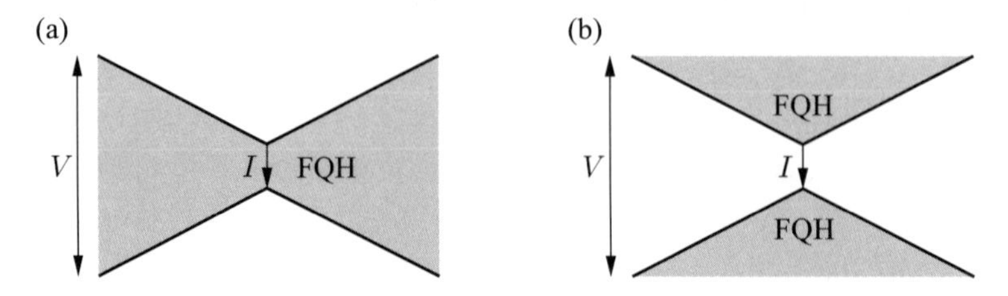

FIG. 7.18. (a) Quasiparticle tunneling between two edges of the same FQH state. (b) Electron tunneling between two edges of the different FQH states.

The following two situations need to be considered separately: (i) two edges of FQH liquids are separated by an insulator; and (ii) two edges are separated by an FQH liquid. In case (i), only an electron can tunnel between the edges (see Fig. 7.18(b)); in case (ii), the quasiparticles supported by the FQH liquid that separates the two edges can tunnel (see Fig. 7.18(a)). The tunneling operator A is given by $A \propto \Psi_{e1}\Psi_{e2}^\dagger$ for case (i), where Ψ_{e1} and Ψ_{e2} are the electron operators on the two edges. The tunneling operator A has the form $A \propto \Psi_{q1}\Psi_{q2}^\dagger$ for case (ii), where Ψ_{q1} and Ψ_{q2} are the quasiparticle operators on the two edges. The physical properties of the tunneling can be calculated from the correlation of the tunneling operator, which in turn can be expressed as a product of the electron or quasiparticle propagators on the two edges.

Let $g = g_e$ for case (i) and $g = g_q$ for case (ii). Then, at zero temperature, the anomalous exponents of the electron and quasiparticle operators lead to $\langle A(t)A(0)\rangle \propto |t|^{-2g}$, which results in a nonlinear tunneling I–V curve (see Section 3.7.4)

$$I \propto V^{2g-1}$$

The noise spectrum of the tunneling current also contains a singularity at the frequency $f = QV/h$, i.e.

$$S(f) \sim |f - \frac{QV}{h}|^{2g-1} \tag{7.4.49}$$

where V is the voltage difference between the two edges and Q is the electric charge of the tunneling electron or the tunneling quasiparticle. At a finite temperature T, the zero-bias conductance also has the following power law dependence:

$$\sigma \propto T^{2g-2}$$

We see that the anomalous exponents can be easily measured by the tunneling experiments. The noise spectrum further reveals the charges of the tunneling particles.

The exponents g_e and g_q are calculated in Section 7.4.6. To summarize the results in a simple way, it is convenient to divide the FQH edges into the following two classes: (A) all edge excitations move in one direction; and (B) edge excitations propagate in both directions.

For the class A edges, the exponents g_e and g_q are directly related to the statistics of the electrons and quasiparticles. In this case, g_e and g_q are universal and independent of the details of the electron interactions and the edge potentials. Table 7.3 lists some FQH states which support the class A edge states, as well as the corresponding values of the exponents $g_{e,q}$ and the electric charge of the associated particles (note that only the minimum values of g_e and g_q are listed).

TABLE 7.3. (g_e, g_q, Q_q) for some FQH states with the class A edge states.

FQH states	$\nu = 1/m$	$\nu = 2/5$	$\nu = p/(pq+1)$	$(331)_{\nu=1/2}$	$(332)_{\nu=2/5}$
g_e	m	3	$q+1$	3	3
g_q	$1/m$	$2/5$	$p/(pq+1)$	$3/8$	$2/5$
Charge Q_q/e	$1/m$	$2/5$	$p/(pq+1)$	$1/4$	$2/5$

Let us discuss how we obtain the above results for the hierarchical states with the filling fractions $\nu = \frac{p}{pq+1}$ (q = even). These states include the $\nu = 2/5, 3/7, 2/9, ...$ FQH states. The hierarchical states with $\nu = \frac{p}{pq+1}$ are described by the $p \times p$ matrices $K = 1 + qC$ in the basis where $\boldsymbol{q}^\top = (1, 1, ..., 1)$. Here C is the pseudo-identity matrix, namely $C_{IJ} = 1$, $I, J = 1, ..., p$, and $K^{-1} = 1 - \frac{q}{pq+1} C$. Because all of the edge excitations move in the same direction, we have

$$\langle \Psi_{\boldsymbol{l}}^\dagger (x = 0, t) \Psi_{\boldsymbol{l}}(0) \rangle \propto t^{-\lambda_{\boldsymbol{l}}}$$

where $\lambda_{\boldsymbol{l}}$ is given by eqn (7.4.48). The fundamental quasiparticle is given by $\boldsymbol{l}^\top = (1, 0, ..., 0)$ and carries the charge $\frac{1}{pq+1}$. The exponent in its propagator is $\lambda_{\boldsymbol{l}} = 1 - \frac{q}{pq+1}$. The quasiparticle with the smallest exponent is given by $\boldsymbol{l}^\top = (1, ..., 1)$ and carries the charge $\frac{p}{pq+1}$. The exponent is $\lambda_{\boldsymbol{l}} = \frac{p}{pq+1}$, which is less than $1 - \frac{q}{pq+1}$ (note that we have $q \geqslant 2$ and $p \geqslant 1$). Such a quasiparticle (with the charge $\frac{p}{pq+1}$) dominates the tunneling between two edges of the *same* FQH fluid at low energies (see Section 7.4.7).

The electron operators are given by $\Psi_{e,L} = \psi_{\boldsymbol{l}}$, with $\boldsymbol{l}$ satisfying $\sum_I l_I = pq + 1$. The exponent in the propagator is given by $\lambda_{\boldsymbol{l}} = \sum l_I^2 - q(pq + 1)$. The electron operator with a minimum exponent in its propagator is given by $\boldsymbol{l}^\top = (q, ..., q, q + 1)$. The value of the minimum exponent is $\lambda_{\boldsymbol{l}} = q + 1$. Such an electron operator dominates the tunneling between edges of two *different* FQH fluids at low energies.

For the class B edges, g_e and g_q are not universal. Their values depend on the electron interactions and the edge potentials.

Problem 7.4.9.
Derive eqn (7.4.49) using the correlation of the tunneling operator $\langle A(t)A(0)\rangle$ at a finite voltage V.

8

TOPOLOGICAL AND QUANTUM ORDER—BEYOND LANDAU'S THEORIES

In Chapters 3, 4, and 5, we discussed several interacting boson/fermion systems in detail. These simple models illustrate Landau's symmetry-breaking theory (plus RG theory) and Landau's Fermi liquid theory (plus perturbation theory). The two Landau theories explain the behavior of many condensed matter systems and form the foundation of traditional many-body theory. The discussions in the three chapters only scratch the surface of Landau's theories and their rich applications. Readers who want to learn more about Landau's theories (and RG theory and perturbation theory) may find the books by Abrikosov *et al.* (1975), Ma (1976), Mahan (1990), Negele and Orland (1998), and Chaikin and Lubensky (2000) useful.

Landau's theories are very successful and for a long time people could not find any condensed matter systems that could not be described by Landau's theories. For fifty years, Landau's theories dominated many-body physics and essentially defined the paradigm of many-body physics. After so many years, it has become a common belief that we have figured out all of the important concepts and understood the essential properties of all forms of matter. Many-body theory has reached its end and is a more or less complete theory. The only thing to be done is to apply Landau's theories (plus the renormalization group picture) to all different kinds of systems.

From this perspective, we can understand the importance of the fractional quantum Hall (FQH) effect discovered by Tsui *et al.* (1982). The FQH effect opened up a new chapter in condensed matter physics. As we have seen in the last chapter, FQH liquids cannot be described by Fermi liquid theory. Different FQH states have the same symmetry and cannot be described by Landau's symmetry-breaking theory. Thus, FQH states are completely beyond the two Landau theories. The existence of FQH liquids indicates that there is a new world beyond the paradigm of Landau's theories. Recent studies suggest that the new paradigm is much richer than the paradigm of Landau's theories. Chapters 7 to 10 of this book are devoted to the new paradigm beyond Landau's theories.

To take a glimpse at the new paradigm of condensed matter physics, we are going to study quantum rotor systems, hard-core boson systems, and quantum spin systems, in addition to the FQH states. These systems are all strongly-correlated many-body systems. Usually, the strong correlation in these systems will lead to long-range orders and symmetry breaking. However, we have already discussed long-range order and symmetry breaking, using a boson superfluid and a fermion SDW state as examples. So, to study the world beyond Landau's theories, we concentrate on quantum liquid states (such as the FQH states) that cannot be described by long-range order and symmetry breaking.

These quantum liquid states represent new states of matter that contain a completely new kind of order—topological/quantum order. I will show that the new order may have a deep impact on our understanding of the quantum phase and the quantum phase transition, as well as gapless excitations in the quantum phase. In particular, topological/quantum order might provide an origin for light and electrons (as well as other gauge bosons and fermions) in nature.

In this chapter, we will give a general discussion of topological/quantum order to paint a larger picture. We will use FQH states and Fermi liquid states as examples to discuss some basic issues in topological/quantum order. The problems and issues in topological/quantum order are very different from those in traditional many-body physics. It is very important to understand what the problems are before going into any detailed calculations.

8.1 States of matter and the concept of order

- Matter can have many different states (or different phases). The concept of order is introduced to characterize different internal structures in different states of matter.
- We used to believe that different orders are characterized by their different symmetries.

At sufficiently high temperatures, all matter is in the form of a gas. Gas is one of the simplest states. The motion of an atom in a gas hardly depends on the positions and motion of other molecules. Thus, gases are weakly-correlated systems which contain no internal structure. However, as the temperature is lowered the motion of the atoms becomes more and more correlated. Eventually, the atoms form a very regular pattern and a crystal order is developed. In a crystal, an individual atom can hardly move by itself. Excitations in a crystal always correspond to the collective motion of many atoms (which are called phonons). A crystal is an example of a strongly-correlated state.

With the development of low-temperature technology in around 1900, physicists discovered many new states of matter (such as superconductors and superfluids). These different states have different internal structures, which are called different kinds of orders. The precise definition of order involves phase transition. Two states of a many-body system have the same order if we can smoothly change

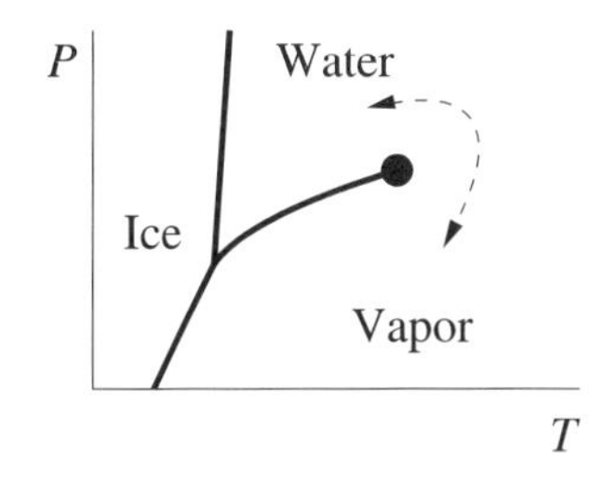

FIG. 8.1. The phase diagram of water.

one state into the other (by smoothly changing the Hamiltonian) without encountering a phase transition (i.e. without encountering a singularity in the free energy). If there is no way to change one state into the other without a phase transition, then the two states will have different orders. We note that our definition of order is a definition of an equivalent class. Two states that can be connected without a phase transition are defined to be equivalent. The equivalent class defined in this way is called the universality class. Two states with different orders can also be said to be two states belonging to different universality classes. According to our definition, water and ice have different orders, while water and vapor have the same order (see Fig. 8.1).

After discovering so many different kinds of order, a general theory is needed to gain a deeper understanding of the states of matter. In particular, we like to understand what makes two orders really different, so that we cannot change one order into the other without encountering a phase transition. The key step in developing the general theory for order and the associated phase and phase transition is the realization that orders are associated with symmetries (or rather, the breaking of symmetries). We find that, when two states have different symmetries, then we cannot change one into the other without encountering a singularity in the free energy (i.e. without encountering a phase transition). Based on the relationship between orders and symmetries, Landau developed a general theory of orders and transitions between different orders (Ginzburg and Landau, 1950; Landau and Lifschitz, 1958). Landau's theory is very successful. Using Landau's theory and the related group theory for symmetries, we can classify all of the 230 different kinds of crystals that can exist in three dimensions. By determining how symmetry changes across a continuous phase transition, we can obtain the critical properties of the phase transition. The symmetry breaking also provides the origin of many gapless excitations, such as phonons, spin waves, etc., which determine the low-energy properties of many systems (Nambu, 1960; Goldstone, 1961). Many of the properties of those excitations, including their gaplessness, are directly determined by the symmetry. Introducing order parameters associated with symmetries,

Ginzburg and Landau developed Ginzburg–Landau theory, which became the standard theory for phase and phase transition. As Landau's symmetry-breaking theory has such a broad and fundamental impact on our understanding of matter, it became a corner-stone of condensed matter theory. The picture painted by Landau's theory is so satisfactory that one starts to have a feeling that we understand, at least in principle, all kinds of orders that matter can have.

8.2 Topological order in fractional quantum Hall states

- The FQH state opened up a new chapter in condensed matter physics, because it was the second state to be discovered by experimentalists that could not be characterized by symmetry breaking and (local) order parameters.
- The FQH states contain a new kind of order—topological order.

However, nature never ceases to surprise us. With advances in semiconductor technology, physicists learnt how to confine electrons on an interface between two different semiconductors, and hence made a two-dimensional electron gas (2DEG). In 1982, Tsui *et al.* (1982) put a 2DEG under strong magnetic fields and discovered a new state of matter—the FQH liquid (Laughlin, 1983). As the temperatures are low and the interaction between the electrons is strong, the FQH state is a strongly-correlated state. However, such a strongly-correlated state is not a crystal, as people had originally expected. It turns out that the strong quantum fluctuations of electrons, due to their very small mass, prevent the formation of a crystal. Thus, the FQH state is a quantum liquid. (A crystal can melt in two ways, namely by thermal fluctuations as we raise temperatures, which leads to an ordinary liquid, or by quantum fluctuations as we reduce the mass of the particles, which leads to a quantum liquid.)

As we have seen in the last chapter, quantum Hall liquids have many amazing properties. A quantum Hall liquid is more 'rigid' than a solid (a crystal), in the sense that a quantum Hall liquid cannot be compressed. Thus, a quantum Hall liquid has a fixed and well-defined density. When we measure the electron density in terms of the filling fraction, defined by

$$\nu = \frac{\text{density of electron}}{\text{density of magnetic flux quanta}},$$

we find that all of the discovered quantum Hall states have densities such that the filling fractions are given exactly by some rational numbers, such as $\nu = 1, 1/3, 2/3, 2/5, \ldots$. Knowing that FQH liquids exist only at certain magical filling fractions, one cannot help but guess that FQH liquids should have some internal orders or 'patterns'. Different magical filling fractions should be due to these different internal 'patterns'. However, the hypothesis of internal 'patterns' appears

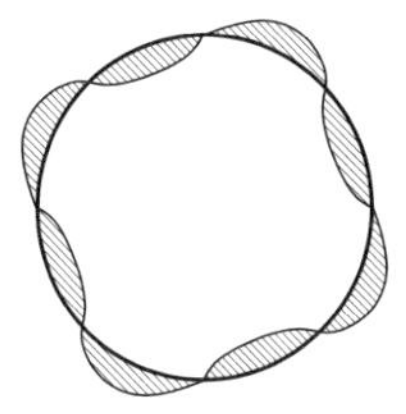

FIG. 8.2. A particle wave on a circle has a quantized wavelength.

to have one difficulty—the FQH states are liquids, and how can liquids have any internal 'patterns'?

To gain some intuitive understanding of the internal order in FQH states, let us try to visualize the quantum motion of electrons in an FQH state. We know that a particle also behaves like a wave, according to quantum physics. Let us first consider a particle moving on a circle with momentum p. Such a particle corresponds to a wave with wavelength $\lambda = h/p$, where h is the Planck constant. Only waves that can fit into the circle are allowed (i.e. the circle must contain an integer number of wavelengths) (see Fig. 8.2). Thus, due to quantum physics, the motion of a particle on a circle is highly restricted (or quantized), and only certain discrete values of momentum are allowed. Such a quantization condition can be viewed in a more pictorial way. We may say that the particle dances around the circle in steps, with a step length given by the wavelength. The quantization condition requires that the particle always takes an integer number of steps to go around the circle.

Now let us consider a single electron in a magnetic field. Under the influence of the magnetic field, the electron always moves along circles (which are called cyclotron motions). In quantum physics, only certain discrete cyclotron motions are allowed due to the wave property of the particle. The quantization condition is such that the circular orbit of an allowed cyclotron motion contains an integer number of wavelengths. We may say that an electron always takes an integer number of steps to go around the circle. If the electron takes n steps around the circle, then we say that the electron is in the nth Landau level. The electrons in the first Landau level have the lowest energy, and the electron will stay in the first Landau level at low temperatures.

When we have many electrons to form a 2DEG, electrons not only do their own cyclotron motion in the first Landau level, but they also go around each other and exchange places. These additional motions are also subject to the quantization condition. For example, an electron must take integer steps to go around another electron. As electrons are fermions, exchanging two electrons introduces a minus sign into the wave function. Also, exchanging two electrons is equivalent to moving one electron half-way around the other electron. Thus, an electron must take half-integer steps to go half-way around another electron. (The half-integer steps introduce a minus sign into the electron wave function.) In other words, an electron must take an odd number of steps to go around another electron. Electrons in an FQH state not only move in a way that satisfies the quantization condition, but they also try to stay away from each other as much as possible, due to the strong Coulomb repulsion and the Fermi statistics between electrons. This means that an electron tries to take more steps to go around another electron, if possible.

Now we see that, despite the absence of crystal order, the quantum motions of electrons in an FQH state are highly organized. All of the electrons in an FQH state dance collectively, following strict dancing rules.

1. All of the electrons do their own cyclotron motion in the first Landau level, i.e. they take one step to go around the circle.
2. An electron always takes an odd number of steps to go around another electron.
3. Electrons try to stay away from each other, i.e. they try to take as many steps as possible to go around another electron.

If every electron follows these strict dancing rules, then only one unique global dancing pattern is allowed. Such a dancing pattern describes the internal quantum motion in the FQH state. It is this global dancing pattern that corresponds to the internal order in the FQH state. Different FQH states are distinguished by their different dancing patterns.

A more precise mathematical description of the quantum motion of electrons outlined above is given by the famous Laughlin wave function (Laughlin, 1983)

$$\Psi_m = \left[\prod (z_i - z_j)^m\right] \mathrm{e}^{-\frac{1}{4l_B^2}\sum |z_i|^2}$$

where m is an odd integer and $z_j = x_j + \mathrm{i} y_j$ is the coordinate of the jth electron. Such a wave function describes a filling fraction $\nu = 1/m$ FQH state. We see that the wave function vanishes as $z_i \to z_j$, so that the electrons do not like to stay close to each other. Also, the wave function changes its phase by $2\pi m$ as we move one electron around another. Thus, an electron always takes m steps to go around another electron in the Laughlin state.

We would like to stress that the internal orders (i.e. the dancing patterns) of FQH liquids are very different from the internal orders in other correlated systems, such as crystals, superfluids, etc. The internal orders in the latter systems can be described by order parameters associated with broken symmetries. As a result, the ordered states can be described by the Ginzburg–Landau effective theory. The internal order in FQH liquids is a new kind of ordering which cannot be described by long-range orders associated with broken symmetries.[51] In 1989, the concept of 'topological order' was introduced to describe this new kind of ordering in FQH liquids (Wen, 1990, 1995).

We would like to point out that topological orders are general properties of any states at zero temperature with a finite energy gap. Non-trivial topological orders not only appear in FQH liquids, but they also appear in spin liquids at zero temperature. In fact, the concept of topological order was first introduced (Wen, 1990) in a study of chiral spin liquids (Kalmeyer and Laughlin, 1987; Khveshchenko and Wiegmann, 1989; Wen *et al.*, 1989). In addition to chiral spin liquids, non-trivial topological orders were also found in anyon superfluids (Chen *et al.*, 1989; Fetter *et al.*, 1989; Wen and Zee, 1991) and short-ranged resonating valence bound states for spin systems (Kivelson *et al.*, 1987; Rokhsar and Kivelson, 1988; Read and Chakraborty, 1989; Read and Sachdev, 1991; Wen, 1991a). The FQH liquid is not even the first experimentally observed state with non-trivial topological orders. That honor goes to the superconducting state discovered in 1911 (Onnes, 1911; Bardeen *et al.*, 1957). In contrast to a common point of view, a superconducting state, with dynamical electromagnetic interactions, cannot be characterized by broken symmetries. It has neither long-range orders nor local order parameters. A superconducting state contains non-trivial topological orders. It is fundamentally different from a superfluid state (Coleman and Weinberg, 1973; Halperin *et al.*, 1974; Fradkin and Shenker, 1979).

It is instructive to compare FQH liquids with crystals. FQH liquids are similar to crystals in the sense that they both contain rich internal patterns (or internal orders). The main difference is that the patterns in the crystals are static, related to the positions of atoms, while the patterns in QH liquids are 'dynamic', associated with the ways that electrons 'dance' around each other. However, many of the same questions for crystal orders can also be asked and should be addressed for topological orders. We know that crystal orders can be characterized and classified by symmetries. Thus, one important question is how do we characterize and classify the topological orders? We also know that crystal orders can be measured by X-ray diffraction. The second important question is how do we experimentally measure the topological orders?

In the following, we are going to discuss topological orders in FQH states in more detail. It turns out that FQH states are quite typical topologically-ordered states. Many other topologically-ordered states share many similar properties with FQH states.

8.2.1 Characterization of topological orders

- Any new concepts in physics must be introduced (or defined) via quantities that can be measured by experiments. To define a physical quantity or concept is to design an experiment.

[51] Although it was suggested that the internal structures of Laughlin states can be characterized by an 'off-diagonal long-range order' (Girvin and MacDonald, 1987), the operator that has long-range order itself is not a local operator. For local operators, there is no long-range order and there are no symmetry-breaking Laughlin states.

- The concept of topological order is (partially) defined by ground-state degeneracy, which is robust against *any* perturbations that can break all of the symmetries.

In the above, the concept of topological order (the dancing pattern) is introduced through the ground-state wave function. This is not quite correct because the ground-state wave function is not universal. To establish a new concept, such as topological order, one needs to find physical characterizations or measurements of topological orders. In other words, one needs to find universal quantum numbers that are robust against any perturbation, such as changes in interactions, effective mass, etc., but which can take different values for different classes of FQH liquids. The existence of such quantum numbers implies the existence of topological orders.

One way to show the existence of topological orders in FQH liquids is to study their ground-state degeneracies (in the thermodynamical limit). FQH liquids have a very special property. Their ground-state degeneracy depends on the topology of space (Haldane, 1983; Haldane and Rezayi, 1985). For example, the $\nu = \frac{1}{q}$ Laughlin state has q^g degenerate ground states on a Riemann surface of genus g. The ground-state degeneracy in FQH liquids is *not* a consequence of the symmetry of the Hamiltonian. The ground-state degeneracy is robust against arbitrary perturbations (even impurities that break all of the symmetries in the Hamiltonian) (Wen and Niu, 1990). The robustness of the ground-state degeneracy indicates that the internal structures that give rise to ground-state degeneracy are universal and robust, hence demonstrating the existence of universal internal structures—topological orders.

To understand the topological degeneracy of FQH ground states, we consider a $\nu = 1/m$ Laughlin state on a torus. We will use two methods to calculate the ground-state degeneracy. In the first method, we consider the following tunneling process. We first create a quasiparticle–quasihole pair. Then we bring the quasiparticle all the way around the torus. Finally, we annihilate the quasiparticle–quasihole pair and go back to the ground state. Such a tunneling process produces an operator that maps ground states to ground states. Such an operator is denoted by U_x if the quasiparticle goes around the torus in the x direction, and U_y if it goes around the torus in the y direction (see Fig. 8.3(a,b)). Then, the four tunnelings in the x, y, $-x$, and $-y$ directions generate $U_y^{-1}U_x^{-1}U_yU_x$ (see Fig. 8.3(c)). We note that the path of the above four tunnelings can be deformed into two linked loops (see Fig. 8.3(d)). The two linked loops correspond to moving one quasiparticle around the other. It gives rise to a phase $\mathrm{e}^{2\mathrm{i}\theta}$, where θ is the statistical angle of the quasiparticle. For the $1/m$ Laughlin state, we have $\theta = \pi/m$. Therefore, we have

$$U_y^{-1}U_x^{-1}U_yU_x = \mathrm{e}^{\,\mathrm{i}2\pi/m}$$

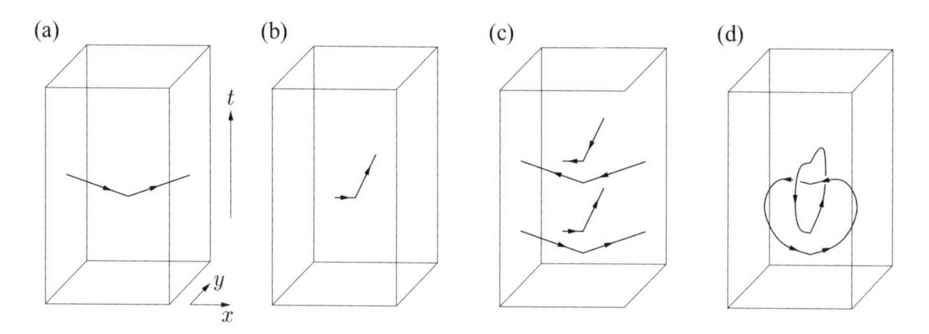

FIG. 8.3. (a) Tunneling in the x direction generates U_x. (b) Tunneling in the y direction generates U_y. (c) The four tunnelings in the x, y, $-x$, and $-y$ directions generate $U_y^{-1}U_x^{-1}U_yU_x$. (d) The above four tunnelings can be deformed into two linked loops.

As $U_{x,y}$ acts within the ground states, the ground states form the representation of the above algebra. The algebra has only one m-dimensional irreducible representation. Thus, the $1/m$ Laughlin state has $m \times$ integer number of degenerate ground states. This approach allows us to see the direct connection between the quasiparticle statistics and the ground-state degeneracy.

The second way to calculate the ground-state degeneracy is to use the effective theory (7.3.11). The degenerate ground states arise from the following collective fluctuations:

$$a_i(t,x,y) = \theta_i(t)/L, \quad i = x,y \tag{8.2.1}$$

where L is the size of the torus in the x and y directions. All of the other fluctuations generate a nonzero 'magnetic' field $b = f_{xy}$ and have a finite energy gap, as one can see from the classical equation of motion. The following Lagrangian describing the dynamics of the collective excitations in eqn (8.2.1) can be obtained by substituting eqn (8.2.1) into eqn (7.3.11):

$$L = -\frac{m}{4\pi}(\dot{\theta}_x\theta_y - \dot{\theta}_y\theta_x) + \frac{1}{2g_1}\dot{\theta}_i^2. \tag{8.2.2}$$

Since the charge of a_μ is quantized as an integer, the gauge transformation $U(x,y)$ that acts on the quasiparticle field, $\psi_q \to U\psi_q$, must be a periodic function on the torus. Thus, the gauge transformation must have the form $U(x,y) = \exp\left(2\pi \mathrm{i}(\frac{nx}{L} + \frac{my}{L})\right)$, where n and m are integers. As the a_μ charge of ψ_q is 1, such a gauge transformation changes the gauge field a_i to $a_i' = a_i - \mathrm{i}U^{-1}\partial_i U$ as follows:

$$(a_x', a_y') = (a_x + \frac{2\pi n}{L}, a_y + \frac{2\pi m}{L}) \tag{8.2.3}$$

Equation (8.2.3) implies that (θ_x, θ_y) and $(\theta_x + 2\pi n, \theta_y + 2\pi m)$ are gauge equivalent and should be identified. The gauge-inequivalent configurations are given by a point on a torus $0 < \theta_i < 2\pi$. As a result, the Lagrangian (8.2.2) describes a particle with unit charge moving on a torus parametrized by (θ_x, θ_y).

The first term in eqn (8.2.2) indicates that there is a uniform 'magnetic' field $B = m/2\pi$ on the torus. The total flux passing through the torus is equal to $2\pi \times m$. The Hamiltonian of eqn (8.2.2) is given by

$$H = \frac{g_1}{2}\left[-(\partial_{\theta_x} - \mathrm{i}A_{\theta_x})^2 - (\partial_{\theta_y} - \mathrm{i}A_{\theta_y})^2\right]. \tag{8.2.4}$$

The energy eigenstates of eqn (8.2.4) form Landau levels. The gap between the Landau levels is of order g_1, which is independent of the size of the system. The number of states in the first Landau level is equal to the number of flux quanta passing through the torus, which is m in our case. Thus, the ground-state degeneracy of the $1/m$ Laughlin state is m.

To understand the robustness of the ground-state degeneracy, let us add an arbitrary perturbation to the electron Hamiltonian. Such a perturbation will cause a change in the effective Lagrangian, $\delta\mathcal{L}(a_\mu)$. The key here is that $\delta\mathcal{L}(a_\mu)$ only depends on a_μ through its field strength. In other words, $\delta\mathcal{L}$ is a function of $\boldsymbol{e}$ and b. As the collective fluctuation in eqn (8.2.1) is a pure gauge locally, $\boldsymbol{e}$ and b, and hence $\delta\mathcal{L}$, do not depend on θ_i. The $\delta\mathcal{L}(a_\mu)$ cannot generate any potential terms $V(\theta_i)$ in the effective theory of θ_i. It can only generate terms that only depend on $\dot{\theta}_i$. Such a correction renormalizes the value of g_1. It cannot lift the degeneracy.

More general FQH states are described by eqn (7.3.18). When K is diagonal, the above result implies that the ground-state degeneracy is $\det(K)$. It turns out that, for a generic K, the ground-state degeneracy is also given by $\det(K)$. On a Riemann surface of genus g, the ground-state degeneracy becomes $(\det(K))^g$.

We see that in a compact space the low-energy physics of FQH liquids are very unique. There are only a finite number of low-energy excitations (i.e. the degenerate ground states), yet the low-energy dynamics are non-trivial because the ground-state degeneracy depends on the topology of the space. Such special low-energy dynamics, which depend only on the topology of the space, are described by the so-called topological field theory, which was studied intensively in the high-energy physics community (Witten, 1989; Elitzur *et al.*, 1989; Fröhlich and King, 1989). Topological theories are effective theories for FQH liquids, just as the Ginzburg–Landau theory is for superfluids (or other symmetry-broken phases).

The dependence of the ground-state degeneracy on the topology of the space indicates the existence of some kind of long-range order (the global dancing pattern mentioned above) in FQH liquids, despite the absence of long-range correlations for all local physical operators. In some sense, we may say that FQH liquids contain hidden long-range orders.

8.2.2 Classification of topological orders

- It is important to understand the mathematical framework behind topological order. (Just like it is important to understand group theory—the mathematical framework behind the symmetry-breaking order.)
- The understanding of the mathematical framework will allow us to classify all of the possible topological orders.

A long-standing problem has been how to label and classify the rich topological orders in FQH liquids. We are able to classify all crystal orders because we know that the crystal orders are described by a symmetry group. However, our understanding of topological orders in FQH liquids is very poor, and the mathematical structure behind topological orders is unclear.

Nevertheless, we have been able to find a simple and unified treatment for a class of FQH liquids—abelian FQH liquids (Blok and Wen, 1990a,b; Read, 1990; Fröhlich and Kerler, 1991; Fröhlich and Studer, 1993). Laughlin states represent the simplest abelian FQH states that contain only one component of incompressible fluid. More general abelian FQH states with a filling fraction such as $\nu = 2/5, 3/7, ...$ contain several components of incompressible fluids, and have more complicated topological orders. The topological orders (or the dancing patterns) in the abelian FQH state can also be described by the dancing steps. The dancing patterns can be characterized by an integer symmetric matrix K and an integer charge vector $\boldsymbol{q}$. An entry of $\boldsymbol{q}$, q_i, is the charge (in units of e) carried by the particles in the ith component of the incompressible fluid. An entry of K, K_{ij}, is the number of steps taken by a particle in the ith component to go around a particle in the jth component. In the $(K, \boldsymbol{q})$ characterization of FQH states, the $\nu = 1/m$ Laughlin state is described by $K = m$ and $\boldsymbol{q} = 1$, while the $\nu = 2/5$ abelian state is described by $K = \begin{pmatrix} 3 & 2 \\ 2 & 3 \end{pmatrix}$ and $\boldsymbol{q} = \begin{pmatrix} 1 \\ 1 \end{pmatrix}$.

All of the physical properties associated with the topological orders can be determined in terms of K and $\boldsymbol{q}$. For example, the filling fraction is simply given by $\nu = \boldsymbol{q}^\top K^{-1} \boldsymbol{q}$ and the ground-state degeneracy on the genus g Riemann surface is $\det(K)^g$. All of the quasiparticle excitations in this class of FQH liquids have abelian statistics, which leads to the name abelian FQH liquids.

The above classification of FQH liquids is not complete. Not every FQH state is described by K-matrices. In 1991, a new class of FQH states—non-abelian FQH states—was proposed (Moore and Read, 1991; Wen, 1991b). A non-abelian FQH state contains quasiparticles with non-abelian statistics. The observed filling fraction $\nu = 5/2$ FQH state (Willett *et al.*, 1987) is very likely to be one such state (Haldane and Rezayi, 1988a,b; Greiter *et al.*, 1991; Read and Green, 2000). Many studies (Moore and Read, 1991; Blok and Wen, 1992; Iso *et al.*, 1992; Cappelli *et al.*, 1993; Wen *et al.*, 1994) have revealed a connection between the topological orders in FQH states and conformal field theories. However, we are still quite far from a complete classification of all possible topological orders in non-abelian states.

8.2.3 Edge excitations—a practical way to measure topological orders

- The edge excitations for FQH states play a similar role to X-rays for crystals. We can use edge excitations to experimentally probe the topological orders in FQH states. In other words, compared to ground-state degeneracy, edge excitations provide a more complete definition of topological orders.

Topological degeneracy of the ground states only provides a partial characterization of topological orders. Different topological orders can sometimes lead to the same ground-state degeneracy. The issue here is whether we have a more complete characterization/measurement of topological orders. Realizing that the topological orders cannot be characterized by local order parameters and long-range correlations of local operators, it seems difficult to find any methods to characterize topological order. Amazingly, FQH states find a way out in an unexpected fashion. The bulk topological orders in FQH states can be characterized/measured by edge excitations (Wen, 1992). This phenomenon of two-dimensional topological orders being encoded in one-dimensional edge states shares some similarities with the holomorphic principle in superstring theory and quantum gravity ('t Hooft, 1993; Susskind, 1995).

FQH liquids as incompressible liquids have a finite energy gap for all of their bulk excitations. However, FQH liquids of finite size always contain one-dimensional gapless edge excitations, which is another unique property of FQH fluids. The structures of edge excitations are extremely rich, which reflects the rich bulk topological orders. Different bulk topological orders lead to different structures of edge excitations. Thus, we can study and measure the bulk topological orders by studying the structures of edge excitations.

As we have seen in the last chapter that, due to the non-trivial bulk topological order, the electrons at the edges of (abelian) FQH liquids form a new kind of correlated state—chiral Luttinger liquids (Wen, 1992). The electron propagator in chiral Luttinger liquids develops an anomalous exponent: $\langle c^\dagger(t,x)c(0)\rangle \propto (x-vt)^{-g}$, $g \neq 1$. (For Fermi liquids, we have $g = 1$.) The exponent g, in many cases, is a topological quantum number which does not depend on detailed properties of the edges. Thus, g is a new quantum number that can be used to characterize the topological orders in FQH liquids. Many experimental groups have successfully measured the exponent g through the temperature dependence of tunneling conductance between two edges (Milliken *et al.*, 1995; Chang *et al.*, 1996), which was predicted to have the form $\sigma \propto T^{2g-2}$ (Wen, 1992). This experiment demonstrates

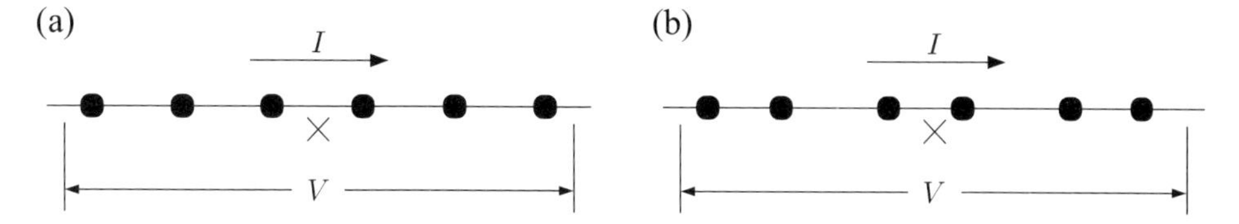

FIG. 8.4. A one-dimensional crystal passing an impurity will generate narrow-band noise in the voltage drop.

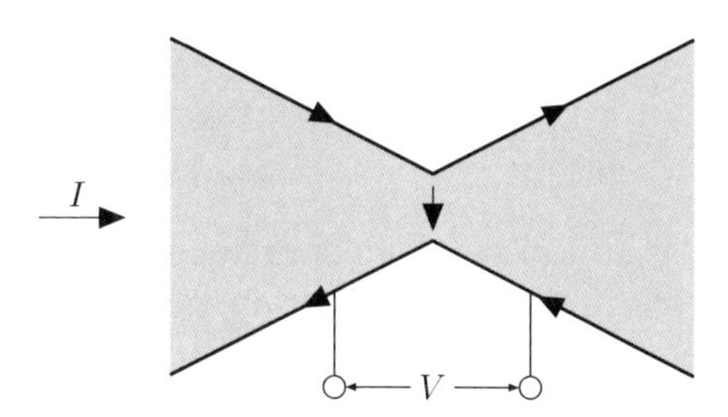

FIG. 8.5. An FQH fluid passing through a constriction will generate narrow-band noises due to the back-scattering of the quasiparticles.

the existence of new chiral Luttinger liquids and opens the door to the experimental study of the rich internal and edge structures of FQH liquids.

The edge states of non-abelian FQH liquids form more exotic one-dimensional correlated systems which have not yet been named. These edge states were found to be closely related to conformal field theories in $1+1$ dimensions (Wen *et al.*, 1994).

We know that crystal orders can be measured by X-ray diffraction experiments. In the following, we would like to suggest that the topological orders in FQH liquids can be measured (in principle) through a noise spectrum in an edge transport experiment. Let us first consider a one-dimensional crystal driven through an impurity (see Fig. 8.4(a)). Due to the crystal order, the voltage across the impurity has a narrow-band noise at a frequency of $f = I/e$ if each unit cell has only one charged particle. More precisely, the noise spectrum has a singularity, i.e. $S(f) \sim A\delta(f - \frac{I}{e})$. If each unit cell contains two charged particles (see Fig. 8.4(b)), then we will see an additional narrow-band noise at $f = I/2e$, so that $S(f) \sim B\delta(f - \frac{I}{2e}) + A\delta(f - \frac{I}{e})$. In this example, we see that the noise spectrum allows us to measure crystal orders in one-dimension. A similar experiment can also be used to measure topological orders in FQH liquids. Let us consider an FQH sample with a narrow constriction (see Fig. 8.5). The constriction induces a back-scattering through quasiparticle tunneling between the two edges. The back-scattering causes a noise in the voltage across the constriction. In the weak-back-scattering limit, the noise spectrum contains singularities at certain frequencies, which allows us to measure the topological orders in the FQH liquids.[52] To be more specific, the singularities in the noise spectrum have the form (see eqn (7.4.49))

$$S(f) \sim \sum_a C_a |f - f_a|^{\gamma_a} \tag{8.2.5}$$

The frequencies and the exponents of the singularities (f_a, γ_a) are determined by the topological orders. For the abelian state characterized by the matrix K and the charge vector $\boldsymbol{q}$, the allowed values of the pair (f_a, γ_a) are given by

$$f_a = \frac{I}{e\nu} \boldsymbol{q}^\top K^{-1} \boldsymbol{l}, \qquad \gamma_a = 2\boldsymbol{l}^\top K^{-1} \boldsymbol{l} - 1 \tag{8.2.6}$$

where $\boldsymbol{l}^\top = (l_1, l_2, ...)$ is an arbitrary integer vector and $\nu = \boldsymbol{q}^\top K^{-1} \boldsymbol{q}$ is the filling fraction. The singularities in the noise spectrum are caused by quasiparticle tunneling between the two edges. The frequency of the singularity f_a is determined by the electric charge of the tunneling quasiparticle Q_q, i.e. $f_a = \frac{I}{e} \frac{Q_q}{e\nu}$. The exponent γ_a is determined by the statistics of the tunneling quasiparticle θ_q, namely $\gamma = 2\frac{|\theta_q|}{\pi} - 1$. Thus, the noise spectrum measures the charge and the statistics of the allowed quasiparticles, which in turn determines the topological orders in FQH states.

8.3 Quantum orders

- Quantum states generally contain a new kind of order—quantum order. Quantum orders cannot be completely characterized by broken symmetries and the associated order parameters.
- Quantum order describes the pattern of quantum entanglements in many-body ground states.
- The fluctuations of quantum order can give rise to gapless gauge bosons and gapless fermions. Quantum order protects the gaplessness of these excitations, just like symmetry protects gapless Nambu–Goldstone bosons in symmetry-breaking states.
- Topological order is a special kind of quantum order in which all excitations have finite energy gaps.

---

[52] The discussion presented here applies only to the FQH states whose edge excitations all propagate in the same direction. This requires, for abelian states, all of the eigenvalues of K to have the same sign.

The topological order, by definition, only describes the internal order of gapped quantum states. Here we make a leap of faith. We will assume that the gap is not important and that gapless quantum states can also contain orders that cannot be described by symmetry and long-range correlations. We will call the non-symmetry-breaking order in quantum ground states the quantum order.

If you believe in this line of thinking, then the only things that need to be done are to show that quantum orders do exist, and to find mathematical descriptions (or symbols) that characterize the quantum orders. We will show that quantum orders do exist in Section 8.3.2 and in Chapter 10. As quantum orders cannot be characterized by broken symmetries and order parameters, we need to develop a new theory to describe quantum orders. At present, we do not have a complete theory that can describe all possible quantum orders. However, in Chapter 9 we manage to find a mathematical object—the projective symmetry group (PSG)—that can describe a large class of quantum orders.

One may ask, why do we need to introduce the new concept of quantum order? What use can it have? To answer such a question, we would like to ask, why do we need the concept of symmetry breaking? Is the symmetry-breaking description useful? Symmetry breaking is useful because it leads to a classification of crystal orders (such as the 230 different crystals in three dimensions), and it determines the structure of low-energy excitations without the need to know the details of a system (such as three branches of phonons from three broken translational symmetries in a solid) (Nambu, 1960; Goldstone, 1961). The quantum order and its PSG description are useful in the same sense; a PSG can classify different quantum states that have the same symmetry (Wen, 2002c), and quantum orders determine the structure of low-energy excitations without the need to know the details of a system (Wen, 2002a,c; Wen and Zee, 2002). The main difference between symmetry-breaking orders and quantum orders is that symmetry-breaking orders generate and protect gapless Nambu–Goldstone modes (Nambu, 1960; Goldstone, 1961), which are scalar bosonic excitations, while quantum orders can generate and protect gapless gauge bosons and gapless fermions. Fermion excitations can even emerge in pure local bosonic models, as long as the boson ground state has a proper quantum order.

One way to visualize quantum order is to view quantum order as a description of the pattern of the quantum entanglement in a many-body ground state. Different patterns of entanglement give rise to different quantum orders. The fluctuations of entanglement correspond to collective excitations above a quantum-ordered state. We will see that these collective excitations can be gauge bosons and fermions.

The concept of topological/quantum order is also useful in the field of quantum computation. People have been designing different kinds of quantum-entangled states to perform different computing tasks. When the number of qubits becomes larger and larger, it is more and more difficult to understand the pattern of quantum entanglements. One needs a theory to characterize different quantum entanglements in many-qubit systems. The theory of topological/quantum order (Wen, 1995, 2002c) is just such a theory. Also, the robust topological degeneracy in topologically-ordered states discovered by Wen and Niu (1990) can be used in fault-tolerant quantum computation (Kitaev, 2003).

In the following, we will discuss the connection between quantum phase transitions and quantum orders. Then we will use the quantum phase transitions in free fermion systems to study the quantum orders there.

8.3.1 Quantum phase transitions and quantum orders

- Quantum phase transitions are defined by the singularities of the ground-state energy as a function of the parameters in the Hamiltonian.

Classical orders can be studied through classical phase transitions. Classical phase transitions are marked by singularities in the free-energy density f. The free-energy density can be calculated through the partition function as follows:

$$f = -\frac{T \ln Z}{V_{\text{space}}}, \quad Z = \int \mathcal{D}\phi \ \mathrm{e}^{-\beta \int \mathrm{d}x \ h(\phi)} \tag{8.3.1}$$

where $h(\phi)$ is the energy density of the classical system and V_{space} is the volume of space.

Similarly, to study quantum orders we need to study quantum phase transitions at zero temperature $T = 0$. Here the energy density of the ground state plays the role of the free-energy density. A singularity in the ground-state-energy density marks a quantum transition. The similarity between the ground-state-energy density and the free-energy density can be clearly seen in the following expression for the energy density of the ground state:

$$\rho_E = \mathrm{i}\frac{\ln Z}{V_{\text{space}-\text{time}}}, \quad Z = \int \mathcal{D}\phi \ \mathrm{e}^{\mathrm{i}\int \mathrm{d}x \mathrm{d}t \ \mathcal{L}(\phi)} \tag{8.3.2}$$

where $\mathcal{L}(\phi)$ is the Lagrangian density of the quantum system and $V_{\text{space}-\text{time}}$ is the volume of space–time. Comparing eqns (8.3.1) and (8.3.2), we see that a classical system is described by a path integral of a positive functional, while a quantum system is described by a path integral of a complex functional. In general, a quantum phase transition, marked by a singularity of the path integral of a complex functional, can be more general than classical phase transitions that are marked by a singularity of the path integral of a positive functional.

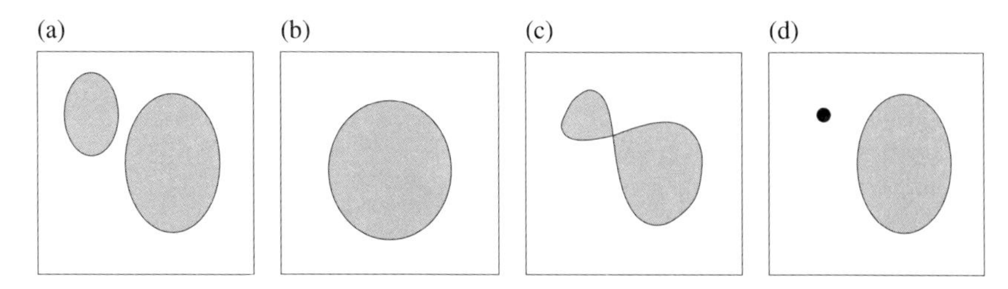

FIG. 8.6. The two sets of oriented Fermi surfaces in (a) and (b) represent two different quantum orders. The two possible transition points between the two quantum orders in (a) and (b) are described by the Fermi surfaces in (c) and (d).

8.3.2 Quantum orders and quantum transitions in free fermion systems

- Free fermion systems contain quantum phase transitions that do not change any symmetry, indicating that free fermion systems contain non-trivial quantum order.
- Different quantum orders in free fermion systems are classified by the topologies of a Fermi surface.

Let us consider a free fermion system with only the translational symmetry and the $U(1)$ symmetry from the fermion number conservation. The Hamiltonian has the form

$$H = \sum_{\langle \boldsymbol{ij} \rangle} \left(c_{\boldsymbol{i}}^\dagger t_{\boldsymbol{ij}} c_{\boldsymbol{j}} + h.c. \right)$$

with $t_{\boldsymbol{ij}}^* = t_{\boldsymbol{ji}}$. The ground state is obtained by filling every negative energy state with one fermion. In general, the system contains several pieces of Fermi surfaces.

To understand the quantum order in the free fermion ground state, we note that the topology of the Fermi surfaces can change in two ways as we continuously change t_{ij}: a Fermi surface can shrink to zero (Fig. 8.6(d)); and two Fermi surfaces can join (Fig. 8.6(c)). When a Fermi surface is about to disappear in a d-dimensional system, the ground-state-energy density has the form

$$\rho_E = \int \frac{\mathrm{d}^d \boldsymbol{k}}{(2\pi)^d} (\boldsymbol{k} \cdot M \cdot \boldsymbol{k} - \mu) \Theta(-\boldsymbol{k} \cdot M \cdot \boldsymbol{k} + \mu) + ...$$

where '...' represents the non-singular contribution and the symmetric matrix M is positive (or negative) definite. We find that the ground-state-energy density has a singularity at $\mu = 0$, i.e. $\rho_E = c\mu^{(2+d)/2}\Theta(\mu) + ...$, where $\Theta(x > 0) = 1$ and $\Theta(x < 0) = 0$. When two Fermi surfaces are about to join, the singularity is still determined by the above equation, but now M has both negative and positive eigenvalues. The ground-state-energy density has a singularity of the form $\rho_E = c\mu^{(2+d)/2}\Theta(\mu) + ...$ when d is odd and $\rho_E = c\mu^{(2+d)/2} \log |\mu| + ...$ when d is even.

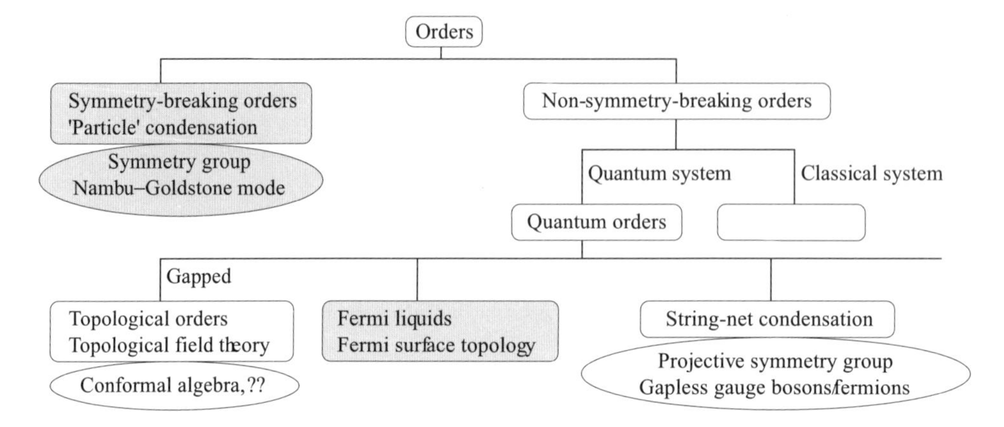

FIG. 8.7. A new classification of orders. The phases in the shaded boxes can be described Landau's theories. Other phases are beyond Landau's theories.

The singularity of the ground-state-energy density at $\mu = 0$ indicates a quantum phase transition. This kind of transition was first studied by Lifshitz (1960). Clearly, there is no change of symmetry across the transition and there is no local order parameters to characterize the phases on the two sides of the transition. This suggests that the two states can only be distinguished by their quantum orders. As the $\mu = 0$ point is exactly the place where the topology of the Fermi surface changes, we find that the topology of the Fermi surface is a 'quantum number' that characterizes the quantum order in a free fermion system (see Fig. 8.6). A change in the topology signals a *continuous* quantum phase transition that changes the quantum order.

Problem 8.3.1.
Consider a two-dimensional spin-1/2 free electron system. As we change the chemical potential μ, the system undergoes a quantum phase transition, as illustrated in Fig. 8.6(d). Find the singular behavior of the spin susceptibility near the transition point μ_c.

8.4 A new classification of orders

- Quantum orders have many classes. FQH states and free fermion systems represent only two of the many classes of quantum orders.

The concept of topological/quantum order allows us to have a new classification of orders, as illustrated in Fig. 8.7. According to this classification, a quantum order is simply a non-symmetry-breaking order in a quantum system, and a topological order is simply a quantum order with a finite energy gap.

From the FQH states and the Fermi liquid states discussed above, we see that quantum order can be divided into several different classes. The FQH states and the free fermion systems only provide two examples of quantum orders. In the next few chapters, we will study quantum orders in some strongly-correlated systems. We will show that these quantum orders belong to a different class, which is closely related to a condensation of nets of strings in the correlated ground state. Chapter 9 studies and classifies this class of quantum order using a projective construction (i.e. the slave-boson approach). Chapter 10 relates the quantum-ordered state studied in Chapter 9 to string-net-condensed states.

9

MEAN-FIELD THEORY OF SPIN LIQUIDS AND QUANTUM ORDER

- The mean-field theory of the spin-liquid state using the slave-boson approach (or projective construction).
- The importance of the gauge interaction and how to apply the mean-field theory (with the gauge interaction) to study the physical properties of the spin liquids.
- The characterization of the spin liquids and the concept of topological order and quantum order.

In this chapter, we are going to develop a mean-field theory for the spin-liquid state. Here by 'spin-liquid state' we mean an insulator with spin–rotation symmetry and with an *odd* number of electrons per unit cell. Usually, a state with an odd number of electrons per unit cell has a half-filled band and is a conductor. Thus, the spin liquids, if they exist, are very unusual states. The spin liquids are so strange that many people do not believe that they can ever exist. Indeed, even now, we do not know of any spin Hamiltonian that can be shown to reliably give rise to a spin-liquid ground state.

For those who believe in the existence of spin liquids, they have to accept or believe the following strange (or fascinating) properties of spin liquids. (i) The excitations in spin liquids always carry fractional quantum numbers, such as neutral spin-1/2, which are impossible to obtain from any collection of electrons. Sometimes excitations can even carry fractional statistics. (ii) Different spin liquids cannot be distinguished by their symmetry properties. Spin liquids are examples of quantum-ordered states. (iii) The ground states of gapped spin liquids always have topological degeneracy, which cannot be related to any symmetry. (iv) Spin liquids always contain certain kinds of gauge fluctuations.

In this chapter, we will assume that spin liquids do exist and develop a mean-field theory for spin liquids. The mean-field theory allows us to understand some physical properties of spin liquids. After including important mean-field fluctuations, we show that some of the mean-field states are stable against these fluctuations and represent real spin-liquid states. As spin liquids are typical states with non-trivial topological/quantum orders, we will use the spin liquids to develop a theory of quantum orders.

9.1 Projective construction of quantum spin-liquid states

In this section, we are going to use the slave-boson approach (or the projective construction) (Baskaran *et al.*, 1987; Affleck *et al.*, 1988; Baskaran and Anderson, 1988; Dagotto *et al.*, 1988; Wen and Lee, 1996; Senthil and Fisher, 2000) to construct two-dimensional spin liquids. The gauge structure discovered by Baskaran and Anderson (1988) in the slave-boson approach plays a crucial role in our understanding of strongly-correlated spin liquids. In addition to the slave-boson approach, one can also use another type of projective construction—the slave-fermion or Schwinger boson approach—to construct spin-liquid states (Arovas and Auerbach, 1988; Read and Sachdev, 1991; Sachdev and Park, 2002). As the slave-fermion approach can only lead to spin liquids with finite energy gaps, here we will concentrate on the slave-boson approach.

9.1.1 Mean-field theory of spin-liquid states

- A mean-field theory of spin liquids can be obtained from a projective construction.
- Mean-field theory and the mean-field ansatz of the π-flux phase.
- Gauge fluctuations of $a_0(\boldsymbol{i})$ impose the constraint.

Let us consider a Hubbard model (5.5.1) on a two-dimensional square lattice. At half-filling, the Hubbard model reduces to the Heisenberg model as follows:

$$H = \sum_{\langle \boldsymbol{ij} \rangle} J_{\boldsymbol{ij}} \boldsymbol{S_i} \cdot \boldsymbol{S_j} \tag{9.1.1}$$

We would like to point out that the above spin-1/2 model can also be viewed as a hard-core boson model, where $|\downarrow\rangle$ corresponds to an empty site and $|\uparrow\rangle$ corresponds to a site occupied by one boson. The above Hamiltonian is hard to solve. So we use a mean-field approximation to understand its physical properties. In the mean-field approximation, we replace one of the $\boldsymbol{S_i}$ by its quantum average $\langle \boldsymbol{S_i} \rangle$ and obtain the following mean-field Hamiltonian:

$$H_{\text{mean}} = \sum_{\langle \boldsymbol{ij} \rangle} J_{\boldsymbol{ij}} \left(\langle \boldsymbol{S_i} \rangle \cdot \boldsymbol{S_j} + \boldsymbol{S_i} \cdot \langle \boldsymbol{S_j} \rangle - \langle \boldsymbol{S_i} \rangle \cdot \langle \boldsymbol{S_j} \rangle \right)$$

The mean-field Hamiltonian H_{mean} is easy to solve and we can obtain the ground state of H_{mean}, namely $|\Phi_{\text{mean}}\rangle$. The only thing we need to do is to choose the values of $\langle \boldsymbol{S_i} \rangle$ carefully, so that they satisfy the so-called self-consistency equation

$$\langle \boldsymbol{S_i} \rangle = \langle \Phi_{\text{mean}} | \boldsymbol{S_i} | \Phi_{\text{mean}} \rangle$$

This is the standard mean-field approach for spin systems.

The problem with the above standard mean-field approach is that it can only be used to study the ordered spin states, because we assume that at the beginning $\langle \boldsymbol{S}_i \rangle \neq 0$. For a long time, it seemed impossible to have a mean-field theory for spin liquids. In 1987, we finally found a strange trick—the slave-boson approach[53]—to do so (Baskaran *et al.*, 1987). To obtain the mean-field ground state of *spin liquids*, we introduce the spinon operators $f_{i\alpha}$, $\alpha = 1, 2$, which are spin-1/2 charge-neutral operators. The spin operator $\boldsymbol{S}_i$ is represented by

$$\boldsymbol{S}_i = \frac{1}{2} f_{i\alpha}^\dagger \boldsymbol{\sigma}_{\alpha\beta} f_{i\beta} \tag{9.1.2}$$

In terms of the spinon operators, the Hamiltonian (9.1.1) can be rewritten as

$$H = \sum_{\langle ij \rangle} -\frac{1}{2} J_{ij} f_{i\alpha}^\dagger f_{j\alpha} f_{j\beta}^\dagger f_{i\beta} + \sum_{\langle ij \rangle} J_{ij} \left(\frac{1}{2} n_i - \frac{1}{4} n_i n_j \right) \tag{9.1.3}$$

Here we have used $\boldsymbol{\sigma}_{\alpha\beta} \cdot \boldsymbol{\sigma}_{\alpha'\beta'} = 2\delta_{\alpha\beta'}\delta_{\alpha'\beta} - \delta_{\alpha\beta}\delta_{\alpha'\beta'}$ and n_i is the number of fermions at site i. The second term in eqn (9.1.3) is a constant and will be dropped in the following discussions. Notice that the Hilbert space of eqn (9.1.3) with four states per site is larger than that of eqn (9.1.1), which has two states per site. The equivalence between eqn (9.1.1) and eqn (9.1.3) is valid only in the subspace where there is exactly one fermion per site. Therefore, to use eqn (9.1.3) to describe the spin state, we need to impose the constraint (Baskaran *et al.*, 1987; Baskaran and Anderson, 1988)

$$f_{i\alpha}^\dagger f_{i\alpha} = 1, \quad f_{i\alpha} f_{i\beta} \epsilon_{\alpha\beta} = 0 \tag{9.1.4}$$

The second constraint is actually a consequence of the first one.

A mean-field ground state at zeroth-order is obtained by making the following approximations. First we replace the constraint (9.1.4) by its ground-state average

$$\langle f_{i\alpha}^\dagger f_{i\alpha} \rangle = 1 \tag{9.1.5}$$

Such a constraint can be enforced by including a *site-dependent* and time-independent Lagrangian multiplier $a_0(\boldsymbol{i})(f_{i\alpha}^\dagger f_{i\alpha} - 1)$ in the Hamiltonian. Second, we replace the operator $f_{i\alpha}^\dagger f_{j\alpha}$ by its ground-state expectation value χ_{ij}, again ignoring their fluctuations. In this way, we obtain the zeroth-order mean-field

---

[53] The slave-boson and the slave-fermion are two very strange and confusing terms. The slave-boson approach has no bosons and the slave-fermion approach has no fermions. This is why I prefer to use projective construction to describe the two approaches.

Hamiltonian

$$H_{\text{mean}} = \sum_{\langle \boldsymbol{ij} \rangle} -\frac{1}{2} J_{\boldsymbol{ij}} \left[(f_{\boldsymbol{i}\alpha}^{\dagger} f_{\boldsymbol{j}\alpha} \chi_{\boldsymbol{ji}} + h.c) - |\chi_{\boldsymbol{ij}}|^2 \right] + \sum_{\boldsymbol{i}} a_0(\boldsymbol{i})(f_{\boldsymbol{i}\alpha}^{\dagger} f_{\boldsymbol{i}\alpha} - 1) \tag{9.1.6}$$

The $\chi_{\boldsymbol{ij}}$ in eqn (9.1.6) must satisfy the self-consistency condition

$$\chi_{\boldsymbol{ij}} = \langle f_{\boldsymbol{i}\alpha}^{\dagger} f_{\boldsymbol{j}\alpha} \rangle \tag{9.1.7}$$

and the site-dependent chemical potential $a_0(\boldsymbol{i})$ is chosen such that eqn (9.1.5) is satisfied by the mean-field ground state. For convenience, we will call $\chi_{\boldsymbol{ij}}$ the 'ansatz' of the spin-liquid state.

Let $\bar{\chi}_{\boldsymbol{ij}}$ be a solution of eqn (9.1.7) and let $\bar{a}(\boldsymbol{i})$ be a solution of eqn (9.1.5). Such an ansatz corresponds to the mean-field ground state. As $\chi_{\boldsymbol{ij}}$ and a_0 do not change under the spin–rotation transformation, the mean-field ground state is invariant under the spin–rotation transformation. If we ignore the fluctuations of $\bar{\chi}_{\boldsymbol{ij}}$ and $\bar{a}(\boldsymbol{i})$, then the excitations around the mean-field state are described by

$$H_{\text{mean}} = \sum_{\langle \boldsymbol{ij} \rangle} -\frac{J_{\boldsymbol{ij}}}{2} \left[(f_{\boldsymbol{i}\alpha}^{\dagger} f_{\boldsymbol{j}\alpha} \bar{\chi}_{\boldsymbol{ji}} + h.c) - |\bar{\chi}_{\boldsymbol{ij}}|^2 \right] + \sum_{\boldsymbol{i}} \bar{a}_0(\boldsymbol{i})(f_{\boldsymbol{i}\alpha}^{\dagger} f_{\boldsymbol{i}\alpha} - 1)$$

We see that the excitations in the zeroth-order mean-field theory are free spinons described by $f_{\boldsymbol{i}\alpha}$. The spinons are spin-1/2 neutral fermions. It is amazing to see that fermionic excitations can emerge from a purely bosonic model eqn (9.1.1).

Now the question is whether we should trust the mean-field result. Should we believe in the existence of spin liquids with spin-1/2 neutral fermionic excitations? One way to check this is to include the fluctuations around the mean-field ansatz, and to see if the fluctuations alter the mean-field result. So, in the following, we will consider the effects of fluctuations.

First, we would like to point out that, if we had included the fluctuations (i.e. the time dependence) of the a_0, then the constraint (9.1.5) would have become the original constraint (9.1.4). To see this, let us consider the path integral formulation of H_{mean}:

$$Z = \int \mathcal{D}f \mathcal{D}[a_0(\boldsymbol{i})] \mathcal{D}\chi_{\boldsymbol{ij}} \, \mathrm{e}^{\mathrm{i} \int \mathrm{d}t \, (\mathcal{L} - \sum_{\boldsymbol{i}} a_0(\boldsymbol{i},t)(f_{\boldsymbol{i}}^{\dagger} f_{\boldsymbol{i}} - 1)}$$

$$\mathcal{L} = \sum_{\boldsymbol{i}} f_{\boldsymbol{i}}^{\dagger} \mathrm{i} \partial_t f_{\boldsymbol{i}} - \sum_{\langle \boldsymbol{ij} \rangle} -\frac{1}{2} J_{\boldsymbol{ij}} \left[(f_{\boldsymbol{i}\alpha}^{\dagger} f_{\boldsymbol{j}\alpha} \chi_{\boldsymbol{ji}} + h.c) - |\chi_{\boldsymbol{ij}}|^2 \right] \tag{9.1.8}$$

We see that the integration of a time-dependent $a_0(\boldsymbol{i}, t)$ produces a constraint

$$\prod_{\boldsymbol{i},t} \delta(f_{\boldsymbol{i}}^{\dagger}(t) f_{\boldsymbol{i}}(t) - 1)$$

which is enforced at every site $\boldsymbol{i}$ and at every time t. We also note that the integration of $\chi_{\boldsymbol{ij}}(t)$ will reproduce the original Hamiltonian (9.1.3). Thus, eqn (9.1.8) is an exact representation of the spin model (9.1.3). The fluctuations in χ_{ij} and $a_0(\boldsymbol{i})$ describe the collective excitations above the mean-field ground state.

Here χ_{ij} has two kinds of fluctuation, the amplitude fluctuations and the phase fluctuations. The amplitude fluctuations have a finite energy gap and are not essential in our discussion. So here we will only consider the phase fluctuations a_{ij} around the mean-field ansatz $\bar{\chi}_{ij}$ as follows:

$$\chi_{ij} = \bar{\chi}_{ij} \mathrm{e}^{-\mathrm{i}a_{ij}} \tag{9.1.9}$$

Including these phase fluctuations and fluctuations of a_0, the mean-field Hamiltonian becomes

$$H = \sum_{\langle ij \rangle} -J_{ij}(f^\dagger_{i\alpha} f_{j\alpha} \bar{\chi}_{ji} \mathrm{e}^{-\mathrm{i}a_{ji}} + h.c) - \sum_{i} a_0(\boldsymbol{i})(f^\dagger_{i\alpha} f_{i\alpha} - 1) \tag{9.1.10}$$

We would like to call eqn (9.1.10) the first-order mean-field Hamiltonian. From eqn (9.1.10), we see that the fluctuations described by a_0 and a_{ij} are simply a $U(1)$ lattice gauge field (see eqn (6.4.9)) (Baskaran and Anderson, 1988; Lee and Nagaosa, 1992). The Hamiltonian is invariant under the gauge transformation

$$a_{ij} \to a_{ij} + \theta_i - \theta_j, \qquad f_i \to f_i \mathrm{e}^{i\theta_i}. \tag{9.1.11}$$

So, in first-order mean-field theory, the excitations are described by spinons coupled to the $U(1)$ gauge field (instead of free spinons in the zeroth-order mean-field theory).

In the usual mean-field theory, we make an approximation to the interacting Hamiltonian to simplify the problem. The Hilbert space is not changed by the mean-field approximation. The projective construction (or slave-boson approach) is very different in this aspect. Not only is the Hamiltonian changed, but the Hilbert space is also changed. This makes the mean-field results from the projective construction not only quantitatively incorrect, but also qualitatively incorrect. For example, the mean-field ground state $|\Psi_{\text{mean}}\rangle$ (the ground state of H_{mean} in eqn (9.1.6)) is not even a valid spin wave function because some sites can have zero or two fermions. However, do not give up. In the usual mean-field theory, we can improve the approximation quantitatively by including fluctuations around the mean-field state. In the above, we see that, in the projective construction, we can improve the approximation qualitatively and recover the original Hilbert space by including gauge fluctuations (a_0, a_{ij}) around the mean-field state. Thus, to obtain even qualitative results from the projective construction, it is important to include, at least, the gauge fluctuations around the mean-field state. In other words, we cut a spin into two halves in the projective construction of spin liquids. It is important to glue them back together to obtain the correct physical results.

According to zeroth-order mean-field theory, the low-energy excitations in the spin liquids are free spinons. From the above discussion, we see that such a result is incorrect. According to first-order mean-field theory, spinons interact via gauge fluctuations. The gauge interaction may drastically change the properties of the spinons. We will return to this problem later.

Problem 9.1.1.
Prove eqn (9.1.3).

9.1.2 To believe or not to believe

- The deconfined phase of first-order mean-field theory leads to new states of matter—quantum-ordered states.
- The projection to the physical spin wave function and the meaning of the $U(1)$ gauge structure.
- Emergent gauge bosons and fermions as fluctuations of entanglements.

According to first-order mean-field theory, the fluctuations around the mean-field ground state are described by gauge fields and fermion fields. Remember that our original model is just an interacting spin model which is a purely bosonic model. How can a purely bosonic model have an effective theory described by gauge fields and fermion fields? This is incredible. Let us examine how we get here. We first split the bosonic spin operator into a product of two fermionic operators (the spinon operators). We then introduce a gauge field to glue the spinons back into a bosonic spin. From this point of view, it looks like the first-order mean-field theory is just a fake theory. It appears that the gauge bosons and the fermions are fake. In the end, all we have are the bosonic spin fluctuations.

However, we should not discard first-order mean-field theory too quickly. It can actually reproduce the above picture of bosonic spin fluctuations if the gauge field is in a confining phase (see Section 6.4.3). In the confining phase, the spinons interact with each other through a linear potential and can never appear as quasiparticles at low energies. The gauge bosons have a large energy gap in the confining phase, and are absent from the low-energy spectrum. The only low energy excitations are the spinon pairs which correspond to the bosonic spin fluctuations. So first-order mean-field theory may be useless, but it is not wrong. It is capable of producing pictures that agree with common sense (although through a long detour).

On the other hand, first-order mean-field theory is also capable of producing pictures that defy common sense when the gauge field is in a deconfined phase. In this case, the spinons and gauge bosons will appear as well-defined quasiparticles. The question is do we believe the picture of deconfined phase? Do we believe the possibility of emergent gauge bosons and fermions from a purely bosonic model? Clearly, the projective construction outlined above is far too formal to convince most people to believe such drastic results.

I have to say that this business of cutting spins into two halves and gluing them back together turns off a lot of people. It is hard to see that any new physical insights and results can possibly be obtained by such a formal manipulation. If any new results do appear from this approach, then many people will attribute them to artifacts of this strange construction rather than true physical properties of the spin liquids. Indeed, the projective construction is very formal, but it is also a gift beyond its time. We now know that what the projective construction really does is to produce a string-net-condensed state. String-net condensation gives rise to a new type of correlated state that has non-trivial quantum orders (see Chapter 10). So, believe me that the striking results from the projective construction can be trusted, once we properly include the effects of gauge fluctuations.

Here let us try to understand the physical picture behind the projective construction, without using the picture of string-net condensation. First, we need to understand how a mean-field ansatz χ_{ij} is connected to a physical spin wave function which has exactly one fermion per site. We know that the mean-field wave state $|\Psi_{\text{mean}}^{(\chi_{ij})}\rangle$ (the ground state of H_{mean}) is not a valid wave function for the spin system, because it may not have one fermion per site. To connect this to a physical spin wave function, we need to include fluctuations of a_0 to enforce the one-fermion-per-site constraint. With this understanding, we may obtain a valid wave function of the spin system, $\Psi_{\text{spin}}(\{\alpha_i\})$, by projecting the mean-field state to the subspace of one fermion per site:

$$\Psi_{\text{spin}}^{(\chi_{ij})}(\{\alpha_i\}) = \langle 0_f | \prod_i f_{i\alpha_i} |\Psi_{\text{mean}}^{(\chi_{ij})}\rangle. \tag{9.1.12}$$

where $|0_f\rangle$ is the state with no f-fermions, i.e. $f_{i\alpha}|0_f\rangle = 0$. Equation (9.1.12) connects the mean-field ansatz (including its fluctuations) to the physical spin wave function. It allows us to understand the physical meaning of the mean-field ansatz and mean-field fluctuations.

For example, the projection (9.1.12) gives the gauge transformation (9.1.11) a physical meaning. The two mean-field ansatz χ_{ij} and $\tilde{\chi}_{ij}$, related by a gauge transformation

$$\tilde{\chi}_{ij} = \mathrm{e}^{\mathrm{i}\theta_i}\chi_{ij}\mathrm{e}^{-\mathrm{i}\theta_j} \tag{9.1.13}$$

give rise to the same projected spin state

$$\Psi_{\text{spin}}^{(\tilde{\chi}_{ij})}(\{\alpha_i\}) = \mathrm{e}^{\mathrm{i}\sum_i \theta_i}\Psi_{\text{spin}}^{(\chi_{ij})}(\{\alpha_i\}) \tag{9.1.14}$$

For different choices of χ_{ij}, the ground states of the H_{mean} in eqn (9.1.6) correspond to different mean-field wave functions $|\Psi_{\text{mean}}^{(\chi_{ij})}\rangle$. After projection, they lead to different physical spin wave functions $\Psi_{\text{spin}}^{(\tilde{\chi}_{ij})}(\{\alpha_i\})$. Thus, we can regard χ_{ij} as labels that label different physical spin states. Equation (9.1.14) tells us that the label is not a one-to-one label, but a many-to-one label. This property is important for us to understand the unusual dynamical properties of the χ_{ij} fluctuations. Using many labels to label the same physical state also makes our theory a gauge theory according to the definition in Section 6.1.1.

Let us consider how the many-to-one property or the gauge structure of χ_{ij} affect its dynamical properties. If χ_{ij} was a one-to-one label of physical states, then χ_{ij} would be like the condensed boson amplitude $\langle\phi(\boldsymbol{x},t)\rangle$ in a boson superfluid or the condensed spin moment $\langle \boldsymbol{S_i}(t)\rangle$ in an SDW state. The fluctuations of χ_{ij} would correspond to a bosonic mode similar to a spin wave mode.[54] However, χ_{ij} does not behave like local order parameters, such as $\langle\phi(\boldsymbol{x},t)\rangle$ and $\langle \boldsymbol{S_i}(t)\rangle$, which label physical states without redundancy. As a many-to-one label, some fluctuations of χ_{ij} do not change the physical state and are unphysical. These fluctuations are called pure gauge fluctuations. For a generic fluctuation $\delta\chi_{ij}$, part of it is physical and the other part is unphysical. The effective theory for χ_{ij} must be gauge invariant. This drastically changes the dynamical properties of the fluctuations. It is this property that makes fluctuations of χ_{ij} behave like gauge bosons, which are very different from the sound mode and the spin wave mode.

We have argued that the phase fluctuations of χ_{ij} correspond to gauge fluctuations. The projective construction (9.1.12) allows us to obtain the physical spin wave function that corresponds to a gauge fluctuation a_{ij}:

$$\Psi_{\text{spin}}^{(a_{ij})} = \langle 0|\prod_{\boldsymbol{i}} f_{\boldsymbol{i}\alpha_i}|\Psi_{\text{mean}}^{(\bar\chi_{ij}\,\mathrm{e}^{\,\mathrm{i}a_{ij}})}\rangle.$$

Similarly, the projective construction also allows us to obtain the physical spin wave function that corresponds to a pair of spinon excitations. We start with the mean-field ground state with a pair of particle–hole excitations. After the projection (9.1.12), we obtain the physical spin wave functions that contain a pair of spinons:

$$\Psi_{\text{spin}}^{\text{spinon}}(\boldsymbol{i}_1,\alpha_1;\boldsymbol{i}_2,\alpha_2) = \langle 0|(\prod_{\boldsymbol{i}} f_{\boldsymbol{i}\alpha_i}) f^\dagger_{\boldsymbol{i}_1\lambda_1} f_{\boldsymbol{i}_2\lambda_2}|\Psi_{\text{mean}}^{(\bar\chi_{ij})}\rangle.$$

If you are not satisfied with the physical picture presented and still do not believe that the projective construction can produce emergent gauge bosons and fermions, then you may go directly to Chapter 10. In Chapter 10, we construct several spin models which can be solved exactly (or quasi-exactly) by the projective construction. These models have emergent $Z_2/U(1)$ gauge structure and fermions. The exactly soluble models reveal the string-net origin of the emergent gauge bosons and fermions. In the rest of this chapter we assume that the projective construction makes sense and study its consequences.

[54] More precisely, the spin wave mode corresponds to scalar bosons. The fluctuations of the local order parameters always give rise to scalar bosons.

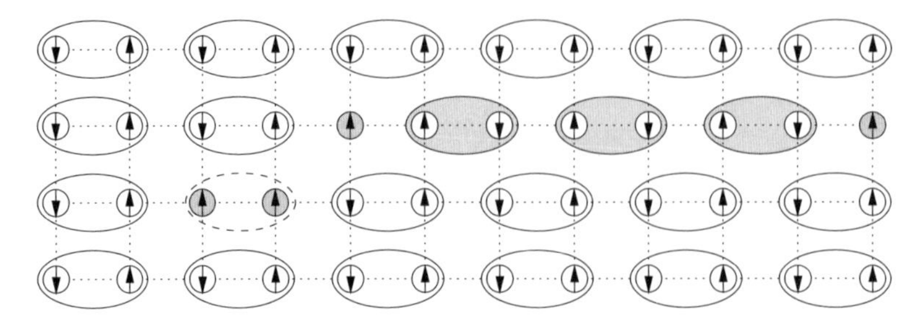

FIG. 9.1. A dimer state is formed by spin-singlet pairs. Changing a dimer into a triplet state creates a spin-1 excitation. Two separated spin-1/2 excitations are connected by a string of displaced dimers.

9.1.3 The dimer state

- The first-order mean-field theory for the dimer state has a confining $U(1)$ gauge interaction. Only spinon-pair bound states appear as physical excitations.

Let us first discuss a specific mean-field ground state, the dimer state (Majumdar and Ghosh, 1969; Affleck *et al.*, 1987; Read and Sachdev, 1989) in the Heisenberg model with nearest-neighbor coupling:

$$H = J_1 \sum_{\boldsymbol{i}} (\boldsymbol{S_i S_{i+x}} + \boldsymbol{S_i S_{i+y}}). \tag{9.1.15}$$

The dimer state is described by the following ansatz:

$$\begin{aligned} \chi_{\boldsymbol{i},\boldsymbol{i}+x} &= \frac{1}{2}(1 + (-)^{i_x}) \\ \text{others} &= 0 \end{aligned} \tag{9.1.16}$$

and $a_0(\boldsymbol{i}) = 0$. One can check that the self-consistency equation (9.1.7) and the constraint (9.1.5) are satisfied by the mean-field ground states of H_{mean} in eqn (9.1.6). The spinon spectrum has no dispersion in the mean-field dimer state:

$$E_k = \pm\frac{1}{2}J_1 \tag{9.1.17}$$

The valence band with $E_k = -\frac{1}{2}J_1$ is completely filled by the spinons. The spin excitations have a finite energy gap in the dimer state. It is also clear that the dimer state breaks the translational symmetry in the x direction.

In zeroth-order mean-field theory, the excitations above the dimer ground state are spin-1/2 spinons. Such a result is obviously wrong because, physically, the dimer state is formed by spin-singlet dimers filling the lattice (see Fig. 9.1). The elementary excitations correspond to spin-triplet dimers, which are bosonic spin-1

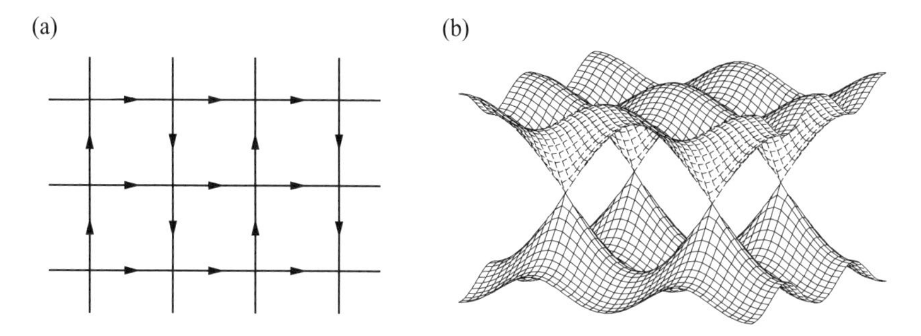

FIG. 9.2. (a) The mean-field ansatz of the π-flux state. Here $\chi_{ij} = i\chi_1$ in the direction of the arrow. (b) The fermion dispersion in the π-flux state. The valence band is filled. The low-energy excitations exist near the four Fermi points, where the valence band and the conduction band touch.

excitations. In first-order mean-field theory, the spinons are coupled to the $U(1)$ gauge field. In $2+1$ dimensions, $U(1)$ gauge theory is confining. Thus, individual spinons are not observable. Only bosonic bound states (which carry integral spin) can appear in the physical spectrum. If we do create two separated spin-1/2 excitations, then, from Fig. 9.1, we see that they are connected by a string of displaced dimers. The string leads to a linear confining interaction between the two spin-1/2 excitations, which agrees with the picture obtained from first-order mean-field theory. We see that first-order mean-field theory gives us qualitatively correct results.

9.1.4 The π-flux state

- After projection, the mean-field π-flux phase gives rise to a translation, a rotation, and a parity-symmetric spin-liquid wave function.
- The low-energy effective theory (the first-order mean-field theory) of the π-flux phase contains gapless Dirac fermions coupled to a $U(1)$ gauge field. The Dirac fermions carry spin-1/2 and are electrically neutral.
- The concepts of stable, marginal, and unstable mean-field states.
- First-order mean-field theory is reliable only for stable mean-field states or marginal mean-field states with weak fluctuations. The π-flux phase is not one of these mean-field states.

The second mean-field state that we are going to study is the π-flux state (Affleck and Marston, 1988; Kotliar, 1988) for the nearest-neighbor Heisenberg model. The π-flux state is given by the ansatz (see Fig. 9.2)

$$\chi_{\boldsymbol{i},\boldsymbol{i}+x} = i\chi_1, \qquad \chi_{\boldsymbol{i},\boldsymbol{i}+y} = i\chi_1(-)^{i_x}, \qquad a_0(\boldsymbol{i}) = 0. \tag{9.1.18}$$

The mean-field Hamiltonian (9.1.6) is equivalent to an electron hopping problem with π flux per plaquette, because around a plaquette $\prod \chi_{ij} \propto \mathrm{e}^{\mathrm{i}\pi}$. The spinon spectrum in the π-flux phase is given by (see Fig. 9.2)

$$E_k = \pm J_1 \chi_1 \sqrt{\sin(k_x)^2 + \sin(k_y)^2} \tag{9.1.19}$$

where $-\frac{\pi}{2} < k_x < \frac{\pi}{2}$ and $-\pi < k_y < \pi$. The value of χ_1 can be obtained by minimizing the mean-field energy

$$-\sum_{\boldsymbol{k}}{}' J_1 \chi_1 \sqrt{\sin(k_x)^2 + \sin(k_y)^2} + J_1 |\chi_1|^2 N_{\text{site}}$$

We find that

$$\chi_1 = \frac{1}{4} \int \frac{\mathrm{d}^2\boldsymbol{k}}{(2\pi)^2} \sqrt{\sin(k_x)^2 + \sin(k_y)^2}$$

In addition to the π-flux phase, there are many other different mean-field ansatz which locally minimize the mean-field energy. However, for the Heisenberg model with only nearest-neighbor coupling J_1, the π-flux phase has the lowest mean-field energy among translationally-invariant spin liquids.

Let us discuss some physical properties of the π-flux phase. First, let us consider the symmetry of the mean-field state. We know that mean-field theory is designed to describe a spin state whose wave function is given by $\Psi_{\text{spin}}(\{\alpha_{\boldsymbol{i}}\})$, where $\alpha_{\boldsymbol{i}} = \pm 1/2$ are the values of S^z at the site $\boldsymbol{i}$. When we say 'the symmetry of the mean-field state', we really mean 'the symmetry of the corresponding spin wave function $\Psi_{\text{spin}}(\{\alpha_{\boldsymbol{i}}\})$'. The corresponding spin wave function is obtained by projecting the mean-field state to the subspace of one fermion per site (see eqn (9.1.12)).

As the mean-field ansatz $\chi_{\boldsymbol{ij}}$ does not depend on the spin orientation, in the mean-field ground state every negative energy level is occupied by a spin-up and a spin-down fermion. Thus, the mean-field state $|\Psi_{\text{mean}}^{(\chi_{ij})}\rangle$ is spin–rotation invariant. As a result, the physical spin wave function obtained by the projection is also spin–rotation invariant.

However, the mean-field ansatz is not invariant under translation in the x direction by one lattice spacing. Therefore, the mean-field state $|\Psi_{\text{mean}}^{(\chi_{ij})}\rangle$ breaks the translational symmetry. This seems to suggest that the physical spin wave function also breaks the translational symmetry. In fact, the physical spin wave function does not break the translational symmetry. This is because the translated ansatz $(\chi_{\boldsymbol{i},\boldsymbol{i}+\boldsymbol{x}}, \chi_{\boldsymbol{i},\boldsymbol{i}+\boldsymbol{y}}) = (-\mathrm{i}\chi_1(-)^{i_x}, \mathrm{i}\chi_1)$ is related to the original ansatz $(\chi_{\boldsymbol{i},\boldsymbol{i}+\boldsymbol{x}}, \chi_{\boldsymbol{i},\boldsymbol{i}+\boldsymbol{y}}) = (\mathrm{i}\chi_1(-)^{i_x}, \mathrm{i}\chi_1)$ by the gauge transformation (9.1.11) with $\mathrm{e}^{\mathrm{i}\theta_i} = (-)^{i_x}$. Therefore, the two ansatz give rise to the same physical spin wave function. The projected spin wave function is invariant under translation. The π-flux phase is a translationally-symmetric spin liquid.

Second, we consider the properties of low-energy excitations. The Fermi surfaces at half-filling (i.e. one fermion per site on average) are points at $(k_x, k_y) = (0, 0)$ and $(0, \pi)$. At the mean-field level, the π-flux state contains gapless spin excitations which correspond to the particle–hole excitations across the Fermi points. In addition to these gapless spin excitations, the mean-field flux state also contains $U(1)$ gauge fluctuations (a_{ij}, a_0). In first-order mean-field theory, the low-energy excitations are described by gapless spinons coupled to the $U(1)$ gauge field.

If we ignore the interaction between the spinons f_i and the gauge fluctuations (a_{ij}, a_0), then the π-flux phase will have gapless neutral spin-1/2 excitations described by fermionic quasiparticles. However, one should not believe this striking result unless it is shown that this result remains valid after the inclusion of the gauge interaction.

Now let us consider the interactions between the spin excitations and the gauge fluctuations. For convenience, let us add the Maxwell term $\frac{1}{8\pi g^2}(v^2 f_{12}^2 - f_{0i}^2)$ to the Hamiltonian (9.1.10), where g is a coupling constant and $f_{\mu\nu}$ is the field strength of a_μ. The original theory corresponds to the $g \to \infty$ limit. The Maxwell term is generated in the process of integrating out the fermions. Let $g(\Lambda)$ be the coupling constant obtained by integrating out the fermions between the energy scales Λ and Λ_0, where Λ_0 is the energy cut-off in the original theory. Using a dimensional analysis, we find that $\mathrm{d}g^{-2}(\Lambda) \sim \mathrm{d}\Lambda/\Lambda^2$ (note that $\mathrm{d}g^{-2}(\Lambda) \propto \mathrm{d}\Lambda$). Thus $g^{-2}(\Lambda) \sim \Lambda^{-1} - \Lambda_0^{-1}$.

Due to the coupling with the gauge field, a spinon creates an 'electric' field f_{0i} of the gauge field (a_0, a_{ij}) (notice that a spinon carries a unit charge of the gauge field). The potential energy between a particle–hole pair in a particle–hole excitation is

$$V(\boldsymbol{r}_1 - \boldsymbol{r}_2) = 2\pi g^2 \ln |\boldsymbol{r}_1 - \boldsymbol{r}_2| \tag{9.1.20}$$

We find that the particle and the hole interact at long distances. To estimate the strength and effect of the interaction, let us compare the kinetic energy $E_K \sim 1/|\boldsymbol{r}_1 - \boldsymbol{r}_2|$ and the potential energy $V(\boldsymbol{r}_1 - \boldsymbol{r}_2)$ of the particle–hole pair. As $g^2 \sim v/|\boldsymbol{r}_1 - \boldsymbol{r}_2|$ (assuming that $\Lambda \sim v/|\boldsymbol{r}_1 - \boldsymbol{r}_2|$), we see that $V/E_K \sim 1$. This implies that the interaction is marginal and we cannot ignore the interaction at low energies. This result suggests that the quantum fluctuations (e.g. the gauge fluctuations) in the π-flux state are important and will drastically change the low-energy properties of the zeroth-order mean-field theory. In this case, zeroth-order mean-field theory (with free spinon excitations) does not provide a reliable picture of the low-energy properties of spin liquids. To obtain reliable low-energy properties of the π-flux state, one has to deal with the coupled system of gapless fermions and the gauge field. The striking properties obtained by ignoring the gauge interaction may not be valid.[55]

We see that the unbelievers of the projective construction are right. The results from zeroth-order mean-field theory are misleading and cannot be trusted. However, the believers are also right. First-order mean-field theory does tell us that the results from zeroth-order mean-field theory are incorrect. So far, the projective construction has not misled us and can be trusted, once we include the proper fluctuations. It is the results from first-order mean-field theory (not zeroth-order mean-field theory) that correspond to the physical properties of spin liquids.

To characterize the importance of the fluctuations around the mean-field state, here we would like to introduce the three concepts of stable, marginal, and unstable mean-field states. In a stable mean-field state, the fluctuations are weak. The interaction induced by the fluctuations vanishes at low energies (i.e. the interactions are irrelevant perturbations). In a marginal mean-field state, the ratios of the interactions and the energy approach finite constants in the low-energy limit. In this case, the interactions are marginal perturbations. In an unstable mean-field state, the ratios of the interactions and the energy diverge in the low-energy limit. Then the interactions are relevant perturbations. The properties of zeroth-order mean-field theories can survive the fluctuations only for stable mean-field states. For unstable mean-field states, the fluctuations will drive a phase transition at low energies and we cannot deduce any physical properties of spin liquids from zeroth-order mean-field theory. In this case, first-order mean-field theory is not useful because it does not help us to deduce the physical properties of the spin system. The π-flux state discussed above is a marginal mean-field state. For marginal mean-field states, first-order mean-field theory can be useful if the ratios of the interaction and the energy are small. In this case, the physical properties of the corresponding spin state can be calculated perturbatively. For the π-flux state, the ratios of the interaction and the energy are of order 1. Thus, it is hard to obtain the low-energy physical properties of the π-flux state from first-order mean-field theory.

From the above discussion, we see that first-order mean-field theory is useful only for stable mean-field states and for marginal mean-field states with weak fluctuations. The key to using mean-field theory to study spin liquids is to find stable mean-field states (see Sections 9.1.6, 9.2.4 and 9.2.6), or marginal mean-field states with weak fluctuations (see Section 9.8).

Problem 9.1.2.
The rotational symmetry of the π-flux state

1. Prove eqn (9.1.14).
2. Show that the π-flux state does not break the $90°$-rotational symmetry.

---

[55] When mean-field fluctuations are weak, a marginal mean-field state can lead to an algebraic spin liquid — an spin liquid with no free quasiparticles at low energies. See Sections 9.9.5 and 9.10.

Problem 9.1.3.
Let $|\Psi_{\text{mean}}\rangle$ be the mean-field ground state for the π-flux ansatz. Calculate $\langle\Psi_{\text{mean}}|\boldsymbol{S_i}\boldsymbol{S_{i+y}}|\Psi_{\text{mean}}\rangle$. Use this result to find the variational mean-field energy $\langle\Psi_{\text{mean}}|H|\Psi_{\text{mean}}\rangle$, where H is the Heisenberg model with nearest-neighbor interactions (see eqn (9.1.15)). Compare the mean-field energy of the π-flux state to the energy of the dimer state.

Problem 9.1.4.
Dirac fermions

1. Show that, in the continuum limit, the low-energy spinons in the π-flux state are described by
$$H = v\sum_{\boldsymbol{k}\sim 0} \lambda^\dagger_{\alpha,\boldsymbol{k}}(k_x\Gamma_x + k_y\Gamma_y)\lambda_{\alpha,\boldsymbol{k}}$$
where $\lambda^\top_{\alpha,\boldsymbol{k}} = (f_{\alpha,\boldsymbol{k}}, f_{\alpha,\boldsymbol{k}+\boldsymbol{Q}})$, $\boldsymbol{Q} = (\pi,\pi)$, and $\Gamma_x^2 = \Gamma_y^2 = 1$. Find v and $\Gamma_{x,y}$. In real space, H can be rewritten as $H = \int \mathrm{d}^2\boldsymbol{x}\; v\lambda^\dagger_\alpha(-\mathrm{i}\Gamma_x\partial_x - \mathrm{i}\Gamma_y\partial_y)\lambda_\alpha$.

2. The corresponding Lagrangian $\mathcal{L} = \mathrm{i}\lambda^\dagger_\alpha\partial_t\lambda_\alpha - v\lambda^\dagger_\alpha(-\mathrm{i}\Gamma_x\partial_x - \mathrm{i}\Gamma_y\partial_y)\lambda_\alpha$ can be rewritten as
$$\mathcal{L} = \mathrm{i}\bar\lambda_\alpha(\gamma^0\partial_t + v\gamma^x\partial_x + v\gamma^y\partial_y)\lambda_\alpha$$
where $\bar\lambda_\alpha = \lambda^\dagger_\alpha\gamma^0$, $\gamma_0^2 = 1$, and $(\Gamma_x,\Gamma_y) = (-\gamma^0\gamma^x, -\gamma^0\gamma^y)$. Find $\gamma^{0,x,y}$. The fermion described by $\mathcal{L}$ is called the massless Dirac fermion.

3. Show that $(\gamma^0,\gamma^x,\gamma^y)$ satisfies the following Dirac algebra in $1+2$ dimensions:
$$\{\gamma^\mu,\gamma^\nu\} = \eta^{\mu\nu}, \qquad \mu,\nu = 0, x, y,$$
where $\eta^{\mu\nu}$ is a diagonal matrix with $\eta^{00} = 1$ and $\eta^{xx} = \eta^{yy} = -1$.

4. A massive Dirac fermion is described by the Lagrangian
$$\mathcal{L} = \mathrm{i}\bar\lambda_\alpha(\gamma^0\partial_t + v\gamma^x\partial_x + v\gamma^y\partial_y)\lambda_\alpha + m\bar\lambda_\alpha\lambda_\alpha.$$
Find the solutions of the corresponding equation of motion $\mathrm{i}(\gamma^0\partial_t + v\gamma^x\partial_x + v\gamma^y\partial_y + m)\lambda_\alpha = 0$. Show that the fermion has dispersion $\omega_{\boldsymbol{k}} = \sqrt{v^2\boldsymbol{k}^2 + m^2}$.

5. *Coupling to the $U(1)$ gauge field*
Use the procedure described in Section 3.7.1 to find the Lagrangian of the Dirac fermion that is minimally coupled to a $U(1)$ gauge field. Make sure that the resulting Lagrangian has a $U(1)$ gauge invariance.

9.1.5 How to kill gapless $U(1)$ gauge bosons

- How to obtain a stable mean-field state: give gauge fluctuations a finite energy gap via the Chern–Simons term and/or the Anderson–Higgs mechanism.
- The stable spin liquids always contain electrically-neutral spin-1/2 spinons with only short-ranged interactions between them.

We have seen that the gapless $U(1)$ gauge bosons interact with gapless fermions strongly in the π-flux state, even down to zero energy. It is hard to obtain the properties of the first-order mean-field theory (i.e. the fermion–gauge coupled system) for the π-flux phase. Although the first-order mean-field theory does not mislead us, it seems too complicated to be useful. Now the question is whether it is possible to construct a stable mean-field state in which mean-field fluctuations interact weakly at low energies. In this case, the properties of the first-order mean-field theory can be obtained easily.

As the first-order mean-field theory must contain, at least, the gauge fluctuations to enforce the one-fermion-per-site constraint, one way to have a stable mean-field state is to give gauge bosons an energy gap. The gapped gauge boson can only mediate a short-range interaction between the fermions. We know how to handle fermions with a short-range interaction, based on our experience with Landau Fermi liquid theory.

First let us clarify what we mean by 'give gauge bosons an energy gap'. We know that gauge bosons are simply fluctuations of the mean-field ansatz χ_{ij}. The dynamics of these fluctuations depends on the mean-field ansatz. So, by 'give gauge bosons an energy gap' we mean to find the mean-field ansatz such that the collective mode fluctuations of the ansatz are gapped.

To motivate our search for a stable mean-field ansatz, let us first consider in what ways a $U(1)$ gauge boson can gain an energy gap. In fact, in Sections 3.7.5 and 4.4, we have already encountered two ways in which a $U(1)$ gauge boson can be gapped. The first way is the Anderson–Higgs mechanism, where the gauge bosons couple to condensed charge bosons. The dynamics of the coupled system is described by

$$\mathcal{L} = \frac{1}{8\pi g^2}(\mathbf{e}^2 - b^2) + c_1 a_0^2 - c_2 \boldsymbol{a}^2$$

in $2+1$ dimensions and in the continuum limit (see eqn (3.7.17)), where c_1 and c_2 come from the condensed charge bosons. The second way is via the Chern–Simons term. If the mean-field ansatz is such that the filled band has a nonzero Hall conductance, then the dynamics of $U(1)$ gauge bosons will be described by

$$\mathcal{L} = \frac{1}{8\pi g^2}(\mathbf{e}^2 - b^2) + \frac{K}{4\pi}\epsilon^{\mu\nu\lambda} a_\mu \partial_\nu a_\lambda. \tag{9.1.21}$$

In Problem 9.1.5 we will find that the Chern–Simons term gives the gauge boson a nonzero energy gap. The key to finding a stable mean-field state is to find a mean-field ansatz that realizes one of the above two mechanisms.

In both cases, the gauge field can only mediate short-range interactions. As a consequence, the spinons are not confined. The quasiparticles above the spin liquid are described by free spinons which carry spin-1/2 and zero electric charge. In the presence of the Chern–Simons term, the spinons can even have fractional statistics.

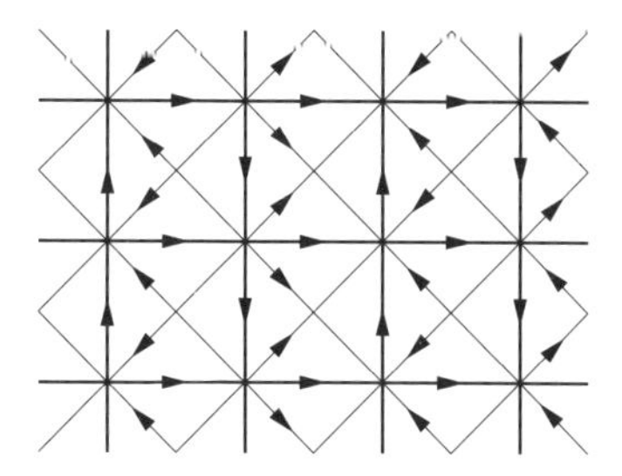

FIG. 9.3. The mean-field ansatz of a chiral spin state. Here χ_{ij} is $\mathrm{i}\chi$ in the direction of the arrow.

Problem 9.1.5.
Show that the fluctuations described by eqn (9.1.21) have a finite energy gap. (You may do so by calculating the classical equation of motion.)

9.1.6 Chiral spin state

- The mean-field chiral spin ansatz gives rise to a translationally- and rotationally-symmetric spin-liquid state. However, the time-reversal and parity symmetries are broken.
- The mean-field chiral spin state is stable. The chiral spin liquid obtained from the mean-field theory does represent a stable quantum phase of real spin systems.
- The chiral spin liquid contains fractionalized excitations—spinons. The spinons carry spin-1/2 and no electric charge, and have fractional statistics.

In this section, we will discuss a mean-field state which realizes the second mechanism discussed above. The mean-field state is described by an ansatz χ_{ij}, where χ_{ij} is complex and generates flux. The spinons described by eqn (9.1.6) behave as if they are moving in a magnetic field. When the flux has a correct commensuration with the spinon density (which is one fermion per site), an integral number of Landau levels (or, more precisely, Landau bands, due to the lattice) are completely filled. In this case, the effective lattice gauge theory contains a Chern–Simons term due to the finite Hall conductance of the filled Landau level. We will call such a stable mean-field state a chiral spin state (Wen *et al.*, 1989; Khveshchenko and Wiegmann, 1989).

The simplest chiral spin state is given by the following ansatz (see Fig. 9.3):

$$
\begin{aligned}
\chi_{\boldsymbol{i},\boldsymbol{i}+x} &= \mathrm{i}\chi_1, & \chi_{\boldsymbol{i},\boldsymbol{i}+y} &= \mathrm{i}\chi_1(-)^{i_x}, & a_0(\boldsymbol{i}) &= 0, \\
\chi_{\boldsymbol{i},\boldsymbol{i}+\boldsymbol{x}+\boldsymbol{y}} &= -\mathrm{i}\chi_2(-)^{i_x}, & \chi_{\boldsymbol{i},\boldsymbol{i}+\boldsymbol{x}-\boldsymbol{y}} &= \mathrm{i}\chi_2(-)^{i_x} & &
\end{aligned}
\tag{9.1.22}
$$

The above χ_{ij} induce π flux for each square and $\frac{1}{2}\pi$ flux for each triangle. The π-flux phase discussed in the last section is somewhat special. As π flux is equivalent to $-\pi$ flux, the π-flux phase respects time-reversal symmetry (T) and parity (P). However, the $\frac{1}{2}\pi$-flux phase described by eqn (9.1.22) is not equivalent to the $-\frac{1}{2}\pi$-flux phase. Under T or P, the $\frac{1}{2}\pi$ flux changes into $-\frac{1}{2}\pi$ flux, and χ_2 in eqn (9.1.22) changes into $-\chi_2$. Therefore, the chiral spin state with nonzero χ_2 breaks T and P spontaneously. Using the identity

$$E_{123} \equiv \boldsymbol{S}_1 \cdot (\boldsymbol{S}_2 \times \boldsymbol{S}_3) = 2\mathrm{i}(\hat{\chi}_{12}\hat{\chi}_{23}\hat{\chi}_{31} - \hat{\chi}_{13}\hat{\chi}_{32}\hat{\chi}_{21}), \qquad \hat{\chi}_{\boldsymbol{ij}} = f_{\boldsymbol{i}}^\dagger f_{\boldsymbol{j}} \quad (9.1.23)$$

we find that $\langle E_{123}\rangle \sim \mathrm{Im}(\chi_{12}\chi_{23}\chi_{31})$ is nonzero in the chiral spin state. Notice that E_{123} is odd under T or P; thus, E_{123} can be regarded as the T- and P-breaking order parameter.

After specifying the chiral spin state, we would like to know which spin Hamiltonian supports the chiral spin state. Let us consider the following frustrated spin Hamiltonian:

$$H = J_1 \sum_{\boldsymbol{i}} (\boldsymbol{S_i}\cdot \boldsymbol{S_{i+x}} + \boldsymbol{S_i}\cdot \boldsymbol{S_{i+y}}) + J_2 \sum_{\boldsymbol{i}} (\boldsymbol{S_i}\cdot \boldsymbol{S_{i+x+y}} + \boldsymbol{S_i}\cdot \boldsymbol{S_{i+x-y}}) \quad (9.1.24)$$

We would like to know when the mean-field Hamiltonian (9.1.6) supports the mean-field chiral spin state. In the mean-field Hamiltonian (9.1.6), $J_{\boldsymbol{ij}}$ is equal to J_1 for the nearest neighbor and J_2 for the second-nearest neighbor.

When $\chi_2 = 0$, the spinon spectrum determined by the mean-field Hamiltonian is given by eqn (9.1.19). The conduction band and the valence band touch at the points $(k_x, k_y) = (0,0)$ and $(0, \frac{\pi}{a})$. When $\chi_2 \neq 0$, an energy gap between the conduction band and the valence band is opened. The mean-field spinon spectrum is given by

$$E_{\boldsymbol{k}}^{\pm} = \pm\sqrt{J_1^2\chi_1^2(\sin(k_x)^2 + \sin(k_y)^2) + J_2^2\chi_2^2(\cos(k_x + k_y) + \cos(k_x - k_y))^2} \quad (9.1.25)$$

The mean-field ground state is obtained by filling the valence band. Due to the energy gap, the spin–spin correlation is short-ranged. Using the ground-state wave function, we can calculate $\langle f_{\alpha i}^\dagger f_{\alpha j}\rangle$ to check the self-consistency condition (9.1.7). We find that, when $J_2/J_1 < 0.49$, the self-consistency equation only supports one solution with $\chi_1 \neq 0$ and $\chi_2 = 0$, i.e. the π-flux phase. When $J_2/J_1 > 0.49$, eqn (9.1.7) also supports a second solution with $\chi_2 \neq 0$. The second solution is found to have a lower mean-field energy. In this case, T and P are spontaneously broken.

Notice that the mean Hamiltonian for the chiral state is equivalent to the problem of electron hopping in a magnetic field. The coupling between the slave fermions and the gauge field a_μ is identical to the coupling between the electrons and the electromagnetic field. Thus, one expects that the slave-fermion system described by eqn (9.1.10) has a phenomenon similar to the Hall effect. The 'Hall effect' in this case implies that an 'electric' field of the a_μ gauge field induces a spinon current in the transverse direction:

$$j_x = \sigma_{xy} e_y, \quad e_y = \partial_t a_y - \partial_y a_0 \tag{9.1.26}$$

where σ_{xy} is the Hall conductance. There is a theorem stating that the Hall conductance of a filled band is always quantized as an integer times $1/2\pi$ (Thouless *et al.*, 1982; Avron *et al.*, 1983). In our case, this implies that $\sigma_{xy} = 2n/2\pi$, where the factor 2 comes from the spins. The value of the 'Hall' conductance can be obtained using the method discussed in Section 4.4. We find that

$$\sigma_{xy} = 2/2\pi. \tag{9.1.27}$$

The simplest way to understand this result is to note that there is one flux quantum for every spin-up spinon, and one flux quantum for every spin-down spinon. Thus, both spin-up and spin-down spinons have a fill fraction of $\nu = 1$. After turning off the lattice potential, the valence band in the mean-field chiral spin state becomes the first Landau level. Both the spin-up and spin-down slave fermions in the valence band contribute $1/2\pi$ to the 'Hall' conductance. Therefore, the total 'Hall' conductance is given by eqn (9.1.27).

The effective action for the gauge fluctuations is obtained by integrating out the spinons in eqn (9.1.10). Using the relationship between the 'Hall' conductance and the Chern–Simons term, we can easily write down the effective action in the continuum limit as follows:

$$S = \int \mathrm{d}^3x \frac{1}{2} \sigma_{xy} a_\mu \partial_\nu a_\lambda \epsilon_{\mu\nu\lambda} + \frac{1}{8\pi g^2} (\mathbf{e}^2 - v^2 b^2) + \ldots \tag{9.1.28}$$

Here g^2 in eqn (9.1.28) is of the order of the spinon gap, and v is of order $1/aJ$, the typical spinon velocity. Using the effective Lagrangian, we can calculate the low-energy dynamical properties of the gauge fluctuations.

Due to the nonzero 'Hall' conductance σ_{xy}, we can change the spinon density *without* creating spinons in the conduction band or creating holes in the valence band. This property is important in understanding the properties of the quasiparticles. Let us slowly turn on the flux of the a_μ field, $\Phi = \int \mathrm{d}^2x\, b$. If the flux is distributed in a large region, the the 'magnetic' field b in the flux is small. In this case, the energy gap between the valence band and the conduction band remains finite and all of the energy levels in the valence band are filled by a spin-up and a spin-down spinon. Turning on the flux induces a circular 'electric' field e_θ, which in turn generates a spinon current in the radial direction $\hat{\boldsymbol{r}}$ due to the nonzero σ_{xy}. Thus, some charges are accumulated near the origin. We find that the total number of induced spinons is given by

$$N \equiv \sum_i (\langle f^\dagger_{\alpha i} f_{\alpha i}\rangle - 1) = -\sigma_{xy}\Phi = -\frac{\Phi}{\pi} \tag{9.1.29}$$

We would like to stress that the flux only changes the spinon density. It does not induce any spin quantum number. No matter how many spinons are induced by the flux, the flux tube always carries zero spin because every energy level in the valence band is filled by a spin-up and a spin-down spinon.

Now we are ready to discuss the quantum number of the quasiparticle excitations in the chiral spin state. The simplest excitation in the mean-field chiral spin state can be obtained by adding a spinon into the conduction band. However, this excitation is not physical because the additional spinon in the conduction band violates the constraint (9.1.5). To satisfy the constraint, we can add flux to change the spinon density in the valence band. From the above discussion, we see that the extra spinon density arising from the spinon in the conduction band can be cancelled by introducing π flux (see eqn (9.1.29)). Therefore, the physical quasiparticles in the chiral spin state are spinons *dressed* by π flux. The dressed spinons carry spin-1/2 because the flux cannot induce any spin quantum numbers. However, as a bound state of a charge and flux (note that the spinon carries a unit charge of the a_μ gauge field), the spinon has fractional statistics. Exchanging two dressed spinons is equivalent to moving one spinon half-way around the other, which induces a phase $\theta = \frac{1}{2}q\Phi$. Here $q = 1$ is the charge of the spinon and $\Phi = \pi$ is the flux bounded to the spinon. Hence the statistical angle of the dressed spinon is $\theta = \frac{\pi}{2}$. Particles with such statistics are half-way between bosons (with $\theta = 0$) and fermions (with $\theta = \pi$), and are called semions.

To understand the low-energy dynamics of the spinons, we would like to derive the effective Lagrangian for the spinons. First let us ignore the gauge fluctuations by setting $a_\mu = 0$. In this case, a spinon in the conduction band or the valence band is described by the dispersion given in eqn (9.1.25). Here E^+_k (E^-_k) has two minima (maxima) at $(0,0)$ and $(0,\pi/a)$. Therefore, in the continuum limit, we have four species of spinons described by the following effective Lagrangian:

$$\mathcal{L} = \sum_{I=1,2}\left[-\mathrm{i} f^\dagger_{I\alpha}\partial_t f_{I\alpha} + \frac{1}{2m_s} f^\dagger_{I\alpha}\partial_i^2 f_{I\alpha} - \mathrm{i}\bar{f}^\dagger_{I\alpha}\partial_t \bar{f}_{I\alpha} + \frac{1}{2m_s}\bar{f}^\dagger_{I\alpha}\partial_i^2 \bar{f}_{I\alpha}\right] \tag{9.1.30}$$

Here $f_{1\alpha}$ and $f_{2\alpha}$ correspond to the spinons near the two minima of the conduction band, and $\bar{f}_{1\alpha}$ and $\bar{f}_{2\alpha}$ correspond to the holes near the maxima of the valence band. When $a_\mu \neq 0$, the coupling between the spinons and the gauge field can be obtained by replacing ∂_μ in eqn (9.1.30) by $\partial_\mu \pm i a_\mu$. Such a form of coupling is determined by the requirement of gauge invariance. After including the coupling to the gauge field, the total effective Lagrangian has the form

$$\mathcal{L} = \sum_{I=1,2}\left[-\mathrm{i}f^\dagger_{I\alpha}(\partial_t - \mathrm{i}a_0)f_{I\alpha} + \frac{1}{2m_s}f^\dagger_{I\alpha}(\partial_i - \mathrm{i}a_i)^2 f_{I\alpha}\right]$$
$$+ \sum_{I=1,2}\left[-\mathrm{i}\bar{f}^\dagger_{I\alpha}(\partial_t + \mathrm{i}a_0)\bar{f}_{I\alpha} + \frac{1}{2m_s}\bar{f}^\dagger_{I\alpha}(\partial_i + \mathrm{i}a_i)^2 \bar{f}_{I\alpha}\right]$$
$$- \frac{2}{4\pi}a_\mu\partial_\nu a_\lambda\epsilon_{\mu\nu\lambda} + \frac{1}{8\pi g^2}(\mathbf{e}^2 - v^2 b^2) \tag{9.1.31}$$

From the equation of motion $\frac{\delta\mathcal{L}}{\delta a_0} = 0$, we find that

$$n_1 + n_2 = -\frac{1}{\pi}b \tag{9.1.32}$$

where $n_I = \sum_\alpha (f^\dagger_{I\alpha}f_{I\alpha} - \bar{f}^\dagger_{I\alpha}\bar{f}_{I\alpha}), I = 1, 2$, is the density of the spinons. Equation (9.1.32) tells us that a spinon is dressed by π flux, which agrees with the previous results. The statistics of the spinons can also be directly calculated from eqn (9.1.31).

We would like to point out that, according to eqn (9.1.31), there are two kinds of quasiparticles for each fixed I, namely the spin-1/2 spinons in the conduction band and the spin-1/2 holes in the valence band. This result is *incorrect*. For each fixed value of I there should only be one kind of quasiparticles, namely the spin-1/2 spinons. The spinons in the conduction band and hole in the valence band give rise to the same spinon after the projection (Affleck *et al.*, 1988; Dagotto *et al.*, 1988). This over-counting problem can be resolved after realizing that the chiral spin ansatz actually has an $SU(2)$ gauge structure (see Problem 9.2.4).

Problem 9.1.6.
Find the equations that determine χ_1 and χ_2 in the ansatz (9.1.22) of the chiral spin state by minimizing the mcan-field energy (9.1.6).

Problem 9.1.7.
Prove eqn (9.1.25).

Problem 9.1.8.
Prove eqn (9.1.27) by using the result of Section 4.4.

9.2 The $SU(2)$ projective construction

9.2.1 The hidden $SU(2)$ gauge structure

- The projective construction has a hidden $SU(2)$ gauge structure.

- The mean-field ansatz $(U_{ij}, a_0^l(i))$ is a many–to–one label of physical spin liquids. The $SU(2)$ gauge-equivalent ansatz labels the same physical wave function.
- The invariance of a physical wave function under a symmetry transformation only requires the invariance of the corresponding mean-field ansatz up to a gauge transformation.

9.2.1.1 *Spinon-pair condensation*

In the last section, we constructed a stable mean-field state using a Chern–Simons term. In this section, we would like to consider another class of stable mean-field state due to the Anderson–Higgs mechanism. In order for the Anderson–Higgs mechanism to work, we first need a boson that carries a_μ charge. Secondly, the charged bosons should have proper dynamics, so that they condense. In the mean-field theory obtained from the projective construction, we do not have bosons that carry a_μ charge; but we have fermions that carry a_μ charge. So we can make charged bosons from pairs of charged fermions and let those fermion pairs condense. We see that the Anderson–Higgs mechanism can be achieved through fermion-pair condensations.[56] In the following, we will include the fermion-pair condensation in our mean-field ansatz. We hope that these mean-field ansatz will represent stable mean-field states.

Remember that, in terms of the spinon operators, the Hamiltonian (9.1.1) can be rewritten as

$$H = \sum_{\langle ij \rangle} -\frac{1}{2} J_{ij} \left(f_{i\alpha}^\dagger f_{j\alpha} f_{j\beta}^\dagger f_{i\beta} + \frac{1}{2} f_{i\alpha}^\dagger f_{i\alpha} f_{j\beta}^\dagger f_{j\beta} \right) \tag{9.2.1}$$

where we have added proper constant terms $\sum_i f_{i\alpha}^\dagger f_{i\alpha}$ to obtain the above result.

To obtain the mean-field Hamiltonian that contains fermion-pair condensation, we replace both the operators $f_{i\alpha}^\dagger f_{j\beta}$ and $f_{i\alpha} f_{i\beta}$ by their ground-state expectation value

$$\begin{aligned} \eta_{ij}\epsilon_{\alpha\beta} &= -2\langle f_{i\alpha} f_{j\beta} \rangle, & \eta_{ij} &= \eta_{ji}, \\ \chi_{ij}\delta_{\alpha\beta} &= 2\langle f_{i\alpha}^\dagger f_{j\beta} \rangle, & \chi_{ij} &= \chi_{ji}^\dagger. \end{aligned} \tag{9.2.2}$$

The χ_{ij} term was included in the previous mean-field ansatz. The η_{ij} term is new and describes fermion-pair condensation. We also replace the constraint (9.1.4) by

[56] If the fermions are electrons, then the fermion-pair condensed state is a BCS superconducting state.

its ground-state average

$$\langle f_{i\alpha}^\dagger f_{i\alpha} \rangle = 1, \qquad \langle f_{i\alpha} f_{i\beta} \epsilon_{\alpha\beta} \rangle = 0 \tag{9.2.3}$$

Such a constraint can be enforced by introducing *site-dependent* and time-independent Lagrangian multipliers $a_0^l(\boldsymbol{i})$, $l = 1, 2, 3$. In this way, we obtain the following zeroth-order mean-field Hamiltonian that includes fermion-pair condensation:

$$\begin{aligned} H_{\text{mean}} = & \sum_{\langle \boldsymbol{ij} \rangle} -\frac{3}{8} J_{\boldsymbol{ij}} \left[(\chi_{\boldsymbol{ji}} f_{\boldsymbol{i}\alpha}^\dagger f_{\boldsymbol{j}\alpha} + \eta_{\boldsymbol{ij}} f_{\boldsymbol{i}\alpha}^\dagger f_{\boldsymbol{j}\beta}^\dagger \epsilon_{\alpha\beta} + h.c) - |\chi_{\boldsymbol{ij}}|^2 - |\eta_{\boldsymbol{ij}}|^2 \right] \\ & + \sum_{\boldsymbol{i}} \left[a_0^3 (f_{\boldsymbol{i}\alpha}^\dagger f_{\boldsymbol{i}\alpha} - 1) + [(a_0^1 + \mathrm{i} a_0^2) f_{\boldsymbol{i}\alpha} f_{\boldsymbol{i}\beta} \epsilon_{\alpha\beta} + h.c.] \right] \end{aligned} \tag{9.2.4}$$

Here $\chi_{\boldsymbol{ij}}$ and $\eta_{\boldsymbol{ij}}$ in eqn (9.1.6) must satisfy the self-consistency condition (9.2.2) and the site-dependent fields $a_0^l(\boldsymbol{i})$ are chosen such that eqn (9.2.3) is satisfied by the mean-field ground state. Such $\chi_{\boldsymbol{ij}}$, $\eta_{\boldsymbol{ij}}$, and a_0^l give us a mean-field solution.

For the Heisenberg model with nearest-neighbor spin coupling, see eqn (9.1.15), the following mean-field ansatz with fermion-pair condensation is a solution of the mean-field equation (9.2.2):

$$\begin{aligned} \chi_{\boldsymbol{i},\boldsymbol{i}+\boldsymbol{x}} &= \chi, & \chi_{\boldsymbol{i},\boldsymbol{i}+\boldsymbol{y}} &= \chi, \\ \eta_{\boldsymbol{i},\boldsymbol{i}+\boldsymbol{x}} &= \eta, & \eta_{\boldsymbol{i},\boldsymbol{i}+\boldsymbol{y}} &= -\eta, \\ a_0^l &= 0. \end{aligned} \tag{9.2.5}$$

Such an ansatz actually corresponds to a d-wave BCS state. We will call it a d-wave state.[57] Due to the fermion-pair condensation, we expect that the mean-field state described by the above ansatz contains no gapless $U(1)$ gauge bosons due to the Anderson–Higgs mechanism.

Well, despite that everything seems to fit together very well and the physical picture seems to be very reasonable, the above result turns out to be incorrect. Where did we make a mistake? The physical picture and reasoning presented here are correct. The mistake turns out to be a mathematical one. We have claimed that the mean-field theory (9.1.10) (or, more generally, eqn (9.2.4)) has a $U(1)$ gauge structure, such that the two ansatz related by the $U(1)$ gauge transformation (9.1.13) correspond to the same physical spin wave function (see eqn (9.1.14)) after the projection. In fact, the mean-field theory has a larger $SU(2)$ gauge structure. Two ansatz related by the $SU(2)$ gauge transformation correspond to the same

[57] We note that the fermion-pairing order parameter $\eta_{\boldsymbol{ij}}$ changes sign under a 90° rotation. This is similar to a wave function that carries an angular momentum of 2. This is why we call the state the d-wave state.

physical spin wave function. The different gauge structure has a profound consequence on our understanding of the properties of mean-field states. In particular, the fluctuations in χ_{ij}, η_{ij}, and $a_0^l(i)$ describe $SU(2)$ gauge fluctuations.

9.2.1.2 *The $SU(2)$ formulation*

To understand the $SU(2)$ gauge structure in the mean-field Hamiltonian (9.2.4) and in the constraints (9.1.4) (Affleck *et al.*, 1988; Dagotto *et al.*, 1988), we introduce a doublet

$$\psi = \begin{pmatrix} \psi_1 \\ \psi_2 \end{pmatrix} = \begin{pmatrix} f_\uparrow \\ f_\downarrow^\dagger \end{pmatrix} \tag{9.2.6}$$

and a matrix

$$U_{ij} = \begin{pmatrix} \chi_{ij}^\dagger & \eta_{ij} \\ \eta_{ij}^\dagger & -\chi_{ij} \end{pmatrix} = U_{ji}^\dagger \tag{9.2.7}$$

Using eqns (9.2.6) and (9.2.7), we can rewrite eqns (9.2.3) and (9.2.4) as follows:

$$\langle \psi_i^\dagger \tau^l \psi_i \rangle = 0, \qquad l = 1, 2, 3 \tag{9.2.8}$$

$$H_{\text{mean}} = \sum_{\langle ij \rangle} \frac{3}{8} J_{ij} \left[\frac{1}{2} \text{Tr}(U_{ij}^\dagger U_{ij}) - (\psi_i^\dagger U_{ij} \psi_j + h.c.) \right] + \sum_i a_0^l \psi_i^\dagger \tau^l \psi_i \tag{9.2.9}$$

where τ^l, $l = 1, 2, 3$, are the Pauli matrices. From eqn (9.2.9), we can clearly see that the Hamiltonian is invariant under a local $SU(2)$ transformation W_i:

$$\begin{aligned} \psi_i &\to W_i \psi_i \\ U_{ij} &\to W_i U_{ij} W_j^\dagger \end{aligned} \tag{9.2.10}$$

The $SU(2)$ gauge structure actually originates from eqn (9.1.2). Here $SU(2)$ is the most general transformation between the spinons that leaves the physical spin operator unchanged. Those transformations become the gauge transformation, since the physical Hamiltonian is a function of the spin operators.

9.2.1.3 *The meaning of the $SU(2)$ gauge transformation*

Just like the $U(1)$ gauge transformation, the $SU(2)$ gauge transformation has the following meaning: two ansatz related by a $SU(2)$ gauge transformation correspond to the same physical spin wave function. To see this, we note that the

f-fermion states and the ψ-fermion states at each site have the following relation:

$$|0_f\rangle = \psi_2^\dagger|0_\psi\rangle, \qquad f_\uparrow^\dagger f_\downarrow^\dagger|0_f\rangle = \psi_1^\dagger|0_\psi\rangle,$$
$$f_\downarrow^\dagger|0_f\rangle = |0_\psi\rangle, \qquad f_\uparrow^\dagger|0_f\rangle = \psi_1^\dagger\psi_2^\dagger|0_\psi\rangle$$

where $|0_\psi\rangle$ is the state with no ψ-fermion, i.e. $\psi_a|0_\psi\rangle = 0$. Thus, the physical one-f-fermion-per-site states correspond to states with even numbers of ψ-fermions per site. These even-ψ-fermion-per-site states are local $SU(2)$ singlet states (i.e. they are an $SU(2)$ singlet on every site). The empty ψ-fermion state corresponds to a down-spin and the doubly-occupied ψ-fermion state corresponds to an up-spin. With this understanding, we may obtain a valid wave function of the spin system by projecting the mean-field state to the even-ψ-fermion-per-site subspace. Let $\boldsymbol{i}_1, \boldsymbol{i}_2, ...$ be the locations of the up-spins. The physical spin wave function can be written as a function of $\boldsymbol{i}_1, \boldsymbol{i}_2, ...$, i.e. $\Psi_{\text{spin}}(\boldsymbol{i}_1, \boldsymbol{i}_2, ...)$. As $\boldsymbol{i}_1, \boldsymbol{i}_2, ...$ are the only sites with two ψ-fermions, we find that

$$\Psi_{\text{spin}}(\{\boldsymbol{i}_n\}) = \langle 0_\psi| \prod_n \psi_{1,\boldsymbol{i}_n}\psi_{2,\boldsymbol{i}_n}|\Psi_{\text{mean}}^{(U_{ij})}\rangle. \tag{9.2.11}$$

As $\langle 0_\psi|$ and $\psi_{1,\boldsymbol{i}}\psi_{2,\boldsymbol{i}}$ are invariant under the local $SU(2)$ transformation, the two ansatz $U_{\boldsymbol{ij}}$ and $U'_{\boldsymbol{ij}}$ related by an $SU(2)$ gauge transformation $U'_{\boldsymbol{ij}} = W_{\boldsymbol{i}}U_{\boldsymbol{ij}}W_{\boldsymbol{j}}^\dagger$ are just two different labels which label the *same physical state*:

$$\langle 0_\psi| \prod_n \psi_{1,\boldsymbol{i}_n}\psi_{2,\boldsymbol{i}_n}|\Psi_{\text{mean}}^{(U_{ij})}\rangle = \langle 0_\psi| \prod_n \psi_{1,\boldsymbol{i}_n}\psi_{2,\boldsymbol{i}_n}|\Psi_{\text{mean}}^{(W_iU_{ij}W_j^\dagger)}\rangle \tag{9.2.12}$$

The relationship between the mean-field state and the physical spin wave function (9.2.11) allows us to construct transformations of a physical spin wave function from those of the corresponding mean-field ansatz. For example, the mean-field state $|\Psi_{\text{mean}}^{(U'_{ij})}\rangle$ with translated ansatz $U'_{\boldsymbol{ij}} = U_{\boldsymbol{i}-\boldsymbol{l},\boldsymbol{j}-\boldsymbol{l}}$ produces a translated physical spin wave function after the projection.

It is obvious that the translationally-invariant ansatz will lead to a translationally-invariant physical spin wave function. However, the translational symmetry of the physical wave function after projection does not require the translational invariance of the corresponding ansatz. The physical state is translationally symmetric if and only if the translated ansatz $U'_{\boldsymbol{ij}}$ is gauge equivalent to the original ansatz $U_{\boldsymbol{ij}}$. We see that the gauge structure can complicate our analysis of symmetries, because the physical spin wave function $\Psi_{\text{spin}}(\{\alpha_{\boldsymbol{i}}\})$ may have more symmetries than the mean-field state $|\Psi_{\text{mean}}^{(U_{ij})}\rangle$ before projection.

9.2.1.4 *Spin rotation invariance*

We note that both of the components of ψ carry spin-up. Thus, the spin–rotation symmetry is not explicit in our formalism and it is hard to tell whether eqn (9.2.9) describes a spin–rotation-invariant state or not. In fact, for a general U_{ij} satisfying $U_{ij} = U_{ji}^\dagger$, eqn (9.2.9) may not describe a spin–rotation-invariant state. However, if U_{ij} has the form

$$U_{ij} = \chi_{ij}^\mu \tau^\mu, \qquad \mu = 0, 1, 2, 3,$$
$$\chi_{ij}^0 = \text{imaginary}, \qquad \chi_{ij}^l = \text{real}, \qquad l = 1, 2, 3, \tag{9.2.13}$$

then eqn (9.2.9) will describe a spin–rotation-invariant state. This is because the above U_{ij} can be rewritten in the form of eqn (9.2.7). In this case, eqn (9.2.9) can be rewritten as eqn (9.2.4), where the spin–rotation invariance is explicit. In eqn (9.2.13), τ^0 is the identity matrix and $\tau^{1,2,3}$ are the Pauli matrices.

9.2.1.5 *Variational approach*

We would like to remark that, in addition to the self-consistency equation (9.2.2), there is another way to obtain the mean-field solutions. We can view the mean-field ground state of H_{mean} (see eqn (9.2.9)), namely $|\Psi_{\text{mean}}^{(U_{ij})}\rangle$, as a trial wave function and U_{ij} as variational parameters. Introducing

$$(\tilde{U}_{ij})_{\alpha\beta} \equiv -2\langle\Psi_{\text{mean}}^{(U_{ij})}|\psi_{i,\alpha}\psi_{j,\beta}^\dagger|\Psi_{\text{mean}}^{(U_{ij})}\rangle$$

we find that the mean-field energy for $|\Psi_{\text{mean}}^{(U_{ij})}\rangle$ is given by

$$E_{\text{mean}}(\{U_{ij}\}) = -\sum_{\langle ij\rangle} \frac{3}{16} J_{ij} \operatorname{Tr}(\tilde{U}_{ij}^\dagger \tilde{U}_{ij}) \tag{9.2.14}$$

Here $E_{\text{mean}}(\{U_{ij}\})$ is a functional of U_{ij} which has $SU(2)$ gauge invariance:

$$E_{\text{mean}}(\{U_{ij}\}) = E_{\text{mean}}(\{W_i U_{ij} W_j^\dagger\})$$

The mean-field solution U_{ij} can now be obtained by minimizing E_{mean}.

9.2.1.6 *First-order mean-field theory*

To obtain the first-order mean-field theory, we start with the zeroth-order mean-field theory described by the mean field Hamiltonian

$$H_{\text{mean}} = \sum_{\langle ij\rangle} -\frac{3}{8} J_{ij} (\psi_i^\dagger \bar{U}_{ij} \psi_j + h.c.) + \sum_i a_0^l \psi_i^\dagger \tau^l \psi_i \tag{9.2.15}$$

where $\bar{U}_{ij}$ is the mean-field solution, which satisfies the self-consistency condition

$$\bar{\chi}_{ij} = \langle\psi_{i\alpha}^\dagger \psi_{j\alpha}\rangle, \quad \bar{\eta}_{ij} = -\langle\psi_{i\alpha}\psi_{i\beta}\epsilon_{\alpha\beta}\rangle \tag{9.2.16}$$

The a_0^l in eqn (9.2.15) are chosen such that eqn (9.2.3) is satisfied. The important fluctuations around the mean-field ground state are the following 'phase' fluctuations of U_{ij}:

$$U_{ij} = \bar{U}_{ij} e^{i a_{ij}} \tag{9.2.17}$$

where $a_{ij} = a^l_{ij} \tau^l$ is a 2×2 traceless hermitian matrix. Due to the $SU(2)$ gauge structure, these fluctuations are the $SU(2)$ gauge fluctuations. We must include these fluctuations in order to obtain qualitatively correct results for spin-liquid states at low energies. This leads us to the first-order mean-field theory

$$H_{\text{mean}} = \sum_{\langle ij \rangle} -\frac{3}{8} J_{ij} (\psi^\dagger_i \bar{U}_{ij} e^{i a^l_{ij} \tau^l} \psi_j + h.c.) + \sum_i a^l_0 \psi^\dagger_i \tau^l \psi_i \tag{9.2.18}$$

which describes spinons coupled to $SU(2)$ lattice gauge fields.

We would like to point out that the mean-field ansatz of the spin liquids U_{ij} can be divided into two classes: the unfrustrated ansatz where U_{ij} only link an even lattice site to an odd lattice site; and the frustrated ansatz where U_{ij} are nonzero between two even sites and/or two odd sites. An unfrustrated ansatz has only pure $SU(2)$ flux through each plaquette, while a frustrated ansatz has $U(1)$ flux of a multiple of $\pi/2$ through some plaquettes in addition to the $SU(2)$ flux.

Problem 9.2.1.
Use eqn (9.2.2) to obtain eqn (9.2.4) from eqn (9.2.1).

Problem 9.2.2.
The equivalence of the π-flux state and the d-wave state

1. Show that the π-flux state (9.1.18) is described by the $SU(2)$ link variables

$$U_{i,i+x} = -i\chi_1 \tau^0, \quad U_{i,i+y} = -i\chi_1 (-)^{i_x} \tau^0$$

2. Show that, when $\eta = \chi$, the d-wave state (9.2.5) is described by the $SU(2)$ link variables

$$U_{i,i+x} = \chi\tau^3 + \chi\tau^1, \quad U_{i,i+y} = \chi\tau^3 - \chi\tau^1$$

3. Show that, when χ and χ_1 are related in a certain way, the above two ansatz are gauge equivalent. Find an $SU(2)$ gauge transformation W_i that changes one ansatz to the other. (Hint: Consider the $SU(2)$ gauge transformation $(i\tau^1)^{i_x}(i\tau^3)^{i_y}$.) Find the relation between χ and χ_1.

4. Show that, when χ and χ_1 are related in this way, the π-flux state and the d-wave state have an identical spinon spectrum.

9.2.2 Dynamics of the $SU(2)$ gauge fluctuations

- The Anderson–Higgs mechanism can be realized through the condensation of $SU(2)$ gauge flux without using Higgs bosons.
- Collinear $SU(2)$ flux breaks the $SU(2)$ gauge structure down to a $U(1)$ gauge structure.

• Non-collinear $SU(2)$ flux breaks the $SU(2)$ gauge structure down to a Z_2 gauge structure. In this case, all of the $SU(2)$ gauge bosons gain a gap.

Just like the $U(1)$ projective construction, to obtain a stable mean-field state in the $SU(2)$ projective construction, we need to find ways to kill the gapless $SU(2)$ gauge bosons. The $SU(2)$ Chern–Simons term is one way to give $SU(2)$ gauge bosons a nonzero energy gap. In the following, we will discuss how to use the Anderson–Higgs mechanism to kill the gapless $SU(2)$ gauge bosons.

We know that $U(1)$ gauge bosons do not carry their own gauge charge. So we need additional charged bosons to implement the Anderson–Higgs mechanism. In contrast, $SU(2)$ gauge bosons themselves carry nonzero gauge charges. Thus, we do not need additional Higgs bosons to realize the Anderson–Higgs mechanism. The $SU(2)$ gauge bosons are capable of killing themselves.

To see how $SU(2)$ gauge bosons commit suicide, we consider a lattice $SU(2)$ gauge theory. The lattice $SU(2)$ gauge field is given by the link variables $U_{ij} \in SU(2)$. The first-order mean-field theory (9.2.18) is an $SU(2)$ lattice gauge theory. The energy of a configuration is a function of U_{ij}, i.e. $E(U_{ij})$. The energy is invariant under the $SU(2)$ gauge transformation

$$E(\tilde{U}_{ij}) = E(U_{ij}), \qquad \tilde{U}_{ij} = W_i(U_{ij})W_j, \qquad W_i \in SU(2) \tag{9.2.19}$$

To understand the dynamics of the lattice $SU(2)$ gauge fluctuations, we write $U_{ij} = \bar{U}_{ij}\mathrm{e}^{\,\mathrm{i}a^l_{ij}\tau^l}$, where the 2×2 matrices a_{ij} on the links describe the gauge fluctuations. The energy can now be written as $E(\bar{U}_{ij}, \mathrm{e}^{\,\mathrm{i}a^l_{ij}\tau^l})$. To see whether the $SU(2)$ gauge fluctuations gain an energy gap or not, we need to examine whether $E(\bar{U}_{ij}, \mathrm{e}^{\,\mathrm{i}a^l_{ij}\tau^l})$ contains a mass term $(a^l_{ij})^2$, or not, in the small-a^l_{ij} limit.

To understand how the mean-field ansatz $\bar{U}_{ij}$ affects the dynamics of the gauge fluctuations, it is convenient to introduce the following loop variable of the mean-field solution:

$$P(C_i) = \bar{U}_{ij}\bar{U}_{jk}...\bar{U}_{li} \tag{9.2.20}$$

Here $P(C_i)$ is called the $SU(2)$ flux through the loop C_i given by $i \to j \to k \to .. \to l \to i$, with base point i. We will also call it the $SU(2)$-flux operator. The loop variable corresponds to gauge field strength in the continuum limit. Under the gauge transformations, $P(C_i)$ transforms as follows:

$$P(C_i) = W_i P(C_i) W_i \tag{9.2.21}$$

We note that the $SU(2)$ flux has the form $P(C) = \chi^0(C)\tau^0 + \mathrm{i}\chi^l(C)\tau^l$. Thus, when $\chi^l \neq 0$, the $SU(2)$ flux has a sense of direction in the $SU(2)$ space which is indicated by χ^l. From eqn (9.2.21), we see that the local $SU(2)$ gauge transformations rotate the direction of the $SU(2)$ flux. As the direction of the $SU(2)$ flux for loops with different base points can be rotated independently by the local $SU(2)$

gauge transformations, it is meaningless to directly compare the directions of the $SU(2)$ flux for different base points. However, it is quite meaningful to compare the directions of the $SU(2)$ flux for loops with the same base point. We can divide different $SU(2)$ flux configurations into the following three classes, based on the $SU(2)$ fluxes through loops with the *same* base point: a trivial $SU(2)$ flux where all $P(C) \propto \tau^0$; a collinear $SU(2)$ flux where all of the $SU(2)$ fluxes point in the same direction; and a non-collinear $SU(2)$ flux where the $SU(2)$ fluxes for loops with the same base point are in different directions.

First, let us consider an ansatz $\bar{U}_{\boldsymbol{ij}}$, with trivial $SU(2)$ flux for all of the loops. The $SU(2)$ flux is invariant under the $SU(2)$ gauge transformation. We can choose a mean-field ansatz (by performing gauge transformations) such that all $\bar{U}_{\boldsymbol{ij}} \propto \tau^0$. In this case, the gauge invariance of the energy implies that

$$E(\bar{U}_{\boldsymbol{ij}}, \mathrm{e}^{\mathrm{i}a^l_{\boldsymbol{ij}}\tau^l}) = E(\bar{U}_{\boldsymbol{ij}}, \mathrm{e}^{\mathrm{i}\theta^l_{\boldsymbol{i}}\tau^l}\mathrm{e}^{\mathrm{i}a^l_{\boldsymbol{ij}}\tau^l}\mathrm{e}^{-\mathrm{i}\theta^l_{\boldsymbol{j}}\tau^l}). \tag{9.2.22}$$

As a result, none of the mass terms $(a^1_{\boldsymbol{ij}})^2$, $(a^2_{\boldsymbol{ij}})^2$, and $(a^3_{\boldsymbol{ij}})^2$ are allowed in the expansion of E.[58] Thus, the $SU(2)$ gauge fluctuations are gapless and appear at low energies. As the ansatz $\bar{U}_{\boldsymbol{ij}} \propto \tau^0$ is invariant under the global $SU(2)$ gauge transformation, we say that the $SU(2)$ gauge structure is not broken.

Second, let us assume that the $SU(2)$ flux is collinear. This means that the $SU(2)$ fluxes for different loops with the same base point all point in the same direction. However, the $SU(2)$ fluxes for loops with different base points may still point in different directions (even for the collinear $SU(2)$ fluxes). Using the local $SU(2)$ gauge transformation, we can always rotate the $SU(2)$ fluxes for different base points into the same direction, and we can pick this direction to be the τ^3 direction. In this case, all of the $SU(2)$ fluxes have the form $P(C) \propto \chi^0(C) + \mathrm{i}\chi^3(C)\tau^3$. We can choose a mean-field ansatz (by performing $SU(2)$ gauge transformations) such that all of the $\bar{U}_{\boldsymbol{ij}}$ have the form $\mathrm{i}\,\mathrm{e}^{\mathrm{i}\phi_{\boldsymbol{ij}}\tau^3}$. As the ansatz is invariant under the global $U(1)$ gauge transformation $\mathrm{e}^{\mathrm{i}\theta\tau^3}$, but not $\mathrm{e}^{\mathrm{i}\theta\tau^{1,2}}$, we say that the $SU(2)$ gauge structure is broken down to a $U(1)$ gauge structure. The gauge invariance of the energy implies that

$$E(\bar{U}_{\boldsymbol{ij}}, \mathrm{e}^{\mathrm{i}a^l_{\boldsymbol{ij}}\tau^l}) = E(\bar{U}_{\boldsymbol{ij}}, \mathrm{e}^{\mathrm{i}\theta_{\boldsymbol{i}}\tau^3}\mathrm{e}^{\mathrm{i}a^l_{\boldsymbol{ij}}\tau^l}\mathrm{e}^{-\mathrm{i}\theta_{\boldsymbol{j}}\tau^3}). \tag{9.2.23}$$

When $a^{1,2}_{\boldsymbol{ij}} = 0$, the above reduces to

$$E(\bar{U}_{\boldsymbol{ij}}, \mathrm{e}^{\mathrm{i}a^3_{\boldsymbol{ij}}\tau^3}) = E(\bar{U}_{\boldsymbol{ij}}, \mathrm{e}^{\mathrm{i}(a^3_{\boldsymbol{ij}}+\theta_{\boldsymbol{i}}-\theta_{\boldsymbol{j}})\tau^3}). \tag{9.2.24}$$

We find that the mass term $(a^3)^2$ is incompatible with eqn (9.2.24). Thereforc, at least the gauge boson a^3 is massless. How about the a^1 and a^2 gauge bosons? Let

---

[58] Under the gauge transformation $\mathrm{e}^{\mathrm{i}\theta^1_{\boldsymbol{i}}\tau^1}$, $a^1_{\boldsymbol{ij}}$ transforms as $a^1_{\boldsymbol{ij}} = a^1_{\boldsymbol{ij}} + \theta^1_{\boldsymbol{i}} - \theta^1_{\boldsymbol{j}}$. The mass term $(a^1_{\boldsymbol{ij}})^2$ is not invariant under such a transformation and is thus not allowed.

$P_A(\boldsymbol{i})$ be the $SU(2)$ flux through a loop with base point $\boldsymbol{i}$. If we assume that all of the gauge-invariant terms that can appear in the energy function do appear in the energy function, then $E(U_{\boldsymbol{ij}})$ will contain the following gauge-inariant term:

$$E = a\mathrm{Tr}[P_A(\boldsymbol{i})U_{\boldsymbol{i},\boldsymbol{i}+\boldsymbol{x}}P_A(\boldsymbol{i}+\boldsymbol{x})U_{\boldsymbol{i}+\boldsymbol{x},\boldsymbol{i}}] + ... \tag{9.2.25}$$

If we write $U_{\boldsymbol{i},\boldsymbol{i}+\boldsymbol{x}}$ as $\chi \mathrm{e}^{\mathrm{i}\phi_{ij}\tau^3}\mathrm{e}^{\mathrm{i}a^l_x\tau^l}$, use the fact that $U_{\boldsymbol{i},\boldsymbol{i}+\boldsymbol{x}} = -U^\dagger_{\boldsymbol{i}+\boldsymbol{x},\boldsymbol{i}}$ (see eqn (9.2.13)), and expand to order $(a^l_x)^2$, then eqn (9.2.25) becomes

$$E = -\frac{1}{2}a\chi^2\mathrm{Tr}([P_A, a^l_x\tau^l]^2) + ... \tag{9.2.26}$$

We see from eqn (9.2.26) that the mass terms for a^1 and a^2 are generated if $P_A \propto \tau^3$.

To summarize, we find that, if the $SU(2)$ flux is collinear, then the ansatz is invariant under the $U(1)$ rotation $\mathrm{e}^{\mathrm{i}\theta\boldsymbol{n}\cdot\boldsymbol{\tau}}$, where $\boldsymbol{n}$ is the direction of the $SU(2)$ flux. As a result, the $SU(2)$ gauge structure is broken down to a $U(1)$ gauge structure. The corresponding mean-field state will have gapless $U(1)$ gauge fluctuations.

Third, we consider the situation where the $SU(2)$ flux is non-collinear. In the above, we have shown that an $SU(2)$ flux P_A can induce a mass term of the form $\mathrm{Tr}([P_A, a^l_x\tau^l]^2)$. For a non-collinear $SU(2)$ flux configuration, we can have two $SU(2)$ fluxes, P_A and P_B, pointing in different directions. The mass term will also contain a term $\mathrm{Tr}([P_B, a^l_x\tau^l]^2)$. In this case, the mass terms for all of the $SU(2)$ gauge fields, namely $(a^1_{\boldsymbol{ij}})^2$, $(a^2_{\boldsymbol{ij}})^2$, and $(a^3_{\boldsymbol{ij}})^2$, will be generated. All of the $SU(2)$ gauge bosons will gain an energy gap (see Section 3.7.5) and the mean-field state described by the ansatz will be a stable state. As the ansatz is invariant under the Z_2 transformation $W_{\boldsymbol{i}} = -\tau^0$, but not other more general global $SU(2)$ gauge transformations, the non-collinear $SU(2)$ flux breaks the $SU(2)$ gauge structure down to a Z_2 gauge structure. So we may guess that the low-energy effective theory is a Z_2 gauge theory. In Sections 9.2.4 and 9.3, we will study the low-energy properties of states with non-collinear $SU(2)$ fluxes. We will show that the low-energy properties of such states, such as the existence of a Z_2 vortex and ground-state degeneracy, are indeed identical to those of a Z_2 gauge theory. So we will call the mean-field state with non-collinear $SU(2)$ fluxes a mean-field Z_2 state.

In a mean-field Z_2 state, all of the gauge fluctuations are gapped. These fluctuations can only mediate short-range interactions between spinons. The low-energy spinons interact weakly and behave like free spinons. Therefore, including mean-field fluctuations does not qualitatively change the properties of the mean-field state. The mean-field state is stable at low energies.

A stable mean-field spin-liquid state implies the existence of a real physical spin liquid. The physical properties of the stable mean-field state apply to the physical spin liquid. If we believe these two statements, then we can study the properties of a physical spin liquid by studying its corresponding stable mean-field state. As the spinons are not confined in mean-field Z_2 states, the physical spin liquid derived from a mean-field Z_2 state contains neutral spin-1/2 fermions as its excitation. This is a very striking result, after realizing that the spin model is a purely bosonic model.

Problem 9.2.3.
(a) Show that the ansatz (9.2.5) is an $SU(2)$-collinear state.
(b) Find the $SU(2)$ gauge transformation that transforms U_{ij} into the form $\mathrm{e}^{\,\mathrm{i}\phi_{ij}\tau^3}$. (Hint: Try $W_i = (\mathrm{i}\tau^3)^{i_x+i_y}$.)

Problem 9.2.4.
(a) Show that the chiral spin ansatz U_{ij} obtained from eqn (9.1.22) is invariant under the global $SU(2)$ gauge transformation $U_{ij} = WU_{ij}W^\dagger$.
(b) Show that the chiral spin ansatz U_{ij} with $a_0^l = 0$ satisfies the constraint (9.2.8).
(c) Show that the ansatz described by eqn (9.1.22) and the following translationally-invariant ansatz:

$$
\begin{aligned}
u_{i,i+x} &= -\chi\tau^3 - \chi\tau^1, & u_{i,i+y} &= -\chi\tau^3 + \chi\tau^1, \\
u_{i,i+x+y} &= \eta\tau^2, & u_{i,i-x+y} &= -\eta\tau^2, \\
a_0^l &= 0,
\end{aligned}
$$

are gauge equivalent. In terms of the f-fermions (see eqns (9.2.4) and (9.2.7)), the above ansatz describes a $d_{x^2-y^2} + \mathrm{i}d_{xy}$ 'superconducting' state.
As the $SU(2)$ gauge structure is unbroken, the low-energy effective theory for the chiral spin state is actually an $SU(2)$ Chern–Simons theory (of level 1).

9.2.3 Spin liquids from translationally-invariant ansatz

- Uniform RVB state, π-flux state and staggered-flux state.
- The symmetry of projected physical wave functions.
- The low energy $SU(2)$ or $U(1)$ gauge fluctuations.

As an application of the $SU(2)$ projective construction, we will study a spin liquid obtained from a translationally-invariant ansatz

$$U_{i+l,j+l} = U_{ij}, \qquad a_0^l(i) = a_0^l,$$

Such a ansatz will lead to a physical wave function with translational symmetry.[59] First, let us introduce

$$u_{ij} = \frac{3}{8} J_{ij} U_{ij} = u^{\mu}_{ij} \tau^{\mu}$$

where $u^{1,2,3}_{l}$ are real, and u^0_{l} is imaginary. Within zeroth-order mean-field theory, the spinon spectrum is determined by the Hamiltonian (see eqn (9.2.9))

$$H = -\sum_{\langle ij \rangle} \left(\psi^{\dagger}_{i} u_{ij} \psi_{j} + h.c. \right) + \sum_{i} \psi^{\dagger}_{i} a^{l}_{0} \tau^{l} \psi_{i} \tag{9.2.27}$$

In $\boldsymbol{k}$ space, we have

$$H = -\sum_{\boldsymbol{k}} \psi^{\dagger}_{\boldsymbol{k}} (u^{\mu}(\boldsymbol{k}) - a^{\mu}_{0}) \tau^{\mu} \psi_{\boldsymbol{k}}$$

where $\mu = 0, 1, 2,$

$$u^{\mu}(\boldsymbol{k}) = \sum_{\boldsymbol{l}} u^{\mu}_{\boldsymbol{i},\boldsymbol{i}+\boldsymbol{l}} \mathrm{e}^{\,\mathrm{i}\boldsymbol{l}\cdot\boldsymbol{k}},$$

$a^0_0 = 0$, and N is the total number of sites. The fermion spectrum has two branches and is given by

$$E_{\pm}(\boldsymbol{k}) = u^0(\boldsymbol{k}) \pm E_0(\boldsymbol{k})$$

$$E_0(\boldsymbol{k}) = \sqrt{\sum_{l} (u^l(\boldsymbol{k}) - a^l_0)^2} \tag{9.2.28}$$

The constraints can be obtained from $\partial E_{\text{ground}}/\partial a^l_0 = 0$ and have the form

$$N \langle \psi^{\dagger}_{i} \tau^l \psi_{i} \rangle = \sum_{\boldsymbol{k}, E_{-}(\boldsymbol{k})<0} \frac{u^l(\boldsymbol{k}) - a^l_0}{E_0(\boldsymbol{k})} - \sum_{\boldsymbol{k}, E_{+}(\boldsymbol{k})<0} \frac{u^l(\boldsymbol{k}) - a^l_0}{E_0(\boldsymbol{k})} = 0 \tag{9.2.29}$$

which allow us to determine a^l_0, $l = 1, 2, 3$.

Let us try the following simple ansatz[60] (Baskaran *et al.*, 1987; Affleck and Marston, 1988; Kotliar and Liu, 1988):

$$a^l_0 = 0, \qquad u_{\boldsymbol{i},\boldsymbol{i}+\boldsymbol{x}} = \chi \tau^3 + \eta \tau^1, \qquad u_{\boldsymbol{i},\boldsymbol{i}+\boldsymbol{y}} = \chi \tau^3 - \eta \tau^1. \tag{9.2.30}$$

First, we would like to show that the corresponding spin liquid has all of the symmetries.

---

[59] Here we will distinguish between the invariance of an ansatz and the symmetry of an ansatz. We say that an ansatz has translational invariance when the ansatz itself does not change under a translation. We say that an ansatz has translational symmetry when the physical spin wave function obtained from the ansatz has translational symmetry. Due to the gauge structure, a translationally-non-invariant ansatz can have translational symmetry.

[60] The ansatz is actually the d-wave ansatz (9.2.5) introduced earlier.

To study the symmetry of the corresponding physical spin state, we need to obtain its wave function $|\Psi_{\text{phy}}\rangle$. Using the ansatz, we can obtain the mean-field ground state $|\Psi_{\text{mean}}\rangle$ of the mean-field Hamiltonian (9.2.27). The physical wave function $|\Psi_{\text{phy}}\rangle$ is obtained by projecting into the subspace with even numbers of ψ-fermions per site, i.e. $|\Psi_{\text{phy}}\rangle = \mathcal{P}|\Psi_{\text{mean}}\rangle$.

The ansatz is already invariant under the two translations $T_x : \boldsymbol{i} \to \boldsymbol{i} + \boldsymbol{x}$ and $T_y : \boldsymbol{i} \to \boldsymbol{i} + \boldsymbol{y}$. The ansatz is also invariant under the two parity transformation $P_x : x \to -x$ and $P_y : y \to -y$. The $P_{xy} : (x, y) \to (y, x)$ parity transformation changes $u_{\boldsymbol{i},\boldsymbol{i}+\boldsymbol{x}} \to u_{\boldsymbol{i},\boldsymbol{i}+\boldsymbol{y}}$ and $u_{\boldsymbol{i},\boldsymbol{i}+\boldsymbol{y}} \to u_{\boldsymbol{i},\boldsymbol{i}+\boldsymbol{x}}$. The transformed ansatz

$$a_0^l = 0, \qquad u_{\boldsymbol{i},\boldsymbol{i}+\boldsymbol{x}} = \chi\tau^3 - \eta\tau^1, \qquad u_{\boldsymbol{i},\boldsymbol{i}+\boldsymbol{y}} = \chi\tau^3 + \eta\tau^1.$$

is different from the original ansatz. However, note that a uniform gauge transformation $G_{P_{xy}}(\boldsymbol{i}) = \mathrm{i}\tau^3$ changes $u_{\boldsymbol{ij}}$ to $G_{P_{xy}} u_{\boldsymbol{ij}} G_{P_{xy}}^\dagger$. It induces the changes $(\tau^1, \tau^2, \tau^3) \to (-\tau^1, -\tau^2, \tau^3)$. Thus the transformed ansatz is gauge equivelant to the original ansatz through a gauge transformation $G_{P_{xy}}$. In other words, the P_{xy} parity transformation, when followed by a gauge transformation $G_{P_{xy}}$, leaves the ansatz unchanged. As a result, the physical wave function is invariant under the P_{xy} parity transformation. A $90°$ rotation R_{90} is generated by $P_x P_{xy}$. The physical wave function also has $90°$-rotational symmetry. The above ansatz also has time-reversal symmetry, because the time-reversal transformation $u_{\boldsymbol{ij}} \to -u_{\boldsymbol{ij}}$ followed by a gauge transformation $G_T(\boldsymbol{i}) = \mathrm{i}\tau^2$ leaves the ansatz unchanged.

To summarize, the ansatz is invariant under the following combined symmetry and gauge transformations: $G_x T_x$, $G_y T_y$, $G_{P_x} P_x$, $G_{P_y} P_y$, $G_{P_{xy}} P_{xy}$, and $G_T T$. The associated gauge transformations are given by

$$\begin{aligned} G_x &= \tau^0, & G_y &= \tau^0, & G_T &= \mathrm{i}\tau^2, \\ G_{P_x} &= \tau^0, & G_{P_y} &= \tau^0, & G_{P_{xy}} &= \mathrm{i}\tau^3. \end{aligned} \tag{9.2.31}$$

So the spin liquid described by the ansatz (9.2.30) has all of the symmetries. We will call such a state a symmetric spin-liquid state.

We note that the link variables $u_{\boldsymbol{ij}}$ in eqn (9.2.30) point to different directions in the $\tau^{1,2,3}$ space, and thus are not collinear. As a result, the ansatz is not invariant under the uniform $SU(2)$ gauge transformations. One might expect that the $SU(2)$ gauge structure is broken down to a Z_2 gauge structure and that the ansatz (9.2.30) describes a Z_2 spin liquid. In fact, this reasoning is incorrect. As we pointed out in Section 9.2.2, to determine if the $SU(2)$ gauge structure is broken down to a Z_2 gauge structure, we need to check the collinearness of the $SU(2)$ *flux*, not the collinearness of the $SU(2)$ link variables.

When $\chi = \eta$ or when $\eta = 0$, the $SU(2)$ fluxes P_C for all of the loops are trivial, i.e. $P_C \propto \tau^0$. In this case, the $SU(2)$ gauge structure is not broken and the $SU(2)$ gauge fluctuations are gapless (in the weak-coupling limit). The spinon in the spin liquid described by $\eta = 0$ has the spectrum $E_{\boldsymbol{k}} = \pm 2|\chi(\cos(k_x) + \cos(k_y))|$, which has a large Fermi surface. We will call this state the $SU(2)$-gapless state (this state is called the uniform RVB state in the literature). The state with $\chi = \eta$ has gapless spinons only at isolated $\boldsymbol{k}$ points, as indicated by the spinon spectrum $E_{\boldsymbol{k}} = \pm 2\sqrt{\chi^2 + \eta^2}\sqrt{\cos^2(k_x) + \cos^2(k_y)}$ (see Fig. 9.2). We will call such a state an $SU(2)$-linear state to stress the linear dispersion $E \propto |\boldsymbol{k}|$ near the Fermi points. (Such a state is called the π-flux state in the literature and in Section 9.1.1.) The low-energy effective theory for the $SU(2)$-linear state is described by massless Dirac fermions (the spinons) coupled to the $SU(2)$ gauge field (see Problem 9.1.4).

After proper gauge transformations, the $SU(2)$-gapless ansatz can be rewritten as

$$u_{\boldsymbol{i},\boldsymbol{i}+\boldsymbol{x}} = \mathrm{i}\chi, \qquad u_{\boldsymbol{i},\boldsymbol{i}+\boldsymbol{y}} = \mathrm{i}\chi, \tag{9.2.32}$$

and the $SU(2)$-linear ansatz as

$$u_{\boldsymbol{i},\boldsymbol{i}+\boldsymbol{x}} = \mathrm{i}\chi, \qquad u_{\boldsymbol{i},\boldsymbol{i}+\boldsymbol{y}} = \mathrm{i}(-)^{i_x}\chi. \tag{9.2.33}$$

In these forms, the $SU(2)$ gauge structure is explicitly unbroken, because $u_{\boldsymbol{ij}} \propto \mathrm{i}\tau^0$. We can easily see that all of the $SU(2)$ fluxes $P_{\boldsymbol{ij}\ldots\boldsymbol{k}}$ are trivial.[61] The ansatz are also invariant under the uniform $SU(2)$ gauge transformations. Spinons in both the $SU(2)$-gapless and the $SU(2)$-linear ansatz have a finite interaction at low energies. So the two mean-field states are marginal mean-field states. It is not clear if the mean-field states correspond to real physical spin liquids or not.

When $\chi \neq \eta$ and $\chi, \eta \neq 0$, the flux P_C is non-trivial. However, the $SU(2)$ fluxes are collinear. In this case, the $SU(2)$ gauge structure is broken down to a $U(1)$ gauge structure. The gapless spinons still only appear at isolated $\boldsymbol{k}$ points, as one can see from the spinon spectrum $E_{\boldsymbol{k}} = \pm 2\sqrt{\chi^2(\cos k_x + \cos k_y)^2 + \eta^2(\cos k_x - \cos k_y)^2}$. We will call such a state the $U(1)$-linear state. (This state is called the staggered-flux state and/or the d-wave pairing state in the literature.) After a proper gauge transformation, the $U(1)$-linear state can also be described by the ansatz (see Problem 9.2.3)

$$u_{\boldsymbol{i},\boldsymbol{i}+\boldsymbol{x}} = \mathrm{i}\chi - (-)^{\boldsymbol{i}}\eta\tau^3, \qquad u_{\boldsymbol{i},\boldsymbol{i}+\boldsymbol{y}} = \mathrm{i}\chi + (-)^{\boldsymbol{i}}\eta\tau^3, \tag{9.2.34}$$

where the $U(1)$ gauge structure is explicit.[62] The low-energy effective theory is described by massless Dirac fermions (the spinons) coupled to the $U(1)$ gauge field. Again the $U(1)$-linear state is a marginal mean-field state, due to the finite interactions induced by the gapless $U(1)$ gauge field at low energy.

---

[61] Under the projective-symmetry-group classification that will be discussed later, the $SU(2)$-gapless ansatz (9.2.32) is labeled by SU2An0 and the $SU(2)$-linear ansatz (9.2.33) is labeled by SU2Bn0 (see eqn (9.4.33)).

[62] Under the projective-symmetry-group classification, such a state is labeled by U1Cn01n (see eqns (9.4.25)–(9.4.29).

9.2.4 A stable Z_2 spin liquids from translationally-invariant ansatz

- A stable spin-liquid states—Z_2 spin-liquid states.
- The Z_2 spin liquid contains fractionalized excitations—spinons. The spinons carry spin-1/2 and no electric charge, and have fermionic statistics.

Having learnt the recipe for constructing stable mean-field states, in this section, we will study a stable mean-field state that does not break any symmetry. The ansatz has the following form (Wen, 1991a)

$$u_{i,i+x} = u_{i,i+y} = -\chi\tau^3, \qquad a_0^{2,3} = 0, \quad a_0^1 \neq 0,$$
$$u_{i,i+x+y} = \eta\tau^1 + \lambda\tau^2, \qquad u_{i,i-x+y} = \eta\tau^1 - \lambda\tau^2 \tag{9.2.35}$$

To show that eqn (9.2.35) describes a stable mean-field state, we only need to show that it generates non-collinear $SU(2)$ fluxes. Around two triangles with the same base point, we obtain the following $SU(2)$ flux:

$$u_{i,i+x}u_{i+x,i+y}u_{i+y,i} = -(\eta\tau^1 - \lambda\tau^2)\chi^2$$
$$u_{i,i+y}u_{i+y,i-x}u_{i-x,i} = -(\eta\tau^1 + \lambda\tau^2)\chi^2$$

We see that, if $\chi\eta\lambda \neq 0$, then the $SU(2)$ fluxes are non-collinear and the ansatz in eqn (9.2.35) describes a stable mean-field state, or more precisely a Z_2 state.

As the ansatz is invariant under a translation, the physical spin state is also invariant under a translation. Thus, the physical wave function $|\Psi_{\text{phy}}\rangle$ has translational symmetry. Under the P_x parity transformation, the ansatz is changed to

$$u_{i,i+x} = u_{i,i+y} = -\chi\tau^3, \qquad a_0^{2,3} = 0, \quad a_0^1 \neq 0,$$
$$u_{i,i+x+y} = \eta\tau^1 - \lambda\tau^2, \qquad u_{i,i-x+y} = \eta\tau^1 + \lambda\tau^2 \tag{9.2.36}$$

(note that P_x changes $u_{i,i+x}$ to $u_{i,i-x} = u_{i,i+x}^\dagger$ for a translationally-invariant ansatz). The transformed ansatz is gauge equivalent to the original ansatz under an $SU(2)$ gauge transformation $G_x(i) = i\tau^1(-)^i$.[63] Although the P_x-transformed ansatz is different to the original ansatz, both ansatz lead to same physical wave function after projection. The physical spin state has a P_x-parity symmetry.

We also note that the above combined transformation G_xP_x changes (a_0^2, a_0^3) to $(-a_0^2, -a_0^3)$ if they are not zero. As the mean-field Hamiltonian has P_x symmetry, the mean-field ground-state energy for nonzero (a_0^2, a_0^3) has the property

---

[63] We define $(-)^i = (-)^{i_x+i_y}$. The gauge transformation $W_i = i\tau^1$ transforms eqn (9.2.36) to

$$u_{i,i+x} = u_{i,i+y} = \chi\tau^3, \qquad a_0^{2,3} = 0, \quad a_0^1 \neq 0,$$
$$u_{i,i+x+y} = \eta\tau^1 + \lambda\tau^2, \qquad u_{i,i-x+y} = \eta\tau^1 - \lambda\tau^2$$

Then the gauge transformation $W_i = (-)^i$ transforms the above to eqn (9.2.35).

$E_{\text{ground}}(a_0^2, a_0^3) = E_{\text{ground}}(-a_0^2, -a_0^3)$. Therefore, $a_0^2 = a_0^3 = 0$ always satisfy the constraint $\langle \psi_{\boldsymbol{i}}^\dagger \tau^2 \psi_{\boldsymbol{i}} \rangle = N_{\text{site}}^{-1} \partial E_{\text{ground}} / \partial a_0^2 = 0$ and $\langle \psi_{\boldsymbol{i}}^\dagger \tau^3 \psi_{\boldsymbol{i}} \rangle = N_{\text{site}}^{-1} \partial E_{\text{ground}} / \partial a_0^3 = 0$ for our ansatz. We only need to adjust a_0^1 to satisfy $\langle \psi_{\boldsymbol{i}}^\dagger \tau^1 \psi_{\boldsymbol{i}} \rangle = 0$.

The action of the P_y parity transformation is the same as the action of P_x. So the physical wave function also has P_y-parity symmetry. Under the time-reversal transformation, T changes $(u_{\boldsymbol{ij}}, a_0) \to (-u_{\boldsymbol{ij}}, -a_0)$. The time-reversal transformation followed by a gauge transformation G_T will leave the ansatz (9.2.35) unchanged, if we choose $G_T = \mathrm{i}\tau^3(-)^{\boldsymbol{i}}$. Thus, $|\Psi_{\text{phy}}\rangle$ has time-reversal symmetry. The ansatz is invariant under the $(x, y) \to (y, x)$ parity P_{xy}. So the physical wave function has P_{xy}-parity symmetry. We see that the ansatz (9.2.35) is invariant under following the combined transformations: $G_x T_x$, $G_y T_y$, $G_{P_x} P_x$, $G_{P_y} P_y$, $G_{P_{xy}} P_{xy}$, and $G_T T$, with the gauge transformations

$$
\begin{aligned}
G_x &= \tau^0, & G_y &= \tau^0, & G_T &= \mathrm{i}(-)^{\boldsymbol{i}}\tau^3, \\
G_{P_x} &= \tau^0, & G_{P_y} &= \tau^0, & G_{P_{xy}} &= \mathrm{i}\tau^3
\end{aligned}
\tag{9.2.37}
$$

The ansatz describes a spin liquid with all the symmetries. We call such a state a Z_2-symmetric spin-liquid state.

The spinon spectrum in the Z_2 state is given by

$$
\begin{aligned}
E_\pm(\boldsymbol{k}) &= \pm\sqrt{\epsilon_1^2 + \epsilon_2^2 + \epsilon_3^2} \\
\epsilon_1 &= 2\chi(\cos k_x + \cos k_y) \\
\epsilon_2 &= 2\eta[\cos(k_x + k_y) + \cos(k_x - k_y)] + a_0^1 \\
\epsilon_3 &= 2\lambda[\cos(k_x + k_y) - \cos(k_x - k_y)]
\end{aligned}
$$

We find that the spinons are fully gapped. We will call the state (9.2.35) a Z_2-gapped spin liquid.[64] The Z_2-gapped spin liquid corresponds to the short-ranged resonating valence bound (sRVB) state proposed by Kivelson *et al.* (1987) and Rokhsar and Kivelson (1988).

As eqn (9.2.35) is a stable ansatz, the gauge fluctuations are all gapped and only cause a short-range interaction between spinons. As the spinons are gapped, the low-energy excitations of the Z_2-gapped state correspond to dilute gases of spinons. These spinons behave like free fermions due to the short range of the interaction. Therefore, the excitations in the Z_2-gapped state are free neutral spin-1/2 fermions.

[64] Here we would like to mention that, under the projective-symmetry-group classification that will be discussed later in Section 9.4.3, the state (9.2.35) is labeled by the projective symmetry group Z2Axx0z.

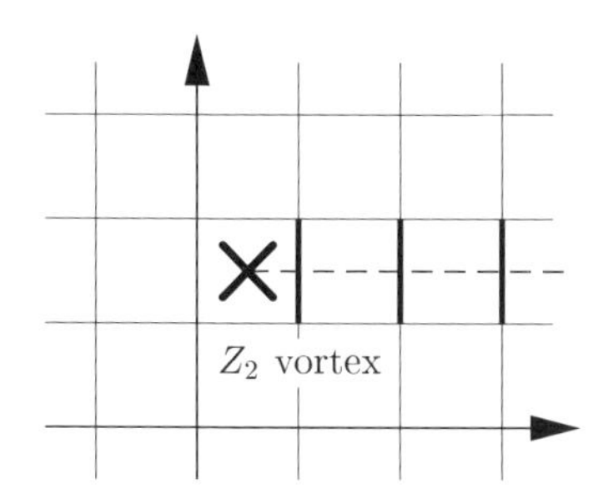

FIG. 9.4. A Z_2 vortex is created by flipping the sign of U_{ij} on the links indicated by the thick lines. Here $\Theta_{ij} = 1$ except on the links indicated by the thick lines, where $\Theta_{ij} = -1$.

9.2.5 A Z_2 vortex in Z_2 spin liquids

- A Z_2 spin-liquid state contains a Z_2 vortex excitation that carries no spin and no electric charge.
- Binding a Z_2 vortex to a spinon changes the spinon statistics from bosonic to fermionic, or from fermionic to bosonic.

In addition to the spinons, there is another type of quasiparticle excitation in the Z_2 spin-liquid state (Wen, 1991a). This excitation appears as a topological soliton in the mean-field theory and corresponds to a π flux of the Z_2 gauge field. So we will call the new excitation a Z_2 vortex. It is created by flipping the signs of u_{ij} along a string of links (see Fig. 9.4). It is described by the following ansatz:

$$\tilde{u}_{ij} = \bar{u}_{ij}\Theta_{ij} \tag{9.2.38}$$

where Θ_{ij} is illustrated in Fig. 9.4. The Z_2 vortex is located at the end of the string. Note that, away from the Z_2 vortex, $\tilde{u}_{ij}$ is locally gauge equivalent to $\bar{u}_{ij}$. Thus, the string of u_{ij} with a flipped sign does not cost any energy and is unobservable. The ansatz $\tilde{u}_{ij}$ actually describes a local excitation at the end of the string.

Since u_{ij} determines the hopping amplitude of the spinons, as a spinon hops around the Z_2 vortex, an additional minus sign will be induced. Therefore, the Z_2 vortex behaves like a π flux to the spinons. According to the discussions in Section 7.1.2, we find that binding a Z_2 vortex to a spinon changes the statistics of the spinon from fermionic to bosonic. Thus, Z_2 spin liquids contain neutral spin-1/2 excitations with both bosonic and fermionic statistics.

Problem 9.2.5.
Show that the $SU(2)$ flux through a loop C for the Z_2-vortex ansatz in eqn (9.2.38) is the same as the $SU(2)$ flux through the same loop for the ground-state ansatz $\bar{u}_{ij}$, except when the loop C encloses the vortex. As the mean-field energy is a function of the $SU(2)$ flux, the above result indicates that the string of the sign-flipped u_{ij}s does not cost energy.

9.2.6 Stable gapless Z_2 spin liquids

- There are many different stable spin-liquid states—Z_2 spin-liquid states. Some of them can have gapless spinon excitations.
- These states have exactly the same symmetry. There is no way to use symmetry to characterize these different spin-liquid states.

In addition to eqn (9.2.35), we can write down another ansatz (Wen, 2002c) for the Z_2-symmetric spin liquid:

$$u_{\boldsymbol{i},\boldsymbol{i}+\boldsymbol{x}} = \mathrm{i}\eta\tau^0 - \chi(\tau^3 - \tau^1), \quad u_{\boldsymbol{i},\boldsymbol{i}+\boldsymbol{y}} = \mathrm{i}\eta\tau^0 - \chi(\tau^3 + \tau^1), \quad a_0^l = 0, \tag{9.2.39}$$

with χ and η nonzero. The above ansatz becomes the $SU(2)$-gapless spin liquid if $\chi = 0$, and the $SU(2)$-linear spin liquid if $\eta = 0$. To show that the ansatz in eqn (9.2.39) describes a symmetric spin liquid, we must show that it is invariant under the following combined symmetry and gauge transformations: G_xT_x, G_yT_y, $G_{P_x}P_x$, $G_{P_y}P_y$, $G_{P_{xy}}P_{xy}$, and G_TT. We find that, if we choose the gauge transformations to be

$$G_x = \tau^0, \qquad G_y = \tau^0, \qquad G_T = (-)^{\boldsymbol{i}}\tau^0,$$
$$G_{P_x} = \mathrm{i}(-)^{i_x}\frac{\tau^1+\tau^3}{\sqrt{2}}, \quad G_{P_y} = \mathrm{i}(-)^{i_y}\frac{\tau^1-\tau^3}{\sqrt{2}}, \quad G_{P_{xy}} = \mathrm{i}\tau^3, \tag{9.2.40}$$

then the symmetry transformations followed by the corresponding gauge transformation will leave the ansatz unchanged.

Using the time-reversal symmetry, we can show that the vanishing a_0^l in our ansatz (9.2.39) do indeed satisfy the constraint (9.2.29). This is because $a_0^l \to -a_0^l$ under the combined time-reversal transformation G_TT. As the ansatz with $a_0^l = 0$ is invariant under G_TT, the mean-field ground-state energy for nonzero a_0^l satisfies $E_{\text{ground}}(a_0^l) = E_{\text{ground}}(-a_0^l)$. Thus, $\partial E_{\text{ground}}/\partial a_0^l = 0$ when $a_0^l = 0$. In fact, any ansatz which only has links between two non-overlapping sub-lattices (i.e. the unfrustrated ansatz) is time-reversal symmetric if $a_0^l = 0$. For such ansatz, including the ansatz (9.2.39), a vanishing a_0^l satisfies the constraint (9.2.29).

The spinon spectrum is given by (see Fig. 9.10(a))

$$E_\pm = 2\eta(\sin(k_x) + \sin(k_y)) \pm 2|\chi|\sqrt{2\cos^2(k_x) + 2\cos^2(k_y)}$$

The spinons have two Fermi points and two small Fermi pockets (for small η). The $SU(2)$ flux is non-trivial. Furthermore, the $SU(2)$ fluxes P_{C_1} and P_{C_2} do not commute, where $C_1 = \boldsymbol{i} \to \boldsymbol{i}+\boldsymbol{x} \to \boldsymbol{i}+\boldsymbol{x}+\boldsymbol{y} \to \boldsymbol{i}+\boldsymbol{y} \to \boldsymbol{i}$ and $C_2 = \boldsymbol{i} \to \boldsymbol{i}+\boldsymbol{y} \to \boldsymbol{i}-\boldsymbol{x}+\boldsymbol{y} \to \boldsymbol{i}-\boldsymbol{x} \to \boldsymbol{i}$ are two loops with the same base point. The non-collinear $SU(2)$ fluxes break the $SU(2)$ gauge structure down to a Z_2 gauge structure. We will call the spin liquid described by eqn (9.2.39) a $Z2$-gapless spin liquid.[65] The low-energy effective theory is described by massless Dirac fermions and fermions with small Fermi surfaces, coupled to the Z_2 gauge field. As there are no gapless gauge bosons, the gauge fluctuations can only mediate short-range interactions between spinons. The short-range interaction is irrelevant at low energies and the spinons are *free* fermions at low energies. Thus, the mean-field Z_2-gapless state is a stable state. The stable mean-field state suggests the existence of the second real physical spin liquid (in addition to the Z_2-gapped spin liquid discussed in Section 9.2.4). The physical properties of such a spin liquid can be obtained from mean-field theory. We find that excitations in the Z_2-gapless spin liquid are described by neutral spin-1/2 fermionic quasiparticles!

The third Z_2-symmetric spin liquid can be obtained from the following frustrated ansatz (Balents *et al.*, 1998; Senthil and Fisher, 2000):

$$\begin{aligned} &a_0^3 \neq 0, \quad a_0^{1,2} = 0, \\ &u_{i,i+x} = \chi\tau^3 + \eta\tau^1, \qquad u_{i,i+y} = \chi\tau^3 - \eta\tau^1, \\ &u_{i,i+x+y} = +\gamma\tau^3, \qquad u_{i,i-x+y} = +\gamma\tau^3. \end{aligned} \tag{9.2.41}$$

The ansatz has translational, rotational, parity, and time-reversal symmetries. It is invariant under the combined transformations $G_{x,y}T_{x,y}$, $G_{P_x,P_y,P_{xy}}P_{x,y,xy}$, and $G_T T$, with

$$\begin{aligned} &G_x = \tau^0, \qquad G_y = \tau^0, \qquad G_T = i\tau^2, \\ &G_{P_x} = \tau^0, \qquad G_{P_y} = \tau^0, \qquad G_{P_{xy}} = i\tau^3. \end{aligned} \tag{9.2.42}$$

When $a_0^3 \neq 0$, $\chi \neq \pm\eta$, and $\chi\eta \neq 0$, we find that $a_0^l\tau^l$ does not commute with the $SU(2)$-flux operators, such as the $SU(2)$ through a square $P(C_{\text{square}})$.[66] Thus, the ansatz breaks the $SU(2)$ gauge structure down to a Z_2 gauge structure. The spinon spectrum is given by (see Fig. 9.8(a))

$$\begin{aligned} E_\pm &= \pm\sqrt{\epsilon^2(\boldsymbol{k}) + \Delta^2(\boldsymbol{k})} \\ \epsilon(\boldsymbol{k}) &= 2\chi(\cos(k_x) + \cos(k_y)) + 2\gamma(\cos(k_x + k_y) + \cos(k_x - k_y)) + a_0^3 \\ \Delta(\boldsymbol{k}) &= 2\eta(\cos(k_x) - \cos(k_y)) + a_0^3 \end{aligned}$$

which is gapless only at the four $\boldsymbol{k}$ points with a linear dispersion. Thus, the spin liquid described by eqn (9.2.41) is a Z_2-linear spin liquid.[67]

---

[65] The Z_2-gapless spin liquid is one of the Z_2 spin liquids classified in Section 9.4.3. Its projective symmetry group is labeled by Z2Ax12(12)n (see Section 9.4.3 and eqn (9.4.16)).

[66] Here $\tau^l a_0^l(\boldsymbol{i})$ can be treated as an $SU(2)$ flux through a loop of zero size.

[67] The Z_2-linear spin liquid is described by the projective symmetry group Z2A0032 or, equivalently, Z2A0013 (see Section 9.4.3).

The Z_2-gapped ansatz (9.2.35), the Z_2-gapless ansatz (9.2.39) and the Z_2-linear ansatz (9.2.41) give rise to three spin liquid states after the projection. The three states have exactly the same symmetry. So we cannot use symmetries to distinguish the three states. To find a new set of quantum numbers to characterize the three different spin liquids, we note that although the three Z_2 spin liquids have the same symmetry, their ansatz are invariant under the symmetry translations followed by *different* gauge transformations (see eqns (9.2.37), (9.2.40) and (9.2.42)). So the invariant groups of the mean-field ansatz for the three spin liquids are different, and we can use such invariant groups to characterize different spin liquids. More detailed and systematic discussion will be given in Section 9.4.

Problem 9.2.6.
Show that the ansatz in eqn (9.2.39) is invariant under the combined transformations $G_x T_x$, $G_y T_y$, $G_{P_x} P_x$, $G_{P_y} P_y$, $G_{P_{xy}} P_{xy}$, and $G_T T$ if the gauge transformations are given by eqn (9.2.40).

Problem 9.2.7.
Show that the chiral spin state has translational and 90°-rotational symmetries. Show that the chiral spin state also has the combined symmetries TP_x, TP_y, and TP_{xy}.

9.2.7 Remarks: the time-reversal transformation in mean-field theory

Under the time-reversal transformation (or the spin-reversal transformation), $\boldsymbol{S}_{\boldsymbol{i}} \to -\boldsymbol{S}_{\boldsymbol{i}}$. As we only consider spin–rotation-invariant states in this chapter, for convenience we will define a T transformation as a combination of a time-reversal and a spin–rotation transformation: $(S^x, S^y, S^z) \to (S^x, -S^y, S^z)$. We will loosely call it a time-reversal transformation.

Such a transformation can be generated by

$$\psi \to \psi' = -\mathrm{i}\tau^2\psi^* \tag{9.2.43}$$

Let us use an operator $\mathcal{T}$ to represent the above transformation as $\mathcal{T}\psi\mathcal{T}^{-1} = -\mathrm{i}\tau^2\psi^*$. Note that $\mathcal{T}$ changes i to $-$i, i.e. $\mathcal{T}^{-1}\mathrm{i}\mathcal{T} = -\mathrm{i}$. In terms of the f-fermion (9.2.6), eqn (9.2.43) becomes

$$f \to \mathcal{T}f\mathcal{T}^{-1} = -\mathrm{i}\tau^2 f^*$$

From $\boldsymbol{S} = \frac{1}{2}f^\dagger\boldsymbol{\sigma}f$, we find that $\mathcal{T}$ changes $\boldsymbol{S}$ to

$$\mathcal{T}\boldsymbol{S}\mathcal{T}^{-1} = -\frac{1}{2}f^\top\boldsymbol{\sigma}f^* = \frac{1}{2}f^\dagger\boldsymbol{\sigma}^\top f = (S^x, -S^y, S^z)$$

In mean-field theory, eqn (9.2.43) induces the following transformation to the mean-field ansatz:

$$\begin{aligned} U_{\boldsymbol{ij}} &\to U'_{\boldsymbol{ij}} = (-\mathrm{i}\tau^2)U^*_{\boldsymbol{ij}}(\mathrm{i}\tau^2) = -U_{\boldsymbol{ij}} \\ a_0^l(\boldsymbol{i})\tau^l &\to a_0'^l(\boldsymbol{i})\tau^l = (-\mathrm{i}\tau^2)(a_0^l(\boldsymbol{i})\tau^l)^\top(\mathrm{i}\tau^2) = -a_0^l(\boldsymbol{i})\tau^l \end{aligned} \tag{9.2.44}$$

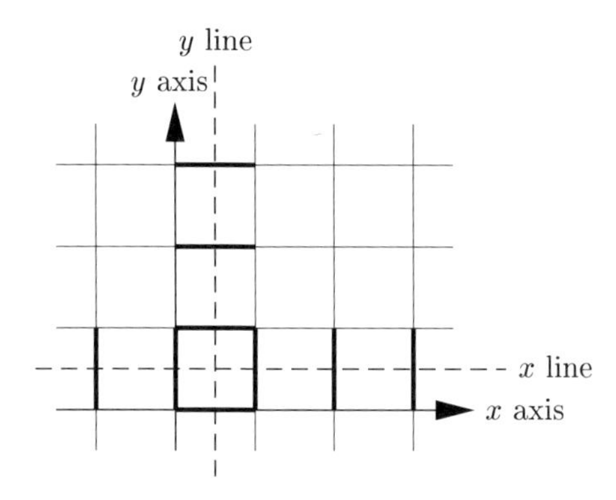

FIG. 9.5. The links crossing the x line and the y line get an additional minus sign.

The above describes how the mean-field ansatz transforms under the T transformation. Both the spin Hamiltonian (9.1.1) and the mean-field Hamiltonian (9.2.4) are invariant under the time-reversal transformation.

Problem 9.2.8.
Find out how U_{ij} and $a_0^l(\boldsymbol{i})$ transform under the real time-reversal transformation $\boldsymbol{S} \to -\boldsymbol{S}$.

9.3 Topological orders in gapped spin-liquid states

- Topological degeneracy of spin-liquid ground states.
- The existence of robust topological degenerate ground states implies the existence of topological order.

We have constructed the Z_2-gapped state and the chiral spin state using mean-field theory. Both of the spin-liquid states have a finite energy gap in all of their excitations and respect the translational symmetry. We would like to point out that the chiral spin liquid and the Z_2-gapped spin liquid studied here contain new types of orders which cannot be characterized by the symmetries and by the order parameters associated with broken symmetries. This is quite obvious for the Z_2-gapped state because it does not break any symmetry. The chiral spin liquid spontaneously breaks the time-reversal and parity symmetries, and has a corresponding local order parameter to describe such a symmetry breaking. As we will see later, the chiral spin liquid contains additional structures that cannot be described by symmetry breaking. As both the chiral spin liquid and the Z_2-gapped spin liquid have a finite energy gap, the new type of orders are topological orders.

As pointed out in Section 8.2.1, one way to show non-trivial topological orders in the Z_2-gapped state and in the chiral spin state is to show that the two states have topologically-degenerate ground states. Let us first consider the Z_2-gapped state and its ground-state degeneracy.

We put the Z_2-gapped state on a torus, with even numbers of sites in both the x and y directions. We consider the following four ansatz constructed from a mean-field solution $\bar{u}_{\boldsymbol{ij}}$:

$$u_{\boldsymbol{ij}}^{(m,n)} = (-)^{ms_x(\boldsymbol{ij})}(-)^{ns_y(\boldsymbol{ij})}\bar{u}_{\boldsymbol{ij}} \tag{9.3.1}$$

where $m, n = 0, 1$. Here $s_{x,y}(\boldsymbol{ij})$ have values 0 or 1, with $s_x(\boldsymbol{ij}) = 1$ if the link $\boldsymbol{ij}$ crosses the x line (see Fig. 9.5) and $s_x(\boldsymbol{ij}) = 0$ otherwise. Similarly, $s_y(\boldsymbol{ij}) = 1$ if the link $\boldsymbol{ij}$ crosses the y line and $s_y(\boldsymbol{ij}) = 0$ otherwise.

We note that $u_{\boldsymbol{ij}}^{(m,n)}$ with different m and n are *locally* gauge equivalent because, on an infinite system, the change $u_{\boldsymbol{ij}} \to (-)^{ms_x(\boldsymbol{ij})}(-)^{ms_y(\boldsymbol{ij})}u_{\boldsymbol{ij}}$ can be generated by an $SU(2)$ gauge transformation $u_{\boldsymbol{ij}} \to W_{\boldsymbol{i}} u_{\boldsymbol{ij}} W_{\boldsymbol{j}}^\dagger$, where $W_{\boldsymbol{i}} = (-)^{m\Theta(i_x)}(-)^{n\Theta(i_y)}$, and $\Theta(n) = 1$ if $n > 0$ and $\Theta(n) = 0$ if $n \leqslant 0$. As a result, the different ansatz $u_{\boldsymbol{ij}}^{(m,n)}$ have the same $SU(2)$ flux through each plaquette.

For a large system, the energy is a local function of $u_{\boldsymbol{ij}}$, and satisfies $F(u_{\boldsymbol{ij}}) = F(\tilde{u}_{\boldsymbol{ij}})$ if $u_{\boldsymbol{ij}}$ and $\tilde{u}_{\boldsymbol{ij}}$ are gauge equivalent. Therefore, the energies for different $u_{\boldsymbol{ij}}^{(m,n)}$s are the same (in the thermodynamic limit). On the other hand, $u_{\boldsymbol{ij}}^{(m,n)}$ with different m and n are not gauge equivalent in the global sense. A spinon propagating all the way around the torus in the x (or y) direction obtains a phase $\mathrm{e}^{\,\mathrm{i}m\pi}$ (or $\mathrm{e}^{\,\mathrm{i}n\pi}$). Therefore, $u_{\boldsymbol{ij}}^{(mn)}|_{m,n=0,1}$ describe different orthogonal degenerate ground states (see also Problem 9.3.1). In this way, we find that the Z_2-gapped state has four degenerate ground states on a torus (Read and Chakraborty, 1989; Wen, 1991a). From the phase that the spinon obtained by going around the torus, we see that m and n label the number of π fluxes going through the two holes on the torus (see Fig. 9.6). The four degenerate ground states correspond to different ways of threading π flux through the two holes. Such a picture can be generalized to a higher-genus Riemann surface. The degenerate ground states can be constructed by having zero or one unit of the π flux through the $2g$ holes on the genus g Riemann surface (see Fig. 9.6). This leads to 2^{2g} degenerate ground states on a genus g Riemann surface. This pattern of degeneracies on different Riemann surfaces is characteristic of the Z_2 gauge theory. The degeneracies and the presence of the Z_2-vortex excitation indicate, in a physical way, that the low-energy effective theory of the Z_2-gapped state is indeed a Z_2 gauge theory (see Problem 9.3.1). After we include the spinons, the first-order mean-field theory for the Z_2-gapped state describes fermions coupled to the Z_2 gauge theory. The physical properties of the Z_2-gapped state can be obtained from such an effective theory.

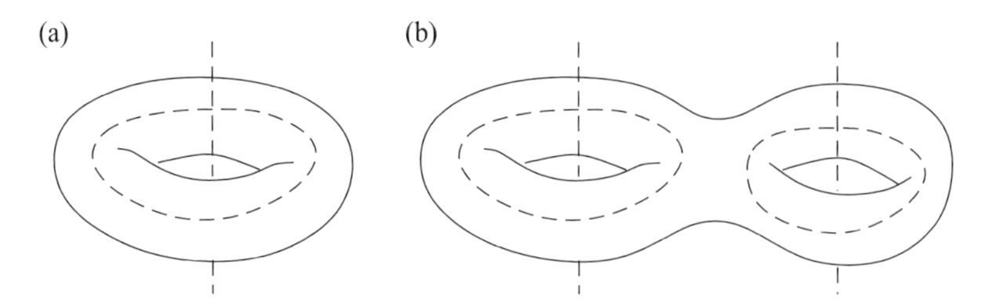

FIG. 9.6. There are (a) $2g = 2$ holes in a $g = 1$ Riemann surface (a torus), and (b) $2g = 4$ holes in a $g = 2$ Riemann surface. We can thread zero flux or π flux through these holes. This leads to $2^{2g} = 4$ or 16 degenerate ground states for the Z_2-gapped spin liquid.

On a finite compactified lattice, the only way for the system to tunnel from one ground state to another is through the following tunneling process. At first, a pair of the Z_2 vortices is created. One of the Z_2 vortices propagates all the way around the torus and then annihilates the other Z_2 vortex. Such a process effectively adds a unit of the π flux to the hole of the torus and changes m or n by 1. The different ground states cannot tunnel into each other through any local fluctuations because of the flux conservation. As a direct consequence of this result, the energy split between different ground states on a finite lattice is expected to be of order $\mathrm{e}^{-L/\xi}$, where $1/\xi$ is related to the finite energy gap of the Z_2 vortex.

In mean-field theory, the degeneracy of ground states is a consequence of the gauge invariance. The gauge invariance remains exact, even after we include the following arbitrary perturbation to the original spin Hamiltonian:

$$\delta H = \sum \delta J_{ij} \boldsymbol{S}_i \cdot \boldsymbol{S}_j + ... \tag{9.3.2}$$

where δH may break translational symmetry, rotational symmetry, etc. Thus, the above arguments are still valid and the mean-field ground states remain four-fold degenerate even for the modified Hamiltonian. We expect this result to be valid even beyond mean-field theory. The ground-state degeneracy of the Z_2-gapped state cannot be changed by any perturbations, as long as the perturbations are weak enough not to close the energy gap of the Z_2 vortex. Therefore, the ground-state degeneracy is a universal quantum number that characterizes different phases of the spin-liquid states. This also shows that a non-trivial topological order exists in the Z_2-gapped state.

Now let us consider the topological orders in the chiral spin state. Again, we would like to calculate the ground-state degeneracy of the chiral spin state on, say, a torus. The easiest way to calculate the ground-state degeneracy is to use the effective Chern–Simons theory (9.1.28), which describes the dynamics of an a_μ collective mode. The ground-state degeneracy of $U(1)$ Chern–Simons theory has been discussed for FQH states. Repeating the calculation in Section 8.2.1, we find that the ground-state degeneracy is 2. We would like to point out that the two-fold degeneracy is for one sector of the T- and P-breaking ground states, say the sector with $\langle \boldsymbol{S}_1 \cdot (\boldsymbol{S}_2 \times \boldsymbol{S}_3)\rangle > 0$. The other sector with $\langle \boldsymbol{S}_1 \cdot (\boldsymbol{S}_2 \times \boldsymbol{S}_3)\rangle < 0$ also has two degenerate ground states. Thus, the total ground-state degeneracy is $4 = 2 \times 2$. One factor of 2 comes from the T and P breaking, and the other factor of 2 comes from the gauge fluctuations.

The two-fold ground-state degeneracy arising from the gauge fluctuations is due to the non-trivial topological orders in the chiral spin state. This degeneracy depends on the topology of the space. On a genus g Riemann surface, the gauge fluctuations give rise to 2^g degenerate ground states. The degeneracy is again directly related to the gauge structure and can be shown to be robust against arbitrary local perturbations of the spin Hamiltonian. Just like the FQH state, the ground-state degeneracy of the chiral spin state is closely related to the fractional statistics of the spinons. In general, one can show that, if the spinons have fractional statistics $\theta = \frac{p\pi}{q}$, then the chiral spin state will have q^g-fold degenerate ground states on a genus g Riemann surface (Wen, 1990). Also, as a result of non-trivial topological order, the chiral spin state has gapless chiral edge excitations (Wen, 1992).

To see the usefulness of the concept of topological order, let us consider the following physical question: what is the difference between the chiral spin state and the Z_2-gapped state? One may immediately say that the two states have different symmetries. However, if we modify our Hamiltonians to break the T and P symmetries, then the two states will have the same symmetries. In this case, we can still ask whether the two states are the same or not in the sense of whether one state can be continuously deformed into the other without phase transitions. When the T and P symmetries are explicitly broken, the chiral spin state has two degenerate ground states, while the Z_2-gapped state still has four degenerate ground states on the torus. Therefore, the chiral spin state and the Z_2-gapped state are different, even when they have the same symmetries. The two states differ by having different ground-state degeneracies and different topological orders.

Problem 9.3.1.

Z_2 gauge structure in states with non-collinear $SU(2)$ fluxes Let $u_{ij} = s_{ij}\bar{u}_{ij}$, where $s_{ij} = \pm 1$ and $\bar{u}_{ij}$ is an ansatz with non-collinear $SU(2)$ fluxes. Let $|\{s_{ij}\}\rangle$ be the projection of $|\Psi_{\text{mean}}^{(u_{ij})}\rangle$ (see eqn (9.2.11)), namely

$$|\{s_{ij}\}\rangle = \langle 0_\psi| \prod_n \psi_{1,i_n}\psi_{2,i_n} |\Psi_{\text{mean}}^{(u_{ij})}\rangle.$$

(a) Show that, like the states in Z_2 gauge theory, $|\{s_{ij}\}\rangle = |\{\tilde{s}_{ij}\}\rangle$ if s_{ij} and $\tilde{s}_{ij}$ are related by a Z_2 gauge transformation.

(b) Show that the four $u_{ij}^{(m,n)}$ defined in eqn (9.3.1) (with $\bar{u}_{ij}$ being a generic ansatz with non-collinear $SU(2)$ fluxes) are not $SU(2)$ equivalent. (Hint: Consider the $SU(2)$ fluxes $P(C_1)$, $P(C_2)$, and $P(C_3)$ through the three loops $C_{1,2,3}$ with the same base point. Here

C_1 and C_2 are two small loops such that $[P(C_1), P(C_2)] \neq 0$, and C_3 is a big loop around the torus.)

9.4 Quantum orders in symmetric spin liquids

- Quantum orders are non-symmetry-breaking orders in quantum phases.
- Quantum orders apply to both gapped quantum states and gapless quantum states.

In this section, we will develop a theory of quantum order based on the invariant group of the mean-field ansatz. We will call the invariant group projective symmetry group. Our theory allows us to characterize and classify hundreds of spin-liquid states that all have the *same* symmetry. In contrast to the ground state degeneracy of topogogical order, the projective symmetry group can characterize both gapped spin liquids and gapless spin liquids.

We have seen that there can be many different spin liquids with the *same* symmetry. Many of these spin liquids occupy a finite region in phase space and represent stable quantum phases. So here we are facing a similar situation to that in the quantum Hall effect, namely that there are many different quantum phases that cannot be distinguished by symmetries and order parameters. The quantum Hall liquids have finite energy gaps. We can use ground state degeneracy or more generally the topological order to describe the internal order of these distinct states. Here we can still] use topological order to describe the internal orders of gapped spin liquids. However, we also have many other stable quantum spin liquids that have gapless excitations.

To describe internal orders in gapless quantum spin liquids (as well as gapped spin liquids), in Chapter 8, we have introduced a new concept—quantum order—that describes the non-symmetry-breaking orders in quantum phases (see Fig. 8.7). The key point in introducing quantum orders is that quantum phases, in general, cannot be completely characterized by broken symmetries and local order parameters. This point is illustrated by quantum Hall states and by the stable spin liquids constructed in the last few sections. However, to make the concept of quantum order useful, we need to find concrete mathematical characterizations of the quantum orders. As quantum orders are not described by symmetries and order parameters, we need to find a completely new way to characterize them. Wen (2002c) proposed the use of the projective symmetry group to characterize quantum (or topological) orders in quantum spin liquids. The projective symmetry group is motivated by the following observation from the section 9.2.6: the ansatz for different symmetric spin liquids are invariant under the same symmetry transformations followed by *different* gauge transformations. We can use these different gauge transformations to distinguish between different spin liquids with the same symmetry. In the next two sections, we will introduce the projective symmetry group in a general and formal setting.

9.4.1 Quantum orders and universal properties

- To define a new type of order is to find a new kind of universal property.
- (A class of) quantum orders can be described by projective symmetry groups.
- A projective symmetry group is the invariant group (or the 'symmetry' group) of a mean-field ansatz.

We know that to define a quantum phase is to find the universal properties of the ground state wave functions. We can use the universal properties to group ground state wave functions into universality classes, such that the wave functions in each class have the same universal properties. These universality classes will correspond to quantum phases.

However, it is very difficult to find universal properties of many-body wave functions. Let us consider the quantum orders (or the universal classes) in free fermion systems to gain some intuitive understanding of the difficulty. We know that a free fermion ground state is described by an anti-symmetric wave function of N variables. The anti-symmetric function has the form of a Slater determinant, namely $\Psi(x_1, ..., x_N) = \det(M)$, where the matrix elements of M are given by $M_{mn} = \psi_n(x_m)$ and ψ_n are single-fermion wave functions. The first step to finding quantum orders in free fermion systems is to find a reasonable way to group the Slater-determinant wave functions into classes. This is very difficult to do if we only know the real-space many-body function $\Psi(x_1, ..., x_N)$. However, if we use the Fourier transformation to transform the real-space wave function to a momentum-space wave function, then we can group different wave functions into classes according to their Fermi surface topologies. To really prove that the above classification is reasonable and corresponds to different quantum phases, we need to show that, as a ground-state wave function changes from one class to another, the ground-state energy always has a singularity at the transition point. This leads to our understanding of quantum orders in free fermion systems (see Section 8.3.2). The Fermi surface topology is a universal property that allows us to classify different fermion wave functions and characterize different quantum phases of free fermions. Here we would like to stress that, without the Fourier transformation, it is very difficult to see Fermi surface topologies from the real-space many-body function $\Psi(x_1, ..., x_N)$.

To understand the quantum orders in spin liquids, we need to find a way to group the spin-liquid wave functions $\Psi_{\text{spin}}(\boldsymbol{i}_1, ..., \boldsymbol{i}_{N_\uparrow})$ into classes, where $\boldsymbol{i}_m$ label the positions of up-spins. What is missing here is the corresponding 'Fourier' transformation. Just like the topology of a Fermi surface, it is very difficult to see universal properties (if any) directly from the real-space wave function. At the moment, we do not know how to classify the spin-liquid wave functions Ψ_{spin}.

However, we do not want to just give up. Motivated by the projective construction, here we consider a simpler problem to get things started. We limit ourselves to a subclass of many-body wave functions that can be obtained from the ansatz $(u_{ij}, a_0^l \tau^l)$ via eqn (9.2.11). Instead of looking for the universal properties of generic many-body wave functions, we try to find universal properties of the many-body wave functions in the subclass. As the many-body wave functions in the subclass are labeled by the ansatz $(u_{ij}, a_0^l \tau^l)$, the universal properties of the wave functions are actually the universal properties of the ansatz, which greatly simplifies the problem. Certainly, one may object that the universal properties of the ansatz (or the subclass of wave functions) may not be the universal properties of the spin quantum phase. This is indeed the case for some ansatz. However, if the mean-field state described by the ansatz $(u_{ij}, a_0^l \tau^l)$ is stable against any fluctuations (i.e. no fluctuations around the mean-field state can cause any infra-red divergence), then the mean-field state faithfully describes a spin quantum state and the universal properties of the ansatz will be the universal properties of the corresponding spin quantum phase. This completes the link between the properties of the ansatz and the properties of physical spin liquids.

So what should be the universal property of an ansatz? Motivated by Landau's theory for symmetry-breaking orders, here we would like to propose that the invariance group (or the 'symmetry' group) of an ansatz is a universal property of the ansatz. Such a group will be called the projective symmetry group (PSG). We will argue later that certain PSGs are indeed universal properties of quantum phases, and these PSGs can be used to characterize quantum orders in quantum spin liquids.

9.4.2 Projective symmetry groups

- A projective symmetry group is an extension of a symmetry group.
- Projective symmetry groups classify all of the mean-field phases.
- A projective symmetry group characterizes the quantum order in a physical spin-liquid state only when the corresponding mean-field state is stable.

Let us give a detailed definition of a PSG. A PSG is a property of an ansatz. It is formed by all of the transformations that keep the ansatz unchanged. Each transformation (or each element in the PSG) can be written as a combination of a symmetry transformation U (such as a translation) and a gauge transformation

G_U. The invariance of the ansatz under its PSG can be expressed as follows:

$$G_U U(u_{ij}) = u_{ij}, \tag{9.4.1}$$
$$U(u_{ij}) \equiv u_{U(i),U(j)}, \quad G_U(u_{ij}) \equiv G_U(i) u_{ij} G_U^\dagger(j), \quad G_U(i) \in SU(2),$$

for each $G_U U \in PSG$.

Every PSG contains a special subgroup, which will be called the invariant gauge group (IGG). The IGG (denoted by $\mathcal{G}$) for an ansatz is formed by all of the pure gauge transformations that leave the ansatz unchanged:

$$\mathcal{G} = \{W_i | W_i u_{ij} W_j^\dagger = u_{ij}, W_i \in SU(2)\} \tag{9.4.2}$$

If we want to relate the IGG to a symmetry transformation, then the associated transformation is simply the identity symmetry transformation.

If the IGG is non-trivial, then, for a fixed symmetry transformation U, there can be many gauge transformations G_U such that $G_U U$ will leave the ansatz unchanged. If $G_U U$ is in the PSG of u_{ij}, then $G G_U U$ will also be in the PSG if and only if $G \in \mathcal{G}$. Thus, for each symmetry transformation U, the different choices of G_U have a one-to-one correspondence with the elements in the IGG. From the above definition, we see that the PSG, the IGG, and the symmetry group (SG) of an ansatz are related as follows:

$$SG = PSG/IGG$$

This relation tells us that a PSG is a projective representation or an extension of the symmetry group.[68]

Certainly, the PSGs for two gauge-equivalent ansatz u_{ij} and $W(i)u_{ij}W^\dagger(j)$ are related. From $WG_U U(u_{ij}) = W(u_{ij})$, where $W(u_{ij}) = W(i)u_{ij}W^\dagger(j)$, we find that

$$WG_U U W^{-1} W(u_{ij}) = WG_U W_U^{-1} U W(u_{ij}) = W(u_{ij})$$

where $W_U \equiv UWU^{-1}$ is given by $W_U(i) = W(U(i))$. Thus, if $G_U U$ is in the PSG of the ansatz u_{ij}, then $(WG_U W_U)U$ is in the PSG of the gauge-transformed ansatz $W(i)u_{ij}W^\dagger(j)$. We see that the gauge transformation G_U associated with the symmetry transformation U is changed in the following way:

$$G_U(i) \to W(i) G_U(i) W^\dagger(U(i)) \tag{9.4.3}$$

after the gauge transformation $W(i)$.

As the PSG is a property of an ansatz, we can group all of the ansatz sharing the same PSG together to form a class. We claim that such a class is a universality

[68] More generally, we say that a group PSG is an extension of a group SG if the group PSG contains a normal subgroup IGG such that PSG/IGG = SG.

class that corresponds to a quantum phase.[69] It is in this sense that we say that quantum orders are characterized by PSGs.

It is instructive to compare the PSG characterization of the quantum orders with the symmetry characterization of the symmetry-breaking orders. We know that a symmetry-breaking order can be described by its symmetry properties. Mathematically, we say that a symmetry-breaking order is characterized by its symmetry group. Similarly, the quantum orders are also characterized by groups. The difference is that the quantum orders are characterized by PSGs—the extensions of the symmetry group. We see that using the projective symmetry group to describe a quantum order is conceptually similar to using the symmetry group to describe a symmetry-breaking order. We also see that quantum states with the same symmetry can have many different quantum orders, because a symmetry group can usually have many different extensions.

In addition to characterizing symmetry-breaking orders, the symmetry description of a symmetry-breaking order is also very useful because it allows us to obtain many universal properties, such as the number of gapless Nambu–Goldstone modes, without knowing the details of the system. Similarly, knowing the PSG of a quantum order also allows us to obtain the low-energy properties of a quantum system without knowing its details. In Section 9.10, we will show that the PSG can produce and protect low-energy gauge fluctuations. In fact, the gauge group of the low-energy gauge fluctuations is simply the IGG of the ansatz. This generalizes the results obtained in Section 9.2.2. In addition to gapless gauge bosons, the PSG can also produce and protect gapless fermionic excitations and their crystal momenta. These gapless fermions can even be produced in a purely bosonic model. We see that gapless gauge bosons and gapless fermions can have a unified origin—the quantum orders.

Here we would like to stress that PSGs really classify different mean-field phases. A continuous mean-field phase transition is always associated with a change of PSG. Thus, strictly speaking, the above discussions about PSGs and their physical implications only apply to mean-field theory. However, some mean-field states are stable against fluctuations. Then the PSGs for these stable mean-field states can be applied to the corresponding physical spin liquids, and they describe the real quantum orders in those spin liquids. On the other hand, there are mean-field states that are unstable against the fluctuations. In this case, the corresponding PSGs may not describe any real quantum orders.

[69] More precisely, such a class is formed by one or several universality classes that correspond to quantum phases. (A more detailed discussion of this point is given in Section 9.9.)

9.4.3 Classification of symmetric Z_2 spin liquids

- We have classified all of the mean-field ansatz whose PSG is a Z_2 extension of the lattice symmetry group (SG), i.e. $PSG/Z_2 = SG$.
- There are over 100 symmetric Z_2 spin liquids.
- The concepts of an invariant PSG and an algebraic PSG.

As an application of the PSG theory of quantum orders in spin liquids, we would like to classify the PSGs associated with translational transformations, where, for simplicity, we restrict the IGG $\mathcal{G}$ to be Z_2. That is, we want to find all of the Z_2 extensions of the translation group.

When $\mathcal{G} = Z_2$, it contains only two elements—the gauge transformations G_1 and G_2:

$$\mathcal{G} = \{G_1, G_2\}, \qquad G_1(\boldsymbol{i}) = \tau^0, \quad G_2(\boldsymbol{i}) = -\tau^0.$$

In order for the IGG of an ansatz to be $\mathcal{G} = Z_2$, the ansatz must generate non-collinear $SU(2)$ fluxes. Otherwise, the IGG will be larger than Z_2. The non-collinear $SU(2)$ fluxes in an ansatz break the $SU(2)$ gauge structure down to a Z_2 gauge structure. The corresponding spin liquid is a Z_2 spin liquid. We see that the Z_2 extensions of the translation group physically classify all of the Z_2 spin liquids that have only translational symmetry.

Consider a Z_2 spin liquid with translational symmetry. Its PSG, as a Z_2 extension of the translation group, is generated by the four elements $\pm G_x T_x$ and $\pm G_y T_x$, where

$$T_x(u_{\boldsymbol{ij}}) = u_{\boldsymbol{i}-\boldsymbol{x},\boldsymbol{j}-\boldsymbol{x}}, \qquad T_y(u_{\boldsymbol{ij}}) = u_{\boldsymbol{i}-\boldsymbol{y},\boldsymbol{j}-\boldsymbol{y}}.$$

Due to the translational symmetry of the ansatz, we can choose a gauge in which all of the $SU(2)$-flux operators of the ansatz are translationally invariant. That is, $P_{C_1} = P_{C_2}$ if the two loops C_1 and C_2 are related by a translation. We will call such a gauge a uniform gauge.

Under the transformation $G_x T_x$, an $SU(2)$-flux operator P_C based at $\boldsymbol{i}$ transforms as $P_C \to G_x(\boldsymbol{i}')P_{T_x C}G_x^\dagger(\boldsymbol{i}') = G_x(\boldsymbol{i}')P_C G_x^\dagger(\boldsymbol{i}')$, where $\boldsymbol{i}' = T_x\boldsymbol{i}$ is the base point of the translated loop $T_x(C)$. We see that translational invariance of P_C in the uniform gauge requires that $G_x(\boldsymbol{i}')P_C G_x^\dagger(\boldsymbol{i}') = P_C$ for any loop C. Also, G_y satisfies a similar condition. As different $SU(2)$-flux operators based at the same base point do not commute, $G_{x,y}(\boldsymbol{i})$ at each site can only take one of the two values $\pm\tau^0$.

We note that a site-dependent gauge transformation of the form $W(\boldsymbol{i}) = \pm\tau^0$ does not change the translationally-invariant property of the $SU(2)$-flux operators. Thus, we can use such gauge transformations to further simplify $G_{x,y}$ through eqn (9.4.3). First, we can choose a gauge to make (see Problem 9.4.1)

$$G_y(\boldsymbol{i}) = \tau^0. \tag{9.4.4}$$

We note that a gauge transformation satisfying $W(\boldsymbol{i}) = W(i_x)$ does not change the condition $G_y(\boldsymbol{i}) = \tau^0$. We can use such kinds of gauge transformation to make

$$G_x(i_x, i_y = 0) = \tau^0. \tag{9.4.5}$$

As the translations in the x and y directions commute, $G_{x,y}$ must satisfy (for any ansatz, Z_2 or not Z_2)

$$G_x T_x G_y T_y (G_x T_x)^{-1} (G_y T_y)^{-1} = G_x T_x G_y T_y T_x^{-1} G_x^{-1} T_y^{-1} G_y^{-1} \in \mathcal{G}. \tag{9.4.6}$$

This means that

$$G_x(\boldsymbol{i}) G_y(\boldsymbol{i} - \boldsymbol{x}) G_x^{-1}(\boldsymbol{i} - \boldsymbol{y}) G_y(\boldsymbol{i})^{-1} \in \mathcal{G} \tag{9.4.7}$$

For Z_2 spin liquids and due to eqn (9.4.4), eqn (9.4.7) reduces to

$$G_x(\boldsymbol{i}) G_x^{-1}(\boldsymbol{i} - \boldsymbol{y}) = +\tau^0 \tag{9.4.8}$$

or

$$G_x(\boldsymbol{i}) G_x^{-1}(\boldsymbol{i} - \boldsymbol{y}) = -\tau^0 \tag{9.4.9}$$

When combined with eqns (9.4.5) and (9.4.4), we find that eqns (9.4.8) and (9.4.9) become

$$G_x(\boldsymbol{i}) = \tau^0, \qquad G_y(\boldsymbol{i}) = \tau^0 \tag{9.4.10}$$

and

$$G_x(\boldsymbol{i}) = (-)^{i_y} \tau^0, \qquad G_y(\boldsymbol{i}) = \tau^0 \tag{9.4.11}$$

Thus, there are only two gauge-inequivalent Z_2 extensions of the translation group. The two PSGs are generated by $G_x T_x$ and $G_y T_y$, with $G_{x,y}$ given by eqns (9.4.10) and (9.4.11).[70] Under the PSG classification, there are only two types of Z_2 spin liquids which have *only* translational symmetry and no other symmetries. The ansatz that is invariant under the PSG (9.4.10) has the form

$$u_{\boldsymbol{i},\boldsymbol{i}+\boldsymbol{m}} = u_{\boldsymbol{m}} \tag{9.4.12}$$

and the one that is invariant under the PSG (9.4.11) has the form

$$u_{\boldsymbol{i},\boldsymbol{i}+\boldsymbol{m}} = (-)^{m_y i_x} u_{\boldsymbol{m}} \tag{9.4.13}$$

Through the above example, we see that the PSG is a very powerful tool. It can lead to a complete classification of (mean-field) spin liquids with prescribed symmetries and low-energy gauge structures.

[70] We would like to remark that $G_x T_x$ and $G_y T_y$ satisfy the usual translational algebra $G_x T_x G_y T_y = G_y T_y G_x T_x$ if G_x and G_y are given by eqn (9.4.8). When G_x and G_y are given by eqn (9.4.9), $G_x T_x$ and $G_y T_y$ satisfy the magnetic translational algebra $G_x T_x G_y T_y = -G_y T_y G_x T_x$.

In the above, we have studied Z_2 spin liquids which have *only* translational symmetry and no other symmetries. We found that there are only two types of such spin liquids. However, if spin liquids have more symmetries, then there can be many more types. Wen (2002c) obtained a classification of symmetric Z_2 spin liquids using the PSG. The classification was obtained by noticing that the gauge transformations $G_{x,y}$, $G_{P_x,P_y,P_{xy}}$, and G_T must satisfy certain algebraic relations similar to eqn (9.4.7). Solving these algebraic relations and factoring out gauge-equivalent solutions, we find that there are 272 different Z_2 extensions of the symmetry group generated by the translational, parity, and time-reversal transformations $\{T_{x,y}, P_{x,y,xy}, T\}$. Here we will just list the 272 Z_2 PSGs. These PSGs are generated by $(G_xT_x, G_yT_y, G_TT, G_{P_x}P_x, G_{P_y}P_y, G_{P_{xy}}P_{xy})$. The PSGs can be divided into two classes. The first class is given by

$$\begin{aligned} G_x(\boldsymbol{i}) &= \tau^0, & G_y(\boldsymbol{i}) &= \tau^0 \\ G_{P_x}(\boldsymbol{i}) &= \eta_{xpx}^{i_x}\eta_{xpy}^{i_y} g_{P_x} & G_{P_y}(\boldsymbol{i}) &= \eta_{xpy}^{i_x}\eta_{xpx}^{i_y} g_{P_y} \\ G_{P_{xy}}(\boldsymbol{i}) &= g_{P_{xy}} & G_T(\boldsymbol{i}) &= \eta_t^{\boldsymbol{i}} g_T \end{aligned} \tag{9.4.14}$$

and the second class is given by

$$\begin{aligned} G_x(\boldsymbol{i}) &= (-)^{i_y}\tau^0, & G_y(\boldsymbol{i}) &= \tau^0 \\ G_{P_x}(\boldsymbol{i}) &= \eta_{xpx}^{i_x}\eta_{xpy}^{i_y} g_{P_x} & G_{P_y}(\boldsymbol{i}) &= \eta_{xpy}^{i_x}\eta_{xpx}^{i_y} g_{P_y} \\ G_{P_{xy}}(\boldsymbol{i}) &= (-)^{i_x i_y} g_{P_{xy}} & G_T(\boldsymbol{i}) &= \eta_t^{\boldsymbol{i}} g_T \end{aligned} \tag{9.4.15}$$

Here the three ηs can independently take the two values ± 1. The gs have 17 different choices, which are given in Table 9.1. Thus, there are $2 \times 17 \times 2^3 = 272$ different PSGs. They can potentially lead to 272 different types of symmetric Z_2 spin liquids on a two-dimensional square lattice.

To label the 272 PSGs, we propose the following scheme:

$$Z2A(g_{px})_{\eta_{xpx}}(g_{py})_{\eta_{xpy}} g_{pxy}(g_t)_{\eta_t}, \tag{9.4.16}$$

$$Z2B(g_{px})_{\eta_{xpx}}(g_{py})_{\eta_{xpy}} g_{pxy}(g_t)_{\eta_t}. \tag{9.4.17}$$

The label Z2A... corresponds to the case in eqn (9.4.14), and the label Z2B... corresponds to the case in eqn (9.4.15). A typical label will look like Z2A$\tau_+^1\tau_-^2\tau^{12}\tau_-^3$. We will also use an abbreviated notation. An abbreviated notation is obtained by replacing $(\tau^0, \tau^1, \tau^2, \tau^3)$ or $(\tau_+^0, \tau_+^1, \tau_+^2, \tau_+^3)$ by $(0, 1, 2, 3)$, and $(\tau_-^0, \tau_-^1, \tau_-^2, \tau_-^3)$ by (n, x, y, z). For example, Z2A$\tau_+^1\tau_-^0\tau^{12}\tau_-^3$ can be abbreviated to Z2A$1n(12)z$.

These 272 different Z_2 PSGs are, strictly speaking, the so-called algebraic PSGs. The algebraic PSGs are defined as extensions of symmetry groups. The algebraic PSGs are different from the invariant PSGs, which are defined as a collection of all transformations that leave an ansatz $u_{\boldsymbol{ij}}$ invariant. Although an

TABLE 9.1. The 17 choices of g_{Pxy}, g_{P_x}, g_{P_y}, and g_T for the Z_2 PSG. Here $\tau^{ab} = (\tau^a + \tau^b)/\sqrt{2}$ and $\tau^{a\bar{b}} = (\tau^a - \tau^b)/\sqrt{2}$.

g_{Pxy}	g_{P_x}	g_{P_y}	g_T	g_{Pxy}	g_{P_x}	g_{P_y}	g_T
τ^0	τ^0	τ^0	τ^0	τ^0	$i\tau^3$	$i\tau^3$	τ^0
$i\tau^3$	τ^0	τ^0	τ^0	$i\tau^3$	$i\tau^3$	$i\tau^3$	τ^0
$i\tau^3$	$i\tau^1$	$i\tau^1$	τ^0	τ^0	τ^0	τ^0	$i\tau^3$
τ^0	$i\tau^3$	$i\tau^3$	$i\tau^3$	τ^0	$i\tau^1$	$i\tau^1$	$i\tau^3$
$i\tau^3$	τ^0	τ^0	$i\tau^3$	$i\tau^3$	$i\tau^3$	$i\tau^3$	$i\tau^3$
$i\tau^3$	$i\tau^1$	$i\tau^1$	$i\tau^3$	$i\tau^1$	τ^0	τ^0	$i\tau^3$
$i\tau^1$	$i\tau^3$	$i\tau^3$	$i\tau^3$	$i\tau^1$	$i\tau^1$	$i\tau^1$	$i\tau^3$
$i\tau^1$	$i\tau^2$	$i\tau^2$	$i\tau^3$	$i\tau^{12}$	$i\tau^1$	$i\tau^2$	$i\tau^0$
$i\tau^{12}$	$i\tau^1$	$i\tau^2$	$i\tau^3$				

invariant PSG must be an algebraic PSG, an algebraic PSG may not be an invariant PSG. This is because certain algebraic PSGs have the following property: *any* ansatz u_{ij} that is invariant under an algebraic PSG G_1 may actually be invariant under a larger PSG G_2. In this case, the original PSG G_1 cannot be an invariant PSG of the ansatz. The invariant PSG of the ansatz is really given by the larger PSG G_2. If we limit ourselves to the spin liquids constructed through the ansatz u_{ij}, then we should drop the algebraic PSGs which are not invariant PSGs. This is because these algebraic PSGs do not characterize mean-field spin liquids.

We find that, among the 272 algebraic Z_2 PSGs, at least 76 of them are not invariant PSGs. Thus, the 272 algebraic Z_2 PSGs can lead to at most 196 possible Z_2 spin liquids.

9.4.4 A Z_2 and a $U(1)$ projective symmetry group and their ansatz

- We can find the most general ansatz for a given PSG. The common physical properties shared by these ansatz are universal properties associated with the PSG.

As an application of the Z_2 PSG, let us construct the most general ansatz that is invariant under the Z2A0013 PSG:

$$
\begin{aligned}
G_x(\boldsymbol{i}) &= \tau^0, & G_y(\boldsymbol{i}) &= \tau^0 \\
G_{P_x}(\boldsymbol{i}) &= \tau^0 & G_{P_y}(\boldsymbol{i}) &= \tau^0 \\
G_{P_{xy}}(\boldsymbol{i}) &= i\tau^1 & G_T(\boldsymbol{i}) &= i\tau^3
\end{aligned}
\tag{9.4.18}
$$

From the invariance under G_xT_x and G_yT_y and because G_x and G_y are trivial, the Z2A0013 ansatz is directly translationally invariant and has the form $u_{\boldsymbol{i},\boldsymbol{i}+\boldsymbol{m}} = u^\mu_{\boldsymbol{m}}\tau^\mu$. The key to obtaining the most general Z2A0013 ansatz is to find out which $u^\mu_{\boldsymbol{m}}$ must vanish. The time-reversal transformation $G_T T$ and the 180° rotation $G_{P_x} P_x G_{P_y} P_y$ are important for this purpose. The invariance under $G_T T$ requires that $-u_{\boldsymbol{m}} = \tau^3 u_{\boldsymbol{m}} \tau^3$. Thus, only $u^{1,2}_{\boldsymbol{m}}$ can be nonzero. The invariance under $G_{P_x} P_x G_{P_y} P_y$ requires that $u_{\boldsymbol{m}} = u_{-\boldsymbol{m}}$, which is always satisfied by $u_{\boldsymbol{m}} = u^1_{\boldsymbol{m}} \tau^1 + u^2_{\boldsymbol{m}} \tau^2$ because $u^{1,2}_{\boldsymbol{m}}$ are real and $u_{-\boldsymbol{m}} = u^\dagger_{\boldsymbol{m}}$. Thus, the 180° rotation does not impose additional conditions. Other transformations relate different $u^{1,2}_{\boldsymbol{m}}$s. The invariance under $G_{P_{xy}} P_{xy}$ requires that $u_{P_{xy}\boldsymbol{m}} = \tau^1 u_{\boldsymbol{m}} \tau^1$, where $P_{xy}\boldsymbol{m} = P_{xy}(m_x, m_y) = (m_y, m_x)$. Therefore, the most general Z2A0013 ansatz has the form

$$
\begin{aligned}
u_{\boldsymbol{m}} &= u^1_{\boldsymbol{m}} \tau^1 + u^2_{\boldsymbol{m}} \tau^2 \\
u^{1,2}_{P_x \boldsymbol{m}} &= u^{1,2}_{\boldsymbol{m}}, \quad u^{1,2}_{P_y \boldsymbol{m}} = u^{1,2}_{\boldsymbol{m}} \\
u^1_{P_{xy} \boldsymbol{m}} &= u^1_{\boldsymbol{m}}, \quad u^2_{P_{xy} \boldsymbol{m}} = -u^2_{\boldsymbol{m}}
\end{aligned}
$$

After replacing (τ^3, τ^1) in eqn (9.2.41) by (τ^1, τ^2) and identifying a_0 with $u_{\boldsymbol{m}=0}$, we find that the ansatz (9.2.41) for a Z_2-linear spin liquid is a special case of the above ansatz. Thus, the PSG of the ansatz (9.2.41) is the Z2A0013 PSG, and the corresponding spin liquid is a Z2A0013 spin liquid.

In addition to the Z_2-symmetric spin liquids studied above, there can be symmetric spin liquids whose low-energy gauge group is $U(1)$ or $SU(2)$. Such $U(1)$- or $SU(2)$-symmetric spin liquids are classified by PSGs whose IGG is $U(1)$ or $SU(2)$. We call these PSGs $U(1)$ and $SU(2)$ PSGs. The $U(1)$ and $SU(2)$ PSGs have been calculated by Wen (2002c). Section 9.4.5 summarizes those results.

One of the $U(1)$ PSGs is given by

$$
\begin{aligned}
G_x &= g_3(\theta_x) \mathrm{i}\tau^1, & G_y &- g_3(\theta_y) \mathrm{i}\tau^1, \\
G_{P_x} &= (-)^{i_x} g_3(\theta_{px}), & G_{P_y} &= (-)^{i_y} g_3(\theta_{py}), \\
G_{P_{xy}} &= g_3(\theta_{pxy}) \mathrm{i}\tau^1, & G_T &= (-)^{\boldsymbol{i}} g_3(\theta_t).
\end{aligned}
\tag{9.4.19}
$$

where

$$
g_a(\theta) \equiv \mathrm{e}^{\mathrm{i}\theta\tau^a}.
$$

The IGG is formed by $\{g_3(\phi) | \phi \in [0, 2\pi)\}$. Such a PSG is labeled by U1Cn01n. It is obtained from eqn (9.4.27) by choosing $\eta_{xpx} = -1$, $\eta_{ypx} = +1$, $\eta_{pxy} = +1$, and $\eta_t = -1$.

Let us obtain the most general ansatz that is invariant under the U1Cn01n PSG. To be invariant under the IGG, the ansatz must satisfy $g_3(\theta) u_{\boldsymbol{ij}} g_3^\dagger(\theta) = u_{\boldsymbol{ij}}$ for any θ. Thus, the $u_{\boldsymbol{ij}}$ must have the form $u_{\boldsymbol{ij}} = \mathrm{i} u^0_{\boldsymbol{ij}} \tau^0 + u^3_{\boldsymbol{ij}} \tau^3$. The invariance under under the translations $G_x T_x$ and $G_y T_y$ requires $u^0_{\boldsymbol{ij}}$ to be translationally invariant and $u^3_{\boldsymbol{ij}}$ to change sign under the two translations $T_{x,y}$. Thus, the $u_{\boldsymbol{ij}}$ have the form $u_{i,i+m} = \mathrm{i}u^0_m\tau^0 + (-)^i u^3_m\tau^3$. The invariance under the time-reversal transformation $G_T T$ requires that $-u_m = (-)^m u_m$. We find that $u^{0,3}_m = 0$ if m = even.[71] The invariance under the 180° rotation $G_{P_x}P_xG_{P_y}P_y$ requires that $u_{-i,-i-m} = u^\dagger_{-i-m,-i} = (-)^m u_{i,i+m}$, which implies that $-\mathrm{i}u^0_m\tau^0 + (-)^{i+m}u^3_m\tau^3 = (-)^m(\mathrm{i}u^0_m\tau^0 + (-)^i u^3_m\tau^3)$. This leads to $u^0_m = -(-)^m u^0_m$ and $u^3_m = u^3_m$, and provides no new constraint on $u^{0,3}_m$. The hermitian relation $u^\dagger_{i,i+m} = u_{i+m,i}$ implies that $u^0_m = -u^0_{-m}$ and $u^3_m = (-)^m u^3_{-m}$. Collecting the above conditions and including the action of $G_{P_x}P_x$, etc., we find that the most general U1Cn01n ansatz has the form

$$
\begin{aligned}
u_{i,i+m} &= \mathrm{i}u^0_m\tau^0 + (-)^i u^3_m\tau^3\\
u^{0,3}_m &= 0, \text{ if } m = \text{even}, \qquad u^{0,3}_m = -u^{0,3}_{-m}\\
u^{0,3}_{P_x m} &= (-)^{m_x}u^{0,3}_m, \qquad u^{0,3}_{P_y m} = (-)^{m_y}u^{0,3}_m\\
u^0_{P_{xy}m} &= u^0_m, \qquad u^3_{P_{xy}m} = -u^3_m
\end{aligned}
$$

The $U(1)$-linear ansatz (the staggered-flux phase) (9.2.34) is a special case of the above ansatz. Thus, the staggered-flux phase (9.2.34) is a U1Cn01n state.

We can use a gauge transformation $W_i = (\mathrm{i}\tau^1)^i$ to make the above ansatz translationally invariant:

$$
\begin{aligned}
u_{i,i+m} &= \tilde{u}^1_m\tau^1 + \tilde{u}^2_m\tau^2\\
\tilde{u}^{1,2}_m &= 0, \qquad \text{for } m = \text{even},\\
G_x(i) &= g_3((-)^i\theta_x), \qquad G_y(i) = g_3((-)^i\theta_y),\\
G_{P_x}(i) &= g_3((-)^i\theta_{px}), \qquad G_{P_y}(i) = g_3((-)^i\theta_{py}),\\
G_{P_{xy}}(i) &= ig_3((-)^i\theta_{pxy})\tau^1, \qquad G_T(i) = (-)^i g_3((-)^i\theta_t);
\end{aligned}
\tag{9.4.20}
$$

where we have also listed the gauge transformations in the transformed PSG. The IGG now has the form $IGG = \{g_3((-)^i\phi)|\phi \in [0, 2\pi)\}$. If we replace (τ^1, τ^2) by (τ^3, τ^1), then one can show that the above ansatz corresponds to a d-wave state in terms of the f-fermions (see eqns (9.2.4) and (9.2.30)).

The spinon spectrum for the ansatz (9.4.20) is given by

$$
E_\pm(k) = \pm\sqrt{\left[\sum_{m=\text{odd}} \tilde{u}^1_m \cos(m\cdot k)\right]^2 + \left[\sum_{m=\text{odd}} \tilde{u}^2_m \cos(m\cdot k)\right]^2}
$$

We note that, for odd m, $\cos(m\cdot k) = 0$ at $k = (\pm\frac{\pi}{2}, \pm\frac{\pi}{2})$. Thus, $E_\pm(k) = 0$ at $k = (\pm\frac{\pi}{2}, \pm\frac{\pi}{2})$. The U1C$n01n$ spin-liquid state always has at least four

[71] Here m is even (odd) if $m_x + m_y$ is even (odd).

branches of gapless spinons at $\boldsymbol{k} = (\pm\frac{\pi}{2}, \pm\frac{\pi}{2})$. The gapless spinons and their crystal momenta $(\pm\frac{\pi}{2}, \pm\frac{\pi}{2})$ are universal properties of the U1Cn01n spin-liquid state. The gaplessness and the crystal momenta of the spinons are protected by the U1Cn01n PSG. This is a striking result. As a purely bosonic model, the U1Cn01n spin-liquid state does not break any symmetry. However, it contains a special quantum order that guarantees the existence of gapless *fermions*! As the gapless spinons at $\boldsymbol{k} = (\pm\frac{\pi}{2}, \pm\frac{\pi}{2})$ is a universal property of the U1Cn01n state, we can use it to detect the U1Cn01n quantum order in experiments.

Problem 9.4.1.
Find the gauge transformation that transforms $G_y(\boldsymbol{i}) = \Theta(\boldsymbol{i})\tau^0$ to $G_y(\boldsymbol{i}) = \tau^0$ through eqn (9.4.3). Here $\Theta(\boldsymbol{i})$ is an arbitrary function with only the two values ± 1.

Problem 9.4.2.
(a) Find the gauge transformations $G_{x,y}$, $G_{P_x,P_y,P_{xy}}$, and G_T for the Z2Azz13 PSG from eqn (9.4.14).
(b) Find the most general ansatz that is invariant under the Z2Azz13 PSG. We will show later that such an ansatz always has gapless fermion excitations.

Problem 9.4.3.
Prove eqn (9.4.20). (Hint: See eqn (9.4.3).)

9.4.5 Remarks: classification of symmetric $U(1)$ and $SU(2)$ spin liquids

- A classification of all of the mean-field ansatz whose PSG is a $U(1)$ or an $SU(2)$ extension of a lattice symmetry group, i.e. $PSG/U(1) = SG$ or $PSG/SU(2) = SG$.

A classification of the $U(1)$ and the $SU(2)$ PSG is given by Wen (2002c). In the following, we only summarize the results. The PSGs that characterize mean-field symmetric $U(1)$ spin liquids can be divided into the four types U1A, U1B, U1C, and U1$_n^m$. There are the following 24 type-U1A PSGs:

$$
\begin{aligned}
G_x &= g_3(\theta_x), & G_y &= g_3(\theta_y),\\
G_{P_x} &= \eta_{ypx}^{i_y} g_3(\theta_{px}), & G_{P_y} &= \eta_{ypx}^{i_x} g_3(\theta_{py})\\
G_{P_{xy}} &= g_3(\theta_{pxy}),\ \ g_3(\theta_{pxy})i\tau^1, & G_T &= \eta_t^{i} g_3(\theta_t)|_{\eta_t=-1},\ \ \eta_t^{i} g_3(\theta_t)i\tau^1
\end{aligned}
\tag{9.4.21}
$$

and

$$
\begin{aligned}
G_x &= g_3(\theta_x), & G_y &= g_3(\theta_y),\\
G_{P_x} &= \eta_{xpx}^{i_x} g_3(\theta_{px})i\tau^1, & G_{P_y} &= \eta_{xpx}^{i_y} g_3(\theta_{py})i\tau^1\\
G_{P_{xy}} &= g_3(\theta_{pxy}),\ \ g_3(\theta_{pxy})i\tau^1, & G_T &= \eta_t^{i} g_3(\theta_t)|_{\eta_t=-1},\ \ \eta_t^{i} g_3(\theta_t)i\tau^1
\end{aligned}
\tag{9.4.22}
$$

We will use U1A$a_{\eta_{xpx}} b_{\eta_{ypx}} c d_{\eta_t}$ to label the 24 PSGs. Here a, b, c, and d are associated with G_{P_x}, G_{P_y}, $G_{P_{xy}}$, and G_T, respectively. They are equal to τ^1 if the corresponding G

contains a τ^1, and equal to τ^0 otherwise. A typical notation looks like U1A$\tau^1_-\tau^1\tau^0\tau^1_-$, which can be abbreviated as U1A$x10x$.

There are also the following 24 type-U1B PSGs:

$$
\begin{aligned}
G_x &= (-)^{i_y} g_3(\theta_x), & G_y &= g_3(\theta_y), \\
G_{P_x} &= \eta_{ypx}^{i_y} g_3(\theta_{px}), & G_{P_y} &= \eta_{ypx}^{i_x} g_3(\theta_{py}) \\
(-)^{i_x i_y} G_{P_{xy}} &= g_3(\theta_{pxy}),\ \ g_3(\theta_{pxy})\,\mathrm{i}\tau^1 & G_T &= \eta_t^{\boldsymbol{i}} g_3(\theta_t)|_{\eta_t=-1},\ \ \eta_t^{\boldsymbol{i}} g_3(\theta_t)\,\mathrm{i}\tau^1
\end{aligned}
\tag{9.4.23}
$$

and

$$
\begin{aligned}
G_x &= (-)^{i_y} g_3(\theta_x), & G_y &= g_3(\theta_y), \\
G_{P_x} &= \eta_{xpx}^{i_x} g_3(\theta_{px})\,\mathrm{i}\tau^1, & G_{P_y} &= \eta_{xpx}^{i_y} g_3(\theta_{py})\,\mathrm{i}\tau^1, \\
(-)^{i_x i_y} G_{P_{xy}} &= g_3(\theta_{pxy}),\ \ g_3(\theta_{pxy})\,\mathrm{i}\tau^1 & G_T &= \eta_t^{\boldsymbol{i}} g_3(\theta_t)|_{\eta_t=-1},\ \ \eta_t^{\boldsymbol{i}} g_3(\theta_t)\,\mathrm{i}\tau^1
\end{aligned}
\tag{9.4.24}
$$

We will use U1B$a_{\eta_{xpx}} b_{\eta_{ypx}} c d_{\eta_t}$ to label the 24 PSGs.

The 60 type-U1C PSGs are given by

$$
\begin{aligned}
G_x &= g_3(\theta_x)\,\mathrm{i}\tau^1, & G_y &= g_3(\theta_y)\,\mathrm{i}\tau^1, \\
G_{P_x} &= \eta_{xpx}^{i_x}\eta_{ypx}^{i_y} g_3(\theta_{px}), & G_{P_y} &= \eta_{ypx}^{i_x}\eta_{xpx}^{i_y} g_3(\theta_{py}) \\
G_{P_{xy}} &= \eta_{pxy}^{i_x} g_3(\eta_{pxy}^{\boldsymbol{i}}\frac{\pi}{4} + \theta_{pxy}), & G_T &= \eta_t^{\boldsymbol{i}} g_3(\theta_t)|_{\eta_t=-1},\ \ \eta_{pxy}^{i_x} g_3(\theta_t)\,\mathrm{i}\tau^1
\end{aligned}
\tag{9.4.25}
$$

$$
\begin{aligned}
G_x &= g_3(\theta_x)\,\mathrm{i}\tau^1, & G_y &= g_3(\theta_y)\,\mathrm{i}\tau^1, \\
G_{P_x} &= \eta_{xpx}^{i_x} g_3(\theta_{px})\,\mathrm{i}\tau^1, & G_{P_y} &= \eta_{xpx}^{i_y}\eta_{pxy}^{\boldsymbol{i}} g_3(\theta_{py})\,\mathrm{i}\tau^1, \\
G_{P_{xy}} &= \eta_{pxy}^{i_x} g_3(\eta_{pxy}^{\boldsymbol{i}}\frac{\pi}{4} + \theta_{pxy}), & G_T &= \eta_t^{\boldsymbol{i}} g_3(\theta_t)|_{\eta_t=-1},\ \ \eta_{pxy}^{i_x}\eta_t^{\boldsymbol{i}} g_3(\theta_t)\,\mathrm{i}\tau^1
\end{aligned}
\tag{9.4.26}
$$

$$
\begin{aligned}
G_x &= g_3(\theta_x)\,\mathrm{i}\tau^1, & G_y &= g_3(\theta_y)\,\mathrm{i}\tau^1, \\
G_{P_x} &= \eta_{xpx}^{i_x}\eta_{ypx}^{i_y} g_3(\theta_{px}), & G_{P_y} &= \eta_{ypx}^{i_x}\eta_{xpx}^{i_y} g_3(\theta_{py}), \\
G_{P_{xy}} &= g_3(\theta_{pxy})\,\mathrm{i}\tau^1, & G_T &= \eta_t^{\boldsymbol{i}} g_3(\theta_t)|_{\eta_t=-1}
\end{aligned}
\tag{9.4.27}
$$

$$
\begin{aligned}
G_x &= g_3(\theta_x)\,\mathrm{i}\tau^1, & G_y &= g_3(\theta_y)\,\mathrm{i}\tau^1, \\
G_{P_x} &= \eta_{xpx}^{i_x}\eta_{ypx}^{i_y} g_3(\theta_{px}), & G_{P_y} &= \eta_{ypx}^{i_x}\eta_{xpx}^{i_y} g_3(\theta_{py}), \\
G_{P_{xy}} &= g_3(\eta_{pxy}^{\boldsymbol{i}}\frac{\pi}{4} + \theta_{pxy})\,\mathrm{i}\tau^1, & G_T &= \eta_{pxy}^{i_x}\eta_t^{\boldsymbol{i}} g_3(\theta_t)\,\mathrm{i}\tau^1
\end{aligned}
\tag{9.4.28}
$$

and

$$
\begin{aligned}
G_x &= g_3(\theta_x)\mathrm{i}\tau^1, & G_y &= g_3(\theta_y)\mathrm{i}\tau^1, \\
G_{P_x} &= \eta_{xpx}^{i_x} g_3(\theta_{px})\mathrm{i}\tau^1, & G_{P_y} &= \eta_{xpx}^{i_y}\eta_{pxy}^{\boldsymbol{i}} g_3(\theta_{py})\mathrm{i}\tau^1, \\
G_{P_{xy}} &= g_3(\eta_{pxy}^{\boldsymbol{i}}\frac{\pi}{4} + \theta_{pxy})\mathrm{i}\tau^1, & G_T &= \eta_t^{\boldsymbol{i}} g_3(\theta_t)|_{\eta_t=-1},\ \eta_t^{\boldsymbol{i}}\eta_{pxy}^{i_x} g_3(\theta_t)\mathrm{i}\tau^1
\end{aligned}
\tag{9.4.29}
$$

which will be labeled by U1C$a_{\eta_{xpx}}b_{\eta_{ypx}}c_{\eta_{pxy}}d_{\eta_t}$.

The type-U1^m_n PSGs have not been classified. However, we do know that, for each rational number $m/n \in (0,1)$, there exists at least one mean-field symmetric spin liquid of type U1^m_n. The ansatz is given by

$$
u_{\boldsymbol{i},\boldsymbol{i}+\boldsymbol{x}} = \chi\tau^3, \qquad u_{\boldsymbol{i},\boldsymbol{i}+\boldsymbol{y}} = \chi g_3(\frac{m\pi}{n}i_x)\tau^3 \tag{9.4.30}
$$

It has $\pi m/n$ flux per plaquette. Thus, there are infinitely many U1^m_n spin liquids.

We would like to point out that the above 108 U1A, U1B, and U1C PSGs are algebraic PSGs. They are only a subset of all possible algebraic $U(1)$ PSGs. However, they do contain all of the invariant $U(1)$ PSGs of type U1A, U1B, and U1C. We find that 46 of the 108 PSGs are also invariant PSGs. Thus, there are 46 different mean-field $U(1)$ spin liquids of type U1A, U1B, and U1C. Their ansatz and labels are given in Wen (2002c).

To classify the symmetric $SU(2)$ spin liquids, we find that there are eight different $SU(2)$ PSGs, which are given by

$$
\begin{aligned}
G_x(\boldsymbol{i}) &= g_x, & G_y(\boldsymbol{i}) &= g_y, \\
G_{P_x}(\boldsymbol{i}) &= \eta_{xpx}^{i_x}\eta_{xpy}^{i_y} g_{P_x}, & G_{P_y}(\boldsymbol{i}) &= \eta_{xpy}^{i_x}\eta_{xpx}^{i_y} g_{P_y} \\
G_{P_{xy}}(\boldsymbol{i}) &= g_{P_{xy}}, & G_T(\boldsymbol{i}) &= (-)^{\boldsymbol{i}} g_T,
\end{aligned}
\tag{9.4.31}
$$

and

$$
\begin{aligned}
G_x(\boldsymbol{i}) &= (-)^{i_y} g_x, & G_y(\boldsymbol{i}) &= g_y, \\
G_{P_x}(\boldsymbol{i}) &= \eta_{xpx}^{i_x}\eta_{xpy}^{i_y} g_{P_x}, & G_{P_y}(\boldsymbol{i}) &= \eta_{xpy}^{i_x}\eta_{xpx}^{i_y} g_{P_y} \\
G_{P_{xy}}(\boldsymbol{i}) &= (-)^{i_x i_y} g_{P_{xy}}, & G_T(\boldsymbol{i}) &= (-)^{\boldsymbol{i}} g_T
\end{aligned}
\tag{9.4.32}
$$

where the gs are in $SU(2)$. We introduce the following notation:

$$
\begin{aligned}
&\text{SU2A}\tau^0_{\eta_{xpx}}\tau^0_{\eta_{xpy}} \\
&\text{SU2B}\tau^0_{\eta_{xpx}}\tau^0_{\eta_{xpy}}
\end{aligned}
\tag{9.4.33}
$$

to denote the above 8 PSGs. Here SU2A$\tau^0_{\eta_{xpx}}\tau^0_{\eta_{xpy}}$ is for eqn (9.4.31) and SU2B$\tau^0_{\eta_{xpx}}\tau^0_{\eta_{xpy}}$ is for eqn (9.4.32). We find that only 4 of the 8 $SU(2)$ PSGs, namely SU2A$[n0, 0n]$ and SU2B$[n0, 0n]$, lead to $SU(2)$-symmetric spin liquids. The SU2A$n0$ state is the uniform RVB state (9.2.32) and the SU2B$n0$ state is the π-flux state (9.2.33). The other two $SU(2)$ spin liquids are labeled by SU2A$0n$, namely

$$
\begin{aligned}
u_{\boldsymbol{i},\boldsymbol{i}+2\boldsymbol{x}+\boldsymbol{y}} &= +\mathrm{i}\chi\tau^0, & u_{\boldsymbol{i},\boldsymbol{i}-2\boldsymbol{x}+\boldsymbol{y}} &= -\mathrm{i}\chi\tau^0, \\
u_{\boldsymbol{i},\boldsymbol{i}+\boldsymbol{x}+2\boldsymbol{y}} &= +\mathrm{i}\chi\tau^0, & u_{\boldsymbol{i},\boldsymbol{i}-\boldsymbol{x}+2\boldsymbol{y}} &= +\mathrm{i}\chi\tau^0,
\end{aligned}
\tag{9.4.34}
$$

and by SU2B0n, namely

$$u_{i,i+2x+y} = +\,\mathrm{i}(-)^{i_x}\chi\tau^0, \qquad u_{i,i-2x+y} = -\,\mathrm{i}(-)^{i_x}\chi\tau^0,$$
$$u_{i,i+x+2y} = +\,\mathrm{i}\chi\tau^0, \qquad u_{i,i-x+2y} = +\,\mathrm{i}\chi\tau^0 \tag{9.4.35}$$

The above results give us a classification of $U(1)$- and $SU(2)$-symmetric spin liquids at the mean-field level. We would like to point out that the $U(1)$ and the $SU(2)$ mean-field states are not stable mean-field states. Some of the $U(1)$ and the $SU(2)$ mean-field states are marginal. So, when compared to the Z_2 mean-field states, it is less clear if the $U(1)$ and the $SU(2)$ mean-field states correspond to any physical $U(1)$- or $SU(2)$-symmetric spin liquids for spin-1/2 models. Physical $U(1)$ or $SU(2)$ states may exist only in the large-N limit (see Section 9.8).

9.5 Continuous phase transitions without symmetry breaking

- The continuous phase transition between quantum phases is governed by the following principle. Let PSG_1 and PSG_2 be the PSGs of the two quantum phases on the two sides of a transition, and let PSG_{cr} be the PSG that describes the quantum critical state. Then $PSG_1 \subseteq PSG_{cr}$ and $PSG_2 \subseteq PSG_{cr}$.
- The above principle applies to both symmetry-breaking transitions and transitions that do not change any symmetry.

After classifying mean-field symmetric spin liquids, we would like to know how these symmetric spin liquids are related to each other. In particular, we would like to know which spin liquids can change into each other through a *continuous* phase transition. At the mean-field level, this problem can be completely addressed by the PSG. The idea is that the PSG is just the 'symmetry' group of the mean-field states. The mean-field phase transitions can be described by the change of PSGs. Just like the symmetry-breaking phase transition, a mean-field state with a PSG of PSG_1 can change to a mean-field state with a PSG of PSG_2 via a continuous (mean-field) phase transition if and only if $PSG_2 \subset PSG_1$ or $PSG_1 \subset PSG_2$.

To understand this result, let us assume that the mean-field state described by PSG_1 has an ansatz u_{ij}. Its neighbor has an ansatz $u_{ij} + \delta u_{ij}$, where δu_{ij} is a small perturbation. Assume the perturbation changes the PSG to a different one PSG_2. As δu_{ij} is a perturbation with an unfixed strength, both u_{ij} and δu_{ij} must be invariant under PSG_2 in order for the $u_{ij} + \delta u_{ij}$ to be invariant under PSG_2. Therefore, $PSG_2 \subset PSG_1$.

Using the above result, we can obtain the symmetric spin liquids in the neighborhood of some important symmetric spin liquids using the following procedure. We start with a symmetric spin liquid u_{ij} with a PSG of PSG_1. Here PSG_1 is an extension of a symmetry group SG by IGG_1, i.e. $PSG_1/IGG_1 = SG$. We then find all of the subgroups of PSG_1 that are extensions of the same symmetry group SG. Let denote these subgroups by PSG_2. Then PSG_2 must have a normal subgroup IGG_2 such that $PSG_2/IGG_2 = SG$. Using those subgroups, we can constrcuct all of the neighboring mean-field states that have the same symmetry.

After lengthy calculations in Wen (2002c), all of the mean-field symmetric spin liquids around the $U(1)$-linear state U1Cn01n in eqn (9.2.34), the $SU(2)$-gapless state SU2An0 in eqn (9.2.32), and the $SU(2)$-linear state SU2Bn0 in eqn (9.2.33) were found. It was shown, for example, that, at the mean-field level, the $U(1)$-linear spin liquid U1Cn01n can continuously change into the 8 different Z_2 spin liquids Z2A0013, Z2Azz13, Z2A001n, Z2Azz1n, Z2B0013, Z2Bzz13, Z2B001n, and Z2Bzz1n.

Let us discuss a simple example to demonstrate the above result. The ansatz

$$u_{\boldsymbol{i},\boldsymbol{i}+\boldsymbol{x}} = \chi\tau^1 - \eta\tau^2, \qquad u_{\boldsymbol{i},\boldsymbol{i}+\boldsymbol{y}} = \chi\tau^1 + \eta\tau^2, \tag{9.5.1}$$
$$u_{\boldsymbol{i},\boldsymbol{i}+\boldsymbol{x}+\boldsymbol{y}} = -\gamma\tau^1, \qquad u_{\boldsymbol{i},\boldsymbol{i}-\boldsymbol{x}+\boldsymbol{y}} = +\gamma\tau^1, \qquad a_0^l = 0.$$

with $\chi \neq \eta$, describes the U1Cn01n spin liquid (the staggered-flux state) when $\gamma = 0$. That is, the ansatz is invariant under the symmetry transformations followed by the gauge transformations given in eqn (9.4.20). When $\gamma \neq 0$, the ansatz generate non-collinear $SU(2)$ fluxes, which break the $U(1)$ gauge structure down to a Z_2 gauge structure. The ansatz now describes a Z_2 state. For a nonzero γ, the ansatz is no longer invariant under the symmetry transformations followed by the U1Cn01n gauge transformations listed in eqn (9.4.20). It is only invariant under the symmetry transformations followed by a subset of the U1Cn01n gauge transformations:

$$G_x(\boldsymbol{i}) = g_3(0) = \tau^0, \qquad G_y(\boldsymbol{i}) = g_3(0) = \tau^0,$$
$$G_{P_x}(\boldsymbol{i}) = g_3((-)^{\boldsymbol{i}}\pi/2) = \mathrm{i}(-)^{\boldsymbol{i}}\tau^3, \quad G_{P_y}(\boldsymbol{i}) = g_3((-)^{\boldsymbol{i}}\pi/2) = \mathrm{i}(-)^{\boldsymbol{i}}\tau^3,$$
$$G_{P_{xy}}(\boldsymbol{i}) = ig_3(0)\tau^1 = \mathrm{i}\tau^1, \qquad G_T(\boldsymbol{i}) = (-)^{\boldsymbol{i}}g_3((-)^{\boldsymbol{i}}\pi/2) = \mathrm{i}\tau^3.$$

The symmetry transformations followed by the above gauge transformations give us the Z2Azz13 PSG. Thus, as γ changes from zero to nonzero, the U1Cn01n state continuously changes to the Z2Azz13 state.

We would like to stress that the above results about the continuous transitions are valid only at the mean-field level. Some of the mean-field results survive the quantum fluctuations, while others do not. One needs to do a case-by-case study to see which mean-field results can be valid beyond mean-field theory (Mudry and Fradkin, 1994).

One possible effect of quantum fluctuations is to destabilize certain mean-field states. We may assume that the destabilization is via certain relevant perturbations generated by the quantum fluctuations. Under such an assumption, some of the

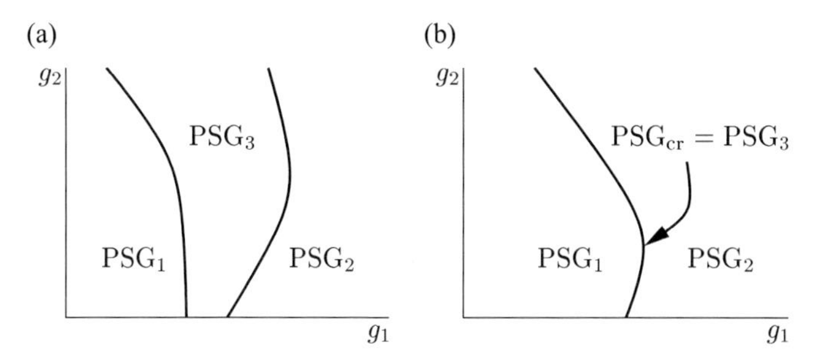

FIG. 9.7. (a) A mean-field phase diagram with three phases characterized by $PSG_{1,2,3}$. The mean-field phase transitions between them are continuous. The PSGs satisfy $PSG_1 \subset PSG_3$ and $PSG_2 \subset PSG_3$. One possible choice for $PSG_{1,2,3}$ is PSG_1 = Z2A0013, PSG_2 = Z2Azz13, and PSG_3 = U1Cn01n. (b) The corresponding physical phase diagram with quantum fluctuations. Here we have assumed that, after including the quantum fluctuations, the state PSG_3 becomes an unstable fixed point with only one relevant perturbation. The mean-field state PSG_3 shrinks into a critical line that describes the transition between the two states PSG_1 and PSG_2. The staggered-flux state U1Cn01n, if unstable, may appear as a critical state.

mean-field stable fixed points become unstable fixed points for real physical systems once quantum fluctuations are included. So some phases in the mean-field phase diagram may shrink into lines that represent critical states at the transition point between two stable quantum phases (see Fig. 9.7). This picture leads to the conjecture stated in the key points at the beginning of this section about the principle that governs the continuous quantum phase transitions.

We would like to stress that all of the above spin liquids have the same symmetry. Thus, the continuous transitions between them, if they exist, represent a new class of quantum continuous transitions which do not change any symmetries (Wegner, 1971; Kosterlitz and Thouless, 1973; Dasgupta and Halperin, 1981; Wen and Wu, 1993; Chen *et al.*, 1993; Senthil *et al.*, 1999; Read and Green, 2000; Wen, 2000).

Problem 9.5.1.
The ansatz

$$u_{i,i+x} = \chi\tau^1 - \eta\tau^2, \qquad u_{i,i+y} = \chi\tau^1 + \eta\tau^2, \qquad a_0^{1,2,3} = 0,$$

$$u_{i,i+2x+y} = u_{i,i-x+2y} = +\lambda\tau^3, \qquad u_{i,i+2x-y} = u_{i,i+x+2y} = -\lambda\tau^3,$$

with $\chi \neq \eta$, describes the U1Cn01n spin liquid (the staggered-flux state) when $\lambda = 0$.
(a) Show that, when $\lambda \neq 0$, the ansatz describes a Z_2 spin liquid.
(b) Find the subgroup of the U1Cn01n PSG that leaves the ansatz invariant.
(c) Find the label of the above Z_2 ansatz.

9.6 The zoo of symmetric spin liquids

- Physical properties of the eight Z_2 spin liquids in the neighborhoods of the staggered-flux state.

In this section, we would like to the study physical properties of the symmetric spin liquids classified in Section 9.4. We have been using projective symmetry to characterize different mean-field states. However, it is not straightforward to see the physical properties of a mean-field state from its PSG. To find the physical properties of a mean-field state, we need to construct the explicit ansatz that is invariant under the corresponding PSG.

However, it is not the easiest thing to list a few hundred symmetric spin liquids, not to mention constructing their ansatz and studying their theory properties one by one. What we want to do here is to study some important spin liquids; but how to determine the importance of a spin liquid?

In the study of high-T_c superconductors, it was found that the $SU(2)$-linear state SU2Bn0 (the π-flux state), the $U(1)$-linear state U1Cn01n (the staggered-flux/d-wave state), and the $SU(2)$-gapless state SU2An0 (the uniform RVB state) are important. They are closely related to some phases observed in high-T_c superconductors. The $SU(2)$-linear, the $U(1)$-linear, and the $SU(2)$-gapless states reproduce the observed electron spectra function for undoped, underdoped, and overdoped samples, respectively. However, theoretically, these spin liquids are unstable at low energies due to the $U(1)$ or $SU(2)$ gauge fluctuations. These states may change into more stable spin liquids in their neighborhood. This motivates us to study these more stable spin liquids in the neighborhood of the $SU(2)$-linear, the $U(1)$-linear, and the $SU(2)$-gapless states. To limit our scope further, here we will mainly study the spin liquids in the neighborhood of the $U(1)$-linear state U1Cn01n.[72]

9.6.1 Symmetric spin liquids around a $U(1)$-linear spin liquid

The U1Cn01n spin liquid (9.2.34 can continuously change into eight different spin liquids that break the $U(1)$ gauge structure down to a Z_2 gauge structure. These eight spin liquids are labeled by different PSGs, despite them all having the same symmetry. In the following, we will study these eight Z_2 spin liquids in more detail. In particular, we would like to find out the spinon spectra in them.

[72] We would like to point out that we will only study symmetric spin liquids here. The U1Cn01n spin liquids may also change into some other states that break certain symmetries. Such symmetry-breaking transitions have actually been observed in high-T_c superconductors (such as the transitions to the anti-ferromagnetic state, the d-wave superconducting state, and the stripe state).

The first one is labeled by Z2A0013 and takes the following form:

$$\begin{aligned} u_{i,i+x} &= \chi\tau^1 - \eta\tau^2, & u_{i,i+y} &= \chi\tau^1 + \eta\tau^2, \\ u_{i,i+x+y} &= +\gamma\tau^1, & u_{i,i-x+y} &= +\gamma\tau^1, \\ a_0^1 &\neq 0, \quad a_0^{2,3} = 0. \end{aligned} \tag{9.6.1}$$

It has the same quantum order as that in the ansatz (9.2.41). The label Z2A0013 tells us the PSG—the 'symmetry' group—of the ansatz. Remember that the PSG for a symmetric spin liquid is generated by the combined symmetry transformation and gauge transformation $\{G_0, G_xT_x, G_yT_y, G_{P_x}P_x, G_{P_y}P_y, G_{P_{xy}}P_{xy}, G_TT\}$. The gauge transformation G_0 generates the IGG of the PSG. The 'Z2' in the label Z2A0013 tells us that IGG $= Z_2$ and $G_0(\boldsymbol{i}) = (-)$. The 'A' in the label tells us that $G_x(\boldsymbol{i}) = G_y(\boldsymbol{i}) = 1$. The next '00' implies that $G_{P_x}(\boldsymbol{i}) = 1$ and $G_{P_y}(\boldsymbol{i}) = 1$. The '1' means that $G_{P_{xy}}(\boldsymbol{i}) = \mathrm{i}\tau^1$ and the '3' means that $G_T(\boldsymbol{i}) = \mathrm{i}\tau^3$. The second ansatz is labeled by Z2Azz13 and takes the following form:

$$\begin{aligned} u_{i,i+x} &= \chi\tau^1 - \eta\tau^2, & u_{i,i+y} &= \chi\tau^1 + \eta\tau^2, \\ u_{i,i+x+y} &= -\gamma\tau^1, & u_{i,i-x+y} &= +\gamma\tau^1, \\ u_{i,i+2x} &= u_{i,i+2y} = 0, & a_0^{1,2,3} &= 0. \end{aligned} \tag{9.6.2}$$

Now the 'zz' in the label tells us that $\eta_{xpx} = \eta_{xpy} = -1$ and $g_{P_x} = g_{Py} = \mathrm{i}\tau^3$. Thus, $G_{P_x}(\boldsymbol{i}) = (-)^{i_x+i_y}\tau^3$ and $G_{P_y}(\boldsymbol{i}) = (-)^{i_x+i_y}\tau^3$ (see eqn (9.4.14)). The third ansatz is labeled by Z2A001n and takes the following form:

$$a_0^l = 0, \tag{9.6.3}$$

$$\begin{aligned} u_{i,i+x} &= \chi\tau^1 + \eta\tau^2, & u_{i,i+y} &= \chi\tau^1 - \eta\tau^2 \\ u_{i,i+2x+y} &= \chi_1\tau^1 + \eta_1\tau^2 + \lambda_1\tau^3, & u_{i,i-x+2y} &= \chi_1\tau^1 - \eta_1\tau^2 - \lambda_1\tau^3, \\ u_{i,i+2x-y} &= \chi_1\tau^1 + \eta_1\tau^2 + \lambda_1\tau^3, & u_{i,i+x+2y} &= \chi_1\tau^1 - \eta_1\tau^2 - \lambda_1\tau^3. \end{aligned}$$

If you are careful, you will find that the label Z2A001n does not appear in our list of 196 Z_2 spin liquids classified in Section 9.4.3 (see Table 9.1). However, the PSG labeled by Z2A001n and the PSG labeled by Z2A003n are gauge equivalent, and the label Z2A003n does appear in our list (see the second row of Table 9.1). In the following, we will call the above spin liquid the Z2A003n state. The fourth ansatz is labeled by Z2Azz1n and takes the following form:

$$\begin{aligned} a_0^l &= 0, \\ u_{i,i+x} &= \chi\tau^1 + \eta\tau^2, & u_{i,i+y} &= \chi\tau^1 - \eta\tau^2 \\ u_{i,i+2x+y} &= \chi_1\tau^1 + \eta_1\tau^2 + \lambda\tau^3, & u_{i,i-x+2y} &= \chi_1\tau^1 - \eta_1\tau^2 + \lambda\tau^3, \\ u_{i,i+2x-y} &= \chi_1\tau^1 + \eta_1\tau^2 - \lambda\tau^3, & u_{i,i+x+2y} &= \chi_1\tau^1 - \eta_1\tau^2 - \lambda\tau^3. \end{aligned} \tag{9.6.4}$$

The above four ansatz have translational invariance. The next four Z_2 ansatz do not have translational invariance because they are all of Z2B type. (However, they still describe translationally-symmetric spin liquids after the projection.) These Z_2 spin liquids are as follows:
Z2B0013:

$$
\begin{aligned}
u_{\boldsymbol{i},\boldsymbol{i}+\boldsymbol{x}} &= \chi\tau^1 - \eta\tau^2, & u_{\boldsymbol{i},\boldsymbol{i}+\boldsymbol{y}} &= (-)^{i_x}(\chi\tau^1 + \eta\tau^2),\\
u_{\boldsymbol{i},\boldsymbol{i}+2\boldsymbol{x}} &= -\gamma_2\tau^1 + \lambda_2\tau^2, & u_{\boldsymbol{i},\boldsymbol{i}+2\boldsymbol{y}} &= -\gamma_2\tau^1 - \lambda_2\tau^2,\\
a_0^1 &\neq 0, \qquad a_0^{2,3} = 0,
\end{aligned}
\tag{9.6.5}
$$

Z2Bzz13:

$$
\begin{aligned}
u_{\boldsymbol{i},\boldsymbol{i}+\boldsymbol{x}} &= \chi\tau^1 - \eta\tau^2, & u_{\boldsymbol{i},\boldsymbol{i}+\boldsymbol{y}} &= (-)^{i_x}(\chi\tau^1 + \eta\tau^2),\\
u_{\boldsymbol{i},\boldsymbol{i}+2\boldsymbol{x}+2\boldsymbol{y}} &= -\gamma_1\tau^1, & u_{\boldsymbol{i},\boldsymbol{i}-2\boldsymbol{x}+2\boldsymbol{y}} &= \gamma_1\tau^1,\\
a_0^{1,2,3} &= 0,
\end{aligned}
\tag{9.6.6}
$$

Z2B001n:

$$
\begin{aligned}
u_{\boldsymbol{i},\boldsymbol{i}+\boldsymbol{x}} &= \chi\tau^1 + \eta\tau^2, & u_{\boldsymbol{i},\boldsymbol{i}+\boldsymbol{y}} &= (-)^{i_x}(\chi\tau^1 - \eta\tau^2),\\
u_{\boldsymbol{i},\boldsymbol{i}+2\boldsymbol{x}+\boldsymbol{y}} &= (-)^{i_x}\lambda\tau^3, & u_{\boldsymbol{i},\boldsymbol{i}-\boldsymbol{x}+2\boldsymbol{y}} &= -\lambda\tau^3,\\
u_{\boldsymbol{i},\boldsymbol{i}+2\boldsymbol{x}-\boldsymbol{y}} &= (-)^{i_x}\lambda\tau^3, & u_{\boldsymbol{i},\boldsymbol{i}+\boldsymbol{x}+2\boldsymbol{y}} &= -\lambda\tau^3,\\
a_0^l &= 0,
\end{aligned}
\tag{9.6.7}
$$

and Z2Bzz1n:

$$
\begin{aligned}
u_{\boldsymbol{i}.\boldsymbol{i}+\boldsymbol{x}} &= \chi\tau^1 + \eta\tau^2, & u_{\boldsymbol{i}.\boldsymbol{i}+\boldsymbol{y}} &= (-)^{i_x}(\chi\tau^1 - \eta\tau^2),\\
u_{\boldsymbol{i}.\boldsymbol{i}+2\boldsymbol{x}+\boldsymbol{y}} &= (-)^{i_x}(\chi_1\tau^1 + \eta_1\tau^2 + \lambda\tau^3), & u_{\boldsymbol{i}.\boldsymbol{i}+-\boldsymbol{x}+2\boldsymbol{y}} &= \chi_1\tau^1 - \eta_1\tau^2 + \lambda\tau^3,\\
u_{\boldsymbol{i}.\boldsymbol{i}+2\boldsymbol{x}-\boldsymbol{y}} &= (-)^{i_x}(\chi_1\tau^1 + \eta_1\tau^2 - \lambda\tau^3), & u_{\boldsymbol{i}.\boldsymbol{i}+\boldsymbol{x}+2\boldsymbol{y}} &= \chi_1\tau^1 - \eta_1\tau^2 - \lambda\tau^3,\\
a_0^l &= 0.
\end{aligned}
\tag{9.6.8}
$$

The spinons are gapless at four isolated points, with a linear dispersion for the first four Z_2 spin liquids given by eqns (9.6.1), (9.6.2), (9.6.3), and (9.6.4) (assuming that a_0^1 is small in eqn (9.6.1)) (see Fig. 9.8). Therefore, the four ansatz describe symmetric Z_2-linear spin liquids. The single-spinon dispersion for the second Z_2 spin liquid Z2Azz13 is quite interesting. It does not have the 90°-rotational symmetry. This is consistent with the 90° symmetry in the ground state, because excitations with odd numbers of spinons can never satisfy the constraint and are not allowed. The spinon dispersion has a 90°-rotational symmetry

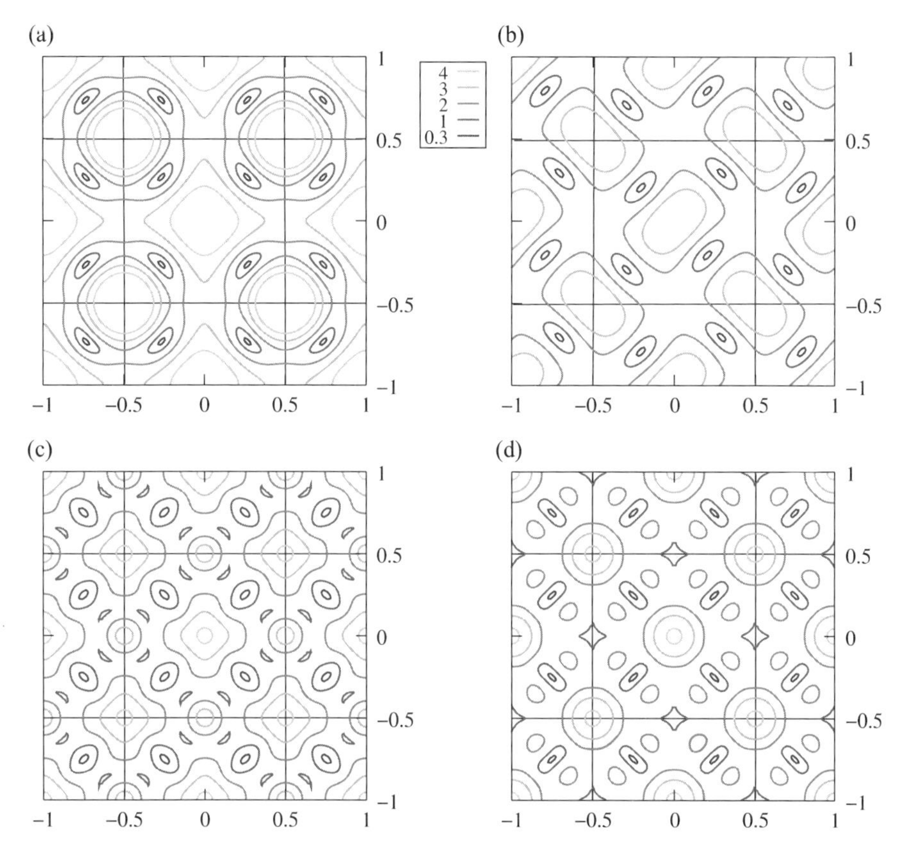

FIG. 9.8. Contour plots of the spinon dispersion $E_+(\boldsymbol{k})$ as a function of $(k_x/2\pi, k_y/2\pi)$ for the Z_2-linear spin liquids: (a) the Z2A0013 state in eqn (9.6.1); (b) the Z2Azz13 state in eqn (9.6.2); (c) the Z2A001n state in eqn (9.6.3); and (d) the Z2Azz1n state in eqn (9.6.4).

around $\boldsymbol{k} = (0, \pi)$, and spectra of excitations with even numbers of spinons have $90°$-rotational symmetry.

We would like to point out that, when a_0^1 is large, eqn (9.6.1) may have a gapped spinon spectrum. Thus, the Z2A0013 state can be a gapped spin liquid or a gapless spin liquid. The other three Z2A states are always gapless and are Z_2-linear states. In Section 9.9, we will show that all of the Z_2-gapped and the Z_2-linear states are stable. The mean-field results for these states can be applied to physical spin liquids.

Next let us consider the ansatz Z2B0013 in eqn (9.6.5). The spinon spectrum for the ansatz (9.6.5) is determined by the eigenvalues of

$$H = -2[\chi\cos(k_x)\Gamma_0 - \eta\cos(k_x)\Gamma_2 - \chi\cos(k_y)\Gamma_1 + \eta\cos(k_y)\Gamma_3] + \lambda\Gamma_4$$

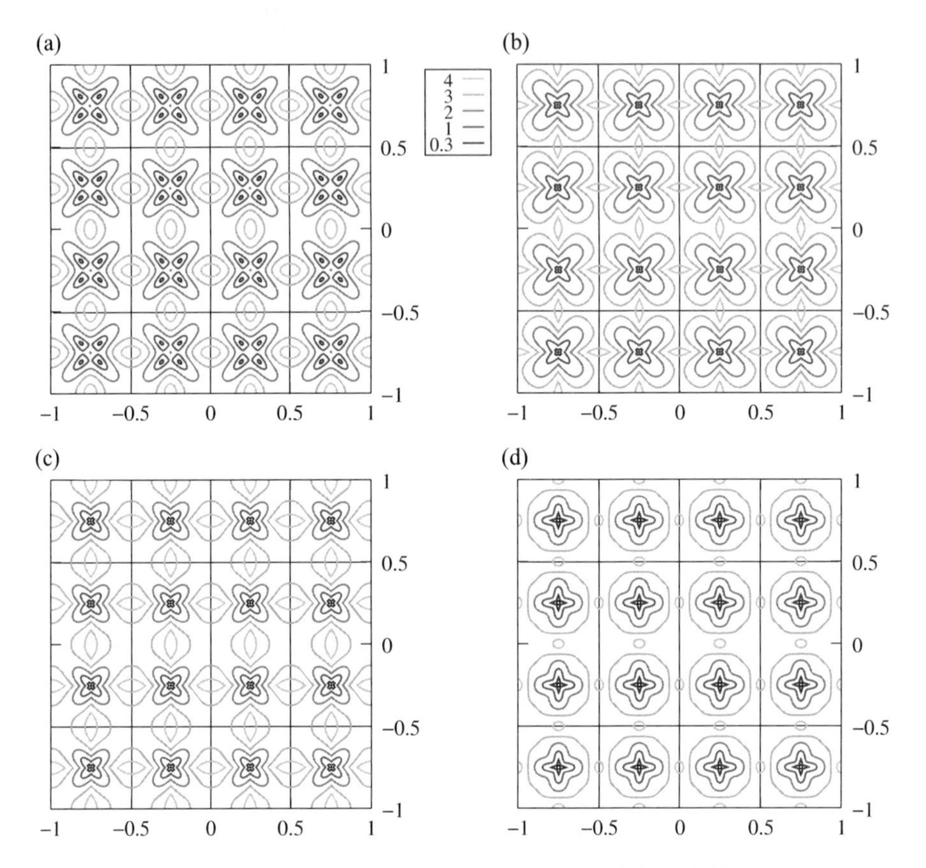

FIG. 9.9. Contour plots of the spinon dispersion $\min(E_1(\boldsymbol{k}), E_2(\boldsymbol{k}))$ as a function of $(k_x/2\pi, k_y/2\pi)$ for the Z_2-linear states: (a) the Z2B0013 state in eqn (9.6.5); (b) the Z2Bzz13 state in eqn (9.6.6); (c) the Z2B001n state in eqn (9.6.7); and (d) the Z2Bzz1n state in eqn (9.6.8).

where $k_x \in (0, \pi)$, $k_y \in (-\pi, \pi)$, and

$$\Gamma_0 = \tau^1 \otimes \tau^3, \qquad \Gamma_1 = \tau^1 \otimes \tau^1, \qquad \Gamma_2 = \tau^2 \otimes \tau^3,$$
$$\Gamma_3 = \tau^2 \otimes \tau^1, \qquad \Gamma_4 = \tau^1 \otimes \tau^0,$$

assuming that $\gamma_{1,2} = \lambda_2 = 0$. The four bands of the spinon dispersion have the form $\pm E_1(\boldsymbol{k})$, $\pm E_2(\boldsymbol{k})$. We find that the spinon spectrum vanishes at eight isolated points near $\boldsymbol{k} = (\pi/2, \pm\pi/2)$ (see Fig. 9.9(a)). Thus, the state Z2B0013 is a Z_2-linear spin liquid.

The spinon spectra of the other three Z2B states can be obtained in a similar way. The spectra are plotted in Fig. 9.9. The three Z2B states are also Z_2-linear states. The gapless spinons appear only at $(\frac{\pi}{2}, \pm\frac{\pi}{2})$.

Knowing the translational symmetry of the above Z2B spin liquid, it seems strange to find that the spinon spectrum is defined only on half of the lattice Brillouin zone. However, this is not inconsistent with translational symmetry in the physical spin liquid, because the single-spinon excitation is not physical. Only two-spinon excitations correspond to physical excitations and their spectrum should be defined on the full Brillouin zone. Now the problem is how to obtain the two-spinon spectrum defined on the full Brillouin zone from the single-spinon spectrum defined on half of the Brillouin zone. Let $|\boldsymbol{k},1\rangle$ and $|\boldsymbol{k},2\rangle$ be the two eigenstates of a single spinon with positive energies $E_1(\boldsymbol{k})$ and $E_2(\boldsymbol{k})$, respectively (here $k_x \in (-\pi/2, \pi/2)$ and $k_y \in (-\pi, \pi)$). The translation by $\boldsymbol{x}$ (followed by a gauge transformation) changes $|\boldsymbol{k},1\rangle$ and $|\boldsymbol{k},2\rangle$ to the other two eigenstates with the same energies as follows:

$$
\begin{aligned}
|\boldsymbol{k},1\rangle &\to |\boldsymbol{k}+\pi\boldsymbol{y},1\rangle \\
|\boldsymbol{k},2\rangle &\to |\boldsymbol{k}+\pi\boldsymbol{y},2\rangle
\end{aligned}
$$

Now we see that the momentum and the energy of two-spinon states $|\boldsymbol{k}_1,\alpha_1\rangle|\boldsymbol{k}_2,\alpha_2\rangle \pm |\boldsymbol{k}_1+\pi\boldsymbol{y},\alpha_1\rangle|\boldsymbol{k}_2+\pi\boldsymbol{y},\alpha_2\rangle$ are given by

$$
\begin{aligned}
E_{2-\text{spinon}} &= E_{\alpha_1}(\boldsymbol{k}_1) + E_{\alpha_2}(\boldsymbol{k}_2) \\
\boldsymbol{k} &= \boldsymbol{k}_1 + \boldsymbol{k}_2, \;\; \boldsymbol{k}_1 + \boldsymbol{k}_2 + \pi\boldsymbol{x}
\end{aligned}
\tag{9.6.9}
$$

Equation (9.6.9) allows us to construct the two-spinon spectrum from the single-spinon spectrum.

9.6.2 A strange symmetric spin liquid around $SU(2)$ spin liquids

There are many types of symmetric ansatz in the neighborhood of the uniform RVB state (or the $SU(2)$-gapless state SU2An0) in eqn (9.2.32) and in the neighborhood of the π-flux state (or the $SU(2)$-linear state SU2Bn0) in eqn (9.2.33). Here we will only consider one of them—the Z_2 spin liquid Z2By1(12)n (note that Z2By1(12)n is gauge equivalent to Z2Bx2(12)n) which takes the following form:

$$
u_{\boldsymbol{i},\boldsymbol{i}+\boldsymbol{x}} = \mathrm{i}\chi\tau^0 + \eta\tau^1, \quad u_{\boldsymbol{i},\boldsymbol{i}+\boldsymbol{y}} = (-)^{i_x}(\mathrm{i}\chi\tau^0 + \eta\tau^2), \quad a_0^{1,2,3} = 0. \tag{9.6.10}
$$

The above ansatz reduces to the ansatz of the $SU(2)$-gapless state SU2An0 when $\chi = 0$, and reduces to the $SU(2)$-linear state SU2Bn0 when $\eta = 0$. After the Fourier transformation, we find that the spinon spectrum is determined by

$$
H = -2\chi\sin(k_x)\Gamma_0 + 2\eta\cos(k_x)\Gamma_2 - 2\chi\sin(k_y)\Gamma_1 + 2\eta\cos(k_y)\Gamma_3
$$

where $k_x \in (-\pi/2, \pi/2)$, $k_y \in (-\pi, \pi)$, and

$$
\Gamma_0 = \tau^0 \otimes \tau^3, \qquad \Gamma_2 = \tau^1 \otimes \tau^3, \qquad \Gamma_1 = \tau^0 \otimes \tau^1, \qquad \Gamma_3 = \tau^2 \otimes \tau^1.
$$

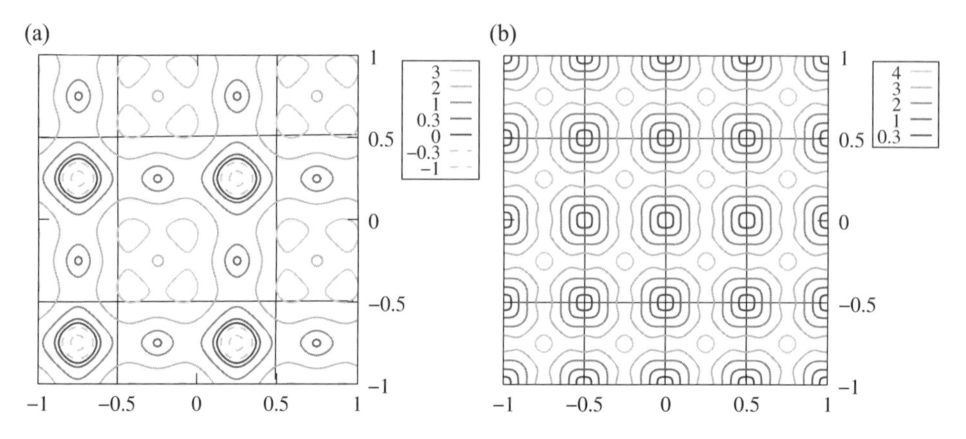

FIG. 9.10. Contour plots of the spinon dispersion $E_+(\boldsymbol{k})$ as a function of $(k_x/2\pi, k_y/2\pi)$ for the Z_2 spin liquids: (a) the Z_2-gapless state Z2Ax2(12)n in eqn (9.2.39); and (b) the Z_2-quadratic state Z2Bx2(12)n in eqn (9.6.10). Despite the lack of rotational and parity symmetries in the single-spinon dispersion in (a), the two-spinon spectrum does have these symmetries.

The spinon spectrum can be calculated exactly and its four branches take the form $\pm E_1(\boldsymbol{k})$ and $\pm E_2(\boldsymbol{k})$. The spinon energy vanishes at the two isolated points $\boldsymbol{k} = (0,0)$ and $(0,\pi)$. Near $\boldsymbol{k} = 0$ the low-energy spectrum is given by (see Fig. 9.10(b))

$$E = \pm\eta^{-1}\sqrt{(\chi^2+\eta^2)^2(k_x^2-k_y^2)^2+4\chi^4k_x^2k_y^2}$$

It is interesting (and strange) to see that the energy does not vanish linearly as $\boldsymbol{k} \to 0$; instead it vanishes like $\boldsymbol{k}^2$! We will call such a state a Z_2-quadratic spin liquid to stress the quadratic $E \propto \boldsymbol{k}^2$ dispersion.

Problem 9.6.1.
The PSG of the ansatz (9.6.3)

1. Find the gauge transformations G_x, G_y, G_{P_x}, etc. for the Z2A001n PSG from the label Z2A001n.
2. Show that the ansatz (9.6.3) is invariant under the Z2A001n PSG.
3. Find the gauge transformation that transforms the Z2A001n PSG to the Z2A003n PSG.

Problem 9.6.2.
Check that, for the spin liquid described by eqn (9.6.10), the $SU(2)$-flux operators (see eqn (9.2.20)) for the loops $\boldsymbol{i} \to \boldsymbol{i}+\boldsymbol{x} \to \boldsymbol{i}+\boldsymbol{x}+\boldsymbol{y} \to \boldsymbol{i}+\boldsymbol{y} \to \boldsymbol{i}$ and $\boldsymbol{i} \to \boldsymbol{i}+\boldsymbol{y} \to \boldsymbol{i}-\boldsymbol{x}+\boldsymbol{y} \to \boldsymbol{i}-\boldsymbol{x} \to \boldsymbol{i}$ do not commute as long as both χ and η are nonzero. Thus, the spin liquid indeed has a Z_2 gauge structure.

Problem 9.6.3.
(a) Show that the ansatz in eqn (9.6.10) reduces to the ansatz of the $SU(2)$-gapless state SU2An0 when $\eta = 0$, and reduces to the $SU(2)$-gapless state SU2Bn0 when $\chi = 0$.
(b) Find the spinon spectrum $\pm E_1$ and $\pm E_2$.

9.7 Physical measurements of quantum orders

- Quantum orders described by PSGs can be measured via spectra of excitations.

After characterizing the quantum orders using the PSG mathematically, we would like to ask how to measure quantum orders in experiments. The quantum orders in gapped states are related to the topological orders. We can use ground-state degeneracy, edge states, quasiparticle statistics, etc. to measure the topological orders (Wen, 1990, 1995; Senthil and Fisher, 2001). The quantum orders in a state with gapless excitations need to be measured differently. In this section, we would like to demonstrate that quantum orders can be measured, in general, by the dynamical properties of gapless excitations. However, not all of the dynamical properties are universal. Thus, we need to identify the universal properties of gapless excitations, before using them to characterize and measure quantum orders. The PSG characterization of quantum orders allows us to obtain these universal properties—we simply need to identify the common properties of gapless excitations that are shared by all of the ansatz with the same PSG.

To demonstrate the above idea, we would like to study the spectrum of two-spinon excitations. We note that spinons can only be created in pairs. Thus, the one-spinon spectrum is not physical. We also note that the two-spinon spectrum includes spin-1 excitations, which can be measured by neutron-scattering experiments.

At a given momentum, the two-spinon spectrum is distributed in one or several ranges of energies. Let $E_{2s}(\boldsymbol{k})$ be the lower edge of the two-spinon spectrum at momentum $\boldsymbol{k}$. In mean-field theory, the two-spinon spectrum can be constructed from the one-spinon dispersion as follows:

$$E_{2\text{-spinon}}(\boldsymbol{k}) = E_{1\text{-spinon}}(\boldsymbol{q}) + E_{1\text{-spinon}}(\boldsymbol{k} - \boldsymbol{q})$$

In Figs 9.11–9.14, we present the mean-field E_{2s} for some simple spin liquids. If the mean-field state is stable against the gauge fluctuations, then we expect that the mean-field E_{2s} should qualitatively agree with the real E_{2s}.

Among our examples, there are eight stable Z_2-linear spin liquids (see Figs 9.11 and 9.12). Using the eight Z_2 spin liquids as examples, we can demonstrate how the universal properties of spin-1 excitations can distinguish different quantum orders.

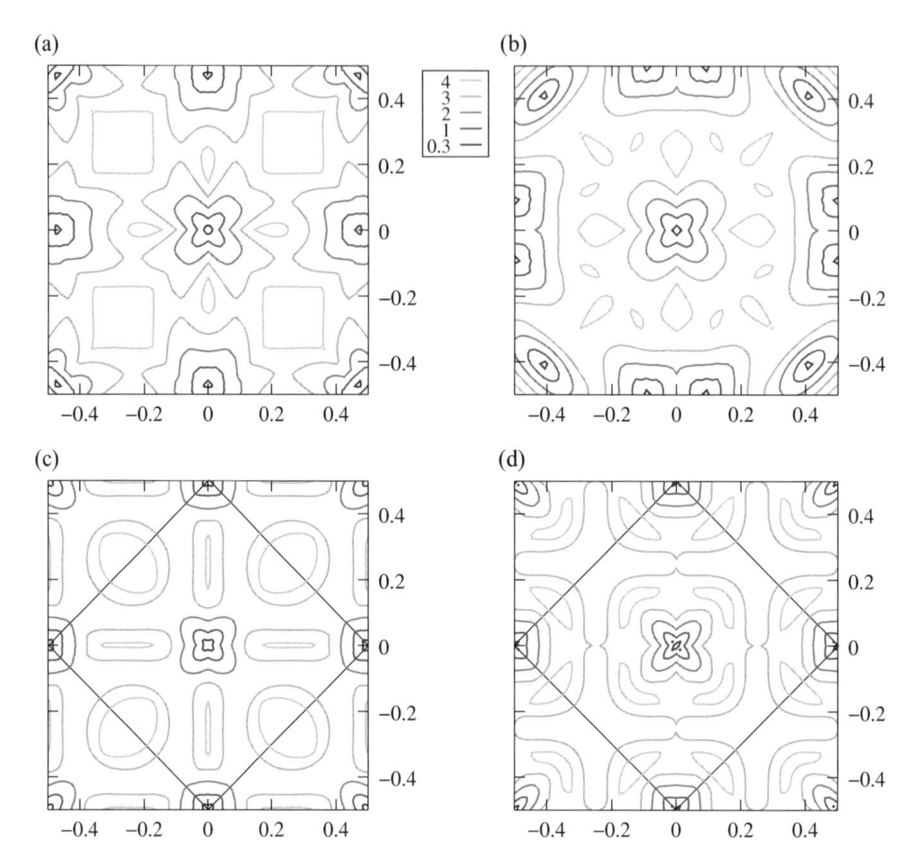

FIG. 9.11. Contour plots of $E_{2s}(\boldsymbol{k})$ as a function of $(k_x/2\pi, k_y/2\pi)$ for the Z_2-linear spin liquids: (a) the Z2A0013 state in eqn (9.6.1); (b) the Z2Azz13 state in eqn (9.6.2); (c) the Z2A001n state in eqn (9.6.3); and (d) the Z2Azz1n state in eqn (9.6.4).

(a) The Z2A spin liquids can be distinguished from the Z2B spin liquids by examining the spectrum of spin-1 excitations. For the Z2B spin liquids, the spin-1 spectrum is periodic in one-quarter of the Brillouin zone (i.e. the spectrum is invariant under $k_x \to k_x + \pi$ and $k_y \to k_y + \pi$). In contrast, the spin-1 spectrum in the Z2A spin liquids does not have such a periodicity.
(b) The periodicity of the spin-1 spectrum can also distinguish the Z2A0013 and Z2Azz13 spin liquids from the Z2A001n and Z2Azz1n spin liquids. The spin-1 spectrum is periodic in one-half of the Brillouin zone (i.e. the spectrum is invariant under $(k_x, k_y) \to (k_x + \pi, k_y + \pi)$) for the Z2A001$n$ and Z2Azz1n spin liquids. Also, the Z2A001n and Z2Azz1n spin liquids have gapless spin-1 excitations at exactly (π, π), $(\pi, 0)$, and $(0, \pi)$.
(c) The Z2A0013 and Z2Azz13 spin liquids can be distinguished by examining the

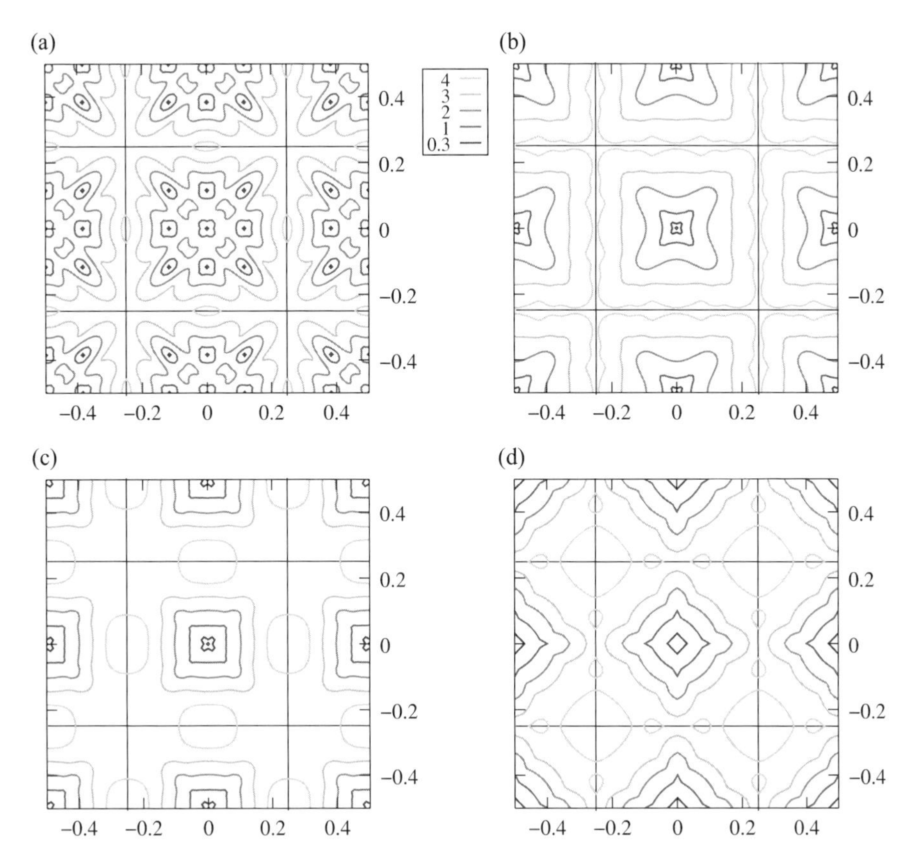

FIG. 9.12. Contour plots of $E_{2s}(\boldsymbol{k})$ as a function of $(k_x/2\pi, k_y/2\pi)$ for the Z_2-linear spin liquids: (a) the Z2B0013 state in eqn (9.6.5); (b) the Z2Bzz13 state in eqn (9.6.6); (c) the Z2B001n state in eqn (9.6.7); and (d) the Z2Bzz1n state in eqn (9.6.8). The spectra are periodic in one-quarter of the Brillouin zone.

gapless spin-1 excitations near $(\pi, 0)$ and $(0, \pi)$. The gapless spin-1 excitations in the Z2Azz13 spin liquids are located along the zone boundary.

We see that quantum orders in spin liquids can be measured by neutron-scattering experiments, which probe the spin-1 excitations.

Next, let us discuss the $U(1)$-linear state U1Cn01n (the staggered-flux state). The U1Cn01n state was proposed to describe the pseudo-gap metallic state in underdoped high-T_c superconductors (Wen and Lee, 1996; Rantner and Wen, 2001). The U1Cn01n state naturally explains the pseudo-gap in the underdoped metallic state. As an algebraic spin liquid, the U1Cn01n state also explains the Luttinger-like electron spectral function (Rantner and Wen, 2001; Franz and Tesanovic, 2001) and the enhancement of the (π, π)-spin fluctuations (Kim and

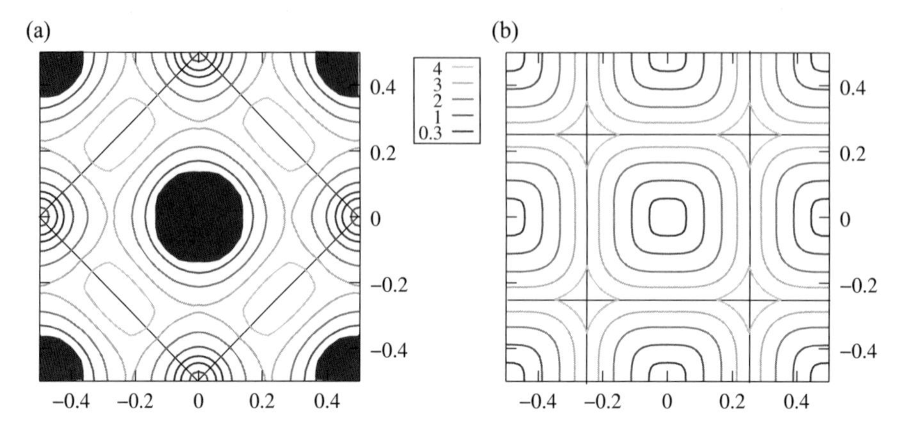

FIG. 9.13. Contour plots of $E_{2s}(\boldsymbol{k})$ as a function of $(k_x/2\pi, k_y/2\pi)$ for (a) the Z_2-gapless state Z2Ax2(12)n in eqn (9.2.39), and (b) the Z_2-quadratic state Z2Bx2(12)n in eqn (9.6.10).

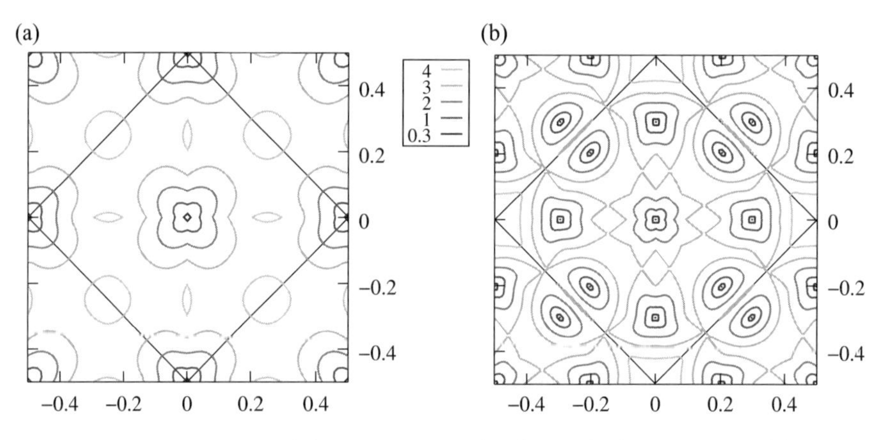

FIG. 9.14. Contour plots of $E_{2s}(\boldsymbol{k})$ as a function of $(k_x/2\pi, k_y/2\pi)$ for the two $U(1)$-linear spin liquids: (a) the U1Cn01n state in eqn (9.2.34) (the staggered-flux phase); and (b) the U1Cn00x state in eqn (9.8.9) in the gapless phase. The spectra are periodic in one-half of the Brillouin zone.

Lee, 1999; Rantner and Wen, 2002) in the pseudo-gap state. From Fig. 9.14, we see that gapless points of the spin-1 excitations in the U1Cn01n state are always at $\boldsymbol{k} = (\pi, \pi)$, $(0, 0)$, $(\pi, 0)$, and $(0, \pi)$. The equal-energy contour for the edge of the spin-1 continuum has the shape of two overlapped ellipses at all of the four $\boldsymbol{k}$ points. Also, the energy contours are not perpendicular to the zone boundary. All of these are the universal properties of the U1Cn01n state. Measuring these properties in neutron-scattering experiments will allow us to determine if the pseudo-gap metallic state is described by the U1Cn01n (the staggered-flux) state or not.

The U1Cn01n state is unstable due to the instanton effect of the $(2+1)$-dimensional $U(1)$ gauge theory. Thus, the U1Cn01n state has to change into some other states, such as the eight Z_2 spin liquids discussed in Section 9.6 or other states not discussed here. From Fig. 9.11(a), we see that the transition from the U1Cn01n state to the Z_2-linear state Z2A0013 can be detected by neutron scattering if one observes the splitting of the node at (π,π) into four nodes at $(\pi\pm\delta,\pi\pm\delta)$, and the splitting of the nodes at $(\pi,0)$ and $(0,\pi)$ into the two nodes at $(\pi\pm\delta,0)$ and $(0,\pi\pm\delta)$. From Fig. 9.11(b), we see that, for the transition from the U1Cn01n state to the Z_2-linear state Z2Azz13, the node at (π,π) still splits into the four nodes at $(\pi\pm\delta,\pi\pm\delta)$. However, the nodes at $(\pi,0)$ and $(0,\pi)$ split differently into the two nodes at $(\pi,\pm\delta)$ and $(\pm\delta,\pi)$. We can also study the transition from the U1Cn01n state to the other six Z_2 spin liquids. We find that the spectrum of spin-1 excitations changes in certain characteristic ways. Thus, by measuring the spin-1 excitation spectrum and its evolution, not only can we detect a quantum transition that does not change any symmetries, but we can also tell which transition is happening.

9.8 The phase diagram of the J_1–J_2 model in the large-N limit

9.8.1 The large-N limit

- The spin-1/2 model can be generalized to the $SP(2N)$ model.
- For the $SP(2N)$ model, the mean-field theory obtained from the $SU(2)$ projective construction has weak fluctuations and is a good approximation.

So far, we have been concentrating on how to characterize, classify, and measure different spin liquids and their quantum orders. We have not discussed how to find physical spin Hamiltonians to realize some of the hundreds of different spin liquids that we have constructed. In this section, we are going to address this issue.

At the mean-field level, it is not very hard to design a spin Hamiltonian that realizes a spin liquid with a given quantum order. It is also not hard to find the mean-field ground state for a given spin Hamiltonian. The real issue is whether we should trust the mean-field results. We have seen that, if the obtained mean-field ground state is unstable (i.e. if the mean-field fluctuations cause diverging interactions at low energies), then the mean-field result cannot be trusted and the mean-field state does not correspond to any real physical spin state. We have also argued that, if the mean-field ground state is stable (if the mean-field fluctuations cause vanishing interactions at low energies), then the mean-field result can be trusted and the mean-field state does correspond to a real physical spin state.

Here we would like to point out that the above statement about stable mean-field states is too optimistic. A 'stable mean-field state' does not have diverging fluctuations at low energies. So it does not have to be unstable. On the other hand, it does not have to be stable either. This is because short-distance fluctuations, if strong enough, can also cause phase transitions and instabilities. Therefore, in order for a mean-field result to be reliable, the mean-field state must be stable (or marginal) *and* the short-distance fluctuations must be weak. As we do not have any small parameters in our spin model, the short-distance fluctuations are not weak, even for stable mean-field states. Due to this, it is not clear if the mean-field results, even for the stable states, can be applied to our spin-1/2 model or not.

In the following, we are going to generalize our spin model to a large-N model. We will show that the mean-field states for the large-N model have weak short-distance fluctuations. Thus, the stable and the marginal mean-field states for the large-N model correspond to real physical states. The mean-field results for the stable or marginal states can be applied to the large-N model.

Let us start with the following path integral representation of the $SU(2)$ mean-field theory:

$$Z = \int \mathcal{D}\psi\mathcal{D}[a_0^l(\boldsymbol{i})]\mathcal{D}U_{\boldsymbol{ij}} \; \mathrm{e}^{\,\mathrm{i}\int \mathrm{d}t\; (L - \sum_{\boldsymbol{i}} a_0^l(\boldsymbol{i},t)\psi_{\boldsymbol{i}}^\dagger \tau^l \psi_{\boldsymbol{i}})}$$

$$L = \sum_{\boldsymbol{i}} \psi_{\boldsymbol{i}}^\dagger \mathrm{i}\partial_t \psi_{\boldsymbol{i}} - \sum_{\langle \boldsymbol{ij}\rangle} \frac{1}{4} J_{ij} \left[\frac{1}{2} \mathrm{Tr} U_{\boldsymbol{ij}} U_{\boldsymbol{ij}}^\dagger - (\psi_{\boldsymbol{i}}^\dagger U_{\boldsymbol{ij}} \psi_{\boldsymbol{j}} + h.c) \right]. \tag{9.8.1}$$

which is obtained from the $SU(2)$ mean-field Hamiltonian (9.2.9), and $U_{\boldsymbol{ij}}$ has the form given in eqn (9.2.13).[73] After integrating out the fermions, we obtain the effective Lagrangian for $U_{\boldsymbol{ij}}$ and a_0^l:

$$Z = \int \mathcal{D}[a_0^l(\boldsymbol{i})]\mathcal{D}U_{\boldsymbol{ij}} \; \mathrm{e}^{\,\mathrm{i}\int \mathrm{d}t\; L_{0,eff}(U_{\boldsymbol{ij}}, a_0^l)}$$

The problem is that the fluctuations of $U_{\boldsymbol{ij}}$ and a_0^l are not weak in the above path integral.

To reduce the fluctuations, we introduce N copies of the fermions $\psi_{\boldsymbol{i}}^a$ and generalize eqn (9.8.1) to the following path integral (Ran and Wen, 2003):

$$Z = \int \mathcal{D}\psi\mathcal{D}[a_0^l(\boldsymbol{i})]\mathcal{D}U_{\boldsymbol{ij}} \; \mathrm{e}^{\,\mathrm{i}\int \mathrm{d}t\; (L - \sum_{\boldsymbol{i}} a_0^l(\boldsymbol{i},t)\psi_{\boldsymbol{i}}^{a\dagger} \tau^l \psi_{\boldsymbol{i}}^a)}$$

$$L = \sum_{\boldsymbol{i}} \psi_{\boldsymbol{i}}^{a\dagger} \mathrm{i}\partial_t \psi_{\boldsymbol{i}}^a - \sum_{\langle \boldsymbol{ij}\rangle} \frac{1}{4} J_{ij} \left[\frac{N}{2} \mathrm{Tr} U_{\boldsymbol{ij}} U_{\boldsymbol{ij}}^\dagger - (\psi_{\boldsymbol{i}}^{a\dagger} U_{\boldsymbol{ij}} \psi_{\boldsymbol{j}}^a + h.c) \right] \tag{9.8.2}$$

where $a = 1, 2, ..., N$. After integrating out the fermions, we obtain

$$Z = \int \mathcal{D}[a_0^l(\boldsymbol{i})]\mathcal{D}U_{\boldsymbol{ij}} \; \mathrm{e}^{\,\mathrm{i}\int \mathrm{d}t\; N L_{0,eff}(U_{\boldsymbol{ij}}, a_0^l)}$$

We see that, in the large-N limit, the fluctuations of $U_{\boldsymbol{ij}}$ and a_0^l are weak and the mean-field approximation is a good approximation. It is also clear that the large-N mean-field theory is an $SU(2)$ gauge theory. The fluctuations of $U_{\boldsymbol{ij}}$ and a_0^l correspond to the $SU(2)$ gauge fluctuations.

In Section 9.2.1, we constructed the mean-field Hamiltonian from the physical spin Hamiltonian. Here we are facing the opposite problem. Knowing the large-N mean-field Hamiltonian

$$H_{\text{mean}} = \sum_{\langle \boldsymbol{ij}\rangle} \frac{1}{4} J_{ij} \left[\frac{N}{2} \mathrm{Tr} U_{\boldsymbol{ij}} U_{\boldsymbol{ij}}^\dagger - (\psi_{\boldsymbol{i}}^{a\dagger} U_{\boldsymbol{ij}} \psi_{\boldsymbol{j}}^a + h.c) \right] + \sum_{\boldsymbol{i}} a_0^l \psi_{\boldsymbol{i}}^{a\dagger} \tau^l \psi_{\boldsymbol{i}}^a \tag{9.8.3}$$

we would like to find the corresponding physical large-N spin model.

[73] We have changed the coefficient $\frac{3}{8}$ to $\frac{1}{4}$ so that integrating out $U_{\boldsymbol{ij}}$ in eqn (9.8.1) will lead to the spin Hamiltonian (9.1.1).

The most important step in constructing the physical model is to find the physical Hilbert space. The physical Hilbert space is a subspace of the fermion Hilbert space. The physical Hilbert space is formed by the $SU(2)$ gauge-invariant states, i.e. the states that satisfy the constraint

$$\psi_{\boldsymbol{i}}^{a\dagger}\tau^l\psi_{\boldsymbol{i}}^a|phy\rangle, \qquad l=1,2,3$$

on every site $\boldsymbol{i}$.

After obtaining the physical Hilbert space on each lattice site, we need to find the physical operators that act within the physical Hilbert space. These physical operators are $SU(2)$ gauge-invariant operators (i.e. the operators that commute with $\psi_{\boldsymbol{i}}^{a\dagger}\tau^l\psi_{\boldsymbol{i}}^a$). Let us write down all of the $SU(2)$ gauge-invariant bilinear forms of ψ for each site as follows:

$$S^{ab+} \equiv \frac{1}{2}\psi_\alpha^{a\dagger}\tilde{\psi}_\alpha^b, \qquad S^{ab-} \equiv \frac{1}{2}\tilde{\psi}_\alpha^{a\dagger}\psi_\alpha^b,$$
$$S^{ab3} \equiv \frac{1}{2}\left(\psi_\alpha^{a\dagger}\psi_\alpha^b - \delta^{ab}\right) = \frac{1}{2}\left(\delta^{ab} - \tilde{\psi}_\alpha^{b\dagger}\tilde{\psi}_\alpha^a\right)$$

where $\tilde{\psi}^a \equiv \mathrm{i}\sigma_2\psi^{a*}$ and the site index has been suppressed. These S operators are the generalization of the spin operators for the spin-1/2 model.

When $N = 1$, the large-N model becomes the spin-1/2 model and the S operator generates the $SU(2)$ spin–rotation group. For $N > 1$, what is the group generated by the S operators? Firstly, let us count the number of different S operators. For S^{ab+} or S^{ab-}, the label is symmetric for a and b, so there are $\frac{N(N+1)}{2}$ different operators of each type. For S^{ab3}, the labels are not symmetric, so there are simply N^2 of them. In total, we have $N(N+1) + N^2 = 2N^2 + N$ different S operators.

One can examine the following commutation relations between the S operators:

$$\left[S^{ab3},\, S^{cd3}\right] = \frac{1}{2}\left(\delta^{bc}S^{ad3} - \delta^{ad}S^{cb3}\right)$$
$$\left[S^{ab3},\, S^{cd+}\right] = \frac{1}{2}\left(\delta^{bc}S^{ad+} + \delta^{bd}S^{ac+}\right)$$
$$\left[S^{ab3},\, S^{cd-}\right] = -\frac{1}{2}\left(\delta^{ad}S^{bc-} + \delta^{ac}S^{bd-}\right)$$
$$\left[S^{ab+},\, S^{cd-}\right] = \frac{1}{2}\left(\delta^{ac}S^{bd3} + \delta^{ad}S^{bc3} + \delta^{bc}S^{ad3} + \delta^{bd}S^{ac3}\right)$$
$$\left[S^{ab-},\, S^{cd-}\right] = 0, \qquad \left[S^{ab+},\, S^{cd+}\right] = 0$$

These are the relations for the $SP(2N)$ algebra. So the S operators are the $2N^2 + N$ generators which generate the $SP(2N)$ group. When $N = 1$, $SP(2)$ is isomorphic to the $SU(2)$ spin–rotation group.

After integrating out $U_{\boldsymbol{ij}}$ and a_0^l in eqn (9.8.2), we obtain the following physical Hamiltonian for the large-N model:

$$H = \sum_{\langle \boldsymbol{ij}\rangle} \frac{J_{\boldsymbol{ij}}}{N}\,\mathbf{S}_{\boldsymbol{i}}^{ab}\cdot\mathbf{S}_{\boldsymbol{j}}^{ba}$$

where

$$S_i^{ab1} = S_i^{ba1} = \frac{1}{2}\left(S_i^{ab+} + S_i^{ab-}\right)$$
$$S_i^{ab2} = S_i^{ba2} = \frac{1}{2i}\left(S_i^{ab+} - S_i^{ab-}\right),$$

and

$$\boldsymbol{S}_i^{ab} = \left(S_i^{ab1}, S_i^{ab2}, S_i^{ab3}\right)$$

We find that H commutes with the $SP(2N)$ generators $\sum_i \boldsymbol{S}_i^{ab}$. Thus, our large-$N$ model has an $SP(2N)$ symmetry. Here we would like to point out that the three components of $\boldsymbol{S}_i^{ab}$ are not actually on the same footing. The first two are symmetric with respect to the a and b labels, but the third one is not.

We know that, when $N = 1$, the physical Hilbert space has two states per site. The two states form an irreducible representation of $SP(2) = SU(2)$. For higher N, the dimensions of the physical Hilbert space per site are

$$\begin{array}{rcccccc} N = & 2 & 3 & 4 & 5 & 6 & \dots \\ \text{Dimension} = & 5 & 14 & 43 & 142 & 429 & \dots. \end{array} \tag{9.8.4}$$

These physical Hilbert spaces turn out to be irreducible representations of the $SP(2N)$ symmetry group. We can label an irreducible representation by its highest-weight state for a particular Cartan basis. The Cartan basis for $SP(2N)$ can be chosen to be the z component spins for each a, i.e.

$$S^{aa3}, \quad a = 1, 2, \dots, N. \tag{9.8.5}$$

Then the highest-weight state in the physical Hilbert space is simply the state with no ψ fermions

9.8.2 The phase diagram of the $SP(2N)$ model

- The mean-field phase diagram is the phase diagram of the $SP(2N)$ model in the large-N limit.
- The different phases in the mean-field theory or in the $SP(2N)$ model can be labeled by PSGs.

From eqn (9.8.3), we see that the $SU(2)$ mean-field theory for the $SP(2N)$ model has the same structure as the $SU(2)$ mean-field theory for the spin-1/2 model. The mean-field states are described by the ansatz $(U_{ij}, a_0^l(i))$ for both models. In particular, the mean-field energy for the $SP(2N)$ model is just N times the mean-field energy of the $SP(2)$ model (i.e. the spin-1/2 model). As a result, the mean-field phase diagram does not depend on N. Just like the spin-1/2 model discussed before, different mean-field states for the $SP(2N)$ model are characterized and classified by PSGs.

Here, we consider a particular $SP(2N)$ model on a square lattice. The model has a nearest-neighbor coupling J_1 and a next-nearest-neighbor coupling J_2. We fix $J_1 + J_2 = 1$.

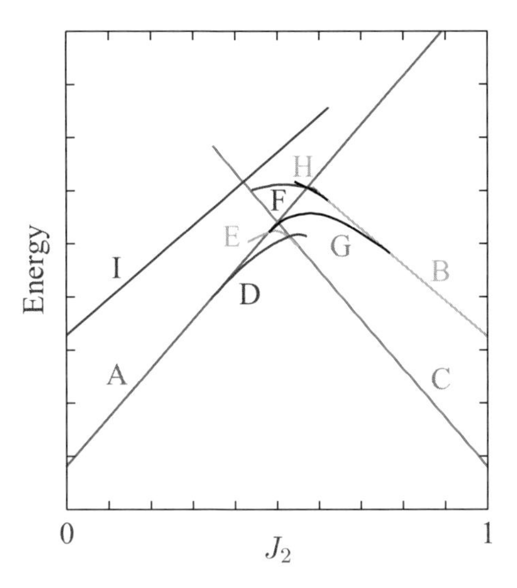

FIG. 9.15. The mean-field energies for the various phases in a J_1–J_2 spin system: 'A' labels the π-flux state (the $SU(2)$-linear state SU2Bn0); 'B' labels the $SU(2) \times SU(2)$-gapless state in eqn (9.8.6); 'C' labels the $SU(2) \times SU(2)$-linear state in eqn (9.8.7); 'D' labels the chiral spin state (an $SU(2)$-gapped state); 'E' labels the $U(1)$-linear state (9.8.8) which breaks $90°$-rotational symmetry; 'F' labels the $U(1)$-gapped state U1C$n00x$ in eqn (9.8.9); 'G' labels the Z_2-linear state Z2Azz13 in eqn (9.6.2); 'H' labels the Z_2-linear state Z2A0013 in eqn (9.6.1); and 'I' labels the uniform RVB state (the $SU(2)$-gapless state SU2An0).

In the large-N limit, the $SU(2)$ mean-field theory (9.8.3) is a good approximation. So we will use the mean-field theory to calculate the phase diagram of the $SP(2N)$ model. The mean-field ansatz that describes the mean-field ground state is calculated by minimizing the mean-field energy. The result is presented in Fig. 9.15. Each curve in Fig. 9.15 represents the mean-field energy of a local minimum. The curve with the lowest energy corresponds to the mean-field ground state. We find that the different ansatz on each curve share the same PSG. This is expected because, as we change J_2, the mean-field energy described by the curve changes in an analytic way. Thus, there is no quantum phase transition as we move from one point of a curve to another point on the same curve. The ansatz on the same curve belong to the same phase and are described by the same PSG.

However, if two curves cross each other, then the crossing point represents a quantum phase transition. This is because the ground-state energy is not analytic at the crossing point. As different curves have different PSGs, we see that a quantum phase transition is characterized by a change in the PSG. In the following, we will discuss different quantum orders (or PSGs) for the mean-field phases in Fig. 9.15.

In Fig. 9.15, the phase A is the π-flux state (the SU2Bn0 $SU(2)$-linear state) given in eqn (9.2.33). The phase B is a state with two independent uniform RVB states on the diagonal links. It has $SU(2) \times SU(2)$ gauge fluctuations at low energies and will be called an $SU(2) \times SU(2)$-gapless state. Its ansatz is given by

$$u_{i,i+x+y} = \chi\tau^3, \qquad u_{i,i+x-y} = \chi\tau^3, \qquad a_0^l = 0. \tag{9.8.6}$$

The phase C is a state with two independent π-flux states on the diagonal links. It has $SU(2)\times SU(2)$ gauge fluctuations at low energies and will be called an $SU(2)\times SU(2)$-linear state. Its ansatz is given by

$$u_{i,i+x+y} = \chi(\tau^3 + \tau^1), \qquad u_{i,i+x-y} = \chi(\tau^3 - \tau^1), \qquad a_0^l = 0 \tag{9.8.7}$$

The phase D is the chiral spin state (9.1.22). The phase E is described by the ansatz

$$\begin{aligned} u_{i,i+x+y} &= \chi_1\tau^1 + \chi_2\tau^2 & u_{i,i+x-y} &= \chi_1\tau^1 - \chi_2\tau^2 \\ u_{i,i+y} &= \eta\tau^3 & a_0^l &= 0 \end{aligned} \tag{9.8.8}$$

which breaks the $90°$-rotational symmetry and is a $U(1)$-linear state. The phase F is described by the following U1C$n00x$ ansatz:

$$\begin{aligned} u_{i,i+x} &= \eta\tau^1 & u_{i,i+y} &= \eta\tau^1 \\ u_{i,i+x+y} &= \chi\tau^3 & u_{i,i+x-y} &= \chi\tau^3 \\ a_0^3 &= \lambda, & a^{1,2} &= 0. \end{aligned} \tag{9.8.9}$$

The U1C$n00x$ state can be a $U(1)$-linear (if a_0^3 is small) or a $U(1)$-gapped state (if a_0^3 is large). The state for the phase F turns out to be a $U(1)$-gapped state. The phase G is described by the Z2A$zz13$ ansatz in eqn (9.6.2), which is a Z_2-linear state. The phase H is described by the Z2A0013 ansatz in eqn (9.6.1) and is also a Z_2-linear state. The phase I is the uniform RVB state (the $SU(2)$-gapless state SU2A$n0$ in eqn (9.2.32)).

From Fig. 9.15, we observe continuous phase transitions (at the mean-field level) between the following pairs of phases: (A,D), (A,G), (B,G), (C,E), and (B,H). The three continuous transitions (B,G), (B,H), and (A,G) do not change any symmetries. We also note that the $SU(2)$ gauge structure in the phase A breaks down to Z_2 in the continuous transition from the phase A to the phase G. The $SU(2) \times SU(2)$ gauge structure in the phase B breaks down to Z_2 in the two transitions (B,G) and (B,H).

9.9 Quantum order and the stability of mean-field spin liquids

- Many gapless mean-field spin liquids can be stable against quantum fluctuations. They can be stable even in the presence of long-range gauge interactions.

We have stressed the importance of the stability of mean-field states against the mean-field fluctuations. Only stable mean-field states and marginal mean-field states have a chance to describe physical spin liquids. We have discussed the ways to obtain stable mean-field states in Sections 9.1.5 and 9.2.2. Here we will discuss the stability of the mean-field states again in the light of quantum orders and their PSG characterization.

9.9.1 The projective symmetry group—a universal property of quantum phases

- The PSG is a universal property of a real quantum phase if the corresponding mean-field state is stable (i.e. has no infra-red divergence). In that case, the PSG describes the quantum order in the quantum phase.

In this section, we would like to show that the PSG can be a universal property of a quantum state, in the sense that it is robust against perturbative fluctuations. Therefore, the PSG, as a universal property, can be used to characterize a quantum phase. Any physical properties that are linked to (or protected by) the PSGs are also universal properties of a phase and can be used to detect and measure quantum order in experiments. In particular, PSGs can protect gapless gauge and fermion excitations (see Section 9.10). Thus, the stability of the PSG also implies the stability of the gapless gauge and fermion excitations.

We know that a mean-field spin-liquid state is characterized by $U_{ij} = \langle \psi_i \psi_j^\dagger \rangle$. If we include perturbative fluctuations around the mean-field state, then we expect U_{ij} to receive perturbative corrections δU_{ij}. Here we would like to argue that the perturbative fluctuations can only change U_{ij} in such a way that U_{ij} and $U_{ij}+\delta U_{ij}$ have the same PSG.

First, we would like to note the following well-known facts. The perturbative fluctuations cannot change the symmetries and the gauge structures. For example, if the ansatz U_{ij} and the Hamiltonian have a symmetry, then δU_{ij} generated by perturbative fluctuations will have the same symmetry. Similarly, the perturbative fluctuations cannot generate the δU_{ij} that, for example, breaks a $U(1)$ gauge structure down to a Z_2 gauge structure.

As both the gauge structure (described by the IGG) and the symmetry group are part of the PSG, it is reasonable to generalize the above observation by saying that, in addition to the IGG and the symmetry group in the PSG, the whole PSG cannot be changed by the perturbative fluctuations.

In fact, the mean-field Hamiltonian and the mean-field ground state are invariant under the transformations in the PSG. Thus, within a perturbative calculation around a mean-field state, the transformations in the PSG behave just like the ordinary symmetry transformations. Therefore, the perturbative fluctuations can only generate δU_{ij} that are invariant under the transformations in the PSG.

As the perturbative fluctuations (by definition) do not change the phase, U_{ij} and $U_{ij} + \delta U_{ij}$ describe the same phase. In other words, we can group U_{ij} into classes (which are called universality classes) such that the U_{ij} in each class are connected by the perturbative fluctuations. Each universality class describes one phase. We see that, if the above argument is true, then the ansatz in a universality class all share the same PSG. In other words, the universality classes or the phases are classified by the PSGs (or quantum orders).

We would like to point out that we have assumed that the perturbative fluctuations have no infra-red divergence in the above discussion. The infra-red divergence implies that the perturbative fluctuations are relevant perturbations. Such diverging corrections may cause phase transitions and invalidate the above argument. Therefore, the above argument and results only apply to stable spin liquids which satisfy the following requirements: (i) all of the mean-field fluctuations have no infra-red divergence; and (ii) the mean-field fluctuations are weak enough at the lattice scale.

In the following, we will discuss the stability of several types of mean-field states. The requirement (ii) can be satisfied through the large-N limit and/or the adjustment of short-range spin couplings in the spin Hamiltonian, if necessary. Here we will mainly consider the requirement (i). We find that, at least in certain large-N limits, many (but not all) mean-field states do correspond to real quantum spin liquids which are stable at low energies.

All spin liquids (with an odd number of electrons per unit cell) studied so far can be divided into four classes. In the following, we will study each class in turn.

9.9.2 Rigid spin liquids

Rigid spin liquids are states in which the spinons and all other excitations are fully gapped. The gap of the gauge field may be produced by the Chern–Simons terms or the Anderson–Higgs mechanism. The gapped gauge field only induces short-range interaction between spinons. As there are no excitations at low energies, the rigid spin liquids are stable states. The rigid spin liquids are characterized by topological orders and they have unconfined neutral spin-1/2 excitations. The low-energy effective theories for rigid spin liquids are topological field theories. The Z_2-gapped spin liquid and the chiral spin liquid are examples of rigid spin liquids.

9.9.3 Bose spin liquids

The U1C$n00x$ state (9.8.9) can be a $U(1)$-gapped spin liquid if a_0^3 in eqn (9.8.9) is large enough. Such a state, at the mean-field level, has gapped spinons and gapless $U(1)$ gauge bosons. We will call it a Bose spin liquid. The dynamics of the gapless $U(1)$ gauge fluctuations are described by the low-energy effective theory

$$\mathcal{L} = \frac{1}{8\pi g^2}(f_{\mu\nu})^2$$

where $f_{\mu\nu}$ is the field strength of the $U(1)$ gauge field. However, in $1 + 2$ dimensions and after including the instanton effect, the $U(1)$ gauge fluctuations will have an infra-red divergence, which leads to an energy gap for the gauge bosons and a confinement for the spinons (Polyakov, 1977). Thus, mean-field $U(1)$-gapped states are not stable in $1+2$ dimensions. Their PSGs may not describe any real quantum orders in physical spin liquids.[74]

9.9.4 Fermi spin liquids

Fermi spin liquids are defined by the following two properties. Firstly, they have gapless excitations that are described by spin-1/2 fermions, and, secondly, these gapless excitations have only short-range interactions between them. The Z_2-linear, Z_2-quadratic, and the Z_2-gapless spin liquids are examples of Fermi spin liquids.

The spinons have a massless Dirac dispersion in Z_2-linear spin liquids. These spinons with short-range interactions are described by the following effective theory in the continuum limit (see Problem 9.1.4):

$$S = \int \mathrm{d}t\mathrm{d}^2x\ [\bar{\psi}\partial_\mu\gamma^\mu\psi + (\bar{\psi}M\psi)^2]$$

In $2+1$ dimensions, ψ has a scaling dimension $[\psi] = 1$, so that the action is dimensionless. Thus, the interaction term $(\psi)^4$ has a scaling dimension $[\psi^4] = 4$, which is bigger than 3. We see that short-range interactions between massless Dirac fermions are irrelevant in $1+2$ dimensions. Thus, Z_2-linear spin liquids are stable states.

Now let us consider the stability of the Z_2-quadratic spin liquid Z2Bx2(12)n in eqn (9.6.10). The spinons have a gapless quadratic dispersion $\omega \propto k^2$ in the Z_2-quadratic spin liquid. In this case, space and time have different scaling dimensionsof $[x^{-1}] \equiv 1$ and $[t^{-1}] = 2$, respectively. From the dimensionless continuum action

$$S = \int \mathrm{d}t\mathrm{d}^2x\ [\psi^\dagger \mathrm{i}\partial_t\psi - c\psi^\dagger(\partial_x)^2\psi]$$

we see that the spinon field ψ has a scaling dimension $[\psi] = 1$ and the four-fermion interaction term has a dimension $[\psi^4] = 4$. So the coupling constant g in the interaction action $S_{\text{int}} = \int \mathrm{d}t\mathrm{d}^2x\ g(\psi^\dagger\psi)^2$ has a scaling dimension $[g] = 0$, because S_{int} always has zero scaling dimension. Thus, unlike the Z_2-linear spin liquid, the short-range interactions between the gapless spinons in the Z_2-quadratic state are marginal in $1+2$ dimensions. Further studies are needed to determine if the higher-order effects of the interaction make the coupling relevant or irrelevant. This will determine the dynamical stability of the Z_2-quadratic spin liquid beyond the mean-field level.

The Z_2-gapless spin liquids are as stable as Fermi liquids in $1+2$ dimensions. If we assume that Fermi liquids are stable in $1+2$ dimensions, then Z_2-gapless spin liquids are also stable.

---

[74] We would like to mention that mean-field $U(1)$-gapped states are stable in $1+3$ dimensions.

9.9.5 Algebraic spin liquids

An algebraic spin liquid is a state with gapless excitations, but none of the gapless excitations are described by free bosonic or fermionic quasiparticles. The $U(1)$-linear spin liquids are examples of algebraic spin liquids. Their low-lying excitations are described by massless Dirac fermions coupled to the $U(1)$ gauge field. As the fermion–gauge coupling is exactly marginal to all orders of perturbation theory (Appelquist and Nash, 1990), the gapless excitations have a finite interaction down to zero energy. As a result, there are neither free bosonic quasiparticles (such as gauge bosons) nor free fermionic quasiparticles (such as spinons) at low energies. This makes the discussion on the stability of these states much more difficult. I refer the interested reader to Rantner and Wen (2002) and Wen (2002c) for a discussion of algebraic spin liquids.

It was shown that, in $2+1$ dimensions, $U(1)$-linear states are stable in certain large-N limits of the spin model where the mean-field fluctuations are weak. The PSG of the $U(1)$-linear states prevents destabilizing counter terms, such as fermion mass terms, from being generated by perturbative fluctuations. This ensures the stability of the $U(1)$-linear states. Thus, at least in the large-N limits, the corresponding algebraic spin liquids exist as phases of physical spin systems.

The existence of the algebraic spin liquid is a very striking phenomenon. According to a conventional wisdom, if bosons/fermions interact at low energies, then the interaction will open an energy gap for these low-lying excitations. This implies that a system can either have free bosonic/fermionic excitations at low energies or have no low-energy excitations at all. The existence of the algebraic spin liquid indicates that such a conventional wisdom is incorrect. It raises the important question of what protects gapless excitations (in particular, when they interact at all energy scales). There should be a 'reason' or 'principle' for the existence of the gapless excitations. In the next section we will show that the PSG prevents the gauge bosons and fermions to obtain mass terms. So *it is the quantum order and the associated PSG that protects the gapless excitations.*

9.10 Quantum order and gapless gauge bosons and fermions

- Quantum order can produce and protect gapless gauge bosons and gapless fermions, just like symmetry breaking can produce and protect gapless Nambu–Goldstone bosons.

Gapless excitations are very rare in nature and in condensed matter systems. Therefore, if we see a gapless excitation, then we would like to ask why it exists. One origin of gapless excitations is spontaneous symmetry breaking, which gives Nambu–Goldstone bosons (Nambu, 1960; Goldstone, 1961). The relationship between gapless excitations and spontaneous symmetry breaking is very important. Due to this relation, we can obtain low-energy physics of a complicated state from its symmetry without knowing the details of the systems. This line of thinking makes Landau's symmetry-breaking theory (Ginzburg and Landau, 1950; Landau and Lifschitz, 1958) for phase and phase transition a very powerful theory with which to study the low-energy properties of a phase. However, spontaneous symmetry breaking is not the only source of gapless excitations. Quantum order and the associated PSGs can also produce and protect gapless excitations. What is striking is that quantum order produces and protects gapless gauge bosons and gapless fermions. Gapless fermions can be produced even from purely bosonic models. In this section, we would like to discuss in some detail the relationship between quantum order (and its PSG) and gapless gauge/fermion excitations (Wen, 2002a,c; Wen and Zee, 2002).

9.10.1 The projective symmetry group and gapless gauge bosons

- The gauge group of gapless gauge bosons (at the mean-field level) is the IGG of the PSG.

The relationship between the gapless gauge fluctuations and quantum order is simple and straightforward. The gauge group for the gapless gauge fluctuations in a quantum-ordered state is simply the IGG in the PSG that describes the quantum order.

To see how quantum order and its PSG produce and protect gapless gauge bosons, let us assume that, as an example, the IGG $\mathcal{G}$ of a quantum-ordered state contains a $U(1)$ subgroup, which is formed by the following constant gauge transformations:

$$\{W_{\boldsymbol{i}} = \mathrm{e}^{\,\mathrm{i}\theta\tau^3} | \theta \in [0, 2\pi)\} \subset \mathcal{G}$$

Next we consider one type of fluctuation around the mean-field solution $\bar{u}_{\boldsymbol{ij}}$, namely $u_{\boldsymbol{ij}} = \bar{u}_{\boldsymbol{ij}} \mathrm{e}^{\,\mathrm{i}a^3_{\boldsymbol{ij}}\tau^3}$. As $\bar{u}_{\boldsymbol{ij}}$ is invariant under the constant gauge transformation $\mathrm{e}^{\,\mathrm{i}\theta\tau^3}$, a spatially-dependent gauge transformation $\mathrm{e}^{\,\mathrm{i}\theta_{\boldsymbol{i}}\tau^3}$ will transform the fluctuation $a^3_{\boldsymbol{ij}}$ to $\tilde{a}^3_{\boldsymbol{ij}} = a^3_{\boldsymbol{ij}} + \theta_{\boldsymbol{i}} - \theta_{\boldsymbol{j}}$. This means that $a^3_{\boldsymbol{ij}}$ and $\tilde{a}^3_{\boldsymbol{ij}}$ label the same physical state and $a^3_{\boldsymbol{ij}}$ corresponds to gauge fluctuations. The energy of the fluctuations has a gauge invariance, i.e. $E(\{a^3_{\boldsymbol{ij}}\}) = E(\{\tilde{a}^3_{\boldsymbol{ij}}\})$. Due to the gauge invariance of the energy, we see that the mass term of the gauge field, $(a^3_{\boldsymbol{ij}})^2$, is not allowed. Therefore, the $U(1)$ gauge fluctuations described by $a^3_{\boldsymbol{ij}}$ will appear at low energies.

If the $U(1)$ subgroup of $\mathcal{G}$ is formed by spatially-dependent gauge transformations as follows:

$$\{W_{\boldsymbol{i}} = \mathrm{e}^{\,\mathrm{i}\theta \boldsymbol{n}_{\boldsymbol{i}}\cdot\boldsymbol{\tau}} | \theta \in [0, 2\pi), |\boldsymbol{n}_{\boldsymbol{i}}| = 1\} \subset \mathcal{G},$$

then we can always use an $SU(2)$ gauge transformation to rotate $\boldsymbol{n}_i$ to the z direction at every site and reduce the problem to the one discussed above. Thus, regardless of whether the gauge transformations in the IGG have spatial dependence or not, the gauge group for low-energy gauge fluctuations is always given by $\mathcal{G}$. As every $U(1)$ subgroup of the IGG corresponds to a gapless $U(1)$ gauge boson, the gauge group of low-energy gauge bosons is given by the IGG, even when the IGG is non-abelian.

We would like to remark that sometimes the low-energy gauge fluctuations not only appear near $\boldsymbol{k} = 0$, but also appear near some other $\boldsymbol{k}$ points. In this case, we will have several low-energy gauge fields, one for each $\boldsymbol{k}$ point. Examples of this phenomenon are given by some ansatz of the $SU(2)$ slave-boson theory discussed in Section 9.8, which have an $SU(2) \times SU(2)$ gauge structure at low energies. We see that the low-energy gauge structure $SU(2) \times SU(2)$ can even be larger than the high-energy gauge structure $SU(2)$. Even for this complicated case where low-energy gauge fluctuations appear around different $\boldsymbol{k}$ points, the IGG still correctly describes the low-energy gauge structure of the corresponding ansatz. If the IGG contains gauge transformations that are independent of the spatial coordinates, then such transformations correspond to the gauge group for gapless gauge fluctuations near $\boldsymbol{k} = 0$. If the IGG contains gauge transformations that depend on the spatial coordinates, then these transformations correspond to the gauge group for gapless gauge fluctuations near nonzero $\boldsymbol{k}$. For example, if $IGG = \{e^{i(-)^{\boldsymbol{i}}\theta\tau^3}\}$, then there will a gapless $U(1)$ gauge boson near $\boldsymbol{k} = (\pi, \pi)$ described by the gauge field $(-)^{\boldsymbol{i}} a^3(\boldsymbol{x})$. Thus, the IGG gives us a unified treatment of all low-energy gauge fluctuations, regardless of their crystal momenta.

In this chapter, we have used the terms Z_2 spin liquids, $U(1)$ spin liquids, $SU(2)$ spin liquids, and $SU(2) \times SU(2)$ spin liquids in many places. Now we can have a precise definition of these low-energy Z_2, $U(1)$, $SU(2)$, and $SU(2) \times SU(2)$ gauge groups. These low-energy gauge groups are simply the IGG of the corresponding ansatz. They have nothing to do with the high-energy gauge groups that appear in the $SU(2)$, $U(1)$, or Z_2 slave-boson approaches. We have also used the terms Z_2 gauge structure, $U(1)$ gauge structure, and $SU(2)$ gauge structure of a mean-field state. Their precise mathematical meaning is again the IGG of the corresponding ansatz. When we say that a $U(1)$ gauge structure is broken down to a Z_2 gauge structure, we mean that an ansatz is changed in such a way that its IGG is changed from the $U(1)$ to the Z_2 group.

9.10.2 The projective symmetry group and gapless fermions

- Sometimes, a PSG guarantees the existence of gapless fermions. In that case, the universality of the PSG implies the universality (or the stability) of the gapless fermions.

To demonstrate the direct connection between the PSG and the gapless fermions, in this section, we are going to study a particular spin liquid whose quantum orders are characterized by Z2Azz13 (Wen and Zee, 2002). The spinon spectrum in a Z2Azz13 spin liquid is given by the spinon hopping Hamiltonian

$$H = \sum_{\boldsymbol{ij}} \psi_{\boldsymbol{i}}^\dagger u_{\boldsymbol{ij}} \psi_{\boldsymbol{j}}$$

One example of the Z2Azz13 ansatz is given in eqn (9.6.2). We note that the ansatz $u_{\boldsymbol{ij}}$ can be viewed as an operator which maps a fermion wave function $\psi(\boldsymbol{i})$ to $\psi'(\boldsymbol{i}) = \sum_{\boldsymbol{j}} u_{\boldsymbol{ij}}\psi(\boldsymbol{j})$. Such an operator, denoted by $\hat{H}$, will be called the Hamiltonian. The eigenvalues of $\hat{H}$ determine the fermion spectrum. The gapless fermions correspond to the zero eigenvalue of the Hamiltonian.

The Z2Azz13 PSG is generated by

$$\begin{aligned}
G_x(\boldsymbol{i}) &= \tau^0, & G_y(\boldsymbol{i}) &= \tau^0 \\
G_{P_x}(\boldsymbol{i}) &= (-)^{\boldsymbol{i}} \mathrm{i}\tau^3 & G_{P_y}(\boldsymbol{i}) &= (-)^{\boldsymbol{i}} \mathrm{i}\tau^3 \\
G_{P_{xy}}(\boldsymbol{i}) &= \mathrm{i}\tau^1 & G_T(\boldsymbol{i}) &= \mathrm{i}\tau^3 \\
G_0(\boldsymbol{i}) &= -\tau^0, & &
\end{aligned} \tag{9.10.1}$$

The combined transformations, such as $G_{P_x} P_x$, can also be viewed as unitary operators acting on $\psi_{\boldsymbol{i}}$. The projective symmetry of the Hamiltonian means that, for example, $G_{P_x} P_x \hat{H} (G_{P_x} P_x)^\dagger = \hat{H}$.

As $G_x = G_y = \tau^0$, the Z2Azz13 ansatz is translationally invariant. In momentum space, the Hamiltonian has the form

$$H(\boldsymbol{k}) = \epsilon^\mu(\boldsymbol{k})\tau^\mu$$

The invariance of the ansatz $u_{\boldsymbol{ij}}$ under $G_T T$ implies that $G_T(\boldsymbol{i}) u_{\boldsymbol{ij}} G_T^\dagger(\boldsymbol{j}) = -u_{\boldsymbol{ij}}$. When expressed in the $\boldsymbol{k}$ space, we have

$$U_T H(k_x, k_y) U_T^\dagger = -H(k_x, k_y), \qquad U_T = \mathrm{i}\tau^3. \tag{9.10.2}$$

where the non-trivial G_T gives rise to a non-trivial U_T. Equation (9.10.2) implies that $\epsilon^{0,3}(\boldsymbol{k}) = 0$ and

$$H(\boldsymbol{k}) = \epsilon^1(\boldsymbol{k})\tau^1 + \epsilon^2(\boldsymbol{k})\tau^2 \tag{9.10.3}$$

The invariance of the ansatz under $G_{P_{xy}} P_{xy}$ gives us

$$U_{P_{xy}} H(k_x, k_y) U_{P_{xy}}^\dagger = H(k_y, k_x), \qquad U_{P_{xy}} = \mathrm{i}\tau^1, \tag{9.10.4}$$

where the exchange of k_x and k_y is generated by P_{xy}. The invariance of the ansatz under $G_{P_x} P_x$ leads to

$$U_{P_x} H(k_x, k_y) U_{P_x}^\dagger = H(-k_x + \pi, k_y + \pi), \qquad U_{P_x} = \mathrm{i}\tau^3, \tag{9.10.5}$$

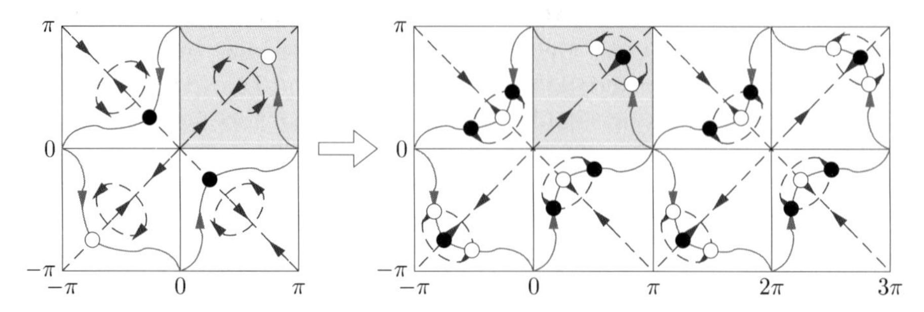

FIG. 9.16. Two patterns of the zeros of $H(\boldsymbol{k})$. The $\epsilon^1 = 0$ and $\epsilon^2 = 0$ lines are represented by solid and dashed lines, respectively. The filled circle are zeros with winding number 1, and the open circles are zeros with winding number -1. The shaded area is one-quarter of the Brillouin zone. When the zero-lines of ϵ^2 move to cross the zero-lines of ϵ^1, a zero with winding number 1 is changed into a zero with winding number -1 plus two zeros with winding number 1.

and the invariance of the ansatz under $G_{P_y}T_{P_y}$ leads to

$$U_{P_y}H(k_x,k_y)U_{P_y}^\dagger = H(k_x+\pi,-k_y+\pi), \qquad U_{P_x} = \mathrm{i}\tau^3. \tag{9.10.6}$$

The momentum shift (π,π) is due to the $(-)^{\boldsymbol{i}}$ term in G_{P_x} and G_{P_y}.

Thus, the nonzero $\epsilon^{1,2}(\boldsymbol{k})$ have the following symmetries:

$$\begin{aligned}
\epsilon^1(k_x,k_y) &= -\epsilon^1(\pi-k_x,\pi+k_y) = -\epsilon^1(\pi+k_x,\pi-k_y) = \epsilon^1(k_y,k_x) \\
\epsilon^2(k_x,k_y) &= -\epsilon^2(\pi-k_x,\pi+k_y) = -\epsilon^2(\pi+k_x,\pi-k_y) = -\epsilon^2(k_y,k_x)
\end{aligned} \tag{9.10.7}$$

In fact, eqns (9.10.3) and (9.10.7) define the most general Z2Azz13 ansatz in the $\boldsymbol{k}$ space.

Equation (9.10.7) allows us to determine $\epsilon^{1,2}(\boldsymbol{k})$ from their values in one-quarter of the Brillouin zone (see Fig. 9.16). In that one-quarter of the Brillouin zone, $\epsilon^2(\boldsymbol{k})$ is anti-symmetric under the interchanging of k_x and k_y. Thus, $\epsilon^2(\boldsymbol{k}) = 0$ when $k_x = k_y$. Equation (9.10.7) also implies that $\epsilon^1(\boldsymbol{k})$ changes sign under a $90°$ rotation around $(0,\pi)$ or $(\pi,0)$, i.e. $\epsilon^1(k_x,k_y+\pi) = -\epsilon^1(k_y,-k_x+\pi)$. Therefore, $\epsilon^1(\boldsymbol{k})$ must vanish at $(0,\pi)$ and $(\pi,0)$. It must also vanish on a line connecting $(0,\pi)$ and $(\pi,0)$ (see Fig. 9.16). As a result, the $\epsilon^1 = 0$ line and the $\epsilon^2 = 0$ line intersect at least once in the quarter of the Brillouin zone. The intersection point is the zero point of the Hamiltonian $H(\boldsymbol{k})$. We see that gapless spinons appear at at least four isolated $\boldsymbol{k}$ points in the Brillouin zone (see Fig. 9.16). The Z2Azz13 symmetry of the ansatz directly leads to gapless spinons.

Two typical distributions of the zeros of $H(\boldsymbol{k})$ may look like those in Fig. 9.16. We see that the zeros have special patterns. To understand which features of the patterns are universal, we would like to study the motion of the zeros as we deform the ansatz. Before doing that, we want to point out that the zeros of $H(\boldsymbol{k})$ have an internal structure that can be characterized by a winding number. As $\boldsymbol{k}$ goes around a zero, the two-dimensional vector $(\epsilon^1(\boldsymbol{k}), \epsilon^2(\boldsymbol{k}))$ draws a loop around $(0,0)$. The winding number of the zero is given by how many times the loop winds around $(0,0)$. The winding number is positive if the loop winds around $(0,0)$ anticlockwise and is negative if the loop winds around $(0,0)$ clockwise. A typical zero has a winding number ± 1. A zero with, say, a winding number 2 can split into two zeros with a winding number 1 once we perturb the Hamiltonian.

For the Z2Azz13 ansatz, the zeros of $H(\boldsymbol{k})$ are given by the intersection of the zero-lines of $\epsilon^1(\boldsymbol{k})$ and $\epsilon^2(\boldsymbol{k})$. As $\epsilon^0(\boldsymbol{k}) = \epsilon^3(\boldsymbol{k}) = 0$, then, as we deform the ansatz, the zeros cannot just appear or disappear. The zeros can only be created or annihilated in groups in a way that conserves the total winding number (see Fig. 9.16). When combined with the symmetry of the ansatz (9.10.7), we find that the following properties are robust against any small perturbations of the Hamiltonian that do not change the PSG: a pattern of $+1$ and -1 zeros along the line $((0,0),(\pi,\pi))$; and (N_+, N_-), the number of $(+1,-1)$ zeros *inside* the triangle $((0,0),(\pi,\pi),(0,\pi))$ (not including the zeros on the sides and corners). We define these properties to be the pattern of the zero's (POZ) for the Z2Azz13 spin liquids. Just like a change in Fermi surface topology, a change in POZ will lead to a singularity in the ground-state energy and signal a phase transition. Thus, the POZ is also a quantum number that characterizes quantum order. We see that the quantum order in the spin liquid is not completely characterized by the PSG. The POZ provides an additional characterization. The combination of the PSG and the POZ provides a more complete characterization of the quantum order.

Problem 9.10.1.
Consider an ansatz of the Z2A0013 state in eqn (9.6.1). You may assume that $\gamma_{1,2} = \lambda = 0$.
(a) Show that changing a_0^1 will cause a change in the POZ. Show that the change in the POZ causes a singularity in the mean-field ground-state energy.
(b) Find the spectrum of low-energy spinons at the transition point.

Problem 9.10.2.
(a) Find $G_{x,y}$, $G_{P_x,P_y,P_{xy}}$, and G_T for the Z2A003n PSG.
(b) Find the most general mean-field Hamiltonian $H(\boldsymbol{k})$ for the Z2A003n state.
(c) Show that the Z2A003n state always has gapless spinons at $\boldsymbol{k} = (\pm\frac{\pi}{2}, \pm\frac{\pi}{2})$.

10

STRING CONDENSATION—AN UNIFICATION OF LIGHT AND FERMIONS

- **What are light and fermions?**
  Light is a fluctuation of nets of condensed strings of arbitrary sizes. Fermions are ends of condensed strings.
- **Where do light and fermions come from?**
  Light and fermions come from the collective motions of condensed string-nets that fill the space.
- **Why do light and fermions exist?**
  Light and fermions exist because our vacuum happens to have string-net condensation.

In Chapter 9, we used the projective construction to construct many quantum-ordered states in two-dimensional spin systems and introduced PSGs to characterize and classify the quantum-ordered states. These quantum-ordered states not only contain gauge bosons, but also contain fermions as their low-energy collective excitations. The emergence of gauge bosons and fermions is a very striking result because the underlying lattice model is a purely bosonic model.

For many years, fermions and gauge bosons were regarded as fundamental and untouchable. When it was suggested that fermions and gauge theory can emerge from a two-dimensional spin liquid (Baskaran *et al.*, 1987; Kivelson *et al.*, 1987; Baskaran and Anderson, 1988), the suggestion was not even surprising because it was greeted with suspicion and disbelief. These doubts were well founded. Even now, it is not clear if two-dimensional spin liquids exist or not. However, this does not imply that the original suggestion was wrong. The calculation, when adapted to the $SU(N)$ spin model (Affleck and Marston, 1988), does lead to well-defined gauge bosons and fermions in three dimensions (see Section 10.7) (Wen, 2002a). Thus, fermions and gauge bosons are not that fundamental and untouchable. They can emerge from certain quantum orders in lowly bosonic models.

However, the projective construction that leads to the emergent gauge bosons and fermions is very formal. It appears that everything relies on the very unreasonable mathematical trick of splitting a spin into two unphysical fermions. It is hard to have any confidence in such an approach. The amazing result of emergent fermions and gauge bosons obtained from such an approach appears to be really unbelievable.

This chapter can be viewed as a confidence builder. We will discuss several exactly soluble spin models which can also be solved by the projective construction. We will show that the result from the projective construction agrees with the exact result for those models. In particular, the exactly soluble models have emergent fermions and a gauge field, as suggested by the projective construction.

However, what is more important is that the exactly soluble model ties the emergence of fermions and the gauge field with a phenomenon—the condensation of nets of closed strings. We find that the quantum-ordered states obtained from the projective construction are actually string-net-condensed states. The PSG that characterizes the quantum order in states obtained from the projective construction really characterizes different string-net condensations. The projective construction and the string-net condensation are just two different ways to describe the same type of quantum order. String-net condensation provides a physical foundation for the formal projective construction.

Trying to connect projective construction and string-net condensation is just a formal motivation for the discussions in this chapter. As fermions and gauge bosons can emerge from string-net condensation, we would like to give the discussions in this chapter a more physical motivation. We would like to ask whether string-net condensation has anything to do with the existence of light and fermions in our universe. In particular, we would like to ask the following questions about light and fermions. What are light and fermions? Where do light and fermions come from? Why do light and fermions exist?

At the moment, the standard answers to the above fundamental questions appear to be 'light is the particle described by a gauge field' and 'fermions are the particles described by anti-commuting fields'. Here, we would like to argue that there is another possible (and, I believe, more physical) answer to the above questions, namely that our vacuum is filled with nets of string-like objects of arbitrary sizes and that these string-nets form a quantum condensed state. According to string-net theory, the light (and other gauge bosons) is a collective vibration of the condensed string-nets, and fermions are the ends of the strings. We see that string-net condensation provides a unified origin of both light and fermions.[75] In other words, string-net condensation unifies light and fermions. If someone says, 'let there be string-net', then we will get both light and fermions.

Before providing evidence for the string-net theory of light and fermions, we would like to first clarify what we mean by 'light exists' and 'fermions exist'. We know that there is a natural mass scale in physics—the Planck mass. The Planck mass is so large that any particle that we see has a mass at least a factor of 10^{16}

[75] Here, by 'string-net condensation' we mean the condensation of nets of string-like objects of arbitrary sizes.

smaller than the Planck mass. So all of the observed particles can be treated as massless when compared with the Planck mass. When we ask why some particles exist, we really mean why are those particles massless (or nearly massless) when compared with the Planck mass. The real issue is what makes certain types of excitation (such as light and fermions) massless (or nearly massless). Why does nature want to have massless excitations at all? Who ordered them?

Secondly, we would like to clarify what we mean by 'the origin of light and fermions'. We know that everything has to come from something. So when we ask, 'where do light and fermions come from?', we have assumed that there are some things simpler and more fundamental than light and fermions. In Section 10.1, we will define local bosonic models which are simpler than models with gauge fields coupled to fermions. We will regard local bosonic models as more fundamental. Such a philosophy will be called the locality principle. We will show that light and fermions can emerge from a local bosonic model if the model contains a condensation of nets of string-like objects in its ground state.

The string-net theory of fermions explains why there is always an even number of fermions in our universe—because a string (or string-net) always has an even number of ends. The string-net theory for gauge bosons and fermions also has an experimental prediction, namely that all fermions must carry certain gauge charges. At first sight, this prediction appears to contradict the known experimental fact that neutrons carry no gauge charges. Thus, one may think that the string-net theory of gauge bosons and fermions has already been falsified by experiments. Here we would like to point out that the string-net theory of gauge bosons and fermions can still be correct if we assume the existence of a new discrete gauge field, such as a Z_2 gauge field, in our universe. In this case, neutrons and neutrinos carry a nonzero charge of the discrete gauge field. Therefore, the string-net theory of gauge bosons and fermions predicts the existence of discrete gauge excitations (such as gauge flux lines) in our universe.

According to the picture of quantum order and string-net condensation, elementary particles (such as photons and electrons) may not be elementary after all. They may be collective excitations of a local bosonic system below the Planck scale. As we cannot do experiments close to the Planck scale, it is hard to determine if photons and electrons are elementary particles or not. In this chapter, through some concrete local boson models on two-dimensional and three-dimensional lattices, we would like to show that the string-net theory of light and fermions is at least self-consistent. The local bosonic models studied here are just a few examples among a long list of local bosonic models that contain emergent unconfined fermions and gauge bosons (Foerster *et al.*, 1980; Kalmeyer and Laughlin, 1987; Wen *et al.*, 1989; Read and Sachdev, 1991; Wen, 1991a; Moessner and Sondhi, 2001; Balents *et al.*, 2002; Ioffe *et al.*, 2002; Motrunich and Senthil, 2002; Sachdev and Park, 2002; Kitaev, 2003; Motrunich, 2003; Wen, 2003c). The ground states of these models all have non-trivial topological/quantum orders.

We would like to remark that, despite some similarity, the above string-net theory of gauge bosons and fermions is different from standard superstring theory. In standard superstring theory, closed strings correspond to gravitons and open strings correspond to gauge bosons. The fermions come from the fermionic fields on the world sheet. All of the elementary particles, including gauge bosons, correspond to different vibrational modes of small strings in superstring theory. In string-net theory, the vacuum is filled with large strings (or nets of large strings, see Fig. 1.2). The massless gauge bosons correspond to the fluctuations of nets of large closed strings, and fermions correspond to the ends of open strings. There are no fermionic fields in string-net theory.

The string-net theory of gauge fluctuations is intimately related to the Wegner–Wilson loop in gauge theory (Wegner, 1971; Wilson, 1974; Kogut, 1979). The relationship between dynamical gauge theory and dynamical Wegner–Wilson-loop theory was suggested and studied by Gliozzi *et al.* (1979) and Polyakov (1979). In Savit (1980), various duality relations between gauge theories and theories of extended objects were reviewed. In particular, some statistical lattice gauge models were found to be dual to certain statistical membrane models (Banks *et al.*, 1977). This duality relationship is directly connected to the relationship between gauge theory and string-net theory in quantum models. The new feature in quantum models is that the ends of strings can sometimes be fermions or anyons (Levin and Wen, 2003).

Here we would like to stress that the string-net theory for the actual gauge bosons and fermions in our universe is only a suggestion at the moment. Although string-net condensation can produce and protect massless photons, gluons, quarks, and other charged leptons, we currently do not know whether string-net condensations can produce neutrinos, which are chiral fermions. We also do not know if string-net condensations can produce the $SU(2)$ gauge field for the weak interaction which couples chirally to the quarks and the leptons. The correctness of string-net condensation in our vacuum depends on the resolution of the above problems.

On the other hand, if we are only concerned with the condensed matter problem of how to use bosons to make artificial light and artificial fermions, then string-net theory and quantum order do provide an answer. To make artificial light and artificial fermions, we simply allow certain strings to condense.

10.1 Local bosonic models

In this chapter, we will only consider local bosonic models. We think that local bosonic models are fundamental because they are really local. We note that a fermionic model is, in general, non-local because the fermion operators at different sites do not commute, even when the sites are well separated. In contrast, local bosonic models are local because different boson operators commute when they are well separated. Due to the intrinsic locality of local bosonic models, we believe that the fundamental theory of nature is a local bosonic model. To stress this point, we will give such a belief the name locality principle.

Let us give a detailed definition of local bosonic models. To define a physical system we need to specify (i) a total Hilbert space, (ii) a definition for a set of local physical operators, and (iii) a Hamiltonian. With this understanding, a local bosonic model is defined to be a model that satisfies the following properties.

(i) The total Hilbert space is a direct product of local Hilbert spaces of finite dimensions.
(ii) Local physical operators are operators acting within a local Hilbert space, or finite products of these operators for nearby local Hilbert spaces. We will also call these operators local bosonic operators because they all commute with each other when far apart.
(iii) The Hamiltonian is a sum of local physical operators.

A spin-1/2 system on a lattice is an example of a local bosonic model. The local Hilbert space is two-dimensional and contains $|\uparrow\rangle$ and $|\downarrow\rangle$ states. The local physical operators are $\sigma^a_{\boldsymbol{i}}$, $\sigma^a_{\boldsymbol{i}}\sigma^b_{\boldsymbol{i}+\boldsymbol{x}}$, etc., where σ^a, $a = 1, 2, 3$, are the Pauli matrices.

A free spinless fermion system (in two or higher dimensions) is not a local bosonic model, despite it having the same total Hilbert space as the spin-1/2 system. This is because the fermion operators $c_{\boldsymbol{i}}$ on different sites do not commute and are not local bosonic operators. More importantly, the fermion hopping operator $c^\dagger_{\boldsymbol{i}}c_{\boldsymbol{j}}$ in two and higher dimensions cannot be written as a local bosonic operator. (However, due to the Jordan–Wigner transformation, a one-dimensional fermion hopping operator $c^\dagger_{i+1}c_i$ can be written as a local bosonic operator. Hence, a one-dimensional fermion system can be a local bosonic model if we exclude $c_{\boldsymbol{i}}$ from our definition of local physical operators.) A lattice gauge theory is not a local bosonic model. This is because its total Hilbert space cannot be a direct product of local Hilbert spaces.

10.2 An exactly soluble model from a projective construction

In this section, we are going to construct an exactly soluble local bosonic model. In the next section, we will show that the model contains string-net condensation and emergent fermions.

10.2.1 Construction of the exactly soluble model

- Exactly soluble models can be constructed by finding mutually-commuting operators.

Usually, the projective construction does not give us exact results. In this section, we are going to construct an exactly soluble model on a two-dimensional square lattice (Kitaev, 2003; Wen, 2003c). Our model has the property that the projective construction give us exact ground states and all of the other exact excited states. The key step in the construction is to find a system of commuting operators.

Let us introduce the four Majorana-fermion operators $\lambda_{\boldsymbol{i}}^{a}$, $a = x, \bar{x}, y, \bar{y}$, satisfying $\{\lambda_{a,\boldsymbol{i}}, \lambda_{b,\boldsymbol{j}}\} = 2\delta_{ab}\delta_{\boldsymbol{ij}}$. We find that the operators

$$\hat{U}_{\boldsymbol{i},\boldsymbol{i}+\hat{\boldsymbol{x}}} = \lambda_{\boldsymbol{i}}^{x}\lambda_{\boldsymbol{i}+\hat{\boldsymbol{x}}}^{\bar{x}}, \qquad \hat{U}_{\boldsymbol{i},\boldsymbol{i}+\hat{\boldsymbol{y}}} = \lambda_{\boldsymbol{i}}^{y}\lambda_{\boldsymbol{i}+\hat{\boldsymbol{y}}}^{\bar{y}}, \qquad \hat{U}_{\boldsymbol{ij}} = -\hat{U}_{\boldsymbol{ji}}, \tag{10.2.1}$$

form a commuting set of operators, i.e. $[\hat{U}_{\boldsymbol{i}_1\boldsymbol{i}_2}, \hat{U}_{\boldsymbol{j}_1\boldsymbol{j}_2}] = 0$. After obtaining a commuting set of operators, we can easily see that, as a function of the $\hat{U}_{\boldsymbol{ij}}$s, the interacting fermion Hamiltonian

$$H = -\sum_{\boldsymbol{i}} V_{\boldsymbol{i}}\hat{F}_{\boldsymbol{i}}, \qquad \hat{F}_{\boldsymbol{i}} = \hat{U}_{\boldsymbol{i},\boldsymbol{i}_1}\hat{U}_{\boldsymbol{i}_1,\boldsymbol{i}_2}\hat{U}_{\boldsymbol{i}_2,\boldsymbol{i}_3}\hat{U}_{\boldsymbol{i}_3,\boldsymbol{i}}, \tag{10.2.2}$$

commutes with all of the $\hat{U}_{\boldsymbol{ij}}$s. Here $\boldsymbol{i}_1 = \boldsymbol{i} + \hat{\boldsymbol{x}}$, $\boldsymbol{i}_2 = \boldsymbol{i} + \hat{\boldsymbol{x}} + \hat{\boldsymbol{y}}$, and $\boldsymbol{i}_3 = \boldsymbol{i} + \hat{\boldsymbol{y}}$. We will call $\hat{F}_{\boldsymbol{i}}$ a Z_2-flux operator.

To obtain the Hilbert space within which the Hamiltonian H in eqn (10.2.2) acts, we group $\lambda^{x,\bar{x},y,\bar{y}}$ into the two complex fermion operators

$$2\psi_{1,\boldsymbol{i}} = \lambda_{\boldsymbol{i}}^{x} + \mathrm{i}\lambda_{\boldsymbol{i}}^{\bar{x}}, \qquad 2\psi_{2,\boldsymbol{i}} = \lambda_{\boldsymbol{i}}^{y} + \mathrm{i}\lambda_{\boldsymbol{i}}^{\bar{y}} \tag{10.2.3}$$

at each site. One can check that $\psi_{1,2}$ satisfy the standard fermion anti-commutation algebra. The two complex fermion operators generate a four-dimensional Hilbert space at each site. The dimension of the total Hilbert space is $4^{N_{\text{site}}}$. The commuting property of the $\hat{U}_{\boldsymbol{ij}}$s and H allows us to solve the interacting fermion system exactly.

To obtain the exact eigenstates and exact eigenvalues of H, let $|\{s_{\boldsymbol{ij}}\}\rangle$ be the common eigenstate of the $\hat{U}_{\boldsymbol{ij}}$ operators with eigenvalue $s_{\boldsymbol{ij}}$. As $(\hat{U}_{\boldsymbol{ij}})^2 = -1$ and $\hat{U}_{\boldsymbol{ij}} = -\hat{U}_{\boldsymbol{ji}}$, we find that $s_{\boldsymbol{ij}}$ satisfies $s_{\boldsymbol{ij}} = \pm\mathrm{i}$ and $s_{\boldsymbol{ij}} = -s_{\boldsymbol{ji}}$. As H is a function of the $\hat{U}_{\boldsymbol{ij}}$s, we find that $|\{s_{\boldsymbol{ij}}\}\rangle$ is also an energy eigenstate of eqn (10.2.2) with energy

$$E = -\sum_{\boldsymbol{i}} V_{\boldsymbol{i}}F_{\boldsymbol{i}}, \qquad F_{\boldsymbol{i}} = s_{\boldsymbol{i},\boldsymbol{i}_1}s_{\boldsymbol{i}_1,\boldsymbol{i}_2}s_{\boldsymbol{i}_2,\boldsymbol{i}_3}s_{\boldsymbol{i}_3,\boldsymbol{i}} \tag{10.2.4}$$

We will call $F_{\boldsymbol{i}}$ the Z_2 flux through the square $\boldsymbol{i}$. When $V_{\boldsymbol{i}} > 0$, the ground state is given by $|\{s_{\boldsymbol{ij}}\}\rangle$, with $s_{\boldsymbol{i},\boldsymbol{i}+\boldsymbol{x}} = s_{\boldsymbol{i},\boldsymbol{i}+\boldsymbol{y}} = i$. For such a state, we have $F_{\boldsymbol{i}} = 1$. To obtain the excited states, we flip the signs of some of the $s_{\boldsymbol{ij}}$ to make some $F_{\boldsymbol{i}} = -1$.

To see if the $|\{s_{\boldsymbol{ij}}\}\rangle$s rcpresent all of the exact eigenstates of H, we need to count the states. Let us assume that the two-dimensional square lattice has N_{site} lattice sites and a periodic boundary condition in both directions. In this case, the lattice has $2N_{\text{site}}$ links. As there are a total of $2^{2N_{\text{site}}}$ different choices of $s_{\boldsymbol{ij}}$

(two choices for each link), the states $|\{s_{ij}\}\rangle$ exhaust all of the $4^{N_{\text{site}}}$ states in the Hilbert space. Thus, the common eigenstates of $\hat{U}_{ij}$ are not degenerate and the above approach allows us to obtain all of the eigenstates and eigenvalues of H. We have solved the two-dimensional interacting fermion system exactly!

We note that the Hamiltonian H can only change the fermion number at each site by an even number. Thus, H acts within a subspace which has an even number of fermions at each site. We will call the subspace the physical Hilbert space. The physical Hilbert space has only two states per site. When restricted to within the physical space, H actually describes a spin-1/2 or a hard-core boson system. To obtain the corresponding spin-1/2 Hamiltonian, we note that

$$\sigma_i^x = \mathrm{i}\lambda_i^y\lambda_i^x, \qquad \sigma_i^y = \mathrm{i}\lambda_i^{\bar{x}}\lambda_i^y, \qquad \sigma_i^z = \mathrm{i}\lambda_i^x\lambda_i^{\bar{x}} \tag{10.2.5}$$

act within the physical Hilbert space and satisfy the algebra of Pauli matrices. Thus, we can identify σ_i^l as the spin operator. Using the fact that

$$(-)^{\psi_{1,i}^\dagger\psi_{1,i}+\psi_{2,i}^\dagger\psi_{2,i}} = \lambda_i^x\lambda_i^y\lambda_i^{\bar{x}}\lambda_i^{\bar{y}} = 1$$

within the physical Hilbert space, we can show that the fermion Hamiltonian (10.2.2) becomes (see Problem 10.2.2)

$$H_{\text{spin}} = -\sum_i V_i\hat{F}_i, \qquad \hat{F}_i = \sigma_i^x\sigma_{i+\hat{x}}^y\sigma_{i+\hat{x}+\hat{y}}^x\sigma_{i+\hat{y}}^y \tag{10.2.6}$$

within the physical Hilbert space.

All of the states in the physical Hilbert space (i.e. all of the states in the spin-1/2 model) can be obtained from the $|\{s_{ij}\}\rangle$ states by projecting into the physical Hilbert space, i.e. $\mathcal{P}|\{s_{ij}\}\rangle$. The projection operator is given by

$$\mathcal{P} = \prod_i \frac{1+(-)^{\psi_{1i}^\dagger\psi_{1i}+\psi_{2i}^\dagger\psi_{2i}}}{2}$$

As $[\mathcal{P}, H] = 0$, the projected state $\mathcal{P}|\{s_{ij}\}\rangle$, if nonzero, is still an eigenstate of H (or H_{spin}) and has the same eigenvalue. After the projection, the exact solution of the interacting fermion model leads to an exact solution of the spin-1/2 model.

We note that, for a system with a periodic boundary condition in both the x and y directions, the product of all links

$$\prod_i (is_{i,i+x})(is_{i,i+y}) = (-)^{\hat{N}_f}, \tag{10.2.7}$$

where $\hat{N}_f = \sum_i(\psi_{1i}^\dagger\psi_{1i} + \psi_{2i}^\dagger\psi_{2i})$, is the total fermion number operator. Thus, the projection of $|\{s_{ij}\}\rangle$ is nonzero only when

$$\prod_i (is_{i,i+x})(is_{i,i+y}) = 1. \tag{10.2.8}$$

The physical states (with even numbers of fermions per site) are invariant under local Z_2 transformations generated by

$$\hat{G} = \prod_i G_i^{\psi_{1,i}^\dagger \psi_{1,i} + \psi_{2,i}^\dagger \psi_{2,i}}$$

where G_i is an arbitrary function with only the two values ± 1. We note that the Z_2 transformations change ψ_{Ii} to $\tilde{\psi}_{Ii} = G_i \psi_{Ii}$ and s_{ij} to $\tilde{s}_{ij} = G_i s_{ij} G_j$. We find that $|\{s_{ij}\}\rangle$ and $|\{\tilde{s}_{ij}\}\rangle$ give rise to the same physical state after projection (if their projection is not zero). Thus $\{s_{ij}\}$ is a many-to-one label of the exact eigenstates of the spin-1/2 system H_{spin}. The projection into the physical Hilbert space with an even number of fermions per site makes our theory a Z_2 gauge theory.

Problem 10.2.1.
(a) Show that the $\hat{U}_{ij}$s in eqn (10.2.1) form a commuting set of operators.
(b) Show that the ψs in eqn (10.2.3) satisfy the standard anti-commutation relation of complex fermions.

Problem 10.2.2.
(a) Show that $2\psi_1^\dagger \psi_1 - 1 = \mathrm{i}\lambda^x \lambda^{\bar{x}}$ and $2\psi_2^\dagger \psi_2 - 1 = \mathrm{i}\lambda^y \lambda^{\bar{y}}$. Show that a state with an even number of fermions at a site is an eigenstate of $\lambda^x \lambda^y \lambda^{\bar{x}} \lambda^{\bar{y}}$ with eigenvalue $+1$.
(b) Prove eqn (10.2.7).
(c) Show that the σ^l in eqn (10.2.5) satisfy the algebra of Pauli matrices within the physical Hilbert space.
(d) Prove that the fermion Hamiltonian (10.2.2) becomes the spin Hamiltonian (10.2.6) within the physical Hilbert space.

Problem 10.2.3.
We have seen that we can obtain a spin-1/2 system from the fermion system by restricting the fermion Hamiltonian H to within the subspace with an even number of fermions per site. We note that the fermion Hamiltonian H also acts within a subspace with odd numbers of fermions per site. Obtain the corresponding spin-1/2 system by restricting the fermion Hamiltonian H to within the odd-number-fermion subspace.

10.2.2 Exact eigenstates and topologically-degenerate ground states

- The ground-state degeneracy is protected by the Z_2 gauge structure and is robust against any local perturbations.

The above Z_2 gauge structure allows us to count the number of physical states obtained from the projection of the $|\{s_{ij}\}\rangle$ states. Again, we assume a periodic boundary condition in both directions. Noting that the constant Z_2 gauge transformation $G_i = -1$ does not change the s_{ij}, we see that there are only $2^{N_{\text{site}}}/2$ distinct s_{ij}s that are gauge equivalent to each other. Among the $4^{N_{\text{site}}}$ number of $|\{s_{ij}\}\rangle$ states, $4^{N_{\text{site}}}/2$ of them satisfy $\prod_i (is_{i,i+x})(is_{i,i+y}) = 1$ (i.e. have even numbers of fermions). Thus, the projection of the $|\{s_{ij}\}\rangle$ states gives us, at most,

$(4^{N_{\text{site}}}/2)/(2^{N_{\text{site}}}/2) = 2^{N_{\text{site}}}$ physical states. On the other hand, the projection of all of the $|\{s_{ij}\}\rangle$ states should give us all of the $2^{N_{\text{site}}}$ spin states. Therefore, different Z_2 gauge-equivalent classes of s_{ij} must lead to independent physical states, so that the projection can recover all of the $2^{N_{\text{site}}}$ spin states. The Z_2 gauge-equivalent class of s_{ij} is a one-to-one label of all of the physical states on a periodic lattice. Thus, we can obtain all of the eigenstates and eigenvalues of H_{spin} on a periodic lattice.

We note that the Z_2 flux F_i is invariant under the Z_2 gauge transformation. Thus, if two configurations s_{ij} and $\tilde{s}_{ij}$ have different Z_2 fluxes F_i and $\tilde{F}_i$, respectively, then s_{ij} and $\tilde{s}_{ij}$ must belong to two different Z_2 gauge-equivalent classes. The two states $|s_{ij}\rangle$ and $|\tilde{s}_{ij}\rangle$ will lead to different physical states after the projection. This result suggests that the projected states (the physical states) are better labeled by the Z_2 flux F_i. It turns out that F_i is not a perfect one-to-one label of the physical states. For a system on an even $\times$ even lattice with periodic boundary conditions, each F_i labels four states (i.e. for each Z_2-flux configuration F_i, there are four and only four physical states that reproduce the same F_i). All of the four physical states with the same F_i have the same energy.

To understand the four-fold degeneracy, let us consider one of the eigenstates that is given by the projection of a $|\{s_{ij}\}\rangle$ state. Other degenerate eigenstate states can be obtained by performing the following two transformations:

$$T_1 : (s_{i,i+x}, s_{i,i+y}) \to (s_{i,i+x}, (-)^{\delta_{i_y}} s_{i,i+y})$$
$$T_2 : (s_{i,i+x}, s_{i,i+y}) \to ((-)^{\delta_{i_x}} s_{i,i+x}, s_{i,i+y})$$

We see that T_1 does not change $s_{i,i+x}$ and only flips the sign of $s_{i,i+y}$ when $i_y = 0$. By construction, T_1 and T_2 do not change the Z_2 flux F_i. If we view the periodic lattice as a torus, then T_1 and T_2 insert π flux through the two holes of the torus (see Fig. 9.6).

On an even$\times$even lattice, the transformations T_1 and T_2 do not change the product $\prod_i (is_{i,i+x})(is_{i,i+y})$. Therefore, the three transformations T_1, T_2, and T_1T_2 generate the other three degenerate states. We see that all of the energy eigenvalues have a four-fold degeneracy. In particular, the spin-1/2 model H_{spin} has four degenerate ground states on an even $\times$ even periodic lattice.

On an even$\times$odd lattice, the situation is a little different. The state generated by T_2 has odd numbers of fermions and does not correspond to any physical spin-1/2 state. Thus, we can only use T_1 to generate the other degenerate state. There are only two degenerate ground states on an even $\times$ odd periodic lattice (generated by T_1). On an odd $\times$ odd lattice, there are also two degenerate ground states generated by T_1T_2.

We note that, locally, the T_1 and T_2 transformations are indistinguishable from the Z_2 gauge transformation. As the physical spin operators are invariant under the Z_2 gauge transformation, they are also invariant under the T_1 and T_2 transformations. Therefore, the degenerate ground states generated by T_1 and T_2 remain degenerate, even after we add an arbitrary local perturbation to our exactly soluble model (10.2.6). The degeneracy of the ground states is a robust topological property, indicating non-trivial topological order in the ground state.

Problem 10.2.4.
Consider eqn (10.2.6) on an $L_x \times L_y$ periodic lattice. We assume that $V_{\boldsymbol{i}} = V$, except on a row of squares where $V_{\boldsymbol{i}} = 0$. In this case, the system can be viewed as being defined on a cylinder with two circular edges in the x direction. Find the ground-state degeneracy of the system. These degenerate ground states can be viewed as edge states. Show that there are $\sqrt{2}$ edge states per edge site. This implies that the edge excitations are described by a Majorana fermion.

10.2.3 The projective symmetry group characterization of ground states

- The $V < 0$ and $V > 0$ ground states of eqn (10.2.6) have different quantum orders.

In this section, we will assume that the $V_{\boldsymbol{i}}$ in the spin-1/2 system (10.2.6) are uniform, i.e. $V_{\boldsymbol{i}} = V$. We have solved eqn (10.2.6) by writing the spin operator as a product of fermion operators, see eqn (10.2.5), i.e. we have solved the spin-1/2 system using the projective construction. It is quite interesting to see that, for the particular spin-1/2 system (10.2.6), the projective construction gives exact results. As the exact ground-state wave function is obtained from the projective construction (i.e. obtained by projecting the free fermion wave function (9.2.11)), we can use the PSG characterization developed in Chapter 9 to characterize the quantum order in the ground state.

The discussion in Chapter 9 is limited to spin–rotation-invariant states. In the following, we will generalize the mean-field theory and the PSG to spin–rotation-non-invariant states. The spin-1/2 model H_{spin} can also be viewed as a hard-core boson model, if we identify the $|\downarrow\rangle$ state as the zero-boson state $|0\rangle$, and the $|\uparrow\rangle$ state as the one-boson state $|1\rangle$. In the following, we will use the boson picture to describe our model.

To use the projective construction to construct quantum-ordered (or entangled) many-boson wave functions, we first introduce the 'mean-field' fermion Hamiltonian

$$H_{\text{mean}} = \sum_{\langle \boldsymbol{ij} \rangle} \left(\psi^\dagger_{I,\boldsymbol{i}} u^{IJ}_{\boldsymbol{ij}} \psi_{J,\boldsymbol{j}} + \psi^\dagger_{I,\boldsymbol{i}} w^{IJ}_{\boldsymbol{ij}} \psi^\dagger_{J,\boldsymbol{j}} + h.c. \right) \tag{10.2.9}$$

where $I, J = 1, 2$. We will use $u_{\boldsymbol{ij}}$ and $w_{\boldsymbol{ij}}$ to denote the 2×2 complex matrices whose elements are $u^{IJ}_{\boldsymbol{ij}}$ and $w^{IJ}_{\boldsymbol{ij}}$, respectively. Let $|\Psi^{(u_{\boldsymbol{ij}}, w_{\boldsymbol{ij}})}_{\text{mean}}\rangle$ be the ground state of the above free fermion Hamiltonian. Then the following many-body boson wave function can be obtained:

$$\Phi^{(u_{ij},w_{ij})}(\boldsymbol{i}_1, \boldsymbol{i}_2...) = \langle 0| \prod_n b(\boldsymbol{i}_n)|\Psi_{\text{mean}}^{(u_{ij},w_{ij})}\rangle \tag{10.2.10}$$

where

$$b(\boldsymbol{i}) = \psi_{1,\boldsymbol{i}}\psi_{2,\boldsymbol{i}} \tag{10.2.11}$$

We note that the physical boson wave function $\Phi^{(u_{ij},w_{ij})}(\{\boldsymbol{i}_n\})$ is invariant under the following $SU(2)$ gauge transformations:

$$(\psi_{\boldsymbol{i}}, u_{\boldsymbol{ij}}, w_{\boldsymbol{ij}}) \to (G(\boldsymbol{i})\psi_{\boldsymbol{i}}, G(\boldsymbol{i})u_{\boldsymbol{ij}}G^\dagger(\boldsymbol{j}), G(\boldsymbol{i})w_{\boldsymbol{ij}}G^\top(\boldsymbol{j}))$$

where $G(\boldsymbol{i}) \in SU(2)$.

The quantum order in the boson wave function $\Phi^{(u_{ij},w_{ij})}(\{\boldsymbol{i}_n\})$ can be characterized by the PSG. The PSG is formed by combined symmetry transformations and the $SU(2)$ gauge transformations that leave the ansatz $(u_{\boldsymbol{ij}}, w_{\boldsymbol{ij}})$ invariant.

Here we would like to point out that the common eigenstates of $\hat{U}_{\boldsymbol{ij}}$, namely $|\{s_{\boldsymbol{ij}}\}\rangle$, are the ground states of the free fermion system

$$\tilde{H}_{\text{mean}} = \sum_{\langle \boldsymbol{ij}\rangle} \left(s_{\boldsymbol{ij}}\hat{U}_{\boldsymbol{ij}} + h.c.\right) \tag{10.2.12}$$

Thus, from the projective construction point of view, $\tilde{H}_{\text{mean}}$ can be viewed as the'mean-field' Hamiltonian H_{mean}. In fact, $\tilde{H}_{\text{mean}}$ is a special case of H_{mean} with

$$-w_{\boldsymbol{i},\boldsymbol{i}+\hat{\boldsymbol{x}}} = u_{\boldsymbol{i},\boldsymbol{i}+\hat{\boldsymbol{x}}} = -\mathrm{i}s_{\boldsymbol{i},\boldsymbol{i}+\hat{\boldsymbol{x}}}(1+\tau^3), \quad -w_{\boldsymbol{i},\boldsymbol{i}+\hat{\boldsymbol{y}}} = u_{\boldsymbol{i},\boldsymbol{i}+\hat{\boldsymbol{y}}} = -\mathrm{i}s_{\boldsymbol{i},\boldsymbol{i}+\hat{\boldsymbol{y}}}(1-\tau^3)$$

The state $|\{s_{\boldsymbol{ij}}\}\rangle$ is equal to the mean-field state $|\Psi_{\text{mean}}^{(u_{ij},w_{ij})}\rangle$ if $s_{\boldsymbol{ij}}$ and $(u_{\boldsymbol{ij}}, w_{\boldsymbol{ij}})$ are related through the above equation. The physical spin wave function is obtained by the projection $\mathcal{P}|\{s_{\boldsymbol{ij}}\}\rangle$ of the mean-field state, which is equivalent to eqn (10.2.10).

Remember that, in the projective construction, we need to choose the ansatz $u_{\boldsymbol{ij}}$ and $w_{\boldsymbol{ij}}$ to minimize the average energy $\langle\Psi_{\text{mean}}^{(u_{ij},w_{ij})}|\mathcal{P}H_{\text{spin}}\mathcal{P}|\Psi_{\text{mean}}^{(u_{ij},w_{ij})}\rangle$. For our spin model H_{spin}, the trivial ground state obtained in this way turns out to be the exact ground state! Also, if we choose a different $s_{\boldsymbol{ij}}$, then the projected state $\mathcal{P}|\Psi_{\text{mean}}^{(u_{ij},w_{ij})}\rangle$ will correspond to an exact eigenstate of H_{spin} with energy (10.2.4). It is in this sense that the projective construction provides an exact solution of H_{spin}.

When $V > 0$, the ground state of our model is given by the Z_2-flux configuration $F_{\boldsymbol{i}} = 1$. To produce such a flux, we can choose $s_{\boldsymbol{i},\boldsymbol{i}+\boldsymbol{x}} = s_{\boldsymbol{i},\boldsymbol{i}+\boldsymbol{y}} = i$. In this case, eqn (10.2.12) becomes eqn (10.2.9) with

$$-w_{i,i+x} = u_{i,i+x} = 1 + \tau^3, \qquad -w_{i,i+y} = u_{i,i+y} = 1 - \tau^3.$$

To obtain the IGG for the above ansatz, we note that u_{ij} is invariant under the constant gauge transformation $G(i) = e^{i\phi\tau^z}$. However, w_{ij} is invariant only when $\phi = 0, \pi$. Thus, IGG $= Z_2$. We find that the low-energy effective theory is a Z_2 gauge theory. As the ansatz is already translationally invariant, the ansatz is invariant under $G_x T_x$ and $G_y T_y$ with the trivial gauge transformations $G_x = G_y = \tau^0$. The PSG for the above ansatz is simply the Z2A PSG in eqn (9.4.10). Thus, the ground state for $V > 0$ is a Z2A state. Note that, for the Z2A PSG, $G_x T_x$ and $G_y T_y$ satisfy the translational algebra $(G_x T_x)^{-1}(G_y T_y)^{-1}(G_x T_x)(G_y T_y) = 1$.

When $V < 0$, the ground state is given by the configuration $F_i = -1$, which can be produced by $s_{i,i+x} = (-)^{i_x} s_{i,i+y} = i$. The ansatz now has the form

$$-w_{i,i+x} = u_{i,i+x} = 1 + \tau^3, \qquad -w_{i,i+y} = u_{i,i+y} = (-)^{i_x}(1 - \tau^3).$$

The ansatz is invariant under the translation $i \to i + x$ followed by the gauge transformation $G_x(i) = (-)^{i_y}$. Its PSG is the Z2B PSG in eqn (9.4.11). Thus, the ground state for $V < 0$ is a Z2B state. For the Z2B PSG, $G_x T_x$ and $G_y T_y$ satisfy the magnetic translational algebra $(G_x T_x)^{-1}(G_y T_y)^{-1}(G_x T_x)(G_y T_y) = -1$. The fermion hopping Hamiltonian H_{mean} describes fermion hopping in the magnetic field with π flux per plaquette. The different PSGs tell us that the $V < 0$ and $V > 0$ ground states have different quantum orders.

10.3 Z_2 spin liquids and string-net condensation on a square lattice

- String-net condensations lead to an emergent Z_2 gauge theory and fermions.

In this section, we are going to discuss the exactly soluble spin-1/2 model (10.2.6) (Kitaev, 2003; Wen, 2003c) from the string-net condensation point of view (Levin and Wen, 2003). The model is one of the simplest models that demonstrates the connection between string-net condensation and emergent gauge bosons and fermions in local bosonic models. The model can also be solved using the slave-boson approach, which allows us to see how the PSG that describes the quantum order is connected to the string-net condensation (see Section 10.4).

10.3.1 Constructing Hamiltonians with closed string-net condensations

- Local boson models with emergent gauge fields can be constructed by making certain string-like objects condense.

Let us first consider an arbitrary spin-1/2 model on a square lattice. The first question that we want to ask is what kind of spin interaction can give rise to a

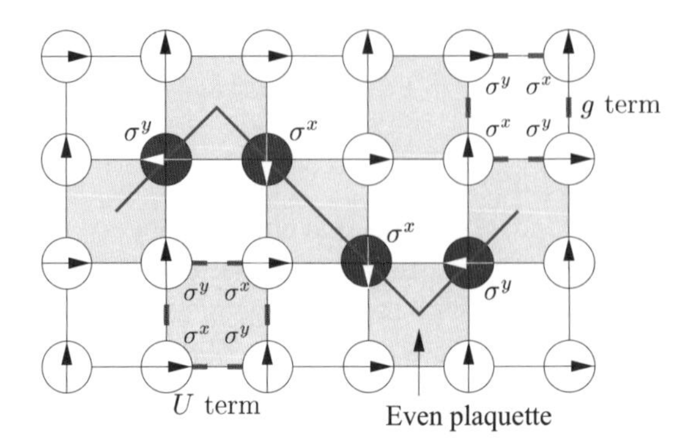

FIG. 10.1. An open-string excitation on top of the ground state of H_J.

low-energy gauge theory. If we believe the connection between gauge theory and string-net theory (Banks *et al.*, 1977; Savit, 1980; Wen, 2003a), then one way to obtain a low-energy gauge theory is to design a spin interaction that allows strong fluctuations of large closed strings, but forbids other types of fluctuations (such as local spin flip, open-string fluctuations, etc.). We hope that the presence of strong fluctuations of large closed strings will lead to the condensation of closed strings of arbitrary sizes, which in turn gives rise to a low-energy gauge theory.

Let us start with

$$H_J = -J \sum_{\text{even}} \sigma^x_{\boldsymbol{i}} - J \sum_{\text{odd}} \sigma^y_{\boldsymbol{i}}$$

where $\boldsymbol{i} = (i_x, i_y)$ labels the lattice sites, $\sigma^{x,y,z}$ are the Pauli matrices, and $\sum_{\text{even}}$ (or $\sum_{\text{odd}}$) is a sum over even sites with $(-)^{\boldsymbol{i}} \equiv (-1)^{i_x+i_y} = 1$ (or over odd sites with $(-)^{\boldsymbol{i}} = -1$). The ground state of H_J, namely $|0\rangle$, has spins pointing in the x direction at even sites and in the y direction at odd sites (see Fig. 10.1). Such a state will be defined as a state with no string.

To create a string excitation, we first draw a string that connects nearest-neighbor *even* squares (see Fig. 10.1). We then flip the spins on the string. Such a string state is created by the following string creation operator (or simply the string operator):

$$W(C) = -\prod_C \sigma^{a_{\boldsymbol{i}}}_{\boldsymbol{i}} \tag{10.3.1}$$

where the product $\prod_C$ is over all of the sites on the string, and $a_{\boldsymbol{i}} = y$ if $\boldsymbol{i}$ is even and $a_{\boldsymbol{i}} = x$ if $\boldsymbol{i}$ is odd. A generic string state has the form

$$|C_1 C_2 ...\rangle = W(C_1)W(C_2)...|0\rangle$$

where C_1, C_2, ... are strings. Such a state will be called a string-net state because strings can intersect and overlap. The operator that creates a string-net, namely

$$W(C_{\text{net}}) = W(C_1)W(C_2)...$$

will be called a string-net operator. The state $|C_1C_2...\rangle$ is an open-string-net state if at least one of the C_i is an open string. The corresponding operator $W(C_{\text{net}})$ will be called an open-string-net operator. If all of the C_i are closed loops, then $|C_1C_2...\rangle$ is a closed-string-net state and $W(C_{\text{net}})$ is a closed-string-net operator.

The Hamiltonian H_J has no string-net condensation because its ground state $|0\rangle$ contains no string-nets, and hence $\langle 0|W(C_{\text{net}})|0\rangle = 0$. To obtain a Hamiltonian with closed-string-net condensation, we need to first find a Hamiltonian whose ground state contains a lot of closed string-nets of arbitrary sizes and does not contain any open strings (or open string-nets).

Let us first write down a Hamiltonian such that closed strings and closed string-nets cost no energy and any open strings cost a large energy. One such Hamiltonian has the form

$$H_U = -U \sum_{\text{even}} \hat{F}_{\boldsymbol{i}}$$

where

$$\hat{F}_{\boldsymbol{i}} = \sigma^x_{\boldsymbol{i}} \sigma^y_{\boldsymbol{i}+\boldsymbol{x}} \sigma^x_{\boldsymbol{i}+\boldsymbol{x}+\boldsymbol{y}} \sigma^y_{\boldsymbol{i}+\boldsymbol{y}} \tag{10.3.2}$$

We find that the no-string state $|0\rangle$ is one of the ground states of H_U (assuming that $U > 0$) with energy $-UN_{\text{site}}$. All of the closed-string states, such as $W(C_{\text{close}})|0\rangle$, are also ground states of H_U because $[H_U, W(C_{\text{close}})] = 0$. Similarly, $[H_U, W(C_{\text{net}})] = 0$ implies that any closed-string state is also a ground state.

Using the commutation relation between the open-string operator $W(C_{\text{open}})$ and $\hat{F}_{\boldsymbol{i}}$, namely

$$\begin{aligned} \hat{F}_{\boldsymbol{i}} W(C_{\text{open}}) &= -W(C_{\text{open}})\hat{F}_{\boldsymbol{i}}, \quad \text{if } C_{\text{open}} \text{ ends at the square } \boldsymbol{i}, \\ \hat{F}_{\boldsymbol{i}} W(C_{\text{open}}) &= +W(C_{\text{open}})\hat{F}_{\boldsymbol{i}}, \quad \text{otherwise}, \end{aligned}$$

we find that the open-string operator flips the sign of $\hat{F}_{\boldsymbol{i}}$ at its two ends. An open-string state $W(C_{\text{open}})|0\rangle$ is also an eigenstate of $\hat{F}_{\boldsymbol{i}}$ and hence an eigenstate H_U, with an energy $-UN_{\text{site}} + 4U$. We see that each end of the open string costs an energy $2U$. We also note that the energy of closed string-nets does not depend on the length of the strings in the net. Thus, the strings in H_U have no tension. We can introduce a string tension by adding the H_J to our Hamiltonian. The string tension will be $2J$ per site (or per segment). We note that any string-net state $|C_1C_2...\rangle$ is an eigenstate of H_U+H_J. Thus, string-nets described by H_U+H_J do not fluctuate, and hence cannot condense. To make strings fluctuate, we need the g-term

$$H_g = -g \sum_{\boldsymbol{p}} W(C_{\boldsymbol{p}}) = -g \sum_{\text{odd}} \hat{F}_{\boldsymbol{i}}$$

where $\boldsymbol{p}$ labels the odd squares and $C_{\boldsymbol{p}}$ is the closed string around the square $\boldsymbol{p}$. When H_g acts on a string-net state, it adds a loop of the string $C_{\boldsymbol{p}}$ to the string-net. Thus the g-term causes the string to fluctuate. The total Hamiltonian of our spin-1/2 model is given by

$$H = H_U + H_J + H_g$$

When $U \gg g \gg J$, the closed strings have little tension and their fluctuations induced by H_g are not limited by the tension. Thus, the ground state contains strong fluctuations of closed strings. In other words, the ground state is filled by closed strings of arbitrary sizes. If $U \gg g$, then it costs too much energy for closed strings to break up into open strings. Thus, when $U \gg g \gg J$, the ground state of our spin-1/2 model contains a condensation of closed strings (or, more precisely, a condensation of closed string-nets). In the following, we will study the physical properties of such a string-net-condensed state.

Problem 10.3.1.
(a) Show that $\hat{F}_i$s satisfy

$$\prod_i \hat{F}_i = 1 \tag{10.3.3}$$

on a lattice with periodic boundary conditions in both the x and y directions. The minus sign in eqn (10.3.2) allows the above simple result.
(b) On an even $\times$ even lattice with periodic boundary conditions, show that the $\hat{F}_i$s satisfy the more strict condition

$$\prod_{i=\text{even}} \hat{F}_i = 1 \tag{10.3.4}$$

10.3.2 String-net condensation and low-energy effective theory

• String-net condensation gives rise to gauge excitations.

Let us first discuss the string-net condensed phase with $U \gg g > 0$ and $J = 0$. When $J = 0$, the model $H_U + H_g$ becomes the model (10.2.6) discussed in the last section, with $V_i = U$ at the even sites and $V_i = g$ at the odd sites. The model is exactly soluble because $[\hat{F}_i, \hat{F}_j] = 0$ (Kitaev, 2003). The eigenstates of $H_U + H_g$ can be obtained from the common eigenstates of $\hat{F}_i$, namely $|\{F_i\}\rangle$, where F_i is the eigenvalue of $\hat{F}_i$. As $\hat{F}_i^2 = 1$, the eigenvalues of $\hat{F}_i$ are simply $F_i = \pm 1$. The energy of $|\{F_i\}\rangle$ is $E = -U \sum_{i=\text{even}} F_i - g \sum_{i=\text{odd}} F_i$.

Note that, for a finite system on an even $\times$ even periodic lattice of size $L_x \times L_y$, $\hat{F}_i$ satisfy the constraints $\prod_{i=\text{even}} \hat{F}_i = 1$ and $\prod_{i=\text{odd}} \hat{F}_i = 1$. Therefore, the F_is are not independent and can only label $2^{L_x L_y}/4$ different configurations. However, the Hilbert space of our spin-1/2 model contains $2^{L_x L_y}$ states. To reproduce the $2^{L_x L_y}$ states, the common eigenstates of $\hat{F}_i$ must be four-fold degenerate. This agrees with the result obtained in the last section.

In the limit $U \gg g$, all of the states containing open strings will have an energy of order U. The low-energy states contain only closed string-nets and satisfy

$$F_{\boldsymbol{i}}|_{\boldsymbol{i}=\text{even}} = 1$$

The different low-energy states are labeled by the $F_{\boldsymbol{i}}$ on odd squares, i.e. $F_{\boldsymbol{i}}|_{\boldsymbol{i}=\text{odd}} = \pm 1$. In particular, the ground state is given by

$$F_{\boldsymbol{i}}|_{\boldsymbol{i}=\text{odd}} = 1.$$

Such a ground state has a closed-string condensation. This is because all of the closed-string operators $W(C_{\text{close}})$ commute with $H_U + H_g$. Hence the ground state $|\Psi_0\rangle$ of $H_U + H_g$ is an eigenstate of $W(C_{\text{close}})$ and satisfies

$$\langle \Psi_0 | W(C_{\text{close}}) | \Psi_0 \rangle = 1.$$

The above equation also implies that the ground state $|\Psi_0\rangle$ is an equal-weight superposition of different closed-string-net configurations.

The low-energy excitations above the ground state can be obtained by flipping $F_{\boldsymbol{i}}$ from 1 to -1 on some odd squares. If we view $F_{\boldsymbol{i}}$ on odd squares as the flux in Z_2 gauge theory, then we find that the low-energy sector of the model is identical to a Z_2 lattice gauge theory, at least for infinite systems. This suggests that the low-energy effective theory of our model is a Z_2 lattice gauge theory.

However, one may object to this result by pointing out that the low-energy sector of our model is also identical to an Ising model with one spin on each odd square. Thus, the low-energy effective theory should be the Ising model. We would like to point out that, although the low-energy sector of our model is identical to an Ising model for infinite systems, the low-energy sector of our model is different from an Ising model for finite systems. For example, on a finite even $\times$ even lattice with periodic boundary conditions, the ground state of our model has a four-fold degeneracy (Kitaev, 2003; Wen, 2003c). The Ising model does not have such a degeneracy, while the Z_2 gauge theory has such a degeneracy. Also, our model contains an excitation that can be identified as a Z_2 charge. Therefore, the low-energy effective theory of our model is a Z_2 lattice gauge theory instead of an Ising model. The $F_{\boldsymbol{i}} = -1$ excitations on odd squares can be viewed as the Z_2 vortex excitations in the Z_2 lattice gauge theory.

To understand the Z_2 charge excitations, we note that, in the string-condensed state, the string costs no energy and is unobservable. Applying the closed-string operator to the ground state results in the ground state itself. As the string is unobservable, a piece of open string behaves like two independent particles. Each end of the open string corresponds to a particle on even squares. The energy required to create such a particle is of order U. Now, let us consider the hopping of one such particle around four nearest-neighbor even squares (see Fig. 10.2). Each hopping step is generated by a small piece of string operator. We see that the product of

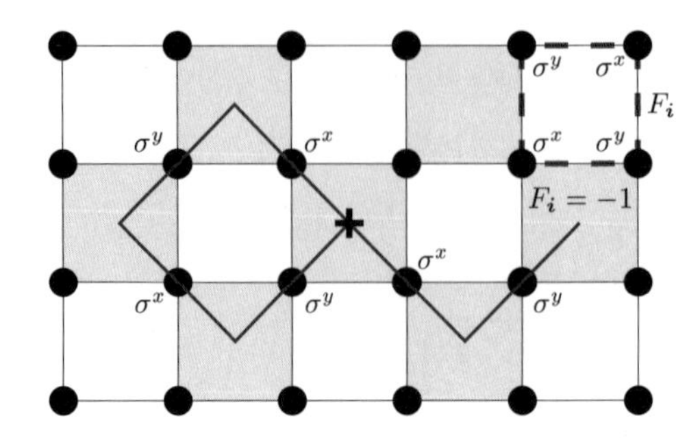

FIG. 10.2. A hopping of the Z_2 charge around four nearest-neighbor even squares.

the four hopping amplitudes is given by the eigenvalue of $\hat{F}_i$ on the odd square in the middle of the four even squares. This is exactly the relationship between charge and flux. Thus, if we identify F_i on odd squares as Z_2 flux, then the ends of strings on even squares will correspond to the Z_2 charges. Due to the closed-string condensation, the ends of open strings are not confined and have only short-ranged interactions between them. Thus, the Z_2 charges behave like point particles with no string attached.

10.3.3 Three types of strings and emergent fermions

- The ends of strings are gauge charges. The ends of certain strings can also be fermions.

In the following, we are going to study several different types of string. To avoid confusion, we will call the strings discussed above T1 strings. The T1 strings connect even squares (see Fig. 10.3(a)).

Just like the Z_2 charges, a pair of Z_2 vortices is also created by an open-string operator. As the Z_2 vortices correspond to flipped $\hat{F}_i$ on *odd* squares, the open-string operator that creates Z_2 vortices is also given by eqn (10.3.1), except that now the product is over a string that connects *odd* squares. We will call such a string a T2 string.

We would like to point out that the reference state (i.e. the no-string state) for the T2 string is different from that of the T1 string. The no-T2-string state is given by $|\tilde{0}\rangle$ with spins pointing in the y direction at even sites and in the x direction at odd sites. As the T1 and T2 strings have different reference states, we cannot have a dilute gas of the T1 strings and the T2 strings at the same time. One can easily check that the T2 string operators also commute with $H_U + H_g$. Therefore, the ground state $|\Psi_0\rangle$, in addition to the condensation of T1 closed strings, also has a condensation of T2 closed strings.

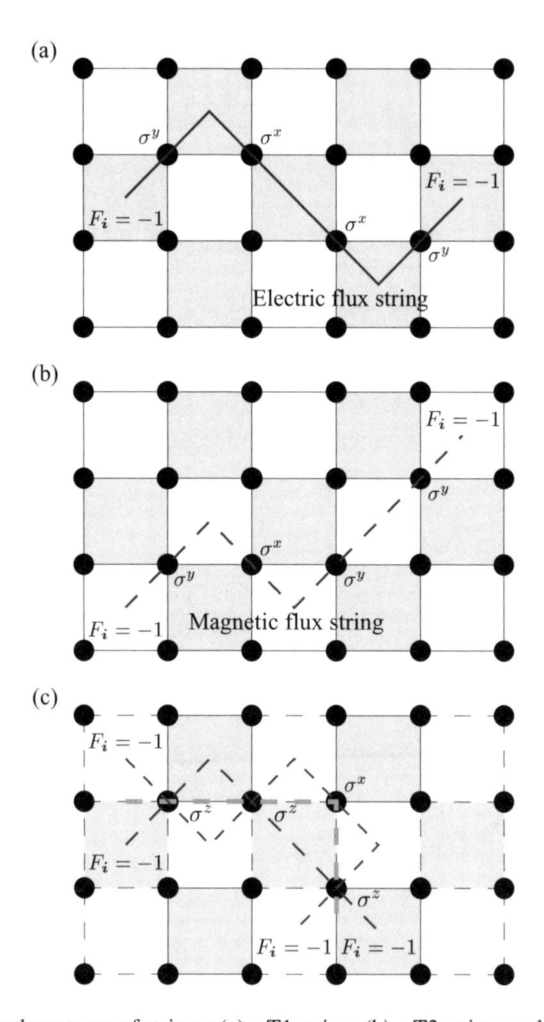

FIG. 10.3. The three types of strings: (a) a T1 string; (b) a T2 string; and (c) a T3 string.

The hopping of a Z_2 vortex is induced by a short T2 open string. As the T2 open-string operators all commute with each other, the Z_2 vortex behaves like a boson. Similarly, the Z_2 charges also behave like bosons. However, T1 open-string operators and T2 open-string operators do not commute. As a result, the ends of the T1 string and the ends of the T2 string have non-trivial mutual statistics. As we have already shown that moving a Z_2 charge around a Z_2 vortex generates a phase π, the Z_2 charges and the Z_2 vortices have semionic mutual statistics.

The T3 strings are defined as bound states of the T1 and T2 strings (see Fig. 10.3(c)). The T3 string operator is given by the product of a T1 string operator and a T2 string operator, and has the form

$$W(C) = \prod_n \sigma_{\boldsymbol{i}_n}^{l_n}$$

where C is a string connecting the mid-point of the neighboring links, and $\boldsymbol{i}_n$ are sites on the string. Here $l_n = z$ if the string does not turn at the site $\boldsymbol{i}_n$; $l_n = x$ or y if the string makes a turn at the site $\boldsymbol{i}_n$; $l_n = x$ if the turn forms an upper-right or lower-left corner; and $l_n = y$ if the turn forms a lower-right or upper-left corner (see Fig. 10.3(c)). The ground state also has a condensation of T3 closed strings. The ends of a T3 string are bound states formed by a Z_2 charge and a Z_2 vortex on the two squares on the two sides of a link (i.e. $F_{\boldsymbol{i}} = -1$ on the two sides of the link). Such bound states are fermions (see Section 7.1.2 and Fig. 7.8). So the fermions live on the links. It is interesting to see that string-net condensation in our model directly leads to a Z_2 gauge structure and three new types of quasiparticle, namely Z_2 charge, Z_2 vortex, and fermions. Fermions, as ends of open T3 strings, emerge from our purely bosonic model!

As the ends of the T1 string are Z_2 charges, the T1 string can be viewed as strings of Z_2 'electric' flux. Similarly, the T2 string can be viewed as strings of Z_2 'magnetic' flux.

10.4 Classifying different string-net condensations by the projective symmetry group

10.4.1 Four classes of string-net condensations

- Different string-net condensations give rise to different quantum-ordered phases.

We have seen that, when $U > 0$, $g > 0$, and $J = 0$, the ground state of our model $H_g + H_U + H_J$ is given by

$$F_{\boldsymbol{i}}|_{\boldsymbol{i}=\text{even}} = 1, \qquad F_{\boldsymbol{i}}|_{\boldsymbol{i}=\text{odd}} = 1.$$

We will call such a phase a Z_2 phase to stress the low-energy Z_2 gauge structure. In the Z_2 phase, the T1 string operator $W_1(C_1)$ and the T2 string operator $W_2(C_2)$ have the following expectation values:

$$\langle W_1(C_1)\rangle = 1, \qquad \langle W_2(C_2)\rangle = 1$$

When $U > 0$, $g < 0$, and $J = 0$, the ground state is given by

$$F_{\boldsymbol{i}}|_{\boldsymbol{i}=\text{even}} = 1, \qquad F_{\boldsymbol{i}}|_{\boldsymbol{i}=\text{odd}} = -1.$$

We see that there is π flux through each odd square. We will call such a phase a Z_2-flux phase. The T1 string operator and the T2 string operator have the following expectation values:

$$\langle W_1(C_1)\rangle = (-)^{N_{\text{odd}}}, \qquad \langle W_2(C_2)\rangle = 1$$

where N_{odd} is the number of odd squares enclosed by the T1 string C_1.

When $U < 0$, $g > 0$, and $J = 0$, the ground state is given by

$$F_{\boldsymbol{i}}|_{\boldsymbol{i}=\text{even}} = -1, \qquad F_{\boldsymbol{i}}|_{\boldsymbol{i}=\text{odd}} = 1.$$

There is a Z_2 charge on each even square. We will call such a phase a Z_2-charge phase. The T1 string operator and the T2 string operator have the following expectation values:

$$\langle W_1(C_1)\rangle = 1, \qquad \langle W_2(C_2)\rangle = (-)^{N_{\text{even}}}$$

where N_{even} is the number of even squares enclosed by the T2 string C_2. Note that the Z_2-flux phase and the Z_2-charge phase, differing only by a lattice translation, are essentially the same phase.

When $U < 0$, $g < 0$, and $J = 0$, the ground state becomes

$$F_{\boldsymbol{i}}|_{\boldsymbol{i}=\text{even}} = -1, \qquad F_{\boldsymbol{i}}|_{\boldsymbol{i}=\text{odd}} = -1.$$

There is a Z_2 charge on each even square and π flux through each odd square. We will call such a phase a Z_2-flux–charge phase. The T1 string operator and the T2 string operator have the following expectation values:

$$\langle W_1(C_1)\rangle = (-)^{N_{\text{odd}}}, \qquad \langle W_2(C_2)\rangle = (-)^{N_{\text{even}}}$$

When $U = g = 0$ and $J \neq 0$, the model is also exactly soluble. The ground state is a simple spin-polarized state (see Fig. 10.1).

From these exactly soluble limits of the $H_U + H_g + H_J$ model, we suggest a phase diagram as sketched in Fig. 10.4. The phase diagram contains four different string-net condensed phases and one phase with no string condensation. All of the phases have the same symmetry and are distinguished only by their different quantum orders.

10.4.2 The projective symmetry group and ends of condensed strings

- The projective symmetry described by the PSG is simply the symmetry of the effective theory for the ends of condensed strings.

From the different $\langle W_1(C_1)\rangle$ and $\langle W_2(C_2)\rangle$, we see that the first four phases have different string-net condensations. However, they all have the same symmetry. This raises an issue. Without symmetry breaking, how do we know that the above four phases are really different phases? How do we know that it is

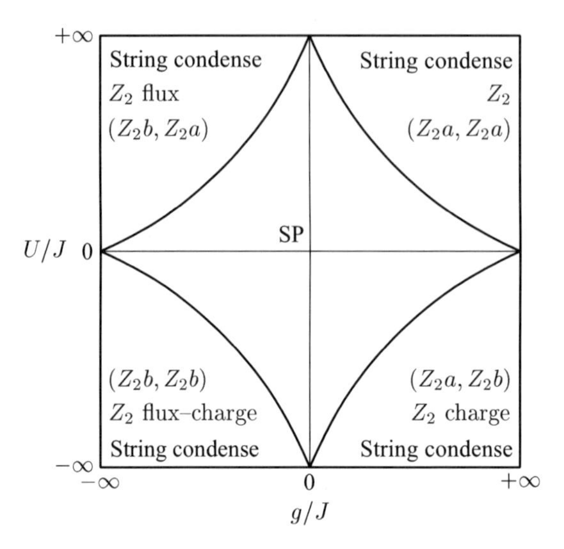

FIG. 10.4. The proposed phase diagram for the $H = H_U + H_g + H_J$ model. Here J is assumed to be positive. The four string-net condensed phases are characterized by a pair of PSGs $(PSG_{\text{charge}}, PSG_{\text{vortex}})$. SP marks a spin-polarized phase.

impossible to change one string-net-condensed state into another without a phase transition?

In the following, we will show that the different string-net condensations can be described by different PSGs (just like different symmetry-breaking orders can be described by different symmetry groups of ground states). In Chapter 9, different quantum orders were introduced via their different PSGs (Wen, 2002b,c). It was shown that the PSGs are universal properties of a quantum phase. Only phase transitions can change PSGs. Thus, the fact that different string-net-condensed states are characterized by different PSGs indicates that these different string-net-condensed states belong to different quantum phases.

The connection between string-net condensation and the PSG also allows us to connect string-net condensation to the quantum order introduced in Chapter 9. In fact, the projective construction discussed in Chapter 9 can be viewed as a way to construct string-net-condensed states.

To see the connection between the PSG and the string-net condensation, we note that, when closed string-nets condense, the ends of the open strings behave like independent particles. Let us consider the two-particle states $|\boldsymbol{p}_1\boldsymbol{p}_2\rangle$ that describe the two ends of a T1 string. Note that the ends of the T1 strings, and hence the Z_2 charges, only live on the even squares. Thus, $\boldsymbol{p}_1$ and $\boldsymbol{p}_2$ only label the even squares. For our model $H_U + H_g$, $|\boldsymbol{p}_1\boldsymbol{p}_2\rangle$ is an energy eigenstate and the

Z_2 charges do not hop. Here we would like to add the term

$$H_t = t\sum_{\boldsymbol{i}}(\sigma^x_{\boldsymbol{i}} + \sigma^y_{\boldsymbol{i}}) + t'\sum_{\boldsymbol{i}}\sigma^z_{\boldsymbol{i}}$$

to the Hamiltonian. The t term $t\sum_{\boldsymbol{i}}(\sigma^x_{\boldsymbol{i}} + \sigma^y_{\boldsymbol{i}})$ makes the Z_2 charges hop among the even squares with a hopping amplitude of order t. The dynamics of the two Z_2 charges are described by the following effective Hamiltonian in the two-particle Hilbert space:

$$H = H(\boldsymbol{p}_1) + H(\boldsymbol{p}_2)$$

where $H(\boldsymbol{p}_1)$ describes the hopping of the first particle $\boldsymbol{p}_1$ and $H(\boldsymbol{p}_2)$ describes the hopping of the second particle $\boldsymbol{p}_2$. Now we can define the PSG of a string-net-condensed state. The PSG is simply the symmetry group of the hopping Hamiltonian $H(\boldsymbol{p})$.

Due to the translational symmetry of the underlying model $H_U + H_g + H_t$, we may naively expect the hopping Hamiltonian of the Z_2 charge $H(\boldsymbol{p})$ to also have translational symmetry in the $\boldsymbol{x}+\boldsymbol{y}$ and $\boldsymbol{x}-\boldsymbol{y}$ directions:

$$\begin{aligned} H(\boldsymbol{p}) &= T^\dagger_{xy}H(\boldsymbol{p})T_{xy}, & T_{xy}|\boldsymbol{p}\rangle &= |\boldsymbol{p}+\boldsymbol{x}+\boldsymbol{y}\rangle \\ H(\boldsymbol{p}) &= T^\dagger_{x\bar{y}}H(\boldsymbol{p})T_{x\bar{y}}, & T_{x\bar{y}}|\boldsymbol{p}\rangle &= |\boldsymbol{p}+\boldsymbol{x}-\boldsymbol{y}\rangle \end{aligned} \tag{10.4.1}$$

If the above is true, then it implies that the PSG is the same as the translational symmetry group.

It turns out that eqn (10.4.1) is too strong. The underlying spin model can have translational symmetry, even when $H(\boldsymbol{p})$ does not satisfy eqn (10.4.1). However, the possible symmetry groups of $H(\boldsymbol{p})$ (the PSGs) are strongly constrained by the translational symmetry of the underlying spin model. In the following, we will explain why the PSG can be different from the symmetry group of the physical spin model, and what conditions the PSG must satisfy in order to be consistent with the translational symmetry of the underlying spin model.

We note that a string always has two ends. Thus, a physical state always has an even number of Z_2 charges. The actions of translation on a two-particle state are given by

$$\begin{aligned} T^{(2)}_{xy}|\boldsymbol{p}_1,\boldsymbol{p}_2\rangle &= |\boldsymbol{p}_1+\boldsymbol{x}+\boldsymbol{y},\boldsymbol{p}_2+\boldsymbol{x}+\boldsymbol{y}\rangle \\ T^{(2)}_{x\bar{y}}|\boldsymbol{p}_1,\boldsymbol{p}_2\rangle &= |\boldsymbol{p}_1+\boldsymbol{x}-\boldsymbol{y},\boldsymbol{p}_2+\boldsymbol{x}-\boldsymbol{y}\rangle \end{aligned}$$

Here $T^{(2)}_{xy}$ and $T^{(2)}_{x\bar{y}}$ satisfy the algebra of translations

$$T^{(2)}_{xy}T^{(2)}_{x\bar{y}} = T^{(2)}_{x\bar{y}}T^{(2)}_{xy} \tag{10.4.2}$$

We note that $T^{(2)}_{xy}$ and $T^{(2)}_{x\bar{y}}$ are direct products of translation operators on the single-particle states. Thus, in some sense, the single-particle translations are square roots of the two-particle translations. The two-particle translational algebra imposes a strong constraint on the single-particle translational algebra.

The most general form of single-particle translations is given by $T_{xy}G_{xy}$ and $T_{x\bar{y}}G_{x\bar{y}}$, where the actions of the operators $T_{xy,x\bar{y}}$ and $G_{xy,x\bar{y}}$ are defined by

$$T_{xy}|\boldsymbol{p}\rangle = |\boldsymbol{p}+\boldsymbol{x}+\boldsymbol{y}\rangle, \qquad T_{x\bar{y}}|\boldsymbol{p}\rangle = |\boldsymbol{p}+\boldsymbol{x}-\boldsymbol{y}\rangle,$$
$$G_{xy}|\boldsymbol{p}\rangle = \mathrm{e}^{\phi_{xy}(\boldsymbol{p})}|\boldsymbol{p}\rangle, \qquad G_{x\bar{y}}|\boldsymbol{p}\rangle = \mathrm{e}^{\phi_{x\bar{y}}(\boldsymbol{p})}|\boldsymbol{p}\rangle.$$

In order for the direct products $T_{xy}^{(2)} = T_{xy}G_{xy} \otimes T_{xy}G_{xy}$ and $T_{x\bar{y}}^{(2)} = T_{x\bar{y}}G_{x\bar{y}} \otimes T_{x\bar{y}}G_{x\bar{y}}$ to reproduce the translational algebra (10.4.2), we only require $T_{xy}G_{xy}$ and $T_{x\bar{y}}G_{x\bar{y}}$ to satisfy

$$T_{xy}G_{xy}T_{x\bar{y}}G_{x\bar{y}} = T_{x\bar{y}}G_{x\bar{y}}T_{xy}G_{xy} \tag{10.4.3}$$

or

$$T_{xy}G_{xy}T_{x\bar{y}}G_{x\bar{y}} = -T_{x\bar{y}}G_{x\bar{y}}T_{xy}G_{xy} \tag{10.4.4}$$

The operators $T_{xy}G_{xy}$ and $T_{x\bar{y}}G_{x\bar{y}}$ generate a group. Such a group is simply the PSG. The two different algebras (10.4.3) and (10.4.4) generate two different PSGs, both of which are consistent with the translation group acting on the two-particle states. We will call the PSG generated by eqn (10.4.3) the Z_2a PSG and the PSG generated by eqn (10.4.4) the Z_2b PSG.

Let us remind the reader of the general definition of a PSG. A PSG is a group. It is an extension of a symmetry group (SG), i.e. a PSG contains a normal subgroup (called the invariant gauge group or IGG) such that

$$PSG/IGG = SG$$

For our case, the SG is the translation group $SG = \{1, T_{xy}^{(2)}, T_{x\bar{y}}^{(2)}, ...\}$. The IGG is formed by the transformations G_0 on the single-particle states that satisfy $G_0 \otimes G_0 = 1$. We find that the IGG is generated by

$$G_0|\boldsymbol{p}\rangle = -|\boldsymbol{p} >$$

The three transformations on the single-particle states, G_0, $T_{xy}G_{xy}$, and $T_{x\bar{y}}G_{x\bar{y}}$, generate the Z_2a or Z_2b PSGs.

We now see that the underlying translational symmetry does not require the single-particle hopping Hamiltonian $H(\boldsymbol{p})$ to have a translational symmetry. It only requires $H(\boldsymbol{p})$ to be invariant under the Z_2a PSG or the Z_2b PSG. When $H(\boldsymbol{p})$ is invariant under the Z_2a PSG, the hopping Hamiltonian has the usual translational symmetry. When $H(\boldsymbol{p})$ is invariant under the Z_2b PSG, the hopping Hamiltonian has a magnetic translational symmetry describing a hopping in a magnetic field with π flux through each odd square.

10.4.3 Projective symmetry groups classify different string-net condensations

- Different string-net condensations have different PSGs. As PSGs are a universal property of the quantum phase, we find that different string-net condensations correspond to different phases.

After understanding the possible PSGs that the hopping Hamiltonian of the ends of strings can have, we are now ready to calculate the actual PSGs. Let us consider two ground states of our model $H_U + H_g + H_t$. One has $F_i|_{i=\text{odd}} = 1$ (for $g > 0$) and the other has $F_i|_{i=\text{odd}} = -1$ (for $g < 0$). Both ground states have the same translational symmetry in the $\boldsymbol{x}+\boldsymbol{y}$ and $\boldsymbol{x}-\boldsymbol{y}$ directions. However, the corresponding single-particle hopping Hamiltonian $H(\boldsymbol{p})$ has different symmetries. For the $F_i|_{i=\text{odd}} = 1$ state, there is no flux through the odd squares and $H(\boldsymbol{p})$ has the usual translational symmetry. It is invariant under the Z_2a PSG. For the $F_i|_{i=\text{odd}} = 1$ state, there is π flux through the odd squares and $H(\boldsymbol{p})$ has a magnetic translational symmetry. Its PSG is the Z_2b PSG. Thus, the $F_i|_{i=\text{odd}} = 1$ state and the $F_i|_{i=\text{odd}} = -1$ state have different orders despite them having the same symmetry. The different quantum orders in the two states can be characterized by their different PSGs.

As mentioned in Section 10.4.1, the above two states have different string-net condensations. For the $\hat{F}_i|_{i=\text{odd}} = 1$ state, the average of the T1 closed-string operator is $\langle W(C)\rangle = 1$; while for the $\hat{F}_i|_{i=\text{odd}} = -1$ state, the average of the T1 closed-string operator is $\langle W(C)\rangle = (-)^{N_{\text{odd}}(C)}$, where $N_{\text{odd}}(C)$ is the number of odd squares enclosed by the T1 closed string C. Thus, we can also say that different string-net condensations can be characterized by their different PSGs.

In the above, we only showed that PSGs can characterize different string-net-condensed states in some particular exactly soluble models. In fact, different PSGs actually characterize different quantum phases in generic models. This is because, as shown in Section 9.9.1, a PSG is a universal property of a phase (Wen, 2002c). The PSG of a string-net-condensed state is robust against any local perturbations of the physical model that do not change the symmetry of the model.

The above discussion also applies to the Z_2 vortex and T2 strings. Thus, the quantum orders in our model are described by the pair of PSGs $(PSG_{\text{charge}}, PSG_{\text{vortex}})$, one for the Z_2 charge and one for the Z_2 vortex. The PSG pairs $(PSG_{\text{charge}}, PSG_{\text{vortex}})$ allow us to distinguish four different string-net-condensed states of the model $H = H_U + H_g + H_t$ (see Fig. 10.5).

10.4.4 Projective symmetry groups for the ends of T3 strings

- For a state with condensations of several different types of strings, the ends of different condensed strings may have different projective symmetries and different PSGs.

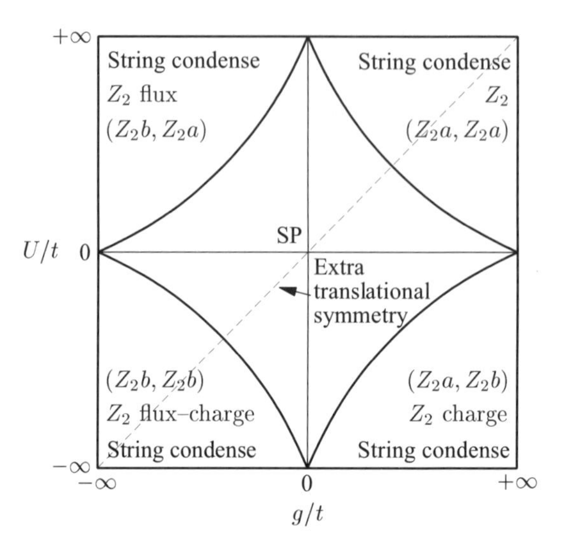

FIG. 10.5. The proposed phase diagram for the $H = H_U + H_g + H_t$ model. Here $t = t'$ is assumed to be positive. The four string-net condensed phases are characterized by the pair of PSGs $(PSG_{\text{charge}}, PSG_{\text{vortex}})$. SP marks a uniform spin-polarized phase. All of the phases have the same symmetry and none of the phase transitions change the symmetry.

- The PSG introduced in Section 9.4.2 corresponds to the PSG of the T3 string.

In this section, we will assume that $U = g = V$ in our model:

$$H_U + H_g + H_t = H_t - V \sum_{\boldsymbol{i}} \hat{F}_{\boldsymbol{i}} \tag{10.4.5}$$

The new model has a larger translational symmetry generated by $\boldsymbol{i} \to \boldsymbol{i} + \boldsymbol{x}$ and $\boldsymbol{i} \to \boldsymbol{i} + \boldsymbol{y}$ (see Fig. 10.5). Note that, when $H_t = 0$, the new model is identical to the spin Hamiltonian (10.2.6) studied in Section 10.2.1. The PSGs for the condensed T1 and T2 strings were studied in the last section. Here we would like to discuss the PSG for the T3 string. As the ends of the T3 strings live on the links, the corresponding single-particle hopping Hamiltonian $H_f(\boldsymbol{l})$ describes fermion hopping between links. Clearly, the symmetry group (the PSG) of $H_f(\boldsymbol{l})$ can be different from that of $H(\boldsymbol{p})$.

Let us define the fermion hopping $\boldsymbol{l} \to \boldsymbol{l} + \boldsymbol{x}$ as the combination of the two hops $\boldsymbol{l} \to \boldsymbol{l} + \frac{\boldsymbol{x}}{2} - \frac{\boldsymbol{y}}{2} \to \boldsymbol{l} + \boldsymbol{x}$ (see Fig. 10.6). The hopping $\boldsymbol{l} \to \boldsymbol{l} + \frac{\boldsymbol{x}}{2} - \frac{\boldsymbol{y}}{2}$ is generated by $\sigma^x_{\boldsymbol{l}+\frac{\boldsymbol{x}}{2}}$ and the hopping $\boldsymbol{l} + \frac{\boldsymbol{x}}{2} - \frac{\boldsymbol{y}}{2} \to \boldsymbol{l} + \boldsymbol{x}$ is generated by $\sigma^y_{\boldsymbol{l}+\frac{\boldsymbol{x}}{2}}$. Thus, the hopping $\boldsymbol{l} \to \boldsymbol{l} + \boldsymbol{x}$ is generated by $\sigma^y_{\boldsymbol{l}+\frac{\boldsymbol{x}}{2}} \sigma^x_{\boldsymbol{l}+\frac{\boldsymbol{x}}{2}}$. Similarly, we define the fermion hopping $\boldsymbol{l} \to \boldsymbol{l} + \boldsymbol{y}$ as the two hops $\boldsymbol{l} \to \boldsymbol{l} + \frac{\boldsymbol{x}}{2} + \frac{\boldsymbol{y}}{2} \to \boldsymbol{l} + \boldsymbol{y}$. It is generated by $\sigma^x_{\boldsymbol{l}+\frac{\boldsymbol{x}}{2}+\boldsymbol{y}} \sigma^y_{\boldsymbol{l}+\frac{\boldsymbol{x}}{2}}$. Under such a definition, a fermion hopping around a

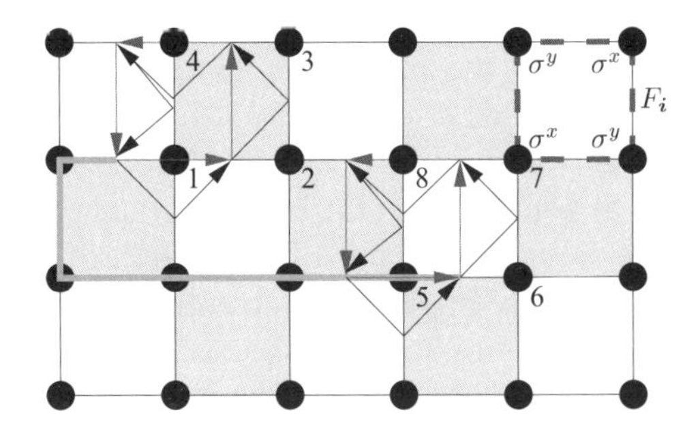

FIG. 10.6. Hoppings of a fermion around a square and around a site.

square $\boldsymbol{l} \to \boldsymbol{l}+\boldsymbol{x} \to \boldsymbol{l}+\boldsymbol{x}+\boldsymbol{y} \to \boldsymbol{l}+\boldsymbol{y} \to \boldsymbol{l}$ is generated by (see Fig. 10.6) $(\sigma_1^y\sigma_4^x)(\sigma_4^x\sigma_4^y)(\sigma_3^x\sigma_2^y)(\sigma_1^y\sigma_1^x) = \sigma_4^y\sigma_3^x\sigma_2^y\sigma_1^x = \hat{F}_1$ or $(\sigma_5^z)(\sigma_5^y\sigma_8^x)(\sigma^z)(\sigma_7^x\sigma_6^y) = -\sigma_5^x\sigma_8^y\sigma_7^x\sigma_6^y = -\hat{F}_5$. The ground state has $F_{\boldsymbol{i}} = \mathrm{sgn}(V)$. However, because the T3 string ends on the link $5 \to 6$, we have $-\hat{F}_5 = \hat{F}_{\boldsymbol{i}} = \mathrm{sgn}(V)$. On the other hand, we have $\hat{F}_1 = \hat{F}_{\boldsymbol{i}} = \mathrm{sgn}(V)$. Therefore, the total amplitude for a fermion hopping around both squares, 1234 and 5678, is given by the sign of V in eqn (10.4.5). When $\mathrm{sgn}(V) = 1$, the fermion sees no flux, while, when $\mathrm{sgn}(V) = -1$, the fermion sees π flux per square. As a result, the generators (T_xG_x, T_yG_y) of the translational symmetries for $H_f(\boldsymbol{l})$ satisfy the algebra of magnetic translation (see eqn (7.2.3))

$$(T_yG_y)^{-1}(T_xG_x)^{-1}T_yG_yT_xG_x = \mathrm{sgn}(V). \tag{10.4.6}$$

In addition to being invariant under (T_xG_x, T_yG_y), $H_f(\boldsymbol{l})$ is also invariant under G_0:

$$G_0|\boldsymbol{l}\rangle = -|\boldsymbol{l}\rangle$$

So (G_0, T_xG_x, T_yG_y) generate the symmetry group—the fermion PSG—of $H_f(\boldsymbol{l})$.

When $\mathrm{sgn}(V) = 1$, T_xG_x and T_yG_y satisfy the translational algebra. The corresponding fermion PSG is the Z2A PSG. When $\mathrm{sgn}(V) = -1$, T_xG_x and T_yG_y satisfy the magnetic translational algebra with π flux per plaquette. The corresponding fermion PSG is the Z2B PSG. We see that the quantum orders in the ground state can also be characterized using the fermion PSG. The quantum order in the $V < 0$ ground state is characterized by the Z2A PSG, and the quantum order in the $V > 0$ ground state is characterized by the Z2B PSG.

In Section 10.2, the spin-1/2 model (10.4.5) (with $t = t' = 0$) was solved using the slave-boson approach by splitting the spin into two fermions. There, it was shown that the fermion hopping Hamiltonian for the $V > 0$ and $V < 0$ states have different symmetries, or are invariant under different PSGs. The different PSGs imply different quantum orders in the $V > 0$ and $V < 0$ ground states. The PSGs obtained in Section 10.2 for the $V > 0$ and $V < 0$ phases agree exactly with the fermion PSGs that we obtained above. This example shows that the PSGs introduced in Section 9.4 to describe the quantum orders are the symmetry groups of the hopping Hamiltonian of the ends of condensed strings. The PSG description and the string-net-condensation description of quantum orders are intimately related.

Here we would like to point out that the PSGs introduced in Section 9.4 are all fermion PSGs. They are only one of many different kinds of PSGs that can be used to characterize quantum orders. In general, a quantum-ordered state may contain condensations of several types of strings. The ends of each type of condensed string will have their own PSG. Finding all types of condensed strings and their PSGs will give us a more complete characterization of quantum orders.

10.5 Emergent fermions and string-net condensation on a cubic lattice

- Emergent fermions in three-dimensional models cannot be understood in terms of attaching flux to a boson. One really needs string-net theory to understand emergent fermions in three dimensions.

In $2+1$ dimensions, a fermion can be obtained by binding π flux to a unit charge (or a Z_2 vortex to a Z_2 charge). This is how we obtain the fermion in our spin-1/2 model $H_U + H_g$. As both the Z_2 vortex and the Z_2 charge appear as ends of open strings, the fermions also appear as ends of strings. We see that we have two theories for emergent fermions in $(2+1)$-dimensional bosonic models. Emergent fermions can be viewed as a bound state of charge and flux, or they can be viewed as ends of strings. However, in $3+1$ dimensions, we cannot change a boson into a fermion by attaching π flux. Thus, one may wonder whether fermions can emerge from $(3+1)$-dimensional local bosonic models. In this section, we are going to study an exactly soluble spin-3/2 model on a cubic lattice. We will study a string-net-condensed state in such a model. We will show that the fermions can still appear as ends of strings. In fact, the string-net theory for emergent fermions is valid in any dimensions.

10.5.1 Exactly soluble spin-3/2 model on a cubic lattice

To construct an exactly soluble model on a three-dimensional cubic lattice, we first introduce the operators $\gamma_{\boldsymbol{i}}^{ab}$, $a, b = x, \bar{x}, y, \bar{y}, z, \bar{z}$, that act within the local Hilbert space at each site. At a site, the γ^{ab} satisfy the following algebra (the site index $\boldsymbol{i}$ is suppressed):

$$\begin{aligned} \gamma^{ab} &= -\gamma^{ba} = (\gamma^{ab})^\dagger, & (\gamma^{ab})^2 &= 1, \\ [\gamma^{ab}, \gamma^{cd}] &= 0, & \gamma^{ab}\gamma^{bc} &= \mathrm{i}\gamma^{ac}, \quad \text{if } a, b, c, d \text{ are all different} \end{aligned} \tag{10.5.1}$$

To show that the algebra is self-consistent, let λ^a be a Majorana-fermion operator labeled by $a = x, \bar{x}, y, \bar{y}, z, \bar{z}$. From the algebra of the Majorana fermion $\{\lambda^a, \lambda^b\} = 2\delta_{ab}$, we can show that

$$\gamma^{ab} = \frac{\mathrm{i}}{2}(\lambda^a_{\boldsymbol{i}}\lambda^b_{\boldsymbol{i}} - \lambda^b_{\boldsymbol{i}}\lambda^a_{\boldsymbol{i}}).$$

satisfy the above algebra.

The above algebra has a four-dimensional representation. We first note that $\gamma^{az} \equiv \gamma^a$, $a = x, \bar{x}, y, \bar{y}$, satisfy the Dirac algebra

$$\{\gamma^a, \gamma^b\} = 2\delta_{ab}, \qquad a, b = x, \bar{x}, y, \bar{y}. \tag{10.5.2}$$

Therefore, we can express them in terms of the four 4×4 Dirac matrices

$$\begin{aligned} \gamma^{xz} &= \gamma^x = \sigma^x \otimes \sigma^x, & \gamma^{\bar{x}z} &= \gamma^{\bar{x}} = \sigma^y \otimes \sigma^x, \\ \gamma^{yz} &= \gamma^y = \sigma^z \otimes \sigma^x, & \gamma^{\bar{y}z} &= \gamma^{\bar{y}} = \sigma^0 \otimes \sigma^y. \end{aligned} \tag{10.5.3}$$

We can also define γ^5 as

$$\gamma^5 = \gamma^x\gamma^{\bar{x}}\gamma^y\gamma^{\bar{y}} = -\sigma^0 \otimes \sigma^z$$

which anti-commutes with γ^a, i.e. $\{\gamma^a, \gamma^5\} = 0$. As $\gamma^{z\bar{z}}$ anti-commutes with $\gamma^{az} \equiv \gamma^a$, $a = x, \bar{x}, y, \bar{y}$, we can identify γ^5 as $\gamma^{z\bar{z}}$. From the algebra (10.5.1), we can express the remaining γ^{ab} as

$$\gamma^{a\bar{z}} = \mathrm{i}\gamma^a\gamma^5, \quad a = x, \bar{x}, y, \bar{y}$$

In this way, we express all of the γ^{ab}, $a, b = x, \bar{x}, y, \bar{y}, z, \bar{z}$, in terms of the Dirac matrices $\gamma^{x,\bar{x},y,\bar{y}}$. One can show that the γ^{ab} defined this way satisfy the algebra (10.5.1).

As the γ^{ab} are four-dimensional, we can say that they act on spin-3/2 states. In terms of $\gamma^{ab}_{\boldsymbol{i}}$, we can write down an exactly soluble spin-3/2 model on a cubic lattice as follows:

$$\begin{aligned} &H_{3D} \\ &= g\sum_{\boldsymbol{i}} \left(\gamma^{yx}_{\boldsymbol{i}}\gamma^{\bar{x}y}_{\boldsymbol{i}+\boldsymbol{x}}\gamma^{\bar{y}\bar{x}}_{\boldsymbol{i}+\boldsymbol{x}+\boldsymbol{y}}\gamma^{x\bar{y}}_{\boldsymbol{i}+\boldsymbol{y}} + \gamma^{zy}_{\boldsymbol{i}}\gamma^{\bar{y}z}_{\boldsymbol{i}+\boldsymbol{y}}\gamma^{\bar{z}\bar{y}}_{\boldsymbol{i}+\boldsymbol{y}+\boldsymbol{z}}\gamma^{y\bar{z}}_{\boldsymbol{i}+\boldsymbol{z}} + \gamma^{xz}_{\boldsymbol{i}}\gamma^{\bar{z}x}_{\boldsymbol{i}+\boldsymbol{z}}\gamma^{\bar{x}\bar{z}}_{\boldsymbol{i}+\boldsymbol{z}+\boldsymbol{x}}\gamma^{z\bar{y}}_{\boldsymbol{i}+\boldsymbol{x}}\right) \\ &= -g\sum_{\boldsymbol{p}} \hat{F}_{\boldsymbol{p}} \end{aligned} \tag{10.5.4}$$

where $\boldsymbol{p}$ labels the squares in the cubic lattice and $\hat{F}_{\boldsymbol{p}}$ is equal to either $-\gamma^{yx}_{\boldsymbol{i}}\gamma^{\bar{x}y}_{\boldsymbol{i}+\boldsymbol{x}}\gamma^{\bar{y}\bar{x}}_{\boldsymbol{i}+\boldsymbol{x}+\boldsymbol{y}}\gamma^{x\bar{y}}_{\boldsymbol{i}+\boldsymbol{y}}$, $-\gamma^{zy}_{\boldsymbol{i}}\gamma^{\bar{y}z}_{\boldsymbol{i}+\boldsymbol{y}}\gamma^{\bar{z}\bar{y}}_{\boldsymbol{i}+\boldsymbol{y}+\boldsymbol{z}}\gamma^{y\bar{z}}_{\boldsymbol{i}+\boldsymbol{z}}$, or $-\gamma^{xz}_{\boldsymbol{i}}\gamma^{\bar{z}x}_{\boldsymbol{i}+\boldsymbol{z}}\gamma^{\bar{x}\bar{z}}_{\boldsymbol{i}+\boldsymbol{z}+\boldsymbol{x}}\gamma^{z\bar{y}}_{\boldsymbol{i}+\boldsymbol{x}}$, depending on the orientation of the square $\boldsymbol{p}$.

It is easy to obtain the eigenvalues of H_{3D} by noticing that

$$[\hat{F}_{\boldsymbol{p}}, \hat{F}_{\boldsymbol{p}'}] = 0, \qquad (\hat{F}_{\boldsymbol{p}})^2 = 1.$$

Let $|\{F_{\boldsymbol{p}}\}, \alpha\rangle$ be the common eigenstates of $\hat{F}_{\boldsymbol{p}}$, i.e. $\hat{F}_{\boldsymbol{p}}|\{F_{\boldsymbol{p}}\}, \alpha\rangle = F_{\boldsymbol{p}}|\{F_{\boldsymbol{p}}\}, \alpha\rangle$, where α labels the possible degeneracy. Then $|\{F_{\boldsymbol{p}}\}, \alpha\rangle$ is also an energy eigenstate with energy $-g\sum_{\boldsymbol{p}} F_{\boldsymbol{p}}$. As $F_{\boldsymbol{p}} = \pm 1$, the ground-state energy is $E_0 = -3|g|N_{\text{site}}$ on a periodic lattice, where N_{site} is the number of lattice sites. The excited states are obtained by flipping the signs of some $F_{\boldsymbol{p}}$, whose energies are E_0 plus integer multiples of $2|g|$.

One subtlety is that the $F_{\boldsymbol{p}}$ are not all independent. If S is the surface of a unit cube, then we have the operator identity

$$\prod_{\boldsymbol{p}\in S} \hat{F}_{\boldsymbol{p}} = 1$$

It follows that $\prod_{\boldsymbol{p}\in S} F_{\boldsymbol{p}} = 1$ for all cubes. If we view $F_{\boldsymbol{p}}$ as the Z_2 flux through the square $\boldsymbol{p}$, then the above constraint implies flux conservation. This means that the spectrum of our model is identical to a Z_2 gauge theory on a cubic lattice. The ground state can be thought of as a state with no flux, i.e. $F_{\boldsymbol{p}} = 1$ for all $\boldsymbol{p}$. The excitations above the ground state are described by flux loops. The elementary excitations correspond to the smallest flux loops where $F_{\boldsymbol{p}} = -1$ for the four squares $\boldsymbol{p}$ adjacent to some link $\langle \boldsymbol{ij}\rangle$. We can think of these excitations as quasiparticles which live on the *links* of the cubic lattice (see Fig. 10.7). As we will see later, such a flux-loop excitation has fermionic statistics.

Problem 10.5.1.
(a) Show that the 4×4 matrices γ^a in eqn (10.5.3) satisfy the Dirac algebra.
(b) Show that the terms in the summation in eqn (10.5.4) all commute with each other. (Hint: You may use the Majorana-fermion representation of γ^{ab}.) This allows us to solve H_{3D} exactly.

Problem 10.5.2.
Solving eqn (10.5.4) using Majorana fermions and the projective construction
(a) Let $\lambda_{\boldsymbol{i}}^a$ be a Majorana-fermion operator, where $a = x, \bar{x}, y, \bar{y}, z, \bar{z}$. For every nearest-neighbor link, define

$$\hat{U}_{\boldsymbol{ij}} = \lambda_{\boldsymbol{i}}^a \lambda_{\boldsymbol{j}}^b \tag{10.5.5}$$

where ab is the pair of indices associated with the link $\boldsymbol{i} \to \boldsymbol{j}$ (see Fig. 10.7). Show that the $\hat{U}_{\boldsymbol{ij}}$s form a commuting set of operators.
(b) Consider the fermion Hamiltonian

$$H = -\sum_{\boldsymbol{p}} g\hat{F}_{\boldsymbol{p}}, \qquad \hat{F}_{\boldsymbol{p}} = \hat{U}_{\boldsymbol{i}_1,\boldsymbol{i}_2}\hat{U}_{\boldsymbol{i}_2,\boldsymbol{i}_3}\hat{U}_{\boldsymbol{i}_3,\boldsymbol{i}_4}\hat{U}_{\boldsymbol{i}_4,\boldsymbol{i}_1} \tag{10.5.6}$$

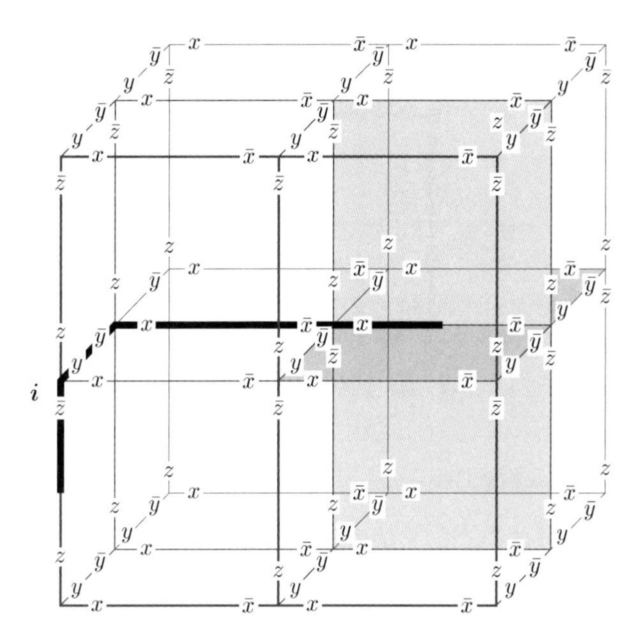

FIG. 10.7. An open string in the cubic lattice and the pairs of indices associated with nearest-neighbor links. An end of an open string flips the sign of $F_{\boldsymbol{p}}$ on the four shaded squares.

where $\sum_{\boldsymbol{p}}$ is a sum over all of the square faces of the cubic lattice. Here $\boldsymbol{i}_1$, $\boldsymbol{i}_2$, $\boldsymbol{i}_3$, and $\boldsymbol{i}_4$ label the four corners of the square $\boldsymbol{p}$. Introduce the complex fermion operators

$$2\psi_{1,\boldsymbol{i}} = \lambda_{\boldsymbol{i}}^{x} + \mathrm{i}\lambda_{\boldsymbol{i}}^{\bar{x}}, \qquad 2\psi_{2,\boldsymbol{i}} = \lambda_{\boldsymbol{i}}^{y} + \mathrm{i}\lambda_{\boldsymbol{i}}^{\bar{y}}, \qquad 2\psi_{3,\boldsymbol{i}} = \lambda_{\boldsymbol{i}}^{z} + \mathrm{i}\lambda_{\boldsymbol{i}}^{\bar{z}},$$

and let $N_{\boldsymbol{i}} = \sum_{a=1,2,3} \psi_{a,\boldsymbol{i}}^{\dagger}\psi_{a,\boldsymbol{i}}$ be the fermion number operator at the site $\boldsymbol{i}$. Show that the fermion Hamiltonian (10.5.6) reduces to the spin-3/2 Hamiltonian (10.5.4) in the subspace with an even number of fermions per site.

(c) Use the procedure in Section 10.2 to find the ground-state energy and the ground-state degeneracy of eqn (10.5.4) on an even $\times$ even $\times$ even lattice with periodic boundary conditions.

10.5.2 String operators and closed-string condensation

To show that the ground state of H_{3D} has a closed-string condensation, we need to find a closed-string operator that commutes with H_{3D}. First, let us construct more general open-string operators.

To construct the string operators in terms of the physical γ^{ab} operator, we note that we can associate each nearest-neighbor link $\boldsymbol{i} \to \boldsymbol{j}$ with a pair of indices ab, as illustrated in Fig. 10.7. For example, the link $\boldsymbol{i} \to \boldsymbol{i}+\boldsymbol{x}$ is associated with $x\bar{x}$, the link $\boldsymbol{i} \to \boldsymbol{i}-\boldsymbol{x}$ with $\bar{x}x$, the link $\boldsymbol{i} \to \boldsymbol{i}+\boldsymbol{y}$ with $y\bar{y}$, etc. The open-string

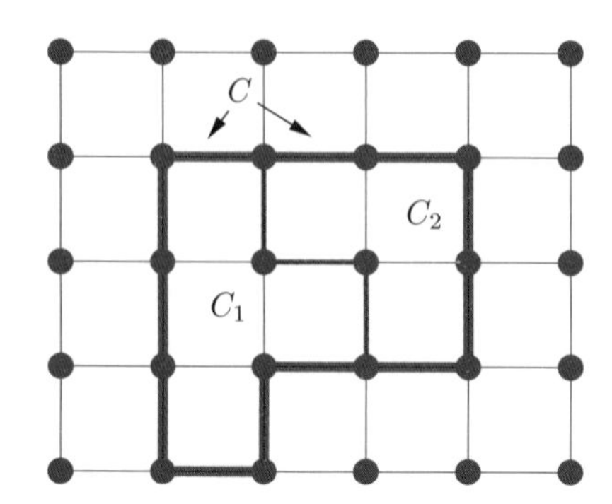

FIG. 10.8. The loop C can be divided into the two loops C_1 and C_2.

operator can then be defined as

$$W(C) = -(-\mathrm{i}\gamma_{\boldsymbol{i}_1}^{a_1 b_1})(-\mathrm{i}\gamma_{\boldsymbol{i}_2}^{a_2 b_2})...(-\mathrm{i}\gamma_{\boldsymbol{i}_n}^{a_n b_n}) \tag{10.5.7}$$

where $b_m a_{m+1}$ is the pair of indices associated with the link $\boldsymbol{i}_m \to \boldsymbol{i}_{m+1}$. For example, the open-string operator for the open string in Fig. 10.7 is given by

$$W(C) = -(-\mathrm{i}\gamma_{\boldsymbol{i}}^{\bar{z}y})(-\mathrm{i}\gamma_{\boldsymbol{i}+\boldsymbol{y}}^{\bar{y}x})(-\mathrm{i}\gamma_{\boldsymbol{i}+\boldsymbol{x}+\boldsymbol{y}}^{\bar{x}x})$$

We would like to stress that an open string C is formed by linking the mid-point of the links through the neighboring sites (see Fig. 10.7). Thus, the ends of open strings live on the links.

The closed-string operator is still given by eqn (10.5.7). However, $\boldsymbol{i}_1$ and $\boldsymbol{i}_n$ are now nearest neighbors, and $b_n a_1$ is the pair of indices associated with the link $\boldsymbol{i}_n \to \boldsymbol{i}_1$.

Let us first discuss some special properties of the closed-string operators that lead to closed-string condensation. Using the Majorana representation of γ^{ab}, we can show that any two closed-string operators commute with each other. As $\hat{F}_{\boldsymbol{p}}$ itself is a closed-string operator, the closed-string operators commute with the spin-3/2 Hamiltonian. Thus, the ground state is also an eigenstate of the closed-string operators and has a closed-string condensation (or, more generally, a closed-string-net condensation).

If we divide a loop C into the two loops C_1 and C_2, as in Fig. 10.8, then the closed-string operators for the three loops are related by

$$W(C) = W(C_1)W(C_2) \tag{10.5.8}$$

Such a relation allows us to express $W(C)$ as a product of the $F_{\boldsymbol{p}}$s and evaluate the amplitude of the condensed closed strings. For the $\hat{F}_{\boldsymbol{p}} = 1$ ground state, we find that

$$\langle W(C_{\text{close}})\rangle = 1$$

For the $\hat{F}_{\boldsymbol{p}} = -1$ ground state, we find that

$$\langle W(C_{\text{close}})\rangle = (-)^{N_p}$$

where N_p is the number of squares on a surface $S_{C_{\text{close}}}$, and $S_{C_{\text{close}}}$ is a surface formed by the faces of the cubes and bounded by the closed string C_{close}.

Problem 10.5.3.
(a) Use the Majorana-fermion representation of $\gamma^{ab} = \mathrm{i}\lambda^a\lambda^b$ to show that the closed-string operator (10.5.7) can be expressed as

$$W(C_{\text{close}}) = \hat{U}_{\boldsymbol{i}_1\boldsymbol{i}_2}\hat{U}_{\boldsymbol{i}_2\boldsymbol{i}_3}...\hat{U}_{\boldsymbol{i}_{n-1}\boldsymbol{i}_n}\hat{U}_{\boldsymbol{i}_n\boldsymbol{i}_1}$$

where $\hat{U}_{ij}$ is defined in eqn (10.5.5).
(b) Show that the closed-string operators (10.5.7) defined for two closed loops commute with each other.
(c) Verify the relation (10.5.8).

10.5.3 Emergent fermions as ends of open strings

In the closed-string condensed state, the ends of the open string become a new type of quasiparticle. When we apply the open-string operator to a ground state, by calculating the commutator between the $F_{\boldsymbol{p}}$ and the open-string operator, we find that the open-string operator only flips the sign of $F_{\boldsymbol{p}}$ near its two ends. As a result, the string itself costs no energy. Only string ends cost energy.

Noting that the string ends live on links and that there are four squares attached to each link, we find that the open-string operator flips the sign of $F_{\boldsymbol{p}}$ on those four squares (see Fig. 10.7). Therefore, each end of an open string corresponds to a small Z_2-flux loop and costs an energy $4|2g| = 8|g|$. These small Z_2-flux loops correspond to quasiparticle excitations above the ground state.

To find the statistics of the Z_2-flux loops, we need to use the following statistical hopping algebra introduced by Levin and Wen (2003) (see Section 4.1.4):

$$\begin{aligned} t_{jl}t_{kj}t_{ji} &= \mathrm{e}^{\mathrm{i}\theta}t_{ji}t_{kj}t_{jl}, \\ [t_{ij}, t_{kl}] &= 0, \quad \text{if } i, j, k, l \text{ are all different,} \end{aligned} \tag{10.5.9}$$

where t_{ji} describes the hopping of particles from link i to link j (note that, here, the ends of strings live on links). It was shown that the particles are fermions if their hopping operators satisfy the algebra (10.5.9) with $\theta = \pi$. If we choose (i, j, k, l)

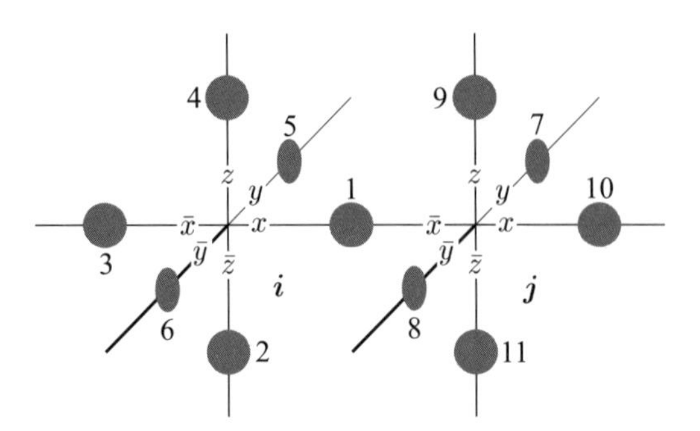

FIG. 10.9. A fermion on link 1 can hop onto the ten different links 2–11. The hopping $1 \to 4$ is generated by γ_i^{zx}, the hopping $1 \to 3$ by $\gamma_i^{\bar{x}x}$, and the hopping $1 \to 7$ by $\gamma_j^{y\bar{x}}$.

to be the links $(2, 1, 3, 4)$ in Fig. 10.9, then the hopping operators (t_{jl}, t_{kj}, t_{ji}) are given by $(\gamma_i^{xz}, \gamma_i^{\bar{x}x}, \gamma_i^{x\bar{z}})$. Equation (10.5.9) becomes

$$\gamma_i^{xz}\gamma_i^{\bar{x}x}\gamma_i^{x\bar{z}} = \mathrm{e}^{\mathrm{i}\theta}\gamma_i^{x\bar{z}}\gamma_i^{\bar{x}x}\gamma_i^{xz}$$

From the Majorana-fermion representation of γ^{ab}, we find that the above equation is valid if $\theta = \pi$. If we choose (i, j, k, l) to be the links $(2, 1, 3, 10)$, then eqn (10.5.9) becomes

$$\gamma_j^{\bar{x}x}\gamma_i^{\bar{x}x}\gamma_i^{x\bar{z}} = \mathrm{e}^{\mathrm{i}\theta}\gamma_i^{x\bar{z}}\gamma_i^{\bar{x}x}\gamma_j^{\bar{x}x}$$

The above equation is again valid with $\theta = \pi$. Other choices of (i, j, k, l) all fall into the above two types. The hoppings of the ends of the strings satisfy the fermion hopping algebra. Hence, the ends of the strings (i.e. the small Z_2-flux loops) are fermions. We see that fermions can emerge from local bosonic models if the bosonic models have condensations of certain types of closed strings.

10.6 The quantum rotor model and $U(1)$ lattice gauge theory

- In higher dimensions, a gauge theory is quite non-trivial. It describes fluctuations of an entangled state. These fluctuations cannot be represented by local degrees of freedom.
- The $U(1)$ gauge fluctuations are fluctuations of closed string-nets in space.
- Gapless $U(1)$ gauge bosons can emerge if the ground state contains a condensation of closed string-nets of arbitrary sizes.

We have studied exactly soluble local bosonic models that have emergent Z_2 gauge fields. In this section, we are going to study the emergence of a continuous

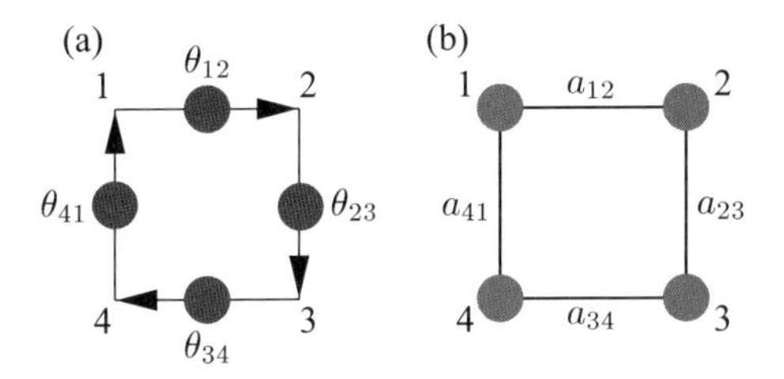

FIG. 10.10. (a) A four-rotor system. (b) A simple lattice gauge theory is described by the lattice gauge field $a_{i,i+1}$ and $a_0(i)$, $i = 1, 2, 3, 4$.

$U(1)$ gauge theory from a quantum rotor model (Foerster *et al.*, 1980; Senthil and Motrunich, 2002; Wen, 2003a). We will also see how string-net condensation and the related quantum order are associated with the emergent $U(1)$ gauge bosons.

This section can also be regarded as a physical way to describe quantum $U(1)$ gauge theory. In my mind, the standard description of gauge theory using gauge potential is unnatural, because the gauge potential itself is unphysical. Here we will show that it is possible to use physical degrees of freedom—nets of closed strings—to formulate the $U(1)$ gauge theory (Banks *et al.*, 1977; Savit, 1980).

10.6.1 A four-rotor system

- Strong quantum fluctuations of certain combinations of rotor angles lead to a low-energy effective gauge theory.

Let us first consider a single rotor described by a particle of mass m moving on a circle $0 \leqslant \theta < 2\pi$. The Lagrangian is

$$L = \frac{1}{2} m \dot{\theta}^2 + K \cos\theta$$

The Hamiltonian has the form

$$H = \frac{(S^z)^2}{2m} - K \cos\theta = \frac{(S^z)^2}{2m} - \frac{K}{2}(a + a^\dagger)$$

where S^z is the angular momentum and $a = \mathrm{e}^{-i\theta}$. If we choose the basis states to be $|n\rangle = (2\pi)^{-1/2} \mathrm{e}^{in\theta}$ with $n =$ integer, then we find that $S^z|n\rangle = n|n\rangle$ and $a|n\rangle = |n-1\rangle$.

To obtain the simplest model that has an emergent low-energy gauge theory, let us consider the following model of four rotors described by $\theta_{\langle 12\rangle}$, $\theta_{\langle 23\rangle}$, $\theta_{\langle 34\rangle}$, and $\theta_{\langle 41\rangle}$ (see Fig. 10.10(a)):

$$H = \sum_i \left(U(S^z_{\langle i-1,i\rangle} - S^z_{\langle i,i+1\rangle})^2 + J(S^z_{\langle i,i+1\rangle})^2 \right), \tag{10.6.1}$$

where we have assumed that $4+1 \sim 1$ and $1-1 \sim 4$. The Hilbert space is spanned by $|n_{\langle 12\rangle} n_{\langle 23\rangle} n_{\langle 34\rangle} n_{\langle 41\rangle}\rangle$, where the integer $n_{i,i+1}$ is the eigenvalue of $S^z_{\langle i,i+1\rangle}$. If $U \gg J$, then the low-energy excitations are described by the $|nnnn\rangle$ states with energy $E = 4Jn^2$. All of the other excitations have energy of order U.

To see the connection with lattice gauge theory, we would like to write down the Lagrangian of our four-rotor model. If we write the Hamiltonian in the form $H = \frac{1}{2} P^\top V P$, where $P^\top = (S^z_{\langle 12\rangle}, S^z_{\langle 23\rangle}, S^z_{\langle 34\rangle}, S^z_{\langle 41\rangle})$ and

$$V = \begin{pmatrix} 4U+2J & -2U & 0 & -2U \\ -2U & 4U+2J & -2U & 0 \\ 0 & -2U & 4U+2J & -2U \\ -2U & 0 & -2U & 4U+2J \end{pmatrix},$$

then the Lagrangian will be

$$L = \frac{1}{2} \dot{\Theta}^\top M \dot{\Theta},$$

where $\Theta^\top = (\theta_{\langle 12\rangle}, \theta_{\langle 23\rangle}, \theta_{\langle 34\rangle}, \theta_{\langle 41\rangle})$ and $M = V^{-1}$.

Obviously, we do not see any sign of gauge theory in the above Lagrangian. To obtain a gauge theory, we need to derive the Lagrangian in another way. Using the path integral representation of H and noting that (θ, S^z) is a canonical coordinate–momentum pair, we find that

$$Z = \int \mathcal{D}S^z \mathcal{D}\theta \; \mathrm{e}^{\mathrm{i}\int \mathrm{d}t \, \left(\sum_i S^z_{\langle i,i+1\rangle} \dot{\theta}_{\langle i,i+1\rangle} - H\right)}$$

Introducing the $a_0(i)$, $i = 1, 2, 3, 4$, field to decouple the U term, we can rewrite the above path integral as

$$Z = \int \mathcal{D}S^z \mathcal{D}\theta \mathcal{D}a_0 \; \mathrm{e}^{\mathrm{i}\int \mathrm{d}t \, \left(\sum_i S^z_{\langle i,i+1\rangle} \dot{\theta}_{\langle i,i+1\rangle} - \tilde{H}(S^z, a_0)\right)}$$

where $\tilde{H} = \sum_i \left(J(S^z_{\langle i,i+1\rangle})^2 + a_0(i)(S^z_{\langle i-1,i\rangle} - S^z_{\langle i,i+1\rangle}) - \frac{a_0^2(i)}{4U} \right)$. After integrating out $S^z_{\langle i,i+1\rangle}$, we obtain

$$Z = \int \mathcal{D}\theta \mathcal{D}a_0 \; \mathrm{e}^{\mathrm{i}\int \mathrm{d}t \; L(\theta, \dot{\theta}, a_0)}$$

where the Lagrangian is given by

$$L = \frac{1}{4J} \sum_i \left((\dot{\theta}_{\langle i,i+1\rangle} + a_0(i) - a_0(i+1))^2 + \frac{a_0^2(i)}{4U} \right)$$

In the large-U limit, we can drop the $a_0^2(i)/4U$ term and obtain

$$L = \frac{1}{4J} \sum_i (\dot{a}_{i,i+1} + a_0(i) - a_0(i+1))^2$$

which is just the Lagrangian of a $U(1)$ lattice gauge theory on a single square with

$$a_{i,i+1} = \theta_{\langle i,i+1\rangle}, \qquad a_{i+1,i} = -\theta_{\langle i,i+1\rangle}$$

as the lattice gauge fields (see Fig. 10.10(b)). One can check that the above Lagrangian is invariant under the following transformation:

$$a_{ij}(t) \to a_{ij}(t) + \phi_j(t) - \phi_i(t), \qquad a_0(i,t) \to a_0(i,t) + \dot{\phi}_i(t) \tag{10.6.2}$$

which is called the gauge transformation. We note that the low-energy wave function $\Psi(a_{12}, a_{23}, a_{34}, a_{41})$ is a superposition of $|nnnn\rangle$ states. All of the low-energy states are gauge invariant, i.e. invariant under the gauge transformation $a_{ij} \to a_{ij} + \phi_j - \phi_i$.

The electric field of a continuum $U(1)$ gauge theory is given by $\boldsymbol{E} = \dot{\boldsymbol{a}} - \boldsymbol{\partial} a_0$ (in the $c = e = 1$ unit). In a lattice gauge theory, the electric field becomes the following quantity defined on the links:

$$E_{ij} = \dot{a}_{ij} - (a_0(j) - a_0(i))$$

We see that our lattice gauge Lagrangian can be written as $L = \frac{1}{4J}\sum_i E^2_{i,i+1}$. Comparing this with the continuum $U(1)$ gauge theory $\mathcal{L} \propto \boldsymbol{E}^2 - \boldsymbol{B}^2$, we see that our Lagrangian contains only the kinetic energy corresponding to $\boldsymbol{E}^2$. A more general lattice gauge theory also contains a potential energy term corresponding to $\boldsymbol{B}^2$.

To obtain a potential energy term, we generalize our rotor model to

$$H = \sum_i \left(U(S^z_{\langle i-1,i\rangle} - S^z_{\langle i,i+1\rangle})^2 + J(S^z_{\langle i,i+1\rangle})^2 + t(\mathrm{e}^{\,\mathrm{i}\theta_{\langle i-1,i\rangle}}\,\mathrm{e}^{\,\mathrm{i}\theta_{\langle i,i+1\rangle}} + h.c.) \right), \tag{10.6.3}$$

We note that $\langle nnn|\mathrm{e}^{\,\mathrm{i}\theta_{\langle i-1,i\rangle}}\,\mathrm{e}^{-\mathrm{i}\theta_{\langle i,i+1\rangle}}|nnn\rangle = 0$. Thus, to first order in t, the new term has no effect at low energies. The low-energy effect of the new term only appears at second order in t.

If we repeat the above calculation with the new term, then we obtain the following Lagrangian:

$$L = \frac{1}{4J}\sum_i \left((\dot{a}_{i,i+1} + a_0(i) - a_0(i+1))^2 - t(\mathrm{e}^{\,\mathrm{i}(a_{i-1,i}+a_{i,i+1})} + h.c.) + \frac{a_0^2(i)}{4U} \right)$$

It is a little more difficult to see in the above Lagrangian why the new term has no low-energy effect at first order in t. Let us concentrate on the fluctuations of the following form:[76]

$$a_{i-1,i} = \phi_i - \phi_{i-1}$$

After integrating out $a_0(i)$, the Lagrangian for the above type of fluctuation has the form $L = \frac{1}{2}\dot{\phi}_i m_{ij}\dot{\phi}_j - \sum_i t(\mathrm{e}^{\,\mathrm{i}(\phi_{i-1}+\phi_{i+1})} + h.c.)$, with $m_{ij} = O(U^{-1})$. We

[76] In lattice gauge theory, such a type of fluctuation is called a pure gauge fluctuation.

see that, in the large-U limit, the above form of fluctuations is fast and strong. As the ϕ_i live on a compact space (i.e. ϕ_i and $\phi_i + 2\pi$ represent the same point), these fast fluctuations all have a large energy gap of order U. We now see that the t term $t\,\mathrm{e}^{\,\mathrm{i}(\phi_{i-1}-\phi_{i+1})}$ averages to zero due to the strong fluctuations of ϕ_i, and has no effect at first order in t. However, at second order in t, there is a term $t^2\prod_{i=1,3}\mathrm{e}^{\,\mathrm{i}(\theta_{\langle i-1,i\rangle}+\theta_{\langle i,i+1\rangle})} = t^2\mathrm{e}^{\,\mathrm{i}\sum_i a_{i-1,i}}$. Such a term does not depend on ϕ_i and does not average to zero. Thus, we expect the low-energy effective Lagrangian to have the form

$$L = \frac{1}{4J}\sum_i\left((\dot{a}_{i,i+1} + a_0(i) - a_0(i+1))^2 + \frac{a_0^2(i)}{4U}\right) + g\cos\Phi \qquad (10.6.4)$$

where $g = O(t^2/U)$ and $\Phi = \sum_i a_{i,i+1}$ is the flux of the $U(1)$ gauge field through the square. The above Lagrangian does not depend on the fast quantum fluctuations and is invariant under the gauge transformations (10.6.2). We see that it is the strong quantum fluctuations of certain combinations of a_{ij} that leads to a gauge-invariant low-energy effective Lagrangian.

To calculate g quantitatively, we would first like to derive the low-energy effective Hamiltonian. If we treat the t term as a perturbation and treat the low-energy states as degenerate states, then, at second order in t, we have

$$\begin{aligned}
&\langle n_1, n_1, n_1, n_1|H_{eff}|nnnn\rangle \\
&= -\frac{t^2}{2U}\langle n_1, n_1, n_1, n_1|\mathrm{e}^{\,\mathrm{i}(\theta_{\langle 34\rangle}+\theta_{\langle 41\rangle})}|n_1, n_1, n, n\rangle\langle n_1, n_1, n, n|\mathrm{e}^{\,\mathrm{i}(\theta_{\langle 12\rangle}+\theta_{\langle 23\rangle})}|nnnn\rangle \\
&\quad + \text{three other similar terms} \\
&= -\frac{2t^2}{U}
\end{aligned}$$

where $n_1 = n + 1$. Thus, the low-energy effective Hamiltonian is

$$H = \sum_i\left(U(S^z_{\langle i-1,i\rangle} - S^z_{\langle i,i+1\rangle})^2 + J(S^z_{\langle i,i+1\rangle})^2\right) - \frac{4t^2}{U}\cos(\Phi).$$

The corresponding Lagrangian is given by eqn (10.6.4) with

$$g = \frac{4t^2}{U}.$$

As discussed earlier, the pure gauge fluctuations have a large energy gap of order U. The low-energy effective theory below U can be obtained by letting $U \to \infty$,

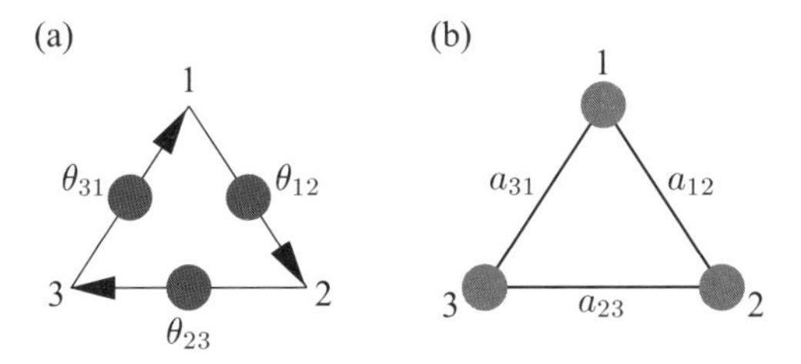

FIG. 10.11. (a) A three-rotor system, and (b) the related lattice gauge theory.

and we obtain

$$L = \frac{1}{4J} \sum_i (\dot{a}_{i,i+1} + a_0(i) - a_0(i+1))^2 + g \cos \Phi \tag{10.6.5}$$

This model was studied in Section 6.4. There it was shown that, when $g = 0$, the energy levels are given by $E_n = 4Jn^2$, which agrees exactly with the energy levels of eqn (10.6.1) at low energies. Hence eqn (10.6.1) is indeed a gauge theory at low energies.

Here we would like to make a remark. What is a gauge theory? Is the four-rotor model really a gauge theory at low energies? The standard definition of gauge theory is a theory with a gauge potential. According to this definition, the four-rotor model can be a gauge theory if we insist on writing its Lagrangian in terms of gauge potentials. On the other hand, the four-rotor model is not a gauge theory because its natural Lagrangian does not contain any gauge potential. We see that the gauge theory is not a well-defined physical concept. Whether a model is a gauge theory or not is not a well-defined question. It is almost like calling the four-rotor model a gauge theory if we label the four sites by (a, b, c, d), and call it a non-gauge theory if we label the four sites by $(1, 2, 3, 4)$. From this point of view, the discussion in the present section is quite formal, with little physical significance.

Problem 10.6.1.
Consider the three-site rotor model described by eqn (10.6.3) with $i = 1, 2, 3$ (see Fig. 10.11).

1. Derive the action for the low-energy lattice gauge theory. You only need to calculate g up to an $O(1)$ coefficient.
2. Calculate the energy levels at low energies, assuming that $U \gg g \gg J$.

10.6.2 A lattice of quantum rotors and artificial light

- A lattice of quantum rotors can be described by a lattice $U(1)$ gauge theory.

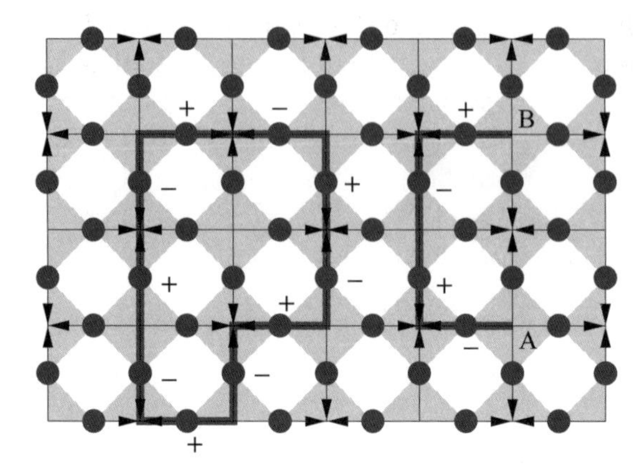

FIG. 10.12. A rotor lattice, a loop representing a low-energy fluctuation, and a pair of charge excitations (A,B). The sum of the angular momenta for rotors at the corners of each shaded square is zero in the large-U limit.

• The deconfined phase of the lattice gauge theory contains artificial light.

After understanding the systems with a few rotors, we are ready to study lattice rotor systems. As an example, we will consider a square lattice where there is one rotor on every link (see Fig. 10.12). We have

$$H = U\sum_{\boldsymbol{i}}\Big(\sum_{\boldsymbol{\alpha}} S^z_{\boldsymbol{i}+\boldsymbol{\alpha}}\Big)^2 + J\sum_{\boldsymbol{i},\boldsymbol{\alpha}}(S^z_{\boldsymbol{i}+\boldsymbol{\alpha}})^2 + \sum_{\boldsymbol{i}}\Big(t_1 \mathrm{e}^{\mathrm{i}\left(\theta_{\boldsymbol{i}+\frac{\boldsymbol{x}}{2}}-\theta_{\boldsymbol{i}+\frac{\boldsymbol{y}}{2}}\right)} \quad (10.6.6)$$

$$+t_2 \mathrm{e}^{\mathrm{i}\left(\theta_{\boldsymbol{i}+\frac{\boldsymbol{y}}{2}}-\theta_{\boldsymbol{i}-\frac{\boldsymbol{x}}{2}}\right)} + t_3 \mathrm{e}^{\mathrm{i}\left(\theta_{\boldsymbol{i}-\frac{\boldsymbol{x}}{2}}-\theta_{\boldsymbol{i}-\frac{\boldsymbol{y}}{2}}\right)} + t_4 \mathrm{e}^{\mathrm{i}\left(\theta_{\boldsymbol{i}-\frac{\boldsymbol{y}}{2}}-\theta_{\boldsymbol{i}+\frac{\boldsymbol{x}}{2}}\right)} + h.c.\Big)$$

Here $\boldsymbol{i} = (i_x, i_y)$ labels the sites of the square lattice, and $\boldsymbol{\alpha} = \pm\frac{\boldsymbol{x}}{2}, \pm\frac{\boldsymbol{y}}{2}$. The U term enforces the constraint that the total angular momentum of the four rotors near the site $\boldsymbol{i}$ is zero. The model can also be generalized to a three-dimensional cubic lattice as follows:

$$H = U\sum_{\boldsymbol{i}}\Big(\sum_{\boldsymbol{\alpha}} S^z_{\boldsymbol{i}+\boldsymbol{\alpha}}\Big)^2 + J\sum_{\boldsymbol{i},\boldsymbol{\alpha}}(S^z_{\boldsymbol{i}+\boldsymbol{\alpha}})^2$$

$$+ \sum_{\boldsymbol{i},\langle\boldsymbol{\alpha}_1\boldsymbol{\alpha}_2\rangle}\Big(t_{\langle\boldsymbol{\alpha}_1\boldsymbol{\alpha}_2\rangle} \mathrm{e}^{\mathrm{i}\left(\theta_{\boldsymbol{i}+\boldsymbol{\alpha}_1}-\theta_{\boldsymbol{i}+\boldsymbol{\alpha}_2}\right)} + h.c.\Big) \quad (10.6.7)$$

Here $\boldsymbol{i} = (i_x, i_y, i_z)$ labels the sites of the cubic lattice, $\boldsymbol{\alpha} = \pm\frac{\boldsymbol{x}}{2}, \pm\frac{\boldsymbol{y}}{2}, \pm\frac{\boldsymbol{z}}{2}$, and the two labels $\boldsymbol{i}+\boldsymbol{\alpha}_1$ and $\boldsymbol{i}+\boldsymbol{\alpha}_2$ label the two nearest-neighbor rotors around the $\boldsymbol{i}$ site.

In the following, we are going to show that, in the limit $U \gg t, J$ and $t^2/U \gg J$, the above two-dimensional and three-dimensional models contain a low-energy collective mode. Such a collective mode is very different from the usual collective modes, such as spin waves and phonons, due to its non-local properties. In fact, it does not correspond to the fluctuations of any local order parameter and one needs to use $U(1)$ gauge theory to describe it. We will call such a collective mode a $U(1)$ gauge fluctuation. The collective mode has an exponentially-small energy gap $\Delta \sim \mathrm{e}^{-C\sqrt{t^2/JU}}$ for the two-dimensional model and is exactly gapless for the three-dimensional model. The collective mode in the three-dimensional model behaves in every way like the light in our universe. Thus, we can also call it artificial light.

What is remarkable about the above models is that they contain no sign of gauge structure at the lattice scale. The $U(1)$ gauge fluctuation and the related gauge structure emerge at low energies. This suggests the possibility that the light in our universe may also appear in a similar fashion. Gauge structure may emerge naturally in strongly-correlated quantum systems. The rotor models that we are going to study in this section are the simplest models that contain artificial light. In the following, for simplicity, we will concentrate on the two-dimensional rotor model. The calculations and the results can be easily generalized to the three-dimensional rotor model.

To understand the dynamics of our two-dimensional rotor system, let us first assume that $J = t = 0$ and $U > 0$. In this case, the Hamiltonian is formed by commuting terms which perform local projections. The ground states are highly degenerate and form a low energy subspace. One of the ground states is the state with $S^z_{\boldsymbol{i}} = 0$ for every rotor. Other ground states can be constructed from the first ground state by drawing an oriented loop in the square lattice, and then alternately increasing or decreasing the angular momenta of the rotors by 1 along the loop (see Fig. 10.12). Such a process can be repeated to construct all of the degenerate ground states. We see that the fluctuations in the projected space are represented by loops. The low energy subspace has some non-local characteristics, despite it being obtained via a local projection. If t and J are nonzero, then the t term will make these loops fluctuate and the J term will give these loops an energy proportional to the loop length. It is clear that, when $U \gg J, t$, the low-energy properties of our system are determined by the fluctuations of the loops. The system can have two distinct phases. When $J \gg t$, the system has only a few small-loop fluctuations. When $J \ll t$, the system has many large-loop fluctuations, which can even fill the whole space. As we will see later, these large-loop fluctuations are actually $U(1)$ gauge fluctuations. We note that the loops can intersect and overlap. In fact, a typical loop looks more like a net of closed strings. In the following, we will call the loop fluctuations the closed-string-net fluctuations to stress the branching structure of the fluctuations.

The degenerate ground states are invariant under local symmetry transformations generated by

$$U(\phi_{\boldsymbol{i}}) = \mathrm{e}^{\mathrm{i}\sum_{\boldsymbol{i}}((-)^{\boldsymbol{i}}\phi_{\boldsymbol{i}}\sum_{\alpha} S^z_{\boldsymbol{i}+\alpha})} \tag{10.6.8}$$

where $(-)^{\boldsymbol{i}} = (-)^{i_x+i_y}$. The above transformation is simply the gauge transformation. Thus, we can also say that the degenerate ground states are gauge invariant.

Using a calculation similar to that in Section 10.6.1, we find that our lattice rotor model can be described by the following low-energy effective Lagrangian in the large-U limit:

$$L = \frac{1}{4J} \sum_{\boldsymbol{i},\boldsymbol{\mu}=\boldsymbol{x},\boldsymbol{y}} [\dot{a}_{\boldsymbol{i},\boldsymbol{i}+\boldsymbol{\mu}} + a_0(\boldsymbol{i}) - a_0(\boldsymbol{i}+\boldsymbol{\mu})]^2 + g\sum_{\boldsymbol{i}} \cos(\Phi_{\boldsymbol{i}}) \tag{10.6.9}$$

Here $a_{\boldsymbol{i},\boldsymbol{i}+\boldsymbol{x}} = (-)^{\boldsymbol{i}}\theta_{\boldsymbol{i}+\frac{\boldsymbol{x}}{2}}$, $a_{\boldsymbol{i},\boldsymbol{i}+\boldsymbol{y}} = (-)^{\boldsymbol{i}}\theta_{\boldsymbol{i}+\frac{\boldsymbol{y}}{2}}$, and $a_{\boldsymbol{i}\boldsymbol{j}} = -a_{\boldsymbol{j}\boldsymbol{i}}$. Also, $\Phi_{\boldsymbol{i}}$ is the $U(1)$ flux through the square at $\boldsymbol{i}$, i.e. $\Phi_{\boldsymbol{i}} = a_{\boldsymbol{i},\boldsymbol{i}+\boldsymbol{x}} + a_{\boldsymbol{i}+\boldsymbol{x},\boldsymbol{i}+\boldsymbol{x}+\boldsymbol{y}} + a_{\boldsymbol{i}+\boldsymbol{x}+\boldsymbol{y},\boldsymbol{i}+\boldsymbol{y}} + a_{\boldsymbol{i}+\boldsymbol{y},\boldsymbol{i}}$. Generalizing the second-order perturbative calculation in Section 10.6.1, we find that

$$g = 2(t_1t_3 + t_2t_4)/U.$$

Equation (10.6.9) describes a $U(1)$ lattice gauge theory.

Here we would like to stress that the Lagrangian (10.6.9), having the form of a $U(1)$ gauge theory, does not necessarily imply that low-energy excitations behave like gauge bosons. This is because quantum fluctuations can be very strong and the intuitive picture from the Lagrangian can be completely misleading. In order to have excitations that behave like gauge bosons, we need to choose J and g so that the Lagrangian (10.6.9) is in the semiclassical limit. To see when the semiclassical limit can be reached, we express the action of the lattice gauge theory in dimensionless form as follows:

$$S = \sqrt{\frac{g}{J}} \int \mathrm{d}\tilde{t} \left(\frac{1}{4} \sum_{\boldsymbol{i},\boldsymbol{\mu}=\boldsymbol{x},\boldsymbol{y}} [\partial_{\tilde{t}} a_{\boldsymbol{i},\boldsymbol{i}+\boldsymbol{\mu}} + \tilde{a}_0(\boldsymbol{i}) - \tilde{a}_0(\boldsymbol{i}+\boldsymbol{\mu})]^2 + \sum_{\boldsymbol{i}} \cos(\Phi_{\boldsymbol{i}}) \right)$$

where $\tilde{t} = \sqrt{gJ}t$ and $\tilde{a}_0 = a_0/\sqrt{gJ}$. We find that the semiclassical limit is reached if $g/J = t^2/JU \gg 1$. In this limit, our model contains $U(1)$ gauge bosons (or an artificial light) as its only low-energy collective excitations.

Due to the instanton effect, a $U(1)$ gauge excitation develops a gap in $1+2$ dimensions (Polyakov, 1977) (see Section 6.3.2). The instanton effect is associated with a change in the $U(1)$ flux Φ from 0 to 2π on a square. To estimate the importance of the instanton effect, let us consider a model with only a single square (i.e. the four-rotor model discussed earlier). Such a model is described by eqn (6.4.11). The instanton effect corresponds to a path $\Phi(t)$ (a time-dependent flux), where Φ

goes from $\Phi(-\infty) = 0$ to $\Phi(+\infty) = 2\pi$. To estimate the instanton action, we assume that

$$\Phi(t) = \begin{cases} 0, & \text{for } t < 0 \\ 2\pi t/T, & \text{for } 0 < t < T \\ 2\pi, & \text{for } T < t \end{cases}$$

The minimal instanton action is found to be

$$S_c = \pi\sqrt{g/J}$$

when $T = \pi/2\sqrt{gJ}$. From the density of the instanton gas on the time axis, $\sqrt{Jg}\mathrm{e}^{-S_c}$, we estimate the energy gap of the $U(1)$ gauge boson to be

$$\Delta_{\text{gauge}} \sim \sqrt{Jg}\mathrm{e}^{-\pi\sqrt{g/J}} \tag{10.6.10}$$

We see that, when $g/J \gg 1$, the gap can be very small and the low-energy fluctuations are very much like a gapless photon. Certainly, the real gapless artificial light only exists in $(3+1)$-dimensional models, such as the three-dimensional rotor model (10.6.7).

The three-dimensional rotor model has the following low-energy effective Lagrangian in the large-U limit:

$$L = \frac{1}{4J} \sum_{\boldsymbol{i},\boldsymbol{\mu}=\boldsymbol{x},\boldsymbol{y},\boldsymbol{z}} [\dot{a}_{\boldsymbol{i},\boldsymbol{i}+\boldsymbol{\mu}} + a_0(\boldsymbol{i}) - a_0(\boldsymbol{i}+\boldsymbol{\mu})]^2 + \sum_{\boldsymbol{p}} g_{\boldsymbol{p}} \cos(\Phi_{\boldsymbol{p}}) \tag{10.6.11}$$

Here $\sum_{\boldsymbol{p}}$ is the summation over all of the squares, and $\Phi_{\boldsymbol{p}}$ is the $U(1)$ flux through a square $\boldsymbol{p}$. Also, $g_{\boldsymbol{p}}$ may depend on the orientation of the square. The value of $g_{\boldsymbol{p}}$ can be tuned by $t_{\langle\boldsymbol{\alpha}_1\boldsymbol{\alpha}_2\rangle}$ in eqn (10.6.7).

10.6.3 String-net theory of artificial light and artificial charge

- A gauge theory is a closed-string-net theory in disguise.
- A confined phase is a phase with dilute small strings. A Coulomb phase is a phase with nets of large closed strings that fill the whole space.
- Gauge bosons (in the Coulomb phase) correspond to fluctuations of large closed string-nets, and gauge charges are end-points of open strings.

As mentioned earlier, the low-energy excitations in our rotor model below U are described by nets of closed strings of increased/decreased S^z. To make this picture more precise, we would like to define a string-net theory on a lattice.

The Hilbert space of the string-net theory is a subspace of the Hilbert space of our rotor model (10.6.6). To construct the string-net Hilbert space, we first need

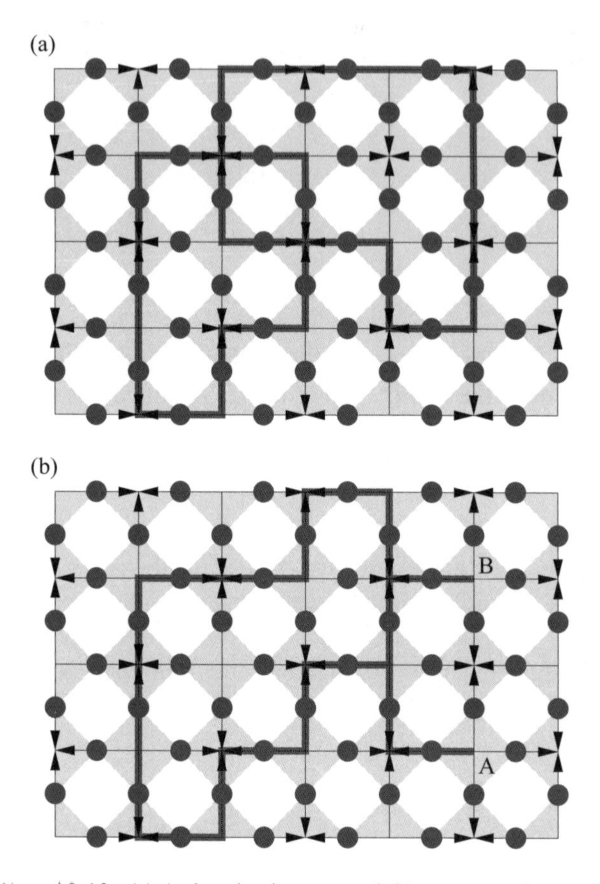

FIG. 10.13. (a) A closed string-net, and (b) an open string-net.

to introduce a string creation operator. A string creation operator is formed by the product of $e^{\pm i\theta_{\langle ij\rangle}}$ operators:

$$U(C) = \prod_{\langle \boldsymbol{ij}\rangle} e^{i(-)^{\boldsymbol{i}}\theta_{\langle \boldsymbol{ij}\rangle}} \tag{10.6.12}$$

where C is a string connecting nearest-neighbor sites in the square lattice, the product $\prod_{\langle \boldsymbol{ij}\rangle}$ is over all of the nearest-neighbor links $\langle \boldsymbol{ij}\rangle$ that form the string, and $\theta_{\langle \boldsymbol{ij}\rangle}$ is the θ field of the rotor on the link $\langle \boldsymbol{ij}\rangle$.

As the string can intersect and overlap, the string looks more like a net. We will call C a string-net and $U(C)$ a string-net operator. If C is formed by a collection of closed strings, then we will call it a closed string-net. If C contains at least an open string, then it will be called an open string-net (see Fig. 10.13).

The string-net Hilbert space contains a state with all $S^z_{\boldsymbol{I}} = 0$. Such a state, by definition, corresponds to a state with no strings. If we apply the closed-string operator (10.6.12) to the $S^z_{\boldsymbol{I}} = 0$ state, then we obtain another state in the string-net Hilbert space. Such a state is formed by $S^z_{\boldsymbol{I}} = \pm 1$ along the closed string C and can be regarded as a state with one closed string C. Other states in the string-net Hilbert space correspond to multiple-string states and are generated by repeatedly applying the closed-string operators (10.6.12) to the $S^z_{\boldsymbol{I}} = 0$ state.

The Hamiltonian of our string-net theory is obtained from the rotor Hamiltonian (10.6.6) restricted in the string-net subspace. It is given by

$$H_{str} = \sum_{\langle \boldsymbol{ij} \rangle} J(S^z_{\boldsymbol{ij}})^2 - \sum_{\boldsymbol{p}} \frac{1}{2}(gW_{\boldsymbol{p}} + h.c.) \tag{10.6.13}$$

where the sum $\sum_{\langle \boldsymbol{ij} \rangle}$ is over all of the nearest-neighbor links $\langle \boldsymbol{ij} \rangle$ in the square lattice, $S^z_{\boldsymbol{ij}}$ is the angular momentum operator for the rotor on the nearest-neighbor links $\langle \boldsymbol{ij} \rangle$, $\sum_{\boldsymbol{p}}$ sums over all of the squares of the square lattice, and $W_{\boldsymbol{p}}$ is the closed-string operator for the closed string around the square $\boldsymbol{p}$. One can check that the above Hamiltonian indeed acts within the string-net Hilbert space. The J term gives strings in the string-net a finite string tension, and the g term, as a string 'hopping' term, causes the string-nets to fluctuate. The J term in eqn (10.6.13) directly comes from the J term in eqn (10.6.6), while the g term comes from the t terms in eqn (10.6.6) at the second order perturbative expansion.

From the construction, it is clear that the string-net Hilbert space is identical to the low-energy Hilbert space of our model (10.6.6), which is formed by states with energy less than U. From our derivation of the effective lattice gauge theory (10.6.9), it is also clear that the string-net Hamiltonian (10.6.13) is directly related to the lattice gauge Lagrangian (10.6.9). In fact, the Hamiltonian of the lattice gauge theory is identical to the string-net Hamiltonian (10.6.13). The $\sum_{\boldsymbol{ij}} J(S^z_{\boldsymbol{ij}})^2$ term in the string-net theory corresponds to the $\frac{1}{4J}\sum_{\langle \boldsymbol{ij} \rangle}[\dot{a}_{\boldsymbol{ij}} + a_0(\boldsymbol{i}) - a_0(\boldsymbol{j})]^2$ term in the gauge theory, and the $\sum_{\boldsymbol{p}} \frac{1}{2} g(W_{\boldsymbol{p}} + h.c.)$ term in the string-net theory corresponds to the $g\sum_{\boldsymbol{p}} \cos(\Phi_{\boldsymbol{p}})$ term in the gauge theory. As $S^z \sim J^{-1}\dot{\theta}_{\langle \boldsymbol{ij} \rangle} = J^{-1}\dot{a}_{\boldsymbol{ij}}$ corresponds to the electric flux along the link, a closed string of increased/decreased S^z corresponds to a loop of electric flux tube. We would like to stress that the above connection between gauge theory and string-net theory not only applies to the two-dimensional model (10.6.6), but it also works for models in any dimensions.

We see that the $U(1)$ gauge theory (10.6.9) is actually a dynamical theory of closed string-nets. Typically, one expects a dynamical theory of closed string-nets to be written in terms of string-nets as in (10.6.13). However, because we are more familiar with field theory, what we did in the last few sections can be viewed as an attempt to describe a string-net theory using a field theory. Through some mathematical trick, we have achieved our goal. We are able to write the string-net theory in the form of a gauge field theory. The gauge field theory is a special field theory in which the field *does not* correspond to physical degrees of freedom and the physical Hilbert space is non-local (in the sense that the total physical Hilbert space cannot be written as a direct product of local Hilbert spaces). The point that we try to make here is that gauge theory (at least the one discussed here) is a closed-string-net theory in disguise. In other words, gauge theory and closed-string-net theory are dual to each other. In fact, if we discretize the time to consider a space–time lattice, then we can find an exact mapping between $U(1)$ lattice gauge theory and a statistical model of membranes in space–time (Banks *et al.*, 1977; Savit, 1980).

In the large-J/g limit (and hence the large-Δ_{gauge} limit), the ground states for both the rotor model and the string-net model are given by $S^z = 0$ for every rotor. In this phase, the closed string-nets or the electric flux tubes do not fluctuate much and have an energy proportional to their length. This implies that the $U(1)$ gauge theory is in the confining phase. In the small-J/g limit, the closed string-nets fluctuate strongly and the space is filled with closed string-nets of arbitrary sizes. According to the calculation in the previous section, we note that the small-J/g phase can also be viewed as the Coulomb phase with gapless gauge bosons. Combining the two pictures, we see that gapless gauge bosons correspond to fluctuations of large closed string-nets.

After relating the closed string-net to artificial light, we now turn to artificial charges. To create a pair of particles with opposite artificial charges for the artificial $U(1)$ gauge field, we need to draw an open string and alternately increase and decrease the S^z of the rotors along the string (see Fig. 10.12). The end-points of the open string, as the end-points of the electric flux tube, correspond to particles with opposite artificial charges. We note that, unlike the rotors, charged particles live on the sites of the square lattice. In the confining phase, the string connecting the two artificial charges does not fluctuate much. The energy of the string is proportional to the length of the string. We see that there is a linear confinement between the artificial charges.

In the small-J/g limit, the large g causes strong fluctuations of the closed string-nets, which leads to gapless $U(1)$ gauge fluctuations. The strong fluctuations of the strings connecting the two charges also changes the linear confining potential to the $\log(r)$ potential between the charges.

To understand the dynamics of particles with artificial charges and to derive the $\log(r)$ potential, let us derive the low-energy effective theory for these charged particles. Let us first assume that $J = t = 0$. A pair of charged particles with opposite unit artificial charges can be created by applying the open-string operator (10.6.12) to the ground state. At the end of the open string, we have $\left(\sum_\alpha S^z_{i+\alpha}\right)^2 = 1$. We find that each charged particle has an energy U which comes from the term $U \sum_i \left(\sum_\alpha S^z_{i+\alpha}\right)^2$. The string itself costs no energy. If we apply the same open-string operator n times, then we create n units of opposite artificial charges at the ends of the open string. The energy of a charge of n units is $n^2 U$.

The gauge theory with charges contains two parts, namely the charges and the strings. Let us first consider only the charges and treat the charged particles as independent particles. In this case, the total Hilbert space of charged particles is formed by the state $|\{n_{\boldsymbol{i}}\}\rangle$, where $n_{\boldsymbol{i}}$ is the number of artificial charges at the site $\boldsymbol{i}$ of the square lattice. Here $|\{n_{\boldsymbol{i}}\}\rangle$ is an energy eigenstate with energy $E = U\sum_{\boldsymbol{i}} n_{\boldsymbol{i}}^2$. Such a system can be described by the Lagrangian

$$L = \sum_{\boldsymbol{i}} \frac{1}{4U}\dot{\varphi}_{\boldsymbol{i}}^2 \tag{10.6.14}$$

where $\varphi_{\boldsymbol{i}}$ is an angular variable. The creation operator of the charged particle of a unit charge is given by $\mathrm{e}^{\,\mathrm{i}\varphi_{\boldsymbol{i}}}$.

The gauge fluctuations are described by the strings and eqn (10.6.9). In the $U(1)$ gauge theory, the ground state is gauge invariant (away from the charges). From eqn (10.6.8), we see that the gauge invariance implies that $\sum_{\boldsymbol{\alpha}} S^z_{\boldsymbol{i}+\boldsymbol{\alpha}} = 0$. Thus, there are no open ends of strings away from the charges.

To find the coupled theory of the charge and the gauge field, we include the fact that the charged particles are always at the ends of the open strings. In other words, the physical state of the gauge theory contains open strings, but the open strings can only end at the charges. Such a constraint can be imposed by the gauge invariance. So we can use the gauge invariance to combine the charge Lagragian (10.6.14) and the gauge Lagragian (10.6.9) together and write down the coupled theory. Using the gauge invariance, we find that the combined Lagrangian has the form (see Problem 6.4.5)

$$L = \sum_{\boldsymbol{i}} \frac{(\dot{\varphi}_{\boldsymbol{i}} + a_0(\boldsymbol{i}))^2}{4U} + \sum_{\boldsymbol{i},\boldsymbol{\mu}=\boldsymbol{x},\boldsymbol{y}} \frac{[\dot{a}_{\boldsymbol{i},\boldsymbol{i}+\boldsymbol{\mu}} + a_0(\boldsymbol{i}) - a_0(\boldsymbol{i}+\boldsymbol{\mu})]^2}{4J} + g\sum_{\boldsymbol{i}} \cos(\Phi_{\boldsymbol{i}})$$

After including the gauge field, the single charge creation operator $\mathrm{e}^{\,\mathrm{i}\varphi_{\boldsymbol{i}}}$ is no longer physical because it is not gauge invariant. The gauge-invariant operator

$$\mathrm{e}^{-\mathrm{i}\varphi_{i_1}}\,\mathrm{e}^{\,\mathrm{i}a_{i_1 i_2}}\,\mathrm{e}^{\,\mathrm{i}a_{i_2 i_3}}\ldots\mathrm{e}^{\,\mathrm{i}a_{i_{N-1} i_N}}\,\mathrm{e}^{\,\mathrm{i}\varphi_{i_N}}$$

always creates a pair of opposite charges, together with the open string connecting the charges. Therefore, open strings always end at the charges. In fact, the above gauge-invariant operator is simply the open-string operator (10.6.12). We also see that the string operator (10.6.12) is closely related to the Wegner–Wilson loop operator (Wegner, 1971; Wilson, 1974; Kogut, 1979).

The t term generates a hopping of charged particles to the next-nearest neighbor in the square lattice. Thus, if $t \neq 0$, then the charged particles will have a non-trivial dispersion. The corresponding Lagrangian is given by

$$L = \sum_{\boldsymbol{i}} \frac{(\dot{\varphi}_{\boldsymbol{i}} + a_0(\boldsymbol{i}))^2}{4U} - \sum_{(\boldsymbol{i}\boldsymbol{j}),a=1,2} t(\mathrm{e}^{\,\mathrm{i}(\varphi_{\boldsymbol{i}} - \varphi_{\boldsymbol{j}} - a_{\boldsymbol{i}\boldsymbol{k}_a} - a_{\boldsymbol{k}_a\boldsymbol{j}})} + h.c.) \tag{10.6.15}$$

where $(\boldsymbol{ij})$ are next-nearest neighbors in the square lattice, and $\boldsymbol{k}_{1,2}$ are the two sites between site $\boldsymbol{i}$ and site $\boldsymbol{j}$. The above Lagrangian also tells us that the charged particles are bosons.

We also note that an increased $S^z_{\boldsymbol{ij}}$ corresponds to two artificial charges at $\boldsymbol{i}$ and $\boldsymbol{j}$. Therefore, each unit of artificial charge carries 1/2 angular momentum! (Note that the total angular momentum $\sum_{\langle \boldsymbol{ij}\rangle} S^z_{\langle \boldsymbol{ij}\rangle}$ is a conserved quantity.)

10.6.4 Physical properties of two-dimensional and three-dimensional rotor systems

- The continuum limit of the lattice gauge theory.
- The speed and the fine-structure constant of artificial light.

To understand the physical properties of the artificial light in the two-dimensional model, let us take the continuum limit of eqn (10.6.9) by writing

$$a_{\boldsymbol{ij}} = \delta \boldsymbol{x}_{\langle \boldsymbol{ij}\rangle} \cdot \boldsymbol{a}(\boldsymbol{x}), \qquad a_0(\boldsymbol{i}) = a_0(\boldsymbol{x}),$$

where $\boldsymbol{a} = (a_x, a_y)$ is a two-dimensional vector field (the vector gauge potential in two dimensions), a_0 corresponds to the potential field, $\boldsymbol{x}$ is near the site $\boldsymbol{i}$, $\delta \boldsymbol{x}_{\langle \boldsymbol{ij}\rangle}$ is the vector that connects the $\boldsymbol{i}$ and $\boldsymbol{j}$ sites in the square lattice, and l is the lattice constant of the square lattice. In the continuum limit, the Lagrangian (10.6.9) becomes

$$L = \int \mathrm{d}^2\boldsymbol{x} \left(\frac{1}{4J}\boldsymbol{e}^2 - \frac{gl^2}{2} b^2 \right) \tag{10.6.16}$$

where $\boldsymbol{e} = \partial_t \boldsymbol{a} - \partial_{\boldsymbol{x}} a_0$ and $b = \partial_x a_y - \partial_y a_x$ are the corresponding artificial electric field and artificial magnetic field, respectively. We see that the velocity of our artificial light is $c_a = \sqrt{2gJl^2/\hbar^2}$. The bandwidth of the artificial light is about $E_a = 2c_a\hbar/l = \sqrt{8gJ}$.

From eqn (10.6.15), we find the continuum Lagrangian that describes the charged particles in the two-dimensional model (in the $U \gg t$ limit) as follows:

$$L = \int \mathrm{d}^2\boldsymbol{x} \sum_{I=1,2} \Big(\phi_I^\dagger (\mathrm{i}\partial_t - a_0 - U)\phi_I - \frac{8tl^2}{2}|(\partial_i + \mathrm{i}a_i)\phi_I|^2$$
$$+ \bar{\phi}_I^\dagger (\mathrm{i}\partial_t + a_0 - U)\bar{\phi}_I - \frac{8tl^2}{2}|(\partial_i - \mathrm{i}a_i)\bar{\phi}_I|^2 \Big)$$

where ϕ_I describe the positively-charged bosons, $\bar{\phi}_I$ describe the negatively-charged bosons, ψ_1 and $\bar{\psi}_1$ describe the charged boson at the even sites of the square lattice, and ψ_2 and $\bar{\psi}_2$ describe the charged boson at the odd sites of the square lattice. It costs energy $2U$ to create a pair of charged bosons. The mass of the bosons is $m = (8tl^2)^{-1}$, and $mc_a^2 = gJ/4t$. We would like to note that the boson velocity $\sim 8tl$ can be larger than the speed of artificial light. The potential energy between a positive and a negative charge is $V(r) = \frac{2J}{\pi}\ln r$.

For the three-dimensional model, the Lagrangian in the continuum limit is given by

$$L = \int d^3\boldsymbol{x} \left(\frac{1}{4Jl}\boldsymbol{e}^2 - \frac{gl}{2}\boldsymbol{b}^2\right) \tag{10.6.17}$$

where $\boldsymbol{e}$ and $\boldsymbol{b}$ are the artificial electric field and artificial magnetic field, respectively, in three dimensions. The speed of artificial light is $c_a = \sqrt{2gJl^2/\hbar^2}$. The above three-dimensional Lagrangian can be rewritten in the more standard form

$$L = \int d^3\boldsymbol{x}\, \frac{1}{8\pi\alpha} \left(\frac{1}{c_a}\boldsymbol{e}^2 - c_a\boldsymbol{b}^2\right) \tag{10.6.18}$$

where $\alpha = \frac{1}{2\pi}\sqrt{J/2g}$ is the artificial fine-structure constant. The mass of the charged boson is $m = (8tl^2)^{-1}$. An artificial atom (a bound state of two positively-charged and negatively-charged bosons) has an energy-level spacing of order $\frac{1}{4}mc_a^2\alpha^2$ and a size of order $2/\alpha m c_a$.

Problem 10.6.2.
Verify eqns (10.6.16) and (10.6.17).

10.7 Emergent light and electrons from an $SU(N_f)$ spin model

- A non-trivial quantum order (i.e. a string-net condensation) in a local bosonic model can lead to *both* massless photons and massless fermions. String-net condensation provides a way to unify light and fermions.
- The masslessness of the photons and fermions is protected by the PSG that characterizes the string-net condensation.

As stressed at the beginning of this chapter, to understand the existence of light and electrons is to understand why these particles are massless (or nearly massless when compared with the Planck mass). For a generic interacting system, massless (or gapless) excitations are rare. If they exist, then they exist for a reason.

We know that symmetry breaking can produce and protect gapless Nambu–Goldstone modes. In Section 9.10, it was proposed that, in addition to symmetry breaking, quantum order and the associated PSGs can also produce and protect gapless excitations. The gapless excitations produced and protected by quantum order can be gapless gauge bosons and/or gapless fermions.

In this section, we are going to study a specific $SU(N_f)$ spin model on a cubic lattice (Wen, 2002a). The ground state of the model has a non-trivial quantum order (i.e. a string-net condensation). As a result, the model has emergent massless $U(1)$ gauge bosons (the artificial light) and massless charged fermions (the artificial electron and proton). In other words, the $SU(N_f)$ spin model has an emergent QED! A world based on such an $SU(N_f)$ spin model will be quite similar to our world; both worlds are formed by charged fermions (i.e. electrons and protons) interacting via Coulomb interactions. A more complicated spin model on a cubic lattice can even lead to an emergent QED and QCD with photons, gluons, leptons, and quarks (Wen, 2003b). Thus, it will not be surprising if someday we find that all of the elementary particles in nature emerge from a generalized spin model (i.e. a local bosonic model).

10.7.1 An $SU(N_f)$ spin model on a cubic lattice

The model is formed by $SU(N_f)$-spins, with one spin at each site of a cubic lattice. Here N_f is even and the lattice sites are labeled by the three integers $\boldsymbol{i} = (i_x, i_y, i_z)$. The states at each site, namely

$$|a_1, a_2, ..., a_{N_f/2}\rangle, \qquad a_1, ..., a_{N_f/2} = 1, 2, ..., N_f$$

form the rank-$N_f/2$ anti-symmetric tensor representation of $SU(N_f)$. The spin operators $S_{\boldsymbol{i}}^{a\bar{b}}$, $a, \bar{b} = 1, 2, ..., N_f$, at each site form the adjoint representation of $SU(N_f)$:

$$\begin{aligned}
[S_{\boldsymbol{i}}^{a\bar{b}}, S_{\boldsymbol{i}}^{c\bar{d}}] &= \delta^{\bar{b}c} S_{\boldsymbol{i}}^{a\bar{d}} - \delta^{a\bar{d}} S_{\boldsymbol{i}}^{c\bar{b}} \\
S_{\boldsymbol{i}}^{a\bar{b}} \delta_{a\bar{b}} &= 0 \\
(S_{\boldsymbol{i}}^{ab})^\dagger &= S_{\boldsymbol{i}}^{ba}
\end{aligned} \tag{10.7.1}$$

where the repeated indices are summed.

Let

$$W_{\boldsymbol{i}} = -N_f^{-4} S_{\boldsymbol{i}+\boldsymbol{x}}^{ab} S_{\boldsymbol{i}+\boldsymbol{x}+\boldsymbol{y}}^{bc} S_{\boldsymbol{i}+\boldsymbol{y}}^{cd} S_{\boldsymbol{i}}^{da}$$

We choose the Hamiltonian for our $SU(N_f)$ model to be (Wen, 2002a)

$$H = g \sum_{\boldsymbol{i}} [W_{\boldsymbol{i}} + h.c] \tag{10.7.2}$$

10.7.2 The ground state of the $SU(N_f)$ model

To understand the ground state, let us introduce the fermion representation of the $SU(N_f)$ spin model (Affleck and Marston, 1988). We first introduce N_f fermion operators ψ_a which form a fundamental representation of $SU(N_f)$. The Hilbert space at each site is then formed by states with $N_f/2$ fermions. The spin operator is given by

$$S^{a\bar{b}} = \psi_a \psi_{\bar{b}}^\dagger - \frac{1}{2}\delta_{a\bar{b}}$$

If we introduce

$$\hat{\chi}_{ij} = N_f^{-1} \psi_{ai}^\dagger \psi_{aj}$$

then W_i is given by

$$W_i = \hat{\chi}_{i,i+x} \hat{\chi}_{i+x,i+x+y} \hat{\chi}_{i+x+y,i+y} \hat{\chi}_{i+y,i} + O(1/N_f)$$

Using the fermion representation, we can express the Hamiltonian (10.7.2) as

$$H = g \sum_p [\prod_\square \hat{\chi}_{ij} + h.c.] + O(N_f^{-1}) \tag{10.7.3}$$

where $\sum_p$ is a sum over all of the squares p, and $\prod_\square$ is a product over all of the four links around the square p; that is, $\prod_\square \hat{\chi}_{ij} = \hat{\chi}_{i_1 i_2} \hat{\chi}_{i_2 i_3} \hat{\chi}_{i_3 i_4} \hat{\chi}_{i_4 i_1}$, with i_1, i_2, i_3, and i_4 being the four sites around the square p.

We note that, in the large-N_f limit, $\hat{\chi}_{ij}$ commute with each other, i.e. $[\hat{\chi}_{ij}, \hat{\chi}_{i'j'}] = O(1/N_f)$. Also, $\hat{\chi}_{ij}$ is a sum of N_f terms. Although each term has strong quantum fluctuations, the sum has little fluctuations. Thus, $\hat{\chi}_{ij}$ behaves like a c-number in the large-N_f limit. That is, for a given state we can replace $\hat{\chi}_{ij}$ by a complex number χ_{ij}. The energy of such a state is $E = g\sum_p [\prod_\square \chi_{ij} + h.c.] + O(N_f^{-1})$. Different states will have different χ_{ij}, and hence different energies.

The ground state of the $SU(N_f)$ model can be obtained by choosing a set of χ_{ij} that minimize $E = g\sum_p [\prod_\square \chi_{ij} + h.c.]$. Here we encounter one difficulty: the χ_{ij} are not independent and we cannot choose their values arbitrarily. For example, we cannot make all of the χ_{ij} to be zero at the same time. We will use the projective construction in Section 9.1.1:[77] to overcome this difficulty, We start with a mean-field Hamiltonian

$$H_{\text{mean}} = -\sum_{\langle ij \rangle} \tilde{\chi}_{ij}^\dagger \psi_i^\dagger \psi_j + \sum_i a_0(i) \psi_i^\dagger \psi_i$$

We choose $a_0(i)$ so that the mean-field ground state of H_{mean}, namely $|\Psi_{\text{mean}}^{\{\tilde{\chi}_{ij}\}}\rangle$, has $N_f/2$ fermions per site on average. A state of the $SU(N_f)$ model can be constructed from the mean-field state $|\Psi_{\text{mean}}^{\{\tilde{\chi}_{ij}\}}\rangle$ by projecting into a subspace with exactly $N_f/2$ fermions per site:

$$|\Psi^{\{\tilde{\chi}_{ij}\}}\rangle = \mathcal{P} |\Psi_{\text{mean}}^{\{\tilde{\chi}_{ij}\}}\rangle$$

We can now choose $\tilde{\chi}_{ij}$ to minimize the average energy $\langle \Psi^{\{\tilde{\chi}_{ij}\}} | H | \Psi^{\{\tilde{\chi}_{ij}\}} \rangle$, and hence obtain the trial energy and the trial wave function of the ground state.

[77] The discussion in Section 9.1.1 is for the $SU(N_f)$ model with $N_f = 2$.

In the large-N_f limit, the projection $\mathcal{P}$ only makes a small change to the wave function $|\Psi_{\text{mean}}^{\{\tilde{\chi}_{ij}\}}\rangle$. Thus, to obtain the ground-state energy, we can perform an easier calculation by minimizing $\langle\Psi_{\text{mean}}^{\{\tilde{\chi}_{ij}\}}|H|\Psi_{\text{mean}}^{\{\tilde{\chi}_{ij}\}}\rangle$ instead.

Here is how we perform the minimization. We first pick a set of $\tilde{\chi}_{\boldsymbol{ij}}$. Then we choose $a_0(\boldsymbol{i})$ so that $\langle\Psi_{\text{mean}}^{\{\tilde{\chi}_{ij}\}}|\psi_{\boldsymbol{i}}^\dagger\psi_{\boldsymbol{i}}|\Psi_{\text{mean}}^{\{\tilde{\chi}_{ij}\}}\rangle = N_f/2$. Then we calculate $\chi_{\boldsymbol{ij}} = N_f^{-1}\langle\Psi_{\text{mean}}^{\{\tilde{\chi}_{ij}\}}|\psi_{\boldsymbol{i}}^\dagger\psi_{\boldsymbol{j}}|\Psi_{\text{mean}}^{\{\tilde{\chi}_{ij}\}}\rangle$. Such $\chi_{\boldsymbol{ij}}$ are the corresponding c-numbers for the operator $\hat{\chi}_{\boldsymbol{ij}}$ discussed above. We see that $\tilde{\chi}_{\boldsymbol{ij}}$ are independent and their values can be chosen arbitrarily. The non-independent $\chi_{\boldsymbol{ij}}$ are functions of $\tilde{\chi}_{\boldsymbol{ij}}$, i.e. $\chi_{\boldsymbol{ij}}(\{\tilde{\chi}_{\boldsymbol{ij}}\})$. The ground state is obtained by minimizing $E = g\sum_{\boldsymbol{p}}[\prod_{\square>}\chi_{\boldsymbol{ij}}(\{\tilde{\chi}_{\boldsymbol{ij}}\}) + h.c.]$ as we change $\tilde{\chi}_{\boldsymbol{ij}}$.

We find that the ground state is obtained by choosing $\tilde{\chi}_{\boldsymbol{ij}}$ as

$$\tilde{\chi}_{\boldsymbol{i},\boldsymbol{i}+\hat{\boldsymbol{x}}} \equiv -\,\mathrm{i}, \qquad \tilde{\chi}_{\boldsymbol{i},\boldsymbol{i}+\hat{\boldsymbol{y}}} \equiv -\,\mathrm{i}(-)^{i_x},$$
$$\tilde{\chi}_{\boldsymbol{i},\boldsymbol{i}+\hat{\boldsymbol{z}}} \equiv -\,\mathrm{i}(-)^{i_x+i_y}, \qquad a_0(\boldsymbol{i}) = 0$$

which lead to the following set of $\chi_{\boldsymbol{ij}}$:

$$\bar{\chi}_{\boldsymbol{i},\boldsymbol{i}+\hat{\boldsymbol{x}}} \equiv -\,\mathrm{i}|\chi|, \qquad \bar{\chi}_{\boldsymbol{i},\boldsymbol{i}+\hat{\boldsymbol{y}}} \equiv -\,\mathrm{i}(-)^{i_x}|\chi|,$$
$$\bar{\chi}_{\boldsymbol{i},\boldsymbol{i}+\hat{\boldsymbol{z}}} \equiv -\,\mathrm{i}(-)^{i_x+i_y}|\chi| \qquad a_0(\boldsymbol{i}) = 0. \tag{10.7.4}$$

The ansatz $\bar{\chi}_{\boldsymbol{ij}}$ has π flux through each square. As a result, $\prod_\square \chi_{\boldsymbol{ij}} = -|\chi|^4$. The minus sign is the reason why the above $\chi_{\boldsymbol{ij}}$ minimize $g\sum_{\boldsymbol{p}}[\prod_\square \chi_{\boldsymbol{ij}} + h.c.]$ (assuming that $g > 0$).

The ground state described by the ansatz $\bar{\chi}_{\boldsymbol{ij}}$ does not break any symmetries. However, the ground state has a non-trivial quantum order. In the following, we will show that the ground state supports massless $U(1)$ gauge bosons and massless charged fermions.

10.7.3 The low-energy dynamics of the $SU(N_f)$ model

The fluctuations of $\chi_{\boldsymbol{ij}}$ describe a collective mode in our $SU(N_f)$ spin model. The dynamics of such a collective mode are described by a classical field theory on a cubic lattice. The fluctuations of $|\chi|$ correspond to a massive excitation and can be ignored. The low-energy fluctuations of the collective mode are described by the phase fluctuations around $\bar{\chi}_{\boldsymbol{ij}}$, i.e. $\chi_{\boldsymbol{ij}} = \bar{\chi}_{\boldsymbol{ij}}\mathrm{e}^{\mathrm{i}a_{ij}}$. As discussed in Section 9.1.1, the fluctuations described by $a_{\boldsymbol{ij}}$ are $U(1)$ gauge fluctuations. The physical wave function for a $U(1)$ gauge fluctuation is given by the projected mean-field state for the deformed ansatz:

$$\mathcal{P}|\Psi_{\text{mean}}^{\{\bar{\chi}_{ij}\mathrm{e}^{\mathrm{i}a_{ij}}\}}\rangle. \tag{10.7.5}$$

The following effective Hamiltonian of a_{ij} is obtained by substituting the above into eqn (10.7.3):

$$H = -2g \sum_{p} |\chi|^4 \cos(\Phi_p) \tag{10.7.6}$$

where

$$\Phi_p = a_{i_1 i_2} + a_{i_2 i_3} + a_{i_3 i_4} + a_{i_4 i_1}$$

and (i_1, i_2, i_3, i_4) are the four sites around the square p. The Hamiltonian (10.7.6) describes a $U(1)$ lattice gauge theory with a_{ij} as the lattice $U(1)$ gauge field and Φ_p as the $U(1)$ flux through the square p.

The $U(1)$ gauge fluctuations (10.7.5) are not the only type of low-energy excitations. The second type of low-energy excitations is given by the projected mean-field state with some particle–hole excitations, such as

$$\mathcal{P}\psi^a_{i_1}\psi^{b\dagger}_{i_2}|\Psi^{\{\bar{\chi}_{ij}\}}_{\text{mean}}\rangle. \tag{10.7.7}$$

These excitations correspond to charged fermions.

To determine the dynamics of the fermion excitations, we write

$$\hat{\chi}_{ij} = \bar{\chi}_{ij} + N_f^{-1} : \psi^\dagger_{a,i}\psi_{a,j} : \tag{10.7.8}$$

where $\bar{\chi}_{ij} = \langle\hat{\chi}_{ij}\rangle$. Substituting eqn (10.7.8) into eqn (10.7.3), we obtain

$$H_f = gN_f^{-1} \sum_{ij} [\chi'_{ij} : \psi^\dagger_j\psi_i : +h.c.] + \sum_i a_0(i) : \psi^\dagger_i\psi_i : +O((:\psi^\dagger\psi:)^2) \tag{10.7.9}$$

where

$$\begin{aligned} \chi'_{i,i+\hat{x}} &\equiv 4\mathrm{i}|\chi|^3, \\ \chi'_{i,i+\hat{y}} &\equiv 4\mathrm{i}(-)^{i_x}|\chi|^3, \\ \chi'_{i,i+\hat{z}} &\equiv 4\mathrm{i}(-)^{i_x+i_y}|\chi|^3. \end{aligned} \tag{10.7.10}$$

We see that eqn (10.7.9) describes fermion hopping in a π-flux phase. The following low-energy effective Hamiltonian for fermions and gauge excitations is obtained by combining eqns (10.7.6) and (10.7.9):

$$\begin{aligned} H_{eff} = {} & gN_f^{-1} \sum_{ij} [\chi'_{ij}\mathrm{e}^{\mathrm{i}a_{ij}} : \psi^\dagger_j\psi_i : +h.c.] + \sum_i a_0(i) : \psi^\dagger_i\psi_i : \\ & - 2g \sum_p |\chi|^4 \cos(\Phi_p) \end{aligned} \tag{10.7.11}$$

In the continuum limit, $-2g\sum_p |\chi|^4 \cos(\Phi_p)$ becomes $\int \mathrm{d}^3x\ (3l_0 g)|\chi|^4 \boldsymbol{B}^2$, where l_0 is the lattice constant of the cubic lattice, and $\boldsymbol{B}$ is the magnetic field strength of the $U(1)$ gauge field. Comparing this with the standard Hamiltonian of $U(1)$ gauge field theory, namely $H = g_1 \boldsymbol{B}^2 + g_2 \boldsymbol{E}^2$ (where $\boldsymbol{E}$ is the electric field of the $U(1)$ gauge field), we find that our $U(1)$ Hamiltonian is missing the kinetic energy term from the electric field. The kinetic energy term appears as the higher-order terms in the $1/N_f$ expansion in our model.

To estimate the value of the coefficient in the $\boldsymbol{E}^2$ term, we note that, for a time-independent gauge potential, $\boldsymbol{E}^2$ has the form $(\partial_{\boldsymbol{x}} a_0)^2$. Such a term is generated from eqn (10.7.11) by integrating out the fermions. The generated term is of order $N_f \frac{1}{gN_f^{-1}|\chi|}(l_0 \partial_{\boldsymbol{x}} a_0)^2$. The first coefficient N_f comes from the fact that each N_f family of the fermions contributes equally to $(l_0 \partial_{\boldsymbol{x}} a_0)^2$. The second coefficient describes the contribution from one family of the fermions. The coefficient has the dimension of inverse energy. Its value can be estimated as the inverse of the fermion bandwidth, i.e. $1/(gN_f^{-1}|\chi|)$. Thus, the Lagrangian for the $U(1)$ gauge field has the form (Marston and Affleck, 1989; Wen, 2002a)

$$\mathcal{L} = \frac{CN_f^2}{gl_0} \boldsymbol{E}^2 - (3l_0 g|\chi|^4) \boldsymbol{B}^2$$

where C is an $O(1)$ constant. Such a Lagrangian describes an artificial light in our $SU(N_f)$ spin model. Comparing the above to the standard Lagrangian (10.6.18), we find that the speed of the artificial light is of order $c_a \sim l_0 g/N_f$ and the fine-structure constant α is of order $1/N_f$.

In momentum space, the fermion hopping Hamiltonian (10.7.9) has the form (note that $a_0 = 0$ for the ground state)

$$H_f = \sum_{\boldsymbol{k}}{}' \Psi_{a,\boldsymbol{k}}^\dagger \Gamma(\boldsymbol{k}) \Psi_{a,\boldsymbol{k}}$$

where

$$\begin{aligned}
\Psi_{a,\boldsymbol{k}}^\top &= (\psi_{a,\boldsymbol{k}}, \psi_{a,\boldsymbol{k}+\boldsymbol{Q}_x}, \psi_{a,\boldsymbol{k}+\boldsymbol{Q}_y}, \psi_{a,\boldsymbol{k}+\boldsymbol{Q}_x+\boldsymbol{Q}_y}), \\
\boldsymbol{Q}_x &= (\pi, 0, 0), \qquad \boldsymbol{Q}_y = (0, \pi, 0), \\
\Gamma(\boldsymbol{k}) &= -8|\chi| N_f^{-1} (\sin(k_x)\Gamma_1 + \sin(k_y)\Gamma_2 + \sin(k_z)\Gamma_3)
\end{aligned} \tag{10.7.12}$$

and $\Gamma_1 = \tau^3 \otimes \tau^0$, $\Gamma_2 = \tau^1 \otimes \tau^3$, and $\Gamma_3 = \tau^1 \otimes \tau^1$. Here $\tau^{1,2,3}$ are the Pauli matrices and τ^0 is the 2×2 identity matrix. The momentum summation $\sum_{\boldsymbol{k}}'$ is over a range $k_x \in (-\pi/2, \pi/2)$, $k_y \in (-\pi/2, \pi/2)$, and $k_z \in (-\pi, \pi)$. As $\{\Gamma_i, \Gamma_j\} = 2\delta_{ij}$, $i, j = 1, 2, 3$, we find that the fermions have the dispersion

$$E(\boldsymbol{k}) = \pm 8g|\chi|^3 N_f^{-1} \sqrt{\sin^2(k_x) + \sin^2(k_y) + \sin^2(k_z)}$$

We see that the dispersion has two nodes at $\boldsymbol{k} = 0$ and $\boldsymbol{k} = (0, 0, \pi)$. Thus, eqn (10.7.9) will give rise to $2N_f$ massless four-component Dirac fermions in the continuum limit.

After including the $U(1)$ gauge fluctuations, the massless Dirac fermions interact with the $U(1)$ gauge field as fermions with unit charge. Therefore, the total effective theory of our $SU(N_f)$ spin model is a QED with $2N_f$ families of Dirac fermions of unit charge. We will call these fermions artificial electrons. The continuum effective theory has the form

$$\mathcal{L} = \bar{\psi}_{I,a} D_0 \gamma^0 \psi_{I,a} + v_f \bar{\psi}_{I,a} D_i \gamma^i \psi_{I,a} + \frac{CN_f^2}{gl_0} \boldsymbol{E}^2 - (3l_0 g|\chi|^4) \boldsymbol{B}^2 + \ldots \tag{10.7.13}$$

where $I = 1, 2$, $D_0 = \partial_t + \mathrm{i}a_0$, $D_i = \partial_i + \mathrm{i}a_i|_{i=1,2,3}$, $v_f = 8l_0 g|\chi|^3/N_f^{-1}$, $\gamma_\mu|_{\mu=0,1,2,3}$ are 4×4 Dirac matrices, and $\bar{\psi}_{I,a} = \psi_{I,a}^\dagger \gamma^0$. Here $\psi_{1,a}$ and $\psi_{2,a}$ are Dirac fermion fields, which form a fundamental representation of $SU(N_f)$. We would like to point out that, although both the speed of the artificial light, c_a, and the speed of the artificial electrons, v_f, are of order $l_0 g/N_f$, the two speeds do not have to be the same in our model. Thus, Lorentz symmetry is not guaranteed.

Equation (10.7.13) describes the low-energy dynamics of the $SU(N_f)$ model in a quantum-ordered phase—the π-flux phase. The fermions and the gauge boson are massless and interact with each other. Here we would like to address an important question: after integrating out high-energy fermion and gauge fluctuations, do the fermions and the gauge boson remain massless? In general, the interaction between massless excitations will generate a mass term for them, unless the masslessness is protected by symmetry, or something else. For our $SU(N_f)$ model, the ground state breaks no symmetry. So we cannot use spontaneously broken symmetry to explain the massless excitations. The massless excitations are protected by the PSG that characterizes the quantum order (or string-net condensation) in the ground state (see Section 9.10 and Wen (2002a)).

10.7.4 Remarks: some historic remarks about gauge theory and Fermi statistics

- There are two ways to view a gauge field, namely as a geometric object of local phase invariance, or as a collective mode of a correlated system.
- The meaning of 'gauge'.
- Gauge fields and fermion fields do not imply gauge bosons and fermions as low-energy quasiparticles.

The first systematic gauge theory was Maxwell's theory for electromagnetism. Although the vector potential A_μ was introduced to express the electric field and the magnetic field, the meaning of A_μ was unclear.

The notion of a gauge field was introduced by Weyl in 1918, who also suggested that the vector potential A_μ is a gauge field. Weyl's idea is motivated by Einstein's theory of gravity and is an attempt to unify electromagnetism and gravity. In Einstein's general relativistic theory, the coordinate invariance leads to gravity. So Weyl thought that the invariance of another geometrical object may lead to electromagnetism. He proposed the scale invariance.

Consider a physical quantity that has a value f. We know that the numerical value f itself is meaningless unless we specify the unit. Let us use ω to denote the unit. The physical quantity is really given by $f\omega$. This is the relativity in scale. Now let us assume that the physical quantity is defined at every point in space (so we are considering a physical field). We would like to know how to compare the physical quantity at different points x^μ and $x^\mu + dx^\mu$. We cannot just compare the numerical values $f(x^\mu)$ and $f(x^\mu + dx^\mu)$ because the unit ω may be different at different points. For the nearby points x^μ and $x^\mu + dx^\mu$, the two units only differ by a factor close to 1. We can express such a factor as $1 + S_\mu dx^\mu$. The difference in the physical quantity at x^μ and $x^\mu + dx^\mu$ is not given by $f(x^\mu + dx^\mu) - f(x^\mu) = \partial_\mu f dx^\mu$, but by $f(x^\mu + dx^\mu)(1 + S_\mu dx^\mu) - f(x^\mu) = (\partial_\mu + S_\mu) f dx^\mu$. Weyl showed that the local scale invariance requires that only the curl of S_μ is physically meaningful, just like only the curl of A_μ is meaningful in Maxwell's theory. Thus, Weyl identified S_μ as the vector potential A_μ. Weyl called the local scale invariance 'Eich Invarianz', which was translated to 'gauge invariance'.

However, Weyl's idea is wrong and the vector potential A_μ cannot be identified as the 'gauge field' S_μ. On the other hand, Weyl was almost right. If we think of our physical field as the amplitude of a complex wave function[78] and the unit ω as a complex phase, i.e. $|\omega| = 1$, then the difference between the amplitudes at different points is given by $(\partial_\mu + \mathrm{i} S_\mu) f dx^\mu$, where the units at different points differ by a factor $(1 + \mathrm{i} S_\mu dx^\mu)$. It is such an S_μ that can be identified as the vector potential. So A_μ should really be called the 'phase field', and 'gauge invariance' should be called 'phase invariance'. However, the old name has stuck.

This part of history is an attempt to give the unphysical vector A_μ some physical (or geometrical) meaning. It views the vector potential as a connection of a fibre bundle. This picture is widely accepted. We now call the vector potential the gauge field, and Maxwell's theory is called gauge theory. However, this does not mean that we have to interpret the vector potential as a geometrical object from the local phase invariance. After all, the phase of a quantum wave function is unphysical.

There is another point of view about the gauge theory. Many thinkers in theoretical physics were not happy with the redundancy of the gauge potential A_μ. It was realized in the early 1970s that one could use gauge-invariant loop operators to characterize different phases of a gauge theory (Wegner, 1971; Wilson, 1974; Kogut and Susskind, 1975). Later, people found that one can formulate the entire gauge theory using closed strings (Banks *et al.*, 1977; Foerster, 1979; Gliozzi *et al.*, 1979; Mandelstam, 1979; Polyakov, 1979; Savit, 1980). These studies revealed the intimate relationship between gauge theories and closed-string theories—a point of view which is very different from the geometrical notion of vector potential.

In a related development in condensed matter physics, people found that gauge fields can emerge from a local bosonic model, if the bosonic model is in certain quantum phases. This phenomenon is also called the dynamical generation of gauge fields. The emergence of gauge fields from local bosonic models has a long and complicated history. The emergent $U(1)$ gauge field was introduced in the quantum-disordered phase of the $(1+1)$-dimensional CP^N model (D'Adda *et al.*, 1978; Witten, 1979). In condensed matter physics, the $U(1)$ gauge field has been found in the slave-boson approach to spin-liquid states of bosonic

[78] The notion of a complex wave function was introduced in 1925, seven years after Weyl's 'gauge theory'.

spin models on a square lattice (Affleck and Marston, 1988; Baskaran and Anderson, 1988). The slave-boson approach not only has a $U(1)$ gauge field, but it also has gapless fermion fields. However, due to the instanton effect and the resulting confinement of the $U(1)$ gauge field in $1+1$ and $1+2$ dimensions (Polyakov, 1975), none of the above gauge fields and gapless fermion fields lead to gauge bosons and gapless fermions that appear as low-energy physical quasiparticles. Even in the large-N limit where the instanton effect can be ignored, the marginal coupling between the $U(1)$ gauge field and the massless Dirac fermions in $2+1$ dimensions destroys the quasiparticle poles in the fermion and gauge propagators. This led to the opinion that the $U(1)$ gauge field and the gapless fermion fields are just an unphysical artifact of the 'unreliable' slave-boson approach. Thus, the key to finding emergent gauge bosons and emergent fermions is not to write down a Lagrangian that contains *gauge fields* and *Fermi fields*, but to show that gauge bosons and fermions actually appear in the physical low-energy spectrum. In fact, for any given physical system, we can always design a Lagrangian with a gauge field of arbitrary choice to describe that system. However, a gauge field in a Lagrangian may not give rise to a gauge boson that appears as a low-energy quasiparticle. Only when the dynamics of the gauge field are such that the gauge field is in the deconfined phase can the gauge boson appear as a low-energy quasiparticle. Thus, many researchers, after the initial findings of D'Adda *et al.* (1978), Witten (1979), Baskaran and Anderson (1988), and Affleck and Marston (1988), have been trying to find the deconfined phase of the gauge field.

In high-energy physics, a $(3+1)$-dimensional local bosonic model with emergent deconfined $U(1)$ gauge bosons was constructed by Foerster *et al.* (1980). It was suggested that light in nature may be emergent. In condensed matter physics, it was shown that, if we break the time-reversal symmetry in a two-dimensional spin-1/2 model, then the $U(1)$ gauge field from the slave-boson approach can be in a deconfined phase due to the appearance of the Chern–Simons term (Khveshchenko and Wiegmann, 1989; Wen *et al.*, 1989). The deconfined phase corresponds to a spin-liquid state of the spin-1/2 model (Kalmeyer and Laughlin, 1987), which is called the chiral spin liquid. A second deconfined phase was found by breaking the $U(1)$ gauge structure down to a Z_2 gauge structure. Such a phase contains a deconfined Z_2 gauge theory (Read and Sachdev, 1991; Wen, 1991a), and is called a Z_2 spin liquid (or a short-ranged RVB state). Both the chiral spin liquid and the Z_2 spin liquid have some amazing properties. The quasiparticle excitations carry spin-1/2 and correspond to *one-half* of a spin flip. These quasiparticles can also carry fractional statistics or Fermi statistics, despite our spin-1/2 model being a purely bosonic model. These condensed matter examples illustrate that both gauge fields and Fermi statistics can emerge from local bosonic models.

We would like to point out that the spin liquids are not the first example of emergent fermions from local bosonic models. The first example of emergent fermions, or, more generally, emergent anyons, is given by the FQH states. Although Arovas *et al.* (1984) only discussed how anyons can emerge from a fermion system in a magnetic field, the same argument can easily be generalized to show how fermions and anyons can emerge from a boson system in a magnetic field. Also, in 1987, in a study of resonating valence bound (RVB) states, emergent fermions (the spinons) were proposed in a nearest-neighbor dimer model on a square lattice (Kivelson *et al.*, 1987; Rokhsar and Kivelson, 1988; Read and Chakraborty, 1989). However, according to the deconfinement picture, the results by Kivelson *et al.* (1987) and Rokhsar and Kivelson (1988) are valid only when the ground state of the dimer model is in the Z_2 deconfined phase. It appears that the dimer liquid on a square lattice with only nearest-neighbor dimers is not a deconfined state (Rokhsar and Kivelson, 1988; Read and Chakraborty, 1989), and thus it is not clear if the nearest-neighbor dimer model on a square lattice (Rokhsar and Kivelson, 1988) has fermionic quasiparticles or not (Read and Chakraborty, 1989). However, on a triangular lattice, the dimer liquid is indeed a Z_2 deconfined state (Moessner and Sondhi, 2001). Therefore, the results of Kivelson *et al.* (1987) and Rokhsar and Kivelson (1988) are valid for the triangular-lattice dimer model, and fermionic quasiparticles do emerge in a dimer liquid on a triangular lattice.

All of the above models with emergent fermions are $(2+1)$-dimensional models, where the emergent fermions can be understood from binding flux to a charged particle (Arovas *et al.*, 1984). Recently, it was pointed out by Levin and Wen (2003) that the key to emergent fermions is a string structure. Fermions can generally appear as ends of open strings in any dimensions. The string picture allows the construction of a $(3+1)$-dimensional local bosonic model that has emergent fermions (Levin and Wen, 2003). Since both gauge bosons and fermions can emerge as a result of string-net condensation, we may say that string-net condensation provides a way to unify gauge bosons and fermions.

Generalizing the bosonic $SU(N)$ spin model on a two-dimensional square lattice (Affleck and Marston, 1988), both gapless deconfined $U(1)$ gauge bosons and gapless fermions were found to emerge from a bosonic $SU(N)$ spin model on a three-dimensional cubic lattice (Wen, 2002a). In $1+3$ dimensions, the two kinds of gapless excitations can be separated because they interact weakly at low energies. The $U(1)$ gauge bosons and gapless fermions behave in every way like photons and electrons. Thus, the bosonic $SU(N)$ spin model not only contains artificial light, but it also contains artificial electrons.

After about one hundred years of gauge theory and Fermi statistics, we are now facing the following questions. What is the origin of the gauge field—geometrical or dynamical? What is the origin of Fermi statistics—given or emergent? In this book, we favor the dynamical and emergent origin of gauge bosons and fermions. The gauge bosons and the Fermi statistics may just be collective phenomena of quantum many-boson systems, and nothing more.

BIBLIOGRAPHY

Abrikosov, A. A., L. P. Gorkov, and I. E. Dzyaloshinski, 1975, *Method of Quantum Field Theory in Statistical Physics* (Dover, New York).

Affleck, I., T. Kennedy, E. H. Lieb, and H. Tasaki, 1987, Phys. Rev. Lett. **59**, 799.

Affleck, I., and J. B. Marston, 1988, Phys. Rev. B **37**, 3774.

Affleck, I., Z. Zou, T. Hsu, and P. W. Anderson, 1988, Phys. Rev. B **38**, 745.

Anderson, P. W., 1963, Phys. Rev. **130**, 439.

Appelquist, T., and D. Nash, 1990, Phys. Rev. Lett. **64**, 721.

Arovas, D., J. R. Schrieffer, and F. Wilczek, 1984, Phys. Rev. Lett. **53**, 722.

Arovas, D. P., and A. Auerbach, 1988, Phys. Rev. B **38**, 316.

Avron, J., R. Seiler, and B. Simon, 1983, Phys. Rev. Lett. **51**, 51.

Balents, L., M. P. A. Fisher, and S. M. Girvin, 2002, Phys. Rev. B **65**, 224412.

Balents, L., M. P. A. Fisher, and C. Nayak, 1998, Int. J. Mod. Phys. B **12**, 1033.

Banks, T., R. Myerson, and J. B. Kogut, 1977, Nucl. Phys. B **129**, 493.

Bardeen, J., L. N. Cooper, and J. R. Schrieffer, 1957, Phys. Rev. **108**, 1175.

Baskaran, G., and P. W. Anderson, 1988, Phys. Rev. B **37**, 580.

Baskaran, G., Z. Zou, and P. W. Anderson, 1987, Solid State Comm. **63**, 973.

Bednorz, J. G., and K. A. Mueller, 1986, Z. Phys. B **64**, 189.

Beenakker, C. W. J., 1990, Phys. Rev. Lett. **64**, 216.

Berry, M. V., 1984, Proc. R. Soc. Lond. **A392**, 45.

Blok, B., and X.-G. Wen, 1990a, Phys. Rev. B **42**, 8133.

Blok, B., and X.-G. Wen, 1990b, Phys. Rev. B **42**, 8145.

Blok, B., and X.-G. Wen, 1992, Nucl. Phys. B **374**, 615.

Brey, L., 1990, Phys. Rev. Lett. **65**, 903.

Buttiker, M., 1988, Phys. Rev. B **38**, 9375.

Caldeira, A. O., and A. J. Leggett, 1981, Phys. Rev. Lett. **46**, 211.

Cappelli, A., C. A. Trugenberger, and G. R. Zemba, 1993, Nucl. Phys. B **396**, 465.

Chaikin, P. M., and T. C. Lubensky, 2000, *Principles of Condensed Matter Physics* (Cambridge University Press).

Chang, A. M., L. N. Pfeiffer, and K. W. West, 1996, Phys. Rev. Lett. **77**, 2538.

Chen, W., M. P. A. Fisher, and Y.-S. Wu, 1993, Phys. Rev. B **48**, 13749.

Chen, Y. H., F. Wilczek, E. Witten, and B. Halperin, 1989, J. Mod. Phys. B **3**, 1001.

Coleman, S., 1975, Phys. Rev. D **11**, 2088.

Coleman, S., 1985, *Aspects of symmetry* (Pergamon, Cambridge).

Coleman, S., and E. Weinberg, 1973, Phys. Rev. D **7**, 1888.

D'Adda, A., P. D. Vecchia, and M. Lüscher, 1978, Nucl. Phys. B **146**, 63.
Dagotto, E., E. Fradkin, and A. Moreo, 1988, Phys. Rev. B **38**, 2926.
Dasgupta, C., and B. I. Halperin, 1981, Phys. Rev. Lett. **47**, 1556.
Eisenstein, J. P., T. J. Gramila, L. N. Pfeiffer, and K. W. West, 1991, Phys. Rev. B **44**, 6511.
Elitzur, S., G. Moore, A. Schwimmer, and N. Seiberg, 1989, Nucl. Phys. B **326**, 108.
Fertig, H., 1989, Phys. Rev. B **40**, 1087.
Fetter, A., C. Hanna, and R. Laughlin, 1989, Phys. Rev. B **39**, 9679.
Floreanini, R., and R. Jackiw, 1988, Phys. Rev. Lett. **59**, 1873.
Foerster, D., 1979, Physics Letters B **87**, 87.
Foerster, D., H. B. Nielsen, and M. Ninomiya, 1980, Phys. Lett. B **94**, 135.
Fradkin, E., and S. H. Shenker, 1979, Phys. Rev. D **19**, 3682.
Franz, M., and Z. Tesanovic, 2001, Phys. Rev. Lett. **87**, 257003.
Fröhlich, J., and T. Kerler, 1991, Nucl. Phys. B **354**, 369.
Fröhlich, J., and C. King, 1989, Comm. Math. Phys. **126**, 167.
Fröhlich, J., and U. M. Studer, 1993, Rev. of Mod. Phys. **65**, 733.
Georgi, H., and S. Glashow, 1974, Phys. Rev. Lett. **32**, 438.
Ginzburg, V. L., and L. D. Landau, 1950, Zh. Ekaper. Teoret. Fiz. **20**, 1064.
Girvin, S. M., and A. H. MacDonald, 1987, Phys. Rev. Lett. **58**, 1252.
Gliozzi, F., T. Regge, and M. A. Virasoro, 1979, Physics Letters B **81**, 178.
Goddard, P., and D. Olive, 1985, Workshop on Unified String Theories, eds. M. Green and D. Gross, (World Scientific, Singapore) , 214.
Goddard, P., and D. Olive, 1986, Inter. J. Mod. Phys. **1**, 303.
Goldstone, J., 1961, Nuovo Cimento **19**, 154.
Green, M. B., J. H. Schwarz, and E. Witten, 1988, *Superstring Theory* (Cambridge University Press).
Greiter, M., X.-G. Wen, and F. Wilczek, 1991, Phys. Rev. Lett. **66**, 3205.
Gross, D. J., J. A. Harvey, E. Martinec, and R. Rohm, 1985, Nucl. Phys. B **256**, 253.
Haldane, F., 1992, Helv. Phys. Acta. **65**, 152.
Haldane, F., and E. Rezayi, 1988a, Phys. Rev. Lett. **60**, E1886.
Haldane, F. D. M., 1983, Phys. Rev. Lett. **51**, 605.
Haldane, F. D. M., and E. H. Rezayi, 1985, Phys. Rev. B **31**, 2529.
Haldane, F. D. M., and E. H. Rezayi, 1988b, Phys. Rev. Lett. **60**, 956.
Halperin, B. I., 1982, Phys. Rev. B **25**, 2185.
Halperin, B. I., 1983, Helv. Phys. Acta **56**, 75.
Halperin, B. I., 1984, Phys. Rev. Lett. **52**, 1583.
Halperin, B. I., T. C. Lubensky, and S. K. Ma, 1974, Phys. Rev. Lett. **32**, 292.
Higgs, P. W., 1964, Phys. Rev. Lett. **12**, 132.
't Hooft, G., 1993, gr-qc/9310026 .

Houghton, A., and J. B. Marston, 1993, Phys. Rev. B **48**, 7790.
Ioffe, L. B., M. V. Feigel'man, A. Ioselevich, D. Ivanov, M. Troyer, and G. Blatter, 2002, Nature **415**, 503.
Iso, S., D. Karabali, and B. Sakita, 1992, Phys. Lett. B **296**, 143.
Jain, J. K., and S. A. Kivelson, 1988a, Phys. Rev. B **37**, 4276.
Jain, J. K., and S. A. Kivelson, 1988b, Phys. Rev. Lett. **60**, 1542.
Jordan, P., and E. Wigner, 1928, Z. Phys. **47**, 631.
Kac, V. G., 1983, *Infinite dimensional Lie algebra* (Birkhauser, Boston).
Kalmeyer, V., and R. B. Laughlin, 1987, Phys. Rev. Lett. **59**, 2095.
Keldysh, L., 1965, Sov. Phys. JETP **20**, 1018.
Khveshchenko, D., and P. Wiegmann, 1989, Mod. Phys. Lett. **3**, 1383.
Kim, D. H., and P. A. Lee, 1999, Annals of Physics **272**, 130.
Kim, Y. B., P. A. Lee, and X.-G. Wen, 1995, Phys. Rev. B **52**, 17275.
Kitaev, A. Y., 2003, Ann. Phys. (N.Y.) **303**, 2.
Kivelson, S. A., D. S. Rokhsar, and J. P. Sethna, 1987, Phys. Rev. B **35**, 8865.
von Klitzing, K., G. Dorda, and M. Pepper, 1980, Phys. Rev. Lett. **45**, 494.
Kogut, J., and L. Susskind, 1975, Phys. Rev. D **11**, 395.
Kogut, J. B., 1979, Rev. Mod. Phys. **51**, 659.
Kosterlitz, J. M., and D. J. Thouless, 1973, J. Phys. C **6**, 1181.
Kotliar, G., 1988, Phys. Rev. B **37**, 3664.
Kotliar, G., and J. Liu, 1988, Phys. Rev. B **38**, 5142.
Landau, L. D., 1937, Phys. Z. Sowjetunion **11**, 26.
Landau, L. D., 1941, Zh. Eksp. Teor. Fiz. **11**, 592.
Landau, L. D., 1956, Sov. Phys. JETP **3**, 920.
Landau, L. D., 1959, Sov. Phys. JETP **8**, 70.
Landau, L. D., and E. M. Lifschitz, 1958, *Statistical Physics - Course of Theoretical Physics Vol 5* (Pergamon, London).
Laughlin, R. B., 1983, Phys. Rev. Lett. **50**, 1395.
Lee, P. A., and N. Nagaosa, 1992, Phys. Rev. B **45**, 5621.
Leinaas, J. M., and J. Myrheim, 1977, Il Nuovo Cimento **37B**, 1.
Levin, M., and X.-G. Wen, 2003, Phys. Rev. B **67**, 245316.
Lifshitz, I. M., 1960, Sov. Phys. JETP **11**, 1130.
Luther, A., 1979, Phys. Rev. B **19**, 320.
Luttinger, J. M., 1963, Journal of Mathematical Physics **4**, 1154.
Ma, S.-K., 1976, *Modern Theory of Critical Phenomena* (Benjamin/Cummings, Reading, MA).
MacDonald, A. H., 1990, Phys. Rev. Lett. **64**, 220.
MacDonald, A. H., P. M. Platzman, and G. S. Boebinger, 1990, Phys. Rev. Lett. **65**, 775.
MacDonald, A. H., and P. Streda, 1984, Phys. Rev. B **29**, 1616.
Mahan, G. D., 1990, *Many-Particle Physics* (Plenum, New York), 2nd

edition.
Majumdar, C. K., and D. K. Ghosh, 1969, J. Math. Phys. **10**, 1388.
Mandelstam, S., 1979, Phys. Rev. D **19**, 2391.
Marston, J. B., and I. Affleck, 1989, Phys. Rev. B **39**, 11538.
Milliken, F. P., C. P. Umbach, and R. A. Webb, 1995, Solid State Comm. **97**, 309.
Moessner, R., and S. L. Sondhi, 2001, Phys. Rev. Lett. **86**, 1881.
Moore, G., and N. Read, 1991, Nucl. Phys. B **360**, 362.
Motrunich, O. I., 2003, Phys. Rev. B **67**, 115108.
Motrunich, O. I., and T. Senthil, 2002, Phys. Rev. Lett. **89**, 277004.
Mudry, C., and E. Fradkin, 1994, Phys. Rev. B **49**, 5200.
Murphy, S. Q., J. P. Eisenstein, G. S. Boebinger, L. N. Pfeiffer, and K. W. West, 1994, Phys. Rev. Lett. **72**, 728.
Nambu, Y., 1960, Phys. Rev. Lett. **4**, 380.
Negele, J. W., and H. Orland, 1998, *Quantum Many-Particle Systems* (Perseus Publishing).
Neto, A. H. C., and E. Fradkin, 1994, Phys. Rev. Lett. **72**, 1393.
Niu, Q., D. J. Thouless, and Y.-S. Wu, 1985, Phys. Rev. B **31**, 3372.
Onnes, H. K., 1911, Comm. Phys. Lab. Univ. Leiden, Nos 119 **120**, 122.
Polchinski, J., 1998, *String Theory* (Cambridge University Press).
Polyakov, A. M., 1975, Phys. Lett. B **59**, 82.
Polyakov, A. M., 1977, Nucl. Phys. B **120**, 429.
Polyakov, A. M., 1979, Phys. Lett. B **82**, 247.
Rammmer, J., and H. Smith, 1986, Rev. Mod. Phys. **58**, 323.
Ran, Y., and X.-G. Wen, 2003, unpublished .
Rantner, W., and X.-G. Wen, 2001, Phys. Rev. Lett. **86**, 3871.
Rantner, W., and X.-G. Wen, 2002, Phys. Rev. B **66**, 144501.
Read, N., 1989, Phys. Rev. Lett. **62**, 86.
Read, N., 1990, Phys. Rev. Lett. **65**, 1502.
Read, N., and B. Chakraborty, 1989, Phys. Rev. B **40**, 7133.
Read, N., and D. Green, 2000, Phys. Rev. B **61**, 10267.
Read, N., and S. Sachdev, 1989, Phys. Rev. Lett. **62**, 1694.
Read, N., and S. Sachdev, 1991, Phys. Rev. Lett. **66**, 1773.
Rokhsar, D. S., and S. A. Kivelson, 1988, Phys. Rev. Lett. **61**, 2376.
Sachdev, S., and K. Park, 2002, Annals of Physics (N.Y.) **298**, 58.
Savit, R., 1980, Rev. Mod. Phys. **52**, 453.
Schwinger, J., 1961, J. Math. Phys. **2**, 407.
Senthil, T., and M. P. A. Fisher, 2000, Phys. Rev. B **62**, 7850.
Senthil, T., and M. P. A. Fisher, 2001, Phys. Rev. Lett. **86**, 292.
Senthil, T., J. B. Marston, and M. P. A. Fisher, 1999, Phys. Rev. B **60**, 4245.
Senthil, T., and O. Motrunich, 2002, Phys. Rev. B **66**, 205104.
Streda, P., J. Kucera, and A. H. MacDonald, 1987, Phys. Rev. Lett. **59**, 1973.

Susskind, L., 1995, J. Math. Phys. **36**, 6377.
Thouless, D. J., M. Kohmoto, M. P. Nightingale, and M. den Nijs, 1982, Phys. Rev. Lett. **49**, 405.
Tomonaga, S., 1950, Prog. Theor. Phys. (Kyoto) **5**, 544.
Trugman, S. A., 1983, Phys. Rev. B **27**, 7539.
Tsui, D. C., H. L. Stormer, and A. C. Gossard, 1982, Phys. Rev. Lett. **48**, 1559.
Wegner, F., 1971, J. Math. Phys. **12**, 2259.
Wen, X.-G., 1990, Int. J. Mod. Phys. B **4**, 239.
Wen, X.-G., 1991a, Phys. Rev. B **44**, 2664.
Wen, X.-G., 1991b, Phys. Rev. Lett. **66**, 802.
Wen, X.-G., 1992, Int. J. Mod. Phys. B **6**, 1711.
Wen, X.-G., 1995, Advances in Physics **44**, 405.
Wen, X.-G., 2000, Phys. Rev. Lett. **84**, 3950.
Wen, X.-G., 2002a, Phys. Rev. Lett. **88**, 11602.
Wen, X.-G., 2002b, Physics Letters A **300**, 175.
Wen, X.-G., 2002c, Phys. Rev. B **65**, 165113.
Wen, X.-G., 2003a, Phys. Rev. B **68**, 115413.
Wen, X.-G., 2003b, Phys. Rev. D **68**, 065003.
Wen, X.-G., 2003c, Phys. Rev. Lett. **90**, 016803.
Wen, X.-G., and P. A. Lee, 1996, Phys. Rev. Lett. **76**, 503.
Wen, X.-G., and Q. Niu, 1990, Phys. Rev. B **41**, 9377.
Wen, X.-G., F. Wilczek, and A. Zee, 1989, Phys. Rev. B **39**, 11413.
Wen, X.-G., and Y.-S. Wu, 1993, Phys. Rev. Lett. **70**, 1501.
Wen, X.-G., Y.-S. Wu, and Y. Hatsugai, 1994, Nucl. Phys. B **422**, 476.
Wen, X.-G., and A. Zee, 1991, Phys. Rev. B **44**, 274.
Wen, X.-G., and A. Zee, 1992a, Phys. Rev. B **46**, 2290.
Wen, X.-G., and A. Zee, 1992b, Phys. Rev. Lett. **69**, 1811.
Wen, X.-G., and A. Zee, 1992c, Phys. Rev. Lett. **69**, 953.
Wen, X.-G., and A. Zee, 1992d, Phys. Rev. Lett. **69**, 3000.
Wen, X.-G., and A. Zee, 1993, Phys. Rev. B **47**, 2265.
Wen, X.-G., and A. Zee, 2002, Phys. Rev. B **66**, 235110.
Wilczek, F., 1982, Phys. Rev. Lett. **49**, 957.
Willett, R., J. P. Eisenstein, H. L. Strörmer, D. C. Tsui, A. C. Gossard, and J. H. English, 1987, Phys. Rev. Lett. **59**, 1776.
Wilson, K. G., 1974, Phys. Rev. D **10**, 2445.
Witten, E., 1979, Nucl. Phys. B **149**, 285.
Witten, E., 1989, Comm. Math. Phys. **121**, 351.
Yang, K., K. Moon, L. Zheng, A. H. MacDonald, S. M. Girvin, D. Yoshioka, and S.-C. Zhang, 1994, Phys. Rev. Lett. **72**, 732.
Zhang, S. C., T. H. Hansson, and S. Kivelson, 1989, Phys. Rev. Lett. **62**, 82.

INDEX